D1291753

24V

MRI and CT
OF THE
MUSCULOSKELETAL
SYSTEM
A Text-Atlas

MRI and CT
OF THE
MUSCULOSKELETAL
SYSTEM
A Text-Atlas

EDITORS

Johan L. Bloem, M.D.

Associate Professor, Department of Radiology
University Hospital Leiden, Leiden, The Netherlands

David J. Sartoris, M.D.

Associate Professor, Department of Radiology
University of California–San Diego, San Diego, California

WILLIAMS & WILKINS
BALTIMORE • HONG KONG • LONDON • MUNICH
PHILADELPHIA • SYDNEY • TOKYO

Editor: Timothy H. Grayson
Managing Editor: Marjorie Kidd Keating
Copy Editor: Susan McColl
Designer: Norman W. Och
Illustration Planner: Kenneth Urban
Production Coordinator: Kathleen C. Millet
Production Service: Lila M. Gardner, Spectrum Publisher Services, Inc.

Printed in the United States of America

Library of Congress Cataloging-in-Publication Data

MRI and CT of the musculoskeletal system : a text-atlas / editors,
  Johan L. Bloem, David J. Sartoris.
     p. cm.—(Recent research in psychology)
  Includes bibliographical references and index.
  ISBN 0-683-00875-7
  1. Musculoskeletal system—magnetic resonance imaging.
2. Musculoskeletal system—Tomography. I. Bloem, Johan L.
II. Sartoris, David J. III. Series.
  [DNLM: 1. Magnetic Resonance Imaging—atlases. 2. Musculoskeletal
System—anatomy & histology—atlases. 3. Musculoskeletal System—
pathology—atlases. 4. Tomography, X-Ray Computed—atlases. WE
17 M939]
RC925.7.M74 1992
616.7'07548—dc20
DNLM/DLC
for Library of Congress
                                   91-15345
                                     CIP

91 92 93 94 95
1 2 3 4 5 6 7 8 9 10

To Els

Hans

To Mom, Dad, and Cyd

D.J.S.

FOREWORD

Musculoskeletal imaging has been profoundly affected by the development of cross-axial computer-based imaging techniques such as magnetic resonance imaging and computed tomography. The impact of these technologies has been profound and pervades virtually all aspects of musculoskeletal radiology today. The evaluation of tumors, all major joints, the marrow space, and the vertebra, are all dramatically augmented through the use of these newer modalities. The incorporation of the new modalities into the active armamentarium of the radiologist, particularly in areas of the musculoskeletal system, represents a somewhat daunting challenge. This includes an understanding of the anatomic features, the morphologic findings which are demonstrable by the new modalities, and the range of useful applications of these.

This text atlas of the musculoskeletal system provides much needed information in a format which is both accessible and comprehensive. The text includes a detailed atlas of normal anatomy, a description of the uses of these techniques in bone marrow disorders, Gaucher's disease, the analysis of cortical bone changes, infection, primary changes in muscle, tumors, and metastatic disease. Detailed sections are included on the applications of MR imaging and CT to the shoulder, wrist, elbow, hand, hip, pelvis, knee, ankle, foot, and spine. The text also includes sections on the temporomandibular joint and craniofacial fractures. The editors have solicited contributions from world leaders in the applications of MR imaging and CT to the musculoskeletal system. The text is abundantly illustrated with state-of-the-art images and numerous illustrations.

Doctors Bloem and Sartoris have developed a most useful text which will serve as a comprehensive reference for those interested in understanding the applications of computed tomography and MR imaging of the musculoskeletal system.

Herbert V. Kressel, M.D.

PREFACE

The rapid advance of magnetic resonance (MR) imaging in diagnosis of musculoskeletal disorders has not only increased our understanding of pathology, but has also changed imaging strategies. The purpose of this book is to provide a comprehensive overview of the current role of MR imaging and computed tomography (CT) in the diagnosis of musculoskeletal disease.

With the exception of the chapters on the temporomandibular joint and the wrist, the first part displays normal anatomy of the musculoskeletal system in an atlas-type format. Its purpose is to provide a quick reference to normal anatomy needed to interpret MR imaging and CT. Five-millimeter thick MR images, made on a Philips 1.5-T system with standard software and surface coils, and some examples of CT arthrography are, together with anatomic line drawings, displayed side by side. Identification of structures is based on in vivo and cadaver studies both in the adult and pediatric age group. Unless stated otherwise axial, coronal, and sagittal planes were used. The anatomic structures are consistently labeled in each chapter, and simple line drawings facilitate orientation. Discussion of anatomy of the temporomandibular joint is tailored to the specific needs demanded by pathology of the temporomandibular joint, as described in Chapter 17. Chapter 4 presents a detailed description of the anatomy of the wrist because it is felt that MR imaging of this complex area will be increasingly important in the near future due to significant hardware and software improvements. It is hoped that this concise, but detailed, display of normal anatomy increases access to the second part of the book for those of us who do not deal with three-dimensional anatomy on a day-to-day basis.

Established and relatively new applications of MR imaging and CT in diagnosis of musculoskeletal disorders are discussed in detail in 34 chapters. The authors were selected because of their significant contributions to specific fields of musculoskeletal imaging. Their observations and opinions are based on extensive experience and on many studies, which have been communicated in recent peer-reviewed articles. The authors focus on examination techniques, applied anatomy, and diagnostic imaging with a clinical perspective.

General and systemic disorders are discussed in the seven chapters comprising the first section. The second section is organized by anatomic subject and deals with facial disorders, shoulder, elbow, wrist and hand, hip and pelvis, knee, ankle and foot, and vertebral column. The whole range of pathology, including trauma, infection, neoplasm, degenerative disease, and congenital and developmental disorders is covered. Two appendices, comprising a glossary and a simple, short introduction to general and technical aspects of MR imaging and CT, may also facilitate use of the book.

The ongoing rapid improvement of MR technology continues to change its role and that of CT in the diagnosis of musculoskeletal disease. Therefore differences of opinion, although channeled to some degree, are not censored by the editors. At this point on the learning curve, we consider it an advantage rather than a disadvantage to be exposed to both concordant and discordant points of view.

We have attempted to present a mixture of anatomic, technical, and clinically oriented information on the use of both MR imaging and CT in the diagnosis of musculoskeletal disorders that may be useful to both radiologists and referring clinicians.

Johan L. Bloem, M.D.
David J. Sartoris, M.D.

ACKNOWLEDGMENTS

Many colleagues, technicians, and secretaries contributed to this project. We especially extend our thanks to the radiology technicians from the various imaging sites and radiologists from Leiden University Hospital, the Netherlands and Thomas Jefferson University Hospital in Philadelphia, Pennsylvania. We gratefully acknowledge the enthusiastic support of the secretarial staff of the department of radiology, and the technicians and artists of the laboratory of anatomy, of Leiden University. The progress of diagnostic imaging would not have been possible without the cooperation of referring clinicians, anatomists, and pathologists. We hope that the remarkable patience and willingness of our patients to cooperate in research projects will be of benefit to many more patients in the near future.

Johan L. Bloem, M.D.
David J. Sartoris, M.D.

CONTRIBUTORS

Mohammed Y. Aabed, M.D., D.M.R.D.
Radiologist, Department of Radiology, Riyadh Armed Forces Hospital, Riyadh, Saudi Arabia

Paul R. Algra, M.D.
Radiologist, Department of Diagnostic Radiology, Free University Hospital, Amsterdam, The Netherlands

Antonio Barile, M.D.
Department of Radiology, University of L'Aquila, S. Maria di Collemaggio Hospital, L'Aquila, Italy

Javier Beltran, M.D.
Associate Professor, Director of Muscular Skeletal Section, Department of Radiology, The Ohio State University Hospitals, Columbus, Ohio

Johan L. Bloem, M.D.
Associate Professor of Radiology, Department of Diagnostic Radiology, University Hospital Leiden, Leiden, The Netherlands

Cees F. A. Bos, M.D.
Orthopaedic Surgeon, Department of Orthopaedics, University Hospital Leiden, Leiden, The Netherlands

John M. Bramble, M.D.
Assistant Professor, Department of Diagnostic Radiology, University of Kansas Medical Center, Kansas City, Kansas

Ferry C. Breedveld, M.D.
Professor and Chairman, Department of Rheumatology, University Hospital Leiden, Leiden, The Netherlands

Anton M. J. Burgers, M.D.
Department of Orthopaedics, University Hospital Leiden, Leiden, The Netherlands

Edwin L. Christiansen, D.D.S., Ph.D.
Professor of Dentistry, Associate Professor of Radiology, Schools of Dentistry and Medicine, Loma Linda University, Loma Linda, California

William F. Conway, M.D., Ph.D.
Associate Professor of Radiology, Co-Director, Musculoskeletal Radiology, Director, Mammography, Medical College of Virginia, Richmond, Virginia

Murray K. Dalinka, M.D.
Professor of Radiology and Orthopaedic Surgery, University of Pennsylvania School of Medicine, Philadelphia, Pennsylvania

Andrew L. Deutsch, M.D.
Associate Clinical Professor of Radiology, University of California, San Diego, San Diego, California; Attending Radiologist, Division of Musculoskeletal Imaging, Cedars-Sinai Medical Center, Los Angeles, California

Joost Doornbos, Ph.D.
Department of Diagnostic Radiology, University Hospital Leiden, Leiden, The Netherlands

Albert Engel, M.D.
Orthopaedic Surgeon, Department of Orthopaedics, University of Vienna, Vienna, Austria

Eva Fascetti, M.D.
Department of Radiology, University of L'Aquila, S. Maria di Collemaggio Hospital, L'Aquila, Italy

Elliott K. Fishman, M.D.
Associate Professor, Director, Computed Body Tomography, The Russell H. Morgan Department of Radiology and Radiological Science, The Johns Hopkins Medical Institutions, Baltimore, Maryland

James L. Fleckenstein, M.D.
Assistant Professor, Department of Radiology, University of Texas Southwestern Medical Center, Dallas, Texas

J. Mark Fulmer, M.D.
Associate Attending Radiologist, Department of Radiology, Baylor University Medical School, Dallas, Texas

Harry K. Genant, M.D.
Professor of Radiology, Medicine, and Orthopaedic Surgery, Department of Radiology, University of California School of Medicine, San Francisco, California

Bernard Ghelman, M.D.
Associate Professor, Department of Radiology, The New York Hospital–Cornell University Medical College; Attending Radiologist, Department of Radiology, The Hospital for Special Surgery, New York, New York

Claus-C. Glüer, Ph.D.
Assistant Adjunct Professor, Department of Radiology, University of California School of Medicine, San Francisco, California

Cooper R. Gundry, M.D.
Assistant Professor of Radiology, University of Vermont College of Medicine; Attending Radiologist, Medical Center Hospital of Vermont, Burlington, Vermont

Maurice C. Haddad, M.D., F.R.C.R.
Consultant Radiologist, Department of Diagnostic Radiology and Imaging, Riyadh Armed Forces Hospital, Riyadh, Saudi Arabia

Paul C. Hajek, M.D.
Department of Radiology, University of Vienna, Vienna, Austria

Steven E. Harms, M.D.
Medical Director of Magnetic Resonance Imaging, Department of Radiology, Baylor University Medical Center, Dallas, Texas

Curtis W. Hayes, M.D.
Assistant Professor, Department of Radiology, Medical College of Virginia, Richmond, Virginia

Charles P. Ho, Ph.D., M.D.
Assistant Clinical Professor of Radiology, University of California, San Diego, School of Medicine, San Diego, California; Medical Director, San Francisco Neuro Skeletal Imaging, Daly City, California

Herma C. Holscher, M.D.
Department of Diagnostic Radiology, University Hospital Leiden, Leiden, The Netherlands

Brian A. Howard, M.D.
Assistant Professor, Department of Radiological Sciences, University of Toronto, Sunnybrook Health Science Center, Toronto, Ontario, Canada

Michele H. Johnson, M.D.
Assistant Professor, Neuroradiology Section, Department of Radiology, Medical College of Virginia, Virginia Commonwealth University, Richmond, Virginia

Ray F. Kilcoyne, M.D.
Professor of Radiology, University of Texas Health Sciences Center, Chief, Radiology Service, Veterans Administration Medical Center, San Antonio, Texas

Cheryl L. Kirby, M.D.
Resident, Department of Radiology, Hospital of the University of Pennsylvania, Philadelphia, Pennsylvania

Herman M. Kroon, M.D.
Radiologist, Department of Diagnostic Radiology, University Hospital Leiden, Leiden, The Netherlands

Sevil Kursunoglu-Brahme, M.D.
Valley Presbyterian Magnetic Resonance Center, Van Nuys, California

S. Howard Lee, M.D.
Clinical Professor of Radiology, University of Medicine and Dentistry of New Jersey and Temple University School of Medicine; Director of Neurosciences and Digital Imaging, Department of Radiology, Muhlenberg Hospital, Plainfield, New Jersey

Te Hua Liu, M.D.
Associate Professor, Chief of Neuroradiology, Department of Diagnostic Imaging, Temple University School of Medicine, Philadelphia, Pennsylvania

Donna Magid, M.D.
Associate Professor, The Russell H. Morgan Department of Radiology and Radiological Science, Department of Orthopaedic Surgery, The Johns Hopkins Medical Institutions, Baltimore, Maryland

Carlo Masciocchi, M.D.
Department of Radiology, University of L'Aquila, S. Maria di Collemaggio Hospital, L'Aquila, Italy

Susan J. F. Meyer, M.D.
Fellow, Department of Radiology, Hospital of the University of Pennsylvania, Philadelphia, Pennsylvania

Jerrold H. Mink, M.D.
Associate Clinical Professor of Radiology, University of California, Los Angeles; Chief, Division of Musculoskeletal Imaging, Cedars-Sinai Medical Center, Los Angeles, California

Donald G. Mitchell, M.D.
Associate Professor of Radiology, Director of Magnetic Resonance Imaging, Thomas Jefferson University, Philadelphia, Pennsylvania

Sheila G. Moore, M.D.
Assistant Professor of Radiology, Department of Diagnostic Radiology and Nuclear Medicine, Stanford University School of Medicine, Stanford, California

Timothy E. Moore, F.R.A.C.R.
Assistant Professor, Department of Radiology, The University of Iowa College of Medicine, Iowa City, Iowa

Mark D. Murphey, M.D.
Assistant Professor, Chief, Section of Musculoskeletal Radiology, Department of Diagnostic Radiology, University of Kansas Medical Center, Kansas City, Kansas

Wim R. Obermann, M.D.
Radiologist, Department of Radiology, University Hospital Leiden, Leiden, The Netherlands

David O'Keefe, M.D., M.R.C.P.I., F.R.C.R.
Skeletal Radiology Fellow, Department of Radiology, Massachusetts General Hospital and Harvard Medical School, Boston, Massachusetts

Roberto Passariello, M.D.
Professor and Chairman, Department of Radiology, University of L'Aquila, S. Maria di Collemaggio Hospital, L'Aquila, Italy

Mayur M. Patel, M.D.
Assistant Professor of Radiology, University of Vermont College of Medicine; Attending Radiologist, Medical Center Hospital of Vermont, Burlington, Vermont

Mini N. Pathria, M.D.
Assistant Professor, Departments of Radiology and Orthopaedics, Case Western Reserve University, University Hospitals of Cleveland, Cleveland, Ohio

Milan Pijl, M.D.
Department of Radiology, University Hospital Leiden, Leiden, The Netherlands

Vijay M. Rao, M.D.
Professor of Radiology and Otolaryngology, Department of Radiology, Thomas Jefferson University, Philadelphia, Pennsylvania

Monique Reijnierse, M.D.
Department of Radiology, University Hospital Leiden, Leiden, The Netherlands

Maximilian F. Reiser, M.D.
Professor and Chairman, Department of Radiology, University of Bonn, Bonn, Germany

Matthew D. Rifkin, M.D.
Professor and Chairman, Department of Radiology, Albany Medical College, Albany, New York

Daniel I. Rosenthal, M.D.
Director, Bone and Joint Radiology, Massachusetts General Hospital; Associate Professor of Radiology, Harvard Medical School, Boston, Massachusetts

Pieter M. Rozing, M.D.
Professor and Chairman, Department of Orthopaedics, University Hospital Leiden, Leiden, The Netherlands

Joel D. Rubenstein, M.D.
Associate Professor, Department of Radiological Sciences, University of Toronto, Sunnybrook Health Science Center, Toronto, Ontario, Canada

Jaap Schipper, M.D.
Department of Diagnostic Radiology, University Hospital Leiden, Leiden, The Netherlands

Heinrich Schüller, M.D.
Department of Radiology, University of Bonn, Bonn, Germany

Hassan S. Sharif, M.D., F.R.C.R.
Senior Consultant, Radiologist—Teaching, Department of Radiology, Riyadh Armed Forces Hospital, Riyadh, Saudi Arabia

Peter Steiger, Ph.D.
Assistant Professor of Radiology, Department of Radiology, University of California School of Medicine, San Francisco, California

Robert M. Steiner, M.D.
Professor of Radiology, Department of Radiology, Thomas Jefferson University, Phialdelphia, Pennsylvania

Anton H. M. Taminiau, M.D.
Department of Orthopaedics, University Hospital Leiden, Leiden, The Netherlands

Edwin van der Linden, M.D.
Department of Radiology, University Hospital Leiden, Leiden, The Netherlands

Jacques A. van Oostayen, M.D.
Department of Radiology, University Hospital Leiden, Leiden, The Netherlands

Ab Verbout, M.D.
Department of Anatomy and Orthopaedics, University Hospital Leiden, Leiden, The Netherlands

Louis H. Wetzel, M.D.
Assistant Professor, Department of Diagnostic Radiology, University of Kansas Medical Center, Kansas City, Kansas

William T. C. Yuh, M.D., M.S.E.E.
Associate Professor, Department of Radiology, The University of Iowa College of Medicine, Iowa City, Iowa

CONTENTS

SECTION ONE
Atlas of Normal Anatomy

SECTION TWO
Pathology

A. General Disorders

SECTION ONE

ATLAS OF
NORMAL ANATOMY

CHAPTER 1

The Temporomandibular Joint

J. Mark Fulmer and Steven E. Harms

Osseous Anatomy
Soft-Tissue Anatomy

Functional Anatomy
Summary

The temporomandibular joint (TMJ) is a complex articulation formed by the skull base, the mandible, and the interposed articular disk. Functional requirements and bilaterality are complicating factors unique to the TMJ. Each joint must provide rotational and translational motion, and must also work in concert with the mirror image contralateral joint. Consequently, diagnostic imaging of the TMJ is best performed when anatomic features of both joints are examined during the functional cycle.

Osseous Anatomy

The skeletal structure of the TMJ is provided by the mandible and the temporal bone. The mandibular condyle supplies a broad, convex articular surface, which is wide in the mediolateral dimension and relatively narrow in the

Abbreviation: TMJ, temporomandibular joint

anterior posterior dimension. The coronoid process, ramus, and body provide points of insertion for the muscles of mastication (Fig. 1.1). The condylar fossa is nestled in the undersurface of the temporal bone directly anterior to the external auditory canal (Fig. 1.2). This concave articular surface houses the mandibular condyle in its resting position. The anterior boundary of the fossa is the convex temporal eminence. Computed tomography (CT) defines the mandible as a high-density cortical margin surrounding the less dense marrow cavity. The petrous portion of the temporal bone displays a uniform high density (Fig. 1.3). Magnetic resonance imaging (MRI) displays compact cortical bone as signal void. The medullary portion of the mandible provides a higher signal due to the presence of marrow fat (Fig. 1.4). The articular surfaces of the condyle and fossa are covered with cartilage. This is not hyaline cartilage, but dense fibrous connective tissue with occasional cartilage cells (1–3). Such distinction is important, because dense fibrous tissue also manifests signal void on MR images while appearing as a thin soft-tissue density on CT images.

Soft-Tissue Anatomy

The keystone of TMJ soft-tissue anatomy is the articular disk. This is a resilient structure composed of dense fibrous tissue rich in proteoglycans (4, 5). The articular disk has three morphologically distinct areas. The central, or intermediate, zone is thin, and is bordered by the thicker

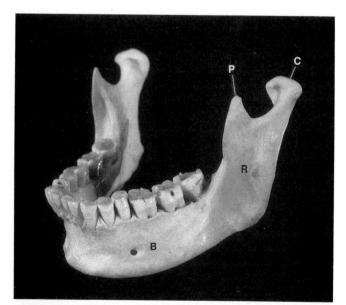

Figure 1.1. Disarticulated mandible. The condyle, *C,* is broad in mediolateral dimension. The coronoid process, *P,* ramus, *R,* and body, *B,* of mandible are labeled.

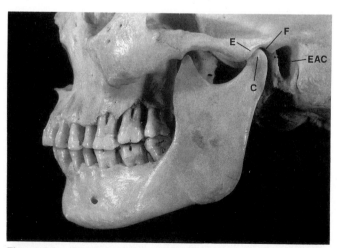

Figure 1.2. Skeletal anatomy of the TMJ. The condyle, *C,* rests in the condylar fossa, *F.* The fossa is bounded anteriorly by the temporal eminence, *E,* and posteriorly by the external auditory canal, *EAC.*

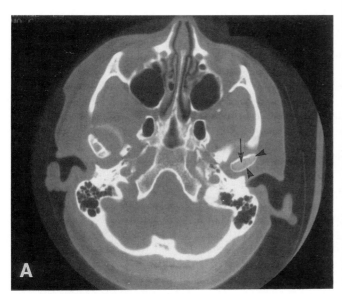

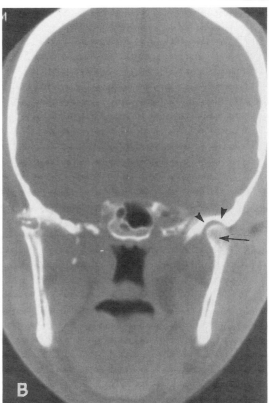

Figure 1.3. **A,** Axial CT image at the level of the TMJ. High-density cortical bone *(arrowhead)* surrounds the dense marrow cavity *(arrow)* of the mandibular condyle. **B,** Direct coronal CT at the level of the TMJ. The left condyle *(arrow)* rests in normal position relative to the dense condylar fossa *(arrowheads)*. Both images demonstrate evidence of previous surgery in the right TMJ.

anterior band and posterior band. Thus, the shape of the disk is reminiscent of an erythrocyte (Fig. 1.5). When viewed in the sagittal plane, the disk has a "bow tie" configuration (Fig. 1.6). The posterior band is slightly thicker than the anterior band. The disk, while firm, is deformable. However, it normally maintains its shape, and does

not act to "smooth out" irregularities in the surface of the condyle or fossa. The disk itself is not innervated, and has no direct vascular supply. The dense fibrous tissue of the disk is displayed as moderately dense soft tissue on CT images. It is more dense than the surrounding musculature and much less dense than the adjacent cortical bone. Direct evaluation of the disk itself may be difficult without specialized manipulation of the CT data (6). Spin-echo MRI sequences display the disk as low signal intensity. Some reports have indicated that the normal disk may contain a small area of high signal intensity when imaged with very short echo time (TE) gradient echo-pulse sequences (7, 8).

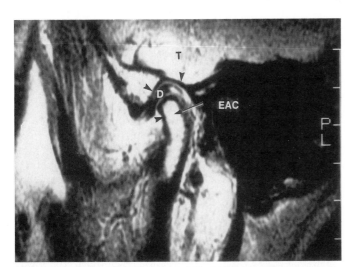

Figure 1.4. Normal saggital T1-weighted (300/16) MR image scan. The dense cortical bone of the condyle and condylar fossa demonstrates signal void *(arrowheads)*. The marrow cavity demonstrates increased signal due to marrow fat *(arrow)*. The articular disk, *D,* also shows low signal. External auditory canals, *EAC,* and temporal lobe, *T,* are labeled.

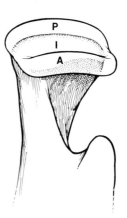

Figure 1.5. The articular disk. The normal disk has an erythroid shape with three distinct zones: the anterior band, *A,* intermediate zone, *I,* and posterior band, *P.*

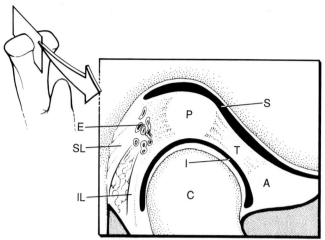

Figure 1.6. Temporomandibular joint anatomy. The following structures are labeled: *A*, anterior band of the disk; *I*, intermediate zone; *T₁*, posterior band; *S*, superior joint space; *I*, inferior joint space; *E*, endothelium-lined vascular spaces in the retrodiskal pad; *SL*, superior retrodiskal lamina; *IL*, inferior retrodiskal lamina; *C*, condyle.

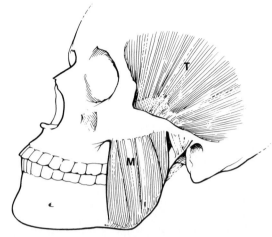

Figure 1.7. Muscles of mastication. The fan-shaped temporalis muscle, *T*, and the masseter muscle, *M*, are shown.

The disk is firmly attached to the condyle along its medial and lateral borders. The undersurface of the disk articulates with the condyle, while the superior surface articulates with the condylar fossa. The posterior band is attached to the tissue posterior to the condyle. This retrodiskal pad is a loosely organized collection of endothelium-lined spaces that is richly innervated. A fibrocartilaginous bilaminar zone attaches the posterior band of the disk to the squamotympanic fissure and to the posterior aspect of the condylar neck. Thus, there is freedom for anterior movement of the disk and condyle, with the bilaminar zone functioning as a limiting mechanism. The fibroelastic bilaminar tissue also serves to maintain the disk in normal position during the functional cycle. The disk is not firmly attached to the temporal bone. The anteromedial surface of the disk is the point of insertion for the superior head of the lateral pterygoid muscle.

The joint itself is defined by the joint capsule, which is attached to the rim of the temporal bone superiorly and to the condylar neck inferiorly. The capsule blends with the disk around its circumference, and contains nerve fibers that supply proprioceptive feedback from the joint. The disk divides the joint into distinct upper and lower compartments. The upper compartment extends anteriorly to the temporal eminence, while the lower compartment is tightly bound by the attachments of the disk to the condyle. The lateral joint capsule is a thick fibrous band termed the temporomandibular ligament. The capsular structures allow predominantly anterior and posterior motion. Lateral and medial motion occurs, but is restricted both by the capsule and by the presence of the contralateral joint.

The muscles of mastication provide motion at the temporomandibular joint. The broad, fan-shaped temporalis muscle originates on the temporal fossa and passes anterior to the TMJ to insert on the coronoid process of the mandible. Originating on the zygomatic arch, the masseter muscle inserts on the mandibular ramus (Fig. 1.7). The medial pterygoid passes from its origin on the lateral pterygoid plate of the sphenoid bone to its insertion on the lower medial ramus (Fig. 1.8). These muscles provide the

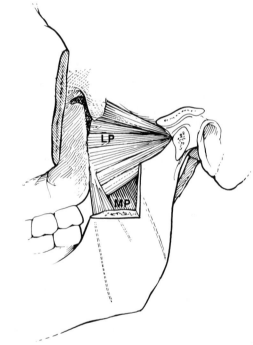

Figure 1.8. Muscles of mastication. The medial pterygoid muscle, *MP*, and the lateral pterygoid muscle, *LP*, are shown.

action of jaw closure. When the medial pterygoids are activated unilaterally, mediolateral mandibular motion is produced. This facilitates the grinding mechanism of mastication, but is not a part of passive jaw function. Opening the jaw is performed primarily by the inferior belly of the lateral pterygoid muscle. The inferior lateral pterygoid originates on the lateral pterygoid plate and inserts on the mandibular neck. The superior belly of the lateral pterygoid inserts on the articular disk, and is active during the power stroke of jaw closure. The superior lateral pterygoid supplies traction on the articular disk when the mandible is closed against resistance (9). This traction is counterbalanced by the pull of the retrodiskal laminae. Suprahyoid muscles (digastric, mylohyoid, and geniohyoid) contribute to jaw opening as well. The masticatory muscles are innervated predominantly by the mandibular branch of the trigeminal nerve (cranial nerve [CN] V), with

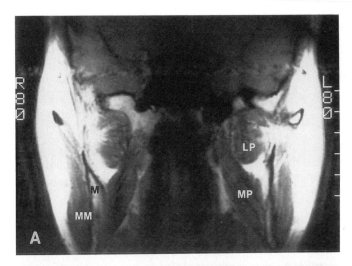

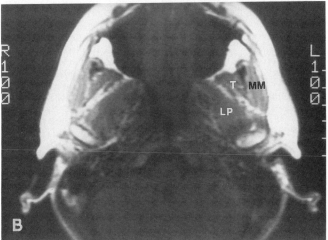

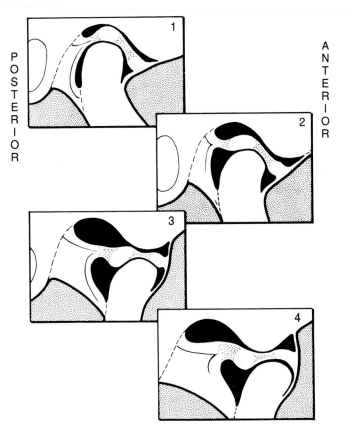

Figure 1.10. Normal TMJ movement during passive opening. In closed mouth position, the posterior band of disk lies directly superior to the condyle. With opening, the condyle rotates on the disk. Further opening results in anterior translation of the disk-condyle complex. With closing, these relations reverse.

Figure 1.9. **A,** Coronal T1-weighted (300/16) MR image of muscles of mastication. Medial pterygoid, *MP,* and lateral pterygoid, *LP,* are shown. The masseter, *MM,* lies outside the mandible, *M.* **B,** Axial T1-weighted (300/16) MR image shows lateral pterygoid, *LP,* masseter, *MM,* and inferior aspect of the temporalis, *T.*

part of the digastric receiving facial innervation (CN VII), and the geniohyoid receiving CN I innervation (10).

Computed tomography displays the masticatory muscles as homogeneous, moderate density. Magnetic resonance imaging demonstrates moderate signal intensity from the muscles of mastication on T1 and proton density-weighted images (Fig. 1.9), and very low intensity on T2-weighted images. The small amount of fat between the fascicles of the muscle provides streaks of increased signal intensity on T1-weighted images that have diminished intensity on the T2-weighted sequences (11). The moderate signal intensity musculature is easily delineated from the low signal intensity disk.

Functional Anatomy

The TMJ is a combined articulation, having both hinge and glide functions. The initial action at the joint is rotational, as the mandibular condyles rotate about a horizontal axis that passes through both joints. This action occurs within the lower joint space. Subsequently, with greater degrees of mouth opening, the condyle and disk translate

forward beneath the articular eminence. Rotation is thus the predominant action of the lower joint space, while translation is the predominant action of the upper joint space (Fig. 1.10). In the resting position, the disk is interposed between the condyle and the fossa, with the posterior band at the most superior portion of the condyle. As the jaw is opened, the disk shifts posteriorly. During the middle portion of the opening cycle the intermediate zone lies superiorly and acts as the articular surface during most of the translational motion. The disk finishes the opening phase by coming to lie with its anterior band between the condyle and the eminence (Fig. 1.11). The forward motion of the disk is probably passive, as the superior head of the lateral pterygoid is inactive during most of the opening cycle. The change in position of the disk is related to the change in the weight-bearing position of the joint. The return of the disk is also passive, but is facilitated by the elastic recoil of the bilaminar zone in the retrodiskal tissues. The bilaminar zone never serves as a force-bearing tissue during normal TMJ function. Since the disk is firmly attached to the condyle by the capsule, the disk and condyle translate as a unit, the disk-condyle complex. Rotation is the only normal motion at the condyle-disk interface. The functional cycle of the temporomandibular joint is a smooth rotation of the condyle, followed by smooth translation of the disk-condyle complex (12, 13). These actions occur bilaterally and symmetrically. Thus, the man-

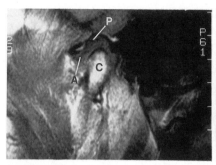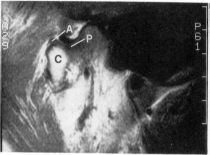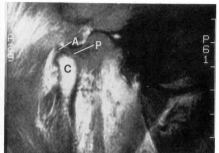

Figure 1.11. Static MR images (300/15) in closed mouth, *A,* mid-cycle, *B,* and open mouth, *C,* positions. Disk movement is demonstrated as the condyle rotates, and the disk-condyle complex translates beneath the eminence. The posterior band, *P,* anterior band, *A,* and condyle, *C,* are labeled.

dible maintains a midline position throughout opening and closing. There should be no palpable or audible "clicks" at the joint, and no limitation of rotation or translation.

Summary

The temporomandibular joint is a complicated structure that acts as the fulcrum for mandibular motion. Modern imaging techniques clearly display the osseous, soft tissue, and functional anatomy of the TMJ. As more thorough anatomic understanding is combined with increased clinical awareness, TMJ disorders can be more accurately diagnosed and properly treated.

References

1. Murphy WA. The temporomandibular joint. In: Resnick D, Niwayama G, eds. Diagnosis of bone and joint disorders. 2nd ed. Philadelphia: W. B. Saunders, 1988:1816–1863.
2. Bell WE. Temporomandibular disorders: classification, diagnosis, management. 2nd ed. Chicago: Year Book Medical Publishers, 1986.
3. Rees LA. The structure and function of the mandibular joint. Br Dent J 1954;96:125–131.
4. Dolwick MF. The temporomandibular joint: normal and abnormal anatomy. In: Helms CA, Katzberg RW, Dolwick MF, eds. Internal derangements of the temporomandibular joints. San Francisco: Radiology Research and Education Foundation, 1983.
5. Scapino RP. Histopathology associated with malposition of the human temporomandibular joint disc. Oral Surg 1983;55:382–397.
6. Christiansen, EL, Thompson JR, Hasso AN, et al. CT number characteristics of malpositioned TMJ menisci: diagnosis with CT number highlighting (blinkmode). Invest Radiol 1987;22:315–321.
7. Helms CA, Kaban LB, McNeill C. Temporomandibular joint: morphology and signal intensity characteristics of the disk at MR imaging. Radiology 1989;172:817–820.
8. Port RB, Mikhael MA, Canzona JE. MR imaging of the temporomandibular joint: correlation with operative findings [abstract]. Radiology 1988;169(P):440.
9. Mahan PE, Wilkinson TM, Gibbs CH, et al. Superior and inferior bellies of the lateral pterygoid muscle EMG activity at basic jaw positions. J Prosth Dent 1983;50:710–719.
10. Thilander B. Innervation of the temporomandibular disc in man. Acta Odontol Scand 1964;22:151–160.
11. Schellhas KP. MR imaging of muscles of mastication. AJR 1989;153:847–855.
12. Isberg-Holm A, Ivarsson R. The movement pattern of the mandibular condyles in individuals with and without clicking. Dentomaxillofac Radiol 1980;9:59–65.
13. Fulmer JM, Harms SE. The temporomandibular joint. Topics Magn Reson Imag 1989;1:75–84.

CHAPTER 2

The Shoulder

Milan Pijl, Edwin van der Linden, and Ab Verbout

Anatomy of the Shoulder

Bone and Cartilage

1 Clavicle
2 (Shaft of) Humerus
3 Epiphyseal Line
4 Greater Tubercle of Humerus
5 Head of Humerus
6 Intertubercular Groove
7 Lesser Tubercle of Humerus
8 Rib
9 Acromion
10 Body of Scapula
11 Coracoid (Process)
12 Glenoid
13 Inferior Angle of Scapula
14 Lateral Border of Scapula
15 Neck of Scapula
16 Spine of Scapula
17 Superior Angle of Scapula
18 Glenoid Labrum
19 Articular (Hyaline) Cartilage
20 Articular Capsule
21 Joint Cavity
22 Acromioclavicular Joint
23 Glenohumeral Joint
24 Axillary Recess
25 Subcoracoid Recess
26 Subscapularis Recess

Muscle

31 Biceps Muscle (Long Head)
32 Coracobrachialis/Biceps Muscle (Short Head)
33 Triceps Muscle (Lateral Head)
34 Triceps Muscle (Long Head)
35 Triceps Muscle (Medial Head)
36 Triceps Muscle
37 Pectoralis Major Muscle
38 Pectoralis Minor Muscle
39 Serratus Anterior Muscle
40 Subclavius Muscle
41 Iliocostal Muscle
42 Latissimus Dorsi Muscle
43 Levator Scapulae Muscle
44 Longissimus Dorsi Muscle
45 Rhomboid Major Muscle
46 Rhomboid Minor Muscle
47 Trapezius Muscle
48 Rotator Cuff
49 Deltoid Muscle
50 Infraspinatus Muscle
51 Subscapularis Muscle
52 Supraspinatus Muscle
53 Teres Major Muscle
54 Teres Minor Muscle
55 Intercostal Muscle
56 Scalenus Medius Muscle

Ligament and Tendon

71 Acromioclavicular Ligament
72 Coracoclavicular Ligament, Conoid
73 Coracoclavicular Ligament, Trapezoid
74 Coracohumeral Ligament

Neurovascular and Miscellaneous

91 Neurovascular Bundle
92 Anterior Circumflex Humeral Vessels
93 Axillary Artery
94 Axillary Vein
95 Brachial Vessels
96 Cephalic Vein
97 Circumflex Scapular Artery and Branches
98 Posterior Circumflex Humeral Vessels
99 Subcutaneous Vessels
100 Subscapular Vessels and Branches
101 Suprascapular Vessels and Branches
102 Thoracodorsal Artery
103 Transverse Cervical (Colli) Artery
104 Dorsal Scapular Artery
105 Intermuscular Septum
106 Lung
107 Lymph Node
108 Quadrangular Space
109 Subacromioclavicular Fat Pad
110 Subcutaneous or Axillary Fat

Abbreviation: t, tendon.

Acknowledgments: Anatomic drawings were made by J. Wetselaar-Whittaker and Bas B. Blankevoort. The assistance of Jacques van Oostaijen, M.D., Department of Radiology, University Hospital Leiden, is gratefully acknowledged.

Figures 2.1–2.4.   Axial MR images.

Figure 2.1.

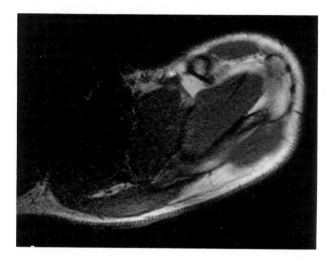

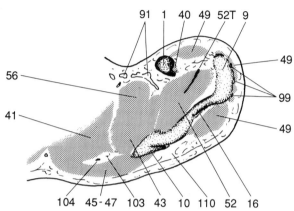

Figure 2.2.

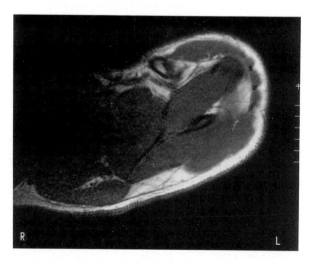

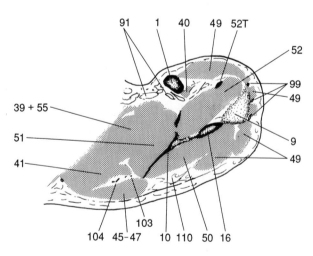

Bone and Cartilage
1 Clavicle
9 Acromion
10 Body of Scapula
16 Spine of Scapula

Muscle
39 Serratus Anterior Muscle

40 Subclavius Muscle
41 Iliocostal Muscle
43 Levator Scapulae Muscle
45 Rhomboid Major Muscle
46 Rhomboid Minor Muscle
47 Trapezius Muscle
49 Deltoid Muscle

50 Infraspinatus Muscle
51 Subcapularis Muscle
52 Supraspinatus Muscle
55 Intercostal Muscle
56 Scalenus Medius Muscle

Neurovascular and Miscellaneous
91 Neurovascular Bundle

99 Subcutaneous Vessels
103 Transverse Cervical (Colli) Artery
104 Dorsal Scapular Artery
110 Subcutaneous or Axillary Fat

Figure 2.3.

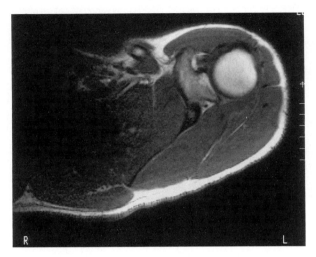

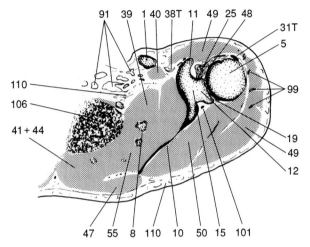

Figure 2.4.

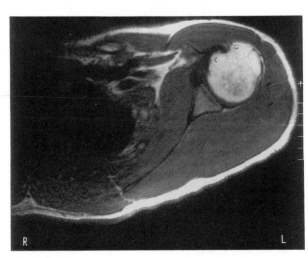

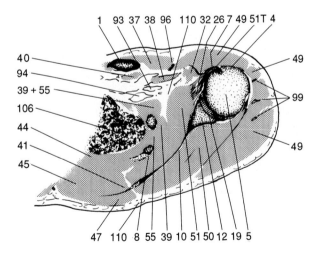

Bone and Cartilage

1 Clavicle
4 Greater Tubercle of Humerus
5 Head of Humerus
7 Lesser Tubercle of Humerus
8 Rib
10 Body of Scapula
11 Coracoid (Process)
12 Glenoid
15 Neck of Scapula
19 Articular (Hyaline) Cartilage
25 Subcoracoid Recess
26 Subscapularis Recess

Muscle

31 Biceps Muscle (Long Head)
32 Coracobrachialis/Biceps Muscle (Short Head)
37 Pectoralis Major Muscle
38 Pectoralis Minor Muscle
39 Serratus Anterior Muscle
40 Subclavius Muscle
41 Iliocostal Muscle
44 Longissimus Dorsi Muscle
45 Rhomboid Major Muscle
47 Trapezius Muscle
48 Rotator Cuff
49 Deltoid Muscle
50 Infraspinatus Muscle
51 Subscapularis Muscle
55 Intercostal Muscle

Neurovascular and Miscellaneous

91 Neurovascular Bundle
93 Axillary Artery
94 Axillary Vein
96 Cephalic Vein
99 Subcutaneous Vessels
101 Suprascapular Vessels and Branches
106 Lung
110 Subcutaneous or Axillary Fat

Figure 2.5. **A** and **B,** CT images made in the supine position after intraarticular injection of 1.5 ml of contrast agent and 10 ml of air. **C,** CT-arthrography displaying the posterior part of the joint with the posterior labrum. Patient is in prone position with arm in external rotation.

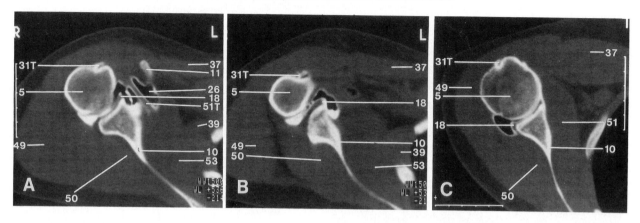

Figures 2.6–2.10. Coronal oblique MR images, made with arm in neutral position.

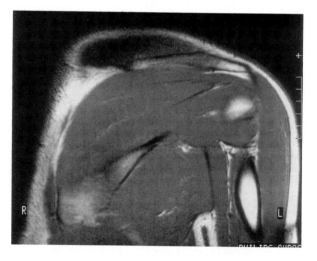

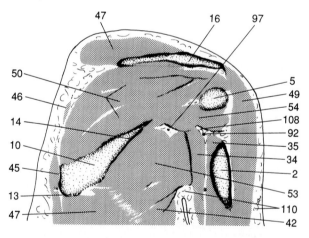

Bone and Cartilage
2 (Shaft of) Humerus
5 Head of Humerus
10 Body of Scapula
11 Coracoid (Process)
13 Inferior Angle of Scapula
14 Lateral Border of Scapula
16 Spine of Scapula

18 Glenoid Labrum
26 Subscapularis Recess

Muscle
31 Biceps Muscle (Long Head)
34 Triceps Muscle (Long Head)
35 Triceps Muscle (Medial Head)
37 Pectoralis Major Muscle
39 Serratus Anterior Muscle

42 Latissimus Dorsi Muscle
45 Rhomboid Major Muscle
46 Rhomboid Minor Muscle
47 Trapezius Muscle
49 Deltoid Muscle
50 Infraspinatus Muscle
51 Subscapularis Muscle
53 Teres Major Muscle
54 Teres Minor Muscle

Neurovascular and Miscellaneous
92 Anterior Circumflex Humeral Vessels
97 Circumflex Scapular Artery and Branches
108 Quadrangular Space
110 Subcutaneous or Axillary Fat

Figure 2.7.

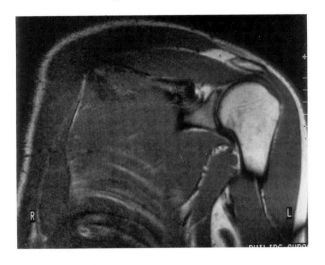
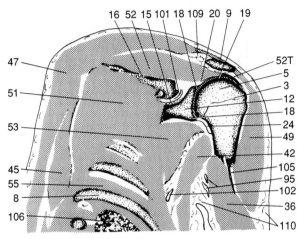

Figure 2.8.

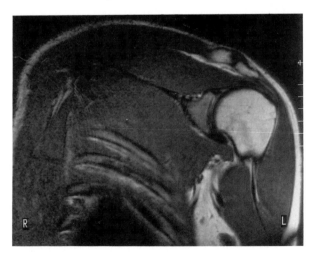
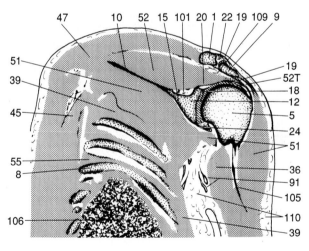

Bone and Cartilage
1 Clavicle
3 Epiphyseal Line
5 Head of Humerus
8 Rib
9 Acromion
10 Body of Scapula
12 Glenoid
15 Neck of Scapula
16 Spine of Scapula
18 Glenoid Labrum
19 Articular (Hyaline) Cartilage
20 Articular Capsule
22 Acromioclavicular Joint
24 Axillary Recess

Muscle
36 Triceps Muscle
39 Serratus Anterior Muscle
42 Latissimus Dorsi Muscle
45 Rhomboid Major Muscle
47 Trapezius Muscle
49 Deltoid Muscle
51 Subscapularis Muscle
52 Supraspinatus Muscle
53 Teres Major Muscle
55 Intercostal Muscle

Neurovascular and Miscellaneous
91 Neurovascular Bundle
95 Brachial Vessels
101 Suprascapular Vessels and
     Branches
102 Thoracodorsal Artery
105 Intermuscular Septum
106 Lung
109 Subacromioclavicular Fat Pad
110 Subcutaneous or Axillary Fat

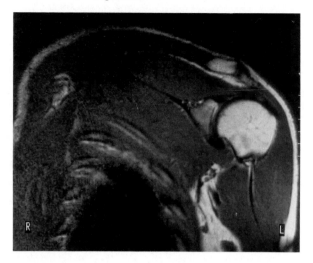

Figure 2.9.

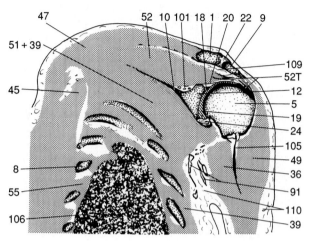

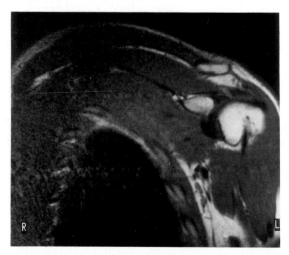

Figure 2.10.

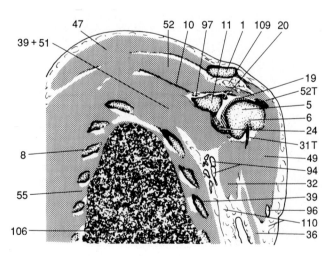

Bone and Cartilage
1 Clavicle
5 Head of Humerus
6 Intertubercular Groove
8 Rib
9 Acromion
10 Body of Scapula
11 Coracoid (Process)
12 Glenoid
18 Glenoid Labrum

19 Articular (Hyaline) Cartilage
20 Articular Capsule
22 Acromioclavicular Joint
24 Axillary Recess

Muscle
31 Biceps Muscle (Long Head)
32 Corcobrachialis/Biceps Muscle
   (Short Head)
36 Triceps Muscle

39 Serratus Anterior Muscle
45 Rhomboid Major Muscle
47 Trapezius Muscle
49 Deltoid Muscle
51 Subscapularis Muscle
52 Supraspinatus Muscle
55 Intercostal Muscle

Neurovascular and Miscellaneous
91 Neurovascular Bundle

94 Axillary Vein
96 Cephalic Vein
97 Circumflex Scapular Artery and
   Branches
101 Suprascapular Vessels and
   Branches
105 Intermuscular Septum
106 Lung
109 Subacromioclavicular Fat Pad
110 Subcutaneous or Axillary Fat

Figures 2.11–2.13. Coronal oblique MR images made with arm in external rotation.

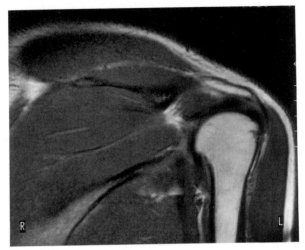

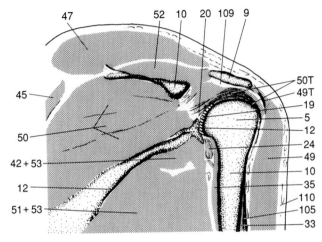

Figure 2.12.

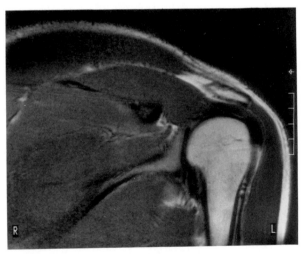

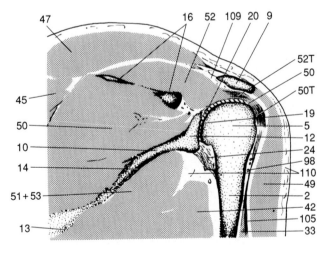

Bone and Cartilage
2   (Shaft of) Humerus
5   Head of Humerus
9   Acromion
10 Body of Scapula
12 Glenoid
13 Inferior Angle of Scapula
14 Lateral Border of Scapula

16 Spine of Scapula
19 Articular (Hyaline) Cartilage
20 Articular Capsule
24 Axillary Recess

Muscle
33 Triceps Muscle
35 Triceps Muscle (Medial Head)

42 Latissimus Dorsi Muscle
45 Rhomboid Major Muscle
47 Trapezius Muscle
49 Deltoid Muscle
50 Infraspinatus Muscle
51 Subscapularis Muscle
52 Supraspinatus Muscle
53 Teres Major Muscle

Neurovascular and Miscellaneous
98   Posterior Circumflex Humeral
      Vessels
105 Intermuscular Septum
109 Subacromioclavicular Fat Pad
110 Subcutaneous or Axillary Fat

Figure 2.13.

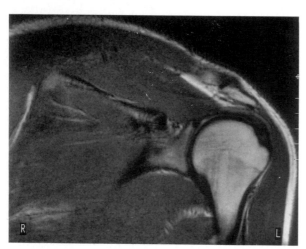

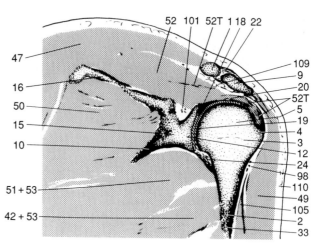

Figures 2.14–2.17. Sagittal oblique MR images.

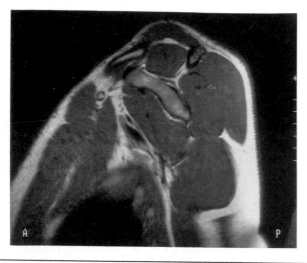

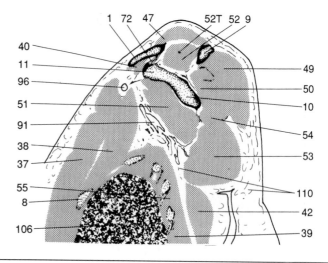

Bone and Cartilage
1 Clavicle
2 (Shaft of) Humerus
3 Epiphyseal Line
4 Greater Tubercle of Humerus
5 Head of Humerus
8 Rib
9 Acromion
10 Body of Scapula
11 Coracoid (Process)
12 Glenoid
15 Neck of Scapula
16 Spine of Scapula
18 Glenoid Labrum
19 Articular (Hyaline) Cartilage
20 Articular Capsule
22 Acromioclavicular Joint
24 Axillary Recess

Muscle
33 Triceps Muscle (Lateral Heaed)
37 Pectoralis Major Muscle
38 Pectoralis Minor Muscle
39 Serratus Anterior Muscle
40 Subclavius Muscle
42 Latissimus Dorsi Muscle
47 Trapezius Muscle
49 Deltoid Muscle
50 Infraspinatus Muscle
51 Subscapularis Muscle
52 Supraspinatus Muscle
53 Teres Major Muscle
54 Teres Minor Muscle
55 Intercostal Muscle

Ligament and Tendon
72 Coracoclavicular Ligament, Conoid

Neurovascular and Miscellaneous
91 Neurovascular Bundle
96 Cephalic Vein
98 Posterior Circumflex Humeral Vessels
101 Suprascapular Vessels and Branches
105 Intermuscular Septum
106 Lung
109 Subacromioclavicular Fat Pad
110 Subcutaneous or Axillary Fat

Figure 2.15.

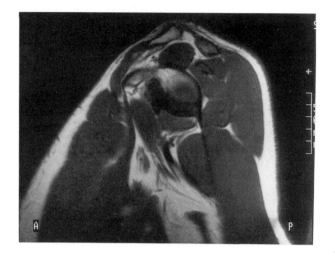

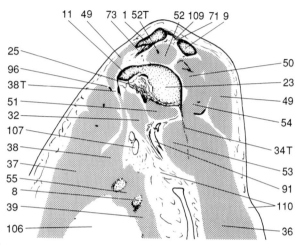

Figure 2.16.

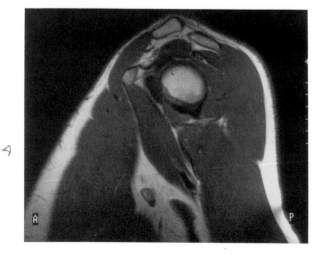

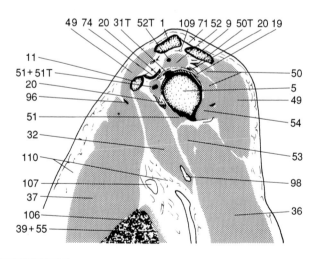

Bone and Cartilage
1   Clavicle
5   Head of Humerus
8   Rib
9   Acromion
11 Coracoid (Process)
19 Articular (Hyaline) Cartilage
20 Articular Capsule
23 Glenohumeral Joint
25 Subcoracoid Recess

Muscle
31 Biceps Muscle (Long Head)
32 Coracobrachialis/Biceps Muscle (Short Head)
34 Triceps Muscle (Long Head)
36 Triceps Muscle
37 Pectoralis Major Muscle
38 Pectoralis Minor Muscle
39 Serratus Anterior Muscle
49 Deltoid Muscle
50 Infraspinatus Muscle
51 Subscapularis Muscle
52 Supraspinatus Muscle
53 Teres Major Muscle
54 Teres Minor Muscle
55 Intercostal Muscle

Ligament and Tendon
71 Acromioclavicular Ligament
73 Coracoclavicular Ligament, Trapezoid
74 Coracohumeral Ligament

Neurovascular and Miscellaneous
91 Neurovascular Bundle
96 Cephalic Vein
98 Posterior Circumflex Humeral Vessels
106 Lung
107 Lymph Node
109 Subacromioclavicular Fat Pad
110 Subcutaneous or Axillary Fat

Figure 2.17.

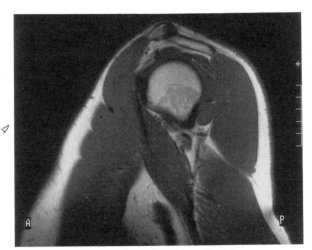

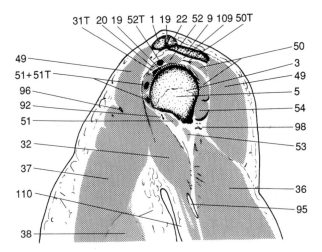

CHAPTER 3

The Elbow

Monique Reijnierse, Herman M. Kroon, and Ab Verbout

Anatomy of the Elbow Joint

Bone and Cartilage

1 Ulna
2 Olecranon
3 Trochlear Notch
4 Coronoid Process
5 Radius
6 Radial Head
7 Humerus
8 Lateral Epicondyle
9 Medial Epicondyle
10 Capitulum
11 Trochlea
12 Joint Fissure
13 Cartilaginous Joint Surface

Muscle

31 Anconeus Muscle
32 Extensor Carpi Ulnaris Muscle
33 Extensor Digitorum Muscle
34 Supinator Muscle
35 Extensor Carpi Radialis Longus Muscle
36 Extensor Carpi Radialis Brevis Muscle

37 Common Head of Extensor Muscles
38 Brachioradialis Muscle
39 Pronator Teres Muscle
40 Flexor Carpi Radialis Muscle
41 Palmaris Longus Muscle
42 Flexor Digitorum Superficialis Muscle
43 Flexor Digitorum Profundus Muscle
44 Flexor Carpi Ulnaris Muscle
45 Brachialis Muscle
46 Biceps (Brachii) Muscle
47 Triceps (Brachii) Muscle
48 Triceps (Brachii) Muscle (Medial Head)
49 Triceps (Brachii) Muscle (Lateral Head)

Ligament and Tendon

71 Lateral Collateral Ligament
72 Medial Collateral Ligament
73 Annular Ligament

Neurovascular and Miscellaneous

91 Radial Artery
92 Radial Recurrent Artery

93 Radial Vein
94 Ulnar Artery
95 Ulnar Recurrent Artery
96 Ulnar Vein
97 Brachial Artery
98 Brachial Vein
99 Median Cubital Vein
100 Cephalic Vein
101 Basilic Vein
102 Communicating Veins
103 Ulnar Nerve
104 Radial Nerve
105 Median Nerve
106 Cutaneous Nerve
107 Cutaneous Vein
108 Antebrachial Cutaneous Nerve
109 Coronoid Fossa
110 Olecranon Fossa
111 Olecranon Bursa
112 Anterior Fat Pad
113 Posterior Fat Pad

Abbreviation: t, tendon.

Acknowledgments: Anatomic drawings were made by Bas
B. Blankevoort and J. Wetselaar-Whittaker.

Figures 3.1–3.6 Axial MR images.

Figure 3.1.

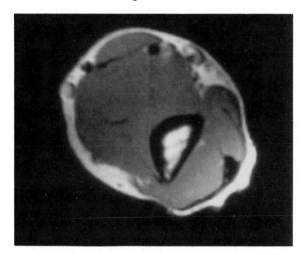

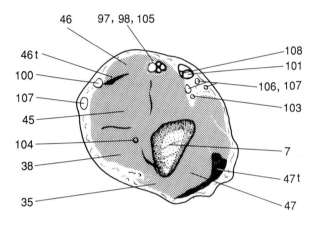

Figure 3.2.

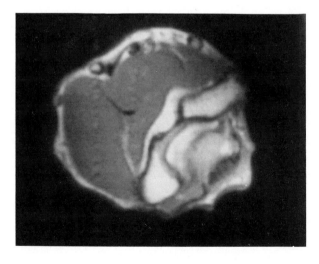

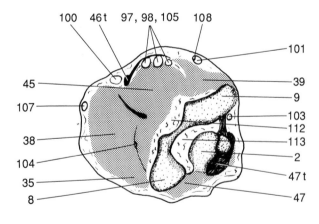

Bone and Cartilage
2 Olecranon
7 Humerus
8 Lateral Epicondyle
9 Medial Epicondyle

Muscle
35 Extensor Carpi Radialis Longus
    Muscle
38 Brachioradialis Muscle
39 Pronator Teres Muscle
45 Brachialis Muscle
46 Biceps (Brachii) Muscle
47 Triceps (Brachii) Muscle

Neurovascular and Miscellaneous
97 Brachial Artery
98 Brachial Vein
100 Cephalic Vein
101 Basilic Vein
103 Ulnar Nerve
104 Radial Nerve

105 Median Nerve
106 Cutaneous Nerve
107 Cutaneous Vein
108 Antebrachial Cutaneous Nerve
112 Anterior Fat Pad
113 Posterior Fat Pad

Figure 3.3.

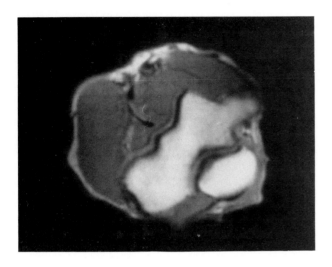

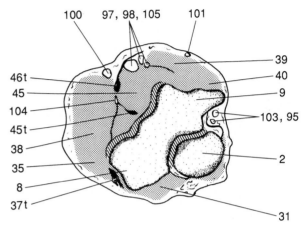

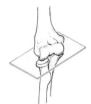

Figure 3.4.

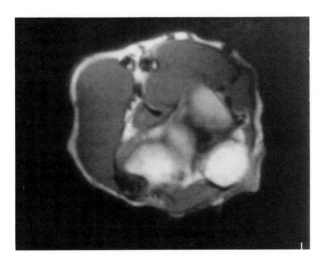

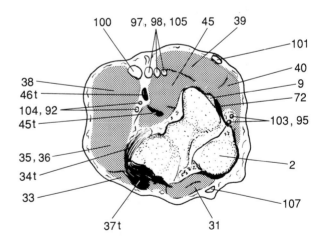

Bone and Cartilage
2 Olecranon
8 Lateral Epicondyle
9 Medial Epicondyle

Muscle
31 Anconeus Muscle
33 Extensor Digitorum Muscle
34 Supinator Muscle

35 Extensor Carpi Radialis Longus Muscle
36 Extensor Carpi Radialis Brevis Muscle
37 Common Head of Extensor Muscles
38 Brachioradialis Muscle
39 Pronator Teres Muscle

40 Flexor Carpi Radialis Muscle
45 Brachialis Muscle
46 Biceps (Brachii) Muscle

Ligament and Tendon
72 Medial Collateral Ligament

Neurovascular and Miscellaneous
92 Radial Recurrent Artery
95 Ulnar Recurrent Artery

97 Brachial Artery
98 Brachial Vein
100 Cephalic Vein
101 Basilic Vein
103 Ulnar Nerve
104 Radial Nerve
105 Median Nerve
107 Cutaneous Nerve

Figure 3.5.

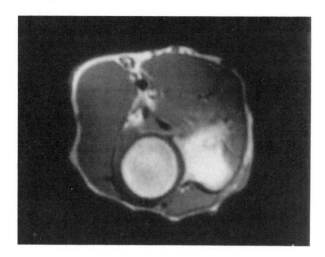

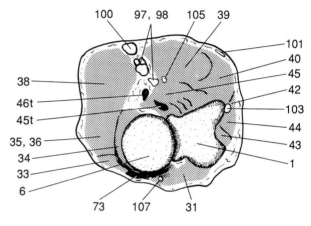

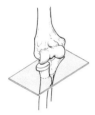

Figure 3.6.

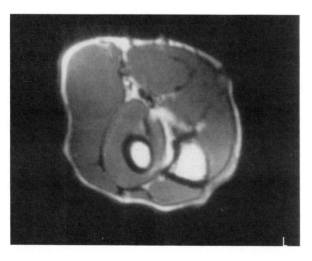

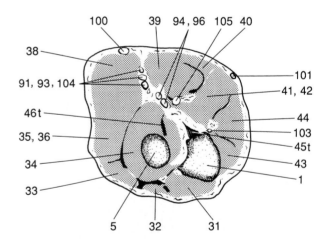

Bone and Cartilage
1 Ulna
5 Radius
6 Radial Head

Muscle
31 Anconeus Muscle
32 Extensor Carpi Ulnaris Muscle
33 Extensor Digitorum Muscle
34 Supinator Muscle
35 Extensor Carpi Radialis Longus Muscle

36 Extensor Carpi Radialis Brevis Muscle
38 Brachioradialis Muscle
39 Pronator Teres Muscle
40 Flexor Carpi Radialis Muscle
41 Palmaris Longus Muscle
42 Flexor Digitorum Superficialis Muscle
43 Flexor Digitorum Profundus Muscle

44 Flexor Carpi Ulnaris Muscle
45 Brachialis Muscle
46 Biceps (Brachii) Muscle

Ligament and Tendon
73 Annular Ligament

Neurovascular and Miscellaneous
91 Radial Artery
93 Radial Vein
94 Ulnar Artery

96 Ulnar Vein
97 Brachial Artery
98 Brachial Vein
100 Cephalic Vein
101 Basilic Vein
103 Ulnar Nerve
104 Radial Nerve
105 Median Nerve
107 Cutaneous Vein

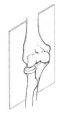

Figures 3.7–3.10. Coronal MR images.

Figure 3.7.

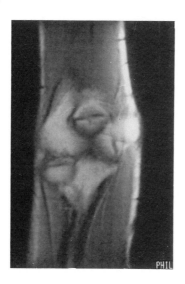
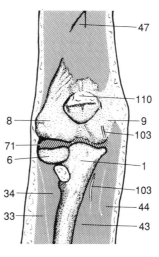

Figure 3.8.

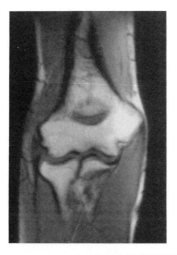
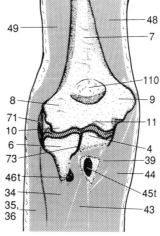

Bone and Cartilage
1 Ulna
4 Coronoid Process
6 Radial Head
7 Humerus
8 Lateral Epicondyle
9 Medial Epicondyle
10 Capitulum
11 Trochlea

Muscle
33 Extensor Digitorum Muscle
34 Supinator Muscle
35 Extensor Carpi Radialis Longus Muscle
36 Extensor Corpi Radialis Brevis Muscle
39 Pronator Teres Muscle
43 Flexor Digitorum Profundus Muscle

44 Flexor Carpi Ulnaris Muscle
45 Brachialis Muscle
46 Biceps (Brachii) Muscle
47 Triceps (Brachii) Muscle
48 Triceps (Brachii) Muscle (Medial Head)
49 Triceps (Brachii) Muscle (Lateral Head)

Ligament and Tendon
71 Lateral Collateral Ligament
73 Annular Ligament

Neurovascular and Miscellaneous
103 Ulnar Nerve
110 Olecranon Fossa

Figure 3.9.

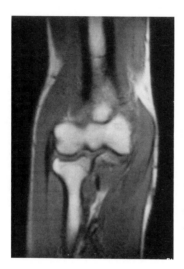

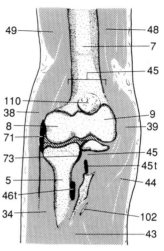

49 — 48
— 7
— 45
110 —
38 — — 9
8 — — 39
71 —
73 — — 45
— 45t
5 — — 44
46t —
34 — — 102
— 43

Figure 3.10.

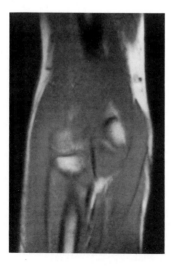

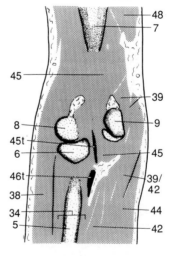

— 48
— 7
45 —
— 39
8 — — 9
45t —
6 — — 45
46t —
— 39/
42
38 —
34 — — 44
5 — — 42

Bone and Cartilage
5   Radius
6   Radial Head
7   Humerus
8   Lateral Epicondyle
9   Medial Epicondyle

Muscle
34 Supinator Muscle
38 Brachioradialis Muscle
39 Pronator Teres Muscle
42 Flexor Digitorum Superficialis
    Muscle
43 Flexor Digitorum Profundus
    Muscle

44 Flexor Carpi Ulnaris Muscle
45 Brachialis Muscle
46 Biceps (Brachii) Muscle
48 Triceps (Brachii) Muscle (Medial
    Head)
49 Triceps (Brachii) Muscle (Lat-
    eral Head)

Ligament and Tendon
71 Lateral Collateral Ligament
73 Annular Ligament

Neurovascular and Miscellaneous
102 Communicating Veins
110 Olecranon Fossa

Figures 3.11–3.14. Sagittal MR images.

Figure 3.11.

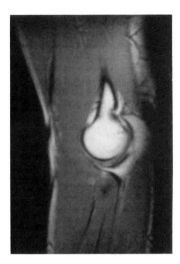 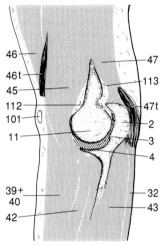

Figure 3.12.

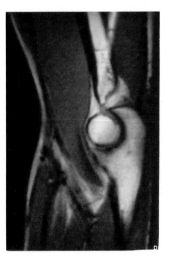 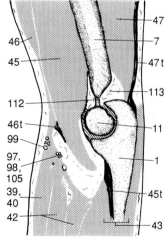

Bone and Cartilage
1 Ulna
2 Olecranon
3 Trochlear Notch
4 Coronoid Process
7 Humerus
11 Trochlea

Muscle
32 Extensor Carpi Ulnaris Muscle
39 Pronator Teres Muscle
40 Flexor Carpi Radialis Muscle
42 Flexor Digitorum Superficialis Muscle
43 Flexor Digitorum Profundus Muscle

45 Brachialis Muscle
46 Biceps (Brachii) Muscle
47 Triceps (Brachii) Muscle

Neurovascular and Miscellaneous
97 Brachial Artery
98 Brachial Vein
99 Median Cubital Vein

101 Basilic Vein
105 Median Nerve
112 Anterior Fat Pad
113 Posterior Fat Pad

Figure 3.13.

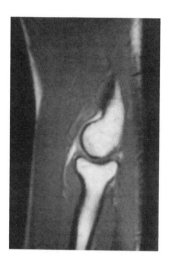
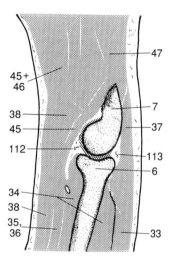

Figure 3.14.

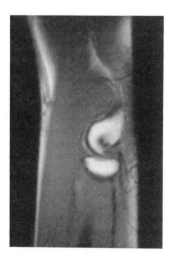
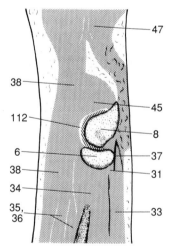

Bone and Cartilage
6   Radial Head
7   Humerus
8   Lateral Epicondyle

Muscle
31   Anconeus Muscle
33   Extensor Digitorum Muscle
34   Supinator Muscle
35   Extensor Carpi Radialis Longus
      Muscle

36   Extensor Carpi Radialis Brevis
      Muscle
37   Common Head of Extensor
      Muscles
38   Brachioradialis Muscle

45   Brachialis Muscle
47   Triceps (Brachii) Muscle

Neurovascular and Miscellaneous
112   Anterior Fat Pad
113   Posterior Fat Pad

CHAPTER 4

The Wrist and Hand

Willem R. Obermann, Monique Reijnierse, and Edwin van der Linden

---

Technique
General Anatomy

Volar Side
Dorsal Side

Technique

For precise anatomical correlation four fresh cadavers were examined with magnetic resonance (MR) imaging and subsequent sectioning. The radiocarpal joint was injected with a silicone rubber compound mixed with barium sulfate powder and methylene blue. One the MR images this compound emits no signal (is black). The hands were fixed in a wooden box in a semipronation position with the third metacarpal parallel or colinear with the radius in the lateral and frontal plane, and embedded in polyurethane foam. The position of the hand in the foam block was verified with radiographs.

Magnetic resonance images were made with a 1.5-T Philips gyroscan (Shelton, Connecticut). A circular surface coil with a diameter of 15 cm was used. Spin-echo (SE) images were made with a repetition time (TR) of 550 to 770 ms and an echo time (TE) of 30 ms. The acquisition matrix was 180×256 when four measurements per data line were taken, and 256×256 when two measurements were used. The field of view was 200 mm. Slices were interleaved, and slice thickness was 3 mm. The phase-encoding gradient was in the left-to-right direction. Subchondral bone may appear thickened in the frequency-encoding direction due to chemical shift artifacts.

The tissue blocks were subsequently deep frozen ($-80°C$) and sectioned with a 0.7-mm-thick bandsaw. The cryosections were 3 to 4 mm thick. The photographs of the cutting surfaces best corresponding to the MR image were selected.

Figures 4.1 to 4.5 show the axial plane from the distal forearm to the proximal metacarpal area of the same specimen. Figures 4.6 to 4.8 show coronal sections of two specimens, and Figures 4.9 to 4.11 show the sagittal sections. The metacarpal region is displayed in Figures 4.12 to 4.14. Eight MR images of a normal volunteer are demonstrated in Figures 4.15 to 4.22.

General Anatomy

The forearm is linked to the carpus by the radiocarpal and ulnocarpal ligaments (extrinsic ligaments). The distal radius and ulna form the distal radioulnar joint, separated from the radiocarpal joint by a discus or triangular pad of fibrocartilage (Figs. 4.2, 4.6, and 4.11). In this joint pronation and supination take place. The carpal bones are also connected to each other by the intercarpal ligaments (intrinsic) and are connected to the metacarpals by the carpometacarpal ligaments. The proximal carpal row, consisting of scaphoid, lunate, triquetrum, and pisiform bones, articulates with the forearm (except the pisiform) and forms with it the radiocarpal joint. The distal carpal row, consisting of the trapezium, trapezoid, capitate, and hamate bones, articulates with the proximal carpal row and forms with it the midcarpal joint. Both joints are responsible for wrist motion. The distal carpal row is connected with the metacarpal bones. The second and third metacarpal bones form a multifaceted joint and interlock with the trapezoid and capitate, which allows almost no movement. The joint surfaces between the fourth and fifth metacarpals and the hamate are flatter, allowing more movement, and the joint surfaces between the first metacarpal and the trapezium are more saddle shaped, allowing a considerable amount of motion. The phalanges are connected with the metacarpal bones by a ligamentous capsular system, the volar side of which consists of a thick reinforcement—the so-called volar plate. Metacarpal phalangeal joints II to V allow flexion and extension and some abduction and adduction, whereas metacarpal phalangeal joint I allows mainly flexion and extension. The phalanges are connected to each other by a capsular ligamentous system in which tendons of the flexor, extensor, interosseous, and lumbrical muscles play a role. These joints also have a volar plate. The origins of the muscles of the carpus and hand are in the distal upper arm and the forearm (extrinsic muscles). There is no insertion at the proximal carpal row, making this an intercalated segment. In the region of the wrist, these muscles are all converted into tendons employed by tendon sheaths. In the hand, intrinsic muscles with their origin and insertion in the distal carpal row, metacarpals, and phalanges (interosseous muscles and thenar and hypothenar muscles) are found. The origin of the lumbrical muscles is in the flexor tendons and their insertions are at the extensor retinaculum. The majority of extrinsic muscle tendons are volar, running into fibrous canals (the major one being the carpal tunnel) covered by a retinaculum. At the ulnar side, the ulnar nerve runs with the ulnar artery, whereas the median nerve runs with the flexor tendons into the carpal tunnel. The nerves have interconnections. The radial artery runs along the volar and radial sides and, with the ulnar artery, forms superficial and deep volar connections at the metacarpal region that are known as the palmar arches. In the palm, the skin and subcutis are connected to each other and to the underlying retinaculum by the palmar aponeurosis, which is a continuation of the palmaris longus tendon and muscle, which has its origin at the humerus.

Volar Side

At the volar side of the dorsal forearm, the flexor muscles of the wrist and hand converge into tendons before entering the carpal region. The quadratus muscle, between the radius and ulna, runs deep to the muscle-tendon transition of the flexors (Figs. 4.10 and 4.11). In the region of the forearm conversion to the carpus, the ulnar nerve runs most medially (ulnarly), accompanied at the lateral side by the ulnar artery (Figs. 4.1 to 4.3). Somewhat volar (superficial) to the ulnar artery and nerve courses the tendon of the flexor carpi ulnaris, which inserts at the pisiform bone (Fig. 4.3). Lateral to these structures and lying deep are the tendons of the flexor digitorum profundus (four in number), close to the radius and radioulnar joint. The tendons of the flexor digitorum superficialis (four in number) run more superficially, with the tendons to digits III and IV being somewhat more superficial than those running to digits II and V (Figs. 4.1 to 4.3). Lateral to the flexor digitorum profundus tendons runs the tendon of the flexor pollicis longus (Figs. 4.1 to 4.5). More superficial to this tendon runs the median nerve, which is distinguished from the tendons by its intermediate signal on magnetic resonance imaging (MRI) and its rather flat form (Figs. 4.1 to 4.5). More distally, the palmar fascia becomes thicker between the pisiform and hamate bones at the ulnar side, and the distal pole of the scaphoid bone and trapezium at the radial side, forming the flexor retinaculum or roof of the carpal tunnel (Figs. 4.3, 4.4, 4.9, and 4.10). The flexor digitorum profundus and superficialis tendons, the flexor pollicis longus tendon, and the median nerve run into a canal in which the median nerve is located superficially at the radial side and the tendon of the flexor pollicis longus is the most laterally located tendon (Figs. 4.3, 4.4, 4.8, and 4.12).

Lateral to the median nerve and more superficial to the flexor pollicis longus runs the flexor carpi radialis tendon (Figs. 4.1 to 4.3). Proximally it remains lateral to the carpal tunnel (Figs. 4.3 and 4.4) and runs at the volar side of the distal pole of the scaphoid, but more distally it runs lateral into the tunnel in a separate canal between the superficial and deep layers of the flexor retinaculum (Figs. 4.3 and 4.4). Distally, this tendon runs through a groove in the trapezium (Fig. 4.4) before it inserts at the base of the second metatarsal bone. The palmaris longus tendon runs most superficially in the subcutis (Figs. 4.1 to 4.3) and the antebrachial fascia and mediovolarly in relation to the median nerve until it blends into the palmar aponeurosis, which is an intricate (complex) fibrous network fixing the skin to the subcutis and the subcutis to the fascia, including the flexor retinaculum (Figs. 4.4, 4.9, 4.10, and 4.12).

The radial artery runs most laterally to the lateral volar border of the distal radius (Fig. 4.1). During its course to the distal location, the artery runs more dorsally along the lateral side of the wrist and crosses the abductor pollicis longus tendon at the level of the waist of the scaphoid (Fig. 4.8). It traverses to the extensor pollicis brevis tendon at the level of the distal pole of the scaphoid (Fig. 4.3) and then courses to the anatomical snuff-box. From here it forms a dorsal branch, the so-called rete carpi dorsale or dorsal arch. At the level of the trapezium the radial artery crosses the extensor pollicis longus tendon deeply and runs along the base of the second metacarpal, again to the palmar location, where it perforates the first interosseous muscle and enters the palm of the hand to the adductor pollicis and lumbrical muscle. There it anastomoses with the deeply lying palmar branch of the ulnar artery to form the deep palmar arch. At the level of the wrist there is also a small superficial branching of the radial artery, which anastomoses with the ulnar artery to form the superficial palmar arch. (The superficial palmar arch provides the main blood supply to the fingers.)

The ulnar artery and nerve run together (Figs. 4.1 to 4.3). At the level of the pisiform bone they are located in a separate space, the so-called space of Guyon (Fig. 4.3). This space is bordered medially by the pisiform bone, volarly by a fascia layer, the so-called volar carpal ligament, and dorsally by the flexor retinaculum, or roof of the carpal tunnel, and by the pisohamate and pisometacarpal ligaments. The space is laterodistally bordered by the hook of the hamate (Fig. 4.4). At the level of the hamate bone the ulnar artery and nerve run somewhat more medially over the pisohamate ligament and subsequently the motor nerve branches (Fig. 4.4), running through the origin of the hypothenar muscles, around the hook of the hamate deep to the metacarpal region, dorsally with respect to the flexor tendons. The sensitive part of the nerve divides on its course distally (Fig. 4.4) over the hypothenar muscles into branches for digits IV and V. The artery also yields a deep branch that forms the deep palmar arterial arch, together with the deep branch of the radial artery, and continues in a distal direction to form, together with the radial artery, the superficial arch. There are many variations in which either the ulnar or the radial artery is dominant. In 80% the superficial palmar arch is complete. The superficial arch provides the main supply of blood for the fingers. Dorsally and volarly at the level of the distal radius and carpus there is also an arterial network (rete carpi dorsale and palmaris) supplied by the radial and ulnar arteries (carpeus dorsalis and palmaris rami). Frequently there is a sensitive communicating branch between the sensitive branch of the ulnar nerve and the median nerve. There are also cutaneous branches originating from the ulnar nerve and from the volar branch of the median nerve. These branches leave their respective nerves at the level of the distal forearm. The volar branch of the median nerve penetrates the flexor retinaculum ulnarly with respect to the flexor carpi radialis tendon, becoming superficial and subsequently dividing.

The flexor digitorum profundus and superficialis tendons of digits II to V run through the carpal canal toward the midcarpal region, where they become paired and carry on to the fingers. At the midlevel of the proximal phalanges, the deep flexor tendon travels through the superficial tendon, inserting at the base of the distal phalanx with the superficial tendon inserting sideways at the middle phalanx. Each tendon has its own synovial sheath, and both are also surrounded by a common synovial sheath enveloped in an intricate ligamentous system. The flexor pollicis longus tendon runs most laterally in the carpal canal. Beyond the carpal tunnel (Figs. 4.3 and 4.4) it runs palmarly along the flexor pollicis brevis muscle, inserting at the base of the distal phalanx of the thumb (Fig. 4.12).

The median nerve, which runs at the level of the distal radius and proximal carpal row, slightly deeper than the tendon of the palmaris longus and more distally palmar and lateral in the carpal tunnel (Figs. 4.1 to 4.5), divides when entering the metacarpal region. A motor branch goes to the thenar muscles and subsequently the nerve divides into sensitive branches to digits I to IV. The sensitive branches are volar with respect to the palmar superficial arch. The motor branch shows many variations. Sometimes the motor branch leaves the nerve at the region of the carpal canal and continues its course deeply to the flexor retinaculum, or perforates the flexor retinaculum to reach the thenar muscles.

The carpus shows a concavity in the region of the carpal tunnel, which is bordered at the ulnar side by the pisiform and hamate bones and at the radial side by the scaphoid bone and trapezium (Figs. 4.3 and 4.4). The floor of the carpal tunnel consists of radiocarpal and intercarpal ligaments. A thick fibrous band, the so-called flexor retinaculum, forms the roof of the carpal tunnel (Figs. 4.3, 4.4, 4.9, and 4.10). At the radial side this ligament is attached to the scaphoid tuberosity, to a part of the trapezium, and at the ulnar side to the pisiform bone and hook of the hamate. In adults the canal is about 4 cm long and is at its narrowest 2.5 cm distal from its proximal border (at the level of the distal carpal row) (Fig. 4.4). The distal border of the flexor retinaculum is approximately at the level of the metacarpal heads and forms the origin of most thenar and hypothenar muscles (Fig. 4.5). The flexor retinaculum has a stabilizing function in the wrist joint. It consists of longitudinal and transverse fibers. The longitudinal fibers are located in the wall of the carpal tunnel and support the stability of the wrist during extension. The transverse fibers lend some support to the stability of the transverse carpal arch. The intercarpal ligaments of the distal carpal row play a major role in this stability.

The hypothenar consists of three muscles (Figs. 4.4, 4.5, 4.8, 4.12, and 4.13). The most superficial and ulnar is the abductor digiti minimi, partly originating at the pisiform bone (Figs. 4.3 and 4.8). Its insertion is at the medial volar side of the base of the proximal phalanx of digit V. The flexor digiti minimi brevis runs deep and lateral to the abductor digiti minimi and originates from the hamalus and flexor retinaculum, also inserting at the mediovolar side of the proximal phalanx (Figs. 4.4 and 4.8). The opponens digiti minimi is deeper than the flexor digiti minimi brevis, has its origin at the hamalus and flexor retinaculum, and inserts at the mediovolar side of metacarpal V (Figs. 4.4 and 4.8).

The thenar consists of four muscles, three at the radial side (the abductor pollicis brevis, opponens pollicis, and flexor pollicis brevis) and one at the ulnar side (the adductor pollicis) (Figs. 4.1, 4.5, 4.8, 4.12, and 4.13). The abduc-

tor pollicis brevis is the most superficial. It has its origin at the flexor retinaculum (Fig. 4.4) and inserts at the lateral side of the proximal phalanx of digit I and the extensor aponeurosis. The flexor pollicis brevis is deeper than the abductor pollicis brevis and ulnar to the opponens pollicis. Proximally the muscle has a superficial part and a deep part. The origin is at the trapezium and flexor retinaculum (Fig. 4.4) for the superficial part and at the trapezoid and capitate bone for the deeper part. Between these two parts is a tunnel through which the flexor pollicis longus tendon courses (Figs. 4.8 and 4.12). The insertion is at the radial sesamoid bone and at the base of the proximal phalanx of digit I. The opponens pollicis muscle lies deeper than the abductor pollicis brevis and proximal to the origin of the superficial part of the flexor pollicis brevis. The insertion is at the radial part of metacarpal I. The adductor pollicis is the deepest muscle of the thenar and consists of an obliquely running head (caput obliquum) and a transversally running head (caput transversum). The origin of the caput obliquum is at the trapezoid, capitate bone, and the base of metacarpal III, and the origin of the caput transversum is at the volar side of the shaft of metacarpal III. The insertion is at the ulnar sesamoid bone and the extensor aponeurosis of digit I.

More muscles are found in the metacarpal region: seven interosseous muscles (four dorsal and three palmar) and four lumbrical muscles (Figs. 4.13 and 4.14). The four lumbrical muscles originate at the radial side of the flexor tendons of the flexor digitorum profundus and insert at the radial side at the extensor aponeurosis, distal to the insertion of the interosseous muscles. The three palmar interosseous muscles have their origin at the radial part of metacarpals IV and V and the ulnar part of metacarpal II and insert at the same side at the extensor aponeurosis of the same digit. Their function is adduction of digits II, IV, and V toward digit III. The four dorsal interosseous muscles each have their origin at metacarpals that face each other, and they insert at the ulnar side of the proximal phalanx of digits III and IV and at the radial side of the proximal phalanx of digits II and III. The insertion is also related to the extensor aponeurosis. Their function is abduction of digits II and IV with respect to digit III, radial and ulnar deviation of digit III, and they assist in the extension of digits II and IV.

Dorsal Side

At the dorsal side of the wrist and hand the tendons of the extrinsic extenders of hand and wrist are the main structures (Figs. 4.1 to 4.5, 4.9, 4.10, and 4.11). In the subcutis are the dorsal cutaneous branch of the ulnar nerve, the cutaneous branch of the radial nerve, with further radial and ulnar branches for innervating most of the skin. The antebrachial fascia in the region of the wrist shows a thickening known as the extensor retinaculum (Figs. 4.2 to 4.4, and 4.10). This retinaculum is attached to the lateral and dorsal sides of the radius and to the dorsal and medial sides of the ulnar head and runs in a volar direction to attach to the flexor carpi ulnaris tendon sheath, the pisiform bone, the triquetrum, and the base of metacarpal V (Figs. 4.2 and 4.3). The extensor retinaculum and flexor retinaculum are fibroosseously connected to each other at the level of the pisiform bone (Fig. 4.3). In this way both ligaments form an extraarticular loop that contributes to the stability of the wrist joint. Proximally, the extensor retinaculum is attached volarly to the volar antebrachial fascia, and distally the extensor retinaculum transforms into a thin dorsal fascia running over the metacarpals and containing the extensor tendons. The extensor retinaculum can be divided into supra- and infratendinium retinacula, which are located above and below the extensor tendons, respectively. Between these two layers are septa that divide this space into six compartments through which the tendons run (Figs. 4.1 and 4.2). These septa are firmly connected to the longitudinally oriented bony prominences at the dorsal and lateral surfaces of the distal radius (Figs. 4.1 and 4.2). The bony surface of the radius also shows smooth concave grooves between the bony ridges for good tendon fitting and guidance (Figs. 4.1 and 4.2).

The first compartment is lateral and contains the abductor pollicis longus tendon and dorsal to it the extensor pollicis brevis tendon (Figs. 4.1 and 4.2). The second compartment is laterodorsal and contains the extensor carpi radialis longus tendon and medial to it the extensor carpi radialis brevis tendon (Figs. 4.1 and 4.2). This compartment is medially bordered by a large bony ridge or prominence called the dorsal tubercle or Lister's tubercle. The third compartment medial to Lister's tubercle contains the extensor pollicis longus tendon (Figs. 4.1 and 4.2). The fourth compartment contains the extensor indicis proprius and extensor digitorum communis tendons. This compartment has its medial border at the level of the distal forearm just at the radial border of the distal radioulnar joint (Figs. 4.1 and 4.2). The fifth compartment contains the extensor digiti quinti (minimi) tendon and shows a relation with the distal radioulnar joint (Figs. 4.1 and 4.2). The sixth compartment contains the extensor carpi ulnaris tendon, running dorsomedially into a groove in the ulnar head and dorsally of the styloid process (Figs. 4.1, 4.2, and 4.6). In the first, second, and fourth compartments thin septa can also separate the tendons. The fifth compartment is a completely fibrous tunnel in which the fibers are attached to the dorsoulnar border of the distal radius, the capsule of the distal radioulnar joint, the triangular fibrocartilage complex, and the capsule of the wrist joint. The sixth compartment is an exception. The tendon sheath is formed by the retinaculum infratendinium, which is separated from the retinaculum supratendinium by loosely woven tissue. The tendon sheath is firmly fixed to the underlying articular disk. Proximal to this tendon sheath transverse fibers of the retinaculum supratendinium attach to the groove in which the tendon runs, and form an extension of the ulnar wall of the tendon sheath. These transverse fibers are strengthened by longitudinal fibers, the linea jugata, running from the styloid process proximal to the antebrachial fascia and to the extensor retinaculum.

The tendons of the first compartment run to the first ray. The most volarly running abductor pollicis tendon inserts laterally at the base of metacarpal I, the trapezium, and the thenar muscles. This tendon has multiple tendon slips. The extensor pollicis brevis tendon inserts laterodorsally at the base of the proximal phalanx of digit I and has

extensions into the extensor aponeurosis. At the level of the scaphoid bone and trapezium these tendons form the lateral border of the *tabatière anatomique* (the anatomical snuff-box). During their course in a distal direction, the radial artery crosses these tendons deeply on its way to its dorsal distal location (Fig. 4.3).

The extensor carpi radialis longus tendon runs lateral to the tendon of the extensor carpi radialis brevis in the second compartment. The longus inserts at the dorsolateral side of the base of metacarpal II and the brevis inserts at the dorsolateral side of the base of metacarpal III (Fig. 4.5). The dorsal branch of the radial artery, the rete carpi dorsale, runs deeper than these tendons to its ulnar location. The tendon of the extensor pollicis longus runs into the third compartment medial to Lister's tubercle and crosses the second compartment superficially at the radiocarpal level in an oblique course (Fig. 4.3) to the extensor aponeurosis of digit I. This tendon forms the medial border of the anatomical snuff-box and also crosses the radial artery superficially (Fig. 4.4). The fourth compartment contains the four tendons of the extensor digitorum communis muscle and the tendon of the extensor indicis. After leaving the compartment, the tendons separate on their way to digits II to V (Figs. 4.4 and 4.5). The tendon of digit IV is connected with the tendons of digits III and V by obliquely running tendinous slips, the so-called juncturae tendinum. Usually a filamentous connection is also found between the tendon running to digit II and that to digit III. The tendons insert at the level of the metacarpophalangeal joints at the extensor aponeurosis. The extensor indicis tendon inserts at the ulnar side of the extensor aponeurosis of digit II. Deep in the fourth compartment at the lateral side run the posterior interosseous artery and nerve. The nerve divides at the level of the scapholunate joint. The extensor digiti quinti tendon usually forms two slips and courses through the fifth compartment to digit V, where it inserts at the ulnar side at the extensor aponeurosis. The extensor carpi ulnaris tendon courses through the sixth compartment to insert at the dorsoulnar side of the base of metacarpal V. The tendon sheath is connected to the carpus by the ulnocarpal complex, or triangular fibrocartilage complex (Figs. 4.2 and 4.6).

Joints and Ligaments

The distal radioulnar joint (DRUJ) is an articulation in which the ulnar head rotates in the opposite ulnar incisure or notch of the radius (Figs. 4.1 and 4.6). This joint is separated from the radiocarpal joint by a discus (Figs. 4.2, 4.6, and 4.11). The discus, or triangular fibrocartilage, is part of the triangular fibrocartilage complex (TFCC), which is a cartilaginous ligamentous system at the ulnar side of the joint that attaches the ulna to the radius and to the carpus. This complex can be subdivided into the discus proper, the dorsal and volar radioulnar ligaments as part of the most dorsal and volar parts of the discus (Fig. 4.2), respectively, the volar ulnolunate and ulnotriquetral ligaments (meniscus homolog) (Fig. 4.11) and the thinner dorsal ulnolunate and ulnotriquetral ligaments (Fig. 4.11), and at least the floor of the extensor carpi ulnaris tendon sheath (Figs. 4.1 and 4.2). The ulnar attachment of the TFCC is located mainly at the region of the base of the ulnar styloid process (Figs. 4.2 and 4.6). The discus attachment to the radius contains the whole ulnar border of the radius just as it reaches the level of the radiocarpal joint lining (Figs. 4.2, 4.6, and 4.11). The discus has a somewhat triangular form when seen from above, with the apex at the base of the ulnar styloid process (Fig. 4.2). It is at its thickest ulnarly and dorsally and at its thinnest centrally, near the attachment to the radius, which results in a biconcave configuration (Fig. 4.11).

The carpus is formed by eight bones. Four constitute the first carpal row: from lateral to medial these are the scaphoid, lunate, and triquetrum, and the pisiform bone volar to the triquetrum. The distal carpal row consists of the four other bones; from lateral to medial they are the trapezium, trapezoid, capitate, and hamate. The wrist joint consists of the radiocarpal and midcarpal joints, both of which contribute approximately equally to the extension-flexion and ulnoradial deviation motion. The supporting ligaments of the wrist can be divided into the extrinsic (running from forearm to carpus or carpus to metacarpus) and the intrinsic ligaments (running between the carpal bones). The volar ligaments are thicker than the dorsal ligaments. The intrinsic ligaments are located dorsally, volarly, or intraarticularly between bones (interosseous ligaments). The extrinsic volar ligaments are the radioscaphocapitate as the most lateral, running from the laterovolar side of the radial styloid to the scaphoid, to which it is loosely attached, and to the capitate (Figs. 4.8, 4.9, and 4.10). The radiolunotriquetral ligament runs somewhat medial to it, to the lunate bone, and continues to the triquetrum (Figs. 4.8, 4.9, and 4.10). Medial to this runs the radioscapholunate ligament from the volar rim of the radius to the scapholunate interosseous ligament (Fig. 4.8). Medial to the volar rim of the radius the short radiolunate ligament runs to the lunate (Figs. 4.8 and 4.10). As part of the TFCC the ulnolunate and ulnotriquetral ligaments are the ulnovolarly running extrinsic ligaments (Fig. 4.11). When viewed volarly, these obliquely running radiocarpal and ulnocarpal ligaments show a reversed V configuration (the proximal V) and form the radial limb of the distal V.

At the dorsal side the dorsal radiocarpal ligament runs from the medial part of the dorsal rim of the radius toward the lunate bone and triquetrum (Fig. 4.10) and the dorsal ulnolunate and ulnotriquetral ligaments as part of the TFCC (Fig. 4.11). The wrist has no real collateral ligaments. What is considered to be the lateral collateral ligament is the most lateral part of the volar radiocarpal ligament, and what is considered to be the ulnar collateral ligament is only a thickening of the medial capsule in relation to the floor of the extensor carpi ulnaris tendon sheath. The main intrinsic volar ligament is the volar intercarpal ligament, which consists of the capitoscaphoid or capitoscaphotrapezium ligament and the capitotriquetral ligament (Figs. 4.3 and 4.10), both running obliquely and giving the impression of a reversed V with the capitate at the apex when seen from a volar direction (the distal V). Furthermore, there are many short and intermediate volar intercarpal and carpometacarpal ligaments. Dorsally, the main intrinsic ligament is the dorsal intercarpal ligament running from the

dorsal aspect of the triquetrum to the scaphoid and trapezium. Also dorsally, there are many intermediate and short intercarpal and carpometacarpal ligaments.

A special group of ligaments are the interosseous ligaments: three between the bones of the distal carpal row, of which the interosseous ligament between capitate and hamate bones is particularly large (Figs. 4.6 and 4.14). The ligaments between capitate and trapezoid bones and between trapezoid and trapezium bones are frequently missing. The proximal carpal row has two interosseous ligaments: between the scaphoid and lunate bones and between the lunate bone and triquetrum (Figs. 4.6 and 4.7). These ligaments are located at the proximal border of those carpal bones, where they follow the convex proximal contour from dorsal to volar to form a continuum with the convex articular surface of the proximal row.

The volarly located pisiform bone articulates with the triquetrum and forms the insertion of the flexor carpi ulnaris tendon and the origin of the abductor digiti minimi muscle. The flexor retinaculum and extensor retinaculum are attached to it (Figs. 4.3 and 4.8). Distally, the pisiform bone is connected with the hamulus and base of metacarpal V by the pisohamate and pisometacarpal ligaments, respectively. The joint of the pisotriquetral joint is capacious and the pisiform bone can also be considered to be a sesamoid bone.

The distal carpal row forms a transverse arch that continues at the metacarpal region. The carpometacarpal (CMC) II and III joints are firmly fixed to the carpals by irregularly formed joints and short thick volar and dorsal carpometacarpal ligaments, whereby metacarpal II articulates with the trapezium and trapezoid, and metacarpal III with the capitate and trapezoid. At these locations flexion and extension are possible only through 1 to 3°. Metacarpal IV has a small, V-like joint surface articulating with the hamate and capitate bones, and metacarpal V has a small, saddle-shaped joint. Both joints have obliquely running carpometacarpal ligaments allowing flexion and extension through 10 to 15° for CMC IV and through 15 to 30° for the CMC V joint. The bases of the metacarpals are also connected to each other by dorsal, volar, and interosseous metacarpal ligaments, and the bases also form an articulation with the neighboring metacarpal bone. The CMC I joint, which is formed by metacarpal I and the trapezium, has a kind of saddle shape that allows singular and compound movements. The capsule contains six ligaments, of which most run obliquely and, at every position, one or more are under tension. The capsule is rather wide and there are no true collateral ligaments. Also found here is an intermetacarpal ligament.

The metacarpophalangeal joints II to V are ball-and-socket joints with flexion-extension and some abduction and adduction movements. These joints contain thick collateral ligaments and accessory collateral ligaments. Volarly, each capsule has a thick, platelike reinforcement, the so-called volar plate (Fig. 4.12), which is also connected with the flexor tendon sheath and dorsally with the sagittal bands of the extensor aponeurosis. The capsules are more capacious than the interphalangeal joints, allowing a certain degree of distraction and rotation. The volar plates become elongated on extension and shorten on flexion, which is not the case at the interphalangeal joints. The volar plates are mutually connected by the transverse metacarpal ligament. The first metacarpophalangeal (MCP) joint shows much interindividual difference—from hinge joint to ball-and-socket joint. The joint is considered to be a transition of a metacarpophalangeal joint to an interphalangeal joint. The joint shows a volar plate with two sesamoid bones. At the ulnar sesamoid bone the adductor pollicis inserts and at the radial sesamoid the flexor pollicis brevis inserts, both muscles acting as stabilizers of the joint. The joint also has collateral and accessory collateral ligaments. The interphalangeal joint of the thumb is a hinge joint with a stout capsule, strengthened by a volar plate, collateral ligaments, and accessory collateral ligaments. The amount of flexion and extension shows a strong interindividual variation, as is shown by the hyperextension possibility. The flexor pollicis longus tendon inserts at the base of the distal phalanx. Sometimes there is a sesamoid bone in this tendon or in the volar plate. The proximal interphalangeal (PIP) joints of digits II to V are also hinge joints allowing only flexion and extension. The joint capsule is reinforced by strong ligaments providing a high degree of stability. At the dorsal side there is no capsule. The extensor aponeurosis at this location is in direct contact with the synovial membrane. There are collateral ligaments and accessory collateral ligaments medially and laterally and a volar plate at the volar side. The proximal insertion of the volar plate is formed by two check ligaments. The distal interphalangeal joint (DIP) is also a hinge joint, with less flexion and more extension possibilities than the PIP joint. The joint capsule is very strong, with reinforcement by collateral and accessory collateral ligaments and by the volar plate. The proximal insertion of the volar plate is at the flexor tendon sheath and the distal fibrocartilaginous part of the plate is at the volar side of the base of the distal phalanx.

Anatomy of Wrist and Hand

Bone and Cartilage

1 Radius
2 Ulna
3 Lister's Tubercle
4 Ulnar Styloid Process
5 Scaphoid Bone
6 Lunate Bone
7 Capitate Bone
8 Hamate Bone
9 Hook of Hamate Bone
10 Partial Volume of Hook of Hamate Bone
11 Triquetrum
12 Pisiform Bone
13 Trapezium Bone
14 Trapezium Tuberosity
15 Trapezoid Bone
16 Metacarpal Bone I
17 Metacarpal Bone II
18 Metacarpal Bone III
19 Metacarpal Bone IV
20 Metacarpal Bone V
21 Base of Metacarpal Bone V
22 Proximal Phalanx Digit I
23 Distal Phalanx Digit I
24 Proximal Phalanx Digit II
25 Proximal Phalanx Digit III
26 Proximal Phalanx Digit IV
27 Proximal Phalanx Digit V
28 Partial Volume of Discus and Ulnar Head

Muscle

31 Palmaris Longus Muscle
32 Flexor Pollicis Longus Muscle
33 Flexor Pollicis Brevis Muscle
34 Flexor Digiti Minimi Brevis and/or Opponens Digiti Minimi Brevis Muscle
35 Flexor Digitorum Muscle
36 Flexor Digitorum Superficialis Muscle
37 Flexor Digitorum Profundus Muscle
38 Flexor Carpi Ulnaris Muscle
39 Flexor Carpi Radialis Muscle
40 Insertion of the Flexor Carpi Ulnaris Tendon
41 Abductor Pollicis Longus Muscle
42 Extensor Pollicis Brevis Muscle
43 Extensor Pollicis Longus Muscle
44 Extensor Carpi Radialis Brevis Muscle
45 Extensor Carpi Radialis Longus Muscle
46 Extensor Digitorum Communis and Extensor Indicis Muscle
47 Extensor Digiti Minimi Muscle
48 Extensor Carpi Ulnaris Muscle
49 Abductor Digiti Minimi Muscle
50 Origin of Abductor Digiti Minimi Muscle
51 Hypothenar Muscles
52 Thenar Muscles
53 Adductor Pollicis Muscle (Fibers)
54 Flexor and/or Opponents Digiti Minimi Brevis
55 Pronator Quadratus Muscle
56 Interosseous Muscle (dorsal)
57 Interosseous Muscle (palmar)
58 Lumbrical Muscle

Ligament and Tendon

68 Volar Ulnotriquetral Ligament
69 Dorsal Intercarpal Ligament
70 Volar Capsule
71 Discus or Triangular Pad of Fibrocartilage
72 Triangular Fibrocartilage Complex (TFCC)
73 Dorsal Radioulnar Ligament
74 Volar Radioulnar Ligament
75 Fibrous Tunnel for the Extensor Digiti Minimi Tendon
76 Fascial or Retinacular Structures
77 Extensor Retinaculum
78 Extensor Retinaculum, Supratendineum Part
79 Extensor Retinaculum, Supra- and Infratendineum Part
80 Flexor Retinaculum as Roof of the Carpal Canal
81 Volar Carpal Ligament
82 Floor of the Carpal Canal, Consisting of Radiocarpal and Intercarpal Ligaments
83 Separate Fibrous Canal for the Flexor Carpi Radialis Tendon
84 Palmar Aponeurosis
85 Scapholunate Interosseus Ligament
86 Dorsal Part of Scapholunate Interosseus Ligament
87 Dorsal Part of Lunotriquetral Interosseus Ligament and Dorsal Ulnocarpal Ligament
88 Triangular Fibrocartilage Complex or Discus with Degenerative Rupture
89 Radial Part of Ruptured Triangular Fibrocartilage
90 Lunotriquetral Interosseus Ligament
91 Radiolunotriquetral Ligament
92 Radioscapholunate Ligament
93 Radioscaphocapitate Ligament
94 Short Radiolunate Ligament
95 Dorsal Radiocarpal Ligament in Dorsal Capsule
96 Short Radiolunate Ligament in Volar Capsule
97 Volar Radiocarpal and Intercarpal Ligaments Inserting at Capitate (Floor of the Carpal Canal)
98 Flexor Retinaculum
99 Flexor Retinaculum with Hypothenar Muscle Origin
100 Flexor Retinaculum with Pisohamate Ligament
101 Volar Ulnolunate Ligament
102 Dorsal Ulnotriquetral Ligament
103 Transverse Fasciculi of the Palmar Aponeurosis
104 Volar Plate
105 Interosseus Ligament Between Capitate and Hamate Bone
106 Space of Guyon
107 Carpal Canal

Neurovascular and Miscellaneous

108 Median Nerve
109 Ulnar Artery
110 Ulnar Nerve
111 Ulnar Nerve, Deep Branch
112 Ulnar Nerve, Superficial Branch
113 Radial Artery
114 Radial Nerve
115 Cutaneous Veins
116 Silicone Barium Contrast in Radiocarpal Joint
117 Sacciform Recess of the Distal Radioulnar Joint with Silicone Barium Contrast
118 Silicone Barium Contrast in Prestyloid Recess
119 Silicone Barium Contrast in Interligamentous Sulcus of the Radiocarpal Joint
120 Silicone Barium Contrast in Proximal Recess of Pisotriquetral Joint
121 Silicone Barium Contrast in Volar Recess of Radiocarpal Joint
122 Silicone Barium Contrast in Distal Sulcus of Radiocarpal Joint

Abbreviations: t, tendon; DRUJ, distal radioulnar joint; TFCC, triangular fibrocartilage complex; CMC, carpometacarpal; MCP, metacarpophalangeal; PIP, proximal interphalangeal; DIP, distal interphalangeal.

Acknowledgments: Anatomic drawings were made by Bas B. Blankevoort and J. Wetselaar-Whittaker.

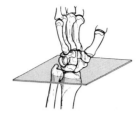

Figure 4.1. Axial section of the distal forearm at the level of the distal radioulnar joint. Of special interest are the extensor tendons running laterally, dorsally, and medially in six fibrous compartments, the arrangement of the flexor tendons with the median nerve volarly, the ulnar artery accompanied by the ulnar nerve volarly at the medial side, and the radial artery volarly at the lateral side.

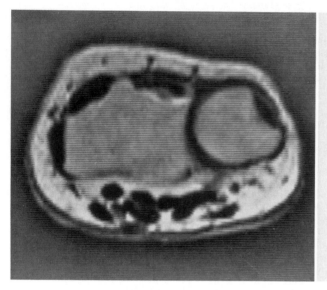
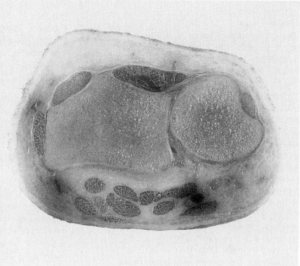

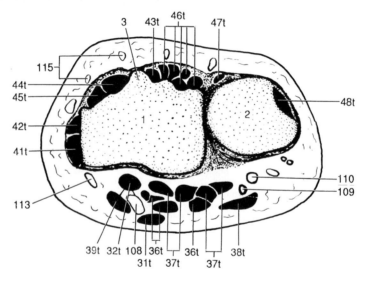

Bone and Cartilage
1   Radius
2   Ulna
3   Lister's Tubercle

Muscle
31 Palmaris Longus Muscle
32 Flexor Pollicis Longus Muscle

36 Flexor Digitorum Superficialis Muscle
37 Flexor Digitorum Profundus Muscle
38 Flexor Carpi Ulnaris Muscle
39 Flexor Carpi Radialis Muscle
41 Abductor Pollicis Longus Muscle

42 Extensor Pollicis Brevis Muscle
43 Extensor Pollicis Longus Muscle
44 Extensor Carpi Radialis Brevis Muscle
45 Extensor Carpi Radialis Longus Muscle
46 Extensor Digitorium Communis and Extensor Indicis Muscle

47 Extensor Digiti Minimi Muscle
48 Extensor Carpi Ulnaris Muscle

Neurovascular and Miscellaneous
108 Median Nerve
109 Ulnar Artery
110 Ulnar Nerve
113 Radial Artery
115 Cutaneous Veins

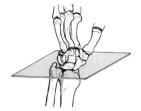

Figure 4.2. Axial section at the level of the articular disk (triangular fibrocartilage). The distal border of the ulnar head has just been cut. The disk is viewed from above.

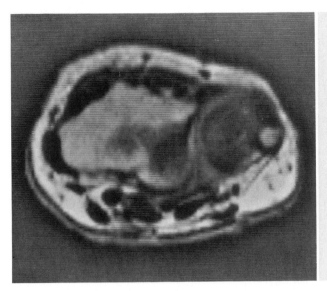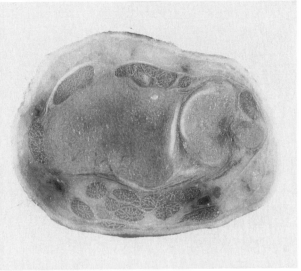

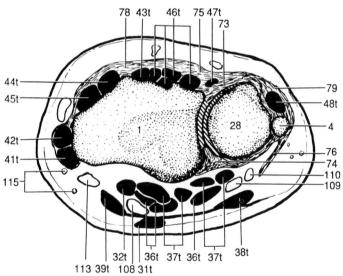

Bone and Cartilage
1 Radius
4 Ulnar Styloid Process
28 Partial Volume of Discus and Ulnar Head

Muscle
31 Palmaris Longus Muscle
32 Flexor Pollicis Longus Muscle
36 Flexor Digitorum Superficialis Muscle
37 Flexor Digitorum Profundus Muscle
38 Flexor Carpi Ulnaris Muscle
39 Flexor Carpi Radialis Muscle
41 Abductor Pollicis Longus Muscle
42 Extensor Pollicis Brevis Muscle
43 Extensor Pollicis Longus Muscle
44 Extensor Carpi Radialis Brevis Muscle
45 Extensor Carpi Radialis Longus Muscle

46 Extensor Digitorum Communis and Extensor Indicis Muscle
47 Extensor Digiti Minimi Muscle
48 Extensor Carpi Ulnaris Muscle

Ligament and Tendon
73 Dorsal Radioulnar Ligament
74 Volar Radioulnar Ligament
75 Fibrous Tunnel for the Extensor Digiti Minimi Tendon
76 Fascial or Retinacular Structures

78 Extensor Retinaculum, Supra-tendineum Part
79 Extensor Retinaculum, Supra- and Infratendineum Part

Neurovascular and Miscellaneous
108 Median Nerve
109 Ulnar Nerve
110 Ulnar Nerve
113 Radial Artery
115 Cutaneous Veins

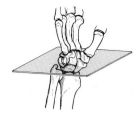

Figure 4.3. Axial section at the midlevel of the carpus. Of special interest are the carpal tunnel with the median nerve volarly, and the space of Guyon between the pisiform bone and the carpal tunnel containing the ulnar nerve and artery. Note also the deep dorsal migration of the radial artery to the tendons.

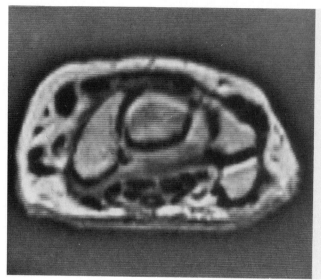

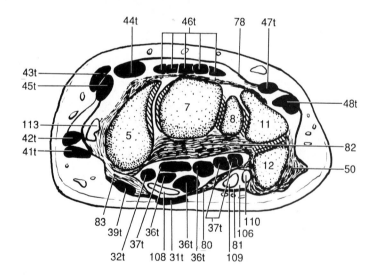

Bone and Cartilage
5 Scaphoid Bone
7 Capitate Bone
8 Hamate Bone
11 Triquetrum
12 Pisiform Bone

Muscle
31 Palmaris Longus Muscle
32 Flexor Pollicis Longus Muscle
36 Flexor Digitorum Superficialis Muscle

37 Flexor Digitorum Profundus Muscle
39 Flexor Carpi Radialis Muscle
41 Abductor Pollicis Longus Muscle
42 Extensor Pollicis Brevis Muscle
43 Extensor Pollicis Longus Muscle
44 Extensor Carpi Radialis Brevis Muscle
45 Extensor Carpi Radialis Longus Muscle

46 Extensor Digitorum Communis and Extensor Indicis Muscle
47 Extensor Digiti Minimi Muscle
48 Extensor Carpi Ulnaris Muscle
50 Origin of Abductor Digiti Minimi Muscle

Ligament and Tendon
78 Extensor Retinaculum, Supratendineum Part
80 Flexor Retinaculum as Roof of the Carpal Canal
81 Volar Carpal Ligament

82 Floor of the Carpal Canal, Consisting of Radiocarpal and Intercarpal Ligaments
83 Separate Fibrous Canal for the Flexor Carpi Radialis Tendon
106 Space of Guyon

Neurovascular and Miscellaneous
108 Median Nerve
109 Ulnar Artery
110 Ulnar Nerve
113 Radial Artery

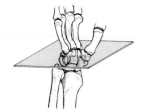

Figure 4.4. Axial section at the level of the distal carpal row. Of special interest are the carpal tunnel at its narrowest point between the hook of hamate and the trapezium tuberosity, the branching of the ulnar nerve, the close relation between the flexor carpi radialis tendon and the trapezium, and the origin of the thenar and hypothenar muscles.

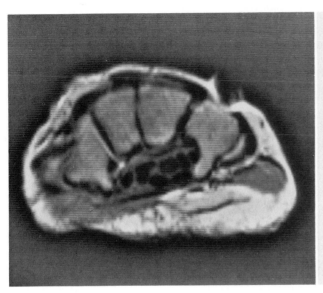
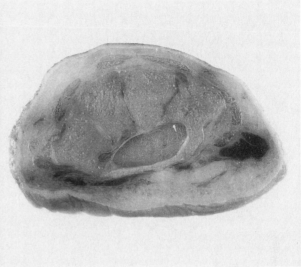

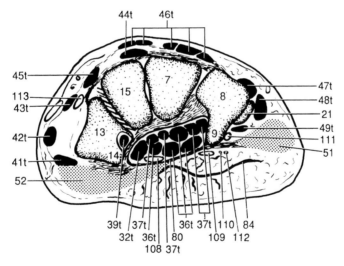

Bone and Cartilage
- 7 Capitate Bone
- 8 Hamate Bone
- 9 Hook of Hamate Bone
- 13 Trapezium Bone
- 14 Trapezium Tuberosity
- 15 Trapezoid Bone
- 21 Base of Metacarpal Bone V

Muscle
- 32 Flexor Pollicis Longus Muscle
- 36 Flexor Digitorum Superficialis Muscle
- 37 Flexor Digitorum Profundus Muscle
- 39 Flexor Carpi Radialis Muscle
- 41 Abductor Pollicis Longus Muscle
- 42 Extensor Pollicis Brevis Muscle
- 43 Extensor Pollicis Longus Muscle
- 44 Extensor Carpi Radialis Brevis Muscle
- 45 Extensor Carpi Radialis Longus Muscle
- 46 Extensor Digitorum Communis and Extensor Indicis Muscle
- 47 Extensor Digiti Minimi Muscle
- 48 Extensor Carpi Ulnaris Muscle
- 49 Abductor Digiti Minimi Muscle
- 51 Hypothenar Muscles
- 52 Thenar Muscles

Ligament and Tendon
- 80 Flexor Retinaculum as Roof of the Carpal Canal
- 84 Palmar Aponeurosis

Neurovascular and Miscellaneous
- 108 Median Nerve
- 109 Ulnar Artery
- 110 Ulnar Nerve
- 111 Ulnar Nerve, Deep Branch
- 112 Ulnar Nerve, Superficial Branch
- 113 Radial Artery

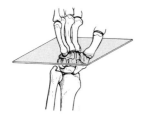

Figure 4.5. Axial section at the level of the bases of the metacarpals at the distal side of the carpal tunnel. Note the origin of the thenar and hypothenar muscles at the flexor retinaculum (roof of the carpal tunnel) and the palmar aponeurosis connecting the cutis and subcutis to the deep fascial structures.

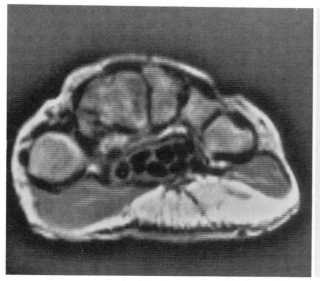
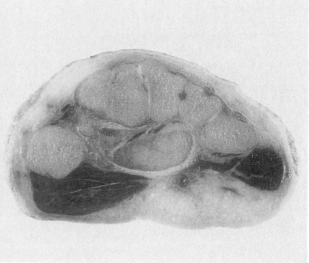

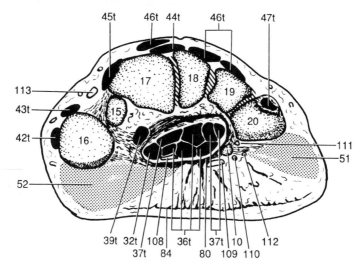

Bone and Cartilage
10 Partial Volume of Hook of Ha-mate Bone
15 Trapezoid Bone
16 Metacarpal Bone I
17 Metacarpal Bone II
18 Metacarpal Bone III
19 Metacarpal Bone IV
20 Metacarpal Bone V

Muscle
32 Flexor Pollicis Longus Muscle
36 Flexor Digitorum Superficialis Muscle
37 Flexor Digitorum Profundus Muscle
39 Flexor Carpi Radialis Muscle
42 Extensor Pollicis Brevis Muscle
43 Extensor Pollicis Longus Mus-cle

44 Extensor Carpi Radialis Brevis Muscle
45 Extensor Carpi Radialis Longus Muscle
46 Extensor Digitorum Communis and Extensor Indicis Muscle
47 Extensor Digiti Minimi Muscle
51 Hypothenar Muscles
52 Thenar Muscles

Ligament and Tendon
80 Flexor Retinaculum as Roof of the Carpal Canal
84 Palmar Aponeurosis

Neurovascular and Miscellaneous
108 Median Nerve
109 Ulnar Artery
110 Ulnar Nerve
111 Ulnar Nerve, Deep Branch
112 Ulnar Nerve, Superficial Branch
113 Radial Artery

Figure 4.6. Coronal section at the dorsal level of the radiocarpal joint. Of special interest are the intermediate signal of the medial part of the TFCC and the low signal of its lateral part. Note also the dorsal part of the scapholunate and lunotriquetral interosseous ligaments.

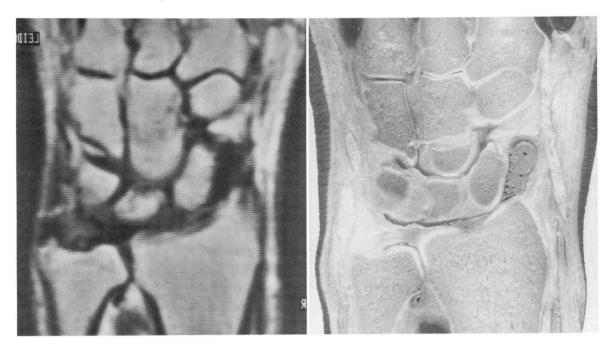

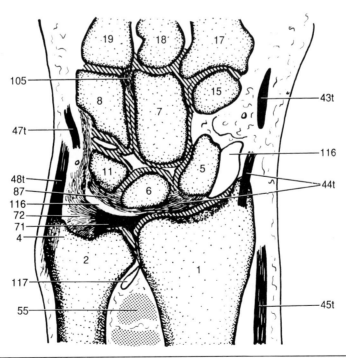

Bone and Cartilage
1 Radius
2 Ulna
4 Ulnar Styloid Process
5 Scaphoid Bone
6 Lunate Bone
7 Capitate Bone
8 Hamate Bone
11 Triquetrum

15 Trapezoid Bone
17 Metacarpal Bone II
18 Metacarpal Bone III
19 Metacarpal Bone IV

Muscle
43 Extensor Pollicis Muscle
44 Extensor Carpi Radialis Brevis Muscle
45 Extensor Carpi Radialis Longus Muscle

47 Extensor Digiti Minimi
48 Extensor Carpi Ulnaris Muscle
55 Pronator Quadratus Muscle

Ligament and Tendon
71 Discus or Triangular of Fibro-cartilage
72 Triangular Fibrocartilage Complex
87 Dorsal Part of Lunotriquetral In-

terosseus Ligament and Dorsal Ulnocarpal Ligament
105 Interosseus Ligament Between Capitate and Hamate Bone

Neurovascular and Miscellaneous
116 Silicon Barium Contrast in Radiocarpal Joint
117 Sacciform Recess of the Distal Radioulnar Joint with Silicone Barium Contrast

Figure 4.7. Coronal section at the middle of the radiocarpal joint. Of special interest are the scapholunate and lunotriquetral interosseous ligaments with an intermediate signal, which forms a continuation of the intermediate signal of the part of the hyaline cartilage of the proximal carpal bones located at the joint side. Note also the disk perforation with an intermediate signal of the degenerated disk.

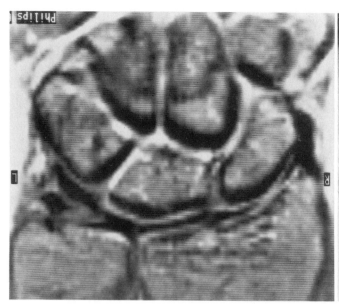
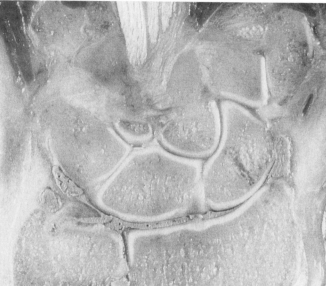

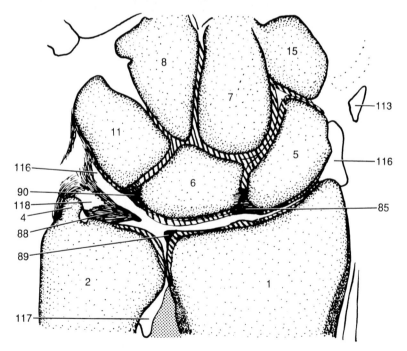

Bone and Cartilage
1 Radius
2 Ulna
4 Ulnar Styloid Process
5 Scaphoid Bone
6 Lunate Bone
7 Capitate Bone
8 Hamate Bone
11 Triquetrum
15 Trapezoid Bone

Ligament and Tendon
85 Scapholunate Interosseus Ligament
88 Triangular Fibrocartilage Complex or Discus with Degenerative Rupture
89 Radial Part of Ruptured Triangular Fibrocartilage
90 Lunotriquetral Interosseus Ligament

Neurovascular and Miscellaneous
113 Radial Artery
116 Silicone Barium Contrast in Radiocarpal Joint
117 Sacciform Recess of the Distal Radioulnar Joint with Silicone Barium Contrast
118 Silicone Barium Contrast in Prestyloid Recess

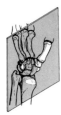

Figure 4.8. Coronal section at the volar side of the wrist. Of special interest are the volar radiocarpal ligaments with an intermediate signal and the flexor tendons running through the carpal tunnel at the level of the pisiform bone, hook of hamate, distal pole of the scaphoid, and trapezium.

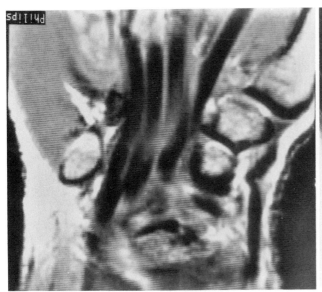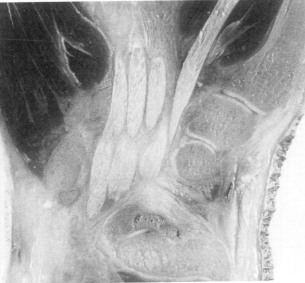

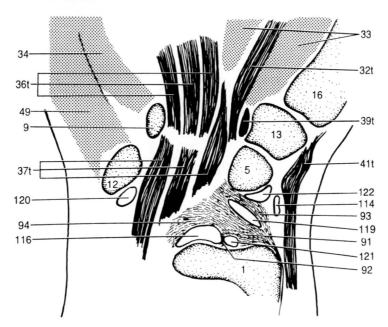

Bone and Cartilage
1   Radius
5   Scaphoid Bone
9   Hook of Hamate Bone
12 Pisiform Bone
13 Trapezium Bone
16 Metcarpal Bone I

Muscle
32 Flexor Pollicis Longus Muscle
33 Flexor Pollicis Boevis Muscle

34 Flexor Digiti Minimi Brevis and/
    or Oponens Digiti Minimi Muscle
    Muscle
36 Flexor Digitorum Superficialis
    Muscle
37 Flexor Digitorum Profundus
    Muscle
39 Flexor Carpi Radialis Muscle
41 Abductor Pollicis Longus Muscle
    cle
49 Abductor Digiti Minimi Muscle

Ligament and Tendon
91 Radiolunotriquetral Ligament
92 Radioscapholunate Ligament
93 Radioscaphocapitate Ligament
94 Short Radiolunate Ligament

Neurovascular and Miscellaneous
114 Radial Nerve
116 Silicone Barium Contrast in Radiocarpal Joint
119 Silicone Barium Contrast in In-

terligamentous Sulcus of the
Radiocarpal Joint
120 Silicone Barium Contrast in
    Proximal Recess of Pisotriquetral Joint
121 Silicone Barium Contrast in Volar Recess of Radiocarpal Joint
122 Silicone Barium Contrast in Distal Sulcus of Radiocarpal Joint

Figure 4.9. Sagittal section at the radial (medial) side of the wrist. Of special interest are the volar radiocarpal ligaments, the flexor and extensor retinaculum, and the flexor and extensor tendons volarly and dorsally, respectively.

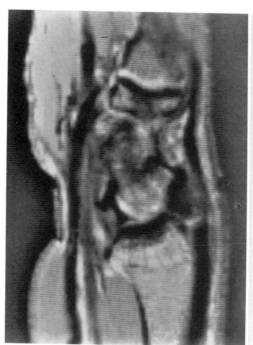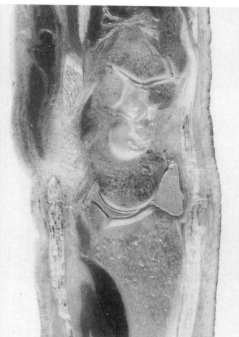

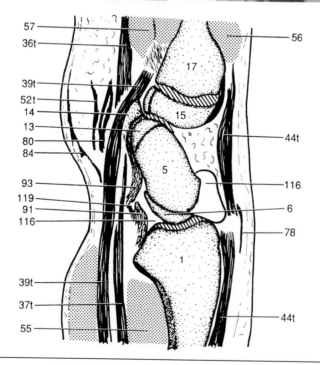

Bone and Cartilage	Muscle		Ligament and Tendon	Neurovascular and Miscellaneous
1 Radius	36 Flexor Digitorum Superficialis Muscle	55 Pronator Quadratus Muscle	84 Palmar Aponeurosis
5 Scaphoid Bone		56 Interosseus Muscle (dorsal)	91 Radiolunotriquetral Ligament
6 Lunate Bone	37 Flexor Digitorum Profundus Muscle	57 Interosseus Muscle (palmar)	93 Radioscaphocapitate Ligament
13 Trapezium Bone		**Ligament and Tendon**	
14 Trapezium Tuberosity	39 Flexor Carpi Radialis Muscle	78 Extensor Retinaculum Supra-tendineum Part	**Neurovascular and Miscellaneous**
15 Trapeziod Bone	44 Extensor Carpi Radialis Brevis Muscle		116 Silicone Barium Contrast in Radiocarpal Joint
17 Metacarpal Bone II		80 Flexor Retinaculum as Roof of the Carpal Canal	119 Silicone Barium Contrast in Ligamentous Sulcus of the Radiocarpal Joint
	52 Thenar Muscles		

Figure 4.10. Sagittal section at the midlevel of the wrist through the capitate and lunate bones. Note the flexor and extensor retinaculum outside the flexor and extensor tendons, and the capsular reinforcement by ligaments.

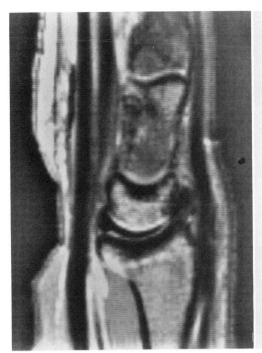 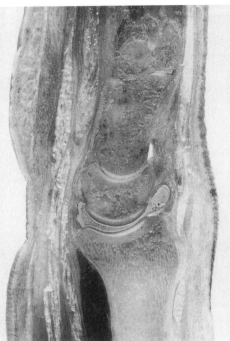

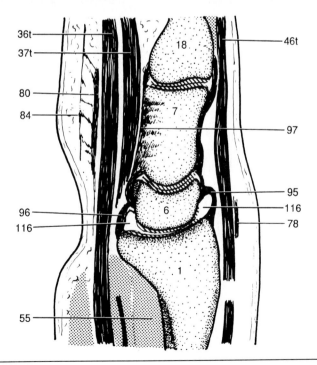

Bone and Cartilage
1 Radius
6 Lunate Bone
7 Capitate Bones
18 Metacarpal Bone III

Muscle
36 Flexor Digitorum Superficialis Muscle
37 Flexor Digitorum Profundus Muscle
46 Flexor Digitorum Communis and Extensor Indicis Muscle
55 Pronator Quadratus Muscle

Ligament and Tendon
78 Extensor Retinaculum, Supratendineum Part
80 Flexor Retinaculum as Roof of the Carpal Canal
84 Palmar Aponeurosis
95 Dorsal Radiocarpal Ligament in Dorsal Capsule

96 Short Radiolunate Ligament in Volar Capsule
97 Volar Radiocarpal and Intercarpal Ligaments Inserting at Capitate (Floor of the Carpal Canal)

Neurovascular and Miscellaneous
116 Silicone Barium Contrast in Radiocarpal Joint

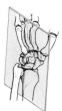

Figure 4.11. Sagittal section at the ulnar (medial) side of the wrist. Of special interest are the low-signal TFCC or discus with its attachments dorsally and volarly and a flexor tendon coursing along the hook of hamate.

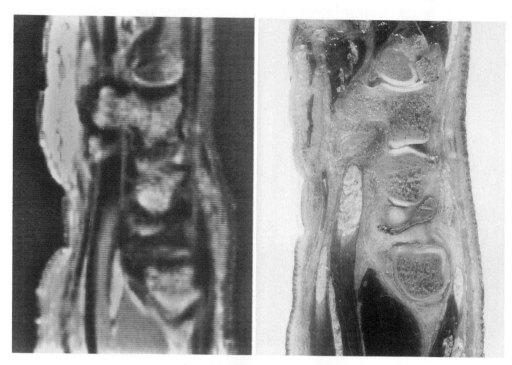

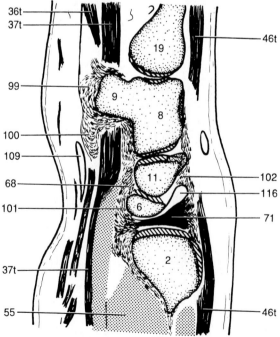

Bone and Cartilage
- 2 Ulna
- 6 Lunate Bone
- 8 Hamate Bone
- 9 Hook of Hamate Bone
- 11 Triquetrum
- 19 Metacarpal Bone IV

Muscle
- 36 Flexor Digitorum Superficialis Muscle
- 37 Flexor Digitorum Profundus Muscle
- 46 Extensor Digitorum Communis and Extensor Indicis Muscle
- 55 Pronator Quadratus Muscle

Ligament and Tendon
- 68 Volar Ulnotriquetral Ligament
- 71 Discus or Triangular pad of Fibrocartilage
- 99 Flexor Retinaculum with Hypothenar Muscle Origin
- 100 Flexor Retinaculum with Pisohamate Ligament
- 101 Volar Ulnolunate Ligament
- 102 Dorsal Ulnolanate Ligament

Neurovascular and Miscellaneous
- 109 Ulnar Artery
- 116 Silicone Barium Contrast in Radiocarpal Joint

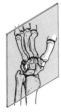

Figure 4.12. Coronal section at the volar side of the wrist and midhand. Of special interest are the flexor tendons, the flexor pollicis longus tendon, the ulnar nerve with branches, and the network of the palmar aponeurosis.

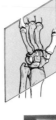

Figure 4.13. Coronal section of the volar side of the wrist and midhand but more dorsally than seen in Fig. 4.12. Of special interest are the lumbrical muscles, the thenar and hypothenar muscles, and the volar plate of the interphalangeal joint of the thumb.

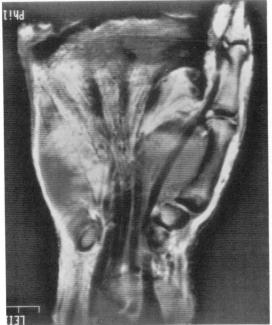

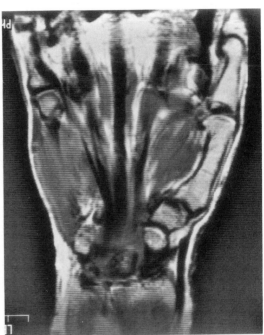

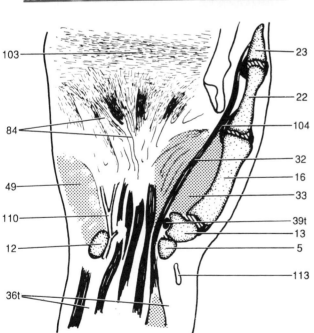

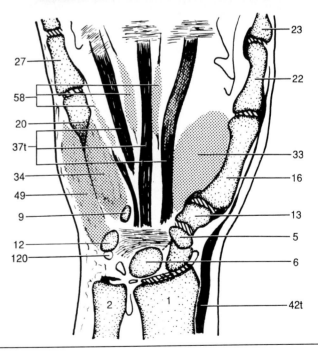

Bone and Cartilage	
1	Radius
2	Ulna
5	Scaphoid Bone
6	Lunate Bone
9	Hook of Hamate Bone
12	Pisiform Bone
13	Trapezium Bone
16	Metacarpal Bone I
20	Metacarpal Bone V

22 Proximal Phalanx Digit I
23 Distal Phalanx Digit I
27 Proximal Phalanx Digit V

Muscle
32 Flexor Pollicis Longus Muscle
33 Flexor Pollicis Brevis Muscle
34 Flexor Digiti Minimi Brevis and/or Opponens Digiti Minimi Brevis Muscle

36 Flexor Digitorum Superficialis Muscle
37 Flexor Digitorum Profundus Muscle
39 Flexor Carpi Radialis Muscle
42 Extensor Pollicis Brevis Muscle
49 Abductor Digiti Minimi Muscle
58 Lumbrical Muscle

Ligament and Tendon
84 Palmar Aponeurosis

103 Transverse Fasciculi fo the Palmar Aponeurosis
104 Volar Plate

Neurovascular and Miscellaneous
110 Ulnar Nerve
113 Radial Artery
120 Silicone Barium Contrast in Proximal Recess of Pisotriquetral Joint

Figure 4.14. Coronal section of the wrist and midhand. Of special interest are the interosseous muscles with their insertions at the phalanges.

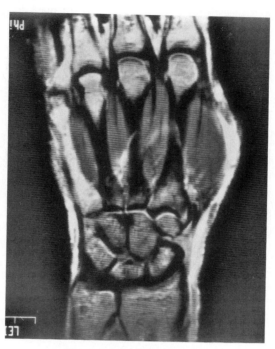

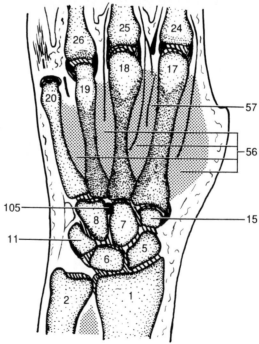

Bone and Cartilage
1 Radius
2 Ulna
5 Scaphoid Bone
6 Lunate Bone
7 Capitate Bone

8 Hamate Bone
11 Triquetrum
15 Trapezoid Bone
17 Metacarpal Bone II
18 Metacarpal Bone III
19 Metacarpal Bone IV

20 Metacarpal Bone V
24 Proximal Phalanx Digit II
25 Proximal Phalanx Digit III
26 Proximal Phalanx Digit IV

Muscle
56 Interosseous Muscle (dorsal)
57 Interosseous Muscle (palmar)

Ligament and Tendon
105 Interosseus Ligament Between
Capitate and Hamate Bone

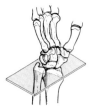

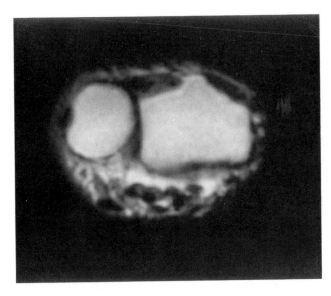

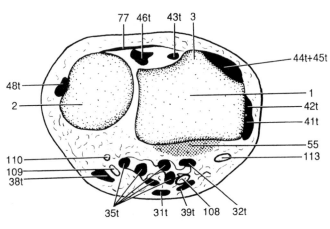

Figure 4.15. Axial section of the distal forearm at the level of the distal radioulnar joint of a normal volunteer.

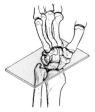

Figure 4.16. Axial section at the midlevel of the carpus of a normal volunteer.

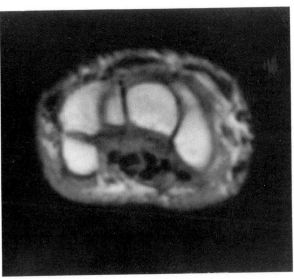

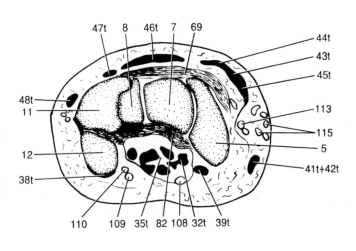

Bone and Cartilage
1 Radius
2 Ulna
3 Lister's Tubercle
5 Scaphoid Bone
7 Capitate Bone
8 Hamate Bone
11 Triquetrum
12 Pisiform Bone

Muscle
31 Palmaris Longus Muscle
32 Flexor Pollicis Longus Muscle
35 Flexor Digitorum Muscle
38 Flexor Carpi Ulnaris Muscle
39 Flexor Carpi Radialis Muscle
41 Abductor Pollicis Longus Muscle
42 Extensor Pollicis Brevis Muscle
43 Extensor Pollicis Longus Muscle

44 Extensor Carpi Radialis Brevis Muscle
45 Extensor Carpi Radialis Longus Muscle
46 Extensor Digitorum Communis and Extensor Indicis Muscle
47 Extensor Digiti Minimi Muscle
48 Extensor Carpi Ulnaris Muscle
55 Pronator Quadratus Muscle

Ligament and Tendon
69 Dorsal Intercarpal Ligament

77 Extensor Retinaculum
82 Floor of the Carpal Canal, Consisting of Radiocarpal and Intercarpal Ligaments

Neurovascular and Miscellaneous
108 Median Nerve
109 Ulnar Artery
110 Ulnar Nerve
113 Radial Artery
115 Cutaneous Veins

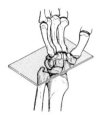

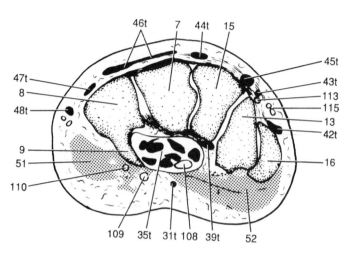

Figure 4.17.   Axial section at the level of the distal carpal row of a normal volunteer.

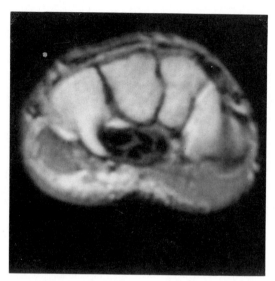

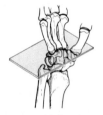

Figure 4.18.   Axial section at the level of the bases of the metacarpals of a normal volunteer.

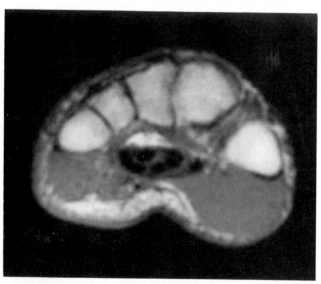

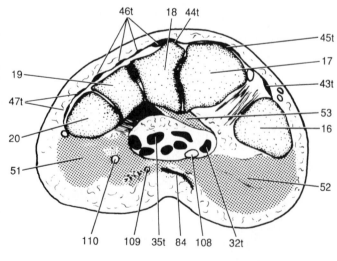

Bone and Cartilage
7   Capitate Bone
8   Hamate Bone
9   Hook of Hamate Bone
13 Trapezium Bone
15 Trapezoid Bone
16 Metacarpal Bone I
17 Metacarpal Bone II
18 Metacarpal Bone III
19 Metacarpal Bone IV
20 Metacarpal Bone V

Muscle
31 Palmaris Longus Muscle
32 Flexor Pollicis Longus Muscle
35 Flexor Digitorum Muscle
39 Flexor Carpi Radialis Muscle
42 Extensor Pollicis Brevis Muscle
43 Extensor Pollicis Longus Muscle
44 Extensor Carpi Radialis Brevis Muscle

45 Extensor Carpi Radialis Longus Muscle
46 Extensor Digitorum Communis and Extensor Indicis Muscle
47 Extensor Digiti Minimi Muscle
48 Extensor Carpi Ulnaris Muscle
51 Hypothenar Muscles
52 Thenar Muscles
53 Abductor Pollicis Muscle (Fibers)

Ligament and Tendon
84 Palmar Aponeurosis

Neurovascular and Miscellaneous
108 Median Nerve
109 Ulnar Artery
110 Ulnar Nerve
113 Radial Nerve
115 Cutaneous Veins

Figure 4.19. Coronal section at the middle of the radiocarpal joint of a normal volunteer.

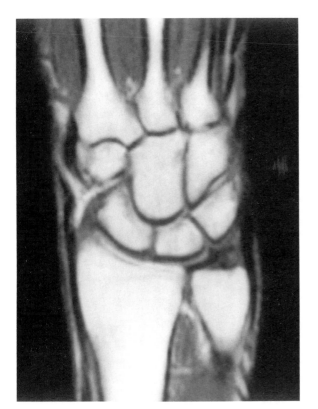

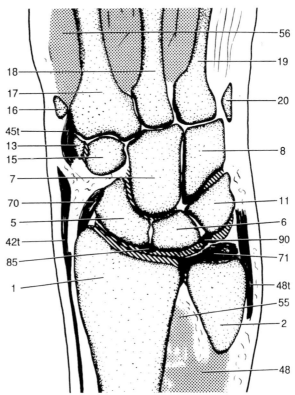

Bone and Cartilage
1 Radius
2 Ulna
5 Scaphoid Bone
6 Lunate Bone
7 Capitate Bone
8 Hamate Bone
11 Triquetrum

13 Trapezium Bone
15 Trapeziod Bone
16 Metacarpal Bone I
17 Metacarpal Bone II
18 Metacarpal Bone III
19 Metacarpal Bone IV
20 Metacarpal Bone V

Muscle
42 Extensor Pollicis Brevis Muscle
45 Extensor Carpi Radialis Longus Muscle
48 Extensor Carpi Ulnaris Muscle
55 Pronator Quadratus Muscle
56 Interosseous Muscle (dorsal)

Ligament and Tendon
70 Volar Capsule
71 Discus or Triangular Pad of Fibrocartilage
85 Scapholunate Interosseus Ligament
90 Lunotriquetral Interosseus Ligament

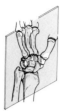

Figure 4.20. Coronal section at the volar side of the wrist of a normal volunteer.

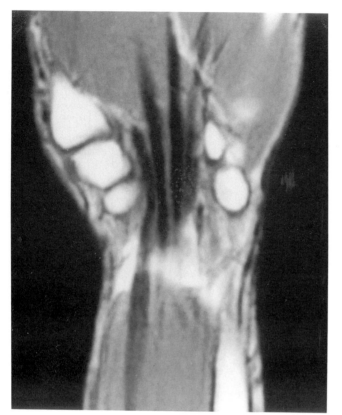

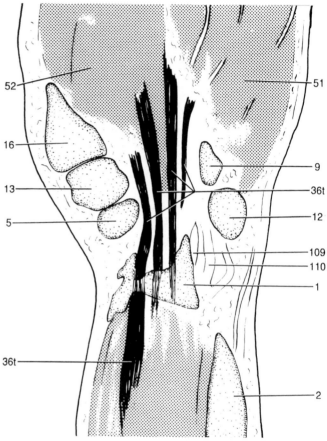

Bone and Cartilage
1   Radius
2   Ulna
5   Scaphoid Bone
9   Hook of Hamate Bone

12   Pisiform Bone
13   Trapezium Bone
16   Metacarpal Bone I

Muscle
36   Flexor Digitorum Superficialis
      Muscle
51   Hypothenar Muscle
52   Thenar Muscle

Neurovascular and Miscellaneous
109   Ulnar Artery
110   Ulnar Nerve

Figure 4.21. Sagittal section at the radial side of the wrist of a normal volunteer.

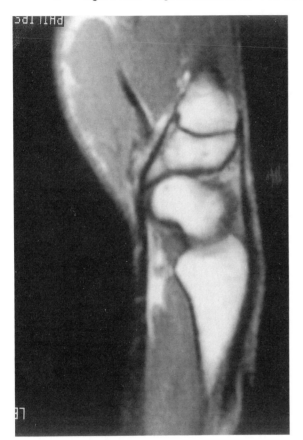 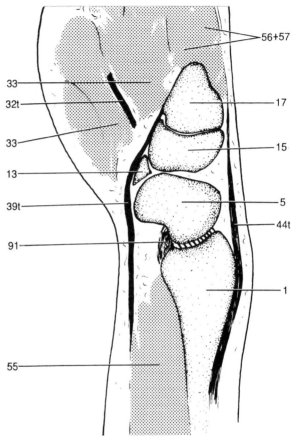

Bone and Cartilage
1 Radius
5 Scaphoid Bone
13 Trapezium Bone

15 Trapeziod Bone
17 Metacarpal Bone II
Muscle
32 Flexor Pollicis Longus Muscle
33 Flexor Pollicis Brevis Muscle

39 Flexor Carpi Radialis Muscle
44 Extensor Carpi Radialis Brevis
    Muscle
55 Pronator Quadratus Muscle
56 Interosseous Muscle (dorsal)

57 Interosseous Muscle (palmar)
Ligament and Tendon
91 Radiolunotriquetral Ligament

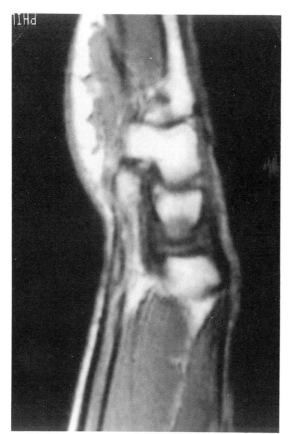 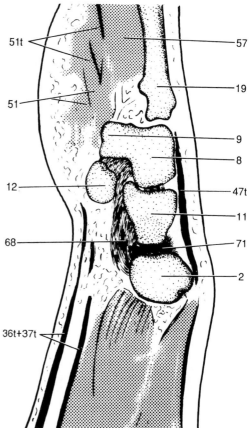

Figure 4.22. Sagittal section at the ulnar side of the wrist of a normal volunteer.

Bone and Cartilage
2 Ulna
8 Hamate Bone
9 Hook of Hamate Bone
11 Triquetrum
12 Pisiform Bone
19 Metacarpal Bone IV

Muscle
36 Flexor Digitorum Superficialis Muscle
37 Flexor Digitorum Profundus Muscle
47 Extensor Digiti Minimi Muscle
51 Hypothenar Muscles
57 Interosseous Muscle (palmar)

Ligament and Tendon
68 Volar Ulnotriquetral Ligament
71 Discus or Triangular Pad of Fibrocartilage

CHAPTER 5

The Adult Pelvis and Hip

Edwin van der Linden and Ab Verbout

Anatomy of the Pelvis and Hip Joint

Bone and Cartilage

1 Femur
2 Head of Femur
3 Neck of Femur
4 Greater Trochanter
5 Lesser Trochanter
6 Fovea of Femoral Head
7 Acetabulum
8 Roof of Acetabulum
9 Iliac Bone
10 Illiac Crest
11 Anterior Iliac Spine
12 Ischial Bone
13 Ischial Tuberosity
14 Ramus of Ischium
15 Ischial Spine
16 Sacral Ala
17 Coccyx Bone
18 Iliosacral Joint
19 Acetabular Rim
20 Pubic Bone
21 Inferior Ramus of Pubis
22 Pubic Symphysis
23 Articular (Hyaline) Cartilage
24 Labrum of Acetabulum
25 Articular Capsule

Muscle

31 Rectus Abdominis Muscle
32 Internal Oblique Abdominal Muscle
33 Transversus Abdominis Muscle
34 Iliacus Muscle
35 Psoas Major Muscle
36 Iliopsoas Muscle
37 Sartorius Muscle
38 Tensor Fasciae Latae Muscle
39 Gluteus Minimus Muscle
40 Gluteus Medius Muscle
41 Gluteus Maximus Muscle
42 Piriformis Muscle
43 Obturator Internus Muscle
44 Obturator Externus Muscle
45 Gracilis Muscle
46 Pectineus Muscle
47 Adductor Brevis Muscle
48 Adductor Magnus Muscle
49 Adductor Longus Muscle
50 Biceps Femoris Muscle
51 Semimembranosus Muscle
52 Semitendinosus Muscle
53 Pyramidalis Muscle
54 Medial or Lateral Hamstrings
55 Inferior Gemellus Muscle
56 Superior Gemellus Muscle
57 Quadratus Femoris Muscle
58 Rectus Femoris Muscle
59 Vastus Lateralis Muscle
60 Vastus Medialis Muscle
61 Vastus Intermedius Muscle
62 Multifidus Muscle
63 Levator Ani Muscle

Ligament and Tendon

71 Transverse Acetabular Ligament
72 Iliofemoral Ligament
73 Lateral or Medial Intermuscular Septum
74 Ligament of Head of Femur (Ligamentum Teres)
76 Acetabulum Fossa (Pulvinar)
77 Sacrospinous Ligament
78 Capsular Ligament
79 Iliotibial Tract

Neurovascular and Miscellaneous

91 Femoral Nerve
92 Sciatic Nerve
93 External Iliac Artery and Vein
94 Femoral Artery and Vein
95 Deep Femoral Artery and Vein
96 Spinal Canal and Cauda Equina
97 Trochanteric Fossa
98 Sacral Root Nerve
99 Inferior Gluteal Vessels
100 Superior Gluteal Vessels

Abbreviation: t, tendon.

Acknowledgments: Anatomic drawings were made by Bas B. Blankevoort and J. Wetselaar-Whittaker.

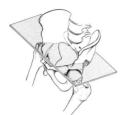

Figures 5.1–5.7. Axial MR images.

Figure 5.1.

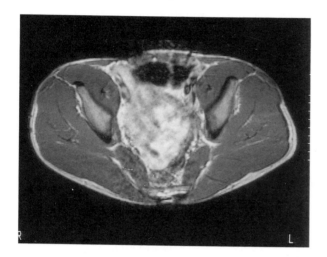

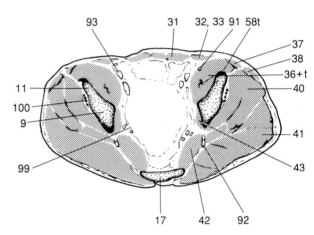

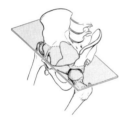

Figure 5.2.

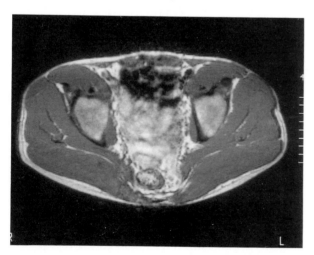

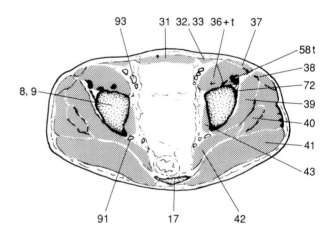

Bone and Cartilage
8   Roof of Acetabulum
9   Iliac Bone
11   Anterior Iliac Spine
17   Coccyx Bone

Muscle
31   Rectus Abdominis Muscle

32   Internal Oblique Abdominal
      Muscle
33   Transversus Abdominis Muscle
36   Iliopsoas Muscle
37   Sartorius Muscle
38   Tensor Fasciae Latae Muscle
39   Gluteus Minimus Muscle

40   Gluteus Medius Muscle
41   Gluteus Maximus Muscle
42   Piriformis Muscle
43   Obturator Internus Muscle
58   Rectus Femoris Muscle

Ligament and Tendon
72   Iliofemoral Ligament

Neurovascular and Miscellaneous
91   Femoral Nerve
92   Sciatic Nerve
93   External Iliac Artery and Vein
99   Sacral Root Nerve
100   Superior Gluteal Vessels

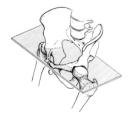

Figure 5.3.

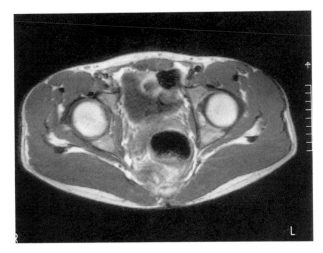

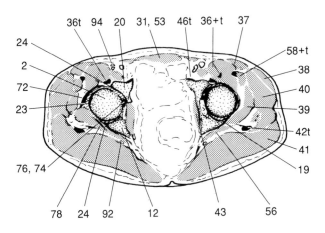

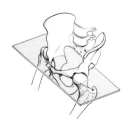

Figure 5.4.

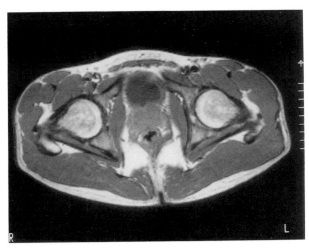

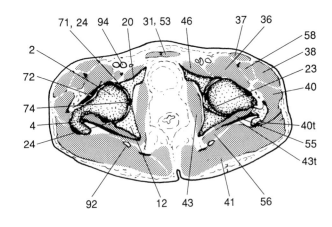

Bone and Cartilage
2   Head of Femur
4   Greater Trochanter
12 Ischial Bone
19 Acetabular Rim
20 Pubic Bone
23 Articular (Hyaline) Cartilage
24 Labrum of Acetabulum

Muscle
31 Rectus Abdominis Muscle
36 Iliopsoas Muscle
37 Sartorius Muscle
38 Tensor Fasciae Latae Muscle
39 Gluteus Minimus Muscle
40 Gluteus Medius Muscle
41 Gluteus Maximus Muscle
42 Piriformis Muscle

43 Obturator Internus Muscle
46 Pectineus Muscle
53 Pyramidalis Muscle
55 Inferior Gemellus Muscle
56 Superior Gemellus Muscle
58 Rectus Femoris Muscle

Ligament and Tendon
71 Transverse Acetabular Ligament

72 Iliofemoral Ligament
74 Ligament of Head of Femur (Ligamentum Teres)
76 Acetabulum Fossa (Pulvinar)
78 Capsular Ligament

Neurovascular and Miscellaneous
92 Sciatic Nerve
94 Femoral Artery and Vein

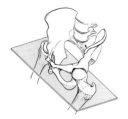

Figure 5.5.

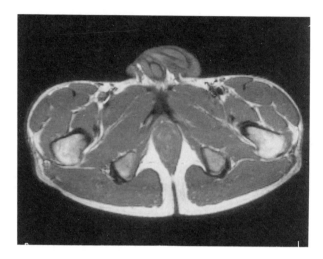

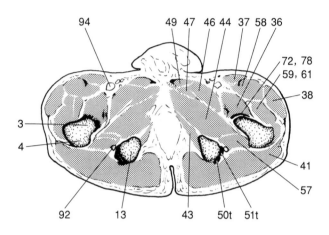

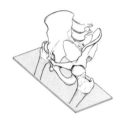

Figure 5.6.

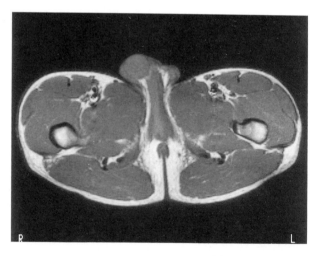

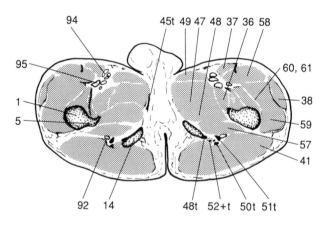

Bone and Cartilage
1   Femur
3   Neck of Femur
4   Greater Trochanter
5   Lesser Trochanter
13 Ischial Tuberosity
14 Ramus of Ischium

Muscle
36 Iliopsoas Muscle
37 Sartorius Muscle
38 Tensor Fasciae Latae Muscle
41 Gluteus Maximus Muscle
43 Obturator Internus Muscle
44 Obturator Externus Muscle
45 Gracilis Muscle
46 Pectineus Muscle

47 Adductor Brevis Muscle
48 Adductor Magnus Muscle
49 Adductor Longus Muscle
50 Biceps Femoris Muscle
51 Semimembranosus Muscle
52 Semitendinosus Muscle
57 Quadratus Femoris Muscle
58 Rectus Femoris Muscle
59 Vastus Lateralis Muscle

60 Vastus Medialis Muscle
61 Vastus Intermedius Muscle

Ligament and Tendon
72 Iliofemoral Ligament
78 Capsular Ligament

Neurovascular and Miscellaneous
92 Sciatic Nerve
94 Femoral Artery and Vein
95 Deep Femoral Artery and Vein

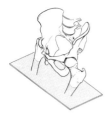

Figure 5.7.

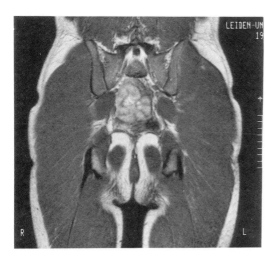

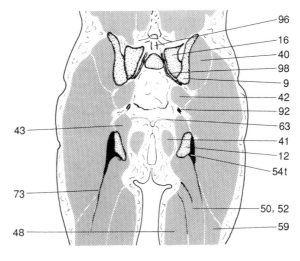

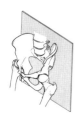

Figures 5.8–5.13.   Coronal MR images.

Figure 5.8.

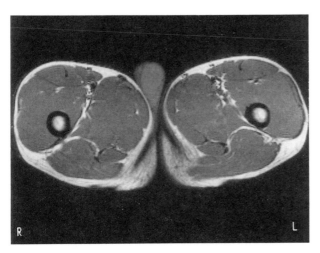

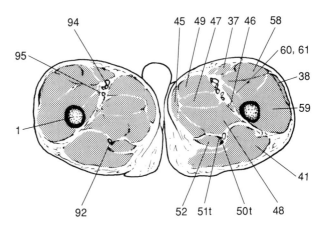

Bone and Cartilage
  1 Femur
  9 Iliac Bone
 12 Ischial Bone
 16 Sacral Ala

Muscle
 37 Sartorius Muscle
 38 Tensor Fasciae Latae Muscle

 40 Gluteus Medius Muscle
 41 Gluteus Maximus Muscle
 42 Piriformis Muscle
 43 Obturator Internus Muscle
 45 Gracilis Muscle
 46 Pectineus Muscle
 47 Adductor Brevis Muscle
 48 Adductor Magnus Muscle
 49 Adductor Longus Muscle

 50 Biceps Femoris Muscle
 51 Semimembranosus Muscle
 52 Semitendinosus Muscle
 54 Medial or Lateral Hamstrings
 58 Rectus Femoris Muscle
 59 Vastus Lateralis Muscle
 60 Vastus Medialis Muscle
 61 Vastus Intermedius Muscle
 63 Levator Ani Muscle

Ligament and Tendon
 73 Lateral or Medial Intermuscular
     Septum

Neurovascular and Miscellaneous
 92 Sciatic Nerve
 94 Femoral Artery and Vein
 95 Deep Femoral Artery and Vein
 96 Spinal Canal and Cauda Equina
 98 Sacral Root Nerve

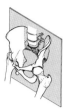

Figure 5.9.

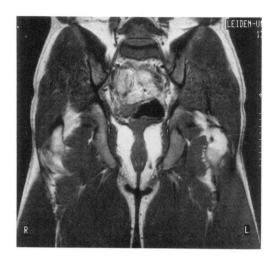

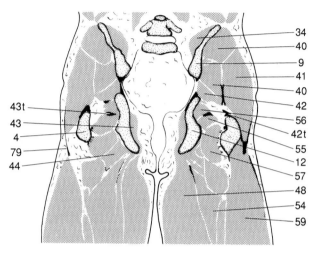

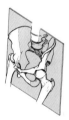

Figure 5.10.

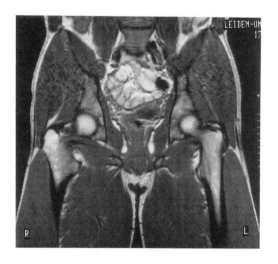

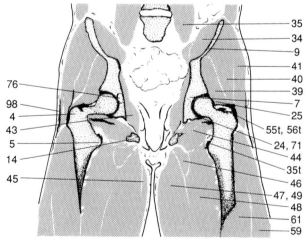

Bone and Cartilage
4   Greater Trochanter
5   Lesser Trochanter
7   Acetabulum
9   Iliac Bone
12 Ischial Bone
14 Ramus of Ischium
24 Labrum of Acetabulum
25 Articular Capsule

Muscle
34 Iliacus Muscle
35 Psoas Major Muscle
39 Gluteus Minimus Muscle
40 Gluteus Medius Muscle
41 Gluteus Maximus Muscle
42 Piriformis Muscle
43 Obturator Internus Muscle
44 Obturator Externus Muscle

45 Gracilis Muscle
46 Pectineus Muscle
47 Adductor Brevis Muscle
48 Adductor Magnus Muscle
49 Adductor Longus Muscle
54 Medial or Lateral Hamstrings
55 Inferior Gemellus Muscle
56 Superior Gemellus Muscle
57 Quadratus Femoris Muscle

59 Vastus Lateralis Muscle
61 Vastus Intermedius Muscle

Ligament and Tendon
71 Transverse Acetabular Ligament
76 Acetabulum Fossa (Pulvinar)
79 Iliotibial Tract

Neurovascular and Miscellaneous
98 Sacral Root Nerve

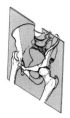

Figure 5.11.

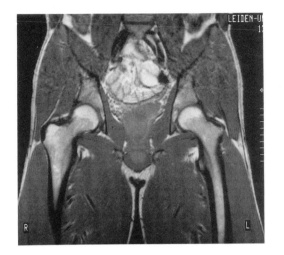
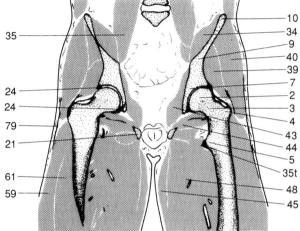

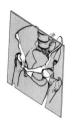

Figure 5.12.

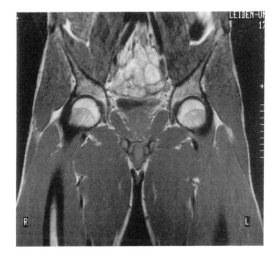
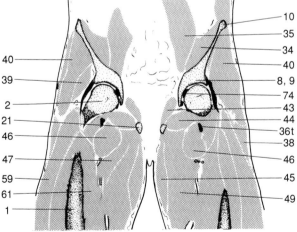

Bone and Cartilage
1　Femur
2　Head of Femur
3　Neck of Femur
4　Greater Trochanter
5　Lesser Trochanter
7　Acetabulum
8　Roof of Acetabulum

9　Iliac Bone
10　Iliac Crest
21　Inferior Ramus of Pubis
24　Labrum of Acetabulum

Muscle
34　Iliacus Muscle
35　Psoas Major Muscle
36　Iliopsoas Muscle

38　Tensor Fasciae Latae Muscle
39　Gluteus Minimus Muscle
40　Gluteus Medius Muscle
43　Obturator Internus Muscle
44　Obturator Externus Muscle
45　Gracilis Muscle
46　Pectineus Muscle
47　Adductor Brevis Muscle

48　Adductor Magnus Muscle
49　Adductor Longus Muscle
59　Vastus Lateralis Muscle
61　Vastus Intermedius Muscle

Ligament and Tendon
74　Ligament of Head of Femur (Lig-
　　amentum Teres)
79　Iliotibial Tract

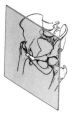

Figure 5.13.

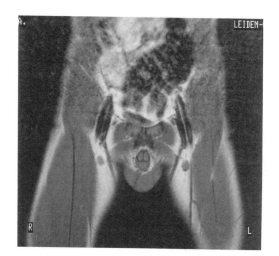 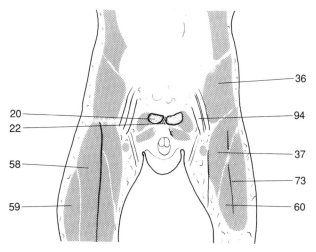

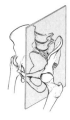

Figures 5.14–5.17. Sagittal MR images.

Figure 5.14.

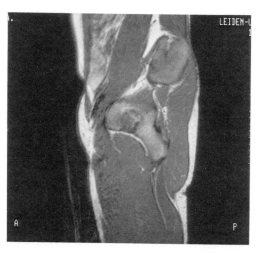 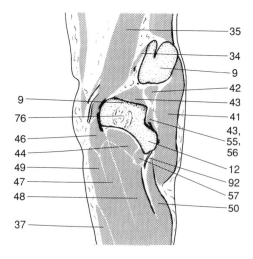

Bone and Cartilage
9 Iliac Bone
12 Ischial Bone
20 Pubic Bone
22 Pubic Symphysis

Muscle
34 Iliacus Muscle
35 Psoas Major Muscle

36 Iliopsoas Muscle
37 Sartorius Muscle
41 Gluteus Maximus Muscle
42 Piriformis Muscle
43 Obturator Internus Muscle
44 Obturator Externus Muscle
46 Pectineus Muscle
47 Adductor Brevis Muscle

48 Adductor Magnus Muscle
49 Adductor Longus Muscle
50 Biceps Femoris Muscle
55 Inferior Gemellus Muscle
56 Superior Gemellus Muscle
57 Quadratus Femoris Muscle
58 Rectus Femoris Muscle
59 Vastus Lateralis Muscle
60 Vastus Medialis Muscle

Ligament and Tendon
73 Lateral or Medial Intermuscular
    Septum
76 Acetabulum Fossa (Pulvinar)

Neurovascular and Miscellaneous
92 Sciatic Nerve
94 Femoral Artery and Vein

Figure 5.15.

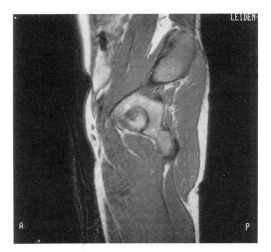

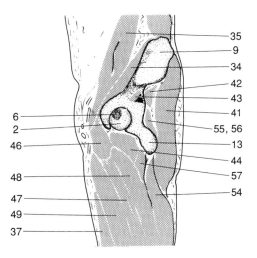

35
9
34
42
43
41
6
2
55, 56
46
13
44
48
57
54
47
49
37

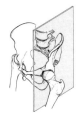

Figure 5.16.

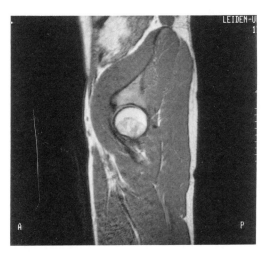

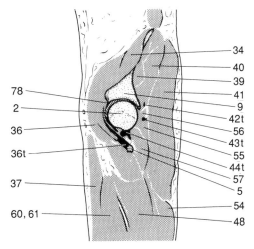

34
40
39
41
78
9
2
42t
36
56
43t
36t
55
44t
57
37
5
54
60, 61
48

Bone and Cartilage
2 Head of Femur
5 Lesser Trochanter
6 Fovea of Femoral Head
9 Iliac Bone
13 Ischial Tuberosity

Muscle
34 Iliacus Muscle
35 Psoas Major Muscle
36 Iliopsoas Muscle
37 Sartorius Muscle
39 Gluteus Minimus Muscle
40 Gluteus Medius Muscle
41 Gluteus Maximus Muscle

42 Piriformis Muscle
43 Obturator Internus Muscle
44 Obturator Externus Muscle
46 Pectineus Muscle
47 Adductor Brevis Muscle
48 Adductor Magnus Muscle
49 Adductor Longus Muscle
54 Medial or Lateral Hamstrings

55 Interior Gemellus Muscle
56 Superior Gemellus Muscle
57 Quadratus Femoris Muscle
60 Vastus Medialis Muscle
61 Vastus Intermedius Muscle

Ligament and Tendon
78 Capsular Ligament

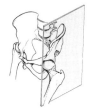

Figure 5.17.

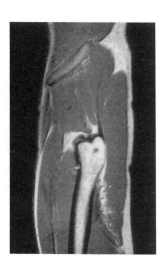

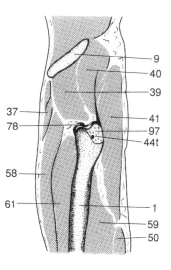

Bone and Cartilage
1 Femur
9 Iliac Bone

Muscle
37 Sartorius Muscle
39 Gluteus Minimus Muscle
40 Gluteus Medius Muscle
41 Gluteus Maximus Muscle

44 Obturator Externus Muscle
50 Biceps Femoris Muscle
58 Rectus Femoris Muscle
59 Vastus Lateralis Muscle
61 Vastus Intermedius Muscle

Ligament and Tendon
78 Capsular Ligament

Neurovascular and Miscellaneous
97 Trochanteric Fossa

CHAPTER 6

The Pediatric Hip

Ab Verbout, Cees F. A. Bos, and Edwin van der Linden

Anatomy of the Pediatric Hip

Bone and Cartilage

1 Femur
2 Head of Femur
3 Chondroepiphyseal Head of Femur
4 Neck of Femur
5 Secondary Ossification Center of Femur
6 Greater Trochanter
7 Lesser Trochanter
8 Acetabulum
9 Cartilaginous Acetabulum
10 Pubic Part of Acetabulum
11 Ischial Part of Acetabulum
12 Labrum of Acetabulum
13 Iliac Bone
14 Ossification Center of Ilium
15 Cartilaginous Ilium
16 Ischial Bone
17 Ramus of Ischium (Ischiopubic Arch)
18 Pubic Bone
19 Inferior Ramus of Pubis (Ischiopubic Arch)
20 Coccyx Bone
21 Growth Plate

Muscle

31 Abdominal Muscles
32 Iliacus Muscle
33 Psoas Major Muscle
34 Iliopsoas Muscle
35 Sartorius Muscle
36 Tensor Fasciae Latae Muscle
37 Gluteus Minimus Muscle
38 Gluteus Medius Muscle
39 Gluteus Maximus Muscle
40 Piriformis Muscle
41 Obturator Internus Muscle
42 Obturator Externus Muscle
43 Gracilis Muscle
44 Pectineus Muscle
45 Adductor Brevis Muscle
46 Adductor Magnus Muscle
47 Biceps Femoris Muscle
48 Semimembranosus Muscle
49 Semitendinosus Muscle
50 Hamstring Muscles
51 Inferior Gemellus Muscle
52 Superior Gemellus Muscle
53 Quadratus Femoris Muscle
54 Rectus Femoris Muscle
55 Vastus Lateralis Muscle
56 Vastus Medialis Muscle
57 Vastus Intermedius Muscle

Ligament and Tendon

71 Sacrotuberous Ligament
72 Ischiofemoral Ligament
73 Ligament of Head of Femur (Ligamentum Teres)
74 Iliofemoral Ligament
75 Ligamentous Capsule
76 Zona Orbicularis
77 Transverse Ligament of Acetabulum
78 Iliotibial Tract

Neurovascular and Miscellaneous

91 Sciatic Nerve
92 Acetabulum Fossa (Pulvinar)
93 Lymph Nodes
94 Femoral Nerve
95 Femoral Artery and Vein
96 Lateral Circumflex Artery
97 Iliac Artery and Vein
98 Superior Gluteal Vessels
99 Inferior Gluteal Vessels

Abbreviation: t, tendon.

Acknowledgments: Anatomic drawings were made by Bas B. Blankevoort and J. Wetselaar-Whittaker.

Figures 6.1–6.2. Axial MR images.

Figure 6.1.

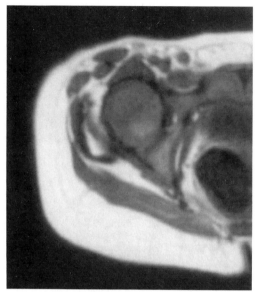

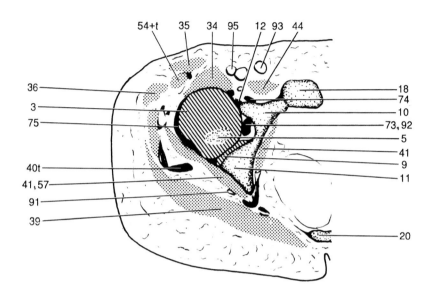

Figure 6.2.

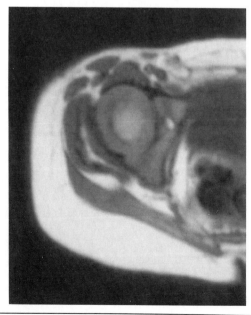

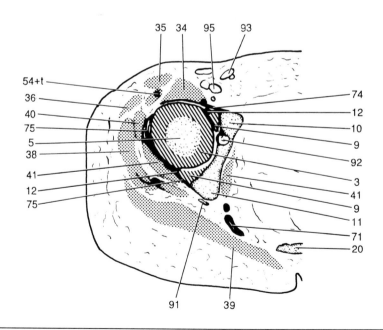

Bone and Cartilage

3 Chondroepiphyseal Head of Femur
5 Secondary Ossification Center of Femur
9 Cartilaginous Acetabulum
10 Pubic Part of Acetabulum
11 Ischial Part of Acetabulum
12 Labrum of Acetabulum
18 Pubic Bone
20 Coccyx Bone

Muscle

34 Iliopsoas Muscle
35 Sartorius Muscle
36 Tensor Fasciae Latae Muscle
38 Gluteus Medius Muscle
39 Gluteus Maximus Muscle
40 Piriformis Muscle
41 Obturator Internus Muscle
44 Pectineus Muscle
54 Rectus Femoris Muscle
57 Vastus Intermedius Muscle

Ligament and Tendon

71 Sacrotuberous Ligament
73 Ligament of Head of Femur (Ligamentum Teres)
74 Iliofemoral Ligament
75 Ligamentous Capsule

Neurovascular and Miscellaneous

91 Sciatic Nerve
92 Acetabulum Fossa (Pulvinar)
93 Lymph Nodes
95 Femoral Artery and Vein

Figures 6.3–6.4.   Coronal MR images.

Figure 6.3.

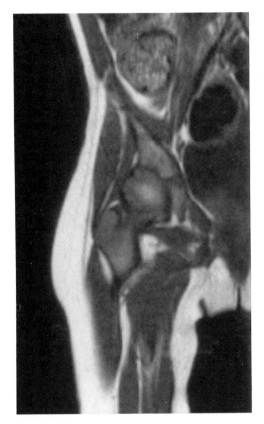

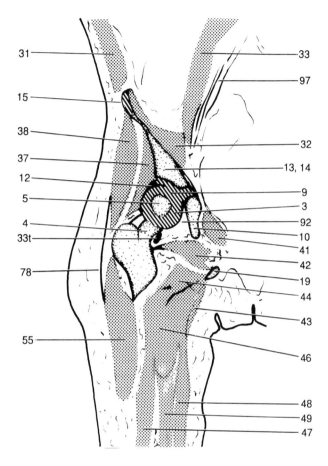

Bone and Cartilage
3 Chondroepiphyseal Head of Femur
4 Neck of Femur
5 Secondary Ossification Center of Femur
9 Cartilaginous Acetabulum
10 Pubic Part of Acetabulum
12 Labrum of Acetabulum
13 Iliac Bone

14 Ossification Center of Ilium
15 Cartilaginous Ilium
19 Inferior Ramus of Pubis (Ischiopubic Arch)

Muscle
31 Abdominal Muscle
32 Iliacus Muscle
33 Psoas Major Muscle
37 Gluteus Minimus Muscle

38 Gluteus Medius Muscle
41 Obturator Internus Muscle
42 Obturator Externus Muscle
43 Gracilis Muscle
44 Pectineus Muscle
46 Adductor Magnus Muscle
47 Biceps Femoris Muscle
48 Semimembranosus Muscle
49 Semitendinosus Muscle

55 Vastus Lateralis Muscle

Ligament and Tendon
78 Iliotibial Tract

Neurovascular and Miscellaneous
92 Acetabulum Fossa (Pulvinar)
97 Iliac Artery and Vein

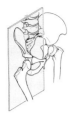

Figure 6.4.

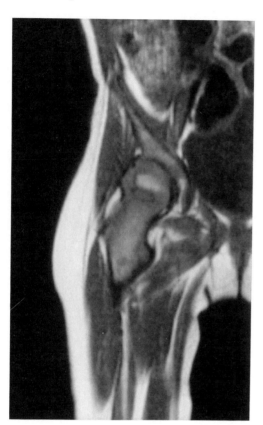

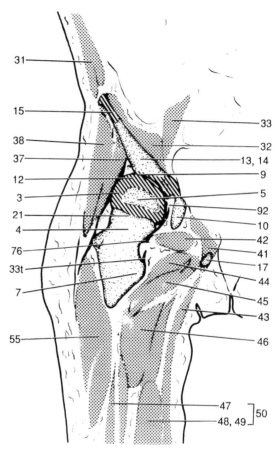

Bone and Cartilage
3 Chondroepiphyseal Head of Femur
4 Neck of Femur
5 Secondary Ossification Center of Femur
7 Lesser Trochanter
9 Cartilaginous Acetabulum
10 Pubic Part of Acetabulum
12 Labrum of Acetabulum

13 Iliac Bone
14 Ossification Center of Ilium
15 Cartilaginous Ilium
17 Ramus of Ischium (Ischiopubic Arch)
21 Growth Plate

Muscle
31 Abdominal Muscles
32 Iliacus Muscle

33 Psoas Major Muscle
37 Gluteus Minimus Muscle
38 Gluteus Medius Muscle
41 Obturator Internus Muscle
42 Obturator Externus Muscle
43 Gracilis Muscle
44 Pectineus Muscle
45 Adductor Brevis Muscle
46 Adductor Magnus Muscle

47 Biceps Femoris Muscle
48 Semimembranosus Muscle
49 Semitendinosus Muscle
50 Hamstring Muscles
55 Vastus Lateralis Muscle

Ligament and Tendon
76 Zona Orbicularis

Neurovascular and Miscellaneous
92 Acetabulum Fossa (Pulvinar)

Figures 6.5–6.6. Sagittal MR images.

Figure 6.5.

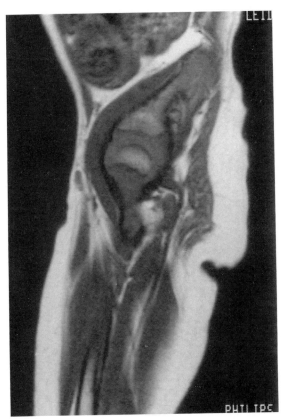

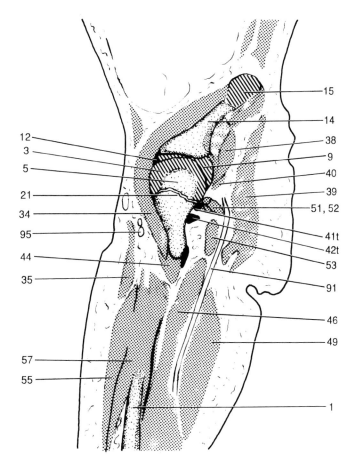

Bone and Cartilage
1 Femur
3 Chondroepiphyseal Head of Femur
5 Secondary Ossification Center of
   Femur
9 Cartilaginous Acetabulum
12 Labrum of Acetabulum

14 Ossification Center of Ilium
15 Cartilaginous Ilium
21 Growth Plate

Muscle
34 Iliopsoas Muscle
35 Sartorius Muscle
38 Gluteus Medius Muscle

39 Gluteus Maximus Muscle
40 Piriformis Muscle
41 Obturator Internus Muscle
42 Obturator Externus Muscle
44 Pectineus Muscle
46 Adductor Magnus Muscle
49 Semitendinosus Muscle
51 Inferior Gemellus Muscle

52 Superior Gemellus Muscle
53 Quadatus Femoris Muscle
55 Vastus Lateralis Muscle
57 Vastus Intermedius Muscle

Neurovascular and Miscellaneous
91 Sciatic Nerve
95 Femoral Artery and Vein

Figure 6.6.

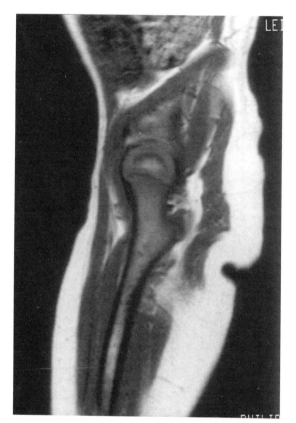

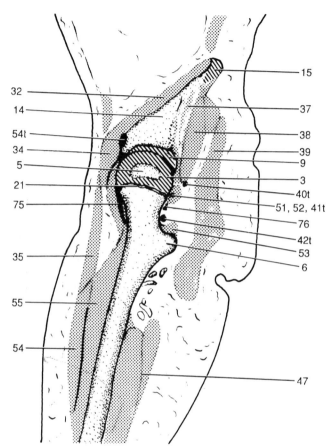

32	15
14	37
54t	38
34	39
5	9
21	3
75	40t
	51, 52, 41t
	76
35	42t
	53
	6
55	
54	
	47

Bone and Cartilage
3 Chondroepiphyseal Head of Femur
5 Secondary Ossification Center of Femur
6 Greater Trochanter
9 Cartilaginous Acetabulum
14 Ossification Center of Ilium
15 Cartilaginous Ilium
21 Growth Plate

Muscle
32 Iliacus Muscle
34 Iliopsoas Muscle
35 Sartorius Muscle
37 Gluteus Minimus Muscle
38 Gluteus Medius Muscle
39 Gluteus Maximus Muscle
40 Piriformis Muscle
41 Obturator Internus Muscle
42 Obturator Externus Muscle
47 Biceps Femoris Muscle
51 Inferior Gemellus Muscle
52 Superior Gemellus Muscle
53 Quadratus Femoris Muscle
54 Rectus Femoris Muscle
55 Vastus Lateralis Muscle

Ligament and Tendon
75 Ligamentous Capsule
76 Zona Orbicularis

CHAPTER 7

The Knee

Edwin van der Linden and Ab Verbout

Anatomy of the Knee Joint

Bone and Cartilage

1　Femur
2　Medial Condyle of Femur
3　Lateral Condyle of Femur
4　Fibula
5　Patella
6　Tibia
7　Articular (Hyaline) Cartilage
8　Lateral Meniscus
9　Medial Meniscus
10　Tibiofibular Joint
11　Tuberosity of Tibia
12　Articular Capsule

Muscle

31　Quadriceps Muscle
32　Vastus Lateralis Muscle
33　Vastus Medialis Muscle
34　Vastus Intermedius Muscle
35　Gastrocnemius Muscle (Lateral Head)
36　Gastrocnemius Muscle (Medial Head)
37　Sartorius Muscle

38　Semimembranosus Muscle
39　Gracilis Muscle
40　Semitendinosus Muscle
41　Biceps Femoris Muscle (Long Head)
42　Biceps Femoris Muscle (Short Head)
43　Plantar Muscle
44　Popliteal Muscle
45　Tibialis Anterior Muscle
46　Soleus Muscle
47　Adductor Magnus Muscle
48　Extensor Digitorum Longus Muscle
49　Peroneus Longus Muscle
50　Pes Anserinus

Ligament and Tendon

71　Iliotibial Tract
72　Anterior Cruciate Ligament
73　Posterior Cruciate Ligament
74　Fibular Collateral Ligament
75　Tibial Collateral Ligament
76　Patellar Ligament
77　Posterior Meniscofemoral Ligament

78　Anterior Meniscofemoral Ligament
79　Transverse Genicular Ligament
80　Lateral Patellar Retinaculum
81　Medial Patellar Retinaculum
82　Oblique Popliteal Ligament
83　Meniscofemoral Ligament

Neurovascular and Miscellaneous

91　Infrapatellar Fat Pad
92　Popliteal Artery and Vein
93　Tibial Nerve
94　Common Peroneal Nerve
95　Great Saphenous Vein
96　Sural Nerve
97　Small Saphenous Vein
98　Middle Genicular Artery and Vein
99　Superior Medial Genicular Artery and Vein
100　Inferior Medial Genicular Artery and Vein
101　Superior Lateral Genicular Artery and Vein
102　Inferior Lateral Genicular Artery and Vein
103　Suprapatellar Fat Pad

Abbreviation: t, tendon.

Acknowledgments: Anatomic drawings were made by Bas B. Blankevoort and J. Wetselaar-Whittaker. The assistance of W. R. Obermann, M.D., in preparing anatomic specimens is gratefully acknowledged.

Figures 7.1–7.8. Axial MR images.

Figure 7.1.

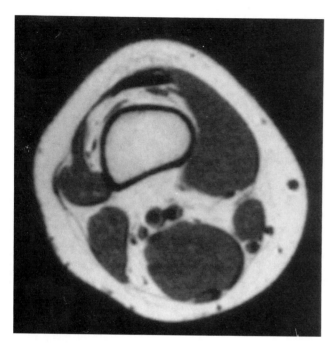

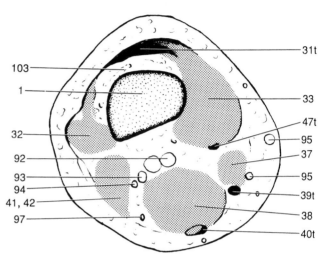

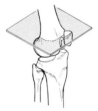

Figure 7.2.

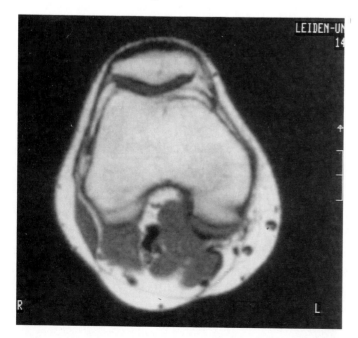

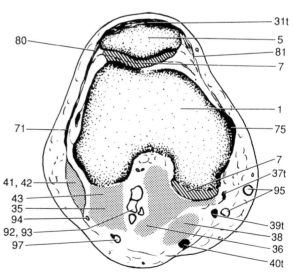

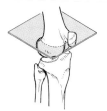

Figure 7.3.

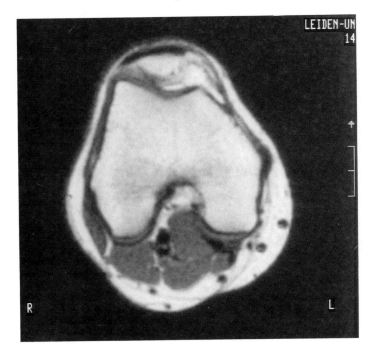

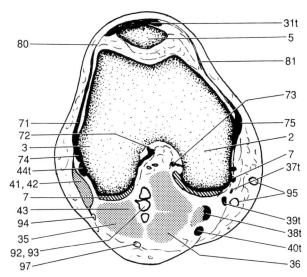

Bone and Cartilage
1　Femur
2　Medial Condyle of Femur
3　Lateral Condyle of Femur
5　Patella
7　Articular (Hyaline) Cartilage

Muscle
31　Quadriceps Muscle
32　Vastus Lateralis Muscle
33　Vastus Medialis Muscle
35　Gastrocnemius Muscle (Lateral Head)
36　Gastrocnemius Muscle (Medial Head)
37　Sartorius Muscle
38　Semimembranosus Muscle
39　Gracilis Muscle
40　Semitendinosus Muscle
41　Biceps Femoris Muscle (Long Head)
42　Biceps Femoris Muscle (Short Head)
43　Plantar Muscle
44　Popliteal Muscle
47　Adductor Magnus Muscle

Ligament and Tendon
71　Iliotibial Tract
72　Anterior Cruciate Ligament
73　Posterior Cruciate Ligament
74　Fibular Collateral Ligament
75　Tibial Collateral Ligament
80　Lateral Patellar Retinaculum
81　Medial Patellar Retinaculum

Neurovascular and Miscellaneous
92　Popliteal Artery and Vein
93　Tibial Nerve
94　Common Peroneal Nerve
95　Great Saphenous Vein
97　Small Saphenous Vein
103　Suprapatellar Fat Pad

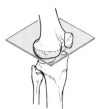

Figure 7.4.

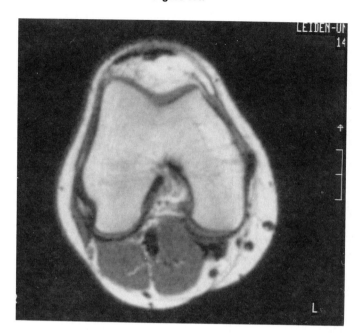

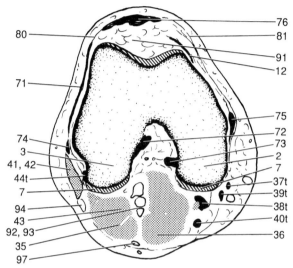

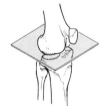

Figure 7.5.

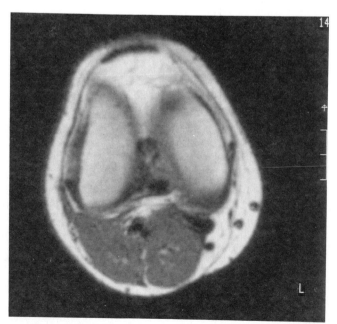

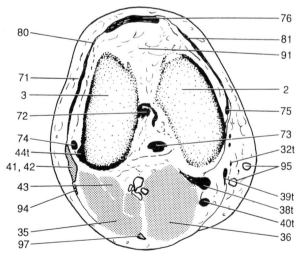

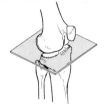

Figure 7.6.

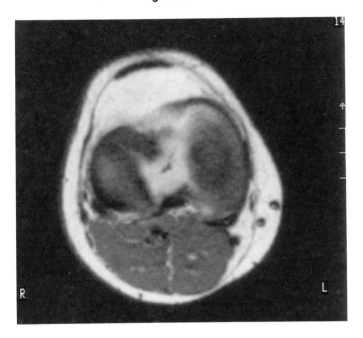

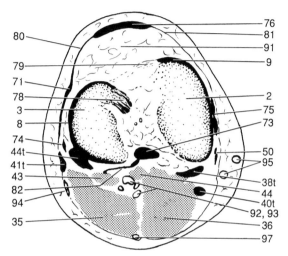

Bone and Cartilage
2 Medial Condyle of Femur
3 Lateral Condyle of Femur
7 Articular (Hyaline) Cartilage
8 Lateral Meniscus
9 Medial Meniscus
12 Articular Capsule

Muscle
32 Vastus Lateralis Muscle
35 Gastrocnemius Muscle (Lateral Head)
36 Gastrocnemius Muscle (Medial Head)
37 Sartorius Muscle
38 Semimembranosus Muscle
39 Gracilis Muscle
40 Semitendinosus Muscle
41 Biceps Femoris Muscle (Long Head)
42 Biceps Femoris Muscle (Short Head)
43 Plantar Muscle
44 Popliteal Muscle
50 Pes Anserinus

Tendon and Ligament
71 Iliotibial Tract
72 Anterior Cruciate Ligament
73 Posterior Cruciate Ligament
74 Fibular Collateral Ligament
75 Tibial Collateral Ligament
76 Patellar Ligament
78 Anterior Meniscofemoral Ligament
79 Transverse Genicular Ligament
80 Lateral Patellar Retinaculum
81 Medial Patellar Retinaculum
82 Oblique Popliteal Ligament

Neurovascular and Miscellaneous
91 Infrapatellar Fat Pad
92 Popliteal Artery and Vein
93 Tibial Nerve
94 Common Peroneal Vein
95 Great Saphenous Vein
97 Small Saphenous Vein

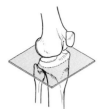

Figure 7.7.

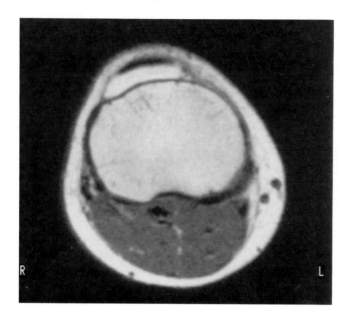

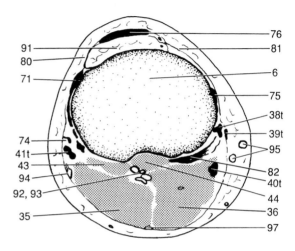

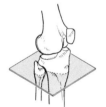

Figure 7.8.

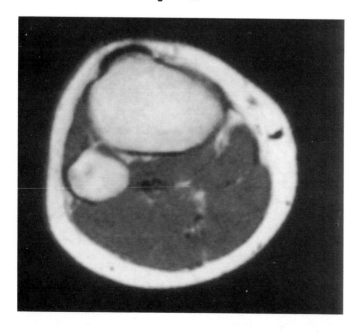

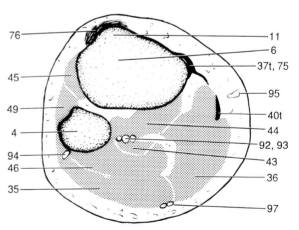

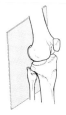

Figures 7.9–7.14.   Coronal MR images.

Figure 7.9.

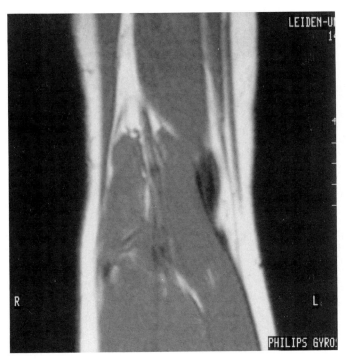

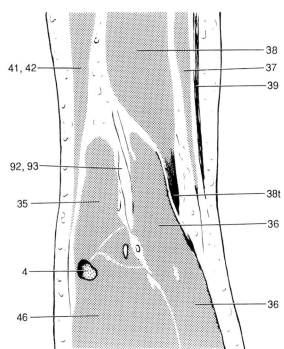

Bone and Cartilage
4 Fibula
6 Tibia
11 Tuberosity of Tibia

Muscle
35 Gastrocnemius Muscle (Lateral Head)
36 Gastrocnemius Muscle (Medial Head)
37 Sartorius Muscle
38 Semimembranosus Muscle
39 Gracilis Muscle
40 Semitendinosus Muscle
41 Biceps Femoris Muscle (Long Head)
42 Biceps Femoris Muscle (Short Head)
43 Plantar Muscle
44 Popliteal Muscle
45 Tibialis Anterior Muscle
46 Soleus Muscle
49 Peroneus Longus Muscle

Tendon and Ligament
71 Iliotibial Tract
74 Fibular Collateral Ligament
75 Tibial Collateral Ligament
76 Patellar Ligament
80 Lateral Patellar Retinaculum
81 Medial Patellar Retinaculum
82 Oblique Popliteal Ligament

Neurovascular and Miscellaneous
91 Infrapatellar Fat Pad
92 Popliteal Artery and Vein
93 Tibial Nerve
94 Common Peroneal Nerve
95 Great Saphenous Vein
97 Small Saphenous Vein

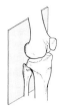

Figure 7.10.

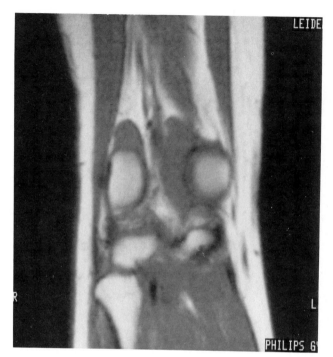

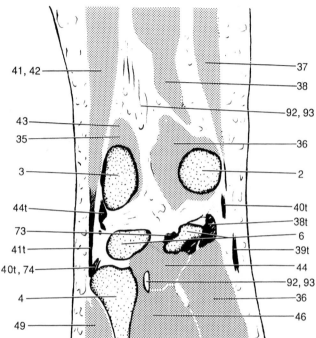

Bone and Cartilage
2 Medial Condyle of Femur
3 Lateral Condyle of Femur
4 Fibula
6 Tibia

Muscle
35 Gastrocnemius Muscle (Lateral Head)

36 Gastrocnemius Muscle (Medial Head)
37 Sartorius Muscle
38 Semimembranosus Muscle
39 Gracilis Muscle
40 Semitendinosus Muscle
41 Biceps Femoris Muscle (Long Head)

42 Biceps Femoris Muscle (Short Head)
43 Plantar Muscle
44 Popliteal Muscle
46 Soleus Muscle
49 Peroneus Longus Muscle

Ligament and Tendon
73 Posterior Cruciate Ligament
74 Fibular Collateral Ligament

Neurovascular and Miscellaneous
92 Popliteal Artery and Vein
93 Tibial Nerve

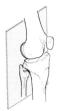

Figure 7.11.

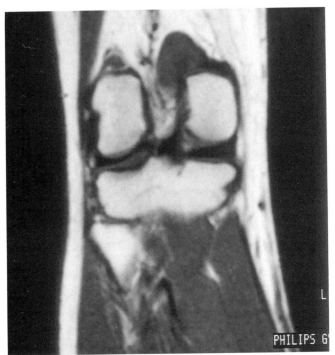

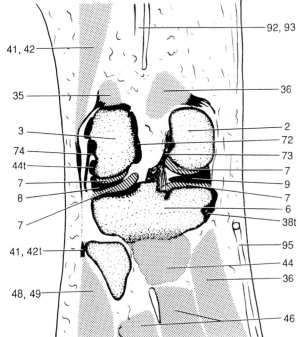

Bone and Cartilage
- 2 Medial Condyle of Femur
- 3 Lateral Condyle of Femur
- 6 Tibia
- 7 Articular (Hyaline) Cartilage
- 8 Lateral Meniscus
- 9 Medial Meniscus

Muscle
- 35 Gastrocnemius Muscle (Lateral Head)
- 36 Gastrocnemius Muscle (Medial Head)
- 38 Semimembranosus Muscle
- 41 Biceps Femoris Muscle (Long Head)
- 42 Biceps Femoris Muscle (Short Head)
- 44 Popliteal Muscle
- 46 Soleus Muscle
- 48 Extensor Digitorum Longus Muscle
- 49 Peroneus Longus Muscle

Ligament and Tendon
- 72 Anterior Cruciate Ligament
- 73 Posterior Cruciate Ligament
- 74 Fibular Collateral Ligament

Neurovascular and Miscellaneous
- 92 Popliteal Artery and Vein
- 93 Tibial Nerve
- 95 Great Saphenous Vein

Figure 7.12.

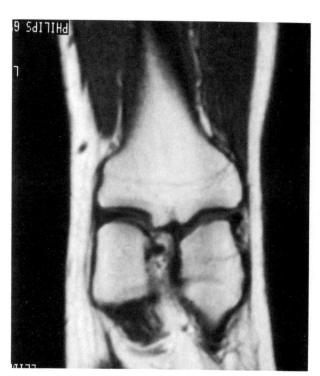

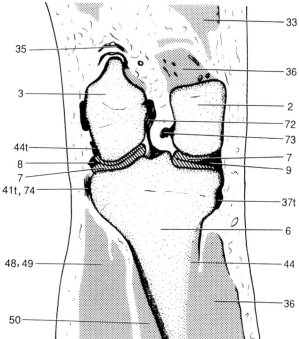

Bone and Cartilage
  2 Medial Condyle of Femur
  3 Lateral Condyle of Femur
  6 Tibia
  7 Articular (Hyaline) Cartilage
  8 Lateral Meniscus
  9 Medial Meniscus

Muscle
  33 Vastus Medialis Muscle
  35 Gastrocnemius Muscle (Lateral
       Head)
  36 Gastrocnemius Muscle (Medial
       Head)

  37 Sartorius Muscle
  41 Biceps Femoris Muscle (Long
       Head)
  44 Popliteal Muscle
  48 Extensor Digitorum Longus
       Muscle
  49 Peroneus Longus Muscle

  50 Pes Anserinus

Ligament and Tendon
  72 Anterior Cruciate Ligament
  73 Posterior Cruciate Ligament
  74 Fibular Collateral Ligament

Figure 7.13.

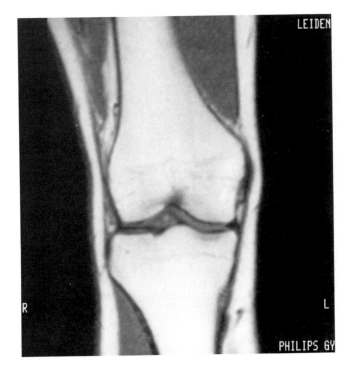

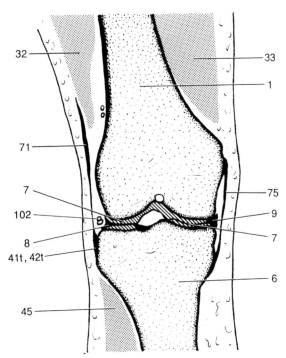

Bone and Cartilage
1 Femur
6 Tibia
7 Articular (Hyaline) Cartilage
8 Lateral Meniscus
9 Medial Meniscus

Muscle
32 Vastus Lateralis Muscle
33 Vastus Medialis Muscle
41 Biceps Femoris Muscle (Long Head)

42 Biceps Femoris Muscle (Short Head)
45 Tibialis Anterior Muscle

Ligament and Tendon
71 Iliotibial Tract

75 Tibial Collateral Ligament

Neurovascular and Miscellaneous
102 Inferior Lateral Genicular Artery and Vein

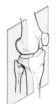

Figure 7.14.

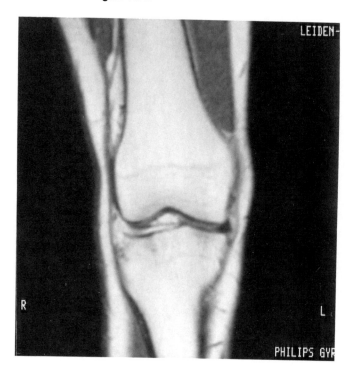

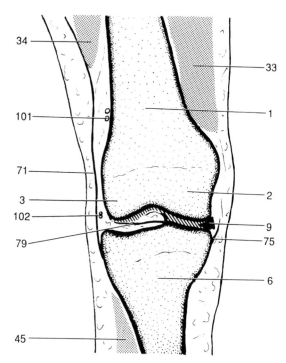

Bone and Cartilage
1 Femur
2 Medial Condyle of Femur
3 Lateral Condyle of Femur

6 Tibia
9 Medial Meniscus

Muscle
33 Vastus Medialis Muscle
34 Vastus Intermedius Muscle

45 Tibialis Anterior Muscle

Ligament and Tendon
71 Iliotibial Tract
75 Tibial Collateral Ligament
79 Transverse Genicular Ligament

Neurovascular and Miscellaneous
101 Superior Lateral Genicular Artery and Vein
102 Inferior Lateral Genicular Artery and Vein

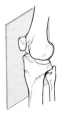

Figure 7.15.

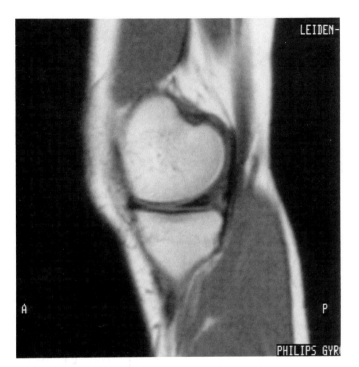

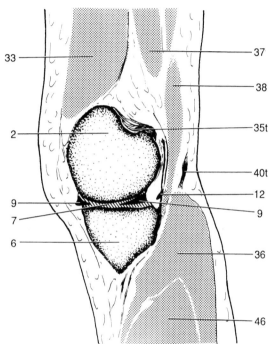

Bone and Cartilage
2   Medial Condyle of Femur
6   Tibia
7   Articular (Hyaline) Cartilage
9   Medial Meniscus

12   Articular Capsule
Muscle
33   Vastus Medialis Muscle
35   Gastrocnemius Muscle (Long Head)

36   Gastrocnemius Muscle (Short Head)
37   Sartorius Muscle

38   Semimembranosus Muscle
40   Semitendinosus Muscle
46   Soleus Muscle

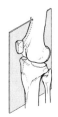

Figure 7.16.

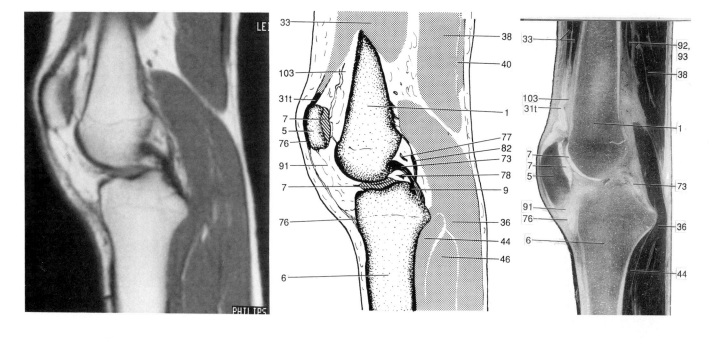

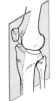

Figure 7.17.

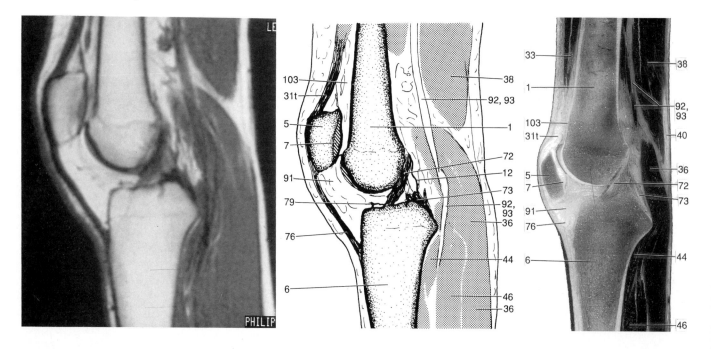

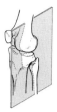

Figure 7.18.

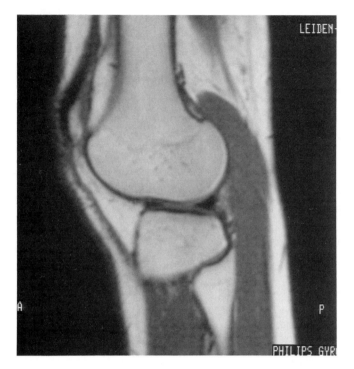

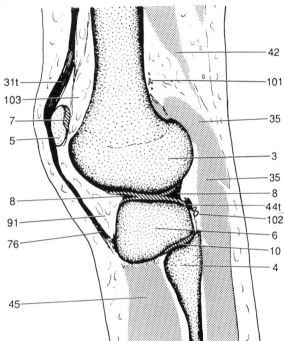

Bone and Cartilage
1   Femur
3   Lateral Condyle of Femur
4   Fibula
5   Patella
6   Tibia
7   Articular (Hyaline) Cartilage
8   Lateral Meniscus
9   Medial Meniscus
10 Tibiofibular Joint
12 Articular Capsule

Muscle
31 Quadriceps Muscle
33 Vastus Medialis Muscle
35 Gastrocnemius Muscle (Lateral Head)
36 Gastrocnemius Muscle (Medial Head)
38 Semimembranosus Muscle
40 Semitendinosus Muscle
42 Biceps Femoris Muscle (Short Head)

44 Popliteal Muscle
45 Tibialis Anterior Muscle
46 Soleus Muscle

Ligament and Tendon
72 Anterior Cruciate Ligament
73 Posterior Cruciate Ligament
76 Patellar Ligament
77 Posterior Meniscofemoral Ligament
78 Anterior Meniscofemoral Ligament

79 Transverse Genicular Ligament
82 Oblique Popliteal Ligament

Neurovascular and Miscellaneous
91 Infrapatellar Fat Pad
92 Popliteal Artery and Vein
93 Tibial Nerve
101 Superior Lateral Genicular Artery and Vein
102 Inferior Lateral Genicular Artery and Vein
103 Suprapatellar Fat Pad

CHAPTER 8

The Foot and Ankle

Edwin van der Linden and Ab Verbout

Anatomy of the Ankle Joint

Bone and Cartilage

1 Tibia
2 Lateral Malleolus
3 Medial Malleolus
4 Fibula
5 Talus
6 Sustentaculum Tali
7 Calcaneus
8 Navicular
9 Medial Cuneiform
10 Lateral Cuneiform
11 Intermediate Cuneiform
12 Cuboid
13 Metatarsal Bone II
14 Metatarsal Bone III
15 Metatarsal Bone IV
16 Metatarsal Bone V
17 Articular (Hyaline) Cartilage

Muscle

31 Extensor Hallucis Longus Muscle
32 Extensor Digitorum Longus Muscle
33 Extensor Hallucis Brevis Muscle
34 Extensor Digitorum Brevis Muscle
35 Peroneus Longus Muscle
36 Peroneus Brevis Muscle
37 Peroneus Tertius Muscle
38 Tibialis Anterior Muscle
39 Tibialis Posterior Muscle
40 Flexor Digitorum Longus Muscle
41 Flexor Digitorum Brevis Muscle
42 Flexor Digiti Minimi Brevis
43 Flexor Hallucis Longus Muscle
44 Flexor Hallucis Brevis Muscle
45 Achilles Tendon
46 Soleus Muscle
47 Quadratus Plantae Muscle
48 Abductor Digiti Minimi Muscle
49 Adductor Hallucis Muscle
50 Abductor Hallucis Muscle
51 Plantar Interosseus Muscle
52 Dorsal Interosseus Muscle
53 Interosseus Muscles
54 Plantar Head of Flexor Digitorum Longus Muscle

Ligament and Tendon

71 Deltoid Ligament
72 Anterior Talofibular Ligament
73 Posterior Talofibular Ligament
74 Posterior Inferior Talofibular Ligament
75 Calcaneofibular Ligament
76 Extensor Retinaculum
77 Talonavicular Ligament
78 Interosseous Ligament
79 Interosseous Membrane
80 Calcaneonavicular Ligament (Spring Ligament)
81 Plantar Aponeurosis
82 Posterior Tibiofibular Ligament

Neurovascular and Miscellaneous

91 Anterior Tibial Artery and Vein
92 Posterior Tibial Artery and Vein
93 Medial Plantar Nerve
94 Lateral Plantar Nerve
95 Sural Nerve
96 Deep Peroneal Nerve
97 Small Saphenus Vein
98 Tibial Nerve
99 Tarsal Sinus

Abbreviation: t, tendon.

Acknowledgments: Anatomic drawings were made by Bas B. Blankevoort and J. Wetselaar-Whittaker. The assistance of W. R. Obermann, M.D., in preparing anatomic specimens is gratefully acknowledged.

Figures 8.1–8.5. Axial MR images.

Figure 8.1.

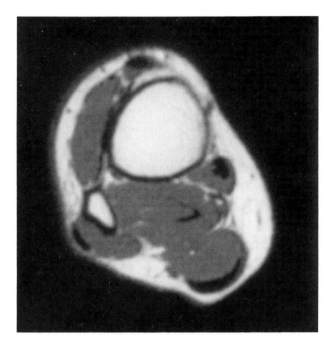

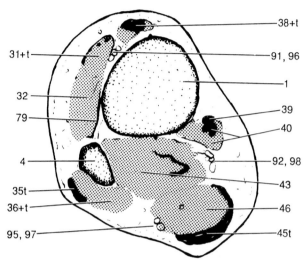

Bone and Cartilage
1   Tibia
4   Fibula

Muscle
31   Extensor Hallucis Longus Muscle

32   Extensor Digitorum Longus Muscle
35   Peroneus Longus Muscle
36   Peroneus Brevis Muscle
38   Tibialis Anterior Muscle
39   Tibialis Posterior Muscle
40   Flexor Digitorum Longus Muscle

43   Flexor Hallicus Longus Muscle
45   Achilles Tendon
46   Soleus Muscle

Ligament and Tendon
79   Interosseous Membrane

Neurovascular and Miscellaneous
91   Anterior Tibial Artery and Vein
92   Posterior Tibial Artery and Vein
95   Sural Nerve
96   Deep Peroneal Nerve
97   Small Saphenus Vein
98   Tibial Nerve

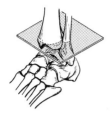

Figure 8.2.

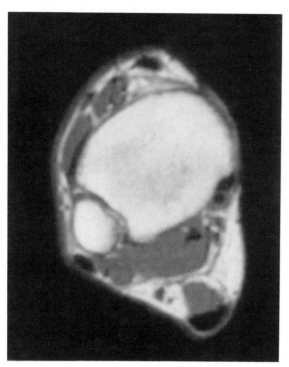

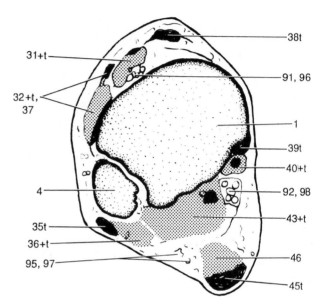

Bone and Cartilage
1 Tibia
4 Fibula

Muscle
31 Extensor Hallucis Longus Muscle

32 Extensor Digitorum Longus Muscle
35 Peroneus Longus Muscle
36 Peroneus Brevis Muscle
37 Peroneus Tertius Muscle
38 Tibialis Anterior Muscle

39 Tibialis Posterior Muscle
40 Flexor Digitorum Longus Muscle
43 Flexor Hallucis Longus Muscle
45 Achilles Tendon
46 Soleus Muscle

Neurovascular and Miscellaneous
91 Anterior Tibial Artery and Vein
92 Posterior Tibial Artery and Vein
95 Sural Nerve
96 Deep Peroneal Nerve
97 Small Saphenus Vein
98 Tibial Nerve

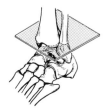

Figure 8.3.

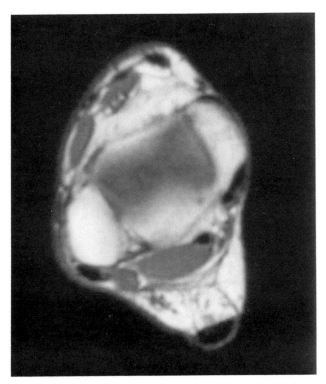

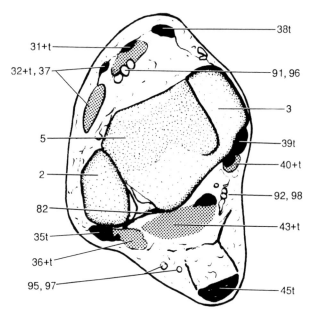

Bone and Cartilage
  2 Lateral Malleolus
  3 Medial Malleolus
  5 Talus

Muscle
 31 Extensor Hallucis Longus Muscle

 32 Extensor Digitorum Longus Muscle
 35 Peroneus Longus Muscle
 36 Peroneus Brevis Muscle
 37 Peroneus Tertius Muscle
 38 Tibialis Anterior Muscle
 39 Tibialis Posterior Muscle
 40 Flexor Digitorum Longus Muscle

 43 Flexor Hallucis Longus Muscle
 45 Achilles Tendon

Ligament and Tendon
 82 Posterior Tibiofibular Ligament

Neurovascular and Miscellaneous
 91 Anterior Tibial Artery and Vein

 92 Posterior Tibial Artery and Vein
 95 Sural Nerve
 96 Deep Peroneal Nerve
 97 Small Saphenus Vein
 98 Tibial Nerve

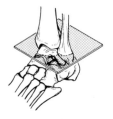

Figure 8.4.

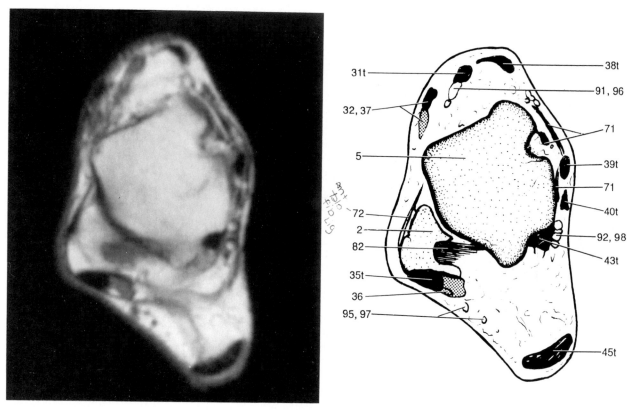

Bone and Cartilage
  2   Lateral Malleolus
  5   Talus

Muscle
 31   Extensor Hallucis Longus Muscle
 32   Extensor Digitorum Longus Muscle
 35   Peroneus Longus Muscle
 36   Peroneus Brevis Muscle
 37   Peroneus Tertius Muscle
 38   Tibialis Anterior Muscle
 39   Tibialis Posterior Muscle
 40   Flexor Digitorum Longus Muscle
 43   Flexor Hallucis Longus Muscle
 45   Achilles Tendon

Ligament and Tendon
 71   Deltoid Ligament
 72   Anterior Talofibular Ligament
 82   Posterior Tibiofibular Ligament

Neurovascular and Miscellaneous
 91   Anterior Tibial Artery and Vein
 92   Posterior Tibial Artery and Vein
 95   Sural Nerve
 96   Deep Peroneal Nerve
 97   Small Saphenus Vein
 98   Tibial Nerve

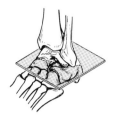

Figure 8.5.

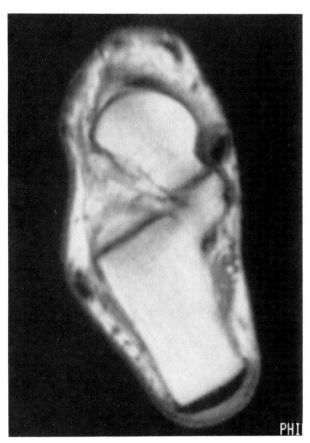

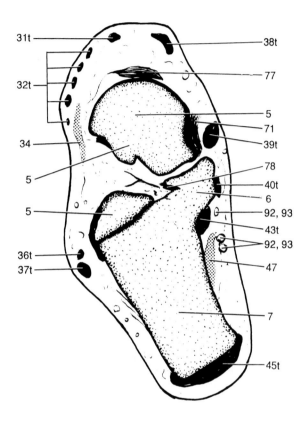

Bone and Cartilage
5 Talus
6 Sustentaculum Tali
7 Calcaneus

Muscle
31 Extensor Hallucis Longus Muscle

32 Extensor Digitorum Longus Muscle
34 Extensor Digitorum Brevis Muscle
36 Peroneus Brevis Muscle
37 Peroneus Tertius Muscle
38 Tibialis Anterior Muscle

39 Tibialis Posterior Muscle
40 Flexor Digitorum Longus Muscle
43 Flexor Hallucis Longus Muscle
45 Achilles Tendon
47 Quadratus Plantae Muscle

Ligament and Tendon
71 Deltoid Ligament
77 Talonavicular Ligament
78 Interosseous Ligament

Neurovascular and Miscellaneous
92 Posterior Tibial Artery and Vein
93 Medial Plantar Nerve

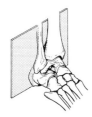

Figures 8.6–8.9. Coronal MR images.

Figure 8.6.

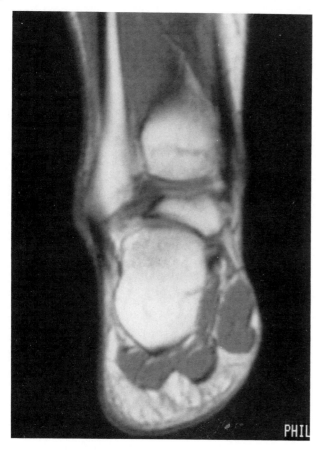

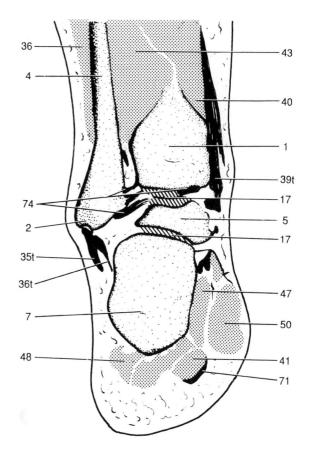

Bone and Cartilage
1 Tibia
2 Lateral Malleolus
4 Fibula
5 Talus

7 Calcaneus
17 Articular (Hyaline) Cartilage
Muscle
35 Peroneus Longus Muscle
36 Peroneus Brevis Muscle

39 Tibialis Posterior Muscle
40 Flexor Digitorum Longus Muscle
41 Flexor Digitorum Brevis Muscle
43 Flexor Hallucis Longus Muscle
47 Quadratus Plantae Muscle

48 Abductor Digiti Minimi Muscle
50 Abductor Hallucis Muscle
Ligament and Tendon
71 Deltoid Ligament
74 Posterior Inferior Talofibular Ligament

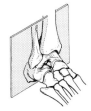

Figure 8.7.

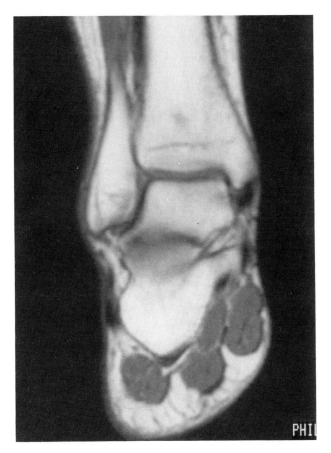

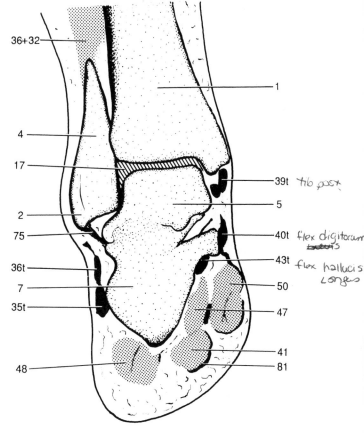

36+32

1

4

17

39t *trb post*

2

5

75

40t *flex digitorum*

36t

43t *flex hallucis longus*

7

50

35t

47

48

41

81

Bone and Cartilage
1   Tibia
2   Lateral Malleolus
4   Fibula
5   Talus
7   Calcaneus

17   Articular (Hyaline) Cartilage

Muscle
32   Extensor Digitorum Longus Muscle
35   Peroneus Longus Muscle
36   Peroneus Brevis Muscle

39   Tibialis Posterior Muscle
40   Flexor Digitorum Longus Muscle
41   Flexor Digitorum Brevis Muscle
43   Flexor Hallucis Longus Muscle
47   Quadratus Plantae Muscle

48   Abductor Digiti Minimi Muscle
50   Abductor Hallucis Muscle

Ligament and Tendon
75   Calcaneofibular Ligament
81   Plantar Aponeurosis

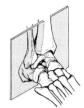

Figure 8.8.

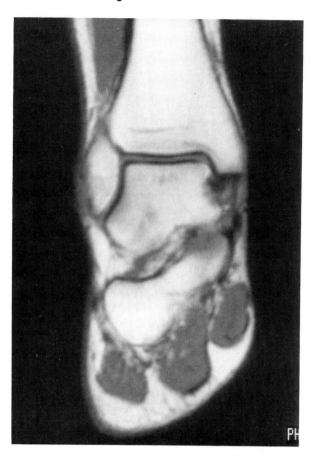

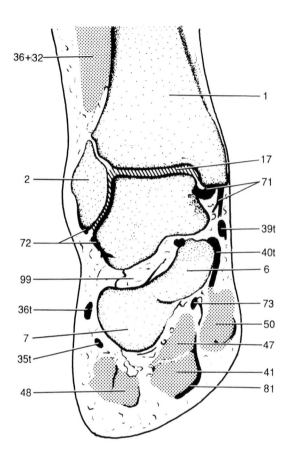

Bone and Cartilage
1   Tibia
2   Lateral Malleolus
6   Sustentaculum Tali
7   Calcaneus
17 Articular (Hyaline) Cartilage

Muscle
32 Extensor Digitorum Longus Muscle
35 Peroneus Longus Muscle
36 Peroneus Brevis Muscle
39 Tibialis Posterior Muscle
40 Flexor Digitorum Longus Muscle

41 Flexor Digitorum Brevis Muscle
47 Quadratus Plantae Muscle
48 Abductor Digiti Minimi Muscle
50 Abductor Hallucis Muscle

Ligament and Tendon
71 Deltoid Ligament

72 Anterior Talofibular Ligament
73 Posterior Talofibular Ligament
81 Plantar Aponeurosis

Neurovascular and Miscellaneous
99 Tarsal Sinus

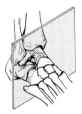

Figure 8.9.

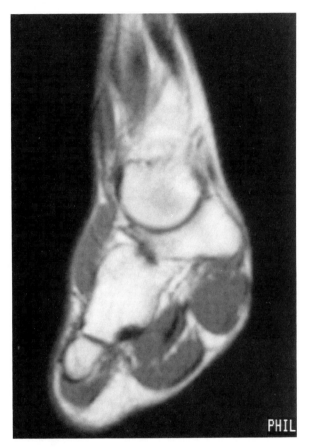

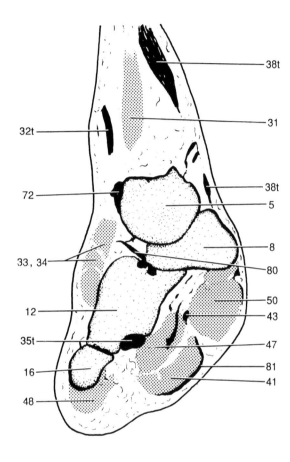

Bone and Cartilage
5 Talus
8 Navicular
12 Cuboid
16 Metatarsal Bone V

Muscle
31 Extensor Hallucis Longus Muscle
32 Extensor Digitorum Longus Muscle
33 Extensor Hallucis Brevis Muscle
34 Extensor Digitorum Brevis Muscle

35 Peroneus Longus Muscle
38 Tibialis Anterior Muscle
41 Flexor Digitorum Brevis Muscle
43 Flexor Hallucis Longus Muscle
47 Quadratus Plantae Muscle
48 Abductor Digiti Minimi Muscle
50 Abductor Hallucis Muscle

Ligament and Tendon
72 Anterior Talofibular Ligament
80 Calcaneonavicular Ligament (Spring Ligament)
81 Plantar Aponeurosis

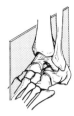

Figures 8.10–8.13. Sagittal MR images.

Figure 8.10.

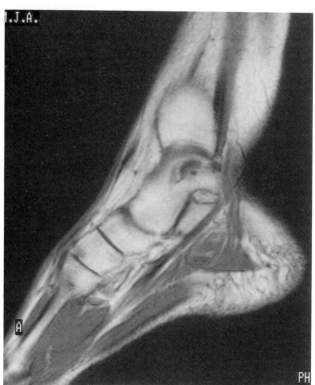

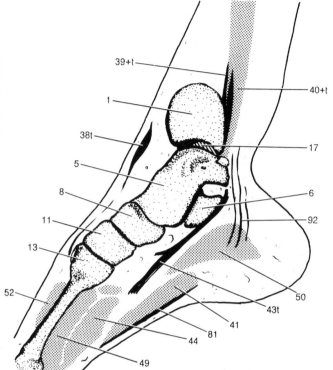

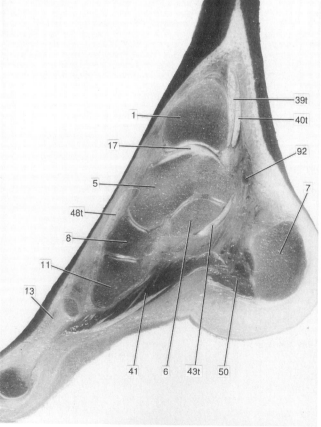

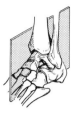

Figure 8.11.

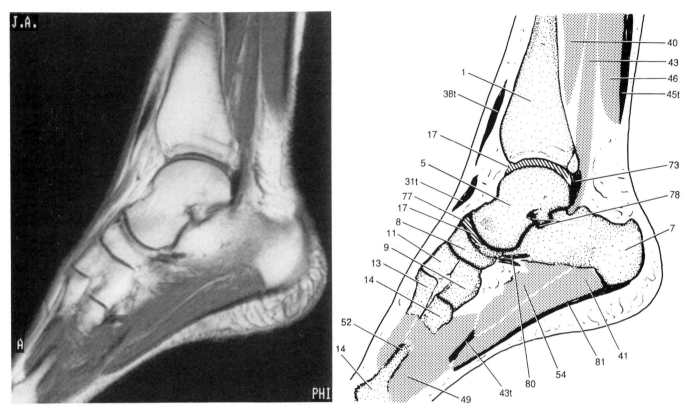

Bone and Cartilage
1 Tibia
5 Talus
6 Sustentaculum Tali
7 Calcaneus
8 Navicular
9 Medial Cuneiform
11 Intermediate Cuneiform
13 Metatarsal Bone II
14 Metatarsal Bone III
17 Articular (Hyaline) Cartilage

Muscle
31 Extensor Hallucis Longus Muscle
38 Tibialis Anterior Muscle
39 Tibialis Posterior Muscle
40 Flexor Digitorum Longus Muscle
41 Flexor Digitorum Brevis Muscle
43 Flexor Hallucis Longus Muscle
44 Flexor Hallucis Brevis Muscle

45 Achilles Tendon
46 Soleus Muscle
48 Abductor Digiti Minimi Muscle
49 Abductor Hallucis Muscle
50 Abductor Hallucis Muscle
52 Dorsal Interosseus Muscle
54 Plantar Head of Flexor Digitorum Longus Muscle

Ligament and Tendon
73 Posterior Talofibular Ligament
77 Talonavicular Ligament
78 Interosseous Ligament
80 Calcaneonavicular Ligament (Spring Ligament)
81 Plantar Aponeurosis

Neurovascular and Miscellaneous
92 Posterior Tibial Artery and Vein

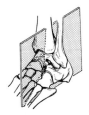

Figure 8.12.

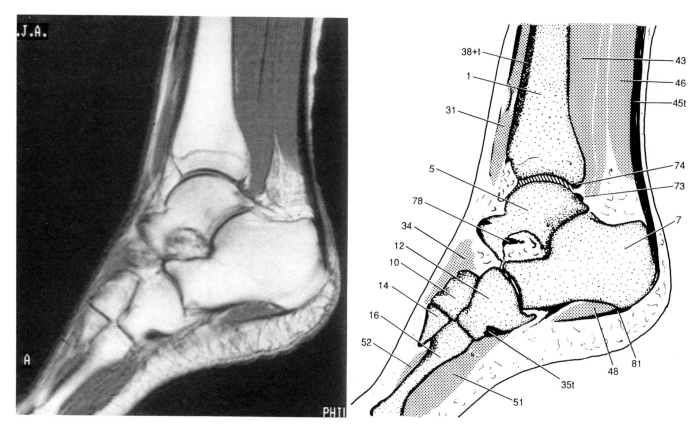

Bone and Cartilage
1 Tibia
5 Talus
7 Calcaneus
10 Lateral Cuneiform
12 Cuboid

14 Metatarsal Bone III
16 Metatarsal Bone V

Muscle
31 Extensor Hallucis Longus Muscle
34 Extensor Digitorum Brevis Muscle

35 Peroneus Longus Muscle
38 Tibialis Anterior Muscle
43 Flexor Hallucis Longus Muscle
45 Achilles Tendon
46 Soleus Muscle
48 Abductor Digiti Minimi Muscle

51 Plantar Interosseus Muscle
52 Dorsal Interosseus Muscle

Ligament and Tendon
73 Posterior Talofibular Ligament
74 Posterior Inferior Talofibular Ligament
78 Interosseous Ligament
81 Plantar Aponeurosis

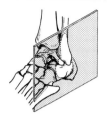

Figure 8.13.

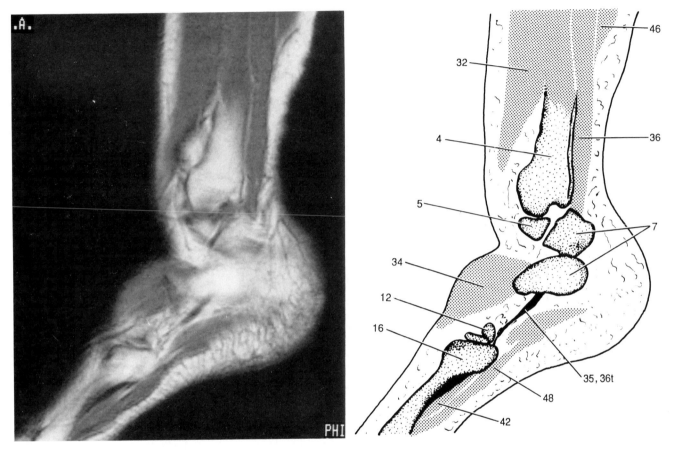

Bone and Cartilage
4 Fibula
5 Talus
7 Calcaneus

12 Cuboid
16 Metatarsal Bone V
Muscle
32 Extensor Digitorum Longus

34 Extensor Digitorum Brevis Muscle
35 Peroneus Longus Muscle
36 Peroneus Brevis Muscle

42 Flexor Digiti Minimi Brevis
46 Soleus Muscle
48 Abductor Digiti Minimi Muscle

CHAPTER 9

The Spine

Edwin van der Linden and Ab Verbout

Anatomy of the Spine

Bone and Cartilage

1   Atlas
2   Anterior Arch of Atlas
3   Posterior Arch of Atlas
4   Axis
5   Dens Axis
6   Vertebral Body
7   Spinous Process
8   Transverse Process
9   Costal Process
10 Uncinate Process
11 Pedicle
12 Lamina
13 Arch
14 Foramen Transversum
15 Facet Joint
16 Inferior Articular Process
17 Superior Articular Process
18 Demifacet for Head of Rib
19 Intervertebral Disk
20 Occipital Condyle
21 Clivus
22 Sternum
23 Mandible
24 Rib
25 Sacrum
26 Iliac Bone
27 Sacral Bone

Muscle

31 Semispinalis Muscle
32 Semispinalis Cervicis Muscle
33 Semispinalis Profunda Muscle
34 Semispinalis Dorsi Muscle
35 Spinalis Thoracis Muscle
36 Splenius Capitis Muscle
37 Sternohyoid Muscle
38 Scalenus Anterior Muscle
39 Scalenus Medius Muscle
40 Sternocleidomastoideus Muscle
41 Trapezius Muscle
42 Prevertebral Muscles
43 Platysma Muscle

44 Multifidus Muscle
45 Longissimus Capitis Muscle
46 Longissimus Cervicis Muscle
47 Longissimus Thoracis Muscle
48 Longissimus Dorsi Muscle
49 Levator Scapulae Muscle
50 Longus Colli Cervicus Muscle
51 Longus Capitis Muscle
52 Zygomaticomandibularis Muscle
53 Masseter Muscle
54 Medial and Lateral Pterygoideus Muscle
55 Gluteal Muscles
56 Digastricus Muscle
57 Rectus Capitus Posterior Minor Muscle
58 Rectus Capitus Posterior Major Muscle
59 Obliquus Capitis Inferior Muscle
60 Iliopsoas Muscle
61 Quadratus Lumborum Muscle
62 Intertransversarius Muscle
63 Iliocostalis Lumborum Muscle
64 Psoas Major Muscle
65 Paraspinal Muscles
66 Erector Spinae Muscle
67 Iliocostal Muscle
68 Intercostal Muscle
69 Latissimus Dorsi Muscle

Ligament and Tendon

71 Anterior Longitudinal Ligament
72 Posterior Longitudinal Ligament
73 Ligamentum Flavum
74 Supraspinous Ligament
75 Interspinous Ligament
76 Thoracolumbar Fascia (Deep Layer)
77 Ligamentum Nuchae
78 Anterior Atlantooccipital Membrane
79 Sacroiliacal Ligament
80 Transverse Ligament of Atlas

Neurovascular and Miscellaneous

91 Canal of Vertebral Artery (Sulcus of Vertebral Artery)
92 Common Carotid Artery
93 External Carotid Artery

94   Internal Carotid Artery
95   Vertebral Venous Plexus
96   Vertebral Artery
97   Internal Jugularis Vein
98   External Jugularis Vein
99   Deep Cervical Artery and Vein
100 Intraforamenal Vein
101 Basivertebral Vein
102 Dorsal Root Ganglion
103 Spinal Nerve (Nerve Root)
104 Cerebellar Tonsil
105 Spinal Cord
106 Subarachnoidal Space with Cerebral Fluid
107 Sympathetic Trunk
108 Intervertebral Foramen
109 Lumbar Nerve
110 Sacral Nerve
111 Cauda Equina
112 Dural Sac
113 Postepidural Fat
114 Epidural Fat
115 Peridural Fat
116 Fovea Dentis
117 Bifurcation of Aorta
118 Inferior Vena Cava
119 Lung
120 Esophagus
121 Trachea
122 Larynx
123 Pharynx
124 Nasopharynx
125 Epiglottis
126 Submandibular Gland
127 Parotid Gland
128 Diaphragm
129 Heart
130 Aorta
131 Azygos Vein
132 Hemiazygos Vein
133 Stomach
134 External Iliac Artery and Vein
135 Internal Iliac Artery and Vein
136 Gluteal Vessels
137 Presacral Vascular Plexus

Abbreviation: t, tendon.

Acknowledgments: Anatomic drawings were made by Bas B. Blankevoort and J. Wetselaar-Whittaker.

Figures 9.1–9.2. Axial MR images.

Figure 9.1.

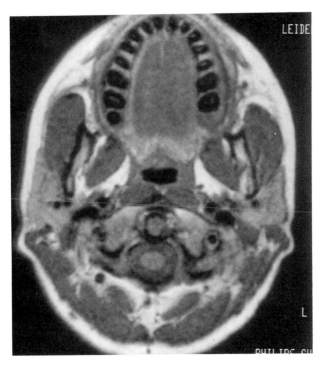

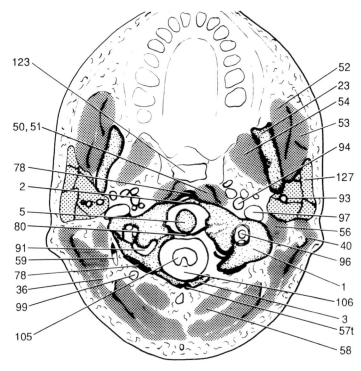

Bone and Cartilage
1 Atlas
2 Anterior Arch of Atlas
3 Posterior Arch of Atlas
5 Dens Axis
23 Mandible

Muscle
36 Splenius Capitis Muscle
40 Sternocleidomastoideus Muscle

50 Longus Colli Cervicus Muscle
51 Longus Capitis Muscle
52 Zygomaticomandibularis Muscle
53 Masseter Muscle
54 Medial and Lateral Pterygoideus Muscle
56 Digastricus Muscle
57 Rectus Capitus Posterior Minor Muscle
58 Rectus Capitus Posterior Major Muscle

59 Obliquus Capitus Inferior Muscle

Ligament and Tendon
78 Anterior Atlantooccipital Membrane
80 Transverse Ligament of Atlas

Neurovascular and Miscellaneous
91 Canal of Vertebral Artery (Sulcus of Vertebral Artery)
93 External Carotid Artery

94 Internal Carotid Artery
96 Vertebral Artery
97 Internal Jugularis Vein
99 Deep Cervical Artery and Vein
105 Spinal Cord
106 Subarachnoidal Space with Cerebral Fluid
123 Pharynx
127 Parotid Gland

Figure 9.2.

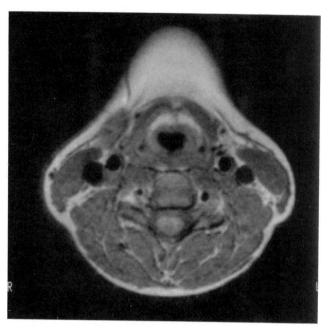

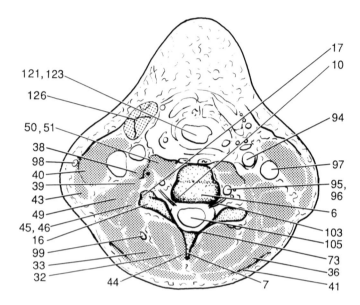

Bone and Cartilage
- 6 Vertebral Body
- 7 Spinous Process
- 10 Uncinate Process
- 16 Inferior Articular Process
- 17 Superior Articular Process

Muscle
- 32 Semispinalis Cervicis Muscle
- 33 Semispinalis Profunda Muscle
- 36 Splenius Capitis Muscle
- 38 Scalenus Anterior Muscle
- 39 Scalenus Medius Muscle
- 40 Sternocleidomastoideus Muscle
- 41 Trapezius Muscle
- 43 Platysma Muscle
- 44 Multifidus Muscle
- 45 Longissimus Capitis Muscle
- 46 Longissimus Cervicis Muscle
- 49 Levator Scapulae Muscle
- 50 Longus Colli Cervicus Muscle
- 51 Longus Capitis Muscle

Ligament and Tendon
- 73 Ligamentum Flavum

Neurovascular and Miscellaneous
- 94 Internal Carotid Artery
- 95 Vertebral Venous Plexus
- 96 Vertebral Artery
- 97 Internal Jugularis Vein
- 98 External Jugularis Vein
- 99 Deep Cervical Artery and Vein
- 103 Spinal Nerve (Nerve Root)
- 105 Spinal Cord
- 121 Trachea
- 123 Pharynx
- 126 Submandibular Gland

Figures 9.3–9.4. Sagittal MR images.

Figure 9.3.

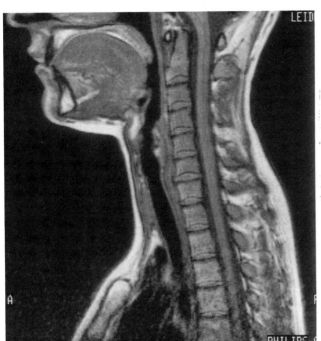

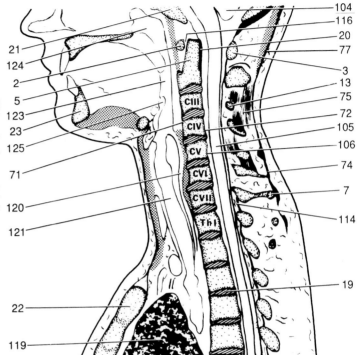

Bone and Cartilage
2 Anterior Arch of Atlas
3 Posterior Arch of Atlas
5 Dens Axis
7 Spinous Process
13 Arch
19 Intervertebral Disk
20 Occipital Condyle

21 Clivus
22 Sternum
23 Mandible

Ligament and Tendon
71 Anterior Longitudinal Ligament
72 Posterior Longitudinal Ligament
74 Supraspinous Ligament

75 Interspinous Ligament
77 Ligamentum Nuchae

Neurovascular and Miscellaneous
104 Cerebellar Tonsil
105 Spinal Cord
106 Subarachnoidal Space with Cerebral Fluid

114 Epidural Fat
116 Fovea Dentis
119 Lung
120 Esophagus
121 Trachea
123 Pharynx
124 Nasopharynx
125 Epiglottis

Figure 9.4.

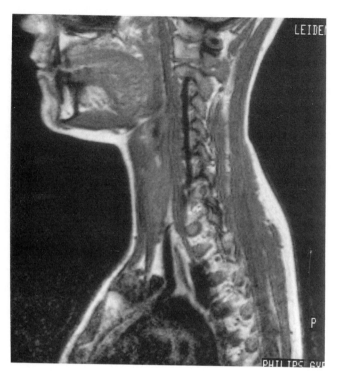

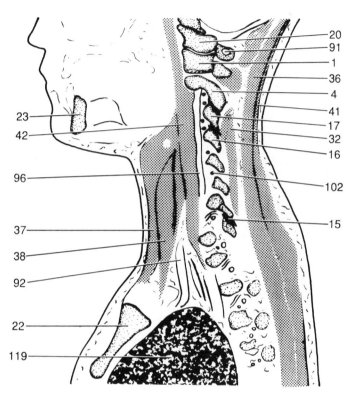

Bone and Cartilage
1 Atlas
4 Axis
15 Facet Joint
16 Inferior Articular Process
17 Superior Articular Process

20 Occipital Condyle
22 Sternum
23 Mandible

Muscle
32 Semispinalis Cervicis Muscle
36 Splenius Capitis Muscle

37 Sternohyoid Muscle
38 Scalenus Anterior Muscle
41 Trapezius Muscle
42 Prevertebral Muscles

Neurovascular and Miscellaneous
91 Canal of Vertebral Artery (Sulcus of Vertebral Artery)
92 Common Carotid Artery
96 Vertebral Artery
102 Dorsal Root Ganglion
119 Lung

Figures 9.5–9.6. Axial MR images.

Figure 9.5.

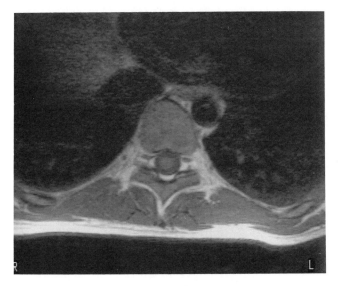

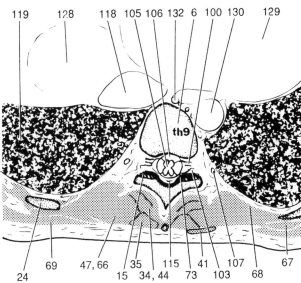

Figure 9.6.

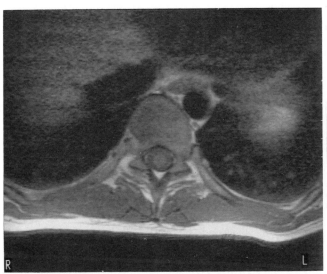

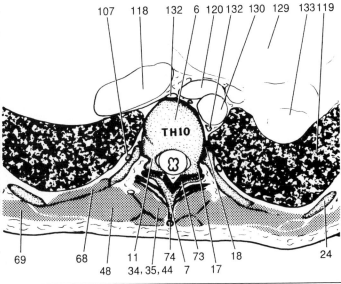

Bone and Cartilage
6	Vertebral Body
7	Spinous Process
9	Costal Process
	Uncimate Process
11	Pedicle
15	Facet Joint
17	Superior Articular Process
18	Demifacet for Head of Rib
24	Rib

Muscle
34	Semispinalis Dorsi Muscle
35	Spinalis Thoracis Muscle
41	Trapezius Muscle
44	Multifidus Muscle
47	Longissimus Thoracis Muscle
48	Longissimus Dorsi Muscle
66	Erector Spinae Muscle
67	Iliocostal Muscle
68	Intercostal Muscle
69	Latissimus Dorsi Muscle

Ligament and Tendon
73	Ligamentum Flavum
74	Supraspinous Ligament

Neurovascular and Miscellaneous
100	Infraforamenal Vein
103	Spinal Nerve (Nerve Root)
105	Spinal Cord
106	Subarachnoidal Space with Cerebral Fluid
107	Sympathetic Trunk

115	Peridural Fat
118	Inferior Vena Cava
119	Lung
120	Esophagus
128	Diaphragm
129	Heart
130	Aorta
132	Hemiazygos Vein
133	Stomach

Figures 9.7–9.8. Sagittal MR images.

Figure 9.7.

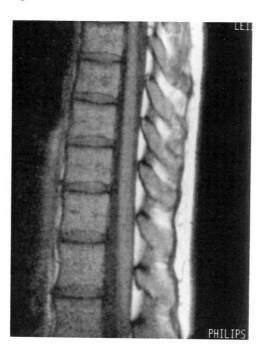
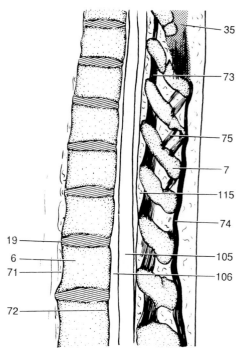

Figure 9.8.

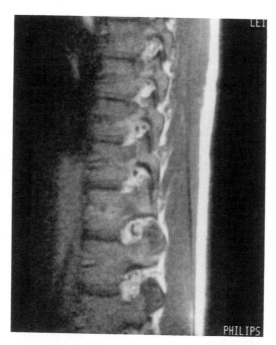
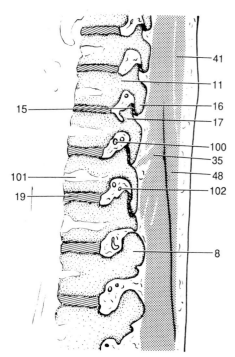

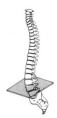

Figures 9.9–9.10. Axial MR images.

Figure 9.9.

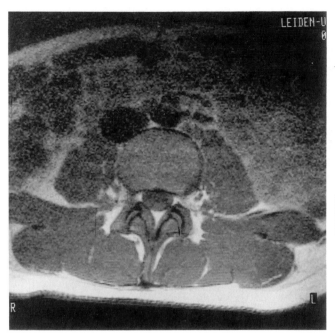

Bone and Cartilage

6	Vertebral Body
7	Spinous Process
8	Transverse Process
11	Pedicle
12	Lamina
15	Facet Joint
16	Inferior Articular Process
17	Superior Articular Process
19	Intervertebral Disk

Muscle

35	Spinalis Thoracis Musle
41	Trapezius Muscle
44	Multifidus Muscle
48	Longissimus Dorsi Muscle
62	Intertransversarius Muscle
63	Iliocostalis Lumborum Muscle
64	Psoas Major Muscle

Ligament and Tendon

71	Anterior Longitudinal Ligament
72	Posterior Longitudinal Ligament
73	Ligamentum Flavum
74	Supraspinous Ligament
75	Interspinous Ligament
76	Thoracolumbar Fascia (Deep Layer)

Neurovascular and Miscellaneous

100	Intraforamenal Vein
101	Basivertebral Vein
102	Dorsal Root Ganglion
105	Spinal Cord
106	Subarachnoidal Space with Cerebral Fulid
109	Lumbar Nerve
111	Cauda Equina
113	Postepidural Fat
115	Peridural Fat
117	Bifurcation of Aorta
118	Inferior Vena Cava

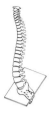

Figure 9.10.

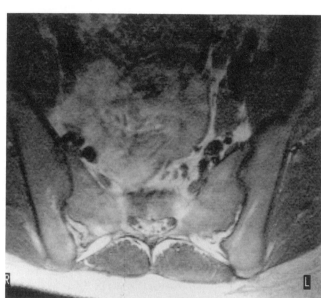

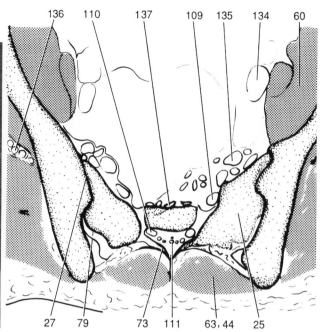

136 110 137 109 135 134 60

27 79 73 111 63, 44 25

Figures 9.11–9.12. Sagittal MR images.

Figure 9.11.

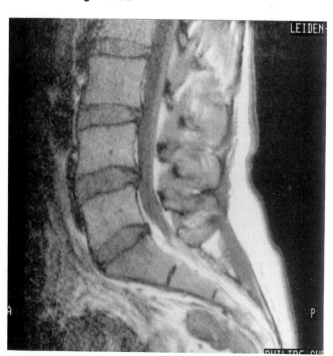

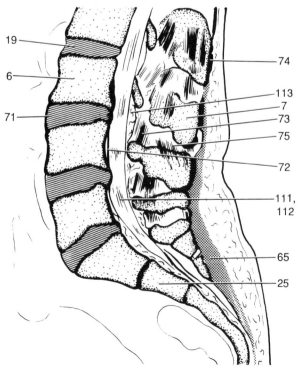

19
6
71

74
113
7
73
75
72
111, 112
65
25

Figure 9.12.

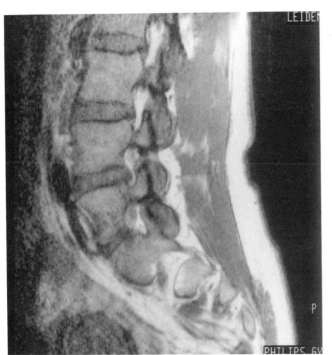

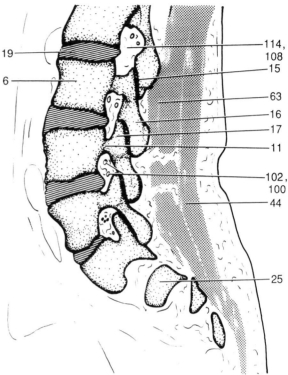

Bone and Cartilage
6   Vertebral Body
7   Spinous Process
11 Pedicle
15 Facet Joint
16 Inferior Articular Process
17 Superior Articular Process
19 Intervertebral Disk
25 Sacrum
27 Sacral Bone

Muscle
44 Multifidus Muscle
60 Iliopsoas Muscle
63 Iliocostalis Lumborum Muscle
65 Paraspinal Muscles

Ligament and Tendon
71 Anterior Longitudinal Ligament
72 Posterior Longitudinal Ligament
73 Ligamentum Flavum

74 Supraspinous Ligament
75 Interspinous Ligament
79 Sacroiliacal Ligament

Neurovascular and Miscellaneous
100 Intraforamenal Vein
102 Dorsal Root Ganglion
108 Intervertebral Foramen
109 Lumbar Nerve
110 Sacral Nerve

111 Cauda Equina
112 Dural Sac
113 Postepidural Fat
114 Epidural Fat
134 External Iliac Artery and Vein
135 Internal Iliac Artery and Vein
136 Gluteal Vessels
137 Presacral Vascular Plexus

SECTION TWO

PATHOLOGY

CHAPTER 10

Magnetic Resonance Imaging of Bone Marrow Disorders

Robert M. Steiner, Donald G. Mitchell, Vijay M. Rao, and Matthew D. Rifkin

Introduction

Since Renade et al. first described an increase in T1 relaxation times of bone marrow aspirates from patients with acute and chronic myelogenous leukemia (1) and Cohen et al. demonstrated magnetic resonance (MR) imaging signal intensity alterations of bone marrow in children with neoplastic disorders (2) MR imaging has evolved into the noninvasive modality of choice for the investigation of bone marrow disorders (3–6).

Although other imaging techniques are helpful in the examination of bone marrow disorders, each has significant limitations. Conventional film radiography has traditionally been the first examination routinely performed when a focal or diffuse bone marrow abnormality is suspected, but there must be considerable destruction of trabecular bone before changes can be visibly detected (7). Computed tomography (CT) has not developed into a primary modality for bone marrow imaging because of poor discrimination between normal and abnormal marrow, particularly in diffuse diseases such as leukemia. Streak and beam-hardening artifacts from cortical bone also obscure marrow detail (8). Radionuclide imaging with technetium-99m (^{99m}Tc)-labeled diphosphonate can assess blood flow and metabolic activity, particularly in cortical bone, but diffuse bone marrow disorders such as leukemia and lymphoma often are not appreciated. Bone marrow can also be labeled with radioisotopes to study hematopoiesis, and phagocytosis can be studied with ^{99m}Tc-labeled sulfur colloid. However, radionuclide studies that portray the distribution of cellular marrow are nonspecific, show ana-

tomic detail poorly, and generate little information about the distribution of fatty marrow (4, 5, 9–11).

Magnetic resonance imaging offers a unique opportunity to investigate the bone marrow because of its ability to clearly distinguish fat from other tissues. Because of this advantage, an understanding of the normal patterns of bone marrow distribution and the reaction of the marrow to the stress of disease can be clearly defined by magnetic resonance imaging and spectroscopy (12–18). As a result, MR imaging has assumed an important role in the detection of a wide variety of bone marrow abnormalities, including diffuse neoplasm (11, 19–22), myeloproliferative (23–25) and marrow-packing disorders (26), chronic anemia (27–30), aplasia (31, 32), avascular necrosis (33–35), infection (36), and trauma (37, 38).

In this chapter, the physiology, anatomy, and the MR imaging patterns of normal bone marrow, the MR methodology required to optimally characterize normal and abnormal bone marrow, and the diagnostic value of MR imaging in diffuse hematologic disorders are emphasized. Attention is also given to some examples of focal marrow injury, including radiation change (39, 40) and marrow edema (41).

Normal Bone Marrow Anatomy and Physiology

After the osseous skeleton, skin, muscle, and fat, the bone marrow is the largest organ of the human body. The hematopoietic function of the bone marrow supplies and regulates circulating platelets and white and red blood cells to meet the body's need for oxygenation, regulation of cell immunity, and coagulation.

The medullary cancellous or spongy bone is composed of primary and secondary trabeculae, which provides a supporting framework for the marrow tissue.

The vascular supply of the marrow is obtained from two major arterial sources, the nutrient and periosteal arteries. One or more nutrient arteries penetrate the bony cortex, enter the medullary cavity, and run parallel to the long axis of the shaft of the bone. The nutrient arteries then branch toward the endosteal surface of the cortex and coalesce at the capillary level with transosteal arterial branches arising from the periosteum. These combined coalescent capillaries widen at the endosteal level of the diaphysis to form an extensive network of sinusoids. The sinusoids then

---

Abbreviations: STIR, short τ inversion recovery pulse sequence; ppm, part per million; LDH, lactate dehydrogenase; SCA, sickle cell anemia; AVN, avascular necrosis.

penetrate into the marrow substance and eventually drain into a central venous channel in the medullary portion of the bone, exiting through a nutrient foramen (3, 5).

Hematopoiesis

Initially, blood cell production occurs exclusively in the yolk sac, which continues to be the source of hematopoiesis until the 6th week of gestation. Afterward, and until the 20th intrauterine week, the liver and the reticuloendothelial system are the primary sites of hematopoiesis. From that time and throughout adult life the bone marrow is the primary site for blood cell synthesis.

The source for blood cell production is the pluripotent stem cell or colony forming unit (42, 43). These cells are present in very small concentrations, with a ratio of 1:2000 nucleated bone marrow cells, and have only recently been isolated. They have the capacity to differentiate into precursor cells of the erythroid, phagocytic or granulocytic, megakaryocytic, and lymphoid cell lines. Pluripotent stem cells may be found in the peripheral blood as well as in the bone marrow itself. This suggests that one role of the pluripotent stem cell is to serve the demands for cell line differentiation and replication throughout the body, not just in the marrow. The presence of the pluripotent stem cell in the peripheral circulation may also explain how engraftment of peripherally introduced marrow tissue during bone marrow transplantation can occur, as well as explain how diffuse bone marrow neoplasms can present for the first time with clinically far advanced disease (42). Disorders affecting the pluripotent stem cell can lead to reversible or permanent stem cell failure, as is found in aplastic anemia, uncontrolled proliferation as exemplified by polycythemia vera, neoplastic transformation including acute lymphocytic leukemia or chronic granulocytic leukemia, and stem cell dysplasia as in paroxysmal nocturnal hemoglobinuria, thalassemia, or sickle cell hemoglobinopathy.

In the adult, the fat cell is the major constituent of bone marrow, accounting for approximately 75% of its total weight. Although the physiologic role of the fat cells of the marrow in hematopoiesis is unclear, during periods of reduced blood cell production the fat cells increase in number and volume. When there is increased demand for hematopoiesis the fat cells atrophy. It is believed that the fat cells provide nutritional support and perhaps growth factors for blood cell production (44). The supporting stroma of the marrow, or reticulin fiber cells, include both non-phagocytic and phagocytic cells or macrophages located within islands of hematopoietic activity (5). Active bone marrow is predominantly cellular or "red." It is composed of granulocytic and red blood cell lines in a ratio of 3:1 in addition to numerous lymphocytes and platelet precursors. In young adults, cellular marrow contains approximately 40% water, 40% fat, and 20% protein (45). The older the patient the greater the fat content of cellular marrow, so that by the age of 70 the composition of cellular marrow is approximately 60% fat, 30% water, and 10% protein. Hematopoietically inactive hypocellular fatty or "yellow" marrow consists of approximately 15% water, 80% fat, and 5% protein (46). Cellular marrow is characteristically as-

sociated with a rich and complex vascular environment while the vascular network in fatty marrow tissue is relatively sparse, with thin vessel walls and capillaries (46, 47).

Marrow Conversion

At birth, almost all marrow is cellular or "red" and is actively involved in blood cell production. Following birth, there is progressive conversion of hematopoietic to fatty marrow (Fig. 10.1). Conversion originates first in the distal phalanges and epiphysis and proceeds in an orderly but often nonuniform fashion from the appendicular skeleton (peripheral) to the axial skeleton (central) (4, 15, 48–50). The marrow of the epiphyses and apophyses of the long bones becomes completely fatty within months after ossification and the growth centers do not appear to participate in active hematopoiesis in the normal child or in the adult (33, 47, 51). Within each bone the diaphyseal marrow converts to fatty marrow, followed by the distal metaphysis and finally by the proximal metaphysial marrow. By the age of 7 years, the marrow of the bones of the feet and hands and the growth centers of the appendicular skeleton contain fatty marrow. By the onset of puberty, there is much fatty replacement in the diaphysis and in the distal metaphysis. By the age of 28, the adult pattern of marrow distribution is reached (Fig. 10.2). Varying amounts of cellular marrow remain in the adult in the vertebral bodies, sternum, ribs, pelvis, skull, and the proximal metaphysis of the humerus and femur (15, 33, 47–51) (Fig. 10.3).

Kricun (49), Ricci et al. (15), and others (6, 12, 50) have emphasized the gradual increase in the fatty component of red marrow with aging. Replacement of cellular marrow in areas surrounded by regions of fatty marrow, such as the intertrochanteric portion of the femur, occurs with aging and at different rates among individuals. Such variations of conversion from cellular to fatty marrow may have important clinical implications. It has been suggested, for

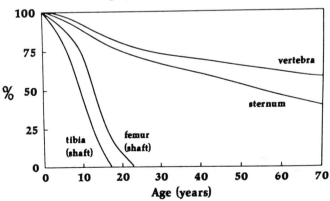

Figure 10.1. Percentage cellularity of red marrow from birth through the eighth decade of life. By the age of 20+ years replacement of cellular marrow by fatty marrow in the tibia and femoral shaft is almost complete. The vertebrae and sternum remain highly cellular throughout life. (Adapted from Custer RP, Ahlfeld FE. J Lab Clin Med 1932;17:960–974.)

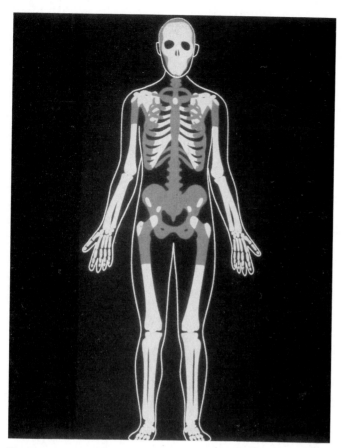

Figure 10.2. Normal adult pattern of marrow distribution. Cellular marrow (grey) is confined mainly to the axial (central) skeleton, including the ribs, proximal femurs and humeri, skull, spine, and pelvis. Fatty marrow (white) predominates in the appendicular skeleton. (Adapted from Custer RP, Ahlfeld FE. J Lab Clin Med 1932;17:960–974.)

when there is an increased demand for hematopoiesis due to destruction or replacement of normal cellular marrow or an increased need because of hemolysis, reconversion of fatty to cellular marrow may occur, often in a matter of hours (49, 54). The reconversion process varies with each skeletal site and with the duration and severity of the stimulus. Reconversion begins in the subendosteal portion of the fatty marrow and is accompanied by hyperemia, capillary proliferation, and sinusoid formation. Red marrow islands will enlarge and zones of normally cellular marrow will become hyperplastic and expand. Because they retain cellular marrow throughout life, flat bones such as the scapula, the sternum, and the spine reconvert more quickly than other structures (39). The process of reconversion occurs in reverse of physiological conversion proceeding from the central to the peripheral skeleton (Fig. 10.5). In the long bones, reconversion will occur first in the proximal femoral and humeral metaphyses, followed by the distal metaphyses, and finally by the diaphyseal portions of the femurs and humeri (Figs. 10.6 and 10.7). Reconversion of fatty marrow in the fibula and tibia and the ulna and radius, as well as in the phalanges, will lag behind more proximal sites of reconversion to cellular marrow (Fig. 10.8). The relatively uniform, symmetrical pattern of reconversion may be modified by the presence of infiltrative disorders such as myelofibrosis, neoplasm, or bone marrow infarction. When the need for increased hematopoiesis is extreme, sites that contain fatty marrow throughout life, such as the epiphysis and apophysis, may convert to cellular marrow (Figs. 10.8 and 10.9). It is possible that initial conversion to fatty marrow may not occur at all in some patients with life-long anemic states such as sickle cell disease (4, 28, 30).

Examination Technique

The MR imaging signal intensity characteristics of bone marrow reflect its major components: fat, water, and the minerals contained in the bony skeleton. Cortical and trabecular bone, both because of their virtual lack of mobile protons and the significant susceptibility effect of trabecular bone and hemosiderin with selected pulse sequences, yield little or no detectable signal (55, 56). Thus, the major contributors to the bone marrow signal are fat and water. However, as the mobile proton densities of fat and water are similar, proton density alone contributes little to the contrast resolution between fat and cellular marrow. Other factors such as T1 and T2 relaxation times, chemical shift effects, and selected pulse sequences provide the needed contrast resolution to successfully separate cellular from fatty and pathologic marrow (4, 56–60).

Spin-Echo Imaging

The use of spin-echo pulse sequences is currently the most common approach to bone marrow imaging. With T1-weighted [short repetition time/echo time (TR/TE)] spin-echo images at 1.5 T fat produces a bright signal intensity because of its relatively short T1 relaxation time (approximate range, 350–500 ms), due to most of the fat protons being in the form of hydrophilic CH₂ groups with efficient

example, that relatively early conversion of the medial portion of the femoral neck may be due to decreased vascular perfusion, decreased temperature, or increased mechanical stress. Mitchell et al. found that there is a greater prevalence of fatty marrow in the intertrochanteric region in patients with nontraumatic avascular necrosis under the age of 50 years compared with age-matched normals, implying that those individuals with more proximal fatty marrow are at increased risk for the development of avascular necrosis, probably due to the relatively poor blood supply of the fatty marrow (51). Varying patterns of fatty deposition along the endplates of vertebral bodies may be due to normal aging (15) or could represent the effect of local ischemia adjacent to degenerative lumbar disks (52). In addition to the normal progression of marrow conversion, isolated islands or foci of cellular marrow may be present in fatty marrow (Fig. 10.4), or fatty marrow foci may be present in regions of predominantly cellular marrow (13, 53) (Fig. 10.15).

Marrow Reconversion

The normal adult requirements for blood cell production are adequately met with available cellular marrow, but

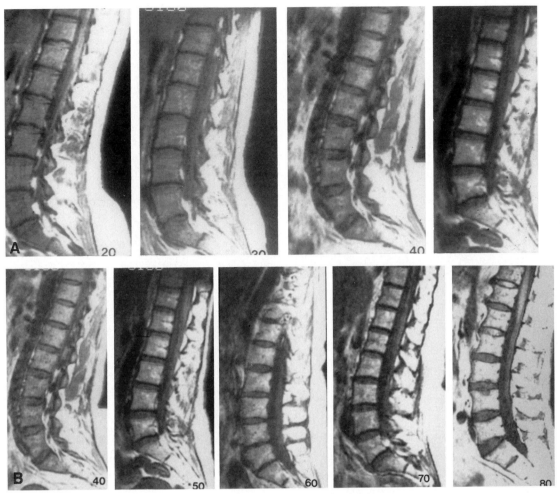

Figure 10.3. Sagittal spin-echo images from the third to ninth decade of the lower dorsal and lumbar spine show the gradual conversion of the vertebral bone marrow from predominantly cellular (intermediate signal intensity) to predominantly fatty marrow (bright signal intensity). (Courtesy of Kenneth Kaplan, M.D., Thomas Jefferson University Hospital, Philadelphia.)

spin-lattice T1 relaxation (3) (Fig. 10.6). With T2-weighted (long TR/TE) spin-echo pulse sequences fatty marrow exhibits less intense signal intensity compared with water, but greater intensity than muscle, because of relatively long T2 relaxation times (Fig. 10.7). Cellular marrow is characterized by an intermediate intensity signal similar to or slightly brighter than muscle on T1-weighted images.

The interaction of water, fat, and protein in cellular marrow is complex and is incompletely understood (3). Although protein has a long T1 relaxation time due to the large size of the protein molecules, in solution protein causes a shortening of the T1 relaxation time. In the case of cellular marrow, the fractional contribution of protein to signal intensity is unclear. The relative contribution of "free" or extracellular water, which has relatively long T1 and T2 relaxation times, and the contribution of intracellular or "bound" water, which has comparatively shorter T1 and T2 relaxation times, to the overall signal intensity of hematopoietic marrow is also unclear. To the contributions of protein and water must be added the contribution of the fat cells of the marrow, which will vary in amount from bone to bone and with age. The increase of the con-

tribution of fat with increasing age is striking, as there is a reduction of red cell mass from 60% in the first decade of life to less than 30% by the age of 90 years (50). The progressive increase in the fat content of cellular marrow explains why the signal intensity of red marrow with T1-weighted images (short TR/TE) in the first year of life is usually less than muscle but is often greater than muscle by adolescence (3, 61). In the elderly, even the vertebral bodies will exhibit the bright signal intensity of fatty marrow. It has been pointed out that when considering the differential diagnosis of marrow reconversion or replacement by neoplasm the signal intensity of the marrow on T1-weighted spin-echo images should be compared with the signal intensity of subcutaneous fat and nearby muscle in the same image. The presence of fat within hematopoietically active cellular marrow will result in an overall signal intensity greater than or equal to muscle but less than that of fat in adult patients. When the marrow is replaced by diffuse neoplasm such as acute leukemia, or where there is extensive reconversion, as in severe hemolytic anemia, the signal intensity closely approximates that of muscle on T1-weighted images because the fat compo-

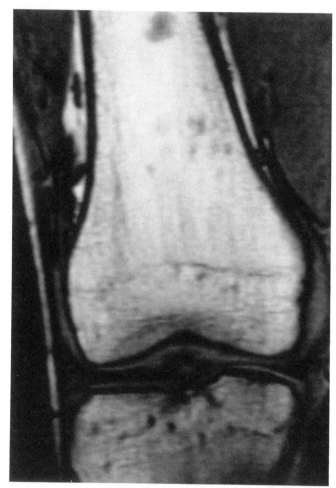

Figure 10.4. Focal islands of low-intensity marrow are scattered throughout the predominantly fatty marrow of the distal metadiaphysis and epiphysis of the femur and the proximal tibia. Similarly, foci of fatty marrow may be seen in regions of predominantly cellular marrow.

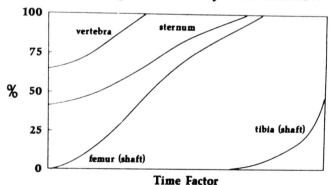

Figure 10.5. Marrow reconversion from fatty to cellular marrow. When requirements for blood cell production increase, reconversion of hypocellular fatty marrow to cellular red marrow will occur. The marrow of the vertebrae and sternum will become hyperplastic and areas of predominantly fatty marrow will reconvert in reverse of normal conversion. Hemolytic anemia and marrow replacement disorders such as myelofibrosis, myeloma, and metastases will stimulate reconversion in areas uninvolved with the underlying disease. (Adapted from Custer RP, Ahlfeld FE. J Lab Clin Med 1932;17:960–974.)

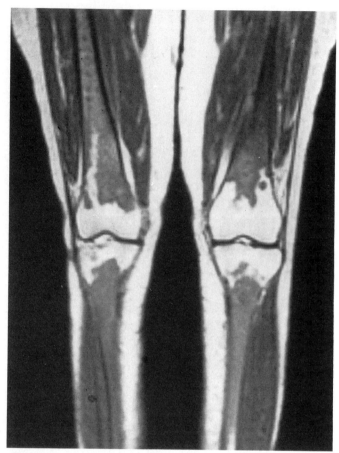

Figure 10.6. Marrow reconversion. An 85-year-old woman with polycythemia vera and myelofibrosis. T1-weighted coronal image (TR/TE 600/20 ms). There is cellular marrow in the metadiaphysis following reconversion from fatty marrow.

nent of the cellular marrow is decreased. As the relative volume of cellular and fatty marrow changes with aging, the mass of the bony skeleton decreases. Although the mineral content of the normal vertebra remains constant up to the third decade, and only small amounts of bone loss occur until the age of 40, by the fifth decade the decrease in trabecular bone is about 1%/year, which is even more pronounced in females during the postmenopausal period. By 75 years, the mineral content of bone is reduced by 40 to 50%. This loss is reflected in reduced cortical thickness and in the number and diameter of the bony trabeculae. The absent bone is replaced with fatty marrow, so that the increased signal seen in fatty marrow with age may reflect both the reduction in cellular mass and the loss of cortical and trabecular bone (14, 62).

Short τ Inversion Recovery Pulse Sequence (STIR)

The short τ inversion recovery sequence is a fat suppression method that has been utilized for bone marrow imaging (63, 64) (Fig. 10.21). In inversion recovery techniques, an initial inverting 180° radio-frequency pulse is followed by a 90° pulse and finally be a second rephasing 180° pulse. The 90° pulse is given following an interval of

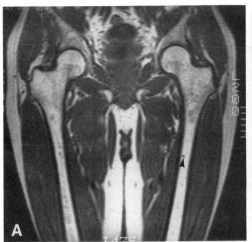
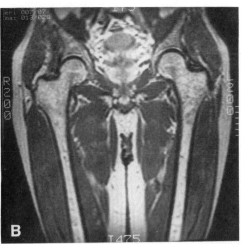

Figure 10.7. Subtle marrow reconversion. A 69-year-old man with polycythemia vera. There is mild reconversion of fatty to cellular marrow in the proximal femurs and small foci of cellular marrow are scattered in the femoral diaphysis *(arrow)*. **(A)** Coronal spin-echo T1-weighted images: TR/TE 600/20 ms; **(B)** TR/TE 1500/20 ms. Contrast resolution between the brighter cellular marrow and bright signal of the fatty marrow is reduced in this image with greater T2 weighting.

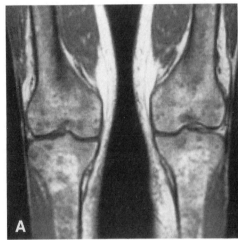
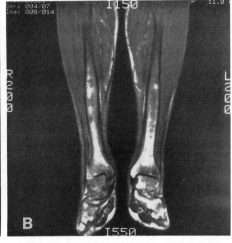

Figure 10.8. Extensive marrow reconversion. There is extensive mottled reconversion involving the lower extremities, including the metatarsal and tarsal bones, in this 70-year-old male with myelofibrosis. **(A)** Coronal image of the foot and lower legs (TR/TE 600/20 ms). **(B)** T1-weighted coronal examination of the knees (TR/TE 600/20 ms). The mottled bright intermediate signal intensity is due to mixed fatty and cellular marrow.

approximately 120 to 170 ms to take advantage of the relatively short T1 relaxation time of fat. At that interval, the fat signal is nulled because there is almost no magnetization of fat protons (63, 65). At the same inversion time (TI), other protons with a relaxation time similar to fat do not emit a signal that can be retrieved by the receiving coil. Tissues with long T1 relaxation times predominate so that normal cellular marrow, and pathologic processes with long T1 relaxation times (including edema, hyperplastic lymph nodes, and tumor) are bright. Tissues that are dark because of low proton density (e.g., bone, air) or because of short T2 relaxation times (fibrosis, hemosiderosis) may mimic fat. Since fat is suppressed, the contrast resolution between predominantly water-laden strictures and fatty marrow is increased. However, because there is an overall reduction in magnetization, the signal-to-noise ratio remains relatively low.

The potential for confusion between normal red marrow and pathologic marrow exists with STIR imaging due to similarities of signal intensity and the reduction in anatomic definition between the poor signal of cortical bone and the nulled signal of fatty marrow (5).

Chemical Shift Imaging

Contrast between fatty marrow and hematopoietic marrow may be enhanced with chemical shift imaging techniques, which take advantage of the differences in resonance frequency of aliphatic and water protons in tissue of approximately 3.5 ppm or 80 to 240 Hz at 0.5 to 1.5 T

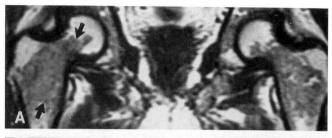

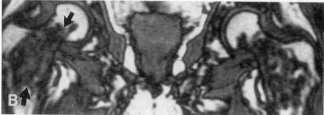

Figure 10.9. Marrow reconversion. **(A)** T1-weighted spin-echo image (TR/TE 700/20 ms) of the proximal femurs demonstrates reconversion of fatty marrow in the femoral neck and proximal shaft. The proximal femoral epiphysis and the greater trochanter remain fatty in this 84-year-old man with polycythemia vera. **(B)** An out-of-phase image (TR/TE 500/22 ms) in the same patient shows the low signal intensity of the expanded region of cellular marrow *(arrows)*.

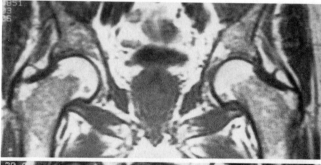

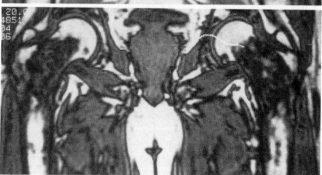

Figure 10.10. Marrow reconversion. An 84-year-old man with myelofibrosis and polycythemia vera. The spin-echo image (top) and Dixon out-of-phase image (bottom) (TR/TE 500/22 ms) show marked decrease in the signal intensity of cellular marrow in out-of-phase images due to relatively balanced contributions of the water and fat in normal cellular marrow. (Reprinted with permission from Steiner RM, Mitchell DG, Rao VM et al. Magn Reson Q 1990;6:17–34.)

(60, 67), echos from where water and fat signals are out of phase are obtained (Figs. 10.9 and 10.10). This occurs by shifting the 180° pulse away from the center of the echo time. The fat and water signals are opposed and the net signal intensity is the magnitude of the difference between the water and fat signal in each pixel (4). This phenomenon strongly reduces the signal from pixels that contain significant amounts of fat and water such as normal cellular marrow. The signal characteristics of hypercellular marrow with high water content and little fat content and predominantly fatty marrow will change very little compared with in-phase spin-echo images. The dark border between the predominantly fat and predominantly water structures seen in out-of-phase imaging is due to cancellation of both water and fat in pixels at the interface of both tissues (66). Measurements from comparable conventional and out-of-phase images can be used to assess the fractions of individual fat and water components within tissue (60). Recently, an enhancement of the Dixon technique, using three separate image acquisitions with phase shifts of 0, $+\pi$, and $-\pi$ between water and fat resonances to provide error-free water and fat images, has been introduced (16, 55, 57, 68). This technique effectively eliminates off-resonance errors due to susceptibility differences, demagnetization, and shim error. Current research with this pulse sequence modification is directly toward characterization of the susceptibility effect of trabecular bone and the paramagnetic effect of hemosiderin on the bone marrow signal.

Fat and Water Suppression Techniques

When it is desirable to obtain "water only" or "fat only" images, a narrow bandwidth saturation pulse can be applied selectively to water or fat prior to the excitation pulse, followed by a "spoiling" or destruction of the resulting transverse magnetization of fat or water so that there is little or no remaining magnetization (Fig. 10.21). Unlike STIR, which alters all signal and contrast in the image, selective peak saturation obliterates only the water or fat signals so that other tissues, including those with T1 relaxation times similar to fat (such as hematoma), and paramagnetic enhanced tissue will not be directly affected. Selective saturation techniques are sensitive to changes in field homogeneity and local susceptibility changes and are compatible with short TR/TE, long TR/short TE, and gradient echo sequences (69–72). Other strategies used to obtain water or fat-only images include selective refocusing of transverse magnetization (73), altered polarity of the section select gradient (74), and a hybrid technique combining a modified phase-sensitive Dixon subtraction method with water or fat spectral peak presaturation (67).

Gradient-Echo Imaging

The role of gradient-refocused echo imaging in the evaluation of bone marrow disorders is currently undergoing investigation and has not been well established (16, 75). Gradient-echo imaging is a fast technique based on the use of a variable flip angle excitation pulse (usually less than 90°) followed by a reversed gradient to refocus the

(33, 51, 60). Standard spin-echo pulse sequences produce images that represent the sum of in-phase signals of fat and water so that the relative percentage of fat and water in bone marrow cannot be easily distinguished. In the method described by Dixon (66) and modified by others

echo and generate a signal. Because of shortened repetition times, imaging time is reduced and most forms of motion artifact are decreased.

Although bone marrow image patterns similar to spin-echo T1- and T2-weighted images can be generated by appropriate selection of TR, TE, and flip angles, gradient-refocused echo imaging has unique characteristics that influence the interpretation of the marrow signal and distinguish it from spin-echo imaging. Most important is dependence on the effective transverse relaxation time (T2*) compared with the T2 dependency of conventional spin-echo sequences. This occurs with gradient-refocused echos because there is no rephasing of local magnetic field inhomogeneities by a 180° pulse. The result is that T2* dephasing is not recovered and dephasing occurs throughout the acquisition time. Additionally, water and adipose tissue may be out of phase, depending on the TE. As a result, gradient refocused echo imaging is especially sensitive to extrinsic and intrinsic magnetic inhomogeneities and chemical shift differences, reducing both fat and water marrow signal in regions with extensive trabecular bone such as the metaphyses of the lower extremities and the vertebral bodies (16, 17).

Magnetic Resonance Imaging Strategy

Of importance in constructing an appropriate imaging protocol for local or diffuse bone marrow disease is inclusion of the proper anatomical parts to encompass the abnormality in question and the selection of the appropriate pulse sequence to provide superior contrast resolution of the disease process.

When diffuse bone marrow disease is suspected, the dorsal and lumbar spine, pelvis, and intertrochanteric areas of the femurs, the predominant sites of cellular marrow until late in life, should be included in the examination as well as the iliac crest, the most common site for biopsy. If the disease process involves the distal extremities additional views of these structures, with contralateral comparative views, should be performed to prevent misinterpretation of normal variations such as low intensity signal islands in fatty marrow. Examination with T1-weighted spin-echo sequences for superior display of fat and cellular marrow patterns should be performed. T2-weighted pulse sequences to accentuate differences between normal and abnormal cellular marrow and fibrosis are an important part of a bone marrow imaging protocol. Additionally, water and fat suppression studies are useful to characterize the range of intensities of predominantly water or fat structures and for purposes of fat and water quantitation.

Our current protocol (4), utilizing a 1.5-T magnetic resonance unit, includes a coronal T1-weighted (TR/TE 300–600/15–17 ms) study of the lumbar spine, pelvis, and upper femurs with 7-mm slice thickness utilizing both in-phase and out-of-phase chemical shift techniques in an attempt to discriminate between normal, hypercellular, and fatty marrow. A T2-weighted (TR/TE 2000, 40/80 ms) coronal examination is obtained of the lumbar spine and upper femurs. A sagittal dorsal and lumbar spine T1-weighted (TR/TE 250–500/20 ms) examination with a 5-mm slice

thickness and a matrix of 256/128 is also performed. An additional T2-weighted (TR/TE 1500–2000/20, 40, 60, 80 ms) spin-echo pulse sequence with fat signal suppression permits T2 estimation of the water fraction. When indicated in the individual case, the distal femurs, tibias, fibulas, and humeri may be examined. The location of the site of iliac crest biopsy, when there is a question of nonrepresentative sampling, may be studied with images of the iliac crest and upper pelvis.

Diffuse and Focal Hematologic Disorders

Disorders that affect bone marrow production may be conveniently divided into four major categories representing conditions involving the pluripotent stem cell. These include reversible or permanent stem cell failure leading to aplastic anemia, uncontrolled proliferation with bone marrow hyperplasia, stem cell dysplasia including the hemoglobinopathies, and malignant transformation (42).

Bone Marrow Stem Cell Failure

Aplastic Anemia

A variety of myelotoxic agents and disease processes cause pancytopenia with decreased erythroid, myeloid, and platelet precursors, resulting in hypoplastic or aplastic anemia. Metastatic cancer and leukemia, granulomatous and viral infections, chemical agents such as organic solvents, and medications including chloramphenicol, ionizing radiation, and chemotherapy have been associated with the development of aplastic anemia. In most cases, however, the etiology is unknown or unclear (42, 76). Patients with aplastic anemia present initially with weakness, pallor, and often infection and bleeding. A bone marrow biopsy is usually diagnostic, showing acellularity or marked hypocellularity with predominantly fatty marrow and areas of fibrosis. But since areas of normal or even increased hematopoiesis may coexist with hypocellular or acellular marrow, iliac crest sampling alone may not always reflect the true state of bone marrow function.

Aplastic anemia may be irreversible but in 30 to 60% of patients transient or long-term improvement will occur, particularly following administration of androgen or immunosuppressive therapy, high-dose steroids, or antithymocytic globulin (76). The presence of normal or increased marrow cellularity in a bone marrow biopsy in the face of peripheral pancytopenia may be a sign of ineffective hematopoiesis of a preleukemic syndrome (77) or may merely be an unrepresentative biopsy sample (31). Although whole-body scintigrams following intravenous injection of indium-111 (^{111}In) are useful indicators of active hematopoiesis, MR imaging offers a global, noninvasive approach for the accurate diagnosis of aplastic anemia and for follow-up of the patient's clinical course.

With MR imaging, hypocellular or aplastic bone marrow is characterized by increased signal intensity with short TR/TE spin-echo pulse sequences due to generalized replacement of hematopoietic by fatty marrow (78) (Fig. 10.11). Increased fatty marrow is most obvious in regions that normally contain predominantly cellular marrow, such

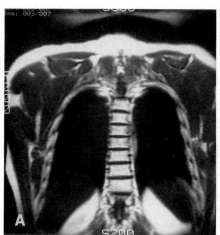
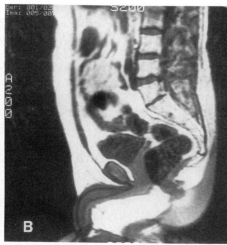
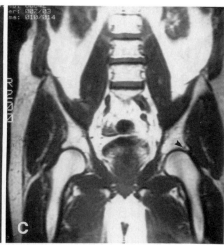

Figure 10.11. A 57-year-old man with aplastic anemia. There is a diffuse increase in signal intensity due to fatty replacement of cellular marrow. **(A)** Coronal dorsal spine and **(B)** sagittal lumbar spine image (TR/TE 600/20 ms). **(C)** In another patient with aplastic anemia, the T1-weighted image of the pelvis, lumbar spine, and upper femurs (TR/TE 600/20 ms) exhibits small focal islands of normal or hyperplastic cellular marrow *(arrow)*. (Reprinted with permission from Steiner RM, Mitchell DG, Rao VM et al. Magn Reson Q 1990;6:17–34.)

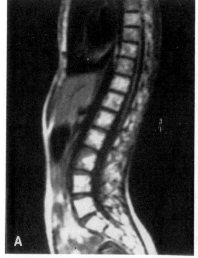
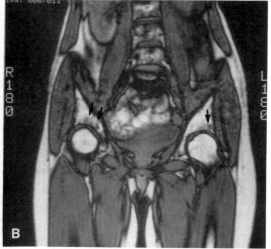

Figure 10.12. A 29-year-old woman with aplastic anemia. Bone marrow biopsy showed marked hypocellularity and abundant fat. **(A)** Sagittal T1-weighted image (TR/TE 500/20 ms) of the dorsal and lumbar spine demonstrates a mixed pattern of fatty and low-intensity marrow, presumably representing regions of hematopoietic activity. **(B)** In the Dixon pulse sequence (TR/TE 500/22 ms) areas of low signal *(arrows)* in this coronal image of the pelvis are due to cancellation of signal from balanced fat and water contributions to the signal indicating cellular marrow. (Reprinted with permission from Steiner RM, Mitchell DG, Rao VM et al. Magn Reson Q 1990;6:17–34.)

as the proximal femurs and spine (Figs. 10.12 and 10.13). Thus depiction of areas that usually contain predominantly cellular marrow have priority in an aplastic anemia protocol. In regions that normally contain predominantly fatty marrow, such as the appendicular skeleton, abnormally increased fatty marrow may be more difficult to appreciate (4, 32, 34, 58).

Quantitative MR imaging studies have been performed to distinguish normal from abnormal fatty marrow in aplastic anemia as well as differences in cellular marrow in other diffuse bone marrow disorders (45, 58, 60, 79). Rosen et al. (60) emphasized that conventional spin-echo imaging of heterogeneous tissue such as bone marrow

represented "bulk" signal intensity reflecting the sum of signals from both water and fat, so that relative changes in the percentage of fat and water cannot be distinguished. However, the calculation of a fat fraction based on in-phase and out-of-phase chemical shift imaging permits improved discrimination between normal bone marrow and the fatty bone marrow found in aplastic anemia when compared with "bulk" T1 relaxation times. This technique may offer an opportunity to assess progression or improvement based on calculated fat factions earlier than is possible with bone marrow aspiration. With treatment, a heterogeneous pattern of mixed cellular and fatty marrow develops in the vertebral body. Enlarging or coales-

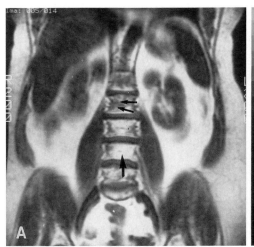

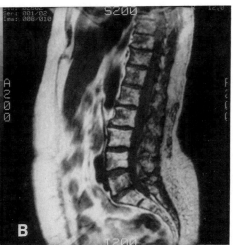

Figure 10.13. Aplastic anemia. **(A)** Coronal and **(B)** sagittal images of the lumbar spine in a 65-year-old male with pancytopenia and a hypocellular bone marrow biopsy (TR/TE 600/20 ms). There is a mot- tled pattern of hematopoietic *(small arrows)* and fatty marrow *(large arrows)*.

cent foci of low signal intensity appear, representing expanded cellular marrow. In one series (32) biopsy of a vertebral body in a patient treated for aplastic anemia exhibited a pattern of mixed fat and cellular marrow. On the other hand, the iliac crest in this same patient exhibited no cellular marrow. This pattern of differential recovery, occurring in all three patients under treatment, may be due to more rapid recovery in the spine because of differences in vascularity. This study suggests that MR imaging has a significant role to play in following clinical response noninvasively in patients with aplastic anemia.

The Effect of Ionizing Radiation

Depletion or extinction of myeloid bone marrow elements with fatty substitution occurs with focused ionizing radiation at therapeutic levels (Fig. 10.14). The extent of marrow depletion is directly related to the radiation dose, fractionation, and the time elapsed since the radiation was administered (39, 40, 80). In addition there are differences in the response of marrow cells to radiation. Mature marrow cells including erythrocytes, granulocytes, and platelets are relatively immune to destruction by ionizing radiation. Immature erythroblasts, myeloblasts, and megakaryoblasts are exquisitely sensitive, as are lymphocytes at any stage of maturity (5).

In a series (39) in which total body irradiation at levels of 1.25 Gy was administered, no change in vertebral signal intensity on T1-weighted images was detected. In the same series, irradiated bone receiving 20 to 30 Gy studied 10 to 23 years after treatment was normal in signal intensity, suggesting that recovery of normal hematopoietic function occurred. A higher level of irradiation in the range of 50 Gy was associated with complete fatty substitution of normal marrow elements in the spine and pelvis up to 9 years after treatment, presumably due to irreversible marrow extinction. In another series (40) irradiated bone was shown by chemical shift MR imaging and dual-energy CT quantitative studies to have a threefold increase in fat content. And, interestingly, there was a 26% decrease in tubercular

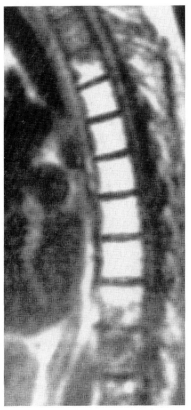

Figure 10.14. Bone marrow extinction. This patient received radiation therapy for metastatic disease of the dorsal spine due to lung cancer. On a T1-weighted sagittal image, increased high-signal marrow is confined to the treatment portal, reflecting cellular marrow depletion with fatty marrow replacement. (Reprinted with permission from Volger JB III, Murphy WA. Radiology 1988;168:679–693.)

bone in the spine measured by dual-energy CT dosimetry. The reduction in bone content suggests that the increase in signal intensity in the irradiated bone on T1-weighted images may be due in part to reduced susceptibility effect because of the decrease in trabecular bony elements as well as to the increased fat.

The changes in MR imaging signal characteristics with ionizing radiation are, in part, related to the elapsed time since treatment. In a recent study (80) of 14 patients with Hodgkin's disease, seminoma, and prostate cancer irradiated from 15 to 30 Gy and followed at monthly intervals for up to 14 months, there was no change in signal intensity during the first 2 weeks after treatment with spin-echo T1-weighted images. There was an increase, however, in signal observed with STIR imaging probably related to early marrow edema and/or necrosis. Between the third and sixth week after irradiation two consistent patterns of change were observed. There was either increased heterogeneity of the vertebral marrow pattern due to mixed fatty and lower intensity marrow, or predominance of high-intensity signal representing fat within the central portion of the vertebral body surrounding the basivertebral veins. After 6 weeks, the heterogeneous marrow pattern became progressively homogeneous with a diffuse, bright signal intensity representing fatty replacement. Alternatively, a bandlike pattern developed within a peripheral region of intermediate signal intensity surrounding a central vertebral zone of bright signal intensity on T1-weighted images. This bandlike pattern may reflect regenerating hematopoietic marrow along the periphery of the vertebral bodies after reconstitution of marrow sinusoids. T2-weighted images during the first 3 weeks following irradiation either showed no change or a subtle increase in signal intensity, suggesting edema and/or necrosis (80).

The published literature suggests the changes in marrow following irradiation are dose related and up to 50 Gy is necessary to induce permanent marrow extinction. These alterations in marrow composition manifested by high signal intensity on T1-weighted images are due to fatty substitution of normal cellular marrow elements. Knowledge of these MR imaging patterns and the possibility of long-term reversibility may be useful in treatment planning as well as in the identification of recurrent disease at or near an irradiated site.

Uncontrolled Stem Cell Proliferation (Myeloproliferative Syndrome)

Excessive proliferation of one or more of the normal bone marrow elements is a commonly found clinical condition. Most often, bone marrow hyperplasia occurs due to a recognizable stimulus such as a granulocytic response to a pyogenic infection or to a transient controlled increase in all cell lines following a hemorrhagic or hemolytic episode. Less often, uncontrolled myeloproliferation of some or all cell lines will occur. A number of acute and chronic clinical disorders associated with uncontrolled proliferation share the same common features and are often associated with myeloid metaplasia at extramedullary sites. They include myelodysplastic syndrome, or preleukemia (77, 81), acute and chronic myelogenous leukemias, polycythemia vera, and myelofibrosis. In this section we will discuss the chronic, relatively benign myeloproliferative syndromes, which include polycythemia vera with myeloid metaplasia and myelofibrosis.

Polycythemia Vera

Polycythemia vera is a rare monoclonal disorder of the pluripotent stem cell characterized by a sustained autonomous increase of the granulocytic, megakaryocytic, and particularly the erythrocytic cell lines occurring in the absence of hypoxic stimulation (54, 82). The peripheral blood cells remain morphologically normal but an increase in red cell mass leads to reduction in blood flow followed by ischemia, thrombosis, liver failure, and hemorrhage. Clinical parameters of severity include increased serum lactate dehydrogenase (LDH), decreased serum cholesterol, and the chronicity of the disease (83, 84). Splenomegaly is a common manifestation of polycythemia vera and the size of the spleen increases with the duration of the disease (82–85). The source of stem cell changes leading to uncontrolled proliferation is not known, although environmental factors such as irradiation and familial propensity have been proposed (83).

The differential diagnosis of polycythemia vera includes other causes of increased red cell mass, including chronic hypoxia, neoplasm with increased erythropoietin production, and "stress" polycythemia (4). The diagnosis of polycythemia vera is established by analysis of the bone marrow biopsy material, which demonstrates cellular hyperplasia in most patients (23). Other patients demonstrate areas of normal cellular or fatty marrow and/or increased reticulin, so that biopsy specimens may not always be representative of the true state of the marrow.

After many years of disease activity the disorder evolves into a "spent" phase in 15% of patients in which hematopoiesis is ineffective and regions of predominantly cellular marrow are replaced by reticulin fibrosis. This condition of postpolycythemic myeloid metaplasia is characterized by myelofibrosis, extramedullary hematopoiesis, and increasing splenomegaly. Although increasing splenomegaly parallels the progression and chronicity of the disease (85), it does not appear to be due to the extramedullary hematopoiesis itself: myelofibrosis may occur independently of polycythemia vera as a sequela to metastatic tumor, Gaucher's disease, infection, or it may be idiopathic (25, 42). With increasing myelofibrosis, pancytopenia with bone marrow failure may develop since myeloid metaplasia cannot compensate for reduced blood cell production. Eventually the patient dies with liver failure, hemorrhage, or infection.

The treatment of polycythemia vera is palliative and includes immunosuppressive therapy, including chlorambucil, hydroxyurea, and radioactive phosphorus during the proliferative phase. Platelets and red blood cell transfusions are administered during the phase of myelofibrosis. Splenectomy may be necessary because of the symptoms of splenomegaly and due to increasing sequestration (42). Acute leukemia, non-Hodgkin's lymphoma, and other malignancies may develop most likely due to the use of immunosuppressive therapy (23).

Magnetic resonance imaging will document the extent of reconversion of fatty to cellular marrow in the spine, femur, humerus, pelvis, and occasionally in the tibia and fibula (Figs. 10.6–10.10). In severe cases, the proximal femoral epiphysis and the apophysis of the greater trochanter will participate in conversion to lower signal in-

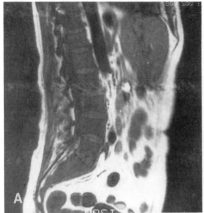
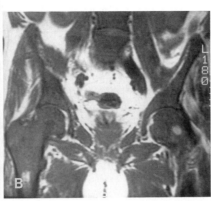
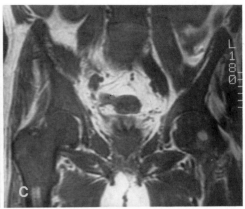

Figure 10.15. Uncontrolled stem cell proliferation disorder. A 56-year-old man with polycythemia vera and myelofibrosis. **(A)** Sagittal T1-weighted image (TR/TE 400/20 ms) of the lumbar spine shows the intermediate signal intensity of increased hematopoietic marrow. **(B)** T1-weighted (TR/TE 600/20 ms) and **(C)** T2-weighted (TR/TE 1500/80 ms) images of the hips demonstrate cellular marrow in the epiphyses, which normally contains fat. The decreased signal on T2-weighted images suggests fibrosis or hemosiderosis due to multiple transfusions. The focus of higher signal in the left femoral neck is a fatty marrow island. (Reprinted with permission from Steiner RM, Mitchell DG, Rao VM et al. Magn Reson Q 1990;6:17–34.)

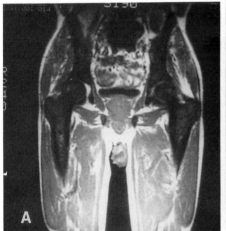
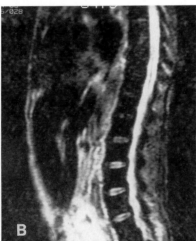
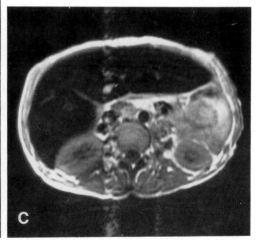

Figure 10.16. Myelofibrosis with probable hemosiderosis in a 63-year-old man. This patient received multiple blood transfusions during the course of his illness. **(A)** There is a diffuse decrease in signal intensity in both the pelvis and femur in this T1-weighted coronal image (TR/TE 600/20 ms). **(B)** In the T2-weighted image (TR/TE 2000/100 ms) of the lumbar spine there is low signal intensity that may be related to the inhomogeneous susceptibility effect of the iron deposition. **(C)** In the same patient an axial projection of the upper abdomen shows the liver to be of low signal intensity due to hemosiderosis. (Reprinted with permission from Steiner RM, Mitchell DG, Rao VM et al. Magn Reson Q 1990;6:17–34.)

tensity marrow as demonstrated on T1-weighted spin-echo pulse sequences (85) (Fig. 10.15). The reconverted cellular marrow is characterized by a signal intensity greater than muscle on both opposed and nonopposed short TR/TE images. T1 relaxation times of this lower signal intensity marrow have been described to be within the range of 690 to 970 ms compared with T1-weighted relaxation times in the normal range of 263 to 266 ms (4, 24, 25). T2 relaxation times are variable in polycythemia vera and have been reported to be elevated and similar to acute leukemia in some patients (79) while remaining within the normal range (24) or decreasing (25) in others. With long TR/TE images the intensity of the signal of the "dark" marrow will vary with the amount of cellular tissue, the extent of reticulin fibro-

sis, and the paramagnetic effect of deposited iron due to multiple transfusions. As a result, on T2-weighted images cellular marrow will be greater, equal, or less intense than fat (24). While normal cellular marrow is much less intense on opposed phase than in-phase images because of signal cancellation between water and fat, hypercellular marrow in patients with polycythemia vera appears similar on both of these sequences because it has little fat. With increased myelofibrosis and siderotic marrow one might expect to find decreased signal in both the opposed and in-phase T1-weighted images as well as in the long TR/TE images (4, 66, 81) (Fig. 10.16). A marked reduction of signal intensity in the liver on T2-weighted images due to tissue iron overload is a clue to the presence of hemosiderosis as a contribution to the lower intensity signal (86–89) (Fig. 10.16).

Malignant Bone Marrow Infiltration and Replacement

Focal or diffuse replacement of normal marrow with tumor cells may be associated with primary neoplasm, metastatic disease, multiple myeloma, leukemia, and lymphoma. Metastatic neoplastic infiltration occurs throughout all marrow-bearing areas, but predominates in the cellular marrow because of its rich vascularity compared with regions of predominantly fatty marrow, where metastases are less common. Since leukemia, lymphoma, and myeloma originate from blood cell precursors in the red marrow, they are most likely to be found in areas with predominantly cellular marrow. In the younger individual, tumor replacement is more likely to be found in the metaphyseal ends of the long bones, where residual cellular marrow is present, rather than in the diaphysis and epiphysis.

Magnetic resonance is exquisitely sensitive to replacement or infiltration of normal marrow by neoplasm, particularly on T1-weighted images because of high contrast resolution between fat and water. In general, tumor cells exhibit long T1 relaxation times due to increased cellularity and more extracellular water (32, 78, 90, 91) yielding a low signal intensity pattern on T1-weighted spin-echo images. This decrease in signal intensity in replacement disorders is for the most part nonspecific, although recently differences in T1 values have been described between hyperplastic disorders such as polycythemia vera and leukemia (92). T2 values are less consistent and vary depending on tissue type and the presence or absence of edema, fibrosis, and necrosis (1, 21). In spite of its lack of specificity, MR is helpful in defining the extent of tumor replacement and in documenting the acute phase of the disease, improvement or lack of improvement following chemotherapy, and progression with relapse (18, 21, 32, 58, 93–98). With remission, fat will replace tumor cells following an interval phase of edema or marrow congestion characterized by long T2 values. Fatty substitution during episodes of remission will be reflected by a decrease in T1 values and an increase in marrow signal intensity. With relapse the marrow again demonstrates reduced signal intensity on low TR/TE images.

Acute Leukemia

Acute leukemia occurs most often in the young and accounts for over one-third of all childhood malignancies. The onset is usually insidious, with low-grade fever, anemia, bruising, petechia, and fatigue. The peripheral blood smear is variable, and there may be pancytopenia with or without circulatory blast cells, a normal white cell count or leukocytosis. The liver and spleen may be enlarged and lymphadenopathy is often present, particularly in acute lymphocytic leukemia.

On short TR/TE MR images, leukemic infiltration is characterized by diffuse replacement of normal marrow with marrow of intermediate to low signal intensity. On long TR/TE images an increase in signal intensity is observed (Figs. 10.17 and 10.18). The abnormality may be focal,

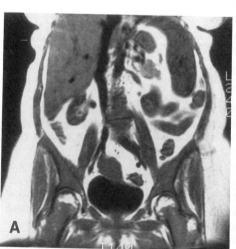
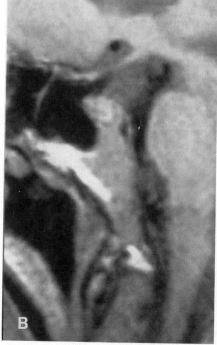
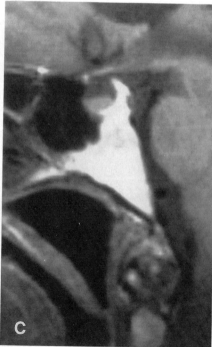

Figure 10.17. Stem cell neoplasm. This 65-year-old woman demonstrated the findings of acute myelogenous leukemia on bone marrow biopsy. **(A)** T1-weighted image (TR/TE 300/15 ms) of the abdomen and pelvis shows replacement of the proximal femoral and pelvic marrow with low-intensity marrow due to tumor infiltration. **(B)** A cranial MR image in the sagittal plane (TR/TE 800/20 ms) shows infiltration of the clivus—normally containing only fatty marrow in this age group. **(C)** The normal pattern of fatty marrow in the clivus of an age-matched asymptomatic patient.

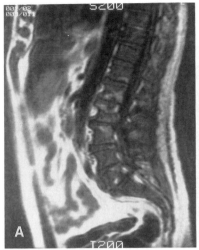 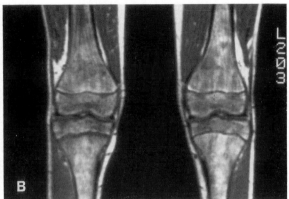

Figure 10.18. **(A)** This 40-year-old woman presented with pancytopenia. The initial bone marrow biopsy was "dry," suggesting aplastic anemia. T1-weighted (TR/TE 600/22 ms) sagittal lumbar spine image suggested almost complete infiltration of the spine with cellular marrow. On the strength of the MR imaging interpretation rebiopsy was performed, showing abnormal cells of acute myelogenous leukemia. Presumably the first biopsy was from an unrepresentative area in this patient with the clinical picture of myelodysplastic syndrome of "preleukemia." **(B)** Another patient, an 8-year-old girl with diffuse infiltration of the distal femurs and proximal tibias with acute lymphocytic leukemia. In a normal subject of this age group, the epiphysis contains fatty marrow, as do the diaphyseal regions on the long bones.

usually in myelocytic leukemia, or diffuse, more common in lymphocytic leukemia (21, 94–96).

Efforts to characterize the abnormal signal patterns found in acute leukemia are highlighted by a study of 30 children (21), in which there was a significant difference in T1 relaxation times between involved vertebral marrow with newly diagnosed acute lymphocytic leukemia and age-matched controls. No significant differences in T1 relaxation time were observed between those children in remission and age-matched controls nor was there any significant difference in T1 relaxation times between those with new disease and relapse. Additionally, there was no overlap in T1 relaxation times between those in remission and controls with those in relapse and new disease. The T2 relaxation times were not significantly different between groups. This study suggests that although an initial bone marrow biopsy is indicated for diagnosis, serial MR imaging is useful to predict the course of the disease without the need for additional biopsies (21).

In another study of adult patients with acute myelocytic and lymphocytic leukemia (59), the T1 relaxation times of lumbar vertebral marrow were two to three times the normal range. On the other hand, there was considerable overlap in T2 values compared with normal values. Following chemotherapy in patients with clinical remission there was a significant reduction in T1 values. In those patients who showed no clinical improvement following chemotherapy, the T1 values remained prolonged. It was concluded that the prolonged T1 relaxation times were due to an increase in total cell number and increased water content of the marrow. Of interest is the fact that during the aplastic phase, following chemotherapy induction when the cellular content of the marrow was only 1%, there were constantly prolonged T1 values. This suggests that the blast cells themselves exhibit prolonged relaxation times independent of the water content of the marrow (81, 91, 96).

Chronic granulocytic leukemia, a malignant line of cells derived from the pluripotent stem cell, is an additional example of a diffuse neoplastic replacement disorder. It is characterized by a slowly increasing granulocytic mass (42). Following a relatively symptom-free chronic phase of 7 to 10 years or more, there is a sudden rapid increase in immature blast cells of the granulocytic series. During this "blast" phase of the disease, there is marked marrow hyperplasia characterized by a diffuse or patchy replacement of fatty marrow by low intensity signal hypercellular marrow similar to other diffuse benign and malignant replacement conditions.

Lymphoma

In lymphoma, bone marrow involvement indicates progression of the disease and may lead to a change in treatment (11, 20, 78, 99, 100). In Hodgkin's disease, unilateral bone marrow biopsy will demonstrate involvement of the iliac crest in 5 to 15% of patients. Bone marrow involvement is found in 25 to 40% of patients with nonHodgkin's lymphoma (20). Although radionuclide examination of bone marrow has the potential to identify areas of involvement, it is limited by low spatial resolution and poor specificity (5). Magnetic resonance imaging may be a more effective method of identifying bone marrow involvement in lymphoma (19, 22). However, in a study of 107 patients in which scintigraphy, MR imaging, and unilateral iliac crest biopsies were compared approximately 50% of the biopsies and either scintigraphy or MR imaging results were in agreement, suggesting that both MR imaging and scintigraphy are helpful as diagnostic tools and represent complementary studies to bone marrow biopsy. An obvious limitation in obtaining correlation between biopsy material and MR imaging is the fact that lymphoma involvement in bone marrow is usually focal (Fig. 10.19) or patchy rather than diffuse, so that bone marrow biopsies

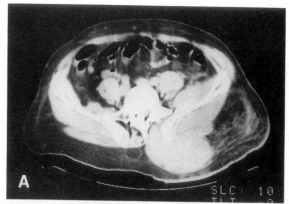

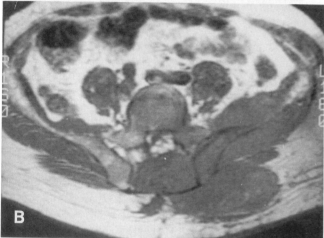

Figure 10.19. Lymphoma. A 60-year-old male with biopsy-proved non-Hodgkin's lymphoma. **(A)** CT shows a soft tissue mass in the gluteal area. There is edema of the soft tissues of the buttocks. **(B)** T1-weighted image (TR/TE 600/20 ms) demonstrates replacement of the fatty marrow by medium signal intensity marrow compatible with tumor or marrow edema.

may be unrepresentative (20). Using MR imaging to localize sites for biopsy will most likely lead to improved correlation in the future.

Characteristically, areas of involvement with lymphoma are hypointense compared with fat and hyperintense compared with muscle on T1-weighted images (22). On T2-weighted images, areas of lymphomatous involvement are hyperintense compared with muscle and isointense with fat (22). Of special interest are areas of bright signal intensity on T2-weighted images within regions of marrow replacement, seen most often in nodular sclerosing Hodgkin's disease. They may represent immature fibrosis within the tumor mass. Although the presence of bright signal intensity in areas of fibrosis appears paradoxical, early fibrosis is characterized by rich vascularity and few collagen fibers so that these areas of bright signal intensity may represent early fibrosis or possibly edema or inflammation surrounding areas of fibrosis (3, 101).

Myeloma

Like leukemia, myeloma exhibits a diffuse or localized decrease in signal intensity on T1-weighted images (Fig. 10.20) (see also Chapter 16). Myeloma is most frequently found in the lower thoracic and lumbar spine, ribs, sternum, skull, and pelvis—regions of richly vascular cellular marrow. Untreated myeloma tends to exhibit increased signal intensity on T2-weighted images, but in those lesions treated with radiation both T1- and T2-weighted pulse sequences show low signal intensity due to osteosclerosis following treatment or due to myelofibrosis. Other myelomatous lesions may exhibit a mixed pattern due to small foci of disease mixed with normal marrow (79). Since scintigraphy and plain film radiography are often normal in multiple myeloma, MR imaging remains the most sensitive imaging alternative to identify areas of involvement (19, 79). As Daffner et al. pointed out in 30 patients with

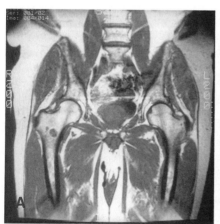

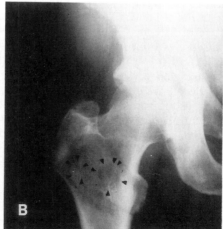

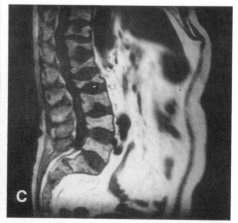

Figure 10.20. Multiple myeloma. **(A)** A 68-year-old man with biopsy-proven myeloma. Focal areas of lower signal intensity on a T1-weighted image (TR/TE 600/20 ms) were present in the right femur. **(B)** Plain film radiographs exhibited focal demineralized lesions in the same area *(arrows)*. **(C)** There is a collapse of multiple vertebral bodies with

punched out areas of abnormality in the upper lumbar and long dorsal spine on this sagittal T1-weighted spin-echo image in another patient with myeloma. (Reprinted with permission from Steiner RM, Mitchell DG, Rao VM et al. Magn Reson Q 1990;6:17–34.)

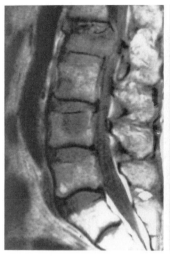
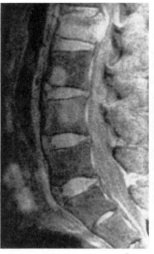
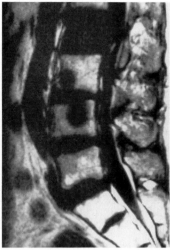
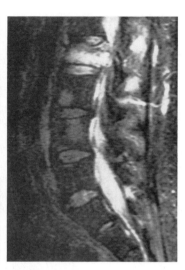

Figure 10.21. Metastatic carcinoma. Four sagittal images of the lumbar spine demonstrate multiple foci of metastatic breast cancer in this 73-year-old woman. There is encroachment of the upper lumbar spinal canal by a collapsed vertebra infiltrated with tumor. **(A)** T1-weighted spin-echo image (TR/TE 800/20 ms); **(B)** with fat saturation and **(C)** with water peak saturation. **(D)** STIR image (TR/TE/TI 1400/43/150 ms).

multiple myeloma, plain film radiographs showed 20 true positives and 10 false negatives. Radionuclide studies showed 6 true positives and 24 false negatives whereas MR imaging yielded 30 true positives out of 30 patients (19). Thus MR imaging appears to be the most sensitive technique for the diagnosis of involvement of the bone marrow with myeloma.

Bone Marrow Metastases

The distribution of hematogenous osseous metastases (see also Chapter 16) is influenced by the vascular anatomy of the bone marrow. Metastases are most commonly seen in the vascular-rich hematopoietic marrow and are less common in the sparsely vascular fatty marrow.

Since metastases often develop first in the bone marrow and then extend to cortical and trabecular bone, they are often detected with MR imaging before enough bone loss occurs to be visible on plain film radiography. On T1-weighted images, metastases are usually focal and are characterized by low signal intensity (Figs. 10.21 and 10.22). With T2-weighted images, metastases are usually high in signal intensity in contrast to hematologic malignancies, such as leukemia and lymphoma, in which the signal intensity on T2-weighted images is variable and often isointense compared with the surrounding marrow. Metastases originating from the breast or prostate may induce an osteoblastic reaction with new bone production. In such lesions, signal loss should be expected in all pulse sequences. Bone marrow metastases are frequently associated with small zones of edema, unlike osteomyelitis (84) or fracture, where a large amount of edema may be present.

Stem Cell Dysplasia–Disorders of Hemoglobin Metabolism

Sickle Cell Anemia

Sickle cell anemia (SCA) is a hereditary hemoglobinopathy characterized by recurrent, acutely painful ischemic crises affecting the bones and joints (30, 102). The

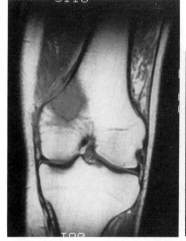
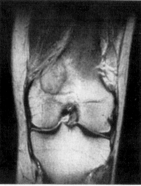

Figure 10.22. Bone marrow metastases in a 42-year-old man with lung cancer. Pain in the left knee prompted the MR imaging study. **(A)** and **(B)** On multiecho T2-weighted spin-echo pulse sequences, the abnormality becomes successively brighter (TR/TE 2000/20–80 ms) relative to surrounding tissue, due to the longer T2 relaxation time of the neoplasm. (Reprinted with permission from Steiner RM, Mitchell DG, Rao VM et al. Magn Reson Q 1990;6:17–34.)

clinical manifestations of vascular obstruction and infarction are the result of intercellular polymerization of deoxygenated hemoglobin S leading to sickling of mature red blood cells. Capillary stasis results because the deformed red blood cells cannot traverse small blood vessels. Hypoxia and tissue damage, especially in the bone marrow and osseous skeleton, follow. A moderate to severe hemolytic anemia is characteristic of SCA, leading to splenomegaly and followed by splenic atrophy due to microinfarction and scarring (54, 103). Metadiaphyseal infarction as well as avascular necrosis of the femoral and humeral heads are common, occurring in the proximal femoral epiphysis in 6 to 62%, depending on the type of sickle cell anemia (7, 42, 104, 105), and in the humeral head in 2 to 17% of patients (106). Because of severe anemia, there is marked expansion of cellular marrow so that in many patients even the epiphyseal centers contain hematopoietic marrow. Many patients develop hemosiderosis because of chronic hemolysis and repeated blood transfusions (20).

The MR imaging findings in patients with sickle cell anemia may be divided into those due to the underlying anemia and those due to marrow ischemia (28–30). Hyperplastic cellular marrow in sickle cell anemia exhibits a signal intensity equal to or greater than muscle but less than fat on both T1- and T2-weighted images (Fig. 10.23). On T1-weighted images there is variable extension of low intensity signal marrow into the epiphysis and throughout

the long bones, replacing fat in a patchy, segmental, or diffuse pattern.

Extension into the epiphysis is seen with frequency in sickle cell anemia. In one series of patients, low-intensity marrow was most common in the tibial shaft and a mixed but predominantly cellular pattern was present in the proximal tibial epiphysis (43) (Fig. 10.24). There was equal distribution of mixed cellular and fatty marrow in the humeral diaphysis. No age-related study describing the age of onset or the extent of marrow reconversion in patients with sickle cell anemia has been published, but since fatty marrow is present in the long bones in many if not most patients with sickle cell anemia, at least some conversion from cellular to fatty marrow occurs in most patients with

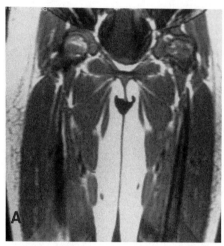
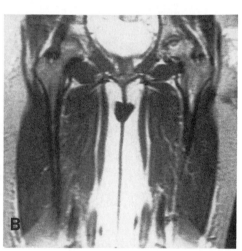

Figure 10.23. Sickle cell anemia. **(A)** Bilateral avascular necrosis of the femoral head in an 18-year-old male with SCA. **(A)** On a T1-weighted (TR/TE 600/20 ms) coronal image there is typical low signal intensity subchondral crescent in an area of mixed fatty and low-intensity marrow. The left femoral head is slightly flattened and there is mixed low-signal and brighter signal intensity in the left femoral epiphysis. The remainder of the femurs are low signal intensity due to reconverted marrow. **(B)** T2-weighted image (TR/TE 2500/80 ms) shows that the cellular marrow is brighter, reflecting the relatively long T2 relaxation time of water. The left crescent is bright and represents an inflammatory zone around a central zone isointense with the remainder of the marrow.

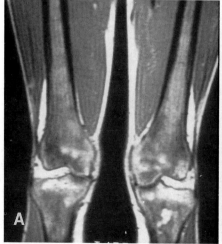
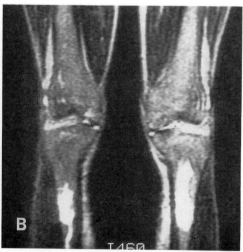
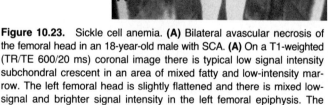

Figure 10.24. A 19-year-old woman with sickle cell disease. **(A)** T1-weighted spin-echo image (TR/TE 600/20 ms) shows the distribution of cellular marrow through the long bones, including epiphyseal involvement. **(B)** T2-weighted images (TR/TE 2000/80 ms) shows medullary zones of bright signal, suggesting acute infarction in both proximal tibias.

sickle cell anemia despite the presence of hemolytic anemia since childhood (3, 28, 30).

The diagnosis of bone marrow infarction complicating SCA has been difficult in the past because of the lack of specific clinical findings or laboratory tests to confirm the diagnosis. Although plain film radiography is useful in documenting the presence of old, partially calcified infarctions in the ribs, long bones, vertebrae, and skull there are no diagnostic plain film findings of acute infarction. Radionuclide studies utilizing ^{99m}Tc sulfur colloid or cyclotron-produced iron-52 (^{52}Fe)-labeled red blood cell precursors detect infarction but both techniques are insensitive and nonspecific. When analyzing radioisotopic scans it is assumed that the erythrocyte precursors distribute in the marrow symmetrically and homogeneously, paralleling reticuloendothelial activity. Since both normal fatty marrow and infarction exhibit reduced activity with radionuclide scans the distribution of fat and cellular marrow may not be symmetrical in sickle cell anemia and a false-negative diagnosis may be made.

Bone marrow infarction occurs most often in fatty marrow, due in large part to its relatively limited vascularity. On the other hand, bone marrow infarction in sickle cell disease is unique in that infarction occurs in areas of nonfatty marrow (Fig. 10.24). Acute infarction in SCA presents typically as a zone of low to intermediate signal intensity on T1-weighted pulse sequences, similar to cellular marrow, but it exhibits high signal intensity on T2-weighted pulse sequences. Infarcts occur most frequently in the femoral head and in the metadiaphyseal portions of the long bones. They are usually sharply marginated, symmetrical, segmental, or serpiginous lesions that most often lie in the center of the cross-section of the bone as seen on axial and coronal images. Liquefaction of a chronic infarct may simulate the MR imaging characteristics of an acute infarction and will be excluded by the absence of a history of recent sickle cell crisis (56). Metaphyseal and diaphyseal infarcts may be difficult to detect on T1-weighted images because they are almost isointense with the surrounding low signal hematopoietic marrow. However, these same regions of abnormality are clearly visualized on T2-weighted images because of their bright signal intensity compared with the intermediate signal intensity of the nearby cellular marrow. In those individuals with hemosiderosis, acute infarction is visible in both T1-weighted and T2-weighted images because of the extremely low signal of the surrounding marrow contrasting with the higher signal of the acute infarction. In an attempt to correlate the presence of acute infarction with the symptoms of sickle cell crisis, Rao et al. found that the presence of acute infarction correlated well with pain in 12 of 14 patient with painful joints (30). Three of five patients with painless joints had evidence of discrete regions of decreased intensity on T1-weighted images, suggesting infarction, but there was no increase in signal in the same areas on T2-weighted images. This finding suggests that these infarctions were not of recent origin or possibly that they represented focal areas of hematopoiesis instead (28, 29).

The diagnosis of avascular necrosis of the femoral head has been correlated carefully with plain film radiography, CT, radionuclide studies, and pathology (33–35, 51) (see also Chapter 26). Avascular necrosis in SCA was described to occur in 15 to 20% of a series of patients with painful sickle cell crisis (30). In this series the femoral epiphysis was characterized by a mixed cellular and fatty marrow pattern in 42% of patients. There was completely fatty marrow in 32%, and 16% exhibited total replacement by hematopoietic marrow (30). Ten percent had evidence of hemosiderotic marrow. With each of these patterns avascular necrosis was found to occur. The magnetic resonance appearance of avascular necrosis (AVN) of the femoral head in patients without sickle cell anemia is often at variance with those patients with sickle cell anemia who develop avascular necrosis. Like the former, a discrete zone of low-intensity signal is the most frequent finding of AVN. On T2-weighted images, however, a pattern of multiple focal low-intensity areas is more typically seen in the femoral head in SCA. The peripheral high signal present in T2-weighted images in most cases of AVN without SCA, representing reactive granulation tissue, is not appreciated in most cases of avascular necrosis with SCA. Joint effusions are also less common in SCA than with avascular necrosis of the femoral head without SCA.

Sickle cell anemia-related medullary bone infarction may be a source of osteomyelitis due to *Staphylococcus aureas* or *Salmonella* species. Signal characteristics of both infection and infarction are similar and may lead to some confusion; however, the presence of cortical involvement, extensive marrow edema beyond the zone of infarction, soft-tissue involvement, and the development of sinus tracts suggest the diagnosis of osteomyelitis. The presence of soft-tissue edema in the muscles of the thigh is common in patients with SCA and if it occurs adjacent to an infarct may be confused with osteomyelitis. If the clinical pictures does not suggest bony inflammation the possibility of an intravenous analgesic injection site should also be considered (4, 30).

Magnetic resonance imaging has become an important imaging examination in patients with SCA to identify the extent of marrow reconversion, the presence or absence of avascular necrosis, and the presence or absence of acute bone marrow infarctions, particularly in patients with a clinical picture of sickle cell crisis. In developing an MR imaging protocol, symptomatic areas are included utilizing T1 and T2 spin-echo pulse sequences. The pelvis, hips, femurs, and lower extremities to the ankles are included to localize patterns of acute and chronic infarction.

Bone Marrow Infarction

The diagnosis of chronic bone marrow infarction is ordinarily established by its typical calcified pattern on plain film radiographs, but early infarction may mimic osteomyelitis or malignancy (107). Magnetic resonance imaging appears to exhibit a sufficiently suggestive pattern to confirm the clinical or plain film diagnosis of early infarction in such cases.

In addition to SCA, bone marrow infarctions are typically found in systemic lupus erythematosus with or without steroid therapy, leukemia, lymphoma, pancreatitis, and following renal transplantation. The MR imaging findings of acute infarction include the presence of a low-signal,

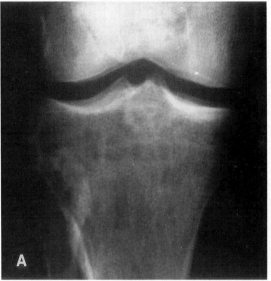 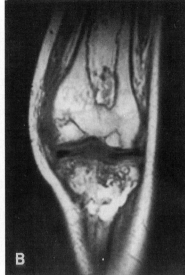 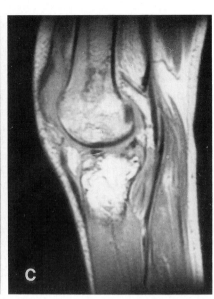

Figure 10.25. Medullary marrow infarction. Multiple bone marrow infarctions are present in the distal femur and proximal tibia in this 33-year-old woman with alcoholic pancreatitis. **(A)** A plain radiograph of the knees shows mottled areas of radiolucency in the distal femur and proximal tibia. **(B)** T1-weighted coronal MR image (TR/TE 600/20 ms) and **(C)** T2-weighted sagittal MR imaging (TR/TE 1500/40 ms) demonstrate multiple serpiginous areas of abnormality. In the T1-weighted image there is a low-intensity zone of inflammation around a central zone isointense with fat in the distal femur. In the T1- and T2-weighted images, the central zone in the proximal tibia is brighter than fat, probably due to subacute hemorrhage.

thin serpentine rim around a central zone that is isointense, or slightly less intense than the nearby fatty marrow on T1-weighted images (Fig. 10.25). This thin dark line may be due to fibrosis or sclerosis surrounding the infarct or to a zone of edema. On T2-weighted images, most of the peripheral zone becomes bright, but an outer rim of low signal may be seen representing sclerosis or chemical shift artifact (33, 35). The center of the infarct typically remains isointense with fat, but central low signal intensity on both T1- and T2-weighted images may develop later on. Identification of the typical pattern of bone marrow infarction with MR imaging initially suspected of osteomyelitis or malignancy with plain films will preclude biopsy and permit the patient to be followed clinically. In the case of AVN, early detection by MR imaging in patients with normal radiographic and scintigraphic examinations allows aggressive treatment with core compression when appropriate.

Bone Marrow Edema

Marrow edema is a nonspecific response to trauma, ischemia, or other stressful stimuli. The pathophysiology of marrow edema is that of hypervascularity with an increase in extracellular water (108, 109). Edema of the bone marrow may be associated with transient osteoporosis (41, 109), femoral capital osteonecrosis (75), interosseous fracture (38), and bone marrow infarction, osteomyelitis (110), or neoplasm. It is uncommon in reflex sympathetic dystrophy.

The MR imaging features of edema are nonspecific. On T1-weighted images there is a decrease in signal intensity compared to fat and an increase in signal intensity on T2-weighted images. The size and character of the area of edema are related to the underlying insult and may be sharply or poorly marginated, geographic, or well defined. In general, edema is extensive when associated with inflammatory processes such as infection, fracture (37), or medullary bruise (38, 108).

Conclusion

Magnetic resonance imaging is a highly sensitive alternative to plain film radiology, CT, and radionuclide studies for the imaging of normal and abnormal marrow and can characterize differences between fatty, cellular, hypercellular, fibrotic, and hemosiderotic marrow. It is useful to depict the extent of the disease and to monitor its clinical course, including the effect of treatment. Magnetic resonance imaging may be helpful in resolving the problem of representative biopsy sampling, i.e., the distribution of disease is often heterogeneous as exemplified by the mixed fatty and cellular patterns of aplastic anemia or polycythemia vera, and the patchy pattern of involvement in lymphoma and metastatic neoplasm. Patterns of disease such as marrow reconversion in epiphyseal and apophyseal marrow have not been previously appreciated. Since MR imaging presents a more global view of the entire bone marrow compartment than does biopsy alone, a better understanding of disease distribution and patterns leading to resolution can be realized. In many respects, MR imaging appears to be ahead of our clinical understanding of diffuse hematologic diseases. Through clinical studies with careful clinical and pathologic correlation, many questions raised by MR imaging may be resolved.

Acknowledgments

The authors would like to thank Maureen Chabot, Tolly Maslen, and Rueben Ybera for typing the manuscript, and Drs. Robert Herfkens and Sheila Moore for their helpful comments.

References

1. Renade SS, Shah S, Advani SH, Kasturi SR. Pulsed nuclear magnetic resonance studies of human bone marrow. Physiol Chem Phys 1977;9:297–299.
2. Cohen MD, Klatte EC, Baehner R, et al. Magnetic resonance imaging of bone marrow disease in children. Radiology 1984;151:715–718.
3. Moore SG, Sebag GH. Primary disorders of bone marrow. In: Cohen MD, Edwards MK, eds. Magnetic resonance imaging in children. Philadelphia: Decker, 1990:765–824.
4. Steiner RM, Mitchell DG, Rao VM, et al. Magnetic resonance imaging of bone marrow: diagnostic value in diffuse hematologic disorders. Magn Reson Q 1990;6:17–34.
5. Vogler JB III, Murphy WA. Bone marrow imaging. Radiology 1988;168:679–693.
6. Weinreb JC. MR imaging of bone marrow: a map would help. Radiology 1990;177:23–24.
7. Sebes JI. Diagnostic imaging of bone and joint abnormalities associated with sickle cell hemoglobinopathies. AJR 1989;152:1153–1159.
8. Helms CA, Cann CE, Brunelle FO, Gilula LA, Chafetz N, Genant HK. Detection of bone marrow metastases using quantitative computed tomography. Radiology 1981;140:745–750.
9. Alavi A, Heyman S, Kim HC. Scintigraphic examination of bone and marrow infarcts in sickle cell disorders. Semin. Roentgenol 1987;22:213–224.
10. Datz FI, Taylor A Jr. The clinical use of radionuclide bone marrow imaging. Semin Nucl Med 1985;15:239–259.
11. Widding A, Smolorz J, Frank M, Linden A, Diehl V, Schicha H. Bone marrow investigation with technetium-99m microcolloid and magnetic resonance imaging in patients with malignant myelo-lympho-proliferative disease. Eur J Nucl Med 1989;15:230–238.
12. Dooms GC, Fisher MR, Hricak H, Richardson M, Crooks LE, Genant HK. Bone marrow imaging: magnetic resonance studies related to age and sex. Radiology 1985;155:429–432.
13. Hajek PC, Baker LL, Goobar JE, et al. Focal fat deposition in axial bone marrow: MR characteristics. Radiology 1987;162:245–249.
14. Jenkins JPR, Stehling M, Sivewright G, et al. Quantitative magnetic resonance imaging of vertebral bodies: A T_1 and T_2 study. Magn Reson Imag 1989;7:17–23.
15. Ricci C, Cova M, Kang YS, et al. Normal age-related patterns of cellular and fatty bone marrow distribution in the axial skeletal: MR imaging study. Radiology 1990;177:83–88.
16. Rosenthal H, Thulborn KR, Rosenthal DI, Kim SH, Rosen BR. Magnetic susceptibility effects of trabecular bone on magnetic resonance imaging of bone marrow. Invest Radiol 1990;25:173–178.
17. Sebag GH, Moore SG. Effect of trabecular bone on the appearance of marrow in gradient-echo imaging of the appendicular skeleton. Radiology 1990;174:855–859.
18. Shah S, Ranade SS, Kasturi SR, Phadke RS, Advani SH. Distinction between normal and leukemic bone marrow by water protons nuclear magnetic resonance relaxation times. Magn Reson Imag 1982;1:23–28.
19. Daffner RH, Lupetin AR, Dash N, Deeb ZL, Sefczek RJ, Schapiro RL. MRI in the detection of malignant infiltration of bone marrow. AJR 1986;146:353–358.
20. Linden A, Zankovich R, Theissen P, Diehl V, Schicha H. Malignant lymphoma: bone marrow imaging versus biopsy. Radiology 1988;173:335–339.
21. Moore SG, Gooding CA, Brasch RC, et al. Bone marrow in children with acute lymphocytic leukemia: MR relaxation times. Radiology 1986;160:237–240.
22. Negendank WG, Al-Katib AM, Karanes C, Smith MR. Lympho-

23. mas: MR imaging contrast characteristics. Radiology 1990;177:209–216.
23. Ellis JT, Peterson P, Geller SA, Rappaport H. Studies of the bone marrow in polycythemia vera and the evolution of myelofibrosis and second hematologic malignancies. Semin Hematol 1986;23(2):144–155.
24. Jensen KE, Grube T, Thomsen C, et al. Prolonged bone marrow T1-relaxation in patients with polycythemia vera. Magn Reson Imag 1988;6:291–292.
25. Lanir A, Aghai E, Simon JS, Lee RGL, Clouse ME. MR imaging in myelofibrosis. J Comput Assist Tomogr 1986;10(4):634–636.
26. Rosenthal DI, Mayo-Smith W, Goodsitt MM, Doppelt HS, Matkin HJ. Bone and bone marrow changes in Gaucher disease. Radiology 1989;170:143–146.
27. Mankad VN, Williams JP, Harpen MD, et al. Magnetic resonance imaging of bone marrow in sickle cell disease. Clinical, hematologic, and pathologic correlations. Blood 1990;75:274–283.
28. Rao VM, Fishman M, Mitchell DG, et al. Painful sickle cell crisis: bone marrow patterns observed with MR imaging. Radiology 1986;161:211–215.
29. Rao VM, Mitchell DG, Steiner RM, et al. Femoral head avascular necrosis in sickle cell anemia: MR characteristics. Magn Reson Imag 1988;6:661–667.
30. Rao VM, Mitchell DG, Rifkin MD, et al. Marrow infarction in sickle cell anemia: correlation with marrow type and distribution by MRI. Magn Reson Imag 1989;7:39–44.
31. Kaplan PA, Asleson RJ, Klassen LW, Duggan MJ. Bone marrow patterns in aplastic anemia: observations with 1.5-T MR imaging. Radiology 1987;164:441–444.
32. Olson DO, Shields AF, Scheurich CJ, Porter BA, Moss AA. Magnetic resonance imaging of bone marrow in patients with leukemia, aplastic anemia and lymphoma. Invest Radiol 1986;21:540–546.
33. Mitchell DG, Joseph PM, Fallon M, et al. Chemical-shift MR imaging of the femoral head; an in vitro study of normal hips and hips with avascular necrosis. AJR 1987;148:1159–1164.
34. Mitchell DG, Rao VM, Dalinka MK, et al. Femoral head avascular necrosis. Correlation of MR imaging, radiographic staging, radionuclide imaging, and clinical findings. Radiology 1987;162:709–715.
35. Mitchell DG, Kressel HY, Arger PH, et al. Avascular necrosis of the femoral head: morphologic assessment by MR imaging with CT. Correl Radiol 1986;161:739–742.
36. Trillet V, Revel D, Combaret V, et al. Bone marrow metastases in small cell lung cancer; detection with magnetic resonance imaging and monoclonal antibodies. Br J Cancer 1989;60(1):83–88.
37. Lee JK, Yao L. Stress fractures: MR imaging. Radiology 1988;169:217–220.
38. Yao L, Lee JK. Occult intraosseous fracture: detection with MR imaging. Radiology 1988;167:749–751.
39. Casamassina F, Ruggiero C, Caramella D, Tinacci E, Villari N, Ruggiero M. Hematopoietic bone marrow recovery after radiation therapy: MRI evaluation. Blood 1989;73:1677–1681.
40. Rosenthal DI, Hayes DW, Rosen B, Mayo-Smith W, Goodsitt MM. Fatty replacement of spinal bone marrow due to radiation: demonstration by dual energy quantitative CT and MR imaging. J Comput Assist Tomogr 1989;13(3):463–465.
41. Bloem JL. Transient osteoporosis of the hip: MR imaging. Radiology 1988;167:753–755.
42. Babior BM, Stossel TP. Hematology: a pathophysiological approach. New York: Churchill Livingstone, 1989.
43. Quesenberry P, Levitt L. Hematopoietic stem cells. N Engl J Med 1979;301:755–760,819–823,868–872.
44. Trubowitz S, Davis S. The bone marrow matrix. In: The human bone marrow: anatomy, physiology and pathophysiology. Boca Raton, Fla.: CRC Press, 1982:43–75.
45. Smith SR, Williams, Davies JM, Edwards RHT. Bone marrow disorders: characterization with quantitative MR imaging. Radiology 1989;72:805–810.
46. Snyder WS, Cook MJ, Nasset ES, Karhausen LR, Howells GP, Tipton IH. Report on the task group on reference man. Oxford: Pergamon Press, 1974:79.
47. Trueta J, Harrison MHM. The normal vascular anatomy of the

femoral head in adult man. J Bone Joint Surg (Br) 1953;351:442–461.

48. Custer RP, Ahlfeld FE. Studies on the structure and function of bone marrow. J Lab Clin Med 1932;17:960–974.

49. Kricun ME. Red-yellow marrow conversion: its effect on the location of some solitary bone lesions. Skel Radiol 1985;14:10–19.

50. Moore SG, Dawson KL. Red and yellow marrow in the femur: age related changes in appearance at MR imaging. Radiology 1990;175:219–223.

51. Mitchell DG, Rao VM, Dalinka M, et al. Hematopoietic and fatty bone marrow distribution in the normal and ischemic hip: new observations with 1.5-T MR imaging. Radiology 1986;161:199–202.

52. Modic MT, Steinberg PM, Ross JS, Masaryk TJ, Carter JR. Degenerative disc disease: assessment of changes in vertebral body marrow with MR imaging. Radiology 1988;166:193–199.

53. Deutsch AL, Mink JH, Rosenfelt FP, Waxman AD. Incidental detection of hematopoietic hyperplasia on routine knee MR imaging. AJR 1989;152:333–336.

54. Guckel F, Brix G, Semmler W, et al. Systemic bone marrow disorders: characterization with proton chemical shift imaging. J Comput Assist Tomogr 1990;14(4):633–642.

55. Glover GH, Schneider E. Three-point Dixon technique for true water/fat decomposition with B^0 inhomogeneity correction. Magn Reson Med 1991 (in press).

56. Mitchell DG, Burk DL Jr, Vinitski S, Rifkin M. The biophysical basis of tissue contrast in extracranial MR imaging. AJR 1987;149:831–837.

57. Lodes CC, Felmlee JP, Ehman RL, et al. Proton MR chemical shift imaging using double and triple phase contrast acquisition methods. J Comput Assist Tomogr 1989;13:855–861.

58. McKinstry CS, Steiner RE, Young AT, Jones L, Swisky D, Aber V. Bone marrow in leukemia and aplastic anemia: MR imaging before, during, and after treatment. Radiology 1987;162:701–707.

59. Richards TL, Davis CA, Barker BR, Bemert WD, Genant HK. Lipid/water ratio measured by phase encoded proton nuclear magnetic resonance spectroscopy. Invest Radiol 1987;22:741–746.

60. Rosen BR, Fleming DM, Kusher DC, et al. Hematologic bone marrow disorders: quantitative chemical shift MR imaging. Radiology 1988;169:799–804.

61. Richards MA, Webb JAW, Jewell SE, et al. In vivo measurement of spin lattice relaxation time (T_1) of bone marrow in healthy volunteers: the effects of age and sex. Br J Radiol 1988;61:30–33.

62. Dunnill MS, Anderson JA. In: Whitehead R, ed. Quantitative histological studies of age changes in bone. J Pathol Bacteriol 1967;94:275–291.

63. Bydder GM, Young IR. MR Imaging: clinical use of the inversion recovery sequence. J Comput Assist Tomogr 1985;9:659–675.

64. Porter BA, Shields AF, Olson DO. Magnetic resonance imaging of bone marrow disorders. Radiol Clin N Am 1986;24:269–288.

65. Field SA, Wehrli. Signa Applications Guide, Vol. I. 4th ed. Milwaukee: General Electric, 1990.

66. Dixon WT. Simple proton spectroscopic imaging. Radiology 1984;153:189–194.

67. Szumowski J, Eisen JK, Vinitski S, Haake PW, Plewes DB. Hybrid methods of chemical-shift imaging. Magn Reson Med 1989;9:388–399.

68. Williams SCR, Horsfield MA, Hall LD. True water and fat MR imaging with use of multiple echo acquisition. Radiology 1989;173:249–253.

69. Keller PJ, Hunter WW, Schmalbrock P. Multisection fat-water imaging with selective presaturation. Radiology 1987;164:539–541.

70. Mitchell DG, Vinitski S, Rifkin MD, Burk DL Jr. Sampling band width and FAT suppression: effects on long term TR/TE MR imaging of the abdomen and pelvis at 1.5T. AJR 1989;153:419–425.

71. Prorok RJ. Signa Applications Guide. 4th ed. Milwaukee: General Electric, 1990.

72. Wismer GL, Rosen BR, Buxton R, Stark DD, Brady TJ. Chemical shift imaging of bone marrow: preliminary experience. AJR 1985;145:1031–1037.

73. Joseph PM. A spin echo chemical shift imaging technique. J Comput Assist Tomogr 1985;9(4):651–658.

74. Gomori JM, Holland GA, Grossman RI, Gefter WB, Lenkinski RE. Fat suppression by section-select gradient reversal on spin-echo MR imaging. Radiology 1988;168:493–495.

75. Turner DA, Templeton AC, Selzer PM, Rosenberg AG, Petasnick JP. Femoral capital osteonecrosis: MR findings of diffuse marrow abnormalities without focal lesions. Radiology 1989;171:135–140.

76. Hotta T, Murate T, Inoue C, et al. Patchy haemopoiesis in long-term remission of idiopathic aplastic anaemia. Eur J Haematol 1990;45:73–77.

77. Koeffler HP, Golde DW. Human preleukemia. Ann Intern Med 1980;93:347–353.

78. Nyman R, Rehn S, Glimelius B, et al. Magnetic resonance imaging in diffuse malignant bone marrow diseases. Acta Radiol 1987;28:199–205.

79. Fruehwald FXJ, Tscholakoff D, Schwaighofer B, et al. Magnetic resonance imaging of the lower vertebral column in patients with multiple myeloma. Invest Radiol 1988;23:193–199.

80. Stevens SK, Moore SG, Kaplan ID. Early and late bone marrow changes after irradiation: MR evaluation. AJR 1990;154:745–750.

81. Jensen KE, Nielsen H, Thomsen C. In vivo measurements of the TI relaxation processes in the bone marrow in patients with myelodysplastic syndrome. Acta Radiol 1989;30:365–368.

82. Wolf BC, Neiman RS. The bone marrow in myeloproliferative and dysmyelopoietic syndromes. Hematol Oncol Clin N Am 1988;2(4):669–694.

83. Murphy S. Polycythemia vera. In: Williams W, Beutler E, Erslev A, Litchman M, eds. Hematology. 4th ed. New York: McGraw Hill, 1989:185–186.

84. Ward HP, Block MH. The natural history of agrogenic myeloid metaplasia (AMM) and a critical evaluation of its relationship with the myeloproliferative syndrome. Medicine 1971;50:357–420.

85. Kaplan K, Mitchell DG, Steiner RM, et al. Polycythemia vera and myelofibrosis: chronic myeloproliferative disease. Spectrum of MRI findings. Soc Magn Reson Med Abstr 1990:18.

86. Kaltwasser JP, Gottschalk R, Schalk KP, Hart LW. Non-invasive quantitation of iron-overload by magnetic resonance imaging. Br J Haematol 1990;74:360–363.

87. Kessing PHL, Falke THM, Steiner RM, Bloem H, Peters A. Magnetic resonance imaging in hemosiderosis. Diagn Imag Clin Med 1985;54:7–10.

88. Leung AWL, Steiner RE, Young IR. NMR imaging of the liver in two cases of iron overload. J Comput Assist Tomogr 1984;8(3):446–449.

89. Siegelman ES, Mitchell DG, Rubin R, et al. Parenchymal vs. reticuloendothelial iron overload in the liver: distinction with MR imaging. Radiology 1991;179:361–366.

90. Kangarloo H, Dietrich RB, Taira TR, et al. MR imaging of bone marrow in children. J Comput Assist Tomogr 1986;10(2):205–209.

91. Thomsen C, Sorensen PG, Karle H, Christoffersen P, Henriksen O. Prolonged bone marrow T1-relaxation in acute leukemia, in vivo tissue characterization by magnetic resonance imaging. Magn Reson Imag 1987;5:251–257.

92. Jensen KE, Jensen M, GrundtVig P, et al. Localized in vivo proton spectroscopy of the bone marrow in patients with leukemia. Magn Reson Imag 1990;8:779–789.

93. Benz-Bohm G, Gross-Fengels WW, Bohndorf K, Guckel C, Berthold F. MRI of the knee region in leukemic children. Part II. Follow-up: responder, non-responder, relapse. Pediatr Radiol 1990;20:272–276.

94. Bohndorf K, Benz-Bohm G, Gross-Fengels W, Berthold F. MRI of the knee region in leukemic children. Part I. Initial pattern in patients with untreated disease. Pediatr Radiol. 1990;20:272–276.

95. Henkelman RM, Messner H, Poon PY, et al. Magnetic resonance imaging for monitoring relapse of acute myeloid leukemia. Leuk Res 1988;12:811–816.

96. Jensen KE, Thomsen C, Henriksen O, et al. Changes in T1 relaxation processes in the bone marrow following treatment in children with acute lymphoblastic leukemia. Pediatr Radiol 1990;20:464–468.

97. Kusnierz-Glaz C, Reiser M, Hagemeister B, Bucher T, Hiddemann W, Van de Loo J. Magnetic resonance imaging. Follow-up in patients with acute leukemia during induction chemotherapy. Haematol Blood Transfus 1990;33:351–356.

98. Pieters R, Van Brenk AI, Veerman AJP. Bone marrow magnetic resonance studies in childhood leukemia. Cancer 1987;60:2994–3000.

99. Dohner H, Guckel F, Knauf W, et al. Magnetic resonance imaging of bone marrow in lymphoproliferative disorders: correlation with bone marrow biopsy. Br J Haematol 1989;73:12–17.

100. Shields AF, Porter BA, Churchley S, Olson DA, Appelbaum FR, Thomas ED. The detection of bone marrow involvement by lymphoma using magnetic resonance imaging. J Clin Oncol 1987;5:225–230.

101. Lee JKT, Glazer HS. Controversy in the MR imaging appearance of fibrosis. Radiology 1990;177:21–22.

102. Papavasiliou C, Trakadas S, Gouliamos A, Vlahos L, Pouliadis G, Phessas P. Magnetic resonance imaging of marrow heterotopia in haemoglobinopathy. Eur J Radiol 1988;8:50–53.

103. Adler DD, Glazer GM, Aisen AM. MRI of the spleen: normal appearance and findings in sickle cell anemia. AJR 1986;147:843–845.

104. Ballas BK, Talacki CA, Rao VM, Steiner RM. The prevalence of avascular necrosis in sickle cell anemia: correlation with alpha-thalassemia. Hemoglobin 1989;13:649–655.

105. Bohrer SP. Bone changes in the extremities in sickle cell anemia. Semin Roentgenol 1987;22:176–185.

106. VanZanten TEG, Stratius Van Eps LW, Golding RP, Valk J. Imaging the bone marrow during a crisis and in chronic forms of sickle cell disease. Clin Radiol 1989;40:486–489.

107. Munk PL, Helms CA, Holt RG. Immature bone infarcts: findings on plain radiographs and MR scans. AJR 1989;152:547–549.

108. Ehman RL. MR imaging of medullary bone. Radiology 1988;167:867–868.

109. Wilson AJ, Murphy WA, Hardy DC, Totty WG. Transient osteoporosis: transient bone marrow edema? Radiology 1988;167:757–760.

110. Ungar E, Moldofsky P, Gatenby R, Hartz W, Broder G. Diagnosis of osteomyelitis by MR imaging. AJR 1988;150:605–610.

CHAPTER 11

Computed Tomography and Magnetic Resonance Imaging in Gaucher's Disease

David O'Keefe and Daniel I. Rosenthal

Introduction

Gaucher's disease is rarely encountered in day-to-day practice; nonetheless it has an importance out of proportion to its rarity. There are several reasons for this. It is relatively more frequent in a particularly susceptible ethnic group. It has been well characterized genetically and biochemically, has marked clinical, biochemical, and radiological changes, and may be viewed as a model of storage and marrow-packing disorders. It has also been successfully treated by replacement of the missing enzyme. The introduction of computed tomography (CT) and particularly magnetic resonance (MR) imaging has led to greater knowledge of the behavior of this disease and of many similar disorders.

In 1882 Philippe C. E. Gaucher reported an adult patient with an enlarged spleen, which in the absence of evidence of leukemia, he ascribed to an epithelioma (1). As more patients gradually came to medical attention, our knowledge of the complex biochemical and genetic defects that constitute Gaucher's disease has grown (2).

Chemistry

Glucocerebroside is a long-chain amino alcohol sphingosine $(CH_3-(CH_2)_{12}-CH=CH-CH(OH)-CH(NH_2)-CH_2OH)$, a long-chain fatty acid, and a molecule of glucose. This is a substance normally found in the body: it is a glycolipid derived from the breakdown of old red and white blood cells. In normal individuals there are between 60 and 280 μg of glucocerebroside per gram of splenic

tissue. In Gaucher's disease the spleen is enormously enlarged and contains more than 200 times the normal concentration of glucocerebroside. Since spleens occasionally weigh up to 10 kg in Gaucher's disease they may contain as much as 2.8 kg of abnormal glycolipid. In the liver there is an even greater increase in concentration of glucocerebroside; up to 23 mg/g. Abnormal accumulation in the brain appears to occur only in infants, in whom the brain is developing rapidly: but there are a few case reports of the biochemical findings in these patients (3). In normal metabolism glucocerebrosidase (glucosylceramidase, EC 3.2.1.45) catalyzes the breakdown of glucocerebroside to ceramide and glucose. Gaucher's disease is now known to be the clinical manifestation of excessive accumulation of glucocerebroside in various organ systems.

Genetics

There are three clinical presentations of the disease: as originally described in adults (type 1, chronic nonneuronopathic), an infantile form (type 2, acute neuronopathic), and an intermediate form that occurs in later adolescence (type 3, subacute neuronopathic). Of the three, type 1 is by far the most frequent. There are more than 4000 patients with the adult form of Gaucher's disease in the United States. Type 1 Gaucher's disease is the most common genetic disorder affecting Ashkenazi Jews and has a reported incidence of 1 in 2500 live births (4). All forms of Gaucher's disease are inherited as an autosomal recessive trait.

Much work has been done to characterize the specific mutations in DNA responsible for the enzyme deficiency states. Several different base pair substitutions may result in type 1 Gaucher's disease. The clinical pattern has been shown to correlate with the genetic make-up of the patient, and no families have been found with more than one type of Gaucher's disease. Thus, of the five genotypes in type 1 Gaucher's disease, those allele-homozygous for a single adenine-to-guanine base mutation (1226/1226) are asymptomatic, heterozygous patients (1226/1448) are more severely affected, and those homozygous for 1448 (a thymine-to-cytosine single base mutation) are most severely affected; usually with a neuronopathic pattern of occur-

Abbreviations: MDP, methylene diphosphonate; QCT, quantitative computed tomography; QMR, quantitative magnetic resonance.

rence (5). These results, although preliminary, go a long way to explain many puzzling features of heterogeneity of Gaucher's disease.

A child of carrier parents has a one-in-four chance of inheriting Gaucher's disease. The type (1 or 2 only) could previously be predicted only when there was an affected child already; the mutation analysis above promises accurate prediction from parental cell cultures and fetal cloned fibroblasts.

Pathology

Glucocerebroside accumulates in the phagocytic cells of the reticuloendothelial system. These cells become distended with elongated lysosomes containing lipid stacked in bilayers. The cells have a foamy, nonvacuolated cytoplasm, with a tissue paper-like appearance, and may reach a diameter of up to 100 μm. Gaucher cells are believed to be altered macrophages derived from the monocyte-histiocyte series. They stain intensely with periodic acid-Schiff reagent. Gaucher cells accumulate in the red pulp of the spleen, sinusoids of the liver, and other tissues, most notably the marrow. Aggregates also occur in the pancreas, lungs, thyroid, and adrenals (6).

Clinical Findings

The clinical findings reflect the pathological process. In children with type 2 Gaucher's disease neurological abnormality predominates and appears to be due to perivascular accumulation of Gaucher cells. Neuronal inflammation and distension is followed by gliosis. There is a pathognomonic clinical triad of trismus, strabismus, and head retraction. (Death is usual before age 2 and results from pulmonary infection or hypoxia (1).) Type 3 Gaucher's disease has an intermediate prognosis; many patients who develop only mild neuropathic disease will have a near-normal prognosis.

Type 1 disease represents almost 80% of cases of Gaucher's disease. Symptoms are present by age 10 in 60% of cases, but some individuals may be asymptomatic until their seventh or eighth decade.

Splenomegaly is the most frequent and striking clinical sign in Gaucher's disease; the sheer size of the spleen leads to hypersplenism with gradual pancytopenia. Splenic infarcts occur and may calcify. Splenic rupture is uncommon (7). Despite the higher concentration of glucocerebroside found in the liver, hepatomegaly is less dramatic than splenomegaly. Sinusoidal deposition of Gaucher cells may result in cirrhosis and portal hypertension, which will exacerbate the signs and symptoms of splenomegaly (8).

Gradually the active red bone marrow becomes infiltrated with Gaucher cells. As the demand for marrow synthetic ability increases, so does conversion of the dormant yellow marrow in the limbs to active red marrow. Eventually, these new areas also become packed by Gaucher cells (9). Marrow packing leads to compromise of the vascular supply to the trabecular bone and bone infarcts. Acute crises that resemble those seen in sickle cell disease may be seen. Because of the clinical and radiographic similarity, trabecular bone infarcts have been called "aseptic os-

teomyelitis." An increased incidence of true osteomyelitis in Gaucher's disease also occurs (10).

The clinical picture is dominated by infarction of the spleen and bone. Fifty to 75% of Gaucher's disease patients develop bone involvement; aseptic necrosis of the femoral head will occur in half of these (11). Renal, lung, and skin infiltration with Gaucher cells occurs but is less common than the above.

Immunological depression is a consequence of the marrow and spleen infiltration, complicated by the decreased synthetic ability of the liver. Susceptibility to infections increases and more unusual manifestations of immune system disturbance occur; monoclonal gammopathy, myeloma, and tumors of the immune system and the CNS occur more frequently in Gaucher's disease than in the general population (1).

A scoring system has been devised by Zimran and colleagues at La Jolla based on thorough clinical, hematological, and biochemical evaluation. They have elegantly demonstrated the close correlation between the severity scoring and DNA mutation (5).

Radiologic Findings

The plain radiographic changes of Gaucher's disease are striking and are basic to an understanding of the changes identified by CT and MR imaging. Seven basic types of findings have been described.

The hallmark of Gaucher's disease is an Erlenmeyer flask deformity, seen most frequently in the distal femur. Failure of remodeling of the metaphyseal bone results in exaggeration of the normal flare. There is no entirely satisfactory explanation for this finding. It is not seen in all patients with Gaucher's disease and is not related to the degree of marrow packing in the femur, as assessed by MR imaging or CT (Fig 11.1).

Infiltrates of Gaucher cells replace the marrow and stimulate osteoclastic resorption of surrounding bone. An osteoblastic response surrounding the deposit may result in zones of alternating lysis and sclerosis. The appearances reflect the slow rate of growth seen in these lesions. Accumulating Gaucher tissue within the bone marrow may produce pressure effects on the internal cortical margin. Occasionally the deposits of Gaucher cells grow extremely large or coalesce to form a pseudotumorous lesion known as a "Gaucheroma" (Figs. 11.2 and 11.3). The appearance may resemble that seen in a low grade chondrosarcoma.

Osteoporosis often occurs in Gaucher's disease, the etiology of which is unclear. It has been attributed to vitamin D deficiency, perhaps as a consequence of hepatic dysfunction (12).

Bone weakened by osteoporosis, infiltration, and marrow packing may sustain a pathological fracture. Interestingly (Fig. 11.3), there is a reported increased likelihood of these occurring following splenectomy (13). Presumably, the availability of the spleen as a reservoir for Gaucher cells confers some degree of protection on the bones. H-shaped vertebral fractures resembling those of hemoglobinopathies occur (these have also been attributed to modeling deformities), as do conventional compression fractures of the vertebrae (14).

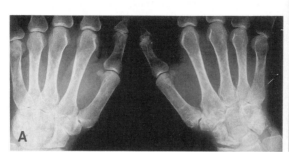

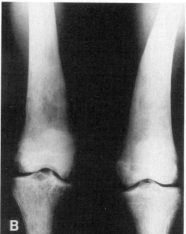

Figure 11.1. Modeling abnormalities. **(A)** Note the undertubulation of the metaphyses of the metacarpals in the hands of this adult with Gaucher's disease. The coarse trabecular pattern reflects osteopenia and marrow packing. **(B)** The lower femur shows a striking example of an Erlenmeyer flask deformity in another patient. The diffuse patchy sclerosis is due to bone marrow infarctions.

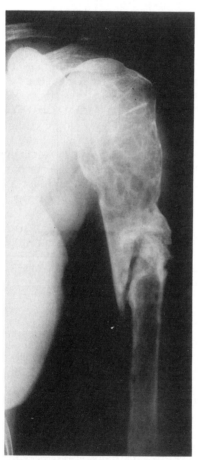

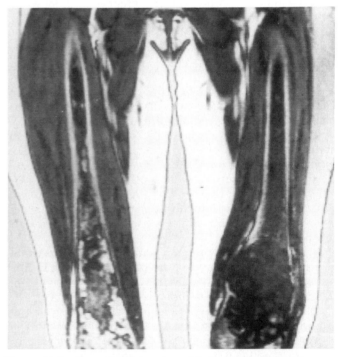

Figure 11.3. Coronal MR imaging of both femurs. Spin-echo sequence (TR 450 ms/TE 20 ms) in a patient with a "Gaucheroma" or expansile tumor-like deposit of Gaucher material in the distal left femur. Note the low signal from the cerebroside deposit contrasting with the high signal from recently infarcted marrow in the right distal femur.

Figure 11.2. Focal Gaucher deposits in the proximal humeri of this patient have led to bilateral pathological fractures (right side not shown) that show nonunion at 12 months.

Trabecular infarcts follow the red marrow infiltrates and are believed to result from compression of the vascular spaces by Gaucher cells (15). Cortical infarcts produce a layer of internal density deep to the endosteal surface, giving a "bone-within-a-bone" appearance. Articular infarcts are a common complication of the condition and result from a similar mechanism (vascular obstruction). They do not appear to occur until the ephiphyses are packed with Gaucher cells (Figs. 11.4 and 11.5).

The alterations in cellular immunity and the presence of infarcted material both in the joint and in the shafts of the long bone predispose patients to the development of infective complications, particularly following surgery or joint

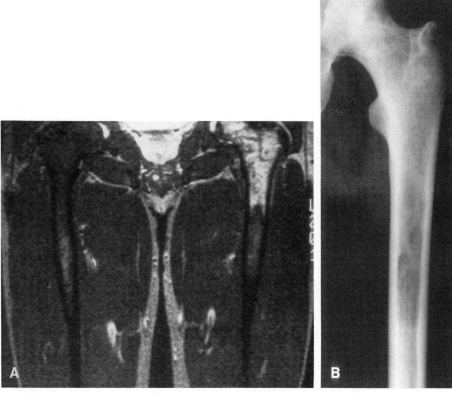

Figure 11.4.   Acute osteonecrosis of the proximal left femur in Gaucher's disease. **(A)** A T2-weighted coronal MR imaging study shows heterogeneous mottling of the marrow, due to Gaucher material deposition. There is a hip joint effusion, and an area of high-intensity signal in the proximal femur. At this time there was clinical evidence of a "Gaucher's disease crisis." The MR imaging high-signal T2-weighted signal is probably due to edema. **(B)** Plain films taken 12 months later demonstrate a slight increase in the density of the proximal femur, particularly in the neck. The focal lucencies of the uninfarcted femoral shaft are more clearly defined at their proximal margins.

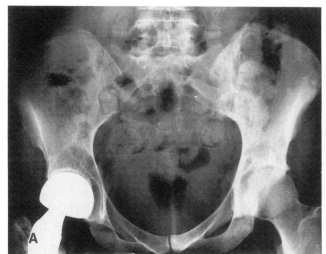
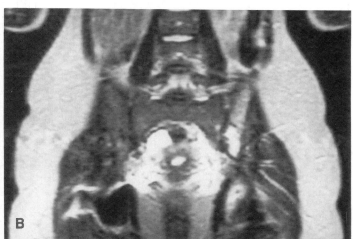

Figure 11.5.   Acute infarction of the left hemipelvis. **(A)** A plain film, and **(B)** T2-weighted (TR 2000 ms/TE 120 ms) coronal images of a female patient with left sacroiliac pain of acute onset show changes of established osteonecrosis in the left hemipelvis. Posterior coronal sections show a region of high T2 signal. There had been a prior right femoral head replacement for osteonecrosis.

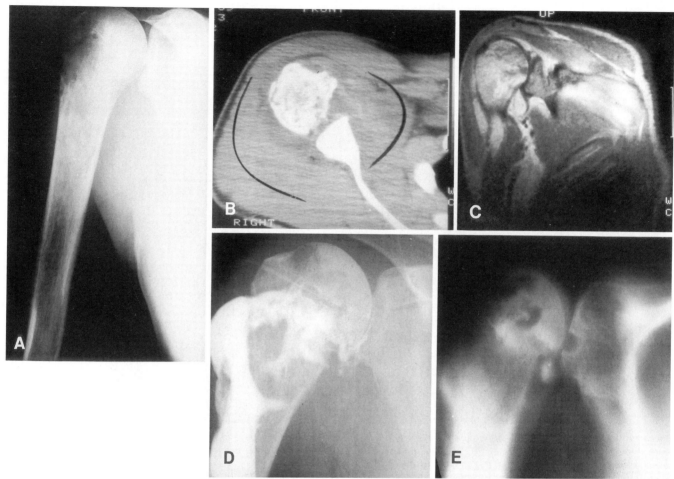

Figure 11.6. Acute infarction complicated by osteomyelitis and septic arthritis. This 23-year-old patient presented with a painful shoulder. Admission radiograph (not shown) demonstrated minimum reduced density of the greater tuberosity. The radiograph was otherwise normal. The primary physician administered intraarticular steroids. **(A)** Two months later changes have occurred, including acute infarction in the proximal humerus and an undisplaced pathological fracture (demonstrated on radiograph (not shown)) of the greater tuberosity. The diaphysis of the humerus shows acute osteoporotic changes with cortical tunneling. **(B)** A CT examination of the shoulder at this time demonstrates a joint effusion. A typical subcortical fracture of osteonecrosis (crescent sign) is visible. **(C)** A coronal proton density image (TR 1500 ms/TE 45 ms) from an MR imaging examination showed a shoulder joint effusion and subarticular changes in the humerus of established infarction. Three months later a sinus developed midway down the arm. A sinogram **(D)** and subsequent tomogram **(E)** through the glenohumeral joint demonstrated communication between a cavity in the proximal humerus and the shoulder. Note the destructive changes affecting the glenohumeral joint.

aspiration (Fig. 11.6). Unusual organisms and atypical radiologic features are commonly found. An inflammatory arthropathy that on aspiration is culture negative is encountered most often as a consequence of aseptic necrosis.

Isotope Scanning

Technitium-99m (^{99m}Tc) methylene dephosphonate (MDP) radionuclide bone scanning may be useful to detect osteonecrosis, pathological fracture, osteomyelitis, and loosening of arthroplasty. Changes are similar to those seen in patients without Gaucher's disease.

^{99m}Tc sulfur colloid has been used to study the location of active synthetic red marrow. Three typical patterns of abnormal uptake found in Gaucher's disease: uniform peripheral extension of red marrow, patchy peripheral extension of red marrow, and late in the disease a diffuse decrease in marrow. These patterns represent abnormal yellow to red marrow conversion, patchy infiltration of infarction of red marrow, and diffuse obliteration of red marrow, respectively.

Studies have shown that Gaucher cells take up inhaled xenon-133m (^{133m}X) avidly. Normal bone marrow shows little or not uptake. In Gaucher's disease three patterns are seen: metaphyseal uptake with epiphyseal sparing, metaphyseal and epiphyseal uptake, and diffuse patchy uptake. Close correlation with MR imaging studies has shown that the first two uptake patterns represents the extent of the marrow infiltration, while the final pattern is found when there is progressive infarction of abnormal Gaucher marrow.

Computed Tomography

Qualitative

Many of the plain film changes of Gaucher's disease are exquisitely defined by CT. Details of the scanning technique depend on the bone being studied. Intravenous contrast enhancement is generally not required. For assessment of visceral disease oral contrast is administered.

There is usually a generalized reduction in bone density. Computed tomography may reveal both cortical thinning and small lucencies in the marrow devoid of recognizable trabeculae. This appearance is similar to that seen in myeloma and amyloid and is presumed to be due to deposits of glucocerebroside.

Focal collections of Gaucher material in the bone may enlarge to a marked degree, the so-called Gaucheroma. These are characterized by CT as expansile, soft-tissue density mass lesions centered in the trabecular bone with a well-defined, occasionally sclerotic margin indicative of a slow rate of growth.

A semiquantitative assessment of the extent of marrow infiltration of the appendicular skeleton can be obtained by measuring the attenuation of the marrow cavity; a positive result in Hounsfield units is abnormal in the appendicular skeleton, reflecting infiltration with Gaucher material or the presence of infarcted bone.

Classical established infarcts result in a sclerotic focus in the trabecular bone with more or less well-defined margins. Over time the sclerosis increases as the infarct matures (Fig. 11.6). Coronal and sagittal reformation of axial CT images may be required to assess the distribution of the lesion within the bone. The early stages of infarction may present a diagnostic problem. Computed tomography and plain films may be normal; MR imaging (see below) is sensitive to the changes of infarction at an earlier stage.

Quantitative

Quantitative CT (QCT) may be used to measure the mineral content of selected sites. Studies in patients with Gaucher's disease are rarely reported; we have reported our experience with an initial cohort of 8 patients (16) and have had an opportunity to perform studies on a further 24. All QCT measurements are performed using a GE 9800 scanner (General Electric Medical Systems, Milwaukee, Wisc.), using methods we have previously described (17). A region of interest is selected in the anterior two-thirds of the L1-L4 vertebral bodies. The same region is maintained at each level for paired sections at 80 and 140 kilovoltage peak (kVp); these values are recorded on the image (Fig. 11.7A). We also measure the trabecular density of the metadiaphyseal region of the distal femur and the proximal tibia (Fig. 11.7B). The fat values required for the calculation are measured on an anthropomorphic phantom with inserts of pure fat in the position of the vertebral body. The Cann-Genant calibration solutions of K_2HPO_4 are used as bone-equivalent markers, and dual-energy calculations are performed to yield values for bone and bone marrow fat content.

There is a moderate reduction in the trabecular bone

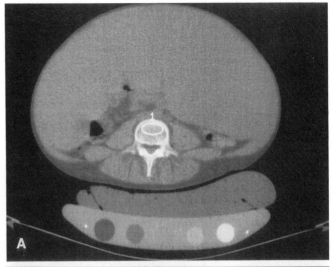

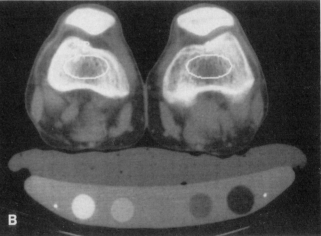

Figure 11.7. Quantitative CT (QCT). This single axial CT image from a QCT examination is at the level of the fourth lumbar vertebra in a 17-year-old patient with massive splenomegaly and hepatomegaly. Note the placement of the measuring cursor on the anterior two-thirds of the vertebral trabecular bone. In this case the bone mineral content was 103 mg/cm³ and no fat was detectable. A normal 17 year old would be expected to have a bone mineral content between 135 and 240 mg/cm³; fat content should be approximately 20 to 30%. **(B)** QCT examination of the distal femoral metaphyses. Note the symmetrical placement of the measuring cursors to encompass as much of the trabecular bone as possible. The peripheral infarct in the right leg is a potential cause of artificially high bone density. In this case there was higher bone density on the right (78 mg/cm³) than on the left (44 mg/cm³). The calculated fat values (right 2%, left 8%) also differ. Normative data for the femora are not yet available.

mineral content of the spine as measured by the single (80 kVp)-energy technique, and a more marked reduction as measured by dual energy. The rate of bone loss with age appears to be faster in Gaucher's disease than in age-matched controls (18).

Striking changes are seen in the percentage of marrow fat found at QCT. Normal control subjects display vertebral fat percentages of approximately 20 to 30%. In Gaucher's disease the fat content is severely reduced. This effect is thought to be due to both the expansion of red marrow and the higher attenuation of the glycolipid that replaces normal fat. It is not yet known whether the amount of fat reduction reflects the severity of disease. Because of the reduced amount of marrow fat, the single-energy bone density measurements are not representative of the true trabecular mineral status. To render them comparable to single-energy values in normal individuals, a downward adjustment of 15 to 30 mg K_2HPO_2 is appropriate to allow for fat usually present within marrow.

Magnetic Resonance Imaging

Technical Factors

Most of the studies we have performed on patients with Gaucher's disease have used a 0.6-T superconducting magnet (Technicare Corporation, Solon, Ohio). Recently we have also used a 1.5-T superconducting magnet (General Electric Medical Systems). Magnetic resonance imaging is performed using three distinctly different routines: an anatomical survey to assess the extent of Gaucher's dis-

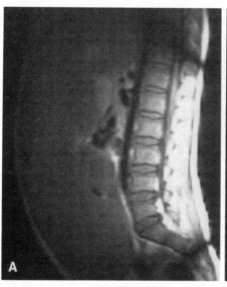

Figure 11.8. Quantitative magnetic resonance (QMR). **(A)** Sagittal in-phase spin-echo (TR 400 ms/TE 25 ms) and **(B)** out-of-phase image sets are obtained, from which region-of-interest measurements are taken to measure fat content. This patient (the same as in Fig. 11.9) shows a "bone within bone" appearance, particularly evident on the out-of-phase imaging sequence *(8B)*. Note also the splenomegaly; the spleen extends to within the pelvis at the midline!

Figure 11.9. **(A)** A more homogeneous appearance of the marrow is more frequently seen in patients who have not undergone episodes of infarction. **(B)** The deposition of Gaucher material is not uniform, exhibiting a coarse, granular appearance.

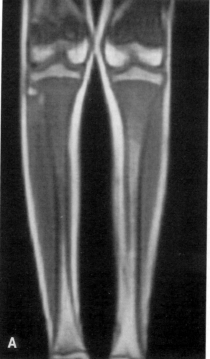

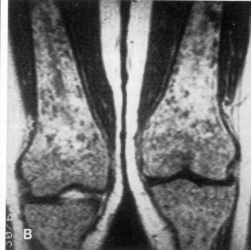

ease for the purpose of staging, a quantitative study of the fat content of the spine also for staging, and a regional anatomical study for evaluation of specific complications.

For the anatomical survey, the pelvis and legs are imaged in their entirety using coronal tomographic 1.5-cm images produced by spin-echo acquisition techniques. The pulse repetition interval (TR) used ranges from 450 to 550 ms, with an echo time (TE) of 20 ms. A sequence of coronal T1-weighted images is obtained that allows differentiation of "normal" from abnormal marrow. Occasionally a sagittal or axial sequence using similar acquisition parameters is used for clarification of abnormalities identified on the coronal sequence.

For analysis of specific regions, sagittal and/or axial images using spin-echo sequences with a TR of 2000 ms and a TE varying from 30 to 80 ms are added. These sequences produce proton density (TR 2000/TE 30, first echo) and predominantly T2-weighted (TR 2000/TE 80, second echo) image sequences.

Quantitative chemical shift imaging of bone marrow fat, with an asymmetric spin-echo technique introducing a dephasing time of 5.3 ms at 0.6 T, can be used to generate opposed phase images from which quantitative studies on the amount of fat are possible (Fig. 11.8).

Findings

The MR imaging characteristics of Gaucher tissue-replaced bone marrow are predominantly due to shortening of the T2 relaxation time. Bone marrow affected by Gaucher's disease gives a weak signal and thus appears dark on both T1- and T2-imaging sequences. The effect is similar on 0.6- and 1.5-T magnets, and differs from most other disease states in which an increase in both T1 and T2 times occurs. The bright signal that arises from normal bone mar-

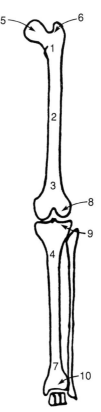

Figure 11.10. Staging of Gaucher's disease by MR imaging of the lower extremities. We have established a 10-stage system based on the sequence observed clinically (9). Coronal T1 images are obtained of the pelvis and entire lower extremities from the patients to be staged. Any reduction in the normally intense T1 signal is regarded as a positive finding. The disease is assigned to a stage based on the highest numbered involved site. The anatomical sites have been numbered in the order in which disease involvement tends to occur.

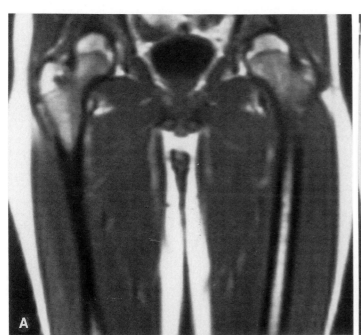

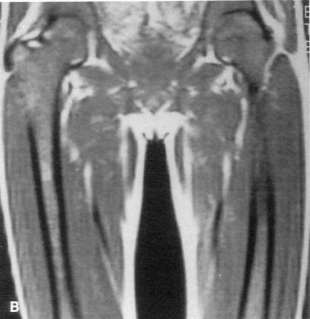

Figure 11.11. **(A)** Stage 4 Gaucher's disease. There is no involvement of the femoral epiphysis or apophysis (TR 500 ms/TE 30 ms) **(B)**

Stage 6 disease. As yet there is no evidence of epiphyseal osteonecrosis, although this patient is now at risk (TR 500 ms/TE 30 ms).

row fat on T1-weighted images contrasts strongly with weak signal from Gaucher tissue and thus the T1-weighted imaging sequences are most useful in assessing the extent of disease.

The MR appearance of Gaucher tissue-replaced marrow may be coarsely granular or homogeneous (Fig. 11.9). In the latter case it may be difficult to distinguish it from normal red marrow. We have developed a staging system for the disease based on the observation that more severely affected individuals show changes in the peripheral appendicular skeleton while less severely affected patients have changes limited to the axial skeleton, and sparing the epiphyses and apophyses (Figs. 11.10 and 11.11). Whether the observed changes are partially due to red marrow may be immaterial to the staging system, since progression of disease precipitates reconversion of yellow to red marrow, as the demand for blood elements increases.

Our staging system has been shown to correlate with the risk of pathological fracture, and osteoarticular necrosis. Recent work using quantitative chemical shift imaging has also confirmed the abnormally decreased amount of marrow fat initially demonstrated by QCT.

Acute bone marrow infarction demonstrates a focal area of bright signal on T2-weighted images (Fig. 11.5). Plain films may be normal, but isotope studies, particularly ^{99m}Tc-MDP, may help by showing focally reduced activity; in this context an infarct is the likeliest diagnosis.

Since Gaucher tissue-replaced marrow is dark on MR imaging, the early stage of osteonecrosis in which signal intensity resembles fat is not observed. More chronic infarcts may develop the surrounding zone of low signal known as the "double-line sign" (19).

Conclusion

It appears probable that in the near future advances in biotechnology will make large quantities of inexpensive glucocerebrosidase available for treatment of these patients. It is even possible that gene-splicing techniques will restore the synthetic capability to the patient. In order to assess the results of such therapies and to optimize their implementation it will be necessary to quantify the extent of bone marrow disease. These new observations using CT and MR imaging represent major advances in this direction.

References

1. Gaucher PCE. De l'epitheliome primitif de la rate, hypertrophie idiopathique de la rate sans lucemie [Thesis]. Paris: 1882.
2. Brady RO, Barranger JA. Glucosylceramide lipidosis: Gaucher's disease. In: Stanbury JB, Wyngaarden JB, Fredrickson JS, Goldstein JL, Brown MS, eds. The metabolic basis of inherited disease. 5th ed. New York: McGraw-Hill, 1983:842–856.
3. Sudo M. Brain glycolipids in infantile Gaucher's disease. J Neurochem 1977;29:379–381.
4. Beutler E. Gaucher disease. In: Goodman RM, Motulsky AG, eds. Genetic disease among Ashkenazi Jews. New York: Raven Press, 1979:157–169.
5. Zimran A, Gross E, West C, Sorge J, Kubitz M, Beutler E. Prediction of the severity of Gaucher's disease by identification of mutations at DNA level. Lancet 1989;ii:349–352.
6. Brady RO, King FM. Gaucher's disease. In: Hers HG, van Hoof F, eds. Lysosomes and storage diseases. New York: Academic Press, 1973:381.
7. Salky B, Kreel I, Gelernt I, Bauer J, Aufses AH Jr. Splenectomy for Gaucher's disease. Ann Surg 1979;190:592–594.
8. James SP, Stromeyer FW, Chang C, Barranger JA. Liver abnormalities in patients with Gaucher's disease. Gastroenterology 1981;80:126–133.
9. Rosenthal DI, Scott JA, Barranger J, et al. Evaluation of Gaucher disease using magnetic resonance imaging. J Bone Joint Surg 1986;68A:802–808.
10. Noyes FR, Smith WS. Bone crises and chronic osteomyelitis in Gaucher's disease. Clin Orthoped Relat Res 1971;79:132–140.
11. Goldblatt J, Sacks S, Beighton P. The orthopaedic aspects of Gaucher's disease. Clin Orthoped 1978;137:208–214.
12. Doppelt S, Osier LK, Neer RM, Barranger J, Mankin HJ. Vitamin D deficiency in patients with Gaucher's disease. J Bone Miner Res 1986;1:113.
13. Rose JS, Grabowski GA, Barnett SH, Desnick RJ. Accelerated skeletal deterioration after splenectomy in Gaucher type I disease. AJR 1982;139:1202–1204.
14. Hansen GC, Gold RH. Central depression of multiple vertebral endplates: A "pathognomic" sign of sickle hemoglobinopathy in Gaucher's disease. AJR 1977;129:343–344.
15. Johnson LC. Histogenesis of avascular necrosis. In: Proceedings of the conference on aseptic necrosis of the femoral head. St. Louis: National Institute of Health, 1964. 55 p.
16. Rosenthal DI, Mayo-Smith W, Goodsitt M, Doppelt S, Mankin HJ. Bone and bone marrow changes in Gaucher disease: Evaluation with quantitative CT. Radiology 1989;170:143–146.
17. Goodsitt MM, Rosenthal DI. Quantitative computed tomography scanning for measurement of bone and bone marrow fat content. Invest Radiol 1987;22:799–809.
18. O'Keeffe D, Rosenthal DI. Unpublished observations, 1990.
19. Mitchell DG, Rao VM, Dalinka MK, et al. Femoral head avascular necrosis: correlation of MR imaging, radiographic staging, radionuclide imaging and clinical findings. Radiology 1987;162:709–715.

CHAPTER 12

Magnetic Resonance Imaging and Computed Tomography of Cortical Bone

Sheila G. Moore

Introduction

Since the first descriptions of skeletal radiographs, analysis of cortical and periosteal remodeling, reaction, and destruction has formed the basis for the interpretation of skeletal abnormalities. While plain films remain the mainstay of initial cortical bone and periosteal analysis (1, 2), loss of 25 to 50% of cortical bone mineral content is required before pathologic processes can be recognized on radiographs. With computed tomography (CT), attenuation differences of 0.5% can be resolved, significantly increasing the sensitivity and specificity of CT for bony analysis (3). Additionally, interpretation of cortical bone and periosteal reaction is unhindered by the presence of overlying tissue (3–6). Although magnetic resonance (MR) appears to have the potential for accurate evaluation of cortical bone (7–12), large prospective studies to fully evaluate the role of MR vs. CT in the evaluation of cortical bone and periosteal reaction have yet to be done.

Cortical Bone

Histology

The histogenesis of cortical bone has been the topic of much study. Bone formation occurs either by transformation of condensed mesenchymal tissue (intramembranous bone formation) or by conversion of cartilage (endochondral bone formation) (13–15). Mineralization of this osteoblast-produced bony matrix results in a network of immature or woven trabeculae, the primary spongiosa (16). The cortical bone derived from this primary spongiosa is formed around, and maintains a relationship to, the haversian system. The haversian system, or osteon, consists of a central haversian canal containing a neurovascular bundle

surrounded by concentric layers of lamellar bone. The conversion of the immature, primary spongiosa to cortical bone is a gradual one; concentric tracts of nonlamellated parallel-fibered bone, termed atypical haversian systems or primary osteons, are remodeled into typical haversian systems (secondary osteons) of lamellated parallel-fibered bone (Fig. 12.1) (15, 17). There are membranes on both the external and internal surfaces of the bone, the fibrovascular periosteum and endosteum, respectively. These membranes retain their osteogenic potential and allow for cortical bone growth and remodeling throughout life.

Branching vessels connect the longitudinally oriented haversian canals to each other. The haversian canal system is supported by interstitial lamellae, which exist in the spaces between the haversian systems (15). The concentric layers of lamellar osseous tissue that surround the central haversian canal each contain collagen fiber bundles and hydroxyapatite crystals, oriented in a specific and complex manner. The orientation of the collagen fiber bundles and

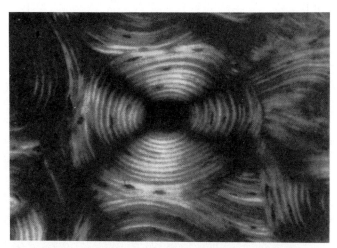

Figure 12.1. Lamellar architecture of cortical bone. Transverse section of a single secondary osteon with high-power polarization optics illustrates the concentric layers of parallel-fibered bone surrounding the central haversian canal containing the neurovascular bundle. The orientation of the collagen fibers differs for each lamella, giving the appearance of concentric rings. (Reprinted with permission from Warwick R, Williams PL, eds. Gray's anatomy. Philadelphia: W. B. Saunders, 1973:219.)

Abbreviations: HU, Hounsfield unit; STIR, short τ inversion recovery.

hydroxyapatite crystals in each lamella differs from that in adjacent lamellae (Fig. 12.1) (15).

Appositional bone growth is the normal, physiologic process through which resorption and formation of cortical bone occurs. It is prominent in the immature skeleton, where bone remodeling leads to marked changes in the size and caliber of bone. This process is less evident in the adult, but is necessary to maintain bone strength and function. Circumferential enlargement of the bone shaft during normal growth and the increase in bone diameter seen in osteoporosis and Paget's disease are examples of appositional bone growth (18).

Computed Tomography and Magnetic Resonance Appearance of Cortical Bone

Several reports have assessed the efficacy of MR and accuracy of MR and CT in evaluating cortical bone (6–10). While initial comparisons of CT and MR imaging for the evaluation of cortical bone suggested that CT was more sensitive and specific than MR (6), recent studies (7–10) indicate that MR may be at least as sensitive to cortical processes as CT examination. A 72% accuracy rate for the MR evaluation of cortical bone was reported in 1987 (8); more recently a study has confirmed that "both MR imaging and CT are very accurate in demonstrating involvement of cortical bone" (9). Anecdotal note was made of one patient in whom MR was more sensitive to tumor invasion of cortical bone than CT, while both MR and CT were falsely negative in assessing cortical bone involvement in another patient (9). This changing estimation of MR efficacy in evaluating cortical bone most likely reflects the early lack of criteria for the MR evaluation of cortical bone; as our knowledge of the normal and abnormal appearance of cortical bone on MR examination increases, the accuracy, sensitivity, and specificity of MR in the evaluation of cortical bone may approach, or even surpass, that of CT.

It is the composition and histologic structure of cortical bone that dictates its appearance on CT and MR exami-nation. On CT examination, compact bone containing hydroxyapatite crystals has a high atomic number and therefore increased attenuation of the X-ray beam (19). The magnitude of this attenuation results in Hounsfield unit (HU) values for cortical bone in the range of +1000; the dense, ossified cortical bone surrounds and can be distinguished from the less dense cancellous portion of bone (Fig. 12.2). Processes that erode or invade cortical bone result in decreased attenuation and appreciable decreased cortical bone density. When present, trabecular bone attenuates the soft-tissue density of marrow; regions of bone with significant amounts of trabecular bone (epiphyses) can be distinguished from regions of bone with a paucity of trabecular bone (diaphyses) (Fig. 12.2). Optimum spatial resolution of cortical bone is achieved by using the bone reconstruction algorithm. Image contrast must be manipulated at the CT console to enhance bony detail. A near-maximum window width of 1000 to 2000 HU and a relatively high window level of 200 to 600 HU will usually result in acceptable images. Apparent destruction of bone on CT images may be artifactually produced by changing window settings on the CT console so that the bone appears very thin.

The criteria for CT evaluation of cortical bone are similar to those of radiographs and plain film tomography; osteopenia is appreciated as decreased attenuation of the X-ray beam when compared to normal, nonosteopenic bone (20). Focal decreased density of cortical bone is seen with intercalation or penetration of the cortex by a pathologic process (e.g., tumor); associated density changes in the marrow and/or surrounding soft tissues will also be seen. Destruction of bone results in loss of visualized cortex, or deformity of the cortical contour; fragments of bone may or may not be present in the surrounding soft tissues. Bony sclerosis and thickening will be appreciated as increased bone density on the CT examination.

On MR examination, cortical bone is best evaluated on transverse images, although sagittal and coronal images may be useful for including or excluding disease suspected on the transverse images. The crystalline, ordered

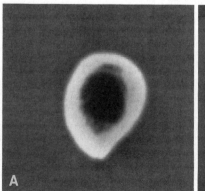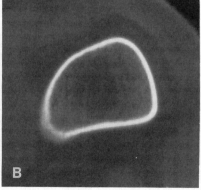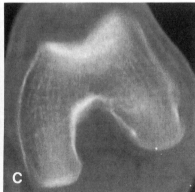

Figure 12.2. CT appearance of normal cortical and trabecular bone. **(A)** Transverse CT image through the midfemur in a 29-year-old woman. Dense cortical bone surrounds the diaphyseal fatty marrow. The diaphysis contains no trabecular bone. Attenuation of surrounding musculature is higher than that of the fatty marrow. **(B)** Transverse image through the distal femoral metaphysis. Increased attenuation cortical bone surrounds the moderately attenuated metaphyseal trabecular bone. Attenuation of the X-ray beam by trabecular bone is appreciated when compared to the nontrabeculated diaphyseal fatty marrow in **(A)**. **(C)** Increased attenuation in the heavily trabeculated distal epiphysis. A rim of normal cortical bone surrounds the marrow cavity.

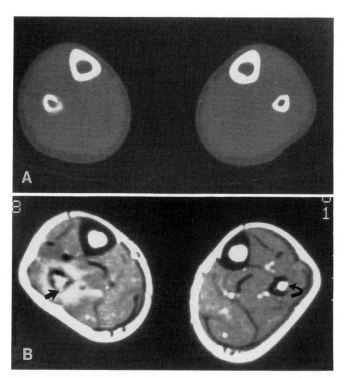

Figure 12.3. Langerhans' cell histiocytosis. **(A)** Transverse CT scan through the midcalf in a 6-year-old boy with Langerhans' cell histiocytosis in the middiaphysis of the right fibula. Decreased density of the medial fibular cortex caused by erosion and intercalation of tumor is appreciated. **(B)** Transverse proton density (TR 1500/20) image through the same region as the CT scan shows thinning and irregularity of the cortical bone, particularly in the medial aspect of the fibula. Intermediate signal intensity within the normally black cortical bone indicates intercalation of tumor *(arrow)*. Extensive increased signal intensity of edema and tumor is noted surrounding the fibula. Note the normal, low signal intensity *(black)* cortical bone in the left tibia and fibula. Chemical shift effect is particularly prominent in the fibula *(curved arrow)*. Corresponding CT image **(A)** confirms the artifactual nature of the apparent lateral cortical thinning on the MR image. (Reprinted with permission from Moore SG, Dawson KL. In: Cohen MD, Edwards MK, eds. Magnetic resonance imaging of children. Philadelphia: B. C. Decker, 1990:825–911.)

structure of lamellar bone does not readily allow free precession of the protons. This, coupled with the low proton density and short T2 relaxation time of cortical bone, results in low (black) cortical signal intensity on MR images (Fig. 12.3) (21, 22). In general, cortical bone abnormality will be appreciated as intermediate signal intensity within the normally black cortical bone on T1-weighted (Fig. 12.3), T2-weighted, short τ inversion recovery (STIR), and fat saturation images. Cortical bone is usually best evaluated by T1-weighted or proton-density spin-echo images (22–24); T2-weighted and STIR images may be useful for subtle areas of cortical bone abnormality, since the increased signal intensity within the cortex may or may not be better appreciated on these sequences (24).

In regions of rapid tapering or expansion of bone, e.g., the metaphyseal regions, there may be partial volume averaging of cortical bone with adjacent muscle or bone marrow, suggesting intermediate signal intensity within the

cortical bone (Fig. 12.4). This partial volume effect is more prominent when slice volume is increased. When partial volume averaging is suspected, that area of cortex should be evaluated on an orthogonal plane (either sagittal or coronal) for persistence of abnormal cortical signal. Confirmation of intermediate signal intensity on the coronal or sagittal image indicates cortical abnormality. Additionally, transverse images of the contralateral side may be helpful in excluding partial volume averaging. The presence of cortical bone chemical shift effect can also result in artifactual abnormality of cortical bone on MR images (Fig. 12.3) (25).

No prospective studies have specifically addressed the significance of abnormal signal intensity within cortical bone on MR examinations. Currently, intermediate signal intensity within the normally black cortical bone is considered abnormal, and may represent cortical "edema" or inflammation, abnormal mineralization of the cortical bone, or intercalation of a pathologic process through the haversian canals. If there is an abnormal marrow or soft-tissue process (e.g., tumor or infection) adjacent to an area of intermediate signal intensity within the cortical bone, then either inflammation (edema) of the cortical bone or involvement of the cortical bone by the pathologic process is likely. Hyperemia due to increased vascularity of a neoplasm may result in cortical osteolysis (26), possibly contributing to abnormal cortical signal.

In the absence of adjacent marrow or soft-tissue abnormality, abnormal signal intensity within the cortical bone may be secondary to aggressive osteoporosis, reactive

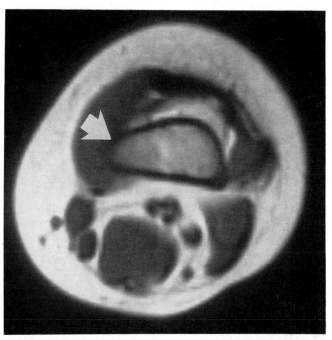

Figure 12.4. Cortical bone. Transverse proton density (TR 2000/20) image through the normal left femur in a 3-year-old girl. While the anterior and posterior cortical signal intensity is normal, intermediate signal intensity within the medial and, to a lesser degree, the lateral aspect of the femur results from partial volume averaging of tapering bone *(arrow)*. This should not be mistaken for abnormal cortical bone signal.

edema, and other inflammatory processes. The intermediate cortical signal intensity seen in these processes may reflect formation of Volkmann's canals. The formation of these additional channels in the cortical bone is provoked by pathologic states associated with neovascularity and bone reabsorption (15, 17). Disease states characterized by rapid bone turnover, such as disuse osteoporosis and reflex sympathetic dystrophy, result in intracortical resorption (27). Intracortical resorption is characterized by the proliferation of resorptive cortical tunnels, leading to an increased amount of cortical blood vessels and connective tissue (27–29). Anecdotally, this may be seen on MR examination as diffuse intermediate signal intensity within the cortical bone (Fig. 12.5). This intracortical resorption is characterized radiographically by prominent longitudinal cortical striations that may also be appreciated on CT examination. Prospective studies establishing specific criteria for and the sensitivity, specificity, and accuracy of MR in the evaluation of cortical bone are needed.

Cortical Bone Evaluation in Musculoskeletal Tumors

Cortical bone involvement has some impact on the staging of musculoskeletal tumors. On MR examinations, endosteal erosion of cortical bone is appreciated as irregular thinning of the endosteal surface of the very low signal intensity cortical bone. The marrow cavity will likely appear expanded, and tumor within the marrow space may deform the endosteal surface of cortical bone (Figs. 12.3 and 12.6). Intercalation of tumor through the cortex via haversian canals is seen as intermediate signal intensity extending either partially or completely through the cortical bone (Figs. 12.3 and 12.6). Some residua of normal cortical outline or signal can usually be recognized (22, 23, 30).

In contrast to intercalation of tumor, cortical breakthrough is seen as interruption of the cortex by solid, abnormal signal intensity marrow or soft-tissue tumor. Tumor will extrude through the cortex, leaving no remnant of normal low signal intensity cortex, and extend directly into an adjacent soft-tissue mass (marrow tumor) (Fig. 12.6), or adjacent marrow (soft-tissue tumor). Cortical breakthrough of tumor into the soft tissues is important to recognize for two reasons: First, a large defect through the cortex will likely decrease the overall chance of patient survival since there will be spread of tumor cells associated with the pathologic fracture and hematoma formation. Second, a biopsy should be performed through a preexisting cortical defect if one is present. While it may occasionally be difficult to distinguish cortical breakthrough and pathologic fracture from diffuse, permeative cortical intercalation on MR images, the requisite evaluation of plain films in conjunction with the MR examination will aid in that determination.

Computed tomography can be used as an adjunctive modality when precise determination of the presence or absence of cortical breakthrough is crucial. Actual breakthrough of tumor through the cortex will be recognized as destruction of cortical bone. Intercalation of tumor through

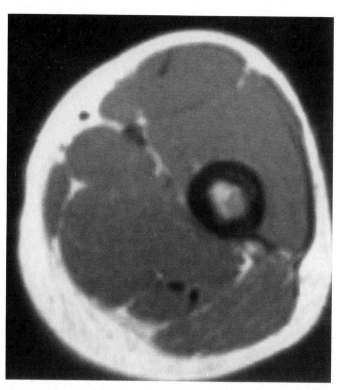

Figure 12.5. Intracortical resorption. Transverse T1-weighted (TR 600/20) image through the left femur of a 6-year-old boy with aggressive disuse osteoporosis. Intermediate signal intensity is present within the cortical bone.

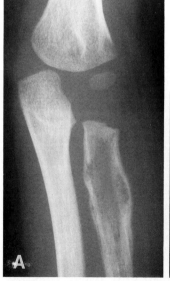
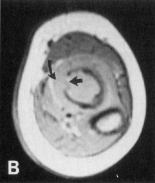

Figure 12.6. *Cortical breakthrough of tumor.* **(A)** Anterior-posterior radiograph of the left forearm in a 2-year-old boy with Langerhans' cell histiocytosis. Endosteal erosion, periosteal reaction, and cortical destruction can be appreciated. **(B)** Transverse proton density (TR 2400, TE 20) image through the proximal radius and ulna shows replacement and expansion of the radial marrow cavity by intermediate signal intensity tumor. Destruction and tumor breakthrough of cortical bone *(arrow)* is appreciated, as is the circumferential soft tissue mass *(curved arrow).*

the cortex will result in decreased cortical attenuation and density.

Periosteal Reaction

Histology

The envelope of tissue surrounding cortical bone, the periosteum, is composed of both an outer "fibrous" layer and an inner, cellular "cambium" layer. The zone between the two layers of active periosteum is a zone of transition (18). This envelope of periosteal tissue separates the cortical bone from the surrounding tissues. The fibrous layer can be replenished from the surrounding periosteal soft tissues, including the fascia, fat, and muscle. Fibrous layer "fibroblasts" are progressively modulated into "preosteoblasts" through nuclear enlargement and acquisition of cytoplasm. Subsequent mitosis and further cell enlargement result in formation of the cambium layer, which contains the osteoid-secreting cells (18). The periosteal membranes contains arterial and capillary vessels that pierce the cortex and enter the medullary canal. These vessels supplement the blood supply to the bone provided by the nutrient artery. The periosteum is continuous about the bone, except in regions of bone (intraarticular) that are covered by cartilage or synovial membrane. Tendons and ligaments that attach to the bone blend with the periosteal membrane at the site of attachment. These entheses can respond to a variety of pathologic stimuli (18).

In the adult, the quiescent periosteum is primarily fibrous, and consists of a single layer of cellular material that has resulted from the fusion of the fibrous and cambium layers. It is firmly attached to the bone. The periosteum in adults responds to bony injury by reverting to distinct fibrous and cambium layers and laying down new bone. In children, the periosteal membrane is thicker, more vascular, and more loosely attached than the adult periosteal membrane. For this reason, the periosteum is more easily lifted from the bone in children than in adults, accounting to some extent for the ease with which the periosteum of infants and children is stimulated to form new bone. The endosteum is less well defined than the periosteum, and is often not identified in the adult. Its primary role likely relates to bone formation in the fetus and infant (15).

While appositional bone growth is a normal process, the term "periosteal reaction" is used to describe a more vigorous deposition of juxtacortical bone in response to stress and disease processes. The quantity of periosteal new bone laid down will depend on the degree of periosteal elevation as well as the aggressiveness of the stimulating process. The configuration of this periosteal reaction will reflect the rate of production and the manner in which the new bone is laid down (18). Mineralization is necessary for appreciation of periosteal reaction on radiographs and CT. This will usually require a period of 10 days to 3 weeks after the initial stimulus, although mineralization of periosteal reaction may occur earlier in children.

The initial phase of periosteal reaction, before mineralization, can best be described as the "cellular" periosteal reaction. The periosteum thickens and differentiates into

the two distinct fibrous and cambium layers. Mineralized periosteal reactions are classified as continuous, interrupted, or complex (Figs. 12.7–12.9) (18, 31). Continuous periosteal reaction may be accompanied by destruction of the endosteal surface of the cortex, giving a "shell" appearance (Fig. 12.7). Continuous periosteal reaction with cortical persistence results in solid, single lamellar, lamellated, and parallel spiculated formation (Fig. 12.7). This subgroup of periosteal reactions results in addition to, rather than replacement of, the original cortex. This does not preclude penetration of the cortex by a marrow process; in fact penetration of the cortex by a marrow process is frequently encountered with the lamellated and parallel spiculated forms. Interrupted periosteal reaction results when an aggressive process breeches a preformed contin-

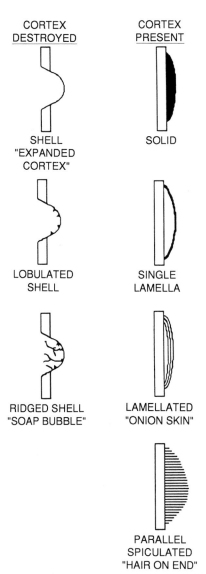

Figure 12.7. Continuous periosteal reaction. Continuous periosteal reaction with *(column 1)* destruction of underlying cortex, and *(column 2)* preservation of cortex. (Adapted from and reprinted with permission from Ragsdale BD, Madewell JE, Sweet DE. Radiol Clin N Am 1981;19:749–784.)

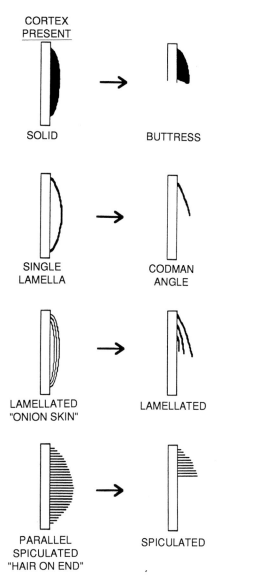

Figure 12.8. Interrupted periosteal reaction. (Adapted from and reprinted with permission, Ragsdale BD, Madewell JE, Sweet DE. Radiol Clin N Am 1981;19:749–784.)

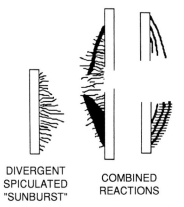

Figure 12.9. Complex periosteal reaction. (Adapted from and reprinted with permission from Ragsdale BD, Madewell JE, Sweet DE. Radiol Clin N Am 1981;19:749–784.)

uous periosteal reaction (Fig. 12.8). The appearance is dictated by the type and amount of underlying, continuous periosteal reaction. Complex periosteal reaction is the most commonly encountered in musculoskeletal tumors (Fig. 12.9).

Computed Tomography and Magnetic Resonance Appearance of Periosteal Reaction

Periosteal reaction is recognized on CT examination when mineralization of the reactive tissue results in increased attenuation of the X-ray beam. The appearance and criteria for interpretation of periosteal reaction on CT are similar to those of plain radiographs; continuous, interrupted, and complex periosteal reactions can all be identified. It is possible to recognize periosteal reaction on reconstructed coronal or sagittal CT images; however, greatest resolution of periosteal reaction will be in the acquired transverse plane.

The MR appearance of periosteal reaction will reflect the underlying histology of the periosteal reaction. Cellular periosteal reaction, characterized by thickening of the periosteum and differentiation of the membrane into two distinct layers, is appreciated as intermediate signal intensity on T1-weighted and increased signal intensity on T2-weighted images. Well-organized lamellar bone is primarily acellular, with low proton density. The protons that are present within lamellar bone are not significantly affected at the field strength and radio frequencies used during clinical proton magnetic resonance since the compact, crystalline structure of lamellar bone does not allow substantial free precession of protons. Therefore, lamellar bone is seen primarily as black or low signal intensity on T1- and T2-weighted spin-echo images as well as STIR and fat saturation images. Tumor bone, and to some degree woven bone in response to more aggressive processes, is structurally disorganized and intermixed with a protein and cellular matrix. This less ordered and more cellular bone is seen as intermediate signal intensity on T1-weighted images and increased on T2-weighted images. It will not be appreciated as low signal intensity "mineralized" bone on MR images even though it appears calcified on plain radiographs and CT images (Fig. 12.10) (11,12). This is similar to the phenomenon seen in Fahr's disease, in which dense calcification embedded in a protein matrix is not appreciated as "calcified" on MR examination (32).

Periosteal reaction on MR images is often best appreciated on transverse images. Anatomic detail is usually best on T1-weighted or proton density images; however, T2-weighted images are useful in the identification of cellular and more subtle periosteal reactions (11, 12, 33). Cortical bone is usually abutted by either increased signal intensity fat or intermediate signal intensity muscle. The presence of normal muscle or fat abutting cortical bone on both T1- and T2-weighted images excludes the possibility of periosteal reaction on the MR examination. Intermediate signal intensity, primarily curvilinear, tissue on T1-weighted images that is contiguous with the cortical bone and increases in signal intensity on T2-weighted images is char-

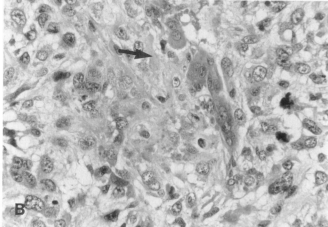

Figure 12.10. Histology of periosteal reaction. **(A)** High-power micrograph of lamellated periosteal reaction *(open arrow)*. **(B)** Tumor bone, while ossified, is disordered and embedded within the cellular matrix of tumor *(arrow)*.

acteristic of the cellular periosteal reaction (Fig. 12.11) (11, 12).

The appearance of periostitis and periosteal reaction in 30 cases of acute experimental *Staphylococcus aureus* osteomyelitis in New Zealand White rabbits has been described (33). The sensitivity, specificity, and accuracy of MR, unenhanced CT, and plain radiographs in detecting periostitis were compared. The sensitivities of MR imaging, CT, and plain film radiography were 100, 77, and 77%, respectively. The specificities were 85, 100, and 100%, while the accuracies were 95, 85, and 85%, respectively. The increased sensitivity and accuracy of MR when compared to CT resulted from identification of nonmineralized periostitis on the MR examination. In four MR cases, "cellular" periostitis could not be distinguished from surrounding soft-tissue edema resulting in a decreased specificity of MR when compared to CT (33).

Continuous Periosteal Reactions

In the shell type of periosteal reaction, osteoclast stimulation by either pressure from an impinging growth or the presence of active hyperemia causes resorption of the endosteal surface of bone (18). At the same time, periosteal new bone on the outer surface of the cortex is formed. If endosteal resorption transiently exceeds periosteal apposition, a thin bony shell will result (Fig. 12.7). The "smooth" shell has an outer contour that is uniform, the result of expansile pressure from benign lesions, i.e., enchondroma, lipoma, fibrous dysplasia, and giant cell tumor. The transverse medullary diameter exceeds the original diameter of bone. The "lobulated" shell results from focal variations in lesion growth rate. The "ridged" shell, or "soap bubble" reaction, occurs when a proliferative lesion slows in growth. The most active lobules of the growth become outlined by ridges. The ridged shell is most commonly seen

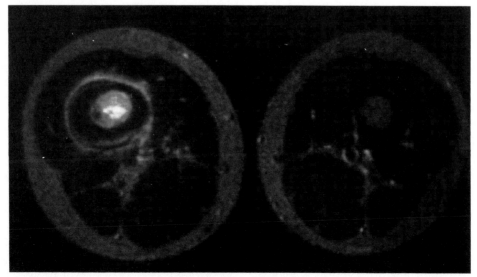

Figure 12.11. Cellular periosteal reaction. Transverse T2-weighted image through the midfemur in an 8-year-old girl with fibrous dysplasia. Cellular periosteal reaction (biopsy proven) is appreciated as curvilinear increased signal intensity border surrounding cortical bone *(arrow)*.

with fibroxanthomas, long-standing giant cell tumors, enchondromas, and occasionally slowly growing malignant processes. On CT examination, this type of periosteal reaction will be seen as a very thin rim of bone surrounding the lesion (Fig. 12.12). The "shell" of periosteal new bone is not usually seen on MR, but may occasionally be seen as a very thin rim of low signal intensity surrounding the abnormal process.

Solid periosteal reaction represents multiple successive layers of new bone applied to the cortex. This leads to cortical thickening (Fig. 12.7). Cortical thickening is usually seen when a chronic, indolent lesion is present in the marrow space, cortex, or adjacent soft tissue resulting in the slow addition of layer after layer of compact lamellar bone. These new layers of lamellar bone become incorporated or fused with the already present lamellar bone. Alternatively, deposition of osseous tissue in the spaces between lamellar layers of bone ("onion skin" periosteal reaction) or between the woven bone network in spiculated periosteal reactions may reduce these reactions to solid, continuous periosteal reactions. Lesions that are often accompanied by this solid periosteal reaction include osteoid osteoma, eosinophilic granuloma, and large enchondromas (particularly those in the intertrochanteric region of the femur) (18).

On CT, this solid periosteal reaction is seen as thickening of the high-attenuation cortical bone. On MR, solid periosteal reaction is usually seen as thick, low signal intensity cortical bone. Intermediate signal intensity within the bone is usually absent, but when present suggests edema (34), aggressive osteoporosis, or involvement of the cortex by an underlying pathologic process such as eosinophilic granuloma. Although changes in attenuation (CT) and signal intensity (MR) are expected when thickened cortical bone contains tumor, there has been one reported case of Ewing sarcoma treated with preoperative therapy that had a "normal" appearance of thickened cortical bone on both CT and MR examinations but viable tumor present within the cortical bone at pathologic examination (9).

Single lamellar periosteal reaction (Fig. 12.7) consists of one sheet of new bone, which on CT and radiographs begins as a faint, radiodense line 1 to 2 mm from the cortical surface. It may or may not join the cortex at the proximal and distal margins. Single lamellar reaction consists of a woven bone network. The initial network of woven bone becomes thickened by surface apposition of lamellar bone (18). On MR examination, single lamellar periosteal reaction can be appreciated as a thin, often curvilinear, low signal intensity line adjacent to cortical bone on both T1- and T2-weighted images (Figs. 12.13–12.15). The periosteal reaction may envelope part, or all, of the cortical bone. The signal intensity of the unossified tissue between the cortical bone and lamellar periosteal reaction is usually intermediate on T1-weighted images and increased on T2-weighted, STIR, and fat saturation images (11, 12, 33).

"Onion skin" periosteal reaction (Fig. 12.7) has traditionally been associated with osteosarcoma, acute suppurative or syphilitic osteomyelitis, Langerhans' cell histiocytosis, and Ewing sarcoma. It may be a transient finding during normal bone growth (35). It is created by concentric planes of lamellar ossification surrounding the cortex

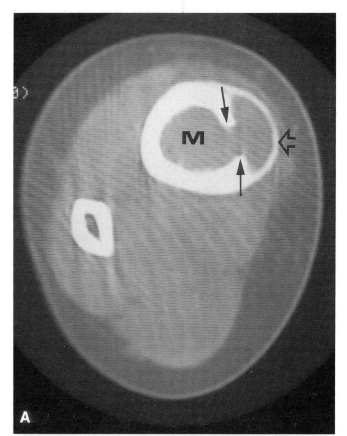

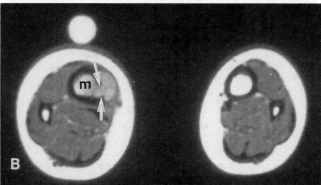

Figure 12.12. Shell-type periosteal reaction. Chronic infection. **(A)** CT scan and **(B)** T1-weighted images through the calves in a child with chronic osteomyelitis. The infectious process has eroded the cortex *(arrows)* and extended into the surrounding soft tissues. "Shell" periosteal reaction surrounding the infectious process is identified on the CT scan; however, the thin rim of lamellated bone cannot be appreciated as a low signal intensity linear rim on the T1-weighted image. (Case courtesy of Mervin Cohen, M.D., Indianapolis, Indiana. Reprinted with permission from Moore SG, Berger P, Stanley P, et al. In: Cohen MD, Edwards MK, eds. Magnetic resonance imaging of children. Philadelphia: B. C. Decker, 1990:913–958.)

(36). These concentric rings of new bone may be caused by permeation of tumor between the layers of lamellated new bone, which separates the layers of lamellated new bone and stimulates further new bone production. However, onion skin periosteal reaction can also be seen in nonneoplastic conditions such as osteomyelitis, pulmonary osteoarthropathy, and stress fractures. The radio-

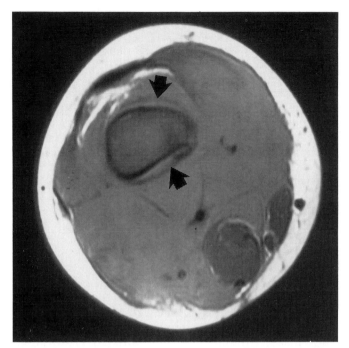

Figure 12.13. Lamellated periosteal reaction. Transverse T1-weighted image through the midthigh in a patient with extensive tumor and curvilinear, low signal intensity lamellated periosteal reaction *(arrows)*. (Courtesy of Dr. G. S. Bisset, III, Cincinnati, Ohio.)

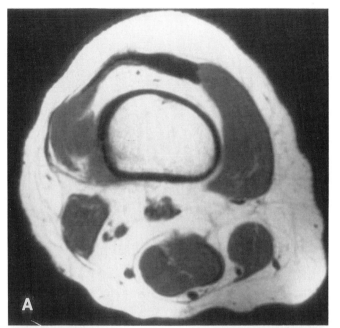

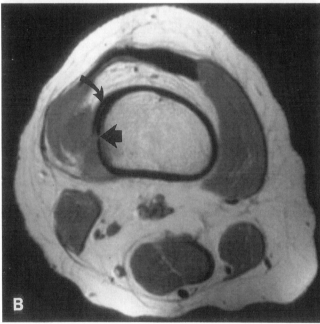

Figure 12.14. Lamellated periosteal reaction. **(A)** T1-weighted (TR 820/20) and **(B)** proton density (TR 2000/20) images through the superior aspect of a femoral, well-differentiated, osteogenic sarcoma. Low signal intensity lamellated periosteal reaction is noted anterolaterally *(curved arrow)*. Intercalation of tumor can be identified *(arrow)*. (Reprinted with permission from Moore SG, Dawson KL. In: Cohen MD, Edwards MK, eds. Magnetic resonance imaging in children. Philadelphia: B. C. Decker, 1990:825–911.)

graphically lucent zones between the layers of new lamellar bone are initially occupied by prominent, dilated vessels and loose connective tissue. These spaces can later be colonized by tumor. In nonneoplastic conditions, lamellated periosteal reaction often follows excessive cortical tunneling in response to active hyperemia.

The initial bone of reactive lamellated periosteal reaction consists of a fine network of woven bone appearing within fibrous tissue. The woven bone is thickened by apposition of lamellar bone. Dynamic refinement of the lamellar layers by osteoclasts (cortical surface) and osteoblasts (soft-tissue surface) results in layers of lamellar bone that are thicker and more mature in histologic structure as the distance from cortical bone increases. The more lamellar outer layers will often contain haversian canals, while the newest layer just under the periosteum will maintain an appearance of woven fiber bone (18). The mineralized layers of lamellar new bone will be visible on CT examination. The radiolucent zones between the mineralized new bone may have some soft-tissue attenuation that reflects the presence of connective tissue, vessels, and/or tumor. On MR examination, the onion skin periosteal reaction is seen as concentric, alternating zones of intermediate and low signal intensity on T1-weighted images, which become alternating zones of intermediate (or bright) and low signal intensity on T2-weighted images (Fig. 12.16). The concentric rings of low signal intensity on both T1- and T2-weighted images correspond grossly and histologically to organized lamellar bone. The zones of intermediate (T1-weighted) and intermediate or bright (T2-weighted) signal intensity between the low signal intensity lamellar bone corresponds histologically to connective tissue, suppera-

tive material, or tumor (11, 12). The innermost layers of woven bone may or may not be appreciated as low signal intensity lamellar bone (11, 12).

Parallel, spiculated "hair on end" periosteal reaction (Fig. 12.7) is the result of a more aggressive process. The texture of individual spicules can be uniform and fine, or they can be lengthy and irregular. They show a gradual reduc-

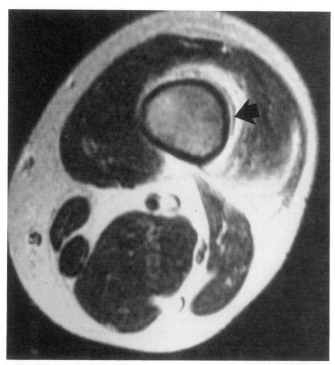

Figure 12.15. Lamellated periosteal reaction. Transverse T2-weighted (TR 2000/70) image through the midthigh in a 15-year-old girl with osteogenic sarcoma. Curvilinear, low signal intensity lamellar periosteal reaction surrounds the lateral aspect of the femur *(arrow)*. Increased signal intensity edema is identified within the vastus lateralis. (Reprinted with permission from Moore SG, Dawson KL. In: Cohen MD, Edwards MK, eds. Magnetic resonance imaging in children. Philadelphia: B. C. Decker, 1990:825–911.)

tion in height in each direction along the shaft from the midzone to the edge of the reaction (18). Histologically, the periosteal new bone has a honeycomb configuration. Tumor may occupy the spaces between the spiculated bone. This pattern of periosteal reaction is often seen in Ewing sarcoma and in the calvarial reaction associated with severe anemias such as thalassemia. When associated with severe anemias, red marrow replaces the loose areolar tissue between the spicules of bone. This pattern has also been reported in myositis, syphilis, and Caffey's disease. The parallel spicules of new bone are appreciated on transverse CT images as radiating spicules of new bone extending from the cortical bone. Depending on the thickness, regularity, and possibly the chronicity of the process, hair on end periosteal reaction may or may not be appreciated on MR images (11, 30, 37). The thick, regular spicules of bone seen in the calvarium of patients with marrow or hematopoietic disorders may be appreciated on MR examination as intermediate signal intensity spicules on both T1- and T2-weighted images, emanating from the cortical bone (Fig. 12.17) (37). However, in tumors such as Ewing sarcoma, the thin strands of spiculated periosteal reaction and new bone formation, with interposed tumor between the honeycomb of bone, may not be appreciated on MR images. The new bone will often be indistinguishable from surrounding abnormal (intermediate signal intensity on T1-weighted images and increased signal intensity on T2-weighted images) tumor.

Interrupted Periosteal Reactions

Interrupted periosteal reaction reflects the structure of the underlying continuous periosteal reaction, and results from tumor involvement of the soft-tissue space within an al-

Figure 12.16. "Onion skin" periosteal reaction. **(A)** Lateral radiograph (right femur) and **(B)** transverse proton density (TR 2000/20) image through the femora in a 3-year-old girl with right femoral Ewing sarcoma. The onion skin periosteal reaction is seen as concentric rings of low and intermediate signal intensity surrounding the cortical bone.

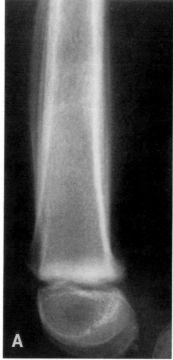
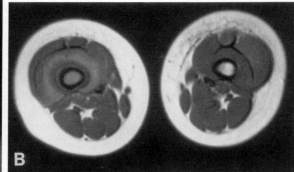

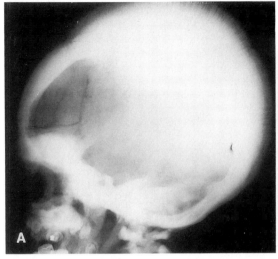

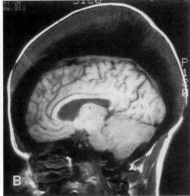

Figure 12.17. "Hair on end" periosteal reaction. **(A)** Lateral view of the skull shows extension of the diploic space and prominent hair on end appearance. **(B)** Midline sagittal T1-weighted (TR 600/TE 20) MR image reveals alternating bands of low and intermediate signal intensity in the diploic space. The bands of low signal intensity correspond to dense trabeculae, while the bands of intermediate signal intensity correspond to hematopoietic marrow. In this case of osteopetrosis, the skull base and inner table are markedly thickened. (Reprinted with permission from Cook PF, Moore SG, Skel Radiol 1991 (submitted).)

ready formed periosteal reaction (Fig. 12.8). There are two ways the continuous periosteal reaction will be interrupted; the tumor may colonize the soft-tissue space within the periosteal reaction, thereby denying space for elaboration of new bone and interrupting the periosteal reaction, or pressure by the tumor can stimulate osteoclast activity, remove the reactive bone, and interrupt the periosteal reaction (18). Computed tomography examination will often show disruption of the attenuated, mineralized cortical bone. Soft-tissue density can be seen emanating through the cortical bone and into the surrounding soft tissue. On MR examination, there is usually interruption of the "normal" black cortical bone by intermediate or bright signal intensity tumor or pathologic tissue. The underlying periosteal reaction that was not destroyed by the pathologic process will be visible at the edges of the lesion on both CT and MR images.

Buttressing occurs when solid periosteal reaction is interrupted, resulting in a solid-appearing wedge of reactive bone at the margins of a (usually) slowly enlarging bone lesion (Fig. 12.9). A thin shell of bone may or may not be formed beyond the region where the solid periosteal reaction was destroyed. Buttressing can also occur at the lateral margins of shell-type periosteal reaction. Histologically, the buttressed bone has a normal, ordered, lamellar configuration; it is seen as low signal intensity on both T1- and T2-weighted MR images (11). The presence of buttressing should prompt suspicion of malignant degeneration in a long-standing, antecedent benign process such as enchondroma or bone cyst (18, 38).

A Codman angle occurs when there is interruption of a single lamellar periosteal reaction. It is commonly seen when an aggressive malignant lesion has broken out of bone, through the periosteum, and into the soft tissues. The side facing the tumor is lucent on the plain films and CT, and tumor can infiltrate into the angle from the open end. On the MR examination, Codman angles are seen as linear, low signal intensity structures on both T1- and T2-weighted images extending away from the cortical bone. The signal intensity of tumor or cellular tissue between the lamellar bone of the Codman angle and the cortical

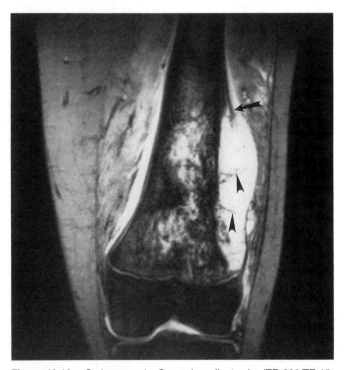

Figure 12.18. Codman angle. Coronal gradient echo (TR 200/TE 15, 30°) image of the distal femur in a child with Ewing sarcoma. The low signal intensity Codman's triangle *(arrow)* and the calcified Sharpey fibers *(arrowheads)* can be clearly identified. (Courtesy of Dr. G. S. Bisset, III, Cincinnati, Ohio. Reprinted with permission from Moore SG, Bisset SG III, Siegel MJ, Donaldson J. Radiology 1991;179:345–360.)

bone is usually intermediate signal intensity on T1-weighted and increased signal intensity on T2-weighted, STIR, and fat saturation images (Fig. 12.18).

Complex Periosteal Reactions

Complex periosteal patterns are commonly seen in bone tumors, especially sarcomas. The divergent spiculated, or "sunburst," pattern is a common sign of malignant os-

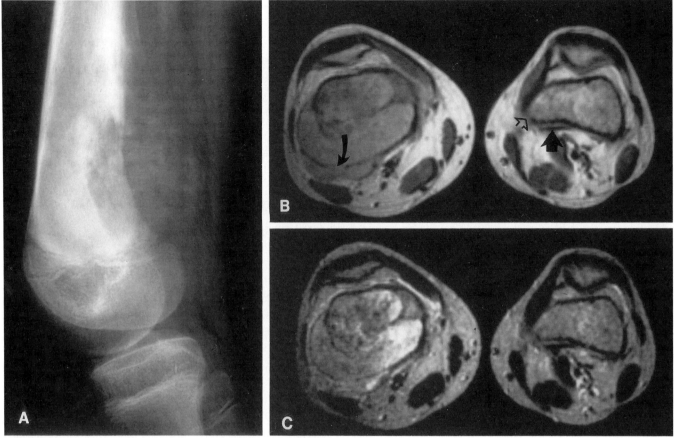

Figure 12.19. Osteogenic sarcoma with complex periosteal reaction. **(A)** Lateral radiograph of the distal femur in an 11-year-old girl with osteogenic sarcoma. Complex periosteal reaction is identified posteriilly. **(B)** Proton density (TR 2000/20) and **(C)** T2-weighted (TR 2000/80) images through the distal femora. The mineralized complex periosteal reaction cannot be appreciated on the MR image; only the posterior soft tissue mass is appreciated. Cortical tumor breakthrough is identified posteriorly; tumor can also be appreciated breaking through the fibrous capsule *(curved arrow).* The low signal intensity fibrous capsule should not be confused for lamellated periosteal reaction. On the left, partial volume averaging of the distal femoral epiphysis gives the appearance of low signal intensity lamellated periosteal reaction *(arrow).* Intermediate signal intensity within the medial cortical bone is also due to partial volume effect *(open arrow).* Cortical abnormalities must be confirmed on saggital or coronal images, particularly in regions of rapid tapering of bone.

Figure 12.20. Partial volume averaging. Transverse proton density (TR 2000/20) image through the proximal femora in a child shows partial volume averaging of the cartilaginous lesser trochanter *(arrow),* giving the appearance of lamellated periosteal reaction.

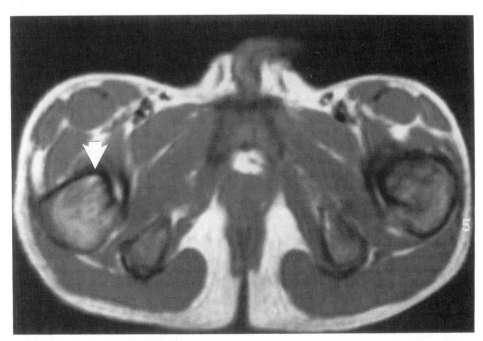

teoid production with only some component of reactive bone (Fig. 12.9). The imaged (CT, radiography) bone will have variable thickness and orientation, pointing toward the epicenter of the marrow space. The individual rays consist of reactive bone, sarcoma bone, or a combination of both (18). The space between the individual rays of new bone is occupied by tumor and tumor products. The reactive bone is most prominent nearest the cortex, since it is oldest at that location. On CT, the calcified complex periosteal reaction can be appreciated within the soft-tissue mass. Most commonly, complex periosteal reaction will not be appreciated as such on the MR image; only the intermediate signal intensity (T1-weighted images) and increased signal intensity (T2-weighted images) soft-tissue mass adjacent to the cortical bone will be seen (Fig. 12.19). The oldest and thickest reactive bone may be seen as linear intermediate signal intensity, emanating from the cortical bone on both T1- and T2-weighted images (11, 12).

Pitfalls

Pitfalls in the evaluation of cortical bone (18) and periosteal reaction include partial volume averaging of the fol-

lowing: surrounding soft tissues in regions of rapid tapering of bone (Fig. 12.19), muscle attachments (especially in the posterior and inferior femur), the vastus lateralis in the femoral introchanteric region, and the cartilaginous epiphyses (Fig. 12.19) and apophyses (Fig. 12.20). This partial volume averaging can result in intermediate signal intensity within or adjacent to the cortical bone, which may masquerade as cortical or periosteal abnormality (Figs. 12.19 and 12.20). Chemical shift artifact is also problematic in the evaluation of cortical bone; chemical shift in the frequency-encoding direction will cause apparent thickening of the cortex on one side of the bone and apparent thinning of the cortex on the opposite side (Fig. 12.3). Reversal of the frequency-encoding direction may be necessary to accurately evaluate cortical bone, especially when a soft-tissue mass abuts the apparently thinned cortical bone (25). Posterior femoral hyperostosis (Fig. 12.21) and normal infantile periostitis can also be mistaken for abnormal periosteal reaction on MR examination. A soft-tissue mass with a low signal intensity fibrous capsule can simulate single lamellated periosteal reaction (Fig. 12.19) (11, 12).

Conclusion

The assessment of cortical bone and periosteal reaction can be pivotal in the differential diagnosis of musculoskeletal lesions. As MR continues to become the imaging modality of choice (after plain films) for many musculoskeletal lesions, knowledge of the appearance, sensitivity, specificity, and accuracy of MR in the evaluation of cortical bone and periosteal reaction becomes increasingly important. However, until prospective studies further define the criteria for cortical and periosteal abnormalities on MR images, CT remains the modality of choice for the evaluation of subtle cortical or periosteal changes and should be used if subtle cortical erosion or periosteal reaction would influence differential diagnosis and/or lesion treatment.

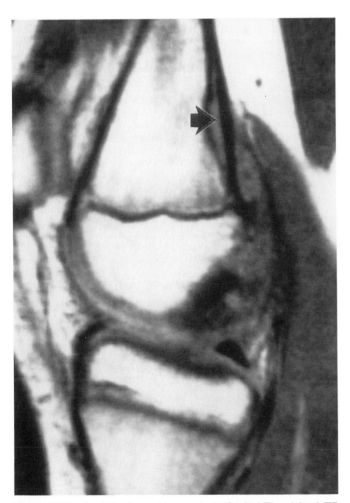

Figure 12.21. Posterior femoral hyperostosis. Sagittal T1-weighted (TR 800/30) image through the knee in an 11-year-old boy with posterior cortical hyperostosis *(arrow)*. This appears as intermediate signal intensity within the normally black cortical bone.

References

1. Coffre, C, Vanel D, Contesso G, et al. Problems and pitfalls in the use of computed tomography for the local evaluation of long bone osteosarcoma: report of 30 cases. Skel Radiol 1985;13:147–153.
2. Sundaram M, McGuire MH. Computed tomography or magnetic resonance for evaluating the solitary tumor or tumor-like lesion of bone? Skel Radiol 1988;17:393–401.
3. Aronberg DJ. Techniques. In: Lee JKT, Sagel SS, Stanley RJ, eds. Computed body tomography. New York: Raven Press, 1983:9–36.
4. Destouet JM, Gilula LA, Murphy WA. Computed tomography of long-bone osteosarcoma. Radiology 1979;131:439–445.
5. Andre¢ M, Resnick D. Computed tomography. In: Resnick D, Niwayama G, eds. Diagnosis of bone and joint disorders. Philadelphia: W. B. Saunders, 1988:143–202.
6. Zimmer WD, Berquist TH, McLeod RA, et al. Bone tumors: magnetic resonance imaging versus computed tomography. Radiology 1985;155:709–718.
7. Petersson H, Gillespy T, Hamlin DJ, et al. Primary musculoskeletal tumors: examination with MR imaging compared with conventional modalities. Radiology 1987;164:237–241.
8. Wetzel LH, Levine E, Murphey MD. A comparison of MR imaging and CT in the evaluation of musculoskeletal masses. Radiographics 1987;7:851–874.
9. Bloem JL, Taminiau AHM, Eulderink F, et al. Radiologic staging

of primary bone sarcoma: MR imaging, scintigraphy, angiography, and CT correlated with pathologic examination. Radiology 1988;169:805–810.

10. Bloem JL, Bluemm RG, Taminiau AHM, et al. Magnetic resonance imaging of primary malignant bone tumors. Radiographics 1987;7:425–445.

11. Moore SG, Berry G. MRI of abnormal cortical bone and periosteal reaction: radiologic and pathologic correlation. Radiology 1991 (submitted).

12. Moore SG, Sebag GH, Dawson KL. MR evaluation of cortical bone and periosteal reaction in bone lesions: pathologic and radiographic correlation. Soc Magn Reson Med [Book of Abstracts] 1989;1:19.

13. Jaffe HL. Metabolic, degenerative, and inflammatory diseases of bones and joints. Philadelphia: Lea & Febiger, 1972.

14. Warwick R, Williams PL. Gary's Anatomy. 35th British ed. Philadelphia: W. B. Saunders, 1973:207.

15. Resnick D, Manolagas SC, Niwayama G. Histogenesis, anatomy and physiology of bone. In: Resnick D, Niwayama G, eds. Diagnosis of bone and joint disorders. Philadelphia: W. B. Saunders, 1988:1940–1974.

16. Jee WSS. The skeletal tissues. In: Weiss, L, Lansing, L, eds. Histology: cell and tissue biology. 5th Ed. New York: Elsevier Biomedical, 1983.

17. Warshawsky H. Embryology and development of the skeletal system. In: Crues RL, ed. The musculoskeletal system. Embryology, biochemistry, physiology. New York: Churchill Livingstone, 1982:33.

18. Ragsdale BD, Madewell JE, Sweet DE. Radiologic and pathologic analysis of solitary bone lesions. Part II. Periosteal reactions. Radiol Clin N Am 1981;19:749–784.

19. Ter-Pogossian MM. Physical principles and instrumentation. In: Lee JKT, Sagel SS, Stanley RJ, eds. Computed body tomography. New York: Raven Press, 1983:1–7.

20. Revak CS. Mineral content of cortical bone measured by computed tomography. J. Comput Assist Tomogr 1980;4:342–350.

21. Mitchell DG, Burk DL Jr, Vinitski S, et al. The biophysical basis of tissue contrast in extracranial MR imaging. AJR 1987;149:831–837.

22. Moore SG, Dawson KL. Tumors of the musculoskeletal system. In: Cohen MD, Edwards MK, eds. Magnetic resonance imaging of children. Philadelphia: B. C. Decker, 1990:825–911.

23. Moore SG. MR precisely evaluates bone tumors: a practical approach to magnetic resonance evaluation of pediatric musculoskeletal tumors. Diagn Imag 1988;10:282–289.

24. Moore SG. MR imaging evaluation of bone lesions: comparison of inversion recovery and spin echo images [Abstract]. Radiology 1988;169(P):191.

25. Dick BW, Mitchell DG, Burk DL, et al. The effect of chemical shift misrepresentation on cortical bone thickness on MR imaging. AJR 1988;151:537–538.

26. Martel W, Abell MR. Radiologic evaluation of soft tissue tumors. A retrospective study. Cancer 1973;32:352.

27. Resnick D, Niwayama G. Osteoporosis. In: Resnick D, Niwayama G, eds. Diagnosis of bone and joint disorders. Philadelphia: W. B. Saunders, 1988:2022–2126.

28. Meena HE. Recognition of cortical bone resorption in metabolic bone disease *in vivo*. Skel Radiol 1977;2:11.

29. Jaworski ZF, Lok E. The rate of ostseosclerosis bone erosion in haversian remodelling sites of adult dog's ribs. Calcif Tissue Res 1972;10:103.

30. Moore SG. Pediatric musculoskeletal imaging. In: Stark DD, Bradley WG, eds. Magnetic resonance imaging. St. Louis: Mosby, 1991 (in press).

31. Edeiken J. New bone production and periosteal reaction. In: Edeiken J, ed. Roentgen diagnosis of diseases of bone. Baltimore: Williams & Wilkins, 1981:11–29.

32. Scotti G, Scialfa G, Tampieri D, et al. MR imaging in Fahr disease. J Comput Assist Tomogr 1985;9:790–792.

33. Spaeth HJ, Chandnani VP, Beltran J, et al. MR imaging detection of early periostitis in acute experimental ostoemyelitis: a comparative study between MR imaging, CT, and plain radiography with pathologic correlation [Abstract]. Radiology 1989;173(P):205; Invest Radiol 1991 (in press).

34. Patel M, Patel U. MR imaging of osteoid osteoma: new observations. Radiology 1990;177(P):146.

35. Hancox NJ, Hay JD, Holden WS, et al. The radiological double contour effect in the long bones of newly born infants. Arch Dis Child 1951;26:543–548.

36. Johnson LC, Vetter H, Putschar WGJ. Sarcoma arising in bone cysts. Virchows Arch. Pathol Anat 1962;335:428–451.

37. Cook PF, Moore SG. Osteopetrosis: MR evaluation of an unusual case with cranial diploid space expansion. Skel Radiol 1991 (submitted).

38. Brunschwig A, Harmon PH. Studies in bone sarcoma. Part III. An experimental and pathological study of role of the periosteum in formation of bone in various primary bone tumors. Surg Gynecol Obstet 1935;60:30–40.

39. Moore SG, Berger P, Stanley P, et al. Infectious, traumatic, mechanical, collagen, and miscellaneous disorders of the musculoskeletal system. In: Cohen MD, Edwards MK, eds. Magnetic resonance imaging of children. Philadelphia: B. C. Decker, 1990:913–958.

40. Moore SG, Bisset SG III, Siegel MJ, Donaldson J. Pediatric musculoskeletal MRI. Radiology 1991;179:345–360.

CHAPTER 13

Magnetic Resonance Imaging and Computed Tomography of Musculoskeletal Infection

Javier Beltran

Introduction

Involvement of bone and soft tissues by an infectious process is a relatively common occurrence that can affect both children and adults. Early diagnosis and treatment are essential to prevent long-term complications of the disease, including growth disturbances, limb deformity, osteolysis, joint ankylosis, osteoarthritis, amyloidosis, and neoplasm (1, 2).

Hematogenous spread, extension from contiguous source, direct implantation, postoperative infection, or a combination of these factors are the main routes of contamination of bones and joints, as described by Resnick and Niwayama (2).

The clinical entity resulting from musculoskeletal infection depends on the location and course of the process. Osteomyelitis indicates infection involving the bone marrow space, with or without extension into the cortical bone. The terms osteitis and periostitis are preferred for those cases in which only the cortical bone or periosteum are involved, without extension into the bone marrow space. The terms acute, subacute (Brodie's abscess), and chronic are used to refer to the clinical stages of osteomyelitis.

Not infrequently soft-tissue infections are associated with osteomyelitis, but they can present also as isolated entities. Soft-tissue infections include septic arthritis, abscess, cellulitis or phlegmon, tenosynovitis, septic bursitis, lymphadenitis, and infectious myositis.

Diagnostic modalities used for evaluating musculoskeletal infections include plain film radiography, with or without magnification, conventional tomography, computed tomography (CT) (3), radionuclide techniques (4),

arthrography, sinus tract injections, and more recently magnetic resonance (MR) imaging (5–9). Each one has its own indications and inherent advantages and limitations.

Plain film radiography and conventional tomography are excellent techniques for detecting bone infection. The main radiographic manifestations of osteomyelitis and soft-tissue infection include periosteal reaction, bone destruction, bone sclerosis, Brodie's abscess, formation of sequestra, involucra, and cloacae, soft-tissue gas, soft-tissue swelling, joint effusion, and sinus tract formation (2). In the vast majority of cases, these findings are sufficient for diagnosis and staging of the infection. The main limitations of plain film radiography include its relatively low sensitivity during the very early stages of the disease, and low specificity when other lesions such as healing fractures, tumors, or neuroarthropathy are among the diagnostic considerations (6, 7).

Radionuclide techniques, including three-phase technetium-99m (^{99m}Tc)-labeled methylene diphosphonate (MDP), gallium-67 (^{67}Ga), and indium-111 (^{111}In)-labeled leukocyte scintigraphy, alone or in combination, are highly sensitive techniques for the demonstration of bone and soft-tissue infection. Lack of specificity, relatively low accuracy in chronic osteomyelitis, and low anatomical resolution are the main drawbacks of radionuclide techniques (10–12).

Arthrography in musculoskeletal infections can be used to verify intraarticular needle position following joint aspiration. Sinus tract injections are helpful in assessing connection of the fistulous tract with the adjacent bone.

Computed tomography and MR imaging can provide additional information in specific clinical situations. The main advantages of these newer techniques over plain film radiography and radionuclide imaging include better soft-tissue contrast resolution, the possibility of multiplanar imaging, and improved assessment of bone marrow involvement (5–8). Relatively low specificity, image degradation due to artifacts originating from metallic implants, and beam-hardening artifacts when using CT are the main limitations of these techniques.

Technique

Computed Tomography

Extremity CT scans can be obtained with a 75-mA, 120-kVp, shaped source filter, and with a normal convolution

Abbreviations (see also glossary): SI, signal intensity; MDP, methylene diphosphonate. STIR, short τ inversion recovery; DTPA, diethylenetriamine pentaacetic acid; HU, Hounsfield unit; GRASS, gradient-recalled acquisition in the steady state.

filter. The scan diameter is zoomed to the specific part of the extremity. A slice thickness of 10 mm can be used to cover large areas. For small areas, 5-mm slice thickness or smaller may be used. Contrast material can be injected intravenously to assess soft-tissue extent. Intrathecal contrast material can be indicated in the evaluation of epidural abscesses (13). The cases illustrated in this chapter were imaged using a variety of CT scanners, including Technicare 2060, Technicare 1440, and General Electric 9800 models. (General Electric, Milwaukee, Wisc.)

Magnetic Resonance Imaging

Spin-echo pulse sequences with T1-weighted (short TR/TE) and T2-weighted (long TR/TE) images in at least two orthogonal planes are adequate for imaging musculoskeletal infections (6, 7). Occasionally oblique planes of section may be necessary to image areas not oriented in the orthogonal planes, like the foot. The short τ inversion recovery (STIR) technique, with an inversion time (TI) of 100 to 110 ms, producing suppression of the signal from the subcutaneous fat and fatty marrow, is also very useful to assess intramedullary and soft-tissue extension of the infection (14).

The surface coil technique is essential to obtain images with high signal and contrast-to-noise ratios. This is especially convenient in small areas, such as the hand and foot. Different types and shapes of surface coils are available from different manufacturers to optimize image quality over the area of interest.

Slice thickness, field of view, and matrix size should be adjusted to the area under study. In general, thick sections (5 to 10 mm), a field of view from 12 to 25 cm, and a matrix of 256×192 to 128 are adequate in most cases. All cases illustrated in this chapter were imaged using high-field superconductive magnets, including a 1.5-T system (Signa, General Electric Co.) or 1.0-T system (Somatom, Iselin, N.J.).

Acute Osteomyelitis

Acute osteomyelitis is typically a disease of childhood, although it can be seen in infants and adults as well. It can involve any bone and any part of the bone, but it occurs most frequently in the long bones and spine. A polyostotic form can be seen in infants. The most common causative organism is *Staphylococcus aureus*. Clinically it presents with the classical symptoms of inflammation, including fever, pain, swelling of the area, redness, and limitation of motion.

The pathophysiologic events occurring during the acute stage of hematogenous osteomyelitis are accurately reflected by the different imaging techniques. Hyperemia, edema, and inflammatory cellular response are the earliest changes taking place within the bone marrow cavity, occurring within hours or days depending on the virulence of the offending organism. At this stage, plain films may be normal or demonstrate regional osteoporosis and loss of the deep soft-tissue fat planes. Bone scintigraphy demonstrates increased uptake, reflecting the local hyperemia.

During this early stage, CT depicts the edematous changes within the bone marrow space, demonstrating increased density (positive Hounsfield units (HUs)) replacing the negative absorption values of the normal fatty bone marrow (Fig. 13.1). As the infection progresses, cancellous bone destruction takes place. Since more than 60% of trabecular bone needs to be destroyed before it can be seen radiographically, plain films still may be negative. Computed tomography can readily demonstrate spongy bone destruction, and on occasion fat-fluid levels, probably due to the release of free fat globules from the adipose cells of the necrotic bone marrow (15) (Fig. 13.2). Intramedullary gas collections can be seen in cases of infections produced by gas-forming organisms (16) (Fig. 13.3).

The earliest MR imaging changes of acute osteomyelitis also reflect the bone marrow edema and hyperemia, with decreased signal intensity (SI) of the bone marrow space

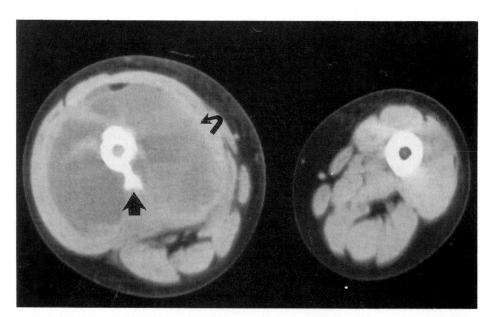

Figure 13.1. Acute osteomyelitis. CT image at midthigh level demonstrates increased density of the bone marrow space in the right femur, compared with the fat density of the normal left femur. Note also the presence of a periosteal reaction *(arrow)* and a large soft-tissue abscess surrounding the femur almost entirely *(curved arrow)*. (Courtesy of V. Chandnani, M.D., University of California, San Diego.)

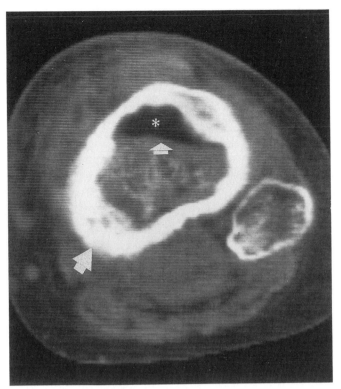

Figure 13.2.   Acute osteomyelitis. CT image at the level of the proximal tibia shows a fat-fluid level within the medullary space *(upper arrow)*. The fatty component (*) had a mean density of −105 HU. Irregularity of the cortical bone and minimal periostitis *(lower arrow)* is also evident. (Courtesy of M. Rafii, M.D. From Rafii M, Firooznia H, Golimbu C, McCauley DL. Radiology 1984;153:493–494.)

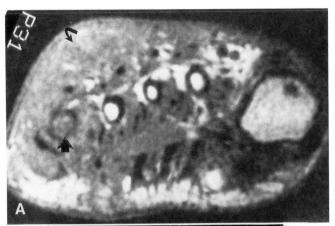

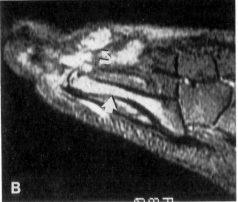

Figure 13.4.   Acute osteomyelitis of the fifth metatarsal in a young diabetic patient. **(A)** Coronal MR image (SE 800/20) shows low signal intensity of the bone marrow of the fifth metatarsal *(arrow)*. **(B)** Sagittal MR imaging (SE 2000/80) along the shaft of the fifth metatarsal demonstrates increased signal intensity of the bone marrow space *(arrow)*. Note the presence of a lobulated abscess in the dorsal soft tissues *(curved arrows* in **A** and **B**). (Reprinted with permission from Beltran J, Noto AM, McGhee RB, Freedy RM, McCalla MS. Radiology 1987;164:448–454.)

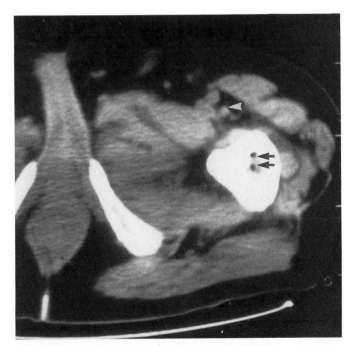

Figure 13.3.   Acute osteomyelitis. CT image through the proximal thigh shows intramedullary gas *(arrows)*. (From Ram PC, Martinez S, Korobkin M, et al. AJR 1981;137:721–723.)

on T1-weighted images and increased SI on T2-weighted images, due to the prolonged T1 and T2 relaxation times of the water-rich inflammatory tissue (6) (Fig. 13.4).

With further extension of the infection, penetration into the cortex through the Haversian canals is facilitated by osteoporosis resulting from the inflammatory response. Once the cortex is penetrated, the periosteum becomes elevated. This may cause interruption of the blood supply to the outer surface of the cortex, and subsequent sequestrum formation in the more chronic stages of the infection. If the infection penetrates the periosteum, soft-tissue abscesses and cellulitis take place.

During this second stage of hematogenous osteomyelitis, plain films readily demonstrate cortical permeation and periosteal reaction. Computed tomography and MR images can also reflect the cortical and periosteal changes as well as the coexisting soft-tissue infection, if present. Lucencies within the cortical bone and calcified periosteal reaction are shown on CT (Figs. 13.1, 13.2, and 13.5).

Involvement of the cortex can be seen on MR images as focal or linear areas of high SI on T2-weighted images (Fig. 13.6). Subperiosteal infection is demonstrated on MR imaging as a halo of high SI on T2-weighted images surrounding the cortex (Fig. 13.6). The halo sign indicates that the periosteum has become elevated, even if plain films or CT fail to demonstrate reactive periosteal calcification. When it becomes thickened or calcified, the periosteum itself may be seen on MR images as another single or multiple low SI ring surrounding the cortex on axial images, on any pulse sequence (Fig. 13.7). Signs of coexistent soft-tissue infection are often present at this stage, with high SI fluid collection on T2-weighted images in the proximity of the bone representing abscesses (Fig. 13.6).

Similar findings of cortical and periosteal infection (osteitis), but without the intramedullary changes, can be seen when the infection occurs as a direct spread from a contiguous source (Fig. 13.8).

Differential diagnosis of the findings demonstrated on CT and MR imaging in acute osteomyelitis includes

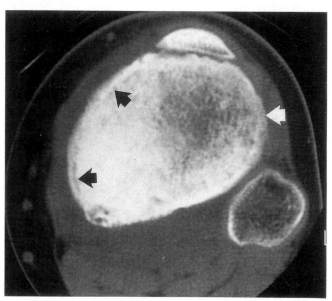

Figure 13.5. Acute osteomyelitis. CT section through the proximal tibia shows cortical permeation and periosteal reaction *(arrows)*. (Courtesy of F. Aparisi, M.D., Hospital la Fe, Valencia, Spain.)

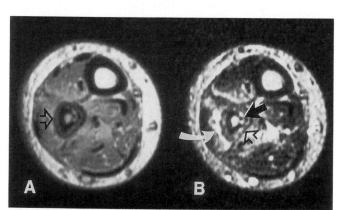

Figure 13.6. Acute osteomyelitis. MR image, axial section **(A,** SE 2000/20; **B,** SE 2000/80) through the tibia and fibula, demonstrates low signal intensity of the bone marrow of the fibula on the proton density image **(A),** and high signal intensity on the T2-weighted image **(B).** A localized hyperintense area on the T2-weighted image within the cortex of the fibula *(arrow* in **B)** represents cortical permeation of the infection. In addition, a hyperintense halo is surrounding the fibular cortex, probably representing infection between the cortex and the periosteum. Outside the halo, a second low signal intensity "ring" is seen *(open arrows* in **A** and **B),** representing the elevated periosteum. A small hyperintense abscess can be seen adjacent to the fibula *(curved arrow* in **B).**

Figure 13.7. Acute osteomyelitis. **(A)** Axial MR image (SE 2000/80) shows multiple rings around the femur, representing multiple layers of periosteal reaction. Note also the hyperintense signal of the bone marrow. **(B)** Radiograph of the femur confirming the presence of a multilayered periostitis ("onion skin").

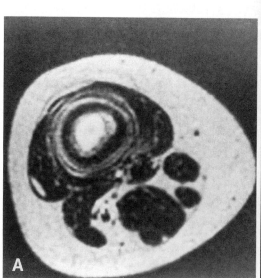
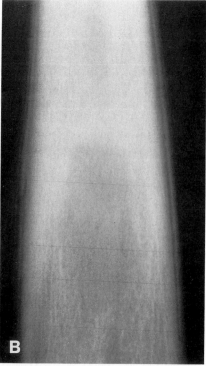

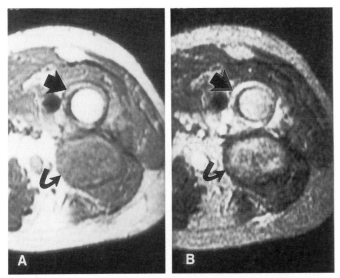

Figure 13.8. Osteitis from contiguous source. **(A)** Axial MR image through the femur (SE 2000/20); **(B)** same plane (SE 2000/80). The signal intensity of the bone marrow is normal in the proton density image **(A)** and the T2-weighted image **(B)**, similar to that of the subcutaneous fat. A hyperintense halo around the femur surrounded by a low signal intensity ring (*arrows* in **A** and **B**) represents periostitis and cortical infection originating from an adjacent soft-tissue abscess (*curved arrows*).

Ewing's sarcoma, when cortical permeation and periosteal reaction are prominent, or any other bone tumor producing increased absorption values of the bone marrow space on CT or SI changes on MR imaging. The relative lack of specificity of these techniques can be improved if one takes into consideration the clinical and laboratory findings.

Subacute Osteomyelitis

Subacute osteomyelitis takes place when the acute infection progresses to intraosseous abscess formation with purulent material contained within an area of trabecular bone destruction. The host bone reacts against the infection, forming a rim of new bone around the abscess. This complex is termed Brodie's abscess, and it is accurately depicted on plain films and CT as a lytic area surrounded by a rim of sclerosis (Fig. 13.9). Absorption values inside the lytic area are within the water range. Not infrequently a periosteal reaction and soft-tissue abscess are associated findings.

The bone marrow surrounding the Brodie's abscess often demonstrates reactive hyperemia, and the internal wall of the abscess is covered by granulation tissue. These changes cannot be seen on plain films or CT, but are well depicted by MR imaging. High SI on T2-weighted images around the area of the abscess reflects the bone marrow hyperemia (Figs. 13.9 and 13.10). Granulation tissue layering the inner wall of the abscess is better demonstrated on proton density weighted images (Fig. 13.11). The high SI of the granulation tissue surrounded by the low SI band of bone sclerosis creates a "double line" effect, not to be confused with the double line sign of avascular necrosis. The contour of a Brodie's abscess is smooth and rounded or oval,

whereas the margins of an area of avascular necrosis tend to be undulating.

Other differential diagnostic considerations of subacute osteomyelitis include eosinophilic granuloma, osteoid osteoma, osteoblastoma, aneurysmal bone cyst, and chondroblastoma. Clinical and laboratory findings help in making the correct diagnosis, but the fluid contents of the lesion and absence of a nidus or internal calcifications, as demonstrated by CT, can help in excluding solid tumors. Aneurysm bone cyst can demonstrate fluid contents, with fluid-fluid levels on CT and MR imaging (17–20), also described in chondroblastoma, telangiectatic osteosarcoma, and giant cell tumor (20). Expanded cortical margin, multiloculated internal configuration, and absence of double line are MR imaging and CT characteristics compatible with aneurysmal bone cyst rather than Brodie's abscess.

Chronic Osteomyelitis

Progression of acute and subacute osteomyelitis to the chronic form can occur when partial or inadequate antibiotic therapy is given, or when the infection is untreated but is produced by a low virulence organism, or in cases of increased host resistance. Interruption of the blood supply to areas of bone leads to necrosis of the cortex and sequestrum formation. Existing purulent material in the bone marrow space and subperiosteum finds its way out through sinus tracts and fistulae. The term *cloaca* is applied to the draining channel through an area of chronic osteomyelitis and surrounding granulation tissue (involucrum). Sequestra can occasionally be extruded through the cloacae and soft tissue-draining sinuses. If the sequestrum is not extruded or removed surgically, it can cause intermittent episodes of exacerbation of the infection, since it often harbors viable organisms. Sclerosing osteomyelitis of Garré refers to a chronic infection that induces marked bone eburnation with intense periosteal reaction, without bone necrosis, purulent exudate, or granulation tissue (2).

An important exception to the findings described in bacterial osteomyelitis is tuberculous infection, which commonly presents in the metaphyseal or epiphyseal regions and often lacks sclerosis, sequestra, or periosteal reaction.

All these manifestations are obvious on conventional tomography in the vast majority of cases, but occasionally small or multiple sequestra may be undetected by conventional radiographic techniques because of the extensive sclerotic bone reaction around them. Computed tomography can reliably demonstrate the exact location and number of sequestra, delineate sinus tracts, and detect the presence of associated soft-tissue abscesses (21) (Fig. 13.12). Computed tomography has been found particularly helpful in identifying areas of chronic osteomyelitis in flat bones with difficult anatomy such as skull, scapula, and pelvis (22).

Experience with MR imaging in chronic osteomyelitis is somewhat limited, but preliminary results indicate that MR imaging is as accurate as CT in demonstrating the different components of chronic osteomyelitis (23). Sequestra are seen as low SI fragments in all pulse sequences, similar to cortical bone (Fig. 13.13). On T2-weighted images,

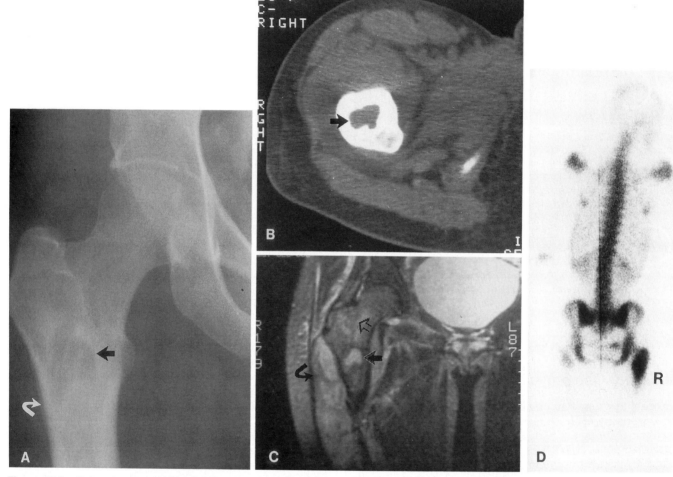

Figure 13.9. Subacute osteomyelitis. Brodie's abscess and soft-tissue abscess. **(A)** Radiograph of the right femur shows a lucency in the subtrochanteric area *(arrow)* and periosteal reaction *(curved arrow).* **(B)** Axial CT demonstrates the Brodie's abscess *(arrow).* **(C)** Coronal MR image (SE 2000/80) shows the hyperintense intraosseous abscess *(arrow),* but also shows a large, soft-tissue abscess adjacent to the femur, in the area of the vastus lateralis muscle *(curved arrow).* Note the high signal intensity of the bone marrow of the proximal femur, probably representing edema *(open arrow).* **(D)** Bone scan with 99mTc-MDP, delayed phase, demonstrates intense uptake in the area of the proximal right femur. (Reprinted with permission from Beltran J, Noto AM, McGhee RB, Freedy RM, McCalla MS. Radiology 1987;164:448–454.)

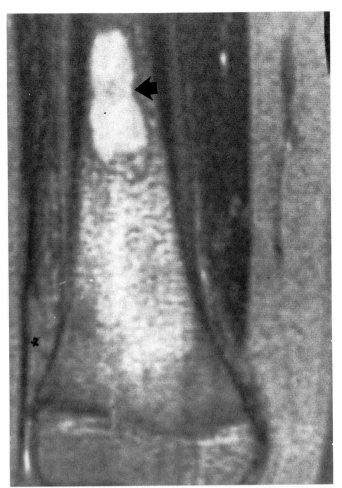

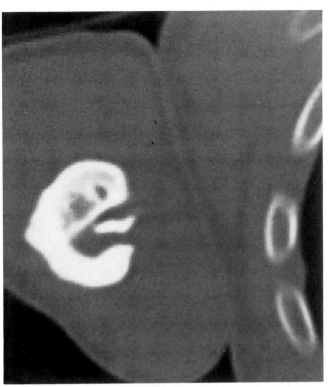

Figure 13.12. Chronic osteomyelitis of the humerus with sequestrum. CT image through the proximal humerus demonstrates a lytic defect surrounded by reactive sclerosis and a dense bone fragment representing the partially extruded sequestrum through a cloaca.

Figure 13.10. Brodie's abscess of the distal femur. Coronal MR image (SE 1800/80) demonstrates a hyperintense abscess *(arrow)*. Below the abscess a large, ill-defined area of high signal intensity probably reflects bone marrow edema. (Courtesy of F. Aparisi, M.D., Hospital la Fe, Valencia, Spain.)

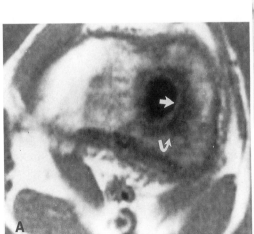
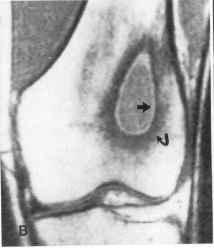
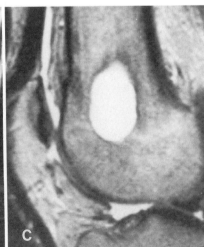

Figure 13.11. Brodie's abscess of the distal femur. **(A)** Axial MR image (SE 500/20); **(B)** coronal MR image (SE 1800/20); **(C)** sagittal MR image (SE 1800/80). An intraosseous abscess shows low signal intensity on **(A)** T1-weighted and **(B)** proton density images, and high signal intensity on a T2-weighted image **(C)**. The abscess is surrounded by a hypointense rim *(curved arrows in **A** and **B**)*. A relatively high signal intensity layer of granulation tissue is seen covering the inner wall of the abscess *(arrows in **A** and **B**)*. This is not seen on the T2-weighted image **(C)** because of the similar signal intensity of the pus inside of the abscess and the granulation tissue. (Courtesy of F. Aparisi, M.D., Hospital la Fe, Valencia, Spain.)

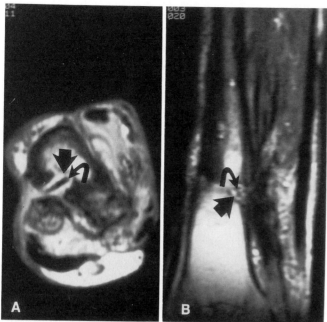

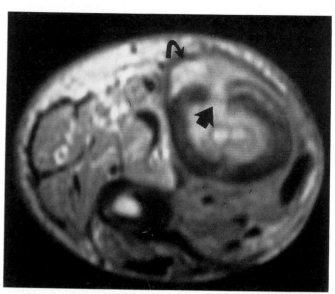

Figure 13.13. Chronic osteomyelitis of the distal tibia. **(A)** Axial MR image (SE 2000/80); **(B)** coronal MR image (SE 2000/80). The site of chronic osteomyelitis is seen as a focal area of high signal intensity on these T2-weighted images, representing the involucrum (*arrows* in **A** and **B**). The hypointense fragment represents the sequestrum (*curved arrows* in **A** and **B**). The contour of the tibia is deformed due to a previous fracture, subsequently infected.

Figure 13.14. Chronic osteomyelitis of the tibia. Axial MR image (SE 2000/80) demonstrates cortical disruption in the anterior aspect of the tibia (*arrow*) due to a draining sinus and a subperiosteal abscess (*curved arrow*). The two rings inside the bone marrow are related to previous intramedullary nailing.

Figure 13.15. Soft-tissue abscess. CT image following intravenous contrast material. Note the water density abscess in the region of the rectus femoris muscle, surrounded by an enhancing rim (*arrow*).

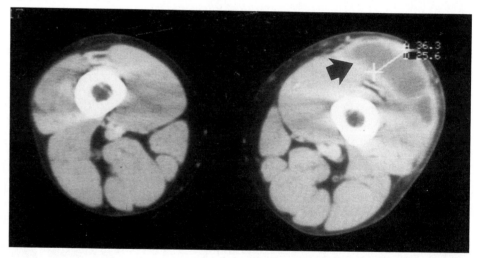

high SI granulation tissue (involucrum) can be well delineated and separated from the surrounding sclerotic bone, which is low in SI in all pulse sequences. Existing soft-tissue abscesses and draining sinuses are also well depicted by MR imaging (Fig. 13.14).

The axial imaging capability of CT and the multiplanar capability of MR imaging are added significant advantages of these techniques over conventional radiography.

Differential diagnosis of chronic osteomyelitis with sequestrum formation is limited and includes osteoid osteoma, eosinophilic granuloma, and rarely fibrosarcoma. Areas of extensive bone sclerosis in chronic osteomyelitis can simulate osteosarcoma, chondrosarcoma, and lym-

phoma. Identification of the radiographic and MR imaging findings of infection as described above, along with clinical and laboratory correlation, should lead to the correct diagnosis of infection in the vast majority of cases.

Soft-Tissue Infections

Direct implantation, hematogenous spread, and spread from contiguous source are the main routes of contamination of the soft tissues. Associated osteomyelitis is often present. Abscess, infected bursitis, lymphadenitis, cellulitis, myositis, septic arthritis, and tenosynovitis are differ-

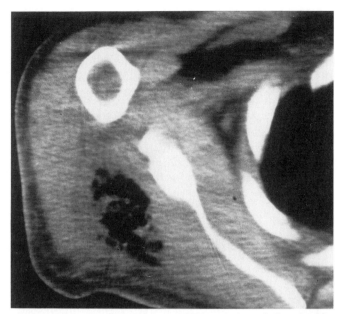

Figure 13.16. Soft-tissue abscess. CT image through the shoulder area demonstrates a cluster of air bubbles in the region of the infraspinatus muscle.

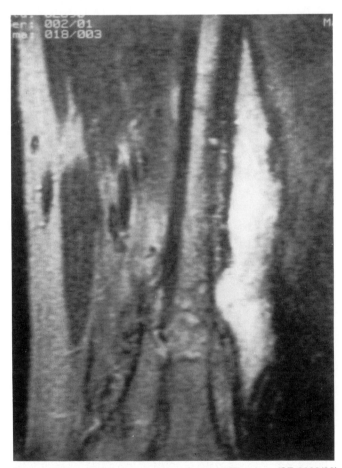

Figure 13.17. Soft-tissue abscess. Coronal MR image (SE 2000/80) demonstrates a large, hyperintense abscess adjacent to the femur in this patient (status postfracture). Note the normal signal intensity of the bone marrow, excluding osteomyelitis. (Reprinted with permission from Beltran J, Noto AM, McGhee RB, Freedy RM, McCalla MS. Radiology 1987;164:448–454.)

ent manifestations of soft-tissue infection. In the majority of cases plain film radiography in soft-tissue infections is normal or nonspecific. Obliteration of normal fat planes, soft-tissue swelling, joint space widening, or presence of air bubbles are the most frequent radiographic manifestations. Bone scintigraphy using three-phase ^{99m}Tc-MDP, ^{67}Ga, and ^{111}In-leukocyte scanning has improved the rate of detection of soft-tissue infection, but lack of specificity and poor spatial resolution are the main limitations of scintigraphic studies (10–12). Since their introduction, cross-sectional imaging techniques with superior soft-tissue contrast resolution have made the diagnosis and staging of soft-tissue infection easier and more accurate.

Abscess and Infected Bursitis

Abscesses are seen on CT as rounded or oval areas of attenuation values within the water range, surrounded by an enhancing rim, if intravenous contrast material is used (Fig. 13.15). If the causative organism is gas producing, air bubbles can be identified inside the fluid collection (Fig. 13.16). On MR imaging, abscesses are shown as low SI fluid collections on T1-weighted images and high SI on T2-weighted images, often surrounded by a low SI rim, which represents the abscess pseudocapsule and probably corresponds to the enhancing rim seen on CT (Figs. 13.17 and 13.18). Not infrequently a central region of low SI on all pulse sequences is identified inside an abscess. This probably represents an area of necrotic tissue and cellular

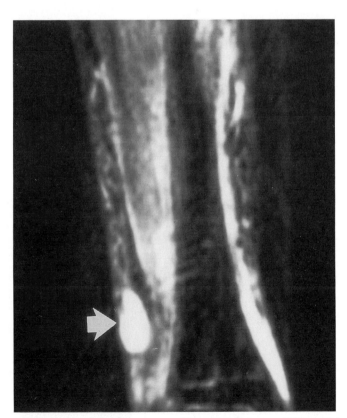

Figure 13.18. Soft-tissue abscess. Coronal MR image (STIR 1500/140/20) through the distal leg demonstrates a hyperintense, small abscess adjacent to the distal fibula *(arrow).*

debris (Fig. 13.19). If air bubbles are present, they are seen on MR images as low SI dots, indistinguishable from calcifications. Plain film or CT correlation is necessary to make the distinction between calcification and air bubbles.

Infected bursae have the same CT and MR imaging characteristics as abscesses, but they can be identified because of their specific anatomical location around the joints (Figs. 13.20 and 13.21). Bursae can become distended by fluid in cases of aseptic inflammation. Distinction between infected or noninfected bursitis cannot be made on the basis of CT or MR imaging, unless gas is present within the fluid collection (Fig. 13.22).

Septic Arthritis and Tenosynovitis

Water-density fluid within a joint space, often associated with narrowing and irregularity of the joint, and soft-tissue swelling are the common manifestations of septic arthritis on CT when the infectious process has already produced articular cartilage destruction (Fig. 13.23). A nonspecific joint effusion may be seen during the early stages of septic arthritis. Air bubbles can also be detected with CT if the infection is produced by gas-forming organisms (Fig. 13.24). Similar findings can also be seen using conventional radiographic techniques. The most signifi-

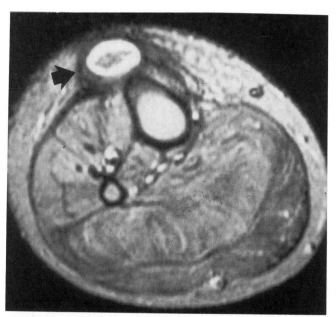

Figure 13.19. Soft-tissue abscess. Axial MR image (SE 2000/80) through the proximal tibia demonstrates a hyperintense soft-tissue abscess surrounded by a thick pseudocapsule *(arrow)*. Note the low signal intensity center, probably related to necrotic debris.

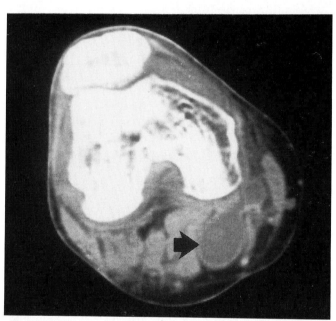

Figure 13.20. Aseptic bursitis. CT image through the knee shows a fluid-distended gastrocnemius-semimembranosus bursa (Baker's cyst) *(arrow)*.

Figure 13.21. Aseptic trochanteric bursitis. **(A)** Coronal MR image (SE 800/20); **(B)** coronal MR image (SE 2000/80) through right hip, demonstrating a lobulated fluid collection adjacent to the greater trochanter *(arrows* in **A** and **B**).

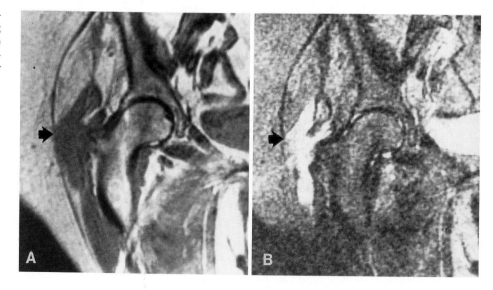

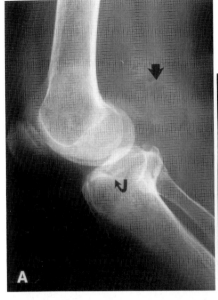

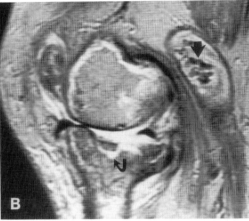

Figure 13.22. Septic arthritis and bursitis. **(A)** Radiograph of the knee demonstrates air bubbles in the popliteal space *(arrow)*. In addition, this diabetic patient had collapse of the tibial plateau *(curved arrow)* due to associated neuroarthropathy. **(B)** Sagittal MR image (SE 2000/80) demonstrates signal intensity changes in the bone marrow and a joint effusion due to septic arthritis. Note tibial plateau collapse *(curved arrow)*, and a distended bursa in the popliteal space, with hypointense dots *(arrow)* representing air bubbles.

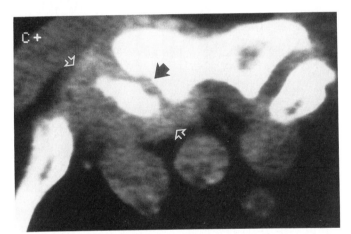

Figure 13.23. Septic arthritis. CT image through the sternoclavicular joints demonstrates irregular articular surface of the right joint *(arrow)*, with capsular distension *(open arrows)*.

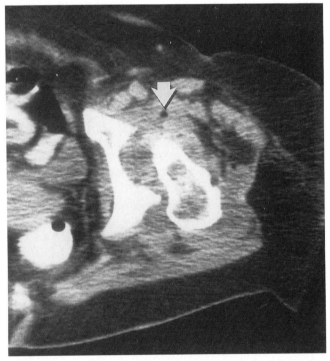

Figure 13.24. Septic arthritis. CT image through the left hip joint demonstrates a joint effusion and a small air bubble in the anterior aspect of the joint capsule *(arrow)*.

cant contribution of CT in the evaluation of joint infection is CT-guided joint aspiration. This has proven a very useful technique in areas with difficult anatomy, such as the sacroiliac joint or disk space (Fig. 13.25).

Magnetic resonance imaging manifestations of septic arthritis parallel those of CT. During the early stages, T2-weighted images show nonspecific high SI fluid distending the capsule (Fig. 13.26). Cartilage destruction, joint space narrowing, and occasionally areas of osteomyelitis with bone erosions can be detected by MR imaging in more advanced cases (Fig. 13.27).

Similarly, CT and MR image manifestations of infectious tenosynovitis are nonspecific. Synovial sheath effusion (Figs. 13.28 and 13.29), often seen in patients with diabetic foot, and abnormal areas of high SI on T2-weighted images within the tendon substance, are signs indicative of synovial and tendon inflammation. Since joint or tendon sheath effusions can occur as a result of aseptic synovitis (e.g., trauma and rheumatoid arthritis), aspiration

of synovial fluid may be necessary to confirm the infectious nature of the effusion (24, 25).

Cellulitis

Cellulitis implies inflammation of the deep subcutaneous tissues, frequently seen in diabetic patients and drug addicts. Streptococci and staphylococci are frequent causative organisms. Inflammatory signs can be seen clinically,

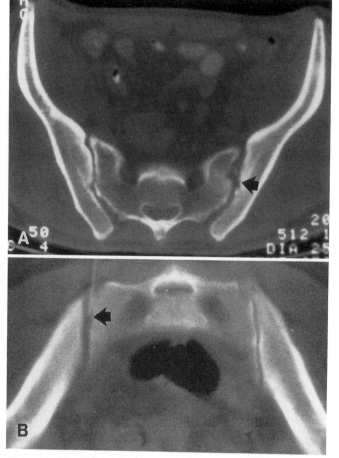

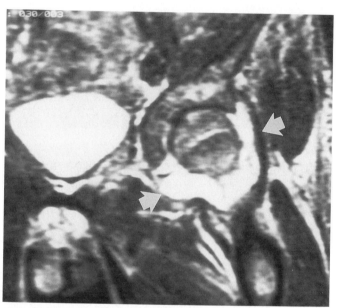

with soft-tissue swelling, redness, and increased temperature. No drainable fluid collection is present on needle aspiration. Indistinct fat planes in the subcutaneous tissues with increased absorption values are seen on CT (Fig. 13.30). On MR imaging, the SI of the fat is altered, showing irregular strands of low SI on T1-weighted images and high SI on T2-weighted images (Fig. 13.31). The main indication to use CT or MR imaging in cellulitis is to exclude a drainable soft-tissue abscess. Ultrasonography can be used as an alternative and inexpensive technique for detection

Figure 13.25. Tuberculous septic arthritis. **(A)** CT image through the sacroiliac joints shows irregularity of the left joint *(arrow).* **(B)** Prone CT image during needle aspiration of the left sacroiliac joint verifies the intraarticular position of the needle *(arrow).*

Figure 13.26. Septic arthritis. Coronal MR image (SE 2000/80) shows nonspecific high signal intensity fluid distending the joint capsule *(arrows).*

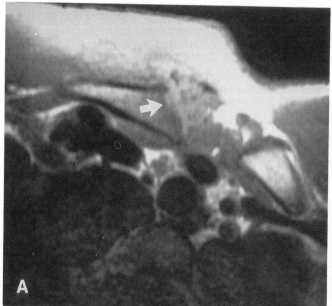

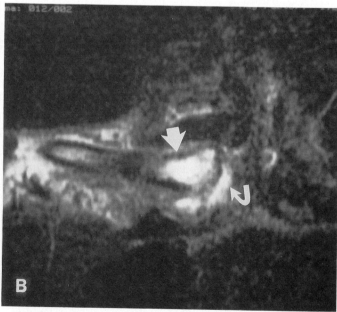

Figure 13.27. Septic arthritis. **(A)** Axial MR image (SE 800/20) through the sternoclavicular joints demonstrates irregular destruction of the articular surface of the right clavicle *(arrow).* **(B)** Coronal MR image (SE

2000/80) shows high signal intensity of the bone marrow space of the proximal right clavicle *(arrow)* due to osteomyelitis, and a hyperintense joint effusion *(curved arrow).*

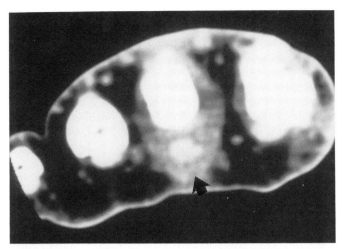

Figure 13.28. Septic tenosynovitis. CT image through the distal metacarpals shows irregular thickening and distension of the synovial sheath of the flexor tendon of the third digit *(arrow)*. (Courtesy of Francisco Aparisi, M.D., Hospital la Fe, Valencia, Spain.)

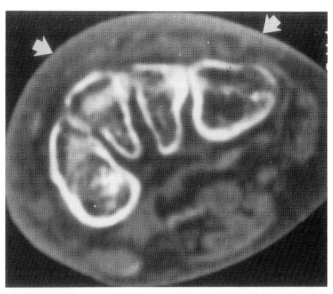

Figure 13.30. Cellulitis. CT image through the metatarsals demonstrates ill-defined, increased density in the subcutaneous tissues of the dorsum of the foot *(arrows)*, replacing the normal fat density.

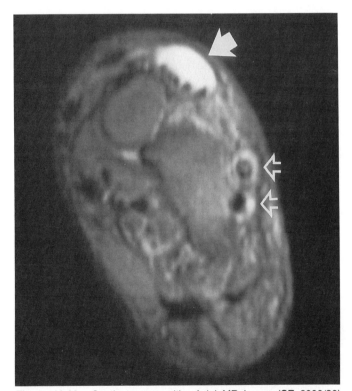

Figure 13.29. Septic tenosynovitis. Axial MR image (SE 2000/80) through the tarsal bones demonstrates high signal intensity fluid distending the synovial sheath of the extensor digitorum tendons *(arrow)* and peroneal tendons *(open arrows)* in this diabetic patient.

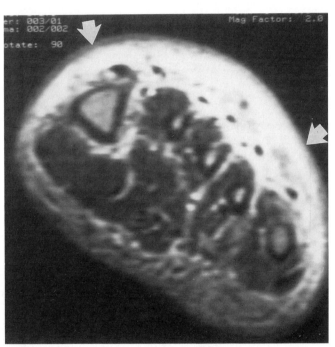

Figure 13.31. Cellulitis. Axial MR image (SE 2000/80) through the metatarsals demonstrates ill-defined, high signal intensity in the dorsum of the foot *(arrows)*.

of abscesses, but it is limited to the evaluation of superficially located lesions, and it does not provide the anatomical resolution achievable with CT or MR imaging.

Myositis

Infectious myositis can be produced by a variety of organisms, including viruses, bacteria, protozoa, and parasites.

Pyomyositis is a specific clinical condition most often due to *Staphylococcus aureus* infection, seen more often in tropical regions, and affecting children and young adults. Thighs and buttocks are more frequently involved. Induration and swelling of the muscle and skin can be found clinically. Purulent fluid collections can occasionally be aspirated (26). Computed tomography can demonstrate increased size of the muscle group involved, with areas of low attenuation representing fluid collections. Computed

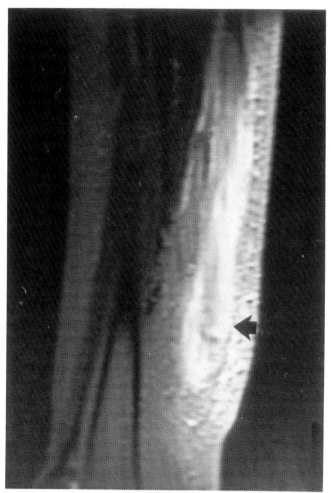

tomography-guided aspiration of these fluid collections is helpful in the diagnosis and management of pyomyositis.

Diffuse increase in SI on T2-weighted images of a muscle or a muscle group can be seen on MR imaging (Fig. 13.32). Since SI changes of the inflamed muscle and fluid collections are similar on MR images, identification of abscesses may be difficult under these circumstances. Infiltrating soft-tissue tumors and muscle necrosis may also demonstrate diffuse increase in SI on T2-weighted images.

Lymphadenitis

Lymphadenitis can be seen accompanying soft-tissue infections. Palpable, enlarged regional lymph nodes can be

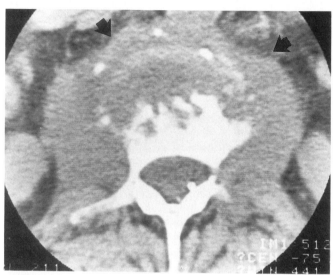

Figure 13.32. Infectious myositis. Sagittal MR image (SE 2000/80) through the thigh demonstrates hyperintensity of the adductor muscles *(arrow)*.

Figure 13.34. Infectious diskitis. CT image at the L2 level demonstrates fragmentation of the end plate, soft-tissue mass *(arrows)*, and obliteration of the epidural fat planes.

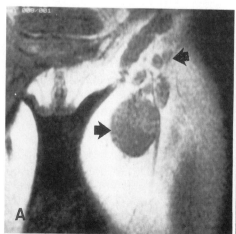
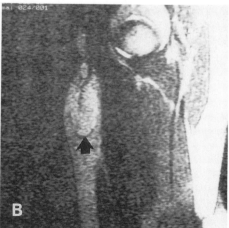

Figure 13.33. Lymphadenitis. **(A)** Coronal MR image (SE 800/20) demonstrates a cluster of enlarged lymph nodes of different sizes in the left inguinal area *(arrows)*. **(B)** Sagittal MR image (SE 2000/80) shows high signal intensity of the large lymph node in this T2-weighted image *(arrow)*.

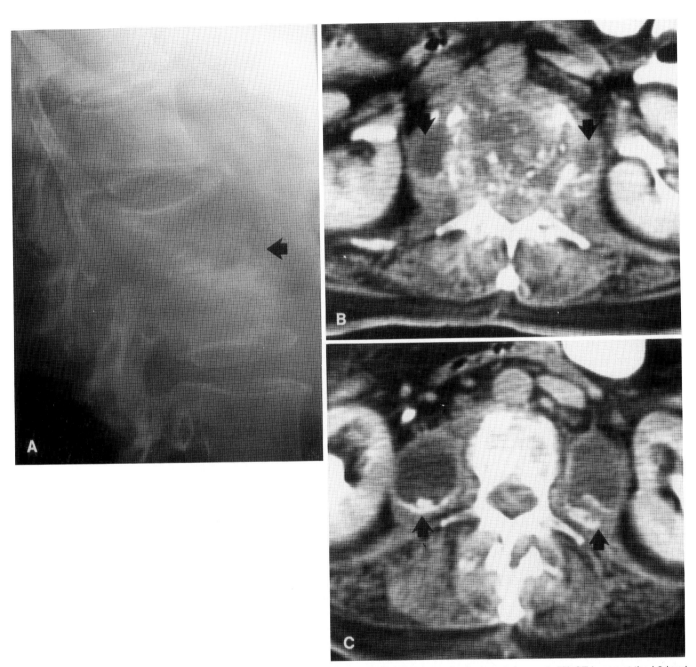

Figure 13.35. Tuberculous spondilitis. **(A)** Radiograph of the lumbar spine demonstrates L2–L3 destruction of the vertebral end plates at this level *(arrow)*. **(B)** CT image at the same level demonstrates destruction of the L2 vertebral body, perivertebral soft-tissue swelling, and water-density psoas abscesses *(arrows)*. **(C)** CT image at the L3 level shows distal extension of the bilateral psoas abscesses, with calcification of their posterior walls *(arrows)*.

detected clinically. Computed tomography and MR imaging can readily demonstrate the presence of lymphadenopathy (Fig. 13.33) but unfortunately differentiation between neoplastic and inflammatory lymph node enlargement is not possible with these techniques (27).

Spinal Infection

Infectious processes involving the spine (see also Chapter 41) can be found in different locations. The most frequent areas of involvement are the disks and vertebral end plates (diskitis), followed by the posterior elements, apophyseal joints, perivertebral soft tissues, and epidural space.

The most common causative organism is *Staphylococcus aureus*, but a variety of other organisms, including streptococci, *Escherichia coli*, *Pseudomonas*, *Klebsiella*, *Salmonella*, and mycobacteria, can be found.

The routes of contamination are not different from other areas of the skeleton, and include hematogenous infection (via arterial blood supply or via venous drainage through the Batson's paravertebral plexus), contamination from an adjacent focus (pharyngeal infection, rectal and pelvic abscesses), or direct implantation (penetrating injuries, needle punctures, surgical intervention) (2).

Hematogenous infection originates in the subchondral area of the end plate and rapidly spreads into the disk and adjacent vertebral end plate. From this location it can disseminate into the subligamentous perivertebral tissues, perforate the ligaments, and reach the paravertebral soft tissues and epidural space, forming abscesses.

Plain films are normal during the early stages, until bone destruction and disk space narrowing occurs. Nonspecific increased uptake can be demonstrated by bone scintigraphy. The earliest described CT finding of disk space infection is decreased attenuation of the disk on axial CT images (28). Once bone and disk destruction is apparent on plain films, CT can confirm the diagnosis and demonstrate associated findings (29). The typical CT manifestations of spinal infection include fragmentation and erosion of the vertebral end plates, anterior paravertebral soft-tissue swelling with obliteration of the vertebral fat planes, obliteration of fat planes in the spinal canal by epidural inflammatory mass or abscess, and paravertebral abscesses, which can become calcified in tuberculous spondilitis (30) (Figs. 13.34 and 13.35).

Other CT findings of spinal infection include destruction of the posterior elements of the neural arch, demonstration of intraspinal air in cases of epidural abscess (31), and intraosseous and intradiskal gas (32). Occasionally, high-resolution CT following myelography may be superior to MR imaging in demonstrating the extent of an epidural abscess (13).

Differential diagnosis of spinal infection is extensive, and includes fractures, postdiskectomy changes, neuropathic arthropathy, and tumors. In one series (33) sensitivity of CT for disk space infection was 63%, with a positive predictive value of only 63%. Complete involvement of the prevertebral soft tissues, diffuse osteolytic destruction, bone and soft-tissue gas, and a process centered in the intervertebral disk space, have been found reliable criteria to identify spinal infection. Tumors have more tendency to

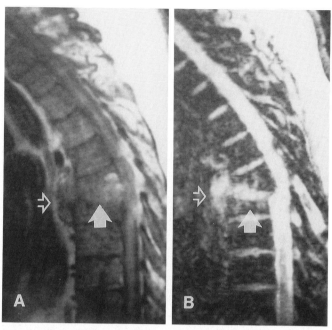

Figure 13.36.   Septic diskitis. **(A)** Midline sagittal MR image (SE 2000/20) at the level of the thoracic spine shows destruction of the T5–T6 end plates, with intermediate signal intensity changes of the bone marrow of the adjacent vertebral bodies *(arrow)*, and an anterior hypointense soft-tissue abscess *(open arrow)*. **(B)** Sagittal MR image at the same level (SE 2000/80) shows abnormally high signal intensity of the involved disk space *(arrow)*, and the anterior paravertebral abscess *(open arrow)*.

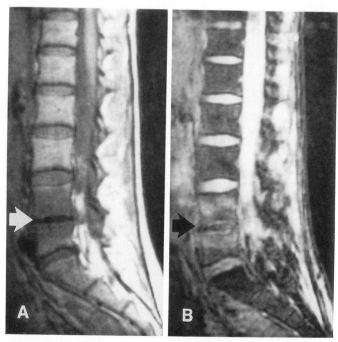

Figure 13.37.   *Aspergillus* diskitis. **(A)** Midline sagittal MR image (SE 800/20) demonstrates collapse and low signal intensity of the L4–L5 disk space and adjacent vertebral bodies *(arrow)*. **(B)** Sagittal MR image at the same level (gradient-recalled echo [GRE] 1000/20, 12° flip angle) shows relatively low signal intensity of the involved disk *(arrow)*, hyperintense L4–L5 vertebral bodies, and lack of paravertebral soft-tissue changes.

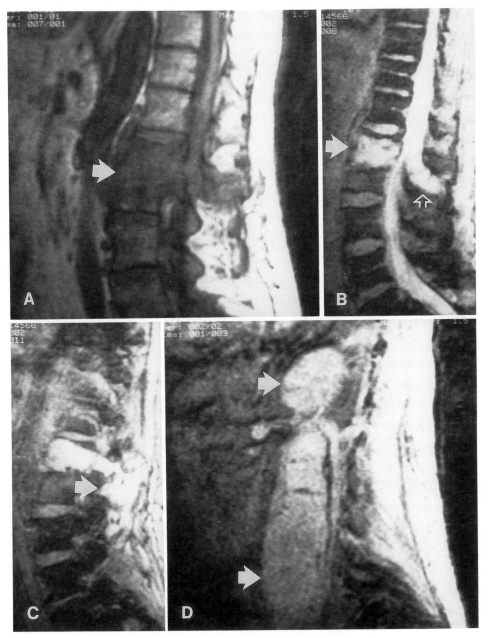

Figure 13.38. Tuberculous spondilitis. **(A)** Midline sagittal section (SE 800/20) shows hypointense, irregular L2–L3 disk space and adjacent vertebral bodies *(arrow).* **(B)** Midline sagittal MR image (GRASS 1000/20, 12° flip angle) at the same level demonstrates high signal intensity of the L2–L3 disk space and vertebral bodies *(arrow),* with extension in the posterior interspinus ligament *(open arrow).* **(C)** Parasagittal MR imaging (GRASS 1000/20, 12° flip angle) shows the posterior extent of the inflammatory changes *(arrow).* **(D)** Parasagittal MR image (GRASS 1000/20, 12° flip angle) demonstrates a large, bilobed, hyperintense psoas abscess *(arrows).*

involve the posterior elements, demonstrate partial or absent prevertebral soft-tissue swelling, and not infrequently show osteoblastic changes (34).

During the early stages of spinal infection, or on the rare occasions when only the vertebral body is infected but no disk infection is present, MR imaging can demonstrate abnormal areas of low SI on T1-weighted images and high SI on T2-weighted images within the bone marrow space of the vertebral body. This finding is indistinguishable from bone tumor or metastasis, unless associated findings such as perivertebral soft-tissue inflammation or abscess are demonstrated. Once the infectious process reaches the disk, and the adjacent vertebral body, a distinct MR imaging pattern can be seen, with increased SI of the disk space and adjacent end plates on T2-weighted images (Fig. 13.36). At this stage of septic diskitis, MR im-

aging is more sensitive than plain films or CT and more specific than bone scintigraphy (35). Indolent fungal infections and tuberculosis cause fewer disk space reactive changes than bacterial septic diskitis and may not demonstrate increased SI of the disk on T2-weighted images (Fig. 13.37).

Other findings reported in tuberculous spondilitis include frequent involvement of the posterior elements and multiplicity of involvement of the vertebral bodies (36) (Fig. 13.38).

In contrast, aseptic diskitis, which is an entity associated with disk degeneration, arthritis, chemonucleolysis, or trauma, demonstrates on MR images a relatively low SI disk space on all pulse sequences, associated with variable vertebral end plate signal changes, depending on the stage of inflammation (Fig. 13.39). Another diagnostic feature of

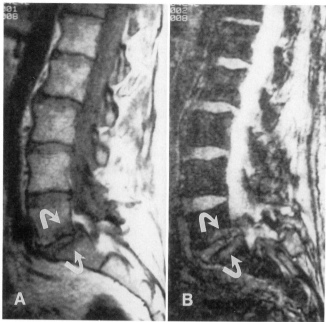

Figure 13.39. Aseptic diskitis. **(A)** Midline sagittal MR image (SE 800/20) shows low signal intensity changes in the bone marrow adjacent to the narrowed L5–S1 disk space *(curved arrows)*. **(B)** Midline sagittal MR image (GRASS 1000/20, 12° flip angle) at the same level demonstrates hyperintense signal of the bone marrow *(curved arrows)*, but low signal intensity of the L5–S1 disk.

aseptic diskitis is the lack of associated soft-tissue abscess, frequently present in septic diskitis.

With further progression of the infection disk space narrowing, irregularity of the end plates, and paravertebral or epidural soft-tissue edema or abscess can be shown on MR imaging. Psoas abscess is shown on MR imaging as a loculated fluid collection with low SI on T1-weighted images and high SI on T2-weighted images (Fig. 13.40). Fungal infections usually have less paraspinal soft-tissue thickening (Fig. 13.37), but tuberculous spondilitis presents with large, often bilateral abscesses (37) (Fig. 13.38). Coronal MR imaging has been found particularly helpful in imaging tuberculous spondilitis (37).

Epidural abscesses represent extension of the infection into the spinal canal, with subsequent compression of the nerve roots and cord. An epidural abscess is shown on MR imaging as a low to intermediate SI mass on T1-weighted images, depending on whether the internal composition of the abscess is liquefied, thick, or fibrous inflammatory tissue. Most of the time, high SI can be seen on T2-weighted images. In cases where the abscess is isointense with the spinal cord, gadolinium-labeled diethylenetriamine pentaacetic acid (DTPA) injection can produce increased SI of the abscess capsule on T1-weighted images, demonstrating the full extent of the lesion (Fig. 13.41). However, other epidural lesions, such as meningiomas or metastasis, can show a similar appearance.

Conclusion

In conclusion, MR imaging and CT are valuable tools to investigate musculoskeletal infections. However, these ex-

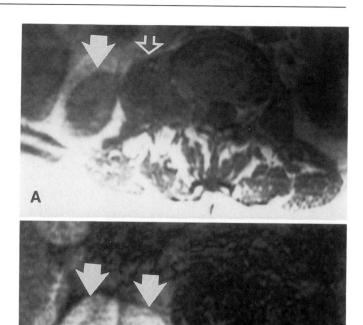

Figure 13.40. Psoas abscess. **(A)** Axial MR image (SE 800/20) at the L3 level demonstrates enlargement of the right psoas muscle *(open arrow)*, and a hypointense mass lateral to the psoas *(arrow)* on this T1-weighted image. **(B)** Axial MR image (GRASS 1000/20, 12° flip angle) at the same level shows better the bilobed abscess extending outside of the psoas *(arrows)*.

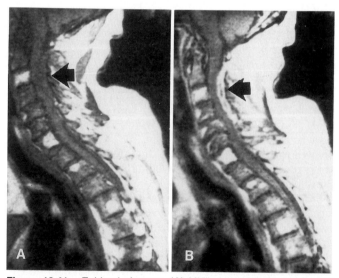

Figure 13.41. Epidural abscess. **(A)** Midline sagittal MR image (SE 800/20) of the cervical spine demonstrates a low signal intensity fluid collection in the posterior epidural space *(arrow)*. High signal intensity lesions within multiple vertebral bodies probably represent hemangiomas. **(B)** Midline sagittal MR image (SE 800/20) at the same level, following gadolinium DTPA injection, shows enhancement of the epidural abscess *(arrow)*, contrasting with the hypointense compressed cord. The low signal intensity line inside of the abscess probably represents necrotic debris.

pensive tests should be used as problem-solving techniques. In the vast majority of cases, conventional techniques including plain films, tomography, and bone scintigraphy provide enough information for diagnosis and treatment of a musculoskeletal infection. Subacute osteomyelitis with Brodie's abscess, chronic osteomyelitis with an obvious sequestrum on plain films, and advanced septic arthritis with cartilage and bone destruction are examples of situations where CT or MR imaging may not provide further significant information. Computed tomography and MR imaging play an important role in the evaluation of osteomyelitis with questionable plain film findings, in the evaluation of soft-tissue infections, or in cases in which conventional imaging demonstrates bone infection but detection of potential extension into the soft tissues is crucial for patient management, such as in diabetic foot or spinal infection.

References

1. Langenskiold A. Growth disturbance after osteomyelitis of femoral condyles in infants. Acta Orthoped Scand 1984;55:1.

2. Resnick D, Niwayama G. Osteomyelitis, septic arthritis, and soft tissue infection: the mechanisms and situations. In: Resnick D, Niwayama G, eds. Diagnosis of bone and joint disorders. 2nd ed. Philadelphia: Saunders, 1988:2524–2618.

3. Arger PH, Coleman BG, Dalinka MK. Computed tomography in orthopedics. Orthoped Clin N Am 1983;14:217–232.

4. Maurer AH, Millmond SH, Knight LC, et al. Infection in diabetic osteoarthropathy: use of indium labeled leukocytes for diagnosis. Radiology 1986;161:221–223.

5. Beltran J, McGhee RB, Shaffer PB, et al. Experimental infections of the musculoskeletal system: evaluation with MR imaging and Tc-99m MDP and Ga-67 scintigraphy. Radiology 1988;167:167–172.

6. Beltran J, Noto AM, McGhee RB, Freedy RM, McCalla MS. Infections of the musculoskeletal system: high field strength MR imaging. Radiology 1987;164:448–454.

7. Berquist TH, Brown ML, Fitzgerald RH, May GR. Magnetic resonance imaging: application in musculoskeletal infection. Magn Reson Imag 1985;3:219–230.

8. Fletcher BD, Scoles PV, Nelson AD. Osteomyelitis in children: detection by magnetic resonance. Radiology 1984;150:57–60.

9. Tang JSH, Gold RH, Bassett LW, Seeger LL. Musculoskeletal infection of the extremities: evaluation with MR imaging. Radiology 1988;166:205–209.

10. Datz FL, Thorne DA. Effect of chronicity of infection on the sensitivity of the In-111 labeled leukocyte scan. AJR 1986;147:809–812.

11. Handmaker H. Acute hematogenous osteomyelitis: has the bone scan betrayed us? Radiology 1980;135:787–789.

12. Yuh WTC, Corson JD, Baraniewski HM, et al. Osteomyelitis of the foot in diabetic patients: evaluation with plain films, 99mTc-MDP bone scintigraphy, and MR imaging. AJR 1989;152:795–800.

13. Donovan Post MJ, Quencer RM, Montalvo BM, Katz BH, Eismont FJ, Green BA. Spinal infection: evaluation with MR imaging and intraoperative US. Radiology 1988;169:765–771.

14. Unger E, Moldofsky P, Gatenby R, et al. Diagnosis of osteomyelitis by MR imaging. AJR 1988;150:605–610.

15. Rafii M, Firooznia H, Golimbu C, McCauley DL. Hematogenous osteomyelitis with fat-fluid level shown by CT. Radiology 1984;153:493–494.

16. Ram PC, Martinez S, Korobkin M, et al. CT detection of intraosseous gas: a new sign of osteomyelitis. AJR 1981;137:721–723.

17. Beltran J, Simon DC, Levy M, et al. Aneurysmal bone cysts: MR imaging at 1.5 T. Radiology 1986;158:689–690.

18. Hudson TM. Fluid levels in aneurysmal bone cysts: a CT feature. AJR 184;142:1001–1004.

19. Hudson TM, Hamlin DJ, Fitzimmons JR. Magnetic resonance imaging of fluid levels in aneurysmal bone cysts and in anticoagulated human blood. Skel Radiol 1985;13:267–270.

20. Munk PL, Helms CA, Johnston J, Steinbach L, Neumann C. MR imaging of aneurysmal bone cysts. AJR 1989;153:99–101.

21. Wing VW, Jeffrey RB, Federle M, Helms CA, Trafton P. Chronic osteomyelitis examined by CT. Radiology 1985;154:171–174.

22. Hernandez RJ, Conway JJ, Poznanski AK, Tachdjian MO, Dias LS, Kelikian AS. The role of computed tomography and radionuclide scintigraphy in the localization of osteomyelitis in flat bones. J Pediatr Orthoped 1985;5:151–154.

23. Quinn SF, Murray W, Clark RA, Cochran C. MR imaging of chronic osteomyelitis. J Comput Assist Tomogr 1988;12:113–117.

24. Edmons ME. The diabetic foot: pathophysiology and treatment. Clin Endocrinol and Metab 1986;15:889–916.

25. Kaufman J, Breeding L, Rosenberg N. Anatomic location of acute diabetic foot infection. Its influence on the outcome of treatment. Am Surg 1987;53:109–112.

26. Yusefzadeh DK, Schumann EM, Mulligan GM, Bosworth DE, Young CS, Pringle KC. The role of imaging modalities in diagnosis and management of lyomyositis, Skel Radiol 1982;8:285–287.

27. Glazer GM, Orringer MB, Chevernet TL, et al. Mediastinal lymph nodes: relaxation time/pathologic correlation and implications in staging of lung cancer with MR imaging. Radiology 1988;168:429–431.

28. Puranen J, Makela J, Lahde S. Postoperative intervertebral discitis. Acta Orthoped Scand 1984;55:461–465.

29. Altman, N, Hardwood-Nash DC, Fitz C, Chuang SH, Armstrong D. Evaluation of the infant spine by direct sagittal computed tomography. Am J Neuroradiol 1985;6:65–69.

30. Golimbu C, Firooznia H, Ralii M. CT of osteomyelitis of the spine. AJR 1984;42:159–163.

31. Kirzner H, Oh YK, Lee SH. Intraspinal air: a CT finding of epidural abscess. AJR 1988;151:1217–1218.

32. Bielecki DK, Sartoris D, Resnick D, Lom KV, Fierer J, Haghighi P. Intraosseous and intradiscal gas in association with spinal infection: report of three cases. AJR 1986;147:83–86.

33. Kopecky KK, Gilmor RL, Scott JA, Edwards MK. Pitfalls of computed tomography in diagnosis of discitis. Neuroradiology 1985;27:57–66.

34. Van Lom KJ, Kellerhouse LE, Pathria MN, et al. Infection versus tumor in the spine: criteria for distinction with CT. Radiology 1988;166:851–855.

35. Modic MT, Feiglin DH, Piraino DW, et al. Vertebral osteomyelitis: assessment using MR. Radiology 1985;157:157–166.

36. Smith AS, Weinstein MA, Mizushima A, et al. MR imaging characteristics of tuberculous spondilitis vs. vertebral osteomyelitis. AJR 1989;153:399–405.

37. deRoos A, van Persijn van Meerten EL, Bloem JL, et al. MRI of tuberculous spondilitis. AJR 1986;146:79–82.

CHAPTER 14

Magnetic Resonance Imaging and Computed Tomography of Skeletal Muscle Pathology

James L. Fleckenstein

Introduction   Pathology
Imaging Techniques   Conclusion

Introduction

Skeletal muscle constitutes up to 40% of the body's weight, is the primary organ of locomotion, and is a major metabolic reservoir. Its anatomic complexity is rivaled only by the variety of diseases that affect it. Presented here is a list classifying the major muscle disorders:

1. Primary neurogenic disorders
   a. Diseases of motor neurons (MNs)
      i. Upper MNs—primary lateral sclerosis
      ii. Upper and lower MNs—amyotrophic lateral sclerosis
      iii. Lower MNs—spinal muscular atrophy
   b. Diseases of the neuromuscular junction
      i. Myasthenia gravis
      ii. Eaton-Lambert syndrome
      iii. Toxic (botulism, tick paralysis)
2. Primary muscle disorders
   a. "Pure" muscular dystrophies
      i. X-Linked (Duchenne, Becker)
      ii. Autosomal recessive (limb-girdle, childhood, congenital)
      iii. Facioscapulohumeral
      iv. Distal
      v. Ocular
      vi. Oculopharyngeal
   b. Muscular diseases with myotonia
      i. Myotonia congenita
      ii. Dystrophic myotonia
      iii. Paramyotonia congenita

   c. Metabolic myopathies
      i. Inherited enzyme defects (e.g., McArdle's disease)
      ii. Endocrinopathies (Cushing's syndrome, thyroid disease)
      iii. Electrolyte disturbances
   d. Inflammatory muscle disease
      i. Infection (pyomyositis, viral polymyositis)
      ii. Idiopathic (polymyositis, dermatomyositis)
      iii. Collagen vascular diseases (lupus)
   e. Trauma
      i. Crush injury
      ii. Burns
      iii. Contusion
      iv. Strains and ruptures
      v. Delayed onset muscle soreness
      vi. Overuse syndromes
      vii. Iatrogenic (intramuscular injection, surgery, etc.)
   f. Mass lesions (see Chapter 15)
      i. Neoplastic (primary and metastatic)
      ii. Vascular (hemangiomas, arteriovenous malformations)
      iii. Hematomas
      iv. Myositis ossificans

While many of the diseases are rare, muscle pain and weakness are among the commonest symptoms in patients. Despite the ubiquity of these symptoms, determining that muscle is the cause of the pain or weakness can be difficult. For these reasons, attention long ago turned to radiologic techniques to help study muscle (1, 2).

Soft-tissue radiography of muscles was first reported in 1928, but only gross assessments of fat deposition could be made (cited by DiChiro (1)). The advent of cross-sectional imaging techniques improved patient evaluation by circumventing the major obstacles of physical examination of the muscles: skin, subcutaneous fat, and adjacent muscles. That this is an important development is suggested by over 2000 cases being reported in the last decade (1–86). Although neurogenic disease and muscular dystrophies have been the primary focus of these studies (4, 6–11, 13, 16, 29–37, 39–41, 43, 47–49, 53, 54, 56, 58, 59, 61–65, 67–73, 75, 78, 79), orthopedic conditions have also been reported (3, 5, 12, 14, 15, 17, 20, 24, 25, 27, 28, 44, 46, 50–

Abbreviations (see also glossary): MN, motor neuron; STIR, short τ inversion recovery; CK, creatine kinase; AIDS, acquired immunodeficiency syndrome; DOMS, delayed onset muscle soreness.

52, 57, 60, 66, 74, 76, 77, 80–82, 84–86). Among the most interesting and constant features in all these disorders is highly focal involvement of individual muscles. Why some muscles are selectively damaged and others are spared is unknown. Patterns of muscle involvement in neuromuscular diseases appear to have some diagnostic specificity (10, 13, 16, 30–37, 41, 46, 49, 58, 61, 62, 67–73, 78) but this has yet to be rigorously quantitated.

Besides providing some diagnostic information on the basis of the pattern of disease, imaging modalities provide the critical ability to avoid common sampling errors in diagnostic tests (8, 19, 33). The ability to precisely localize focal muscle abscess (12, 27, 57, 60, 66, 74, 76, 82, 84, 86), necrosis (5, 28, 52, 80, 81, 86), and hematoma (14, 43) aids in their presurgical evaluation. The high sensitivity of magnetic resonance (MR) imaging to soft-tissue composition allows one to objectify and monitor tissue pathology over time in poorly understood but common clinical syndromes (see below).

Critical factors yet to be completely characterized are the relative sensitivity, specificity, cost effectiveness, and availability of the various modalities in different clinical applications. In the meantime, an understanding of the basic disorders affecting skeletal muscle is important. The purpose of this chapter is to review basic pathological processes that affect the muscles and to discuss relative advantages and disadvantages of the available modalities in evaluating them. Neurogenic and dystrophic causes of muscle pathology will be less emphasized because reviews already exist in the literature (10, 69).

Imaging Techniques

Technical Considerations

Ultrasound imaging of the muscles is best performed with real-time linear array transducers with the highest frequency possible while still adequately penetrating the tissue of interest (35). A water "path" may be placed between the transducer and the skin for evaluation of superficial structures if a higher frequency transducer is not available (77).

The optimal kilovolt peak, milliamp, and reconstruction algorithm for CT scanning of muscles depends on the particular device. A review of different CT techniques has previously been published (10). Computed tomography images shown in this chapter employ the "abdominal" technique available on the Picker 1200SX device (Picker, Cleveland, Ohio) (130 kVp, 385 mA, and 512 × 512 matrix).

For a complete MR imaging evaluation of the muscles, both T1- and T2-weighted examinations are usually necessary because fat is bright on the former and edematous processes are bright on the latter. An alternative approach is to use a fat-suppression technique that is sensitive to edema, such as a T2-weighted chemical shift sequence tuned to water proton frequency or the short τ inversion recovery (STIR) sequence [1500 repetition time (TR), 100 inversion time (TI), 30 echo time (TE)]. Because pathologic processes tend to have increased proton density, and T1 and T2 times, and since each of those positively contribute to signal intensity on STIR, it is highly sensitive to pathol-

ogy. Drawbacks of STIR include prominent motion artifacts, especially at high field (1.5 to 2.0 T), signal to noise is relatively low and scan times are long. However, relatively low resolution images can be obtained with large time savings. This is an appropriate trade-off when the objective is tissue characterization and not detailed anatomic depiction, such as in the large muscles of the thighs. According to this logic, a fast, short τ inversion recovery (fast STIR) sequence was developed that allows the acquisition of 16 contiguous 1-cm slices in slightly over 1 min (23). Motion-generated artifacts are tolerable because they occur in the phase-encoding direction, which can be rotated away from the area of interest on many devices. The sequence is ideal for use as a scout scan, rapidly screening for gross abnormalities. In some situations, only one sequence can be performed (e.g., agitated patients). The fast STIR sequence is helpful in this situation because of its rapidity, sensitivity to edema, and specificity for fat. Currently, comprehensive MR imaging evaluations of the lower or upper extremities require 45 min, with follow-up studies requiring less.

A Toshiba America, MRI Inc. (South San Francisco, Calif.) 0.35-T MR imaging device was used in all cases shown in this chapter.

Advantages and Disadvantages of Imaging Modalities

Before addressing the radiologic appearance of basic pathological processes, we will consider general advantages and disadvantages of the imaging modalities. Greater detail is provided in special circumstances below.

Ultrasound detects gross changes in muscle composition and size (5, 12, 16, 24, 25, 32–37, 44, 46, 47, 52, 53, 62, 66, 67, 77, 79, 82, 84). Ultrasound is the most convenient, fastest, and least expensive method to survey the muscles. Because it is safe, large volumes of muscles can be studied and serial examinations do not pose a health threat. Real-time techniques can assess the muscle during the dynamic process of muscular contraction. This helps differentiate abnormal muscle from hematoma and makes muscle ruptures more easily visualized (24). The major disadvantages are that (*a*) its display of anatomy is inferior to CT and MR imaging and (*b*) it is the least sensitive of the three modalities in detecting myonecrosis (52, 69) and other pathology (69).

Computed tomography provides high-resolution images and is sensitive to alterations in tissue density. It is intermediate between ultrasound and MR imaging in terms of cost and availability. Computed tomography can quantitate muscle cross-sectional area (7, 39, 40) and detect fatty infiltration (6, 8–11, 13, 30, 31, 40, 41, 43, 50, 58, 61–63, 67–69, 71–73, 78), necrosis (15, 28, 52), and inflammation (62, 67), including abscess (15, 27, 51, 57, 60, 74, 76, 84). However, CT has significant shortcomings. First, it is limited in its ability to discriminate between these processes in muscle, because each tends to decrease muscle density. Second, ionizing radiation necessarily limits the volume of muscle that can be studied on serial survey examinations. Third, iodinated contrast may improve detection of infection (52, 74), hematoma (14), or necrosis (80, 81), but does not improve distinction between them. It requires an intravenous injection and places the patient at risk for adverse systemic reactions. Fourth, "beam hardening" artifacts are a major disadvantage as they may mimic or obscure a muscle lesion. These limitations are circumvented by MR imaging.

Magnetic resonance imaging is the most sensitive modality in detecting myonecrosis (52, 69) and other pathology (69). It can be used to quantitate changes in muscle size (22), water compartmentation (18, 42, 55, 64), and fat deposition (65). Unlike CT, MR imaging readily differentiates fat (short T1 relaxation time) from other pathologic processes, such as edema (long T1). Thus, MR imaging offers some specificity that the other tests lack. Magnetic resonance imaging also appears to be safe, lending itself to the survey of large volumes of muscle in populations at risk for muscle disease (21).

The high sensitivity of MR imaging to soft-tissue processes is exemplified not only by its ability to distinguish static pathological abnormalities of muscle but also by its demonstration of dynamic alterations in muscle composition that occur as part of normal muscular exertion. Transient increases in muscle water occur during intense exercise and this was hypothesized as the reason for exercised muscles becoming bright on spin-density and T2-weighted images (17, 18). Magnetic resonance imaging can thus be used to noninvasively monitor muscle recruitment during exercise (see below). The diagnostic potential for the technique was recently shown in a study in which patients with chronic compartment syndromes showed unusual delayed increases in T1 relaxation times after exercise (3).

Because MR imaging offers the greatest sensitivity in evaluating changes in muscle composition, it is the primary modality focused on in this chapter. Disadvantages of MR imaging are its cost, availability, and sensitivity to motion artifacts.

Pathology

In broad terms, neuromuscular disorders can be categorized as those due to dysfunction of neurons, neuro-

muscular junction, and muscle itself (see "Introduction"). Muscular spinal atrophies and muscular dystrophies have been extensively studied (4, 6–11, 13, 16, 29–37, 39–41, 43, 47–49, 53, 54, 56, 59, 61–65, 67–73, 75, 78, 79). More important to the orthopedist are necrotic (15, 20, 21, 25, 28, 38, 46, 52, 86), inflammatory (12, 15, 17, 27, 51, 57, 60, 66, 67, 74, 76, 77, 80–82, 84, 85), hemorrhagic (14, 43), and traumatic (24, 77) processes of skeletal muscles. Imaging modalities in general, and MR imaging in particular, are sensitive to these conditions (52, 69).

In contrast to the wide variety of diseases that may involve skeletal muscles, the response of muscle to disease is relatively restricted, both microscopically and radiographically. Denervation, necrosis, fiber size alteration (i.e., atrophy and hypertrophy), and connective tissue changes (i.e., fatty replacement and fibrosis) are dominant histopathologic findings (45). The radiologic appearances of basic pathological processes of skeletal muscle will now be reviewed.

Denervation

In primary neurogenic disease, denervation is the major histopathologic finding, characterized by fiber type grouping and relatively greater atrophy of type 2 fibers than type 1 (38). Magnetic resonance imaging is the only imaging technique that identifies the subacute phase of denervation of muscles prior to fatty deposition (64, 70). Alterations in muscle during subacute denervation include prolongation of proton T1 and T2 relaxation times (64). This is in contrast to the end-stage appearance, in which T1 shortening is the dominant MR imaging alteration due to fat deposition (see below). In an animal model, prolongation of muscle relaxation times was evident in the first few weeks after denervation. These changes were proportional to a diminution in fiber size and attendant increase in the extracellular water content (64). The MR imaging appearance of end-stage denervation is exemplified by fatty re-

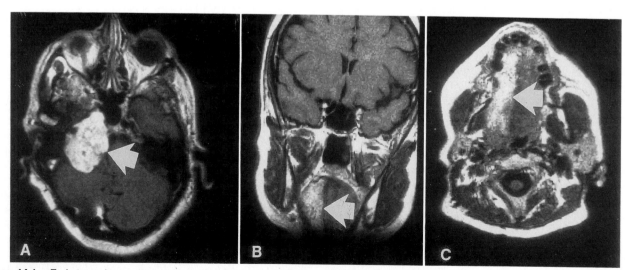

Figure 14.1. End-stage denervation. A large, recurrent glossopharyngeal neuroma impinges on the right hypoglossal nerve at the foramen magnum (*arrow*, **A**). The half of the tongue ipsilateral to the par- alyzed nerve is largely replaced by fat, as shown by high signal intensity *(arrows)* on coronal (**B**) and axial (**C**) (500/40) images.

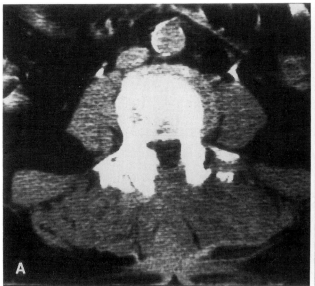

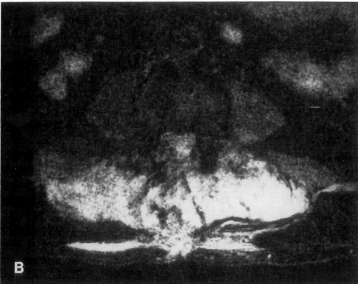

Figure 14.2. Postsurgical changes of paraspinal muscles attributed to denervation. This patient had mutlilevel laminectomies 12 weeks earlier but had persistent back pain at the time of noncontrast CT **(A)** and MR imaging **(B)**. Paraspinal muscle abnormality is underestimated by CT whereas it dominates the STIR image **(B)**. The MR image appearance is nonspecific, however, and it cannot differentiate edema due to myonecrosis from denervation.

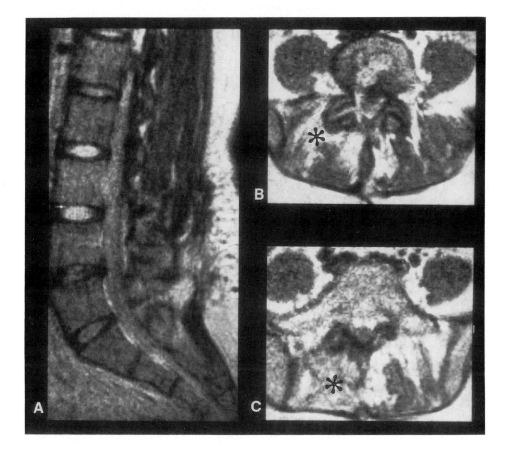

Figure 14.3. Focal atrophy and fatty replacement in the "failed back" syndrome. A sagittal view (500/40) of the lumbar spine **(A)** shows diminution in size and fatty infiltration of the paravertebral muscles. Fatty replacement of the right multifidus muscle (*) begins near the incision site **(B)** but is more extensive near the sacral insertion **(C)**.

placement of half of the tongue, ipsilateral to the side of a tumor impinging on the hypoglossal nerve at the foramen magnum (Fig. 14.1).

Obliteration of the paraspinal muscles has been reported as a complication of spine surgery, with the incidence being 30% in patients with failed back syndromes (50). This was attributed to denervation because the changes tend to occur below the surgical site and extend to the most inferior aspect of the muscle. Infarction due to prolonged retraction is another possible mechanism. The greater sensitivity of MR imaging to this process relative to CT is shown in Figure 14.2. Focal atrophy and fatty replacement are easily seen in the end stage of this process (Fig. 14.3).

Myonecrosis, Muscle Trauma, and Inflammation

Fiber necrosis is a basic histopathologic feature common to many conditions affecting the muscles (44). Serum creatine kinase (CK) elevations indicate myonecrosis but the test has significant limitations. First, it cannot distinguish between a small amount of enzyme release from a large volume of muscle and a large amount of enzyme release from a small volume of tissue. Second, the timing of specimen collection relative to the inciting event is a critical determinant of the enzyme level and therefore apparent degree of injury. Third, the test obviously provides no information as to the location and spatial extent of necrosis. Magnetic resonance imaging has documented that tissue abnormality may persist in necrotic muscles for far longer than serum CK levels (20). Necrotic muscle becomes a surgical problem when central liquefaction or infection are associated. In these situations imaging is especially useful to localize the fluid (5, 15, 25, 28, 46).

On ultrasound, myonecrosis is usually hypoechoic (5, 25, 28, 46) but may be echogenic (52). Compared to surgery, ultrasound was reported to be only 42% sensitive (52). On CT, necrosis may be more readily detected when infection is concomitant (15) or intravenous contrast is administered (80, 81). On noncontrast studies, however, CT was only 62% sensitive compared to surgery (52). In most published cases of imaging myonecrosis, the clinical situation has been that of drug overdose or suspected crush injury. An interesting feature of this syndrome is that some muscles are severely affected while nearby muscles are spared. The pattern of distribution of abnormal muscles does not necessarily conform to that expected on the basis of crush alone. This is exemplified by a case in which CT and MR imaging were both performed (Fig. 14.4). The strikingly focal distribution of muscle necrosis is not specific to drug overdose, however, because similarly focal patterns appear to be the rule in exertional rhabdomyolysis (20, 21).

In contrast to ultrasound and CT, MR imaging is highly sensitive to myonecrosis (100% sensitive, compared to surgery (52)). On MR images, myonecrosis is evident by high signal intensity in the muscle on spin-density, T2-weighted, and STIR sequences but usually is not apparent on T1-weighted spin-echo images. This is in part because prolonged T1 and T2 times have opposing effects on muscle signal intensity when using moderately T1-weighted sequences (18, 20). The high sensitivity of MR imaging to myonecrosis is exemplified by two cases of severe electrical burns (Fig. 14.5).

Postexertion muscle necrosis is similar on MR imaging to that of thermal injury except that little swelling is evident (Figs. 14.6 to 14.8) (20). This is exemplified in the frequent situation of the "week-end athlete," who is untrained, but engages in intense physical activity to which he or she is unaccustomed (Figs. 14.6 and 14.7). Frequent but more mild injuries are also evident on MR imaging in the well-trained athlete (Fig. 14.9). The fact that these frequent but incompletely understood conditions can now be spatially localized should facilitate research probing their pathophysiology (20). Similar statements can be made about exertional muscle necrosis in rare inherited disorders of

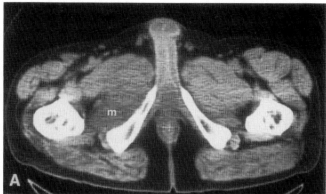
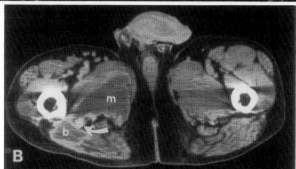
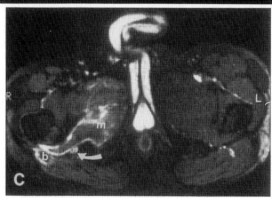

Figure 14.4. Myonecrosis in drug overdose: noncontrast CT of the proximal thighs **(A)** in a 34-year-old man 1 week after attempted suicide by diphenhydramine overdose. Severe rhabdomyolysis and sciatic nerve palsy were present clinically. Faintly decreased density is apparent in the right adductor magnus *(m)* but beam-hardening artifact contributes. A postcontrast CT image **(B)** obtained slightly more inferiorly shows peripheral enhancement about the low-density adductor magnus *(m)*and biceps femoris *(b)* as well. The sciatic nerve enhances and is enlarged *(arrow)*. These findings are readily apparent using the STIR MR imaging sequence **(C)**.

carbohydrate metabolism, e.g., McArdle's disease (myophosphorylase deficiency). This group undergoes myonecrosis with relatively trivial muscular exertion. Magnetic resonance imaging has been used to document the high frequency of rhabdomyolysis occurring in these patients during activities of daily living (Fig. 14.10) (21).

Muscle strains are another form of exertion-related injury, but are not usually considered to be a form of necrosis. However, these injuries have MR imaging features that are indistinguishable from necrosis (Fig. 14.11). Similar to other pathological processes, tissue abnormality occurs fo-

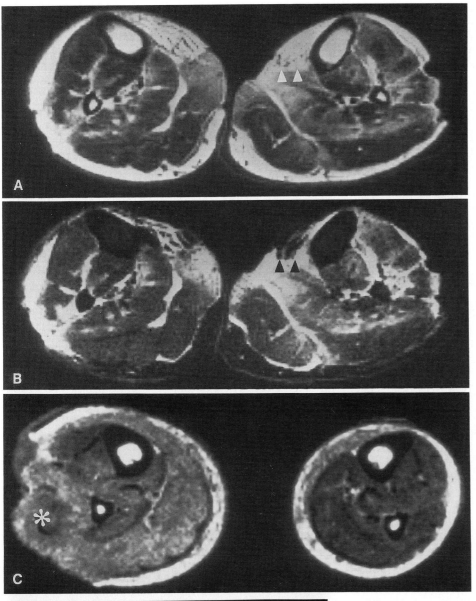

Figure 14.5. Thermal myonecrosis. Axial images of the legs in two patients who grounded high-voltage wires. In this setting, areas of necrosis show a patchy distribution. Note on the moderately T2-weighted image (2000/60) that edema and fat blend imperceptibly (*arrowheads,* **A**). Using the STIR sequence, signal from the fat is suppressed *(dark)* so that edema can be distinguished from fat (*arrowheads,* **B**). Marked swelling of the peroneus longus is evident in the second patient (*, **C**). Both patients underwent subsequent amputation. (Courtesy of J. Hunt, M.D., Department of Surgery, University of Texas Southwestern Medical Center of Dallas.)

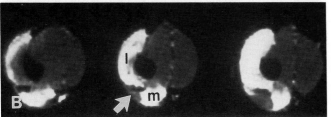
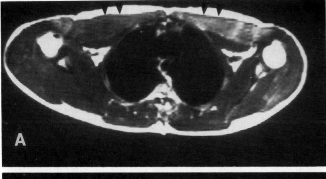

Figure 14.6. Delayed onset muscle soreness (DOMS). A sedentary but heavily muscled chef decided it was time to "get into shape." After 2 days of upper body work he noted soreness in his arms and chest. His urine turned brown on the 4th day, leading him to medical attention. Serial serum creatine kinase levels peaked at 250,000 IU/ml at 7 days. Magnetic resonance imaging was performed on the 11th day, 3 days after all symptoms abated. The entire pectoralis major shows increased signal intensity on a 2000/50 image (*arrows,* **A**). Bilaterally symmetrical alterations of the triceps were seen (**B**). The medial *(m)* and lateral *(l)* heads of the triceps show abnormally increased signal intensity on the STIR sequence but the long head appears normal *(arrow)*. Biopsy and electromyogram (EMG) studies of the upper extremities were deferred pending complete recovery of the muscles by MR imaging criteria.

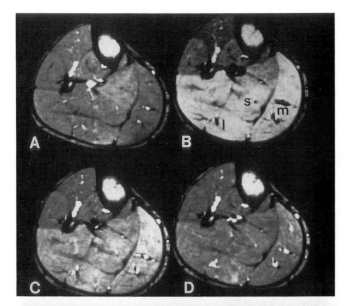

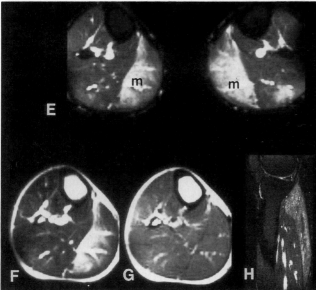

Figure 14.7. Acute **(A–D)** and delayed **(E–H)** effects of calf plantar-flexion exercise on leg muscles. A gradient-echo 500/30, 30° sequence was obtained before **(A)** and during successive 4-min postexercise periods following 5 min of ankle plantarflexion **(B–D)**. This increase in signal intensity with exercise is a normal result and is evident in active muscles, i.e., soleus *(S)* and medial *(M)* and lateral *(L)* gastrocnemius. Twenty-four hours after bilateral exercise, pain and rhabdomyolysis were associated with marked signal intensity abnormality in the medial head of the gastrocnemius. This is shown using STIR **(E** and **H)** and a 2000/60 sequence **(F)** but is not apparent using a 500/30 sequence **(G)**. A sagittal STIR image of the leg shows that the entirety of the muscle is involved **(H)**. Note that muscles used to perform the work **(A–D)** are not necessarily injured by the exercise **(E–H)**. (Reprinted with permission from Fleckenstein JL, Weatherall PT, Parkey RW, Payne JA, Peshock RM. Sports-related muscle injuries: evaluation with MR imaging. Radiology 1988;172:793–798.)

cally in specific muscle groups. For instance, the rectus femoris muscle has been called "the sprinter's muscle" because it may rupture during short explosive bursts of running (Fig. 14.12). In an ultrasound study of 120 muscle

injuries occurring in advanced level athletes, the rectus femoris was the most commonly injured (38%) (24). Magnetic resonance imaging can detect not only the injury acutely, but also after complete clinical resolution, providing improved monitoring of tissue healing over time (Fig. 14.13).

A prospective study of delayed muscle soreness following exercise observed that signal intensity in necrotic muscle initially rose proportional to pain and CK levels, but prolonged relaxation times persisted for up to 4 weeks postexercise (20). This indicates that MR imaging may be used to assess muscle pathology even though considerable time has elapsed since the injurious event. It also implies that radiologists may observe evidence of exertional muscle injury in patients scanned for related or unrelated reasons.

Paraspinal muscle strains are similar in appearance to exertional strains. Magnetic resonance imaging in this setting helps to document the presence, location, and severity of paraspinal strains. This may help to identify patients with muscle injury as a cause of symptoms in "whiplash" and on-the-job injuries (Fig. 14.14).

Muscle pain is also a frequent occurrence in patients with occupation-related muscle overuse syndromes. This is demonstrated by the case of a waitress who had pain in the arm used to hold the serving tray. The pain resulted in inability to work. Magnetic resonance imaging confirmed the presence of focal muscle abnormality (Fig. 14.15). After 4 days of leave from work and treatment with antiinflammatory analgesics, she was free of pain and reported no subsequent episodes during 1 year of follow-up. The muscle lesion is similar in MR imaging appearance to a patient with postmarathon myalgia, in whom the MR imaging finding was small and located near myotendinous junctions (Fig. 14.9). Another example of muscle overuse demonstrable by MR imaging is a patient with recurrent "tennis elbow" (Fig. 14.16).

Inflammatory conditions of the muscles also show necrosis histopathologically. Idiopathic polymyositis, drug reaction polymyositis, and dermatomyositis all present clinically with proximal muscle weakness and markedly elevated serum creatine kinase levels. Magnetic resonance imaging detects acute and chronic changes in these conditions (Fig. 14.17) and can be used to guide biopsies (26, 67). In these conditions tissue alterations occur with striking symmetry but focal involvement remains the rule, resembling other conditions described above. Other inflammatory disorders of muscle that have been studied with MR imaging include orbital pseudotumor and Graves' ophthalmopathy. Swelling and edema of the extraocular muscles are readily apparent on MR images (Fig. 14.18).

Pyomyositis shares the same features as myonecrosis, although swelling and pain may be prominent (Fig. 14.19). Serum creatine kinase levels are usually normal and sepsis is conspicuously absent in pyomyositis, so imaging can play an important role in the diagnosis; *Staphylococcus aureus* is usually the organism responsible (12, 27, 57, 60, 66, 74, 76, 82, 84, 85). Pyomyositis has recently been recognized in association with the acquired immunodeficiency syndrome (AIDS) and constitutes one of the readily treatable infections in AIDS (27, 66, 76).

Iatrogenic muscle trauma is readily shown on MR im-

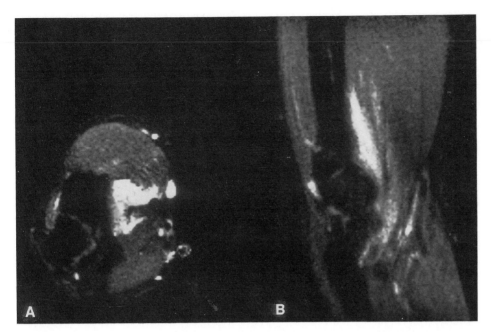

Figure 14.8. Injurious effect of "bicep curls" exercise on the arm. STIR images were obtained 4 hr after a single bout of 50 bicep curls in an untrained subject. Although pain was absent for 24 hr, range of motion was limited during elbow extension at the time the image was obtained. Axial **(A)** and sagittal **(B)** images show alteration in the distal brachialis near its insertion onto the ulna.

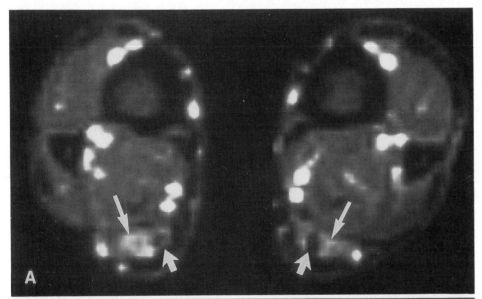

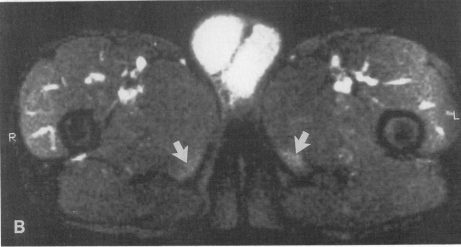

Figure 14.9. Postmarathon myalgia. This marathoner curtailed his training schedule because of recurrent "Achilles tendinitis." The pain resulted in his having to quit halfway through a 30-km long-distance run. STIR image shows that the soleus (*long arrows,* **A**) is bright near its contribution to the Achilles tendon (*short arrows,* **B**). Similarly, small areas of signal abnormality are evident on axial STIR image of the adductor magnus near its origin on the ischium (*arrow,* **B**). This case exemplifies muscle injuries in elite athletes by the paucity of abnormality and the location of lesions near the myotendinous junction. (Reprinted with permission from Fleckenstein JL, Weatherall PT, Parkey RW, Payne JA, Peshock RM. Sports-related muscle injuries: evaluation with MR imaging. Radiology 1988;172:793–798.)

Figure 14.10. Muscle contracture. In McArdle's disease, exertional rhabdomyolysis occurs during relatively minor physical activity. An axial 2000/80 image of the proximal forearm was obtained 24 hr after handgrip-induced contracture **(A)**. Extensive damage is evident in the superficial *(s)* and deep finger flexors *(d)*. A coronal STIR image **(B)** shows the relationship of the injured flexor digitorum profundus *(d)* to the ulna *(u)*. Lateral to the radial head *(r)* a smaller region of necrosis is seen in the extensor carpi radialis *(e)*.

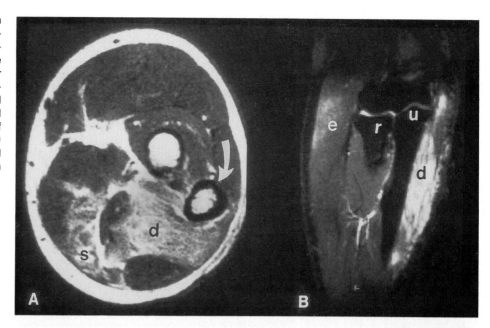

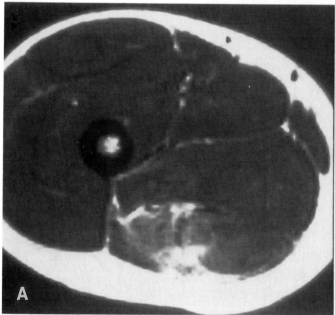

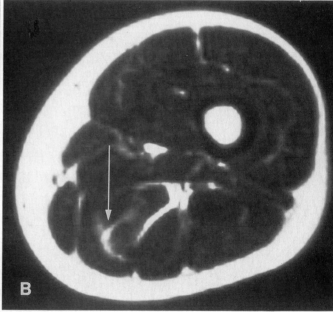

Figure 14.11. Muscle strains in week-end athletes. Two healthy but sedentary patients complained of acute pain while playing intramural softball. Focal areas of high signal intensity are present in the biceps femoris of one patient **(A)** and the semimembranosus *(arrow)* of the other **(B)**. The cause of the high signal intensity is unproven but likely represents edema with or without hemorrhage. (Reprinted with permission from Fleckenstein JL, Weatherall PT, Parkey RW, Payne JA, Peshock RM. Sports-related muscle injuries: evaluation with MR imaging. Radiology 1988;172:793–798.)

aging (17). Intramuscular injections may result in bright signal intensity on T2-weighted or STIR images. Muscle biopsy also leaves a mark, sometimes for long periods of time. This fact can be used to document the site of biopsy and should not be mistaken as a part of the patient's disease process (Fig. 14.20).

Atrophy, Hypertrophy, and Pseudohypertrophy

Alterations in muscle size, i.e., atrophy, hypertrophy, and pseudohypertrophy, may be diffuse but are frequently fo-

cal, with the distribution of abnormalities depending on the specific disease and its chronicity. Atrophy indicates diminution in size, hypertrophy an increase in size, and pseudohypertrophy an apparent increase in size of muscle due to fatty infiltration. These are hallmarks of neuromuscular disease, being present in patients with muscular dystrophies (4, 8–11, 13, 16, 29–33, 35, 37, 40, 41, 43, 49, 53, 54, 56, 58, 61, 63, 67, 69, 71–73, 75, 78), myotonias (4, 10, 11, 16, 41, 47, 67–69, 73), and neurogenic atrophies (4, 6, 10, 11, 16, 30–32, 35, 37, 50, 53, 61, 63, 69, 70, 73, 78).

One of the most striking findings in imaging neuromus-

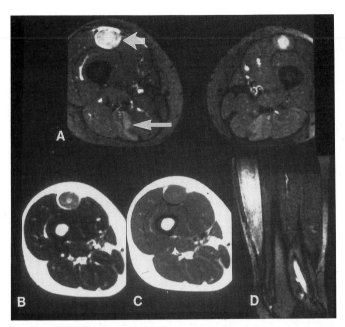

Figure 14.12. Acute thigh pain while running for a ball in a week-end athlete. MR imaging was performed 36 hr after the event **(A–D)**. An axial STIR image shows abnormal signal intensity in the rectus femoris muscle bilaterally, with the right more severe than the left (**A**, *short arrow*). Associated semitendinosus strains are less obvious *(long arrow)*. A 2000/60 spin-echo image of the right thigh does not detect the hamstring tears **(B)**. A 500/30 spin-echo image of the thigh shows minimal abnormal signal intensity in the rectus femoris but the abnormally round shape of the muscle is still evident **(C)**. A sagittal STIR image shows the longitudinal extent of the rectus femoris injury **(D)**. (Reprinted with permission from Fleckenstein JL, Weatherall PT, Parkey RW, Payne JA, Peshock RM. Sports-related muscle injuries: evaluation with MR imaging. Radiology 1988;172:793–798.)

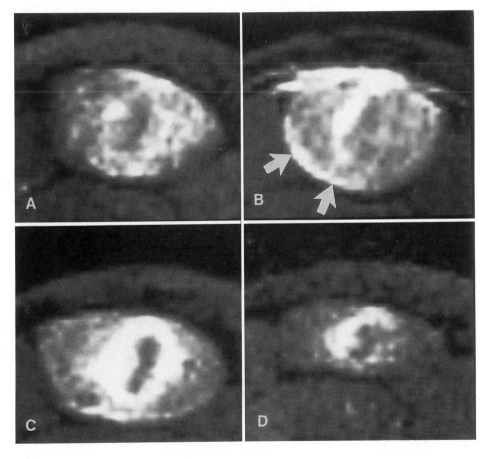

Figure 14.13. Healing of a muscle strain. Serial, axial, STIR images demonstrate the time course of the rectus femoris strain in the patient in Figure 14.12, at 18 hr **(A)**, 36 hr **(B)**, 72 hr **(C)**, and at 12 days after the injury, when he was asymptomatic **(D)**. Note the evolution of concentric heterogeneity over time, with a bright, thin rim of signal intensity adjacent to the intermuscular fascia at 36 hr *(arrows)*. (Reprinted with permission from Fleckenstein JL, Weatherall PT, Parkey RW, Payne JA, Peshock RM. Sports-related muscle injuries: evaluation with MR imaging. Radiology 1988;172:793–798.)

cular disorders is the highly focal distribution of atrophic and hypertrophic muscles in the muscular dystrophies (8–11, 13, 30–35, 40, 41, 43, 56, 61, 67–69, 71–73, 78). In Duchenne's muscular dystrophy there is an apparently ordered pattern of atrophy and fatty replacement of certain muscles and focal sparing and hypertrophy of others (9–11, 30, 31, 41, 67, 72). Early in the course of the disease selective atrophy occurs in the quadriceps, adductor magnus, biceps femoris, and medial gastrocnemius (10, 35, 63, 67, 72, 73). Focal sparing of gracilis and sartorius in the thigh, and tibialis posterior, tibialis anterior, and peronei in the leg, is seen late in the disease (10, 35, 63, 67, 72, 73). In the trunk, the psoas and multifidus are spared while longissimus and iliocostalis are selectively atrophied (30).

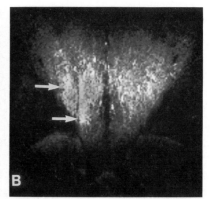

Figure 14.14.   Paraspinal muscular strains in trauma. **(A)** Cervical strain. Coronal STIR image of the neck and shoulders of a woman (made following a motorcycle accident) in whom a brachial plexus injury was suspected clinically. Diffuse injury is obvious in the left scalenus anterior muscle *(arrows)* as well as in the supraspinatus *(arrowheads).* Other areas of bright signal intensity in the neck are due to the normal STIR appearance of blood flowing slowly in veins. (Courtesy of G. Gibson, M.D.) **(B)** Lumbar strain. A construction laborer had severe paraspinous muscle spasm while lifting at work. Coronal STIR shows abnormal signal intensity in the trapezius muscles, worse on the right *(arrow).*

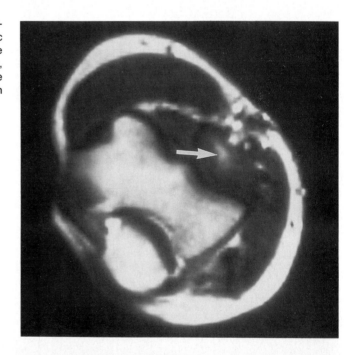

Figure 14.15.   "Waitress elbow." Occupational muscle overuse syndromes are characterized by pain in the extremity, brought on by specific muscular work. A waitress complained of several days of progressive pain near her elbow, which was exacerbated by holding her cocktail tray, ultimately forcing her to stop work. The axial 2000/60 image confirms the presence of tissue abnormality in the distal brachialis consistent with edema *(arrow).*

Figure 14.16.   "Tennis elbow." Because of progressively increasing pain over the lateral humeral epicondyle, this patient quit playing tennis. Axial STIR images **(A and B)** show focal signal abnormality within the proximal extensor digitorum, including where it inserts onto the lateral epicondyle *(arrow).* (Courtesy of A. Miller, M.D. and P. T. Weatherall, M.D., Department of Radiology, University of Texas Southwestern Medical Center, Dallas.)

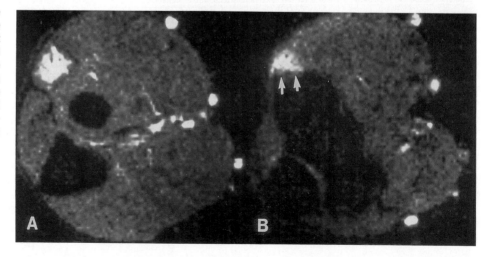

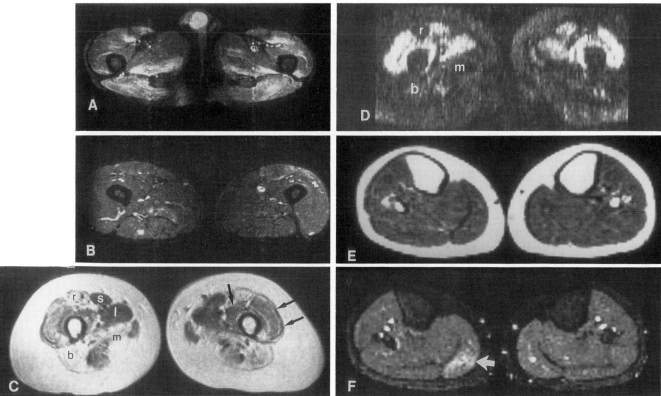

Figure 14.17. Acute and chronic idiopathic polymyositis. A previously healthy 28-year-old man presented with rapidly progressive proximal weakness and rhabdomyolysis. Strikingly symmetrical involvement of proximal thigh muscles **(A)** is evident using STIR while much less abnormality is seen 10 cm more distally **(B)**. A 37-year-old woman was diagnosed with polymyositis 17 years prior to MR imaging. A 2000/30 spin-echo image **(C)** of the thighs shows symmetrical fatty replacement of the rectus femoris *(r)*, adductor magnus *(m)*, and biceps femoris *(b)*. Sparing of the sartorious *(s)*, adductor longus *(l)*, and vasti is apparent *(arrows)*. A 1-min fast STIR image **(D)** at the same level shows suppressed signal intensity where fat predominates. Signal intensity in the remaining muscle is abnormally increased, suggesting ongoing edema/inflammation. A 2000/30 image of the same patient's calves shows no abnormality **(E)**, whereas the STIR sequence **(F)** shows a single focus of increased signal intensity in the medial head of the right gastrocnemius *(arrow)*.

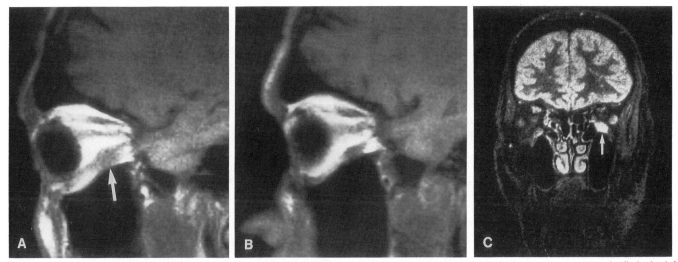

Figure 14.18. Graves' ophthalmopathy. T1-weighted (500/15) sagittal images show thickening of the left inferior rectus muscle *(arrow,* **A**) compared to the normal side **(B)**. A coronal STIR image **(C)** reveals increased signal intensity in multiple muscles, most markedly in the left inferior rectus *(arrow)*.

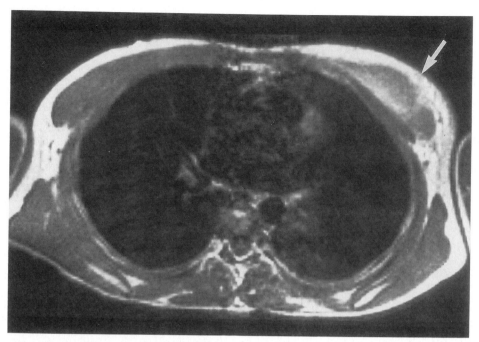

Figure 14.19. Pyomyositis. This patient was shot 1 month before this T2-weighted image was obtained (2000/60). He complained of chronic drainage, swelling, and pain at the site of penetration *(arrow).* High signal intensity in the left pectoralis major muscle is consistent with inflammation. At surgery necrotic muscle was debrided and *Staphylococcus aureus* was cultured.

In contrast to the highly focal distribution of abnormal muscles in dystrophies, spinal muscular atrophies tend to show diffuse muscle atrophy with respect to increased subcutaneous fat (10, 11, 16, 31, 32, 35, 37, 53, 61, 69, 73).

While the peculiar localization of pathology to individual muscles in dystrophies is poorly understood, the potential to make a specific diagnosis based on the pattern of involved muscles is immediately suggested. However, clinical diagnosis of certain conditions, such as Duchenne's dystrophy, is relatively straightforward, so that an important role for imaging studies in its initial diagnosis is yet to be absolutely proved (11, 31, 78). However, imaging can provide prognostic information not obtainable by other means by detecting evidence of subclinical disease. This has been demonstrated using CT to detect asymptomatic carriers of Duchenne's dystrophy (13, 71). Imaging can also monitor progression of diseases such as Duchenne's and the response to therapy (72). Magnetic resonance imaging may prove to be the most sensitive tool for detecting muscle damage in dystrophies, as it has been in inflammatory myopathies (26) and posttraumatic myonecrosis (52); ongoing work is expected to clarify this issue (67). The ability to assess bulk relaxation times (56) of water protons independent of fat protons (54) attests to additional advantages of MR imaging in evaluating dystrophies.

Pattern recognition of gross changes in muscle size on imaging studies in the various disorders offers potential advantages in the diagnosis of muscle diseases, although difficult cases will continue to tax our clinical resources. For example, a patient presented to our institution with an unknown disease characterized by clinical features of Duchenne's muscular dystrophy, "ragged red fibers" on biopsy (suggesting mitochondrial myopathy), and MR imaging findings that argue against Duchenne's dystrophy (Fig. 14.21). Atrophy and fatty replacement of the sartorius, gracilis, and semitendinosus are in contrast to the usual pattern of hypertrophy of these muscles in Duchenne's muscular dystrophy. On the other hand, the MR imaging appearance is highly similar to that published of a patient with inclusion body myositis (65). The case illustrates the difficulty of diagnosing myopathies and the limitations in the available diagnostic technologies.

Experience with imaging studies in the evaluation of inherited myopathies has been limited, but a few positive findings have been reported. For example, MR imaging of patients with phosphofructokinase deficiency revealed that muscles predisposed to acute necrosis are also prone to atrophy and fatty replacement (22). This suggests that myonecrosis and atrophy are causally linked in this disease (Fig. 14.22).

Changes in muscle volume may occur in endocrinopathies. Diffuse muscle atrophy has been detected using CT in patients with Cushing's syndrome, in which glucocorticoid excess is associated with weakness (31). Muscle hypertrophy has been reported in hypothyroidism (2). One might predict that imaging could be useful in detecting exogenous anabolic steroid abuse in athletes but ethical considerations make testing this hypothesis difficult.

Fatty Replacement and Fibrosis

Common mesenchymal abnormalities in muscle lesions include fat deposition and fibrosis. Microscopically, fatty deposits occur early in many pathologic conditions but become apparent on imaging studies in later stages. This was

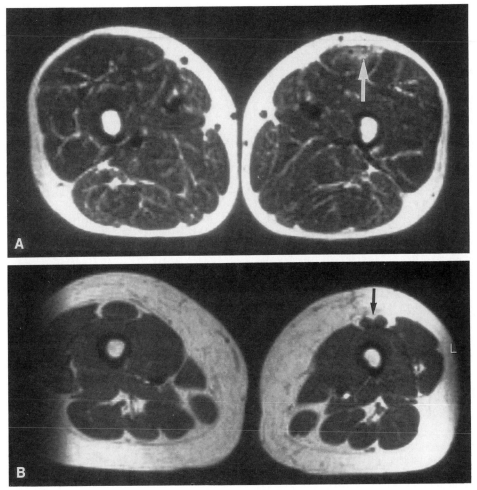

Figure 14.20. Iatrogenic trauma: the muscle biopsy. Axial 2000/60 images of two patients with suspected metabolic myopathy who were biopsied by the same surgeon. The first patient **(A)** was imaged 4 days after the biopsy. The second patient **(B)** was biopsied 2 years before the image. Bright signal intensity in the left rectus femoris muscle in the first patient *(arrow)* and focal diminution in size of the same muscle in the second patient *(arrow)* suggest that the surgeon reliably misses in his attempt to biopsy the vastus lateralis. The biopsy site should be recognized as such, rather than as part of the disease process.

demonstrated above in the case of advanced denervation (Fig. 14.1).

One of the first uses of ultrasound (34) and CT (63) in evaluating muscles was the detection of fatty deposition. Using the criteria of decreased density on CT and increased echogenicity on ultrasound as indicators of fatty replacement, together with the distribution of atrophic and hypertrophic muscles, characteristic patterns suggestive of specific diagnoses were observed. Selective sparing of the sartorius and gracilis is characteristic of Duchenne's muscular dystrophy, whereas in spinal muscular atrophies a patchy pattern of low densities is evident.

Sonographic criteria for fatty replacement (increased echogenicity) and CT criteria (decreased density) are nonspecific. As described, MR imaging can differentiate fat from most other processes because of the unusually short T1 relaxation time of fat. Subacute hematoma and fat may have a similar appearance on MR imaging but can be distinguished using a fat-suppression technique. The clinical circumstances also usually differentiate the two processes.

Recurrent or severe injury, regardless of cause, may re-sult in fibroblast proliferation and subsequent fibrosis. However, only a few cases of fibrosis of muscles have been reported using CT (10) and MR imaging (83).

Conclusion

As modalities before it, MR imaging circumvents traditional obstacles in the clinical evaluation of muscle pathology. Preliminary results suggest that MR imaging is more sensitive and specific than CT or ultrasound in detecting muscle pathology (52). Magnetic resonance imaging is very promising as a means for improving sampling accuracy in diagnostic evaluation of patients with muscle disorders. Muscle denervation, necrosis, atrophy, hypertrophy, and mesenchymal changes are sensitively detected and safely monitored using MR imaging. These applications are relatively new and their clinical efficacy and cost effectiveness remain to be quantitated. While studies address these issues, it remains desirable for clinicians and radiologists alike to be aware of the kind of information available to them using current imaging technology. What is clear is that imaging will continue to shed new light on muscle disorders by improving the sampling accuracy of other diagnostic tests and by noninvasively probing the natural history of incompletely understood disorders.

References

1. Di Chiro G, Nelson KB. Soft tissue radiography of extremities in neuromuscular disease with histological correlations. Acta Radiol 1965;3:65–88.
2. Gay BB, Weens HS. Roentgenologic evaluation of disorders of muscle. Semin Roentgenol 1973;3(1):25–36.
3. Amendola A, Rorabeck CH, Vellett D, Vezina W, Rutt B, Nott L. The use of magnetic resonance imaging in exertional compartment syndromes. Am J Sports Med 1990;18(1):29–34.
4. Barany M, Siegel I, Venkatasubramanian PN, Mok E, Wilbur AC. Human leg neuromuscular diseases: P-31 MR spectroscopy. Radiology 1989;172:503–508.
5. Barloon TJ, Zachar CK, Harkens KL, Honda H. Rhabdomyolysis: computed tomography findings. J Comput Tomog 1988;12(3):193–195.
6. Bertorini TE, Igarashi M. Postpoliomyelitis muscle pseudohypertrophy. Muscle Nerve 1985;8:644–649.
7. Bulcke JA, Termote JL, Palmers Y, Crolla D. Computed tomography of the human skeletal muscular system. Neuroradiology 1979;17:127–136.
8. Bulcke JA, De Meirsman J, Termote JL. The influence of skeletal muscle atrophy on needle electromyography. Electromyogr Clin Neurophysiol 1979;19:269–279.
9. Bulcke JA, Crolla D, Termote JL, Baert A, Palmers Y, Van Den Bergh R. Computed tomography of muscle. Muscle Nerve 1981;4:67–72.
10. Bulcke JAL, Baert AL. Clinical and radiological aspects of myopathies. Heidelberg: Springer-Verlag, 1982.
11. Calo M, Crisi G, Martinelli C, Colombo A, Schoenhuber R, Gibertoni M. CT and the diagnosis of myopathies: preliminary findings in 42 cases. Neuroradiology 1986;28:53–57.
12. Datz FL, Lewis SE, Conrad M, Maravilla A, Parkey W. Pyomyositis diagnosed by radionuclide imaging and ultrasonography. South Med J 1980;73(5):649–651.
13. De Visser M, Verbeeten B Jr. Computed tomographic findings in manifesting carriers of Duchenne muscular dystrophy. Clin Genet 1985;27:269–275.

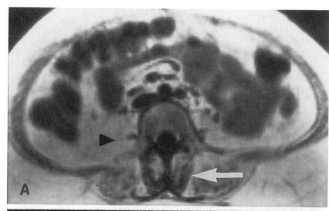
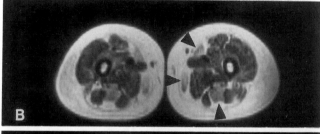
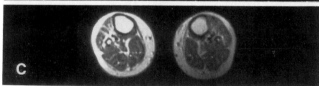

Figure 14.21. Proximal progressive atrophy in an undiagnosed myopathy. Axial T1-weighted (500/17) image of the trunk, thighs, and calves shows focal loss of paraspinal muscles with bilateral sparing of the multifidus (*arrow*, **A**). The psoas muscles are absent from their usual location (*arrowhead*, **A**). In the thigh **(B)** no hypertrophy is apparent, as may be expected in certain dystrophies, but focal fatty replacement of sartorius, gracilis, and the semimembranosus has occurred (*arrows*, from superior to inferior). The leg muscles are relatively preserved **(C)**.

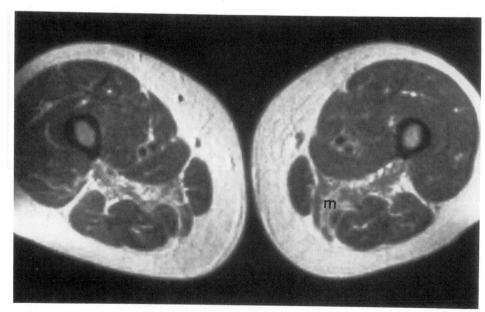

Figure 14.22. Atrophy and fatty replacement in metabolic myopathy. A 48-year-old man with an inherited defect in glycogenolysis (phosphofructokinase deficiency) and recurrent subclinical rhabdomyolysis had MR imaging of the thighs performed as a screening test for muscle abnormalities. A 2000/30 spin-echo image demonstrates fatty deposition and focal diminution of the adductor magnus *(m)* bilaterally. The case suggests that severe muscle necrosis may lead to atrophy and fatty replacement. (Reprinted with permission from Fleckenstein JL, Peshock RM, Lewis SF, Haller RG. Magnetic resonance imaging of muscle injury and atrophy in glycolytic myopathies. Muscle Nerve 1989;12(10):849–855.)

14. Dooms GC, Fisher MR, Hricak H, Higgins CB. MR imaging of intramuscular hemorrhage. J Comput Assist Tomogr 1985;9(5):908–913.

15. Farmlett EJ, Fishman EK, Magid D, Siegelman SS. Computed tomography in the assessment of myonecrosis. Can Assoc Radiol J 1987;38:278–282.

16. Fischer AG, Carpenter DW, Hartlage PL, Carroll JE, Stephens S. Muscle imaging in neuromuscular disease using computerized real-time sonography. Muscle Nerve 1988;11:270–275.

17. Fisher MR, Dooms GC, Hricak H, Reinhold C, Higgins CB. Magnetic resonance imaging of the normal and pathologic muscular system. Magn Reson Imag 1986;4:491–496.

18. Fleckenstein JL, Canby RC, Parkey RW, Peshock RM. Acute effects of exercise on MR imaging of skeletal muscle in normal volunteers. AJR 1988;15:231–237.

19. Fleckenstein JL, Bertocci LA, Nunnally RL, Parkey RW, Peshock RM. Exercise-enhanced MR imaging of variations in forearm muscle anatomy and use: importance in MR spectroscopy. AJR 1989;153:693–698.

20. Fleckenstein JL, Weatherall PT, Parkey RW, Payne JA, Peshock RM. Sports-related muscle injuries: evaluation with MR imaging. Radiology 1988;172:793–798.

21. Fleckenstein JL, Peshock RM, Lewis SF, Haller RG. Magnetic resonance imaging of muscle injury and atrophy in glycolytic myopathies. Muscle Nerve 1989;12(10):849–855.

22. Fleckenstein JL, Haller RG, Lewis SF, Peshock RM. Skeletal muscle size as a determinant of exercise performance: a new application of MRI [Abstract]. Proceedings of the 7th annual meeting of the Society of Magnetic Resonance in Medicine 1988;(1):192.

23. Fleckenstein JL, Archer BT, Barker BA, Vaughan JT, Parkey RW, Peshock RM. Fast, short tau inversion recovery MR imaging. Radiology 1991;179:499–504.

24. Fornage BD, Touche DH, Segal P, Rifkin MD. Ultrasonography in the evaluation of muscular trauma. J Ultrasound Med 1983;2:549–554.

25. Fornage BD, Nerot C. Sonographic diagnosis of rhabdomyolysis. J Clin Ultrasound 1986;14:389–392.

26. Fraser DD, Frank JA, Dalakas M, Miller FW. A comparative study of magnetic resonance imaging, muscle biopsy and clinical assessment in the idiopathic inflammatory myopathies [Abstract]. Proceedings of the 8th annual meeting of the Society of Magnetic Resonance in Medicine 1989;2:701.

27. Gaut P, Wong PK, Meyer RD. Pyomyositis in a patient with the acquired immunodeficiency syndrome. Arch Intern Med 1988;148:1608–1610.

28. Grau A, Pomes J, Davalos A, Gomez E, Sola P, Genis D. Computed tomography in acute alcoholic myopathy. J Comput Tomogr 1988;12:161–164.

29. Grindrod S, Tofts P, Edwards R. Investigation of human skeletal muscle structure and composition by x-ray computerised tomography. Eur J Clin Invest 1983;13:465–468.

30. Hadar H, Gadoth N, Heifetz M. Fatty replacement of lower paraspinal muscles: normal and neuromuscular disorders. AJR 141:895–898, 1983.

31. Hawley RJ, Schellinger D, O'Doherty, DS. Computed tomographic patterns of muscles in neuromuscular diseases. Arch Neurol 1984;41:383–387.

32. Heckmatt JZ, Leeman S, Dubowitz V. Ultrasound imaging in the diagnosis of muscle disease. J Pediatr 1981;101(5):656–660.

33. Heckmatt JZ, Dubowitz V. Diagnostic advantage of needle muscle biopsy and ultrasound imaging in the detection of focal pathology in a girl with limb girdle dystrophy. Muscle Nerve 1985;8:705–709.

34. Heckmatt JZ, Leeman S, Dubowitz V. Ultrasound imaging in the diagnosis of muscle disease. J Pediatr 1982;101(5):656–660.

35. Heckmatt JZ, Pier N, Dubowitz V. Real-time ultrasound imaging of muscles. Muscle Nerve 1988;11:56–65.

36. Heckmatt JZ, Pier N, Dubowitz V. Measurement of quadriceps muscle thickness and subcutaneous tissue thickness in normal children by real-time ultrasound imaging. J Clin Ultrasound 1988;16:171–176.

37. Heckmatt JZ, Pier N, Dubowitz V. Assessment of quadriceps femoris muscle atrophy and hypertrophy in neuromuscular disease in children. J Clin Ultrasound 1988;16:177–181.

38. Herfkens RJ, Sievers R, Kaufman L, et al. Nuclear magnetic resonance imaging of the infarcted muscle: a rat model. Radiology 1983;147:761–764.

39. Horber FF, Hoopeler H, Scheidegger JR, Grunig BE, Howald H, Frey FJ. Impact of physical training on the ultrastructure of mid-thigh muscle in normal subjects and in patients treated with glucocorticoids. J Clin Invest 1987;79:1181–1190.

40. Horikawa H, Konagaya M, Takayanagi T, Otsuji H. Quantitative analysis of muscular wastings of lower limbs in Duchenne muscular dystrophy by computed tomography [Abstract]. Muscle Nerve 1986;9(suppl):242.

41. Jiddane M, Gastaut JL, Pellissier JF, Pouget J, Serratrice G, Salamon G. CT of primary muscle disease. Am J Neuroradiology 1983;4:773–776.

42. Jolesz FA, Schwartz LH, Streter E, et al. Proton NMR of fast and slow twitch muscles and the effects of stimulation [Abstract]. Proceedings of the 5th annual meeting of the Society of Magnetic Resonance in Medicine 1986;2:444–445.

43. Jones DA, Round JM, Edwards RHT, Grindwood SR, Tofts PS. Size and composition of the calf and quadriceps muscles in Duchenne muscular dystrophy. J Neurol Sci 1983;60:307–322.

44. Kaftori JK, Rosenberger A, Pollack S, Fish JH. Rectus sheath hematoma: ultrasonographic diagnosis. AJR 1977;128:283–285.

45. Kakulas BA, Adams RD. Disease of muscle: pathological foundations of clinical myology. 4th ed. Philadelphia: Harper & Row, 1985.

46. Kaplan GN. Ultrasonic appearance of rhabdomyolysis. AJR 1980;134:375–377.

47. Kaschka WP, Druschky KF, Rott HD. Myotonic dystrophy: structural changes visualized by ultrasound. J Neurol 1987;234:122–123.

48. Kaufman LD, Gruber BL, Gerstman DP, Kaell AT. Preliminary observations on the role of magnetic resonance imaging for polymyositis and dermatomyositis. Ann Rheum Dis 1987;46:569–572.

49. Kuriyama M, Hayakawa K, Konishi Y, et al. MR imaging of myopathy. Comput Med Imag Graph 1989;13(4):329–333.

50. Laasonen EM. Atrophy of sacrospinal muscle groups in patients with chronic, diffusely radiating lumbar back pain. Neuroradiology 1984;26:9–13.

51. Lachiewicz PF, Hadler NM. Spontaneous pyomyositis in a patient with Felty's syndrome: diagnosis using computerized tomography. South Med J 1986;79(8):1047–1048.

52. Lamminen AE, Hekali PE, Tiula E, Suramo I, Korhola OA. Acute rhabdomyolysis: evaluation with magnetic resonance imaging compared with computed tomography and ultrasonography. Br J Radiol 1989;62:326–331.

53. Lamminen A, Jaaskelainen J, Rapola J, Suramo I. High-frequency ultrasonography of skeletal muscle in children with neuromuscular disease. J Ultrasound Med 1988;7:505–509.

54. Lamminen A, Tanttu JI, Sepponen RE, Suramo IJI. Magnetic resonance imaging at 0.02 T: evaluation of muscular dystrophies and congenital myopathies [Abstract]. Proceedings of the 8th annual meeting of the Society of Magnetic Resonance in Medicine 1989;2:702.

55. Le Rumeur E, de Certaines J, Toulouse P, Rochcongar P. Water phases in rat striated muscles as determined by T2 proton NMR relaxation times. Magn Reson Imag 1987;5:267–272.

56. Matsumura K, Nakano I, Fukuda N, Ikehira H, Tateno Y, Aoki Y. Proton spin-lattice relaxation time of Duchenne dystrophy skeletal muscle by magnetic resonance imaging. Muscle Nerve 1988;11:97–102.

57. McLoughlin J. CT and percutaneous fine-needle aspiration biopsy in tropical myositis. AJR 1980;134:167–168.

58. Medici M, deTenyi A, Vincent O, Tome F. Muscle computed axial tomography in oculopharyngeal dystrophy [Abstract]. Muscle Nerve 1986;9(suppl):244.

59. Mielke U, Ricker K, Emser W, Boxler K. Unilateral calf enlargement following S1 radiculopathy. Muscle Nerve 1982;5:434–438.

60. Moore DL, Delage G, Labelle H, Gauthier M. Peracute streptococcal pyomyositis. J Pediatr Orthoped 1986;6:232–235.

61. Murphy WA, Totty WG, Carroll JE. MRI of normal and pathologic skeletal muscle. AJR 1986;146:565–574.

62. Naegele M, Reimers CD, Fenzl G, Witt TN, Pongratz DE, Hahn

D. Muscular imaging (CT, MRI, ultrasound) in inflammatory myopathies: differential diagnostic aspects [Abstract]. Magn Reson Imag 1989;7(1):171.

63. O'Doherty DS, Schellinger D, Raptopoulos V. Computed tomographic patterns of pseudohypertrophic muscular dystrophy: preliminary findings. J Comput Assist Tomogr 1977;1:482–486.

64. Polak JF, Jolesz FA, Adams DF. Magnetic resonance imaging of skeletal muscle prolongation of T1 and T2 subsequent to denervation. Invest Radiol 1988;23:365–369.

65. Poon CS, Szumowski J, Plewes DB, Ashby P, Henkelman RM. Fat/water quantitation and differential relaxation time measurement using chemical shift imaging technique. Magn Reson Imag 1989;7:369–382.

66. Raphael SA, Wolfson BJ, Parker P, Lischner HW, Faerber EN. Pyomyositis in a child with acquired immunodeficiency syndrome. Am J Dis Child; 1989;143:779–781.

67. Reimers CD, Naegele M, Fenzl G, Muller W, Witt TN, Hahn D. Muscular imaging (CT, MRI, ultrasound) in degenerative myopathies [Abstract]. Magn Reson Imag 1989;7(1):55.

68. Rickards D, Isherwood I, Hutchinson R, Gibbs A, Cummings WJK. Computed tomography in dystrophia myotonica. Neuroradiology 1982;24:27–31.

69. Rodiek S.O. CT, MR-Tomographie und MR-Spektroskoipie bei Neuromuskularen Erkrankungen. Stuttgart: Enke, 1987.

70. Shabas D, Gerard G, Rossi D. Magnetic resonance imaging examination of denervated muscle. Comput Radiol 1987;11(1):9–13.

71. Stern LM, Caudrey DJ, Clark MS, Perrett LV, Boldt DW. Carrier detection in Duchenne muscular dystrophy using computed tomography. Clin Genet 1985;27:392–397.

72. Stern LM, Caudrey DJ, Perrett LV, Boldt DW. Progression of muscular dystrophy assessed by computed tomography. Dev Med Child Neurol 1984;26:569–573.

73. Termote JL, Baert A, Crolla D, Palmers Y, Bulcke JA. Computed tomography of the normal and pathologic muscular system. Radiology 1980;137:439–444.

74. Tumeh SS, Butler GJ, Maguire JH, Nagel JS. Pyogenic myositis: CT evaluation. J Comput Assist Tomogr 1988;12(6):1002–1005.

75. Uppal G, Dewbury KC, Dennis NR. Carrier detection in DMD. Clin Genet 1986;31(1):62–63.

76. Vartian C, Septimus EJ. Pyomyositis in an intravenous drug user with human immunodeficiency virus [Letter]. Arch Int Med 1988;148:2689.

77. Vincent LM. Ultrasound of soft tissue abnormalities of the extremities. Radiol Clin Am 1988;26(1):131–144.

78. Vliet AM, Thijssen HOM, Joosten E, Merx JL. CT in neuromuscular disorders: a comparison of CT and histology. Neuroradiology 1988;30:421–425.

79. von Rohden L, Wiemann D, Steinbicker V, Krebs P, Köeditz H. Diagnosis of malignant hyperthermia in childhood and adulthood by ultrasound [Abstract]. Paediatric Radiology 1988;18:260.

80. Vukanovic S, Hauser H, Wettstein P. CT localization of myonecrosis for surgical decompression. AJR 1980;135:1298–1299.

81. Vukanovic S, Hauser H, Curati WL. Myonecrosis induced by drug overdose: pathogenesis, clinical aspects and radiological manifestations. Eur J Radiol 1983;3:314–318.

82. Weinberg WG, Dembert ML. Tropical pyomyositis: delineation by gray scale ultrasound. Am J Trop Med Hyg 1984;33(5):930–932.

83. Whyte AM, Lufkin RB, Bredenkamp J, Hoover L. Sternocleidomastoid fibrosis in congenital muscular torticollis: MR appearance. J Comput Assist Tomogr 1989;13(1):163–166.

84. Yeh H, Rabinowitz JG. Ultrasonography of the extremities and pelvic girdle and correlation with computed tomography. Radiology 1982;143:519–525.

85. Yuh WTC, Schreiber AE, Montgomery WJ, Ehara S. Magnetic resonance of pyomyositis. Skel Radiol 1988; 17:190–193.

86. Zagoria RJ, Karstaedt N, Koubek TD. MR imaging of rhabdomyolysis. J Comput Assist Tomogr 1986;10(2):268–270.

CHAPTER 15

Magnetic Resonance Imaging and Computed Tomography of Primary Musculoskeletal Tumors

Johan L. Bloem, Herma C. Holscher, and Anton H. M. Taminiau

Introduction

Until recently, a mutilating surgical procedure has been the most effective method of treating the majority of primary malignant tumors originating in the musculoskeletal system. Surgery not only has a major impact on survival but also on quality of life. Treatment has been improved considerably by effective chemotherapy and by modern operation techniques, often allowing reconstructive and limb salvage procedures instead of amputation or disarticulation (1–8). Only after meticulous preoperative staging is it possible to execute limb salvage surgery, resulting in control of the primary tumor and good residual function (1, 9–20). Other major factors that are important in therapy planning are histologic grade of the tumor, sensitivity to (neo)adjuvant chemotherapy, and patient compliance. Five diagnostic imaging procedures are essential for effective patient management.

1. Detection of tumor and determination of specific diagnosis: High-quality radiographs are as a rule sufficient for detection of tumors. Computed tomography (CT) (vertebral column, pelvis), ultrasound (soft tissue), magnetic resonance (MR) imaging (bone marrow and soft tissue), and bone scintigraphy are sometimes needed to detect tumors when radiographs are negative or equivocal. A radiologic diagnosis is initially made on plain radiographs. At present, CT and MR imaging have not been able to increase specificity substantially, for instance on the basis of T1/T2 measurements (1, 21–25). As will be shown, MR imaging and CT are sometimes able to suggest a specific diagnosis, but this is an exception rather than the rule. A

needle or open biopsy is, with some exceptions (such as fibrous cortical defect or nonossifying fibroma), necessary to make a histologic diagnosis. Histologic and radiologic findings combined are used to make a final diagnosis. Imaging studies are, when staging is needed, performed prior to biopsy and can thus guide the biopsy needle.

2. Local tumor staging: Reconstructive, limb salvage procedures can be performed only when the surgeon is informed in detail about the intra- and extraosseous extension of tumor growth (1, 9, 10, 26). Because this is relevant only in primary tumors we will in our discussion exclude secondary tumors. Magnetic resonance imaging can be used to stage benign tumors. This is, however, of limited clinical importance since these well-contained benign tumors are usually accurately staged by radiographs and CT. The major indication for MR imaging is staging primary malignant musculoskeletal tumors. The accuracy of MR imaging in staging these tumors in relation to conventional modalities such as CT, technetium-99m (^{99m}Tc)-labeled methylene diphosphonate (MDP) scintigraphy, and angiography will be discussed.

3. Monitoring of preoperative (neoadjuvant) chemotherapy: This is important to determine the effect of chemotherapy and to select the proper time for surgery. Magnetic resonance imaging is useful for its ability to monitor the effect of chemotherapy. Accurate monitoring of the effect of chemotherapy may also have an impact on planning postoperative (adjuvant) chemotherapy.

4. Search for disseminated disease: ^{99m}Tc-MDP is the method of choice for screening for osseous metastases, because of its capability to image the whole skeleton. However, MR imaging is superior to planar bone scintigraphy in the detection of metastases in the vertebral column (see also Chapter 16). Since primary musculoskeletal tumors metastasize most frequently to the lungs, CT of the chest is a routine procedure in our patients.

5. Detection of recurrent disease: Clinical examination is not sufficient to detect early recurrent disease, and tumor markers are not available. Imaging studies will be needed to manage the problem of tumor recurrence. The application of imaging studies depends on the histologic grade of the primary tumor, its location, type of margins achieved at surgery, and therapeutic options when recurrence is detected.

Abbreviations: SE, spin echo; DTPA, diethylenetriamine pentaacetic acid; FFE, fast field echo; MDP, methylene diphosphonate.

Technique

When a bone lesion is potentially malignant, imaging studies (to allow local staging) and biopsy are needed. Imaging studies are performed prior to biopsy, because they can be used to plan the biopsy procedure. Furthermore, biopsy induces reactive changes (edema, hemorrhage), which interfere with accuracy of staging.

Magnetic Resonance Imaging Technique

A high field strength is not mandatory. It is possible to stage bone tumors accurately at 0.5 T (1, 10, 11, 27). Image quality and accuracy will, however, improve at higher fields (1 to 1.5 T) because of higher spatial resolution. Whenever possible, surface coils or head coils should be used to increase spatial resolution. Large field of view images can be used to detect skip lesions. It is important to use both T1- and T2-weighted pulse sequences. This is accomplished in our institution by using short repetition time/echo time (TR/TE) spin-echo (SE) pulse sequences (TR 550 ms, TE 20 to 30 ms) for T1 weighting and long TR/TE sequences (TR 1500 to 2000 ms, TE 50 to 100 ms) for T2 weighting at 0.5 or 1.5 T (Philips, Shelton, Conn.) (1, 10, 11).

T1-weighted sequences are used for intramedullary staging and T2-weighted sequences are used for defining soft-tissue extension and cortical involvement. Inversion recovery or short τ (TI) inversion recovery (STIR) images are rarely used, but may occasionally be necessary for detection or intraosseous staging, especially when the short TR/TE SE sequence does not provide enough T_1 weighting. The STIR sequence is an inversion recovery technique designed to accentuate the signal intensity of bulk water of pathologic processes (28, 29). The prolonged T1 and T2 relaxation times of the pathological process are added while signal from fat (bone marrow) is nulled. An example of a STIR sequence is TR 1500 ms, TE 30 ms, and TI 100 to 150 ms.

Routinely, multiple imaging planes are used: the transverse plane (T2 weighted) for soft-tissue extension and sagittal or coronal planes (T1 weighted) for intraosseous extension. Additional planes are used when necessary, for instance the axial plane for intraosseous staging to rule out partial volume effects of cortical bone, which may occur in the sagittal and coronal plane in thin tubular bones in children. Slice thickness varies between 5 and 10 mm and number of excitations will be one to four, depending on TR, TE, type of coil, and field strength used.

Gadolinium-labeled diethylenetriamine pentaacetic acid (DTPA)-enhanced studies are performed following intravenous injection of 0.1 to 0.2 mmol Gd-DTPA/kg body weight with SE technique (TR 250 to 600 ms, TE 20 to 30 ms), and gradient-recalled echo or fast field echo (FFE) imaging (30–33). Enhancement is evaluated on dynamic gradient-recalled echo images with large flip angles to emphasize T1 contrast. We use a flip angle of 80°, TR 100 ms, and TE 9.2 ms. These images are made every 30 sec. Numerous gradient echo sequences, for instance with shorter TE, spoiler gradients, and inversion pulses can be used to increase T1 weighting.

The high density of Gd-DTPA is seen in the renal collecting system on CT scans, if they are made after a Gd-DTPA enhanced MR imaging study (34). Gd-DTPA can be used in patients with a poor renal function, and can be removed from the body with dialysis (35).

Gradient-recalled echo techniques with a small flip angle can also be used to obtain images with proton density or T2* contrast. However, contrast between tumor and muscle on T2-weighted SE images is superior to that on native or enhanced gradient-recalled images (32, 36).

Computed Tomography Technique

Computed tomography is mainly used to evaluate benign lesions. The patient is in the supine position and when the lower extremity is imaged both legs are included in the field of view to allow comparison. The data are stored to allow zoom reconstruction when necessary. Following a scout view, transverse images are made with a slice thickness ranging from 1.5 to 10 mm and an increment of 1 to 20 mm. The slice thickness used depends on the tumor volume and the location of the tumor. An osteoid osteoma located in the cervical spine is, for instance, evaluated with 1.5-mm thick slices, and preferably with overlap, i.e., an increment of 1 mm (Fig. 15.1). We use a matrix size of 215×215 or 512×512 and zoom reconstruction when needed (field of view ranges from 5 to 30 cm). Thus an in-plane resolution of 0.1 to 1.2 mm is obtained. A bone algorithm is used for reconstruction. Images are reviewed with bone and soft-tissue window settings.

Intravenous contrast agents are only used to visualize major vascular bundles when necessary, or to evaluate the vascularization of the lesion under observation.

Magnetic Resonance Imaging and Computed Tomography Characteristics

As most pathologic tissues, primary malignant bone tumors usually have prolonged T1 and T2 relaxation times (1, 21, 22, 25, 26). Therefore osteolytic tumors have a relatively high signal intensity on T2-weighted images and a relatively low signal intensity on T1-weighted images (Fig. 15.2). Osteosclerotic components or calcifications within the tumor have because of their low spin density and short T2 relaxation time a low signal intensity on both T1- and T2-weighted images. Small calcifications detected by plain radiographs or CT often cannot be detected by MR imaging. Liquefaction or lacunae caused by necrosis within a tumor can be identified because T1 and T2 relaxation times are even longer than those of viable tumor.

The use of Gd-DTPA-enhanced T1 weighted images facilitates differentiation between viable tumor tissue, reactive changes (edema), necrosis, and so on (30–33). Identification of the aforementioned tumor components, and especially viable tumor, may be used to guide the biopsy needle (Fig. 15.3).

Only in selected cases does MR imaging allow a specific diagnosis to be made. The inhomogeneity of the tumor, consisting of viable tumor with different histologic components, necrosis, hemorrhage, and reactive changes make it impossible to differentiate histologic types by relaxation

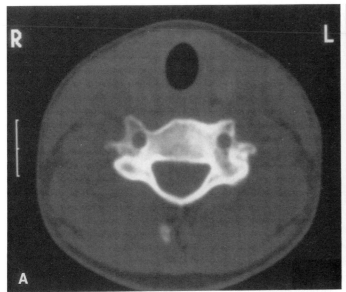

Figure 15.1. Osteoid osteoma of the cervical spine. **(A)** This 3-mm thick slice through the cervical spine shows the abnormality in the left pedicle, but the nidus is not well defined. **(B)** The nidus with central calcification is easily appreciated on this 1.5-mm thick slice.

times. Magnetic resonance imaging increases specificity only when unusual features such as high signal intensity on T1-weighted images, low signal intensity on T2-weighted images, or specific morphology or enhancement patterns are observed.

High Signal Intensity on T1-Weighted Images

Subacute hematoma has a high signal intensity on T1-weighted images (37). The signal intensity of hematoma depends on the sequential degradation of hemoglobin, cell lysis, and field strength of the MR system (38). Interstitial hemorrhage is more diffuse than hematoma and is accompanied by edema, resulting in nonspecific prolonged T1 and T2 relaxation times irrespective of age of the hemorrhage.

In the first hours of the hyperacute phase of hematoma, oxygenated hemoglobin determines the signal intensities. Oxygenated hemoglobin is not paramagnetic, and the signal intensities are not characteristic. Hyperacute hematoma usually has an intermediate signal intensity similar to muscle on T1-weighted sequences and a high signal intensity relative to muscle on T2-weighted sequences.

In the acute phase (first days, less than 1 week) intracellular oxyhemoglobin is converted into intracellular deoxyhemoglobin; this does not affect the signal intensity on T1-weighted images. The signal intensity on T2-weighted images depends on the field strength. At high field the intracellular paramagnetic deoxyhemoglobin causes susceptibility effects that induce shortening of T2-relaxation time, resulting in low signal intensity. A difference in magnetic susceptibility exists across the cell membrane. Protons diffusing through this cell membrane experience a change of field strength resulting in phase shifts that enhance T2 relaxation. The gradient is proportional to the field strength. This phenomenon of rapid decay of T2

magnetization does not occur at low field when spin-echo images are used. Therefore signal intensity of acute hematoma at low or midfield will be similar to that of hyperacute hematoma. The phase shifts may be observed only at low field when gradient-recalled images are obtained. The gradient across the cell membrane disappears after cell lysis has occurred.

Subsequently deoxyhemoglobin is oxidized into intracellular methemoglobin. Because of cell lysis intracellular methemoglobin becomes free methemoglobin. The influence of methemoglobin on signal intensity depends on its concentration; in the subacute phase hematoma will usually display an increase of signal intensity on T1- and T2-weighted images. This high signal intensity is first observed in the periphery of the hematoma (Fig. 15.4). The low signal intensity outer rim, seen on T1- and T2-weighted sequences, represents the presence of hemosiderin-laden macrophages (susceptibility effect).

Hyperacute or acute hematoma may be identified on CT by virtue of an increased density. The increased density is, however, as a rule not as marked as the density of intracerebral hematoma.

Although hemorrhage and hematoma may occur in any tumor or hemophiliac pseudotumor (Fig. 15.5), telangiectatic osteosarcoma and aneurysmal bone cyst are the lesions to think of when hemorrhage is found (1, 26, 39). Telangiectatic osteosarcoma will display malignant features such as indistinct margins and inhomogeneity, whereas aneurysmal bone cyst often contains multiple compartments and displays well-defined margins (Fig. 15.6) (40). Layering or a fluid level, as can be demonstrated with MR imaging and CT (high density in the most dependent part of the lesion), are more often seen in aneurysmal bone cyst than in telangiectatic osteosarcoma. These fluid levels may also be encountered in chondroblastoma, giant cell tumor, especially when a secondary aneurysmal bone cyst is present, and occasionally in other tumors as well (41).

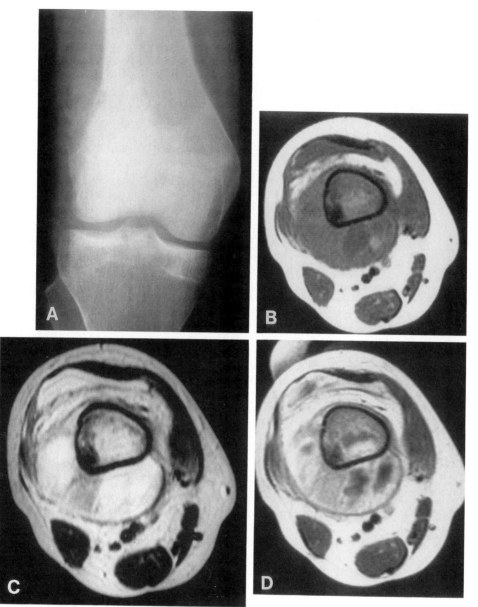

Figure 15.2. **(A)** Classic osteosarcoma of the distal femur. **(B)** Axial T1-weighted image (SE 550/30). Tumor has an inhomogeneous low to intermediate signal intensity. The intraosseous signal void represents osteosclerosis within the tumor. **(C)** T2-weighted image (SE 2000/100). The sclerotic part of the tumor has a persistent low signal intensity as opposed to the inhomogeneous signal intensity of the nonsclerotic part of the tumor. Note also the high signal intensity of joint effusion in the suprapatellar bursa. **(D)** Same pulse sequence as in **(A)**. After administration of Gd-DTPA viable tumor tissue enhances. The intraosseous sclerotic part and the liquefied necrotic areas in the soft-tissue component do not enhance.

Another cause for hematoma is surgical intervention. Since hematoma and edema may be quite extensive following needle or open biopsy, imaging studies are preferably performed prior to invasive procedures.

Round cell tumors (such as Ewing's sarcoma and lymphomas) and other well-vascularized tumors such as metastases from renal cell carcinoma may exhibit a signal intensity on T1-weighted sequences that is higher than usual for bone tumors. However, the signal intensity characteristically is not as high as that of hematoma and does usually not allow a specific diagnosis to be made (Fig. 15.7).

High signal intensity may also be encountered in li-poma (1, 42). On T2-weighted images the signal intensity will remain identical to that of subcutaneous fat (Fig. 15.8). Therefore these lesions may be missed when located in subcutaneous fat. Deep lipomas are either intra- or intermuscular. Intramuscular lipomas may, because of inhomogeneity and indistinct interface with muscle, mimic a malignant tumor. Intraosseous lipomas are displayed on CT as low-density lesions surrounded by a thin sclerotic margin. Osseous ridges and a central calcified nidus may be encountered. Liposarcomas may contain areas consisting of well-differentiated fat; these areas have the same appearance on MR imaging and CT as benign lipomas.

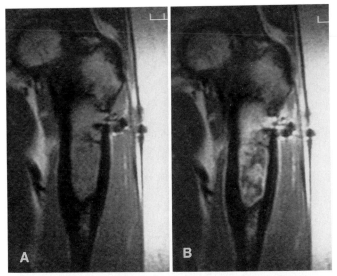

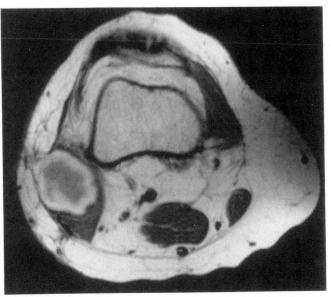

Figure 15.3. MR images of chondrosarcoma obtained after biopsy. **(A)** Susceptibility artifacts due to presence of metallic particles are seen on lateral side as low signal intensity areas with bright halo. Tumor has a low to intermediate signal intensity on this T1-weighted (SE 550/30) image. **(B)** Enhancement of tumor is most pronounced in cephalic part, indicating high cellularity. The poorly enhanced distal part represents area of paucicellular mucoid degeneration. The homogeneous enhancement is atypical, whereas the enhancement of septations and cellular areas in the distal part is typical for cartilage (see text).

Figure 15.4. Subacute hematoma on T1-weighted (SE 550/30) image after tumor resection. The low signal intensity center is surrounded by a high signal intensity in the periphery (methemoglobin). The thin, low signal intensity margin represents hemosiderin.

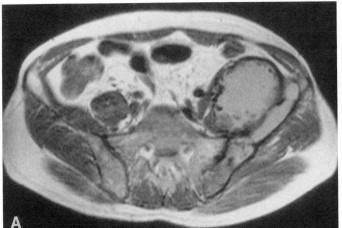

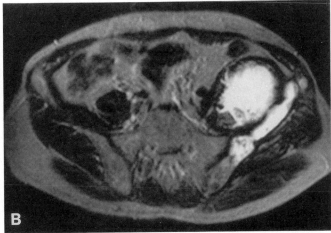

Figure 15.5. Hemophiliac pseudotumor located in the iliac bone and iliopsoas muscle. Signal intensity is relatively high on T1-weighted image (**A**, SE 550/30) and very high on T2-weighted image (**B**, SE 2000/

100). Hemosiderin in the margin has a low signal intensity on both sequences.

Liposarcomas, however, invariably contain large (myxoid liposarcoma) or some (lipoblastic liposarcoma) areas of poorly differentiated fat and mesenchymal tissue, which have the same MR imaging and CT characteristics as other high-grade malignancies (Fig. 15.9) (1, 42).

Hemangiomas more often are single than multiple (43). Histomorphology of hemangiomas in bone does not differ from that in soft tissues. Localized hemangiomas have been classified as capillary, cavernous, arteriovenous, and venous (44). They are typically of the mixed capillary-cavern-

ous type. Usually the capillary (capillaries with sparse stroma) or cavernous (large sinusoidal blood-filled spaces) component is predominant. Soft-tissue hemangiomas are usually found in skin and subcutaneous tissue, but intramuscular hemangiomas do occur (45, 46). The signal intensity of soft tissue or intraosseous hemangioma is variable (43, 45–47). The signal intensity may be high on T1-weighted sequences due to adipose tissue within the lesion and due to slow flow through dilated sinuses and vessels (47). Because of slow flow the serpiginous vascular

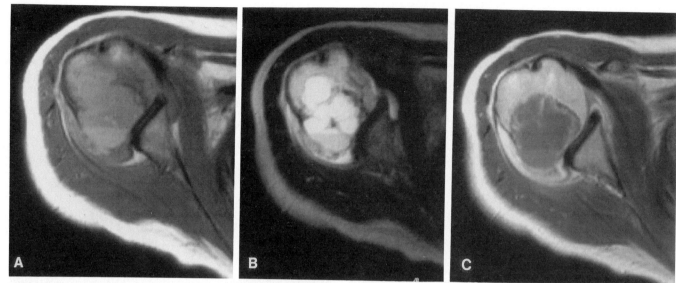

Figure 15.6. Aneurysmal bone cyst of the proximal humerus. **(A)** High signal intensity of hemorrhagic fluid is seen in dependent part of lesion on this T1-weighted (SE 550/30) image. A fluid level is depicted. Compression on infraspinatus muscle is appreciated. **(B)** Edema in the infraspinatus muscle and the ventral loculated part of the lesion exhibit a higher signal intensity than hemorrhagic fluid on this T2-weighted image (SE 2000/100). **(C)** Septations within the lesion enhance after Gd-DTPA injection. The cystic components do not enhance. (Case courtesy of Dr. W. Koops, Dijkzicht Hospital, Rotterdam.)

Figure 15.7. Ewing's sarcoma originating in the femur diaphysis has a signal intensity on this T1-weighted image (SE 550/30) that is slightly higher relative to muscle than is usually found in bone sarcoma. Note the low signal intensity line within the soft-tissue component of the tumor, representing elevated periosteum.

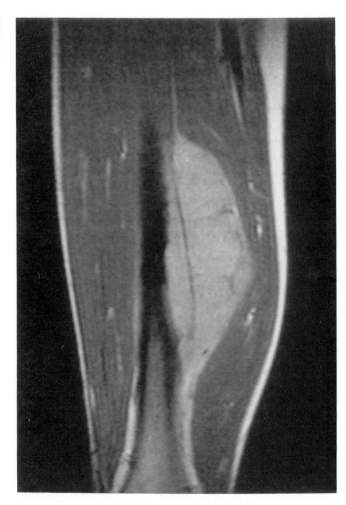

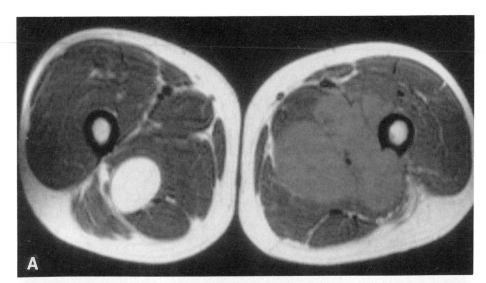

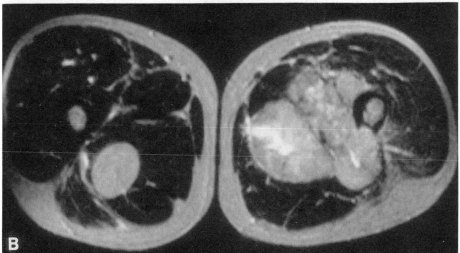

Figure 15.8. Patient analyzed for mass in left leg. The symptomatic malignant fibrous histiocytoma of the left leg has a nonspecific low signal intensity on T1-weighted image (**A,** SE 550/30) and an inhomogeneous high signal intensity on the T2-weighted image (**B,** SE 2000/100). The asymptomatic lipoma of the right leg has a homogeneous signal intensity identical to that of subcutaneous fat on both sequences.

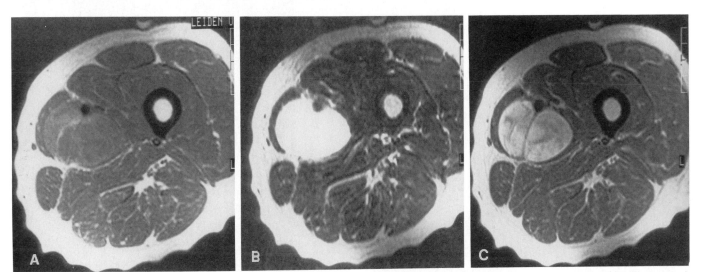

Figure 15.9. A 54-year-old female with a liposarcoma in the adductor compartment of the left leg. **(A)** Contrast between the nonspecific intermediate signal intensity of the liposarcoma and surrounding muscle is poor on this axial T1-weighted (TR 550/TE 30) image. Note the signal void of the large feeding and draining vessels. **(B)** On this T2-weighted image (2000/100) contrast between nonspecific high signal intensity of tumor and muscle is higher than in **(A)** and **(C)**. **(C)** A marked inhomogeneous enhancement after Gd-DTPA injection is easily appreciated. As a result the internal structure of this well-vascularized tumor can be evaluated. Contrast is sufficient but less than on the T2-weighted image. (Reprinted with permission from Bloem JL, Reiser MF, Vanel D. Magn Reson Q 1990;6:136–163.)

channels typically display a high signal intensity on T2-weighted images (Figs. 15.10 and 15.11). Angiography is, because of slow flow in the lesion, inferior to MR imaging in demonstrating venous hemangiomas. Flow void phenomena representing rapid flow are marked in arteriovenous malformations, but some vessels are often also depicted in other hemangiomas.

Vertebral hemangioma is also easily identified on CT because of the thickened trabeculae. Fat, as can be identified on CT or MR imaging, predominates in asymptomatic vertebral hemangiomas. Vascularity is often more pronounced in symptomatic hemangiomas (48). Increased vascularity is indicated by increased density, relative to fat, on CT and decreased signal intensity, relative to fat, on T1-weighted MR images.

Another lesion that may display an intermediate or high signal intensity on T1-weighted images because of adipose components is benign or malignant schwannoma. The signal intensity may also be low on T1-weighted sequences, depending on the amount of fat present (49).

Low Signal Intensity on T2-Weighted Images

A low spin density and short T2 relaxation time, as are present in calcification, osteoid, and bone cement, result in a signal void that is identical to that of cortical bone (Figs. 15.2 and 15.12). These signal void structures are characterized by a high density on CT and radiographs, therefore these MR signal characteristics do not usually increase specificity as compared to radiography and CT.

Hemosiderin deposition is characterized by a low signal intensity on T2-weighted images (37). This may allow a specific diagnosis to be made when combined with other characteristics. Low signal intensity areas within a focal intraarticular lesion with or without focal osseous destruction are indicative for pigmented villonodular synovitis (Fig.

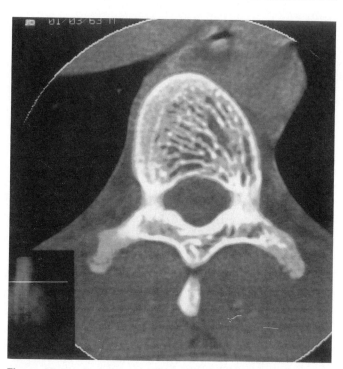

Figure 15.11. Typical coarse trabeculation of hemangioma within vertebral body extending into pedicle and lamina. (Case courtesy of Netherlands Committee on Bone Tumors.)

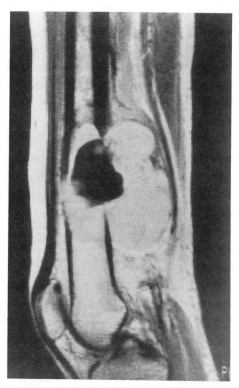

Figure 15.12. T2-weighted (1800/50) sagittal image of fibrosarcoma located in the femur diaphysis. The tumor has a high signal intensity. The signal void in the center is caused by bone cement, which was used to stop bleeding during the biopsy procedure. (Reprinted with permission from Bloem JL, Bluemm RG, Taminiau AHM, et al. Radiographics 1987;7:425–446.)

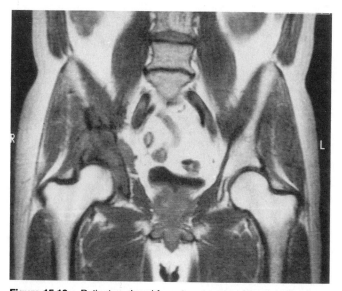

Figure 15.10. Patient analyzed for osteosarcoma of the right iliac bone, coronal T1-weighted image (550/30). The inhomogeneous high signal intensity hemangioma of the third lumbar vertebral body was an incidental finding.

15.13) (50). High signal intensity on T1-weighted images caused by fat-laden macrophages or subacute hematoma may also be encountered in pigmented villonodular synovitis. In disorders such as hemophilia or rheumatoid arthritis diffuse cartilage destruction accompanies hemosiderin deposition (51).

Giant cell tumors of tendon sheath may have the nonspecific signal intensities of ganglia, cysts, or bursae, but typically are characterized by low signal intensity areas on T2-weighted images. This is related to the presence of hemosiderin and fibrous stroma. The distribution along tendons is a feature of both ganglia and giant cell tumors.

Low signal intensity areas on T2-weighted images may also be present in fibromatoses, centrally in sarcoid, fibrous dysplasia, neuroma (Morton neuroma), and others (Fig. 15.14). The signal intensity reflects the amount of collagen and fibrous tissue present (49, 52). Cellular areas will be seen as high signal intensity areas on T2-weighted im-

ages, whereas the paucicellular areas will exhibit a low signal intensity on all pulse sequences.

The fibromatoses have been classified by Enzinger and Weiss as superficial (fascial) fibromatosis and deep (musculoneuronic) fibromatosis (44). The superficial fibromatoses are slow growing, whereas deep fibromatosis, also referred to as aggressive fibromatosis or desmoid tumor, grows rapidly. Aggressive fibromatosis and superficial fibromatosis display the same MR imaging characteristics (53). Usually fibromatosis demonstrates, on T2-weighted images, a heterogeneous, high signal intensity with low signal intensity areas in between. However, both sides of the spectrum (predominantly high or low signal intensity on T2-weighted images) may be encountered. Although aggressive fibromatosis is a benign entity that does not metastasize, local growth is infiltrative and not unlike malignant lesions. This infiltrative growth is reflected in the high recurrence rate and indistinct margins on MR im-

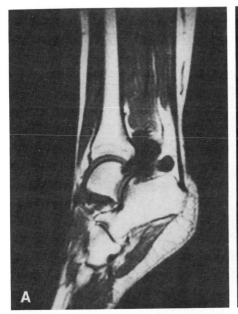
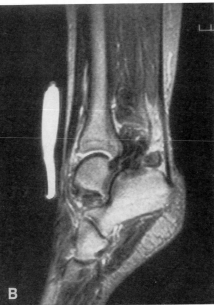

Figure 15.13. Pigmented villonodular synovitis of the tibiotalar joint. **(A)** Sagittal view (550/30). The lobulated lesion located posterior to the joint is characterized by a low signal intensity on this T1-weighted image. The lesion erodes the talus and calcaneus. **(B)** On this T2-weighted image (2000/100) the central part of the lesion exhibits a low signal intensity, indicating a short T2 relaxation time consistent with hemosiderin deposition. The periphery of the lesion has a somewhat higher signal intensity than in **(A)**. (Reprinted with permission from: Bloem JL, Taminiau AHM, Bloem RM. In: Falke THM, ed. Essentials of clinical MRI. The Hague: Martinius Nijhoff, 1988:203–217.)

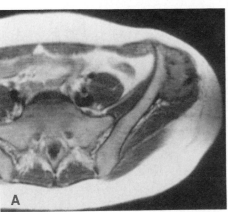
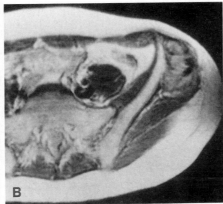
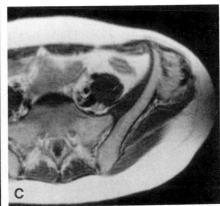

Figure 15.14. Recurrence of aggressive fibromatosis in the gluteal muscle compartment, extending into iliac bone and abdominal wall. **(A)** The lesion has a mixed low and intermediate signal intensity on this T1-weighted image (550/30). **(B)** When T2 contrast is emphasized (2000/100) the cellular component increases in signal intensity, whereas the paucicellular area on the lateral side still has a low signal intensity. **(C)** Increase of signal intensity after Gd-DTPA injection is observed only in the cellular area. Fibrotic paucicellular area does not enhance. Contrast on **(B)** is very similar to that on the enhanced image.

Figure 15.15. Aggressive fibromatosis in the adductor compartment of the left leg has an inhomogeneous, relatively high density on this nonenhanced CT scan. Note marked atrophy of the muscles and cortical thinning on the involved site.

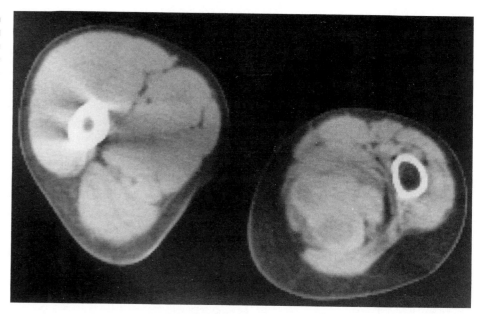

ages. Sometimes well-defined margins, seen on MR images, may falsely suggest the absence of infiltrative growth. The density of aggressive fibromatosis on precontrast CT scans may vary from hyper- to hypodense relative to muscle (Fig. 15.15) (54). Hyperdense areas within the lesion are rather suggestive for fibrous tissue; this may, however, also be seen in and around tumors that trigger a strong desmoplastic reaction, such as lymphomas and some metastatic carcinomas. The cause for hyperdensity on the precontrast scan is not known; it may be related to increased Compton scatter caused by a high electron volume density (54).

Fibrous dysplasia may exhibit a low signal intensity on T2-weighted images. This is caused by the dense collagenous matrix and mineralized fields of woven bone. However, high signal intensity is not unusual, as cartilage and cellular and cystic areas are not infrequently found in fibrous dysplasia (55–58). Benign myxomas are infrequently found in association with fibrous dysplasia. They are usually homogeneous and encapsulated, but may infiltrate musculature. The signal intensity is not characteristic.

Low signal intensity masses located within ligaments (Achilles tendon) representing xanthomas are often encountered in patients suffering from familial hypercholesterolemia. This is caused by deposition of cholesterol (large, relatively immobile molecules) in combination with dense collagen fibers.

The flow void phenomenon, representing rapid flow in vascular channels such as found in arteriovenous malformations, or to a lesser extent in capillary-venous hemangiomas, may also be a cause for low signal intensity on T2-weighted spin-echo images (Fig. 15.16) (59).

Specific Morphology

Morphology, distribution, and localization of disorders can also be used in MR imaging and CT to characterize a le-

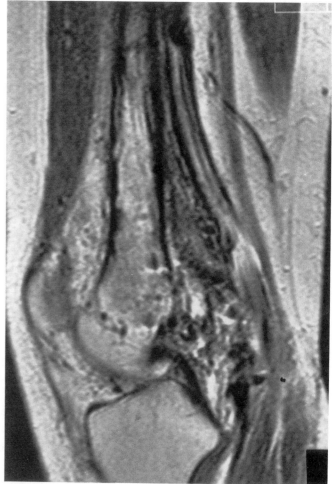

Figure 15.16. The flow void areas within the femur and atrophic soft tissues represent vascular channels with rapid flow in this patient with an arteriovenous malformation.

sion. Indistinct margins and heterogeneity are in favor of malignancy, whereas well-defined margins and homogeneity are more in favor of a benign lesion. There are of course exceptions to these general statements. Aggressive fibromatosis is a benign lesion characterized by infiltrative growth and a very high recurrence rate. Other benign lesions may on occasion be heterogeneous with indistinct margins, just as malignant lesions may be homogeneous with well-defined margins (Fig. 15.17).

The high spatial resolution of CT is sometimes needed to identify typical features of benign lesions. The nidus of osteoid osteoma can consistently be detected with CT, provided a thin (1.5 mm) slice thickness is used. This nidus may not be detected with MR imaging. The extremely thin cortical envelope of aneurysmal bone cyst or giant cell tumor may be differentiated from cortical breakthrough only with the aid of thin CT slices.

Edema, characterized by an increased signal intensity on T2-weighted images, a decreased signal intensity on T1-weighted images, and indistinct margins (feathery appearance), is a poor indicator of malignancy. Extensive soft-tissue and bone marrow edema is frequently encountered in osteoid osteoma, osteoblastoma, and chondroblastoma (Fig. 15.18). Edema in these benign entities is often more pronounced than in malignant tumors (Fig. 15.19). Peritumoral high signal intensity in the soft tissue of patients with primary malignant tumors may represent only edema (73%) or a reactive zone also containing tumor (27%) (60). Intraosseous edema is found in and around many lesions including fracture, bone bruise, necrosis, transient osteoporosis, etc.

Cartilaginous tumors are often characterized by their lobulated morphology, representing cartilage nodules separated by septation (1). Cartilage may have an intermediate to high signal intensity on T1-weighted images and a high signal intensity on T2-weighted images (Figs. 15.20 to 15.22). The cartilage cap is as a rule much better visualized with MR imaging than with CT, unless the cap is

calcified. Changes in thickness of the cartilage cap are easily measured on MR images. Growth in the adult or a thickness of 2 cm or more is by some authors considered to be a sign in favor of malignancy (61, 62). Calcification may be of value in differentiating juxtacortical osteosarcoma from osteochondroma. In juxtacortical osteosarcoma calcification or mineralization of the tumor is most pronounced in the center of the exophytic tumor, whereas calcification in osteochondroma, subperiosteal hematoma, or myositis ossificans is usually most pronounced in the periphery. This is referred to as the zonal sign of CT (Figs. 15.22 and 15.23) (63).

Intraarticular cartilaginous lesions such as dysplasia epiphysialis hemimelica, also called Trevor's disease, may be diagnosed with CT or MR imaging (Fig. 15.24). Calcified cartilaginous bodies in chondromatosis are seen as low signal intensity bodies, whereas the high signal intensity bodies represent bone marrow fat, or noncalcified cartilage particles (Fig. 15.25). Progression of synovial chondromatosis to chondrosarcoma is very rare (Fig. 15.26).

Cortical bone and periosteal reaction can be evaluated not only with CT, but also with MR imaging (see Chapter 12). Normal cortical bone is represented as a signal void area, whereas intracortical lesions show almost invariably areas of increased signal intensity relative to cortex, on T1-weighted, Gd-DTPA enhanced, and T2-weighted images (Figs. 15.27 and 15.28). Likewise, mineralized periosteal reaction is seen as a solid or layered signal void area, whereas the nonmineralized cellular inner layer of the periostium (cambium) has a high signal intensity on T2-weighted images. (See also "Local Tumor Staging" below, and Chapter 12.)

Neurofibroma is easily recognized both on CT and MR imaging when it presents as the classic dumbbell tumor of the vertebral canal. Plexiform neurofibromatosis extends along neural bundles in a lobulated fashion, and is thus also easily recognized. When located in the retroperitoneum or pelvis these plexiform neurofibromas may be quite

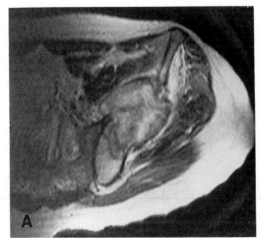 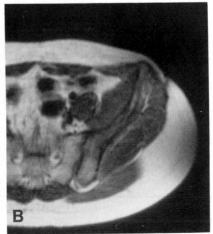

Figure 15.17. Eosinophilic granuloma. **(A)** Inhomogeneous high signal intensity lesion of iliac bone is surrounded by edema in the iliopsoas and gluteal muscle compartment. The ventral margin is not well defined. Open biopsy was needed to allow a diagnosis of eosinophilic granuloma to be made with confidence. **(B)** Following multiple intralesional injections with corticosteroids the lesion is shown to decrease in size on this T1-weighted (550/30) follow-up MR examination made after 3 months. Residual low signal intensity is still present in the iliac bone. No soft-tissue component was seen on the T2-weighted sequence (not shown). At this time the patient was almost completely asymptomatic. Note increase in muscle volume.

Figure 15.18. **(A)** CT scan of the femur at the level of the lesser trochanter shows the nidus with central calcification, allowing a diagnosis of osteoblastoma to be made. Extensive periosteal reaction is present. **(B)** The central calcification is not visualized on the T2-weighted image (2000/100) made at the same level. The tumor, periosteal reaction, and soft-tissue edema beyond the low signal intensity line of periosteum are demonstrated. **(C)** Above the level of the center of the osteoblastoma extensive intra- and extraosseous edema is seen as high signal intensity (2000/100). **(D)** On the T1-weighted image (600/20) at the same level as **(C)**, intraosseous edema and edema outside the elevated periosteum exhibit a low signal intensity. (Reprinted with permission from Kroon HM, Schurmans J. Radiology 1990;175:783–790.)

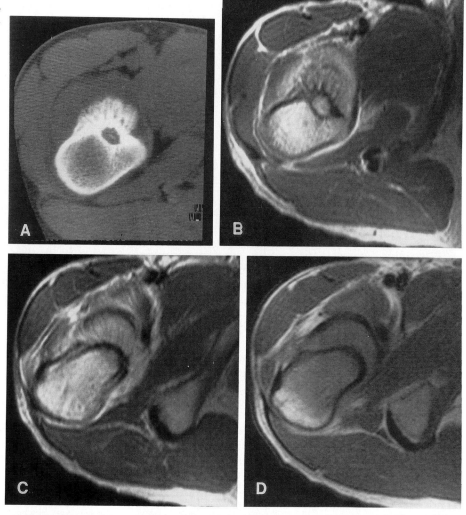

Figure 15.19. Ewing's sarcoma of the humerus. **(A)** The proximal and distal margin are difficult to define on the sagittal T1-weighted image (600/20). **(B)** Following administration of Gd-DTPA, inhomogeneous enhancement of the tumor is observed. Note the homogeneous enhancement of the reactive tissue in the metaphysis. This reactive area did not contain a tumor. Enhancement of soft-tissue extension is seen anteriorly and posteriorly. **(C)** T1-weighted image (600/20) after three cycles of chemotherapy. The tumor is now well defined. The reactive tissue interface in the metaphysis has disappeared and now has a normal signal intensity.

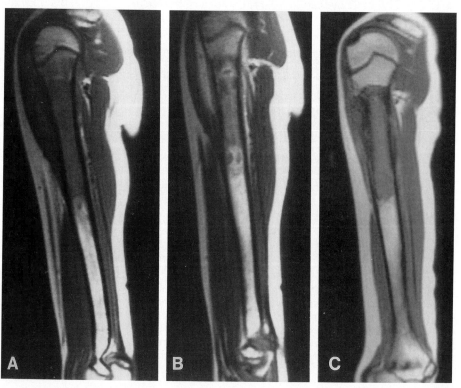

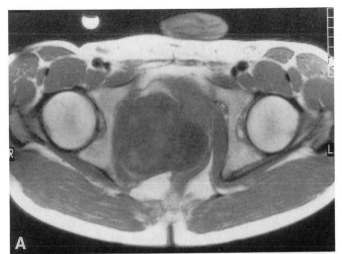

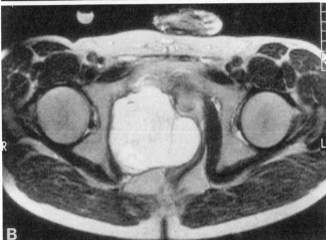

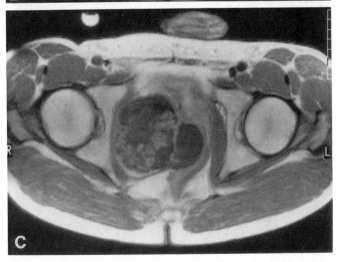

Figure 15.20. Chondrosarcoma, grade I, originating from pubic bone. **(A)** Tumor has a heterogeneous low signal intensity on the axial T1-weighted image (550/30). Within the lesion areas of intermediate to high (relative to muscle) signal intensity represent mucoid degeneration. **(B)** Soft-tissue extension is, due to the high signal intensity, well visualized on the T2-weighted (2000/100) image. The right internal obturator muscle is severely compressed. Although some lobulation is appreciated the internal structure of the lesion is difficult to ascertain. **(C)** On the Gd-DTPA enhanced image (550/30), only the margin, cellular areas, and some septations within the tumor enhance. This septal and curvilinear or ringlike enhancement pattern is typical of cartilaginous tumor. The large mucoid fields do not enhance.

extensive. The signal intensity of neurofibroma is determined by its prolonged T1 and T2 relaxation times: low to intermediate signal intensity on T1-weighted images and a very high signal intensity on T2-weighted images. When present, collagen and fibrous tissue, often centrally located, decrease the signal intensity (49).

Variants and Tumor-Like Lesions

Distribution of areas with abnormal signal intensity are important, not only for differentiating tumors and tumor-like lesions, but also in identifying normal variants. The low to intermediate signal intensity areas of red bone marrow in the axial skeleton and metaphysis and proximal diaphysis of the appendicular skeleton in adults and also in the periphery of the skeleton in adolescents and children on T1-weighted images may be confusing (see also Chapter 10) (64). On STIR sequences, the high signal intensity of red marrow is easily differentiated from absence of signal of the nulled yellow marrow (28, 29, 65, 66). Low signal intensity areas within bone marrow on gradient-recalled echo images may be due to shortening transverse relaxation time related to field inhomogeneity caused by bone trabeculae and cortex, or by susceptibility effect related to iron metabolism in red bone marrow (67).

When reviewing a CT or MR examination, clinical data and radiographs are of paramount importance. Many tumor-like lesions such as Paget's disease, pseudoaneurysms, polymyositis, osteomyelitis, stress fractures, etc. (Figs. 15.29 and 15.30) can be identified in the proper clinical setting. Early osteomyelitis will, in contrast to Ewing's sarcoma, usually display only reactive changes in the surrounding soft tissues. However, a soft-tissue abscess may be encountered (Fig. 15.30).

Paget's disease is characterized by increased size of bone, focal or diffuse decreased signal intensity on T1-weighted images, representing cortical thickening and coarse thickened trabeculae. Increased signal intensity areas representing residual fat are found in between. Fibrovascular marrow in active Paget's disease has a high signal intensity on T2-weighted images (68–70). As in hemangioma, slow flow in vascular spaces within the widened diploic space can be depicted as high signal intensity areas on T1- and T2-weighted images (69). Complications of Paget's disease, such as basilar invagination, spinal stenosis, hemorrhage, liquefactive degeneration, and sarcoma or giant cell tumor, can be diagnosed with MR imaging or CT (68, 69). The MR imaging characteristics of Paget's sarcoma reflect the histology of the tumor and are not different from other high-grade malignant tumors.

Stress fractures present as irregular or bandlike (perpendicular to cortical bone) areas of low signal intensity on T1-weighted images. The high signal intensity seen on T2-weighted or STIR sequences represents edema. Soft-tissue edema is as not a feature of stress fracture. These characteristics are important in differentiating stress fracture from osteomyelitis and Ewing's sarcoma, which invariably demonstrate soft-tissue abnormalities.

Tumor-like lesions may be found in this book in the chapters discussing the specific entities.

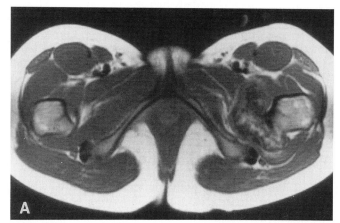

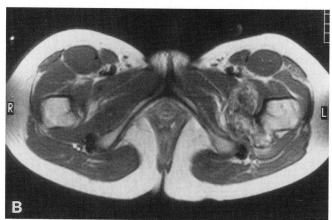

Figure 15.21. **(A)** Osteochondroma of the femur has a heterogeneous signal intensity on the axial T1-weighted image (550/30). The high signal intensity regions represent fatty bone marrow, whereas the low signal intensity areas represent both paucicellular and cellular tumor components. **(B)** After Gd-DTPA administration, a curvilinear enhancement pattern typical of cartilaginous tumor is observed. In this benign osteochondroma the cartilage cap is very small (<1 cm).

Figure 15.22. Osteochondroma of the humerus. **(A)** The central part of this exophytic tumor has a high density, whereas the periphery has a low density. This is unusual for osteochondroma, but not uncommon for juxtacortical osteosarcoma. **(B)** The T2-weighted MR image (2000/100) demonstrates the signal void of the central calcified component. The high signal intensity in the periphery is consistent with a cartilage cap, favoring a diagnosis of osteochondroma. Osteochondroma was found at histologic examination. (Reprinted with permission from Bloem JL, Reiser MF, Vanel D. Magn Reson Q 1990;6:136–163.)

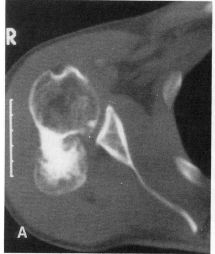

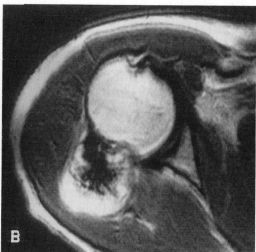

Figure 15.23. CT scan of the femur diaphysis demonstrates an exophytic lesion with low density in the center and high density in the periphery. This is not consistent with juxtacortical osteosarcoma. Clinical history, CT scan, and follow-up were indicative of subperiosteal hemorrhage.

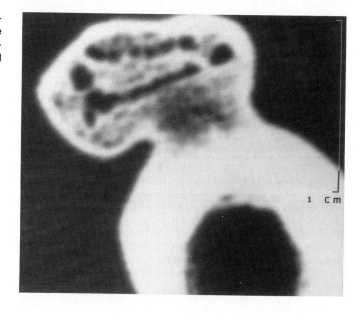

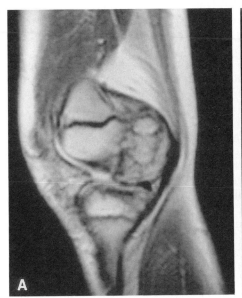
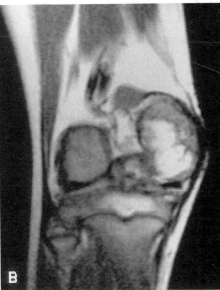

Figure 15.24. Infant with dysplasia epiphysialis hemimelica (Trevor's disease). Sagittal **(A)** and coronal **(B)** T1-weighted images (550/30) demonstrate the intraarticular cartilaginous lesion originating in the medial part of the distal femoral epiphysis. Morphology and signal intensity are identical to those of osteochondroma.

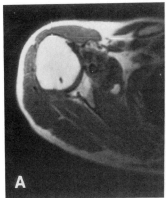
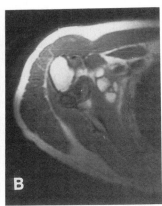

Figure 15.25. Synovial chondromatosis of the shoulder. Axial images made at **(A)** and caudal to the level of the glenoid **(B)**. Cartilaginous bodies seen on the cephalic T1-weighted image **(A)** are, with one exception, calcified and therefore have a low signal intensity. On the caudal slice **(B)** some noncalcified cartilaginous bodies are seen, characterized by a high signal intensity. The high signal intensity represents bone marrow fat or uncalcified cartilage.

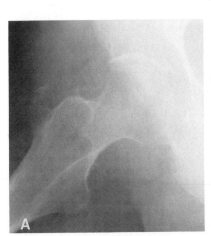
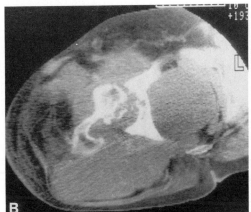
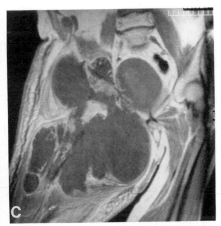

Figure 15.26. **(A)** Synovial chondromatosis of the hip joint was treated in this patient with synovectomy. Patient did not have any symptoms following surgery. Histology showed only synovial chondromatosis, with no signs of tumor. **(B)** Computed tomography was performed 6 years later when patient presented with a large mass and pain. A large soft-tissue mass and osseous destruction are present. **(C)** Coronal Gd-DTPA enhanced MR image (550/30) displays a large tumor with only marginal enhancement. In the cephalic and most distal part a curvilinear enhancement pattern can be appreciated. Biopsy revealed chondrosarcoma.

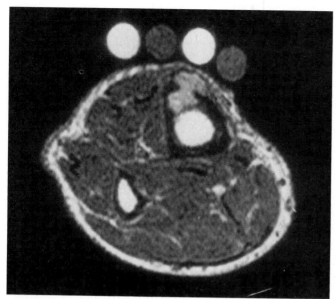

Figure 15.27. Intracortical brown tumor in patient with hyperparathyroidism has a high signal intensity relative to cortical bone on this axial Gd-DTPA-enhanced (550/30) image.

Gd-DTPA-Enhanced Magnetic Resonance Imaging

The pharmacokinetics of Gd-DTPA are similar to iodinated contrast agents (71–78). This paramagnetic contrast agent acts indirectly by facilitating T1 and T2 relaxation processes through alteration of the local magnetic environment. Shortening of both relaxation times is a function of the concentration of the Gd complex (71, 72). Initially an increase of the Gd complex concentration results in a predominantly shortened T1 relaxation time and an ensuing increased signal intensity on T1-weighted images (Fig. 15.2). With an increasing concentration of the Gd complex, the signal intensity drops because of the dominant effect on the T2 relaxation time (dephasing). With the concentration used in clinical practice (0.1 to 0.2 mmol/kg body weight for intravenous administration) the shortening of T1 relaxation time supervenes. Only in areas where a high concentration is reached, such as in the urinary bladder, can a signal void occur. Enhancement of subcutaneous fat (8.5%), bone marrow (4.3%), and cortex is measurable, but usually not discernible on T1-weighted SE images after intravenous administration of Gd-DTPA. Slight enhancement of muscle (13 to 20%) may be measured but is also difficult to observe at visual inspection of the SE images (32).

Erlemann et al. found, with a gradient-echo technique (TR 40 ms, TE 10 ms, flip angle of 90°), on a 1.5-T system, no increase of signal intensity of fat in 51%, and no increase in signal intensity of bone marrow in 56% of 69 patients (79). In 69 patients, no change of the median signal intensity ratio (prior to and following intravenous injection of 0.1 mmol/kg body weight Gd-DTPA) of fat and bone marrow was observed. The upper quartile was only 7% for fat and 8% for bone marrow. However, 95% of patients exhibited an increase of muscle signal intensity on gradient-echo images. The median increase of muscle signal was 20%.

Well perfused, often cellular tumor components are able to accumulate a high concentration of Gd-DTPA, which results, through shortening of the T1 relaxation time, in a high signal intensity on short TR/TE (600/20) images, as opposed to other tumor components, such as sclerosis,

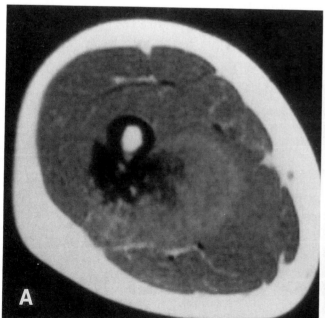
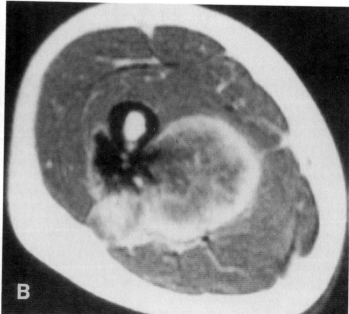

Figure 15.28. **(A)** Paraosseous osteosarcoma of the femur diaphysis has a low signal intensity centrally, and an intermediate signal intensity in the periphery (550/30). **(B)** After injection of Gd-DTPA septations and the periphery enhance (550/30). Only the calcified central core does not enhance. Cartilaginous fields were found in the enhancing periphery of the tumor; the central part exhibited only osteosarcoma.

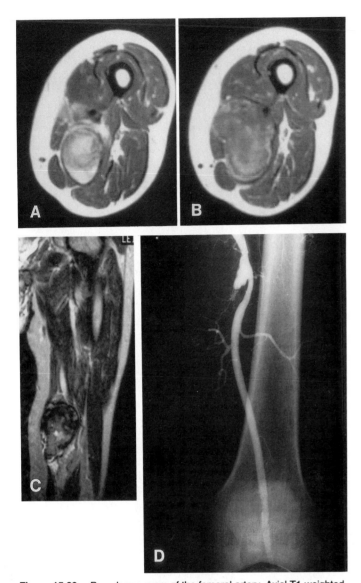

Figure 15.29. Pseudoaneurysm of the femoral artery. Axial T1-weighted (600/20) image displays an inhomogeneous high signal intensity lesion **(A)**. On more cephalic axial images **(B)**, the neurovascular bundle is seen within the periphery of the mass. **(C)** Coronal T2-weighted (2000/100) image. The lesion is inhomogeneous with high and low signal intensity areas. The low signal intensity margin represents the presence of hemosiderin. The findings seen on the T1- and T2-weighted images are consistent with hematoma. **(D)** Arterial phase of angiography demonstrates the origin of the pseudoaneurysm. Tumor vessels were not seen. At surgery the diagnosis of pseudoaneurysm was confirmed.

mature fibrosis, and liquefaction, which are not well perfused and therefore do not show a high signal intensity on the postcontrast images. In general, the presence or absence of homogeneous enhancement is not a very good indicator for malignancy. Still, Gd-DTPA may increase tissue characterization.

Although the intravenous administration of Gd-DTPA may assist in differentiating viable tumor from liquefaction and edema, neovascularity in necrotic areas and immature fibrosis may also enhance. Differentiation between viable

and necrotic tumor is, as a consequence, not always possible at this moment.

Static short TR/TE spin-echo images may suggest a specific diagnosis in selected cases. Well-differentiated cartilaginous tumors containing large paucicellular cartilage fields or cartilage fields with mucoid degeneration demonstrate no or only slight enhancement (30, 32). Only the margin and cellular septations in the periphery and within the tumor enhance (Figs. 15.20 and 15.31). Because of the lobulated gross anatomy of enchondroma and low grade (grade I and II) chondrosarcoma, the postcontrast MR imaging scans of these tumors exhibit a septal or curvilinear (serpiginous) enhancement pattern. The enhancement of benign enchondroma and malignant low-grade chondrosarcoma is identical. An exception is enchondroma with ossification in the margin and center. These inert lesions do not enhance. Only high-grade cartilaginous tumors such as grade III chondrosarcoma and mesenchymal chondrosarcoma exhibit the nonspecific homogeneous or inhomogeneous enhancement displayed by most tumors (Fig. 15.32).

Other cartilage-containing tumors such as osteochondroma or chondroid osteosarcoma may also show areas with curvilinear enhancement (Fig. 15.33). Usually other criteria such as morphology will assist in differentiating these lesions. A chondroid osteosarcoma is less homogeneous than a well-differentiated cartilage tumor and often contains large areas with osteoid, which can be recognized.

The curvilinear enhancement of well-differentiated cartilaginous tumors must be differentiated from thin peripheral enhancement of the pseudocapsule of lesions such as eosinophilic granuloma, solitary bone cyst, etc. No enhancement is found within these lesions.

At our institution we were able to identify all 22 well-differentiated cartilaginous tumors from a group of 100 tumors including osteosarcoma, Ewing's sarcoma, fibrosarcoma, benign tumors, tumor-like conditions, etc. (32). These findings need further confirmation from additional studies before they can be used in clinical practice.

The clinical importance of the ability to identify well-differentiated cartilaginous tumors and their various components lies in the ability to identify cellular areas for biopsy (Fig. 15.3) as well as the option to perform a resection without taking a biopsy, which may facilitate reconstruction of large pelvic tumors.

Other tumors do not exhibit clinically important enhancement patterns. Some differences may be encountered, such as the nonenhancing osteoid components of classic osteosarcoma. These osteoid components can, however, also be identified with conventional radiography, CT, and nonenhanced MR imaging.

The use of gradient-refocused MR imaging in combination with Gd-DTPA may further increase specificity. Erlemann et al. reported a correlation between the slope of enhancement and the biologic activity of the tumor (79). Statistically significant differences were reported between relative signal enhancement and time-dependent signal enhancement of benign and malignant tumors after intravenous administration of Gd-DTPA. When lesions, in which the increase of signal intensity over the baseline value was

Figure 15.30. **(A)** Diffusely increased density is observed in the femur of a patient with osteomyelitis. **(B)** Abnormal low signal intensity in femur and soft tissue are nonspecific. Note the thickened, elevated, nonmineralized periostium. **(C)** High signal intensity representing inflammatory tissue is seen in the femur and soft tissues. Edema is present in the extensor muscle compartment. Only the shape and signal intensity of the soft-tissue mass are more in favor of liquefaction (such as occurs in a soft-tissue abscess) than of a firm tumor mass.

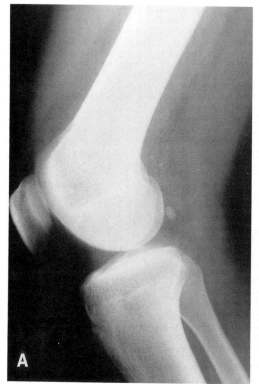

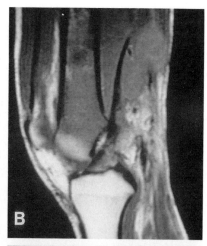

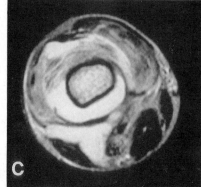

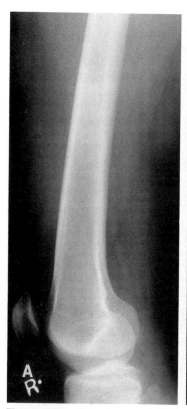

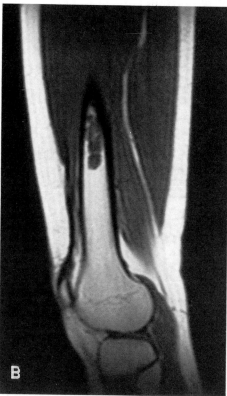

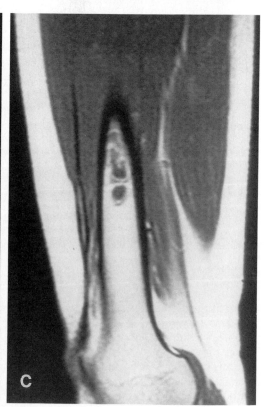

Figure 15.31. Chondrosarcoma in the femur is hard to detect on radiographs in this patient. Subtle calcifications in the diaphysis are hardly discernable. **(B)** Lobulated chondrosarcoma is seen in diaphysis. **(C)** The curvilinear enhancement appreciated on the Gd-DTPA-enhanced image allows a diagnosis of well-differentiated cartilaginous tumor to be made. Histologic diagnosis: chondrosarcoma, grade I.

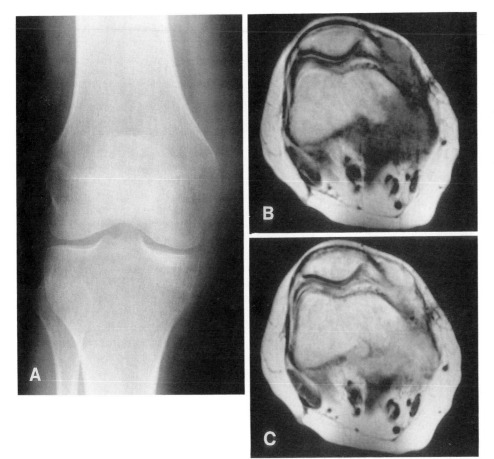

Figure 15.32. **(A)** A 38-year-old female with a mesenchymal chondrosarcoma of the medial femoral condyle. **(B)** Nonspecific intermediate signal intensity of the tumor in the medial condyle on this T1-weighted (550/30) image. A biopsy defect medially in subcutaneous fat can be seen. **(C)** Nonspecific homogeneous enhancement of the cellular tumor is seen on this Gd-DTPA enhanced MR image. Tumor is extending into the medial head of the gastrocnemius and the medial retinaculum. Note enhancement of the synovium. (Reprinted with permission from Bloem JL, Reiser MF, Vanel D. Magn Reson Q 1990;6:136–163.)

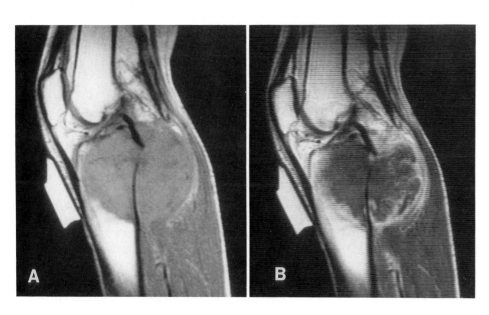

Figure 15.33. Chondroid osteosarcoma of the tibia. **(A)** Nonspecific homogeneous low signal intensity on T1-weighted image (550/30). Posterior cruciate ligament is encased by tumor. **(B)** After Gd-DTPA injection curvilinear enhancement indicative of cartilage is seen in the soft-tissue component. Only some peripheral enhancement is seen in the osseous component. Cartilage was predominant in the soft-tissue component of this osteosarcoma.

less than 30%, were diagnosed as benign, and those of 30% or more malignant, the following results were observed. The sensitivity for detection of malignant tumors was 84% and for benign tumors 72% (79). Peritumoral edema in the soft tissues was reported to exhibit a slower and more gradual increase in signal intensity compared to tumor (79). Thus, a correlation with the pathologic grade or biologic activity was found. These initial studies also need further confirmation. We did, for instance, encounter patients with Ewing's sarcoma and osteosarcoma in whom the intraosseous reactive zone displayed stronger enhancement on delayed spin echo images than the tumor itself (20).

Nonenhanced MR imaging has been proven to be very accurate in local tumor staging (11). As a general rule Gd-complex does not significantly increase the performance of MR imaging in this respect (32). A decrease of the C:N ratio between tumor and bone marrow or fat is observed, when Gd-DTPA-enhanced images are used instead of nonenhanced T1-weighted SE images. Likewise, the signal intensity ratio between tumor and muscle increases up to 60% when Gd-DTPA-enhanced SE images are used instead of the nonenhanced T1-weighted (550/30) SE images. However, the C:N ratio between tumor and muscle usually is lower in Gd-DTPA-enhanced SE or gradient-recalled echo images than in T2-weighted SE images (32, 36).

Local Tumor Staging

Adequate, wide, or radical surgery has been a prerequisite for proper treatment of patients with primary malignant musculoskeletal tumors. The choice of therapy depends on a number of factors such as grade of malignancy, response to neoadjuvant therapy, local tumor extension, and the presence or absence of regional or distant metastases. The surgical staging system developed by Enneking (Table 15.1) takes all of these factors, with the exception of the response to neoadjuvant therapy, into account (9, 80).

The biologic aggressiveness, indicated by the grade, is the key factor in the selection of the surgical margin required to achieve control (9). Four different surgical procedures, which imply four different margins, are recognized:

1. *Intralesional:* This is a debulking procedure (also curettage), in which the surgical plane of dissection is in the lesion itself.
2. *Marginal:* The plane of dissection is in the pseudocapsule or reactive zone.
3. *Wide:* The plane of dissection is in the compartment of the tumor (intracompartmental); a cuff of normal tissue is removed with the tumor.
4. *Radical:* The entire compartment containing the tumor is resected.

Local staging procedures are essential in planning surgery because they establish tumor size and identify the anatomic compartments that contain tumor. The exact anatomic location of tumor is the key factor in selecting how a required tumor-free margin can be accomplished (9, 80,

Table 15.1
Tumor Staging System According to Enneking[a]

Grade	Site
IA	Low grade, intracompartmental
IB	Low grade, extracompartmental
IIA	High grade, intracompartmental
IIB	High grade, extracompartmental
III	Distant metastases present

[a]See Refs. 9 and 80.

81). The compartment analysis used in this chapter is a detailed derivate from Enneking's system. A more detailed analysis was chosen because a more detailed definition of tumor extent is needed when surgery is planned (82, 83). Compartments that are subsequently discussed are bone marrow, cortical bone, muscular compartments, major neurovascular bundles, and joints. The data listed are from a prospective study in which 65 patients with primary malignant musculoskeletal tumors were evaluated with MR imaging, CT, ^{99m}Tc-MDP scintigraphy, angiography, and pathologic/morphologic examination of the resected specimens (11). The MR imaging results of this prospective study have been confirmed at our institution in over 200 patients with malignant primary musculoskeletal tumors.

Bone Marrow Involvement

Definition of bone marrow involvement is essential in determining the level of amputation or resection. Radiography, CT, and bone scintigraphy have some disadvantages. Radiographs do not visualize bone marrow; only advanced disease in epiphyseal and metaphyseal areas can be detected when bony trabeculae are destructed. The transverse imaging plane of CT is a disadvantage in defining tumor that extends in a longitudinal plane. The relationship to the growth plate is difficult to analyze with CT. A second major disadvantage of CT is the beam-hardening artifact, which is a larger problem than in a general CT population because patients with primary malignant bone tumors are usually young and possess a high amount of calcium in their bones. Again, destruction of trabeculae is the most important sign of tumor presence within the marrow cavity (24). Replacement of diaphyseal marrow fat may also be observed, especially when Hounsfield units are measured. ^{99m}Tc-MDP bone scintigraphy offers the advantage of the ideal imaging plane but reactive changes, especially tumor-related hyperemic osteoporosis, result in increased tracer uptake indistinguishable from increased uptake due to the presence of tumor (10, 26, 84, 85). Computed tomography and plain radiographs are also unable to differentiate this hyperemic osteoporosis from tumor-induced osteolysis (1, 10, 11, 26). The lack of anatomical information, especially in relation to the normal uptake in the region of the growth plate, is another disadvantage.

Magnetic resonance imaging offers some crucial advantages: it is a tomographic modality used in the ideal imaging plane. For extremities a plane through the long axis of a long bone is chosen, i.e., sagittal or coronal (Fig. 15.34). Because of the high fat content of bone marrow, normal

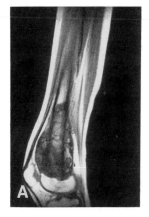

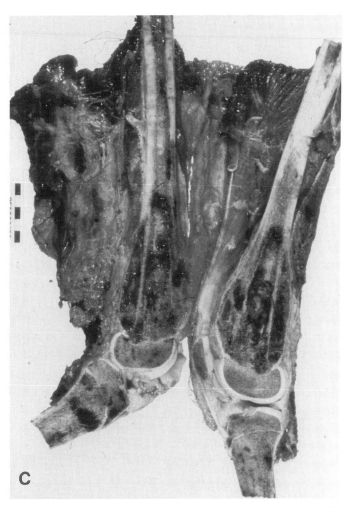

Figure 15.34. **(A)** Sagittal T1-weighted (550/30) image of osteosarcoma. The proximal and distal intraosseous margins are well defined. Signal intensity of epiphysis is normal. **(B)** Lateral view of ^{99m}Tc-MDP bone scintigraphy shows increased tracer uptake in the diaphysis, metaphysis, and epiphysis of the femur. Increased tracer uptake in patella and tibia is considered to be reactive because this is not located within the bone hosting tumor. **(C)** Sagittal section of resected specimen. The discrepancy between MR and bone scintigraphy is solved. The epiphysis is normal. The growth plate is the distal margin of the tumor.

bone marrow can be directly visualized by MR imaging and will have, especially on T1-weighted sequences, a high to intermediate signal intensity. The signal intensity depends on the age of the patient. Magnetic resonance imaging is superior to ^{99m}Tc-MDP bone scanning because tumor-related hyperemic osteoporosis is not depicted on MR imaging and is therefore no source of false positive readings (11, 86). Magnetic resonance imaging has an almost perfect correlation (r = 0.99) with pathologic/morphologic examination, CT has a less substantial correlation (r = 0.93), while ^{99m}Tc-MDP scintigraphy has a weak correlation (r = 0.69) (11). These results are exhibited in Figure 15.35.

The osseous tumor margin may be obscured by the presence of intraosseous edema. Often a double margin can be visualized. The margin nearest to the center of the tumor represents the true tumor margin, whereas the outer margin represents the margin of the edematous reactive zone toward normal bone marrow. Differentiation can be facilitated by using Gd-DTPA (Fig. 15.19). Tumor and reactive edematous zone display different enhancement patterns: a well-vascularized tumor will enhance more than the reactive zone, but usually, for instance in osteosarcoma, the intraosseous reactive zone will enhance more than the tumor itself. Furthermore, enhancement of the reactive edematous zone is, as a rule, more homogeneous than that of tumor. The intraosseous reactive zone is usually not a clinical problem, because it disappears after one or two cycles of chemotherapy. The true tumor margin is then easily identified.

Caution is needed in the diagnosis of skip metastases (Fig. 15.36). Skip metastases, as may be encountered in patients with osteosarcoma and Ewing's sarcoma, are detected with bone scintigraphy unless the size is below the detection threshold. Skip lesions of less than 5 mm may thus pose a diagnostic problem. Imaging of the entire bone on T1-weighted images by using a large field of view may be beneficial.

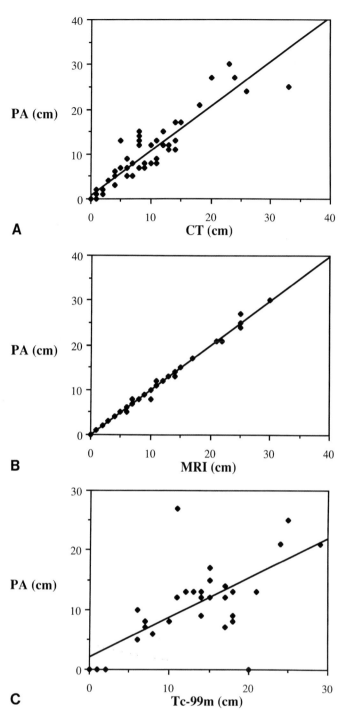

needed, with tomography or CT. Magnetic resonance images are also able to visualize the status of cortex. Similar to calcifications, normal cortical bone has a low signal intensity on T1- and T2-weighted images because of low spin density and short T2 relaxation time. Invasion of cortex by tumor is best shown on T2-weighted images as a disruption of the cortical line and replacement of cortex by high signal intensity tumor (Figs. 15.7, 15.27, and 15.28). The sensitivity and specificity of MR imaging (92 and 99%, respectively) were not found to be significantly higher than the sensitivity and specificity (91 and 98%, respectively) of CT (11). Magnetic resonance imaging is superior to CT for sclerotic osteosarcomas, because the signal intensity of osteosclerotic tumor is still slightly higher than the low signal intensity of normal cortex. The high density of osteosclerotic tumor and cortex may be indistinguishable on CT.

Involvement of Muscular Compartments

Magnetic resonance imaging is significantly superior to CT in identifying muscle compartments containing tumor (Fig. 15.37). In an individual analysis of 600 muscle compartments, we found for MR imaging a sensitivity of 97%, and a specificity of 99%. For CT the sensitivity was 71%, the specificity was 93% (Table 15.2) (11). The superior performance of MR imaging is based on display of tumor relative to muscular compartments with, compared to CT, superior contrast in the ideal imaging plane. Peritumoral edema can be identified because of a slightly different signal intensity as compared to that of the tumor, and especially because of the fading margins of edema as opposed to the distinct margin of the (pseudo)capsule of the tumor. However, accurate delineation of the tumor-edema interface can be difficult. Differences in signal intensity on multiple echos and enhancement following administration of Gd-DTPA may be used to differentiate tumor from edema (see also "Magnetic Resonance Imaging and Computed Tomography Characteristics," above).

Figure 15.35. **(A)** Intraosseous tumor length as measured with CT, correlated with pathomorphological (PA) measurements (in centimeters): $r = 0.93$, $N = 56$. **(B)** Intraosseous tumor length as measured by MR imaging, correlated with pathomorphological (PA) measurements (in centimeters): $r = 0.99$, $N = 61$. **(C)** Intraosseous tumor length as measured with ^{99m}Tc-MDP scintigraphy, correlated with pathomorphological (PA) measurements (in centimeters): $r = 0.69$, $N = 33$. (Reprinted with permission from Bloem JL, Taminiau AHM, Eulderink F, Hermans J, Pauwels EKJ. Radiology 1988;169:805–810.)

Cortical Involvement

Destruction of cortical bone is not a diagnostic problem since it can be evaluated on plain radiographs and, if

Table 15.2
Muscular Involvement:[a] Comparison of CT (57 Patients) and MR Imaging (60 Patients) with Pathomorphology[b]

	CT	MR imaging
Sensitivity		
Mean	71%	97%
95% confidence limits	60–80%	92–99%
Specificity		
Mean	93%	99%
95% confidence limits	86–96%	97–100%
Accuracy	86%	98%
Positive predictive value	81%	98%
Negative predictive value	88%	99%
Prevalence	30%	30%
N	576	600

[a] Muscles of the knee, pelvis, and shoulder were observed. N, Number of muscle compartments.
[b] Reprinted with permission from Bloem JL, Taminiau AHM, Eulderink F, Hermans J, Pauwels EKJ. Radiology 1988; 169:805–810.

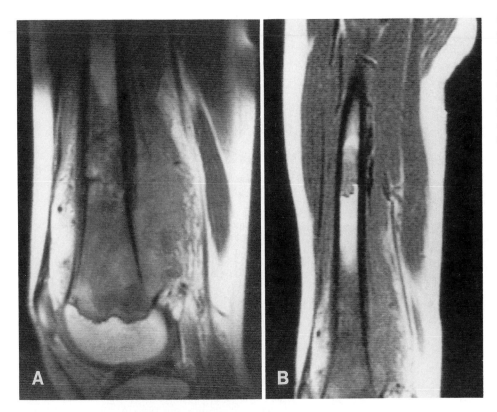

Figure 15.36. **(A)** The osseous proximal and distal margins are visualized on this T1-weighted (550/30) sagittal image, which was made with a surface coil and a field of view of 20 cm. **(B)** This large field of view image, made with the body coil in the same imaging session, demonstrates the need for large field of view images in the detection of skip lesions: a large skip lesion is seen in the diaphysis.

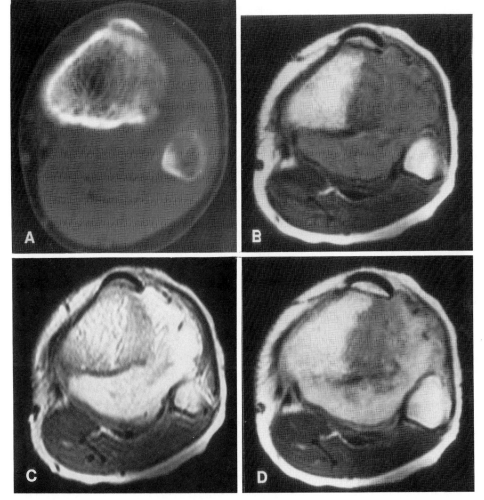

Figure 15.37. **(A)** Cortical destruction, osseous infiltration, and the presence of soft-tissue extension can all be evaluated by CT in this patient with osteosarcoma of the tibia. However, the extent of the soft-tissue component cannot be defined. **(B)** Intraosseous extension is well visualized on the T1-weighted (550/30) sequence. **(C)** Contrast on this T2-weighted sequence (2000/100) between high signal intensity of tumor and low signal intensity of muscle is superior to that of CT, T1-weighted, and Gd-DTPA-enhanced images. **(D)** In this particular patient both intra- and extraosseous extension can be evaluated on this Gd-DTPA-enhanced image. This is, however, not a consistent observation.

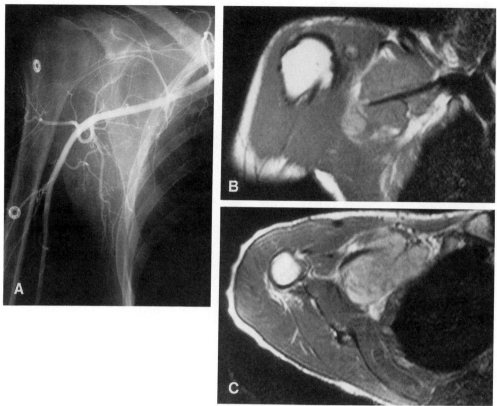

Figure 15.38. Relationship between neurovascular bundle and synovial sarcoma is difficult to ascertain with angiography. Compression of the contrast-filled axillary artery by tumor is seen, but encasement is not demonstrated. **(B)** Coronal T1-weighted image demonstrates encasement of axillary artery by tumor. This was also seen on adjacent slices. **(C)** Axial T2-weighted image (2000/50) confirms encasement of neurovascular bundle.

Vascular Involvement

Vessels are more often displaced than encased by tumor. The relationship between tumor and neurovascular bundle is easily evaluated on T2-weighted images because normal flow in a vessel results in low or absent signal intensity, the so-called flow void phenomenon. The lumen of the vessel may have a higher signal intensity owing to paradoxical enhancement, even-echo rephasing, or turbulence with slow flow caused by compression. Computed tomography (sensitivity 36%, specificity 94%) and MR imaging (sensitivity 92%, specificity 98%) provide more information than angiography (sensitivity 75%, specificity 71%) (11) because large vessels are, especially with MR imaging, well visualized in relation to the tumor. Angiography displays the lumen of the artery and often parts of the tumor having pathologic vascularization. However, it is usually difficult to differentiate displacement from encasement because the outer vessel wall and the tumor margins are not well exhibited with angiography (Fig. 15.38). Furthermore, angiography seldom yields additional information and is thus rarely indicated as a staging procedure. Magnetic resonance imaging is the superior method for displaying the relationships between major vessels and tumor.

Joint Involvement

Computed tomography (sensitivity 94%, specificity 90%) and MR imaging (sensitivity 95%, specificity 98%) are both able to demonstrate joint involvement with high accuracy (11). Joint involvement is sometimes more accurately demonstrated on MR images than by CT, because the articular surfaces may be parallel to the transverse CT plane. Joint effusion with or without hemorrhage, often but not always a secondary sign of a contaminated joint, is readily identified by CT and MR imaging.

Monitoring of Chemotherapy

Spontaneous necrosis of up to 50% is not uncommon in high-grade malignancies. Therefore a histologically good response is defined as the presence of 10% (or less) viable tumor tissue.

A change of tumor volume and signal intensity may correlate with response to chemotherapy (87–90). An increase of tumor volume indicates poor or no response. These tumors usually display a stable, or a slight to moderate increase of signal intensity in the soft-tissue component on the T2-weighted images (Fig. 15.39).

No change or increase in the intraosseous reactive zone is a consistent intraosseous sign of unsatisfactory response (91). This change is easily depicted by visual inspection, and is, together with changes in volume, one of the most important qualitative signs.

A decrease in edema is observed in both good and poor respondents (Fig. 15.19) (91). A thick, well-defined, low signal intensity margin often develops and represents a pseudocapsule consisting of collagen. The development of this collagenous pseudocapsule does not correlate with response.

The good respondents may show a decrease in tumor volume, as well as changing signal intensities (87). The tumor shrinks mainly in the soft tissues, the volume of the intraosseous compartment does not change much, if at all. However, decrease in tumor volume is usually quite pronounced in good respondents with Ewing's sarcoma.

A decrease in signal intensity of the soft-tissue component measured on T2-weighted images is a sign of satisfactory response (Fig. 15.40). These changes are often found

in osteosarcoma and are caused by dehydration, and the development of granulation tissue, fibrosis, and calcification. The decrease in signal intensity is often subtle and is not readily appreciated when no measurements are taken (91). The development of cystic spaces in primary malignant musculoskeletal tumors (92), depicted as a substantial increase in signal intensity on T2-weighted images, may also correlate with a satisfactory response. In osteosarcoma, changes in signal intensities of the intraosseous component or of signal intensities measured on T1-weighted images do not correspond to response (87). In Ewing's sarcoma, an increased intraosseous signal intensity on T2-weighted images may prove to be a reliable sign of good response (87, 93). The high signal intensity reflects the presence of a loose hypocellular myxoid matrix (93).

Exceptions to the described common patterns are encountered. These may be secondary to fracture, hemorrhage, or variation in histology of the tumor under observation; for instance, fibroblastic osteosarcoma will not react in the same way as classic osteosarcoma.

The role of Gd-DTPA enhanced MR imaging also needs further evaluation. Areas of necrosis or residual viable tumor cannot be identified with certainty on nonenhanced MR images, unless cystic spaces develop (92, 94). Although necrotic tissue may develop a new vascular supply, a decrease in enhancement after the intravenous administration of Gd-DTPA, compared to prechemotherapy studies, may indicate vascular deficiency and necrosis. Pilot studies indicate that a rapid increase in signal intensity of the tumor after Gd-DTPA administration is compatible with poor response to chemotherapy, whereas a reduction in signal intensity relative to the prechemotherapy studies indicates satisfactory response (32, 90). The chemotherapy-induced changes in tumor perfusion may also affect intravoxel incoherent motion (95). At this time it is not known whether this correlates with tumor response to chemotherapy.

A noninvasive way to identify good and poor respondents may have an important impact on timing of surgery, selection of chemotherapy, selection of patients for postoperative chemotherapy, and prognosis. The clinical rele-

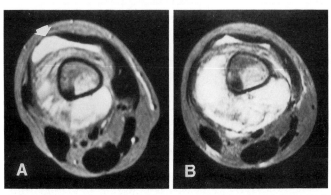

Figure 15.39.   Osteosarcoma of the femur on T2-weighted (2000/100) images. **(A)** Before chemotherapy. Extraosseous component of the tumor has mixed partially high and partially intermediate signal intensity. High signal intensity in the suprapatellar bursa represents joint effusion. **(B)** After treatment. Tumor volume has clearly increased. Signal intensity of extraosseous component has increased also and has become more homogeneous. Histopathology of resected specimen showed poor response. (Reprinted with permission from Holscher HC, Bloem JL, Nooy MA. AJR 1990;154:763–769.)

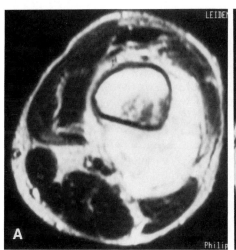

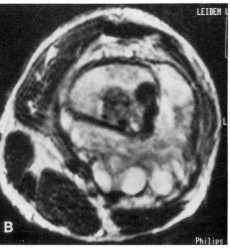

Figure 15.40.   **(A)** Osteosarcoma of the femur. **(B)** T2-weighted (2000/100) MR image before chemotherapy. Extraosseous component of tumor has rather high homogeneous signal intensity; intramedullary component has mixed signal intensity. Signal intensities of both intra- and extraosseous components of tumor have decreased. Signal intensity of tumor has become inhomogeneous. Pathologic response was graded as good.

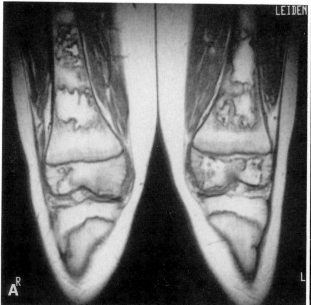 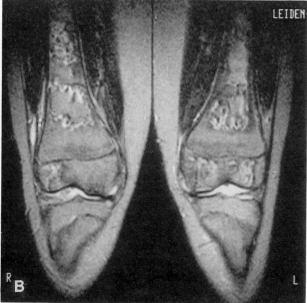

Figure 15.41. Coronal T1-weighted (600/20) **(A)** and T2-weighted (2000/100) **(B)** images of a patient with treated acute leukemia in complete remission presenting with knee pain. The signal intensity characteristics of the lesions and margins, together with the distribution, are consistent with multiple bone infarcts in femurs. On contiguous coronal images (not shown) infarcts were also seen in both tibias. Joint effusion is also present.

vance and encouraging preliminary results warrant further research in this field.

Avascular necrosis can complicate medical treatment of patients with musculoskeletal malignancies (88). This may or may not be accompanied by clinical symptoms. Avascular necrosis is especially, but not exclusively, found in children with leukemia or lymphoma, who are treated with combinations of chemotherapy and steroids (Fig. 15.41).

Recurrence

Detection of recurrent or residual tumor following initial treatment is a challenging problem for diagnostic imaging. A large recurrent tumor mass may be detected at clinical examination, with radiographs or CT. Small recurrent tumors are difficult to define in relation to posttherapy changes caused by surgery, radiation therapy, and chemotherapy. The matter is further complicated when fixation devices have been used in reconstructive surgical procedures. These devices cause streak artifacts on CT and susceptibility artifacts on MR imaging. Even when no hardware is used, susceptibility artifacts, caused by metallic particles left behind after instrumentation, can degrade the MR images.

Despite these drawbacks, CT and especially MR imaging can be used to detect recurrence. A recurrence is very likely (sensitivity 96 to 100%) when following surgery, with or without chemotherapy, a mass is seen on MR imaging that is characterized by a high signal intensity on Gd-DTPA-enhanced or T2-weighted images (Fig. 15.42) (96). Reactive changes, including immature fibrous tissue, may have the same signal intensity characteristics, but are often not space occupying. Cystic masses without tumor will have a high signal intensity on T2-weighted images, but do not enhance after intravenous administration of Gd-DTPA. A

tumor recurrence is very unlikely when, following surgery, no enhancement or a low signal intensity on T2-weighted is found. However, knowledge of the signal intensity of the primary tumor prior to therapy is crucial, especially when the primary tumor is characterized by a low signal intensity on T2-weighted images.

When an equivocal lesion is detected, follow-up studies, biopsy, or other imaging studies such as radiographs or ultrasound may be helpful.

Radiation therapy adds to the confusion because it may induce inflammatory, reactive changes that may be indistinguishable from tumor recurrence. These reactive changes are characterized by a high signal intensity on Gd-DTPA enhanced and T2-weighted images, and may persist for more than a year. However, reactive lesions usually confine themselves to the space available and, unlike tumor recurrence, do not present themselves as space-occupying expanding masses. Differentiation between tumor and reactive changes is not always possible, because the latter are seen in the months following therapy, as well as years later.

The effort put into detection of recurrent disease depends on several factors, such as the therapeutic options when a recurrence is found, the biological grade of the primary tumor, and the degree of success of the surgical procedure, especially in relation to tumor-free margins. When detection of early recurrence is of vital importance, a baseline study obtained 3 to 6 months following surgery may be beneficial.

Conclusion

Currently, MR imaging is the method of choice in local staging of primary malignant musculoskeletal tumors. Additional imaging procedures for local staging, such as an-

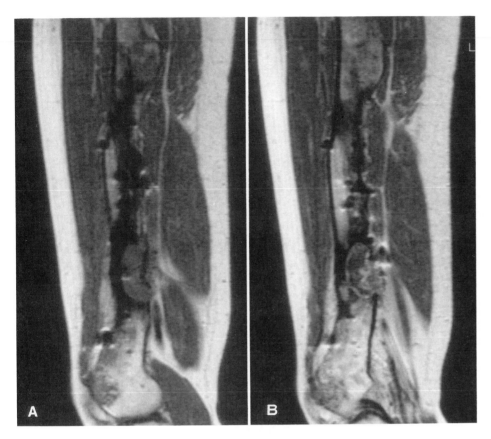

Figure 15.42. Recurrence of chondrosarcoma following resection and reconstruction with use of hardware. **(A)** Sagittal T1-weighted image (600/20) shows many susceptibility artifacts caused by hardware. In the distal diaphysis a low signal intensity area is seen. **(B)** Curvilinear enhancement of this lesion indicative for recurrent chondrosarcoma is depicted on this Gd-DTPA-enhanced image. The second lesion is seen in the proximal diaphysis.

giography and ^{99m}Tc-MDP, are as a rule not indicated: only one imaging study is sufficient. Extension into osseous and soft-tissue compartments is well visualized with T1- and T2-weighted MR imaging. In equivocal cases, additional sequences such as STIR and Gd-DTPA enhanced MR imaging may be of value. When patients are being treated with preoperative chemotherapy, MR imaging studies made prior to and following chemotherapy must be evaluated in the preoperative staging analysis. This is especially important for Ewing's sarcoma, in which the prechemotherapy study may be the only one to display the compartments hosting the primary tumor. Even when response to chemotherapy is excellent, tumor nests may be present in these compartments.

Accurate local tumor staging allows for a high percentage of successful local resections or limb salvage procedures (1, 4, 6–8, 97, 98). Eighty percent of our patients with primary malignant musculoskeletal tumors are being treated with limb salvage procedures, whereas only 20% need ablative surgery. Local recurrences occurred in 9% after ablative surgery and in 5% after resection in 200 of our patients who were treated on the basis of MR imaging. Although local control of these primary malignant tumors is quite feasible, survival depends to a large extent on prevalence and management of disseminated disease.

Benign tumors, which are characteristically much better defined than their malignant counterparts, may be accurately staged with CT. Computed tomography may be useful in imaging benign lesions, especially in complex anatomical regions such as the vertebral column, thoracic wall, and pelvis. Computed tomography has definitely reduced the need for conventional tomography. For detection of disseminated disease, additional studies, such as CT for the detection of lung metastases and ^{99m}Tc-MDP scintigraphy for the detection of osseous metastases and skip lesions or large field-of-view MR imaging, are needed.

Also, MR imaging may, in selected cases, suggest a specific diagnosis. Tissue characterization is increased when Gd-DTPA is used. Magnetic resonance studies with or without Gd-DTPA may thus assist in choosing the appropriate site for biopsy.

Despite some limitations, MR imaging is very well suited as a method to detect recurrent disease.

Finally, the preliminary results of pilot studies indicate that MR imaging may play a role in monitoring the response to chemotherapy. However, further research in this area is needed and is currently being conducted in several institutions.

Acknowledgments

We thank F. Eulderink, M.D., M. A. Nooy, M.D., H. M. Kroon, M.D., M. Geirnaerdt, M.D., E. K. J. Pauwels, Ph.D., J. Beentjes, G. Kracht, and R. van Dijk for their assistance and contributions. Study supported by the Dutch Cancer Foundation, Grants IKW 8589 and IKW 90-21.

References

1. Bloem JL. Radiological staging of primary malignant musculoskeletal tumors; a correlative study of MRI, CT, Tc-99m MDP scintigraphy and angiography. [Thesis]. The Hague, 1988.

2. Bramwell, VHC. Adjuvant chemotherapy in soft tissue and bone sarcomas. In: van Oosterom A, van Unnik J, eds. Management of soft tissue and bone sarcomas. New York: Raven Press, 1985:215–220.
3. Burgers JMV, Voute PA, van Glabbeke M, Kalifa C. Adjuvant treatment for osteosarcomas of the limbs, trial 20781 of the SIOP and the EORTC radiotherapy/chemotherapy group, a preliminary communication. In: van Oosterom, van Unnik, eds. Management of soft tissue and bone sarcomas. New York: Raven Press, 1985.
4. Eilber FR, Morton DL, Eckardt J, Grant T, Weisenburger T. Limb salvage for skeletal and soft tissue sarcomas, multidisciplinary preoperative therapy. Cancer 1984;53:2579–2584.
5. The National Institutes of Health. Limb-sparing treatment of adult soft-tissue sarcomas and osteosarcomas, consensus conference. JAMA 1985;254:1791–1794.
6. Eckhardt JJ, Eilber FR, Dorey FJ, Mirra JM. The UCLA experience in limb salvage surgery for malignant tumors. Orthopedics 1985;8:612–621.
7. Eilber FR, Guiliani AE, Huth JF, Eckhardt JJ. Limb salvage for malignant tumors of bone. Prog Clin Biol Res 1985;201:25–37.
8. Simon MA, Aschliman MA, Thomas N, Mankin HJ. Limb-salvage treatment versus amputation for osteosarcoma of the distal end of the femur. J Bone Joint Surg (Am) 1986;68-A:1331–1337.
9. Enneking WF, Spanier SS, Goodman MA. The surgical staging of musculoskeletal sarcoma. J Bone Joint Surg (Am) 1980;62-A:1027–1030.
10. Bloem JL, Bluemm RG, Taminiau AHM, et al. MRI in staging primary malignant bone tumors. Radiographics 1987;7:425–446.
11. Bloem JL, Taminiau AHM, Eulderink F, Hermans J, Pauwels EKJ. Radiologic staging of primary bone sarcoma: MR imaging, scintigraphy, angiography and CT correlated with pathologic examination. Radiology 1988;169:805–810.
12. Hudson TM, Enneking WF, Hawkins IF. The value of angiography in planning surgical treatment of bone tumors. Radiology 1981;138:283–292.
13. Hudson TM, Schiebler M, Springfield DS, et al. Radiologic imaging of osteosarcoma: role in planning surgical treatment. Skel Radiol 1983;10:137–146.
14. Zimmer WD, Berquist TH, Mcleod RA, et al. Bone tumors: magnetic resonance imaging versus computed tomography. Radiology 1985;155:709–718.
15. Aisen AM, Martell W, Braunstein EM, McMillin KI, Phillips WA, Kling TF. MRI and CT evaluation of primary bone and soft-tissue tumors. AJR 1986;146:749–756.
16. Beltran J, Noto AM, Chakeres DW, Christoforides AJ. Tumors of the osseous spine: staging with MR imaging versus CT. Radiology 1987;162:565–569.
17. Ekelund L, Herrlin K, Rydholm A. Comparison of computed tomography and angiography in the evaluation of soft tissue tumors of the extremities. Acta Radio (Diagn) 1982;23:15–28.
18. Enneking WF, Chew FS, Springfield DS, Hudson TM, Spanier SS. The role of radionuclide bone-scanning in determining the resectability of soft tissue sarcomas. J Bone Joint Surg (Am) 1981;63-A:249–257.
19. Levine E, Lee KR, Neff JR, Maklad JR, Robinson RG, Preston DF. Comparison of computed tomography and other imaging modalities in the evaluation of musculoskeletal tumors. Radiology 1979;131:431–437.
20. Sundaram M, McGuire MH, Herbold DR, Wolverson MK, Heiberg E. Magnetic resonance imaging in planning limb salvage surgery for primary malignant tumors of bone. J Bone Joint Surg (Am) 1986;68-A:809–819.
21. Fisher MR, Dooms GC, Hricak H, Reinhold C, Higgins CB: Magnetic resonance imaging of the normal and pathologic muscular system. Magn Reson Imag 1986;4:491–496.
22. Reiser M, Rupp N, Biehl Th, et al. MR in diagnosis of bone tumors. Eur J Radiol 1985;5:1–7.
23. Totty WG, Murphey WA, Lee JKT. Soft-tissue tumors: MR imaging. Radiology 1986;160:135–141.
24. Zimmer WD, Berquist TH, McLeod RA, et al. Bone tumors: magnetic resonance imaging versus computed tomography. Radiology 1985;155:709–718.
25. Pettersson H, Slone RM, Spanier S, et al. Musculoskeletal tumors: T1 and T2 relaxation times. Radiology 1988;167:783–785.
26. Bloem JL, Falke THM, Taminiau AHM, et al. Magnetic resonance imaging of primary malignant bone tumors. Radiographics 1985;5:853–886.
27. Pettersson H, Gillespy T III, Hamlin DJ, et al. Primary musculoskeletal tumors: examination with MR imaging compared with conventional modalities. Radiology 1987;164:237–241.
28. Dwyer AJ, Frank JA, Sauk VJ, et al. Short TI inversion-recovery pulse sequence: analysis and initial experience in cancer imaging. Radiology 1988;168:827–836.
29. Golfieri R, Baddeley H, Pringle JS, Souhami R. The role of the STIR sequence in magnetic resonance imaging examination of bone tumours. Br J Radiol 1990;63:251–256.
30. Bloem JL, Bluemm RG, Taminiau AHM, Doornbos J. Gadolinium-DTPA enhanced MRI of primary malignant bone tumors [Abstract]. Proceedings of the 6th annual meeting of the Society of Magnetic Resonance in Medicine, New York, 1987:667.
31. Bluemm RG, Doornbos J, Koops W, Bloem JL, te Strake L. Gd-DTPA in fast field echo magnetic resonance imaging. In: Runge VM, Claussen C, et al., eds. Contrast agents in MRI. 1986:177–182.
32. Bloem JL, Reiser MF, Vanel D. Magnetic resonance contrast agents in the evaluation of the musculoskeletal system. Magn Reson Q 1990;6:136–163.
33. Traill MR, Sartoris DJ. Musculoskeletal system. In: Runge VM, ed. Enhanced magnetic resonance imaging. St. Louis: C. V. Mosby, 1989:290–300.
34. Bloem JL, Wondergem J. Gd-DTPA as a contrast agent in computed tomography. Radiology 1989;171:578–579.
35. Haustein J, Lackner K, Krahe Th, et al. The dialysibility of gadolinium-DTPA [Abstract]. Proceedings of the 9th annual meeting of the Society of Magnetic Resonance in Medicine New York, 1990:227.
36. Erlemann R, Vasallo P, Bongartz G, et al. Musculoskeletal neoplasms: fast low-angle shot MR imaging with and without Gd-DTPA. Radiology 1990;176:489–495.
37. Unger EC, Glazer HS, Lee JKT, Ling D. MRI of extracranial hematomas: preliminary observations. AJR 1986;146:403–407.
38. Gomorri JM, Grossman RI. Mechanisms responsible for the MR appearance and evolution of intracranial hemorrhage. Radiographics 1988;8(3):427–440.
39. Wilson DA, Prince JR. MR imaging of hemophilic pseudotumors. AJR 1988;150:349–350.
40. Beltran J, Simon DC, Levy M, et al. Aneurysmal bone cysts: MR imaging and 1.5 T. Radiology 1986;158:689–690.
41. Tsai JC, Dalinka MK, Fallon MD, Zlatkin MB, Kressel HY. Fluid-fluid level: a nonspecific finding in tumors of bone and soft tissue. Radiology 1990;175:779–782.
42. Sundaram M, McGuire MH, Herbold DR, Beshany SE, Fletcher JW. High signal intensity soft tissue masses on T1 weighted pulsing sequences. Skel Radiol 1987;16:30–36.
43. Nelson MC, Stull MA, Teitelbaum GP, et al. Magnetic resonance imaging of peripheral soft tissue hemangiomas. Skel Radiol 1990;19:477–482.
44. Enzinger FM, Weiss SW. Soft tissue tumors. St. Louis: C. V. Mosby, 1983.
45. Cohen EK, Kressel HY, Perosio T, et al. MR imaging of soft tissue hemangiomas: correlation with pathologic findings. AJR 1988;150:1079–1081.
46. Yuh WTC, Kathol MH, Sein MA. Hemangiomas of skeletal muscle: MR findings in five patients. AJR 1986;149:765–768.
47. Ross JS, Masaryk TJ, Modic MT, et al. Vertebral hemangiomas: MR imaging. Radiology 1987;165:165–169.
48. Laredo JD, Assouline E, Gelbert F, Wybier M, Merland JJ, Tubiana JM. Vertebral hemangiomas: fat content as a sign of aggressiveness. Radiology 1990;177:467–472.
49. Mitchell DG, Burk DL, Vinitski S, Rifkin MD. The biophysical basis of tissue contrast in extracranial MR imaging. AJR 1987;149:831–837.
50. Jelinek JS, Kransdorf MJ, Utz JA, et al. Imaging of pigmented villonodular synovitis with emphasis on MR imaging. AJR 1989;152:337–342.
51. Yulish BS, Lieberman JM, Stranjord SE, Bryan PJ, Mulopulos GP, Modic MT. Hemophilic arthropathy: assessment with MR imaging. Radiology 1987;164:759–762.
52. Sundaram M, McGuire MH, Schajowicz F. Soft-tissue masses:

histologic basis for decreased signal (short T2) on T2-weighted MR images. AJR 1987;148:1247–1250.

53. Kransdorf MJ, Jelinek JS, Moser RP, et al. Magnetic resonance appearance of fibromatosis. Skel Radiol 1990;19:495–499.

54. Francis IR, Dorovini-Zis K, Glazer GM, Lloyd RV, Amendola MA, Martel W. The fibromatoses: CT-pathologic correlation. AJR 1986;147:1063–1066.

55. Daffner RH, Kirks DR, Gehweiler JA, Heaston DK. Computed tomography of fibrous dysplasia. AJR 1982;139:943–948.

56. Dominguez R, Saucedo J, Fenstermacher M. MRI findings in osteofibrous dysplasia. Magn Reson Imag 1989;7:567–570.

57. Zeanah WR, Hudson TM, Springfield DS. Computed tomography of ossifying fibroma of the tibia. J Comput Assist Tomogr 1983;7:688–691.

58. Utz JA, Kransdorf MJ, Jelinek JS, Moser RP, Berrey BH. MR appearance of fibrous dysplasia. J Comput Assist Tomogr 1989;13:845–851.

59. Cohen JM, Weinreb JC, Redman HC. Arteriovenous malformations of the extremities: MR imaging. Radiology 1986;158:475–479.

60. Beltran J, Simon DC, Katz W, Weis LD. Increased MR signal intensity in skeletal muscle adjacent to malignant tumors: pathologic correlation and clinical relevance. Radiology 1987;162:251–255.

61. Cohen EK, Kressel HY, Frank TS, et al. Hyaline cartilage origin of bone and soft tissue neoplasms: MR appearance and histologic correlation. Radiology 1988;167:477–481.

62. Hudson TM, Hamlin DJ, Enneking WF, et al. Magnetic resonance imaging of bone and soft tissue tumors: early experience in 31 patients compared with computer tomography. Skel Radiol 1985;13:134–146.

63. Edeiken-Monroe B, Edeiken J, Jacobson HG. Osteosarcoma. Semin Roentgenol 1989;3:153–173.

64. Moore SG, Dawsen KL. Red and yellow marrow in the femur: age-related changes in the appearance at MR imaging. Radiology 1990;175:219–223.

65. Steiner RM, Mitchell DG, Rao VM, et al. Magnetic resonance imaging of bone marrow: diagnostic value in diffuse hematologic disorders. Magn Reson Q 1990;6:17–34.

66. Vogler JB, Murphey WA. Bone marrow imaging. Radiology 1988;168:679–693.

67. Sebag GH, Moore SG. Effect of trabecular bone on the appearance of marrow in gradient-echo imaging of the appendicular skeleton. Radiology 1990;174:855–859.

68. Roberts MC, Kressel HY, Fallon MD, Zlatkin MB, Dalinka MK. Paget disease: MR imaging findings. Radiology 1989;173:341–345.

69. Tjon A, Tam RTO, Bloem JL, Falke THM, et al. Magnetic resonance imaging in Paget disease of the skull. Am J Neuroradiology 1985;6:879–881.

70. Moser RP, Vinh TN, Ros PR, Smirniotopoulos JG, Madewell JE, Berrey BH. Paget disease of the anterior tibial tubercle. Radiology 1987;164:211–214.

71. Felix R, Schörner W, Laniado M, et al. Brain tumors: MR imaging with gadolinium-DTPA. Radiology 1985;156:681–688.

72. Bydder GM. Clinical application of gadolinium-DTPA. In: Stark DS, Bradley WG Jr, eds. Magnetic resonance imaging. 1st ed. St. Louis: C. V. Mosby, 1988:182–200.

73. Slutsky RA, Peterson T, Strich G, Brown JJ. Hemodynamic effects of rapid and slow infusions of manganese chloride and gadolinium-DTPA in dogs. Radiology 1985;154:733–735.

74. Laniado M, Weinman HJ, Schörner W, Felix R, Speck U. First use of gadolinium-DTPA/dimeglumine in man. Physiol Chem Phys Med NMR 1984;16:157–165.

75. Schörner W, Felix R, Laniado M, et al. Prüfung des kernspintomographischen Kontastmittels Gadolinium-DTPA am Menschen: Verträglichkeit, Kontrastbeeinflussung und erste klinische Ergebnisse. Fortschr Röntgenstr 1984;140:493–500.

76. Strich G, Hagan PL, Gerber KH, Slutsky RA. Tissue distribution and magnetic resonance spin lattice relaxation effects of gadolinium-DTPA. Radiology 1985;154:723–726.

77. Runge VM, Clanton JA, Herzer WA, et al. Intravascular contrast agents suitable for magnetic resonance imaging. Radiology 1984;153:171–176.

78. Weinmann H-J, Brasch RC, Press W-R, Wesbey GE. Characteristics of gadolinium-DTPA complex: a potential NMR contrast agent. AJR 1984;142:619–624.

79. Erlemann R, Reiser MF, Peters PE, et al. Musculoskeletal neoplasms: static and dynamic Gd-DTPA enhanced MR imaging. Radiology 1989;171:767–773.

80. Enneking WF. Staging of musculoskeletal neoplasms. Skel Radiol 1985;13:183–194.

81. Enneking WF, Spanier SS, Malawer MM. The effect of anatomic setting on the results of surgical procedures for soft parts sarcoma of the thigh. Cancer 1981;47:1005–1022.

82. Kawaguchi N, Amino K, Matsumoto S, et al. Limiting factors of limb salvage operation for musculoskeletal osteosarcoma. Proceedings of the international symposium on limb salvage in musculoskeletal oncology, Kyoto, 1987:53–54.

83. Rijdholm A, Roser B. Surgical margins for soft tissue sarcoma. J Bone Joint Surg (Am) 1987;69-A:1074–1078.

84. Thrall JH, Geslien GE, Corcoron RJ, Johnson MC. Abnormal radionuclide deposition patterns adjacent to focal skeletal lesions. Radiology 1975;659–663.

85. Chew FS, Hudson TM. Radionuclide bone scanning of osteosarcoma: falsely extended uptake patterns. AJR 1982;139:49–54.

86. Gillespie T III, Manfrini M, Ruggieri P, Spanier SS, Pettersson H, Springfield DS. Staging of intraosseous extent of osteosarcoma: correlation of preoperative CT and MR imaging with pathologic macroslides. Radiology 1988;167:765–767.

87. Holscher HC, Bloem JL, Nooy MA. The value of MR imaging to monitor the effect of chemotherapy on bone sarcomas. AJR 1990;154:763–769.

88. Pan G, Raymond AK, Carasco CH, et al. Osteosarcoma: MR imaging after preoperative chemotherapy. Radiology 1990;174:517–526.

89. Dewhirst MW, Sostman HD, Leopold KA, et al. Soft-tissue sarcomas: MR imaging and MR spectroscopy for prognosis and therapy monitoring. Radiology 1990;174:847–853.

90. Erlemann R, Sciuk J, Bosse A, et al. Response of osteosarcoma and Ewing sarcoma to preoperative chemotherapy: assessment with dynamic and static MR imaging and skeletal scintigraphy. Radiology 1990;175:791–796.

91. Holscher HC, Bloem JL, Vanel D, et al. Quantitative MR imaging changes in osteosarcoma treated with preoperative chemotherapy. Radiology 1990;177(p):221.

92. Picci P, Bacci G, Campanacci M, et al. Histologic evaluation of necrosis in osteosarcoma induced by chemotherapy. Cancer 1985;56:1515–1521.

93. Lemmi MA, Fletcher BD, Marina NM, et al. Use of MR imaging to assess results of chemotherapy for Ewing sarcoma. AJR 1990;155:343–346.

94. Sanchez RB, Quinn SF, Walling A, Estrada J, Greenberg H. Musculoskeletal neoplasms after intraarterial chemotherapy: correlation of MR images with pathologic specimens. Radiology 1990;174:237–240.

95. Tagami T, Sakuma H, Nomura Y, et al. Intravoxel inciherent motion imaging of bone and soft tissue tumors [Abstract]. Proceedings of the 9th annual meeting of the Society of Magnetic Resonance in Medicine New York, 1990:198.

96. Vanel D, Lacombe MJ, Couanet D, Kalifa C, Spielmann M, Genin J. Musculoskeletal tumors: follow-up with MR imaging after treatment with surgery and radiation therapy. Radiology 1987;164:243–245.

97. Sim FH, O'Connor MI, Donati D. Limb salvage in malignant pelvic tumors: a follow-up study [Abstract]. In: Proceedings of the international symposium on limb salvage in musculoskeletal oncology, Kyoto, 1987:22.

98. Simon MA. Limb salvage for osteosarcoma? [Abstract]. In: Proceedings of the international symposium on limb salvage in musculoskeletal oncology, Kyoto, 1987:34–35.

CHAPTER 16

Magnetic Resonance Imaging of Metastatic Disease and Multiple Myeloma

P. R. Algra and J. L. Bloem

Introduction

Metastases

The incidence of cancer continues to rise. For 1990, the estimated total number of new cancers is 1,040,000 and the estimated number of cancer deaths is 510,000 for the United States. Cancer is the second leading cause of death (1). Currently available therapy can now cure approximately 50% of cancer patients. The majority of patients succumb to the effects of metastatic disease. Metastatic neoplasms vastly outnumber primary tumors of the skeleton (2). These facts coupled with increasing financial pressures to optimize the use of diagnostic resources have led third party payers to scrutinize the use of radiologic procedures and radiologists and oncologists to look systematically at the benefits of a variety of staging procedures (3).

Historically, two theories on the pathogenesis of skeletal metastatic disease are noteworthy: the seed and soil theory of Paget ("some tumor cells find certain organs provide a more fertile soil for metastatic growth") (4) and the first station theory of Ewing ("mechanics of circulation") (5).

It is evident that some predisposing factors must prevail to promote the generation of skeletal metastatic disease, considering that the skeleton receives only 10% of the cardiac output, whereas muscle and spleen receive a far greater amount of the cardiac output and yet are rarely affected by metastatic disease (6). This finding supports the seed and soil theory of Paget, e.g., metastases from certain primary tumors have a particular tendency to develop in the bone marrow (7).

The theory explaining the relative frequency of metastatic skeletal disease as a result of mechanics (Ewing) is supported by the presence of slow flow of blood in the sinusoids, the fenestrated membrane of the endothelium lining the vessels, together with other factors promoting the development of metastatic disease (8, 9). Anatomic studies, originally carried out by Batson, suggest a preferential drainage of certain organ sites to the vertebral column, e.g., the breast, prostate (10, 11), colon, and rectum (12) by means of a vertebral venous plexus system (13–16). Later experiments (10) and autopsy studies (17) showed that indeed this venous system can be regarded as an additional pathway for metastases to reach the bone marrow in the vertebral bodies (18, 19).

The distribution of metastases in the skeleton has long been a subject of study for many authors. The distribution of skeletal metastases as assessed by different authors varies widely (20). The reasons for this variability are, among others, patient selection and the techniques used, e.g., gross inspection vs. roentgenologic survey vs. bone scintigraphy. There is general agreement, however, that metastatic disease predominantly occurs in the axial or central skeleton. This is most probably related to the fact that the axial skeleton contains red marrow (21).

The distribution of metastases in the vertebral column is, in order of frequency, lumbar, lower thoracic, thoracic, sacral, and cervical spine. The lower thoracic and lumbar spine are also the most common sites of myelomatous bone lesions (20, 22, 23).

Plain radiographs have a very low yield in depicting skeletal metastases when these are confined to the bone marrow. Experimental studies have shown that only metastases that cause a destruction of more than 50 to 70% of trabecular bone within the vertebral body can be seen on plain radiographs (24, 25). The origin of metastatic deposits is within the bone marrow of the vertebral body. Metastatic tumor may fill the marrow and leave cortical bone intact (26). Detection of metastatic disease can occur at a much earlier stage, when destruction of cortical bone is present. The contours of collapsed vertebral bodies can suggest the nature of the underlying disease (27), but conventional radiography alone is unreliable in the differentiation between benign and malignant causes of vertebral body collapse.

Abbreviations (see also glossary): BS, bone scintigraphy; SE, spin echo, STIR, short τ inversion recovery; GE, gradient echo; CSF, cerebrospinal fluid; DTPA, diethylenetriamine pentaacetic acid; MDP, methylene diphosphonate; FOV, field of view.

Computed tomography (CT) studies show that metastatic deposits in the vertebral column are usually located in the vertebral body. Here marrow is more abundant than in the pedicles and provides a fertile field for the earliest metastatic deposits (28). Computed tomographic findings of metastatic spread to vertebral bodies rather than pedicles correlate well with pathologic findings. The plexus of vertebral veins provides ample blood supply for hematogenous metastases to lodge in vertebral body marrow (13–16). While relative lack of marrow in the pedicles may predispose to radiographic visualization of metastases (29), CT is sensitive enough to discover even earlier changes in the vertebral body.

These findings have been recently corroborated by magnetic resonance (MR) imaging studies (30) and by experimental studies in which tumor cells were injected into the systemic arterial circulation in mice. Tumor cells lodged and grew in the hematopoietic bone marrow of the vertebrae. These experiments also showed that cancer cells invaded the vertebral canal through the foramina of the vertebral veins rather than by destroying cortical bone (31, 32).

The fact that destruction of the vertebral pedicle is far more easily appreciated on plain radiographs than loss of density because of lytic trabecular bone metastases, may have led to the misconception that metastatic disease is located in the pedicle more frequently than in the vertebral body.

Multiple Myeloma

Multiple myeloma is a malignancy composed of plasma cells showing variable degrees of differentiation. It primarily involves bone marrow and is usually multicentric. Patients frequently have bone pain, weakness, fatigue, and a normochromic normocytic anemia. Serum or urine electrophoresis and immunoelectrophoresis are abnormal. The diagnosis is confirmed by bone marrow aspirate and biopsy (33).

In contrast to metastatic disease, multiple myeloma deposits usually do not cause osteoblastic bone responses, therefore bone scintigraphy yields a high false negative rate in multiple myeloma. The lack of osteoblastic response in multiple myeloma is probably related to inhibitory factors produced by the multiple myeloma cells. These factors diminish the proliferation and metabolic activity of osteoblastic cells (34).

Initial evaluation of the spine by plain radiography has been recommended because of the false negative findings of bone scintigraphy in the detection of myelomatous lesions (35, 36).

Magnetic Resonance Imaging Technique

Efficient clinical MR imaging of the vertebral column can be accomplished at various field strengths. Most experience today is gained at field strengths between 0.5 and 1.5 T. A study in which images were obtained with MR scanners operating at different field strengths (37) did not reveal significant differences in diagnostic quality in imaging of the cervical spine. Unless specified differently, images in this chapter were obtained with MR scanners operating at 0.5 or 0.6 T (Philips Gyroscan, Shelton, Conn.; General Electric, Milwaukee, Wisc.).

Long rectangular surface coils (e.g., 10 × 40 cm) with a large field of view (FOV) allow imaging of the entire vertebral column with high signal-to-noise ratio. Coils with a smaller FOV will benefit from a coil holder and marker system that make it possible to move the surface coil without repositioning the patient (38). Switched array coils can change the FOV electronically and the area of the spine imaged without moving the patient (39, 40).

Magnetic resonance imaging of the spine can be performed using several pulse sequences: spin-echo (SE), short τ inversion recovery (STIR), gradient-echo (GE) and chemical shift imaging. Each pulse sequence has its own imaging features and will exhibit normal bone marrow and pathology in a characteristic manner.

Most experience today is gained with the spin-echo, or SE, pulse sequence. In general, T1-weighted SE images will show metastases better than T2-weighted SE images (41). Some authors prefer a combination of T1- and T2-weighted images, as T2-weighted images sometimes reveal additional lesions (42). T2-weighted images can be used for further characterization of the nature of metastases (lytic vs. sclerotic) and to evaluate the arachnoidal space and compressive myelopathy (43). T2-weighted SE images are less useful in evaluating bone marrow changes after radiation therapy than either T1-weighted SE images or STIR images (44).

The STIR pulse sequence produces images characterized by yellow marrow signal intensity being nulled, such that bone marrow appears black. T1 and T2 values other than fat are additive. Lytic metastases, because of their high water content, produce an increase in both T1 and T2 relaxation times. This will enhance the contrast between lytic metastases and bone marrow. Sometimes lytic metastases are better depicted on STIR images than on T1-weighted SE images (45–49). In some cases fatty infiltration after radiation therapy can be seen earlier on STIR images than on T1-weighted SE images (41).

Gradient-echo (GE) imaging offers the opportunity to obtain images in a shorter acquisition time than SE imaging. Image contrast is dependent not only on the instrument settings (repetition and echo times and flip angle) but also on field inhomogeneity, chemical shift, and susceptibility. Bone marrow in contact with trabecular bone results in shortened effective transverse relaxation time (T2*) and a resultant signal loss because of local field inhomogeneities where mineralized matrix interfaces with it. Shortening of T2* is caused by iron deposits in red bone marrow as well. The amount of trabecular bone in addition to marrow composition (red vs. yellow) is especially reflected on GE images (50). When the TR is relatively long (more than 200 ms) or when spoiler pulses are used, T1 weighting is obtained by choosing large flip angles (60° or more). Proton density and T2*-weighting can be achieved by small flip angles (20°) in spoiled GE sequences, or large flip angles in steady-state sequences. Gradient-echo T2*-weighted images define the subarachnoid space well and can increase lesion conspicuousness (51). However, some bone marrow abnormalities may be better seen on T1-

weighted SE than on T2*-weighted GE, so that both techniques are considered complementary (52).

Chemical shift imaging methods depend on the difference in resonance frequency between aliphatic and water protons. This resonance frequency difference is 3.5 ppm or 75 to 150 Hz at 0.5 to 1.0 T and 220 Hz at 1.5 T (53). The echo time determines whether the fat and water protons are in phase or out of phase. Out-of-phase images will increase image contrast in the case of lytic bone marrow metastases (54, 55). Fat suppression with pre-saturation pulses with further increase contrast on T2-weighted and Gd-DTPA enhanced images.

T1 and T2 measurements, or quantitative MR, do not improve diagnostic potential of MR imaging, but may have a role in serial studies in patients undergoing myelosuppressive therapy (46).

Presaturation is a technique for diminishing artifacts caused by blood flow. These artifacts caused by pulsating blood flow in the major blood vessels are superimposed on the vertebral column and can degrade the image. The basic principle is that an extra radio-frequency pulse is applied outside the imaging volume to saturate inflowing spins without altering the signal intensity of stationary tissues within the imaging volume. The inflowing, presaturated spins then appear dark when imaged with the regular pulse sequence. The technique eliminates ghost artifacts from pulsatile flow. Flow compensation and cardiac gating can accomplish cerebrospinal fluid (CSF) flow artifact-free images (56).

Gadolinium-labeled diethylene triamine pentaacetic acid (DTPA) may increase the specificity of MR imaging, as metastatic deposits in vertebral bodies will exhibit diffuse enhancement of signal intensity on T1-weighted images, whereas osteoporotic collapsed vertebral bodies will show bandlike enhancement (57). Some authors, however, found that the use of Gd-DTPA can be disappointing in the differential diagnosis of malignant and benign tissues (58). Leptomeningeal spread of tumor is much more conspicuous on Gd-DTPA enhanced MR images than on nonenhanced images. On the other hand, Gd-DTPA enhanced images may obscure metastatic lesions within the bone marrow, if fat is not suppressed. In multiple myeloma, Gd-DTPA does not increase the number of lesions detected (49, 52, 58–63).

T1- and T2/T2*-weighted surface coil-obtained sagittal images constitute a sufficient routine imaging protocol for the spine. Additional imaging in the axial or coronal planes provides valuable information in the setting of soft-tissue involvement or compressive myelopathy. Gd-DTPA should be reserved for selected cases, such as distinction between malignant and benign collapsed vertebrae, suspicion of leptomeningeal tumor spread, an evaluation of epidural extension of soft-tissue tumor.

Normal Magnetic Resonance Appearance of the Vertebral Column

Bone marrow in the adult shows a high signal intensity on T1-weighted images and an intermediate to high signal intensity on T2-weighted images because of its high lipid content. The variations in signal intensity of bone marrow between vertebral bodies within one individual are small, but the interindividual variations are reported to be large (63). Cortical bone, because of its low percentage of mobile protons, is of low signal intensity on both T1- and T2/T2*-weighted images. The cerebrospinal fluid occupying the subarachnoid space shows a low signal intensity on T1-weighted images and a high signal intensity on T2/T2*-weighted images.

The age-related change in the composition of the bone marrow is described as a successive decrease in cellular red marrow and a corresponding increase in fat content (21, 64, 65, 66). The change in composition is largest during the first decade of life and after the age of 65 years. Thus, as the patient ages, the T1 and T2 relaxation times of the vertebral marrow decrease (64, 67–69). The change in signal intensity of the vertebral bone marrow as a result of aging produces a distinctive peripherally bandlike pattern with multiple small and large areas of high signal intensity fatty marrow (70). The low signal intensity seen on T1-weighted images in younger control subjects may make it difficult to discriminate between T1 relaxation time in healthy persons and T1 relaxation time in patients with infiltrative disease (69).

One study reports that for the age group of 51 years and over, T1 and T2 relaxation times are slightly lower for females than for males. This could be due to the loss of bone and mineral content, which is more rapid and significant in women (67). A recent study did not show differences in signal intensity of bone marrow between sexes (70).

A decrease in T2 relaxation time of the spinal bone marrow has been noted after disuse. This phenomenon is explained by the replacement of hematopoietic marrow by fatty marrow (71).

Metastatic Disease, Multiple Myeloma, and Differential Diagnosis

Bone marrow contains a large quantity of mobile protons within fat and water in cellular hematopoietic and stromal tissue. It is therefore ideally suited for MR imaging. Especially in the case of lytic bone marrow metastases, the high contrast between fat-containing bone marrow and the high concentration of water in the cytoplasm of metastatic cells renders ideal contrast by MR imaging.

When analyzing the MR appearance of metastatic disease in the vertebral column, one can distinguish at least four patterns (72): two focal (focal lytic and focal sclerotic, or a combination of focal lytic and focal sclerotic) and two diffuse (diffuse homogeneous and diffuse inhomogeneous).

Focal lytic metastases show a localized area of decreased signal intensity on T1-weighted images and an increased signal intensity on T2/T2*-weighted images (Fig. 16.1). Focal sclerotic lesions exhibit a low signal intensity on both T1- and T2/T2*-weighted images (Fig. 16.2). In the diffuse

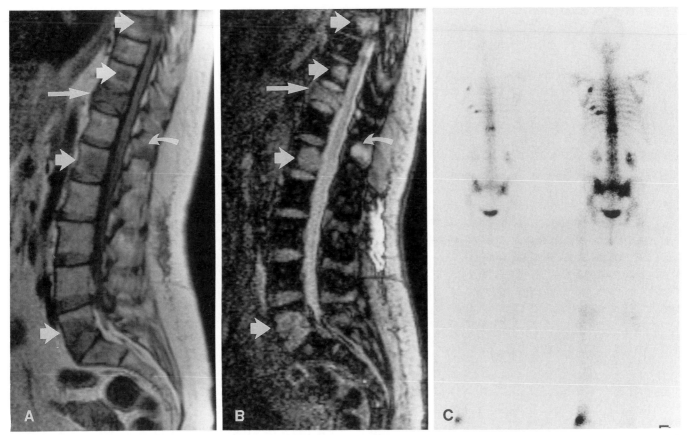

Figure 16.1. Focal lytic metastases. Sagittal T1-weighted SE image **(A),** T2*-weighted gradient-echo image **(B),** and bone scan **(C).** Metastases are seen as focal areas of low signal intensity on T1 and high signal intensity on T2*-weighted images at multiple levels *(short arrows).* Signal intensities are consistent with increased water content. T11 vertebral body shows loss of height and complete loss of normal signal intensity *(long arrow),* representing metastasis as well; there is no evidence of compressive myelopathy. Abnormal signal intensity also shows in T12 spinous process *(curved arrow).* Bone scan of same patient **(C)** shows areas of increased uptake of the radiopharmaceutical in the spine and ribs. Note that MR imaging reveals more metastases.

patterns, the bone marrow of the spine show a diffuse low signal intensity on T1-weighted images and a high signal intensity on T2/T2*-weighted images. These diffuse patterns can present as inhomogeneous irregular (Fig. 16.3) or homogeneous, smooth patterns (Fig. 16.4).

In some cases these abnormalities in signal intensity may indicate benign disease; bone islands will show low signal intensity on both T1-weighted and T2-weighted images and cannot be distinguished from sclerotic metastases by MR imaging.

Multiple myeloma exhibits low signal intensity on T1-weighted images (Fig. 16.5) and high signal intensity on T2-weighted images and cannot be differentiated from lytic metastases. No reliable differentiation is possible between leukemias (myeloproliferative and lymphoproliferative disease) and a variety of benign hematological (see also Chapter 10) disorders by MR imaging: signal intensity reflects the marrow cellularity, as has been shown in patients with diffuse infiltrative diseases (64, 69).

Loss of visibility of the basivertebral vein can be an early sign of diffuse bone marrow disease, such as bone marrow metastasis (Fig. 16.6). Disappearance of the basivertebral vein may be more conspicious than abnormal signal intensity in early bone marrow involvement (73).

Measurement of signal intensity in relation to adjacent normal vertebrae can be helpful in the distinction of malignant and benign lesions (74, 75). Further chararcterization can be accomplished by follow-up studies, which will show a return to normal signal intensity in the case of benign lesions causing collapsed vertebrae (48).

Gd-DTPA-enhanced images are obligatory if leptomeningeal spread is expected. On the other hand, Gd-DTPA may obscure metastatic lesions within the bone marrow as some metastases will become isointense with normal bone marrow (Fig. 16.7) (49, 52, 60–62).

Although the MR characteristics of metastatic disease do not always allow reliable differentiation between benign and malignant disease, MR imaging is more specific than bone scintigraphy in the distinction between benign vs. malignant disease. For instance, by means of MR examination it is usually possible to make a reliable distinction between degenerative bone disease (Figs. 16.8 and 16.9) and malignant infiltration (76–86). Morphologic characteristics of the lesion and adjacent disk must be taken into account. Loss of height of the intervertebral disk can be seen in degenerative disease and in diskitis, whereas the shape and height of the disk is usually preserved in metastatic diseases. In diffuse bone marrow lesions, the in-

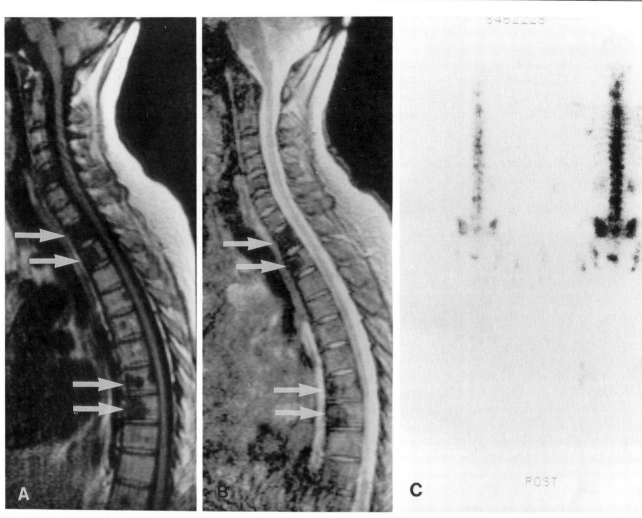

Figure 16.2. Focal sclerotic metastases. Sagittal T1-weighted SE image **(A)**, T2*-weighted gradient-echo image **(B)**, and bone scan **(C)**. Metastatic deposits at multiple levels show a low signal intensity on both T1-weighted and T2*-weighted images *(arrows)*. Signal intensity changes resemble those of cortical bone. MR image is not specific for metastatic disease; bone islands will show the same MR characteristics as sclerotic metastatic disease. Bone scan reveals lesions in the spine, pelvis, and proximal femora.

tervertebral disk may show a relative high signal intensity on T1-weighted SE images (87).

Morphology of vertebral bodies and signal intensity changes of bone marrow, particularly in relation to the vertebral end plates, may assist in differentiating between benign and malignant compression fractures (48, 79, 82, 87–89). In old osteoporotic compression fractures, the signal intensity is typically normal (fat), whereas in malignant compression fractures an abnormal signal intensity due to replacement of bone marrow is seen. In the (sub)acute phase, an inhomogeneous increase of signal intensity on STIR or T2-weighted SE images, or a sharply delineated isointense vertical band of preserved normal (fatty) bone marrow along the dorsal aspect of the compressed body (Fig. 16.10), favors a benign fracture (48, 89, 90). The abnormal signal intensity in benign fractures may have the shape of a horizontal band. Fractures at multiple levels with preservation of normal bone marrow, vertebral body fragmentation, and disk rupture also indicate benign disease. Signs in favor of malignancy are homogeneous signal intensity changes, convex anterior and especially posterior contour, cortical destruction, multiple levels with abnormal signal intensity but without fracture, and abnormal signal intensity in the posterior elements (48, 89, 90). Paraspinal masses are more conspicious in malignant disease, but can be seen in both traumatic and malignant cases (89).

Care must be taken in interpreting MR examinations in recently collapsed vertebral bodies, as they can show signal intensities indistinguishable from metastatic disease (Fig. 16.11) (89, 91–93).

Sensitivity of Magnetic Resonance in the Detection of Vertebral Metastases

Since the introduction of ^{99m}Tc-labeled methylene diphosphonate (MDP) (93), bone scintigraphy is considered the first choice in the detection of skeletal metastatic disease. The major advantages of bone scintigraphy in the

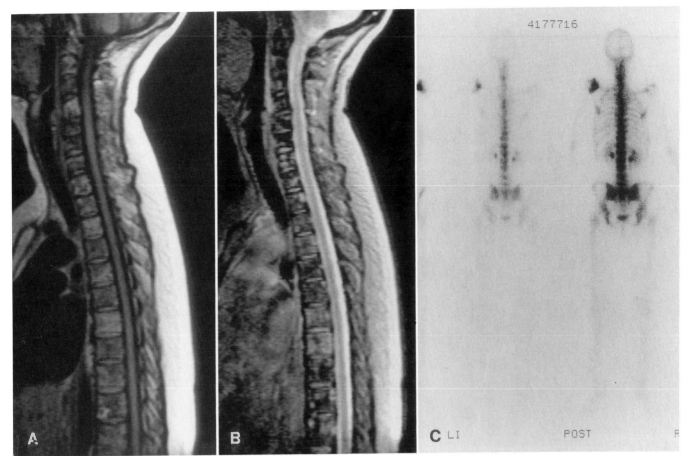

Figure 16.3. Diffuse, inhomogeneous metastases. Sagittal T1-weighted SE image **(A)**, T2*-weighted gradient-echo image **(B)**, and bone scan **(C)**. A metastatic pattern of diffuse, inhomogeneous low signal intensity on the T1-weighted image and a diffuse, inhomogeneous high signal intensity on the T2*-weighted gradient image are shown. High signal intensity of CSF makes it easy to rule out compressive myelopathy. Bone scan shows areas of increased accumulation of radiopharmaceutical in the spine and the left shoulder. MR imaging shows more vertebral metastases than bone scintigraphy; the major advantage of bone scintigraphy is imaging of the entire skeleton in one examination.

diagnosis of skeletal metastases are imaging of the entire skeleton in one examination and its high sensitivity in detecting skeletal metastases. Skeletal metastases are associated with both destruction of bone and new bone formation, which occur simultaneously. New bone is formed in all metastases once there is some bone destruction. This is similar to callus repair after a fracture and may be due to stress to weakened bone (94). Experimental studies suggest that this occurs during osteoclast-mediated bone destruction. Once the destruction is gross and malignant cells surround residual bone, no new bone is formed and no osteoclasts are seen. This may explain the occasional false negatives in skeletal scintigrams when conventional radiographs demonstrate large lytic metastases; at this stage there is no reactive new bone to concentrate the bone-seeking isotope (95).

The specificity of ^{99m}Tc-MDP bone scintigraphy is, in contrast to its sensitivity, very low. Attempts have been made to improve the specificity by additional imaging after 24 hr in combination with the usual 4-hr uptake (or 24 hr: 4 hr ratio) (96) and by obtaining images with a pinhole collimator in order to increase the spatial resolution of bone scintigrams (97).

Several retrospective studies address the sensitivity of MR imaging as compared with bone scintigraphy (42, 45, 98–106). In a prospective study (107) in which MR imaging was compared with bone scintigraphy in the detection of vertebral metastases, it was found that MR imaging is more sensitive in the detection of vertebral metastases than bone scintigraphy (Fig. 16.12). This is not surprising, since MR imaging demonstrates bone marrow per se rather than depending on secondary signs such as new bone formation (Fig. 16.13).

Magnetic resonance imaging in patients with malignant disease, including patients suffering from multiple myeloma, showed abnormal signal intensity in all patients, whereas bone scintigraphy was normal in all patients with multiple myeloma (108, 109). Magnetic resonance imaging, as correlated with gross pathology, correctly depicted myelomatous lesions. In addition, early detection of imminent medullary compression led to successful radiation therapy before symptoms of compressive myelopathy appeared (110).

The high yield of false negative findings by bone scintigraphy in patients with multiple myeloma is well known. This is probably related to the absence of an osteoblastic

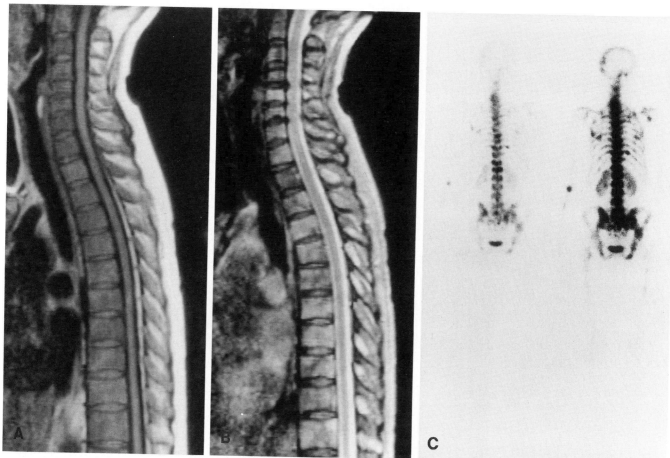

Figure 16.4. Diffuse, homogeneous metastases. Sagittal T1-weighted SE **(A)**, T2*-weighted gradient-echo **(B)** images, and bone scan **(C).** MR imaging shows a diffuse, homogeneous pattern of decreased sig-nal on T1-weighted and increased signal intensity on T2*-weighted im-ages. Bone scan shows multiple hot spots, largely confined to the axial skeleton (spine and pelvis), and lesions in ribs and shoulder.

Figure 16.5. Multiple myeloma. A coronal T1-weighted SE image of the lower lumbar spine and pelvis is shown. Multiple areas of low sig-nal intensity in lumbar spine *(long white arrows),* pelvis *(curved black arrows),* and proximal femora *(short white arrows)* are visible. Lesions are not specific, and cannot be differentiated from focal lytic metas-tases.

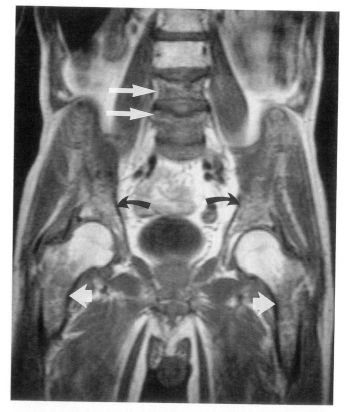

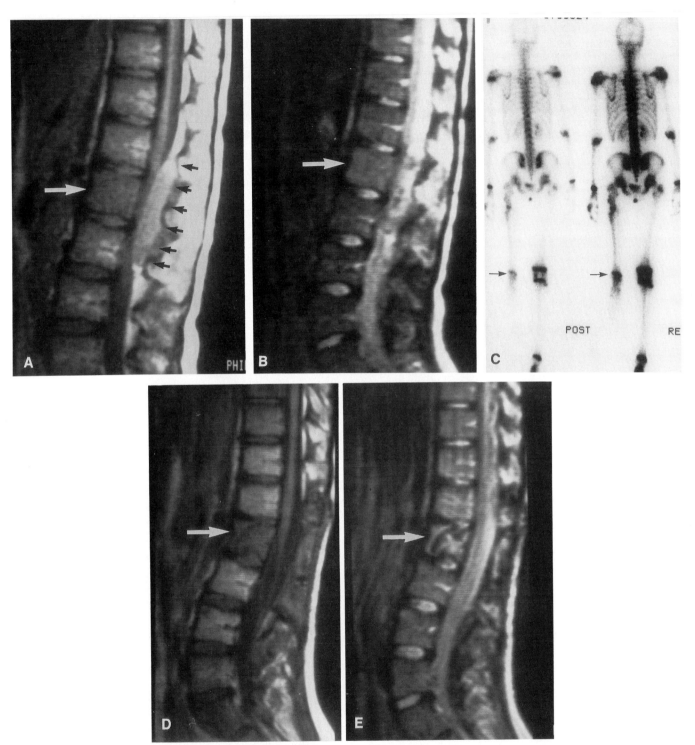

Figure 16.6. Loss of visibility of the basivertebral vein in a patient with metastases from osteosarcoma of the knee. Sagittal T1-weighted SE **(A** and **D)** and T2-weighted SE **(B** and **E)** images of the lumbar spine before and after laminectomy at T12–L2, and bone scan **(C).** Bone marrow of L1 shows a subtle loss of signal intensity on the T1-weighted image **(A)** and a slight increase in signal intensity on the T2-weighted image **(B)** on the preoperative MR examination *(arrows).* The loss of outline of the basivertebral vein is far more conspicious. The large, epidural mass *(short black arrows)* proved to be epidural metastasis at surgery. At the time of initial MR examination, the bone scan **(C)** was negative. The patient was treated with resection and rotation plasty; this accounts for the hot spot at the level of anastomosis of tibia and femur and elevated ankle *(arrow). Follow-up MR imaging* **(D** and **E)** shows laminectomy at T12–L2 and progressive metastatic disease at L1.

Figure 16.7. Gd-DTPA in focal lytic metastases. Sagittal T1-weighted SE images before **(A)** and after **(B)** the administration of Gd-DTPA. Metastatic lesions are seen as areas of low signal intensity on precontrast images. Gd-DTPA was administered in this patient to rule out leptomeningeal spread of metastases. Note that Gd-DTPA increases the signal intensity of the metastases confined to the bone marrow, making them isointense with normal bone marrow *(arrows)*.

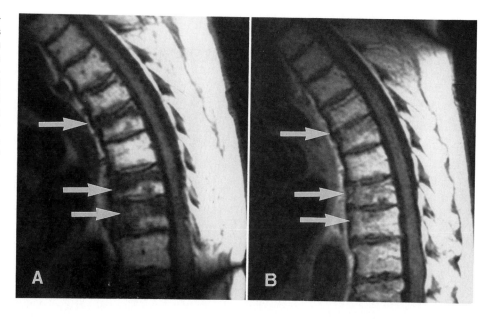

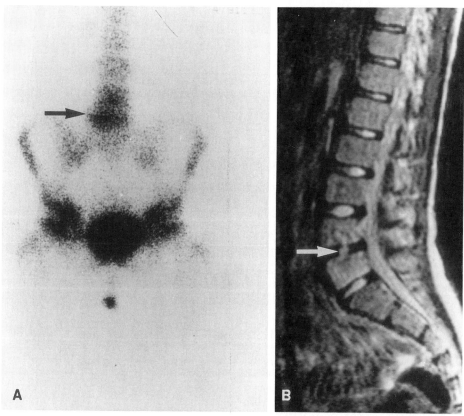

Figure 16.8. Schmorl's node. Coned down view bone scan **(A)** shows a hot spot in the lower lumbar spine *(black arrow)* in a patient with back pain. Differentiation between malignant disease and benign lesion is not possible by bone scintigraphy alone. Sagittal T2-weighted SE image **(B)** shows low signal intensity of the L4–L5 disk that herniates into the adjacent bodies *(white arrow)*. The true nature of the hot spot on bone scan is thus revealed by MR imaging. (Reprinted with permission from Algra PR, Bloem JL, Verboom LJ, et al. Radiographics 1991;11:219–232.)

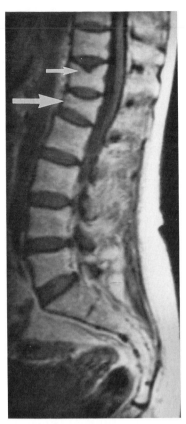

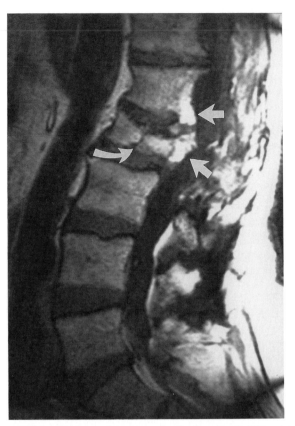

Figure 16.9. Old compression fracture and degenerative disease. Sagittal T1-weighted SE image shows wedge-shaped body of first lumbar vertebra *(large arrow).* Signal intensity of the bone marrow is normal, thus indicating a benign compression fracture. Schmorl's node is visible at T12 *(small arrow).* MR imaging in general is more specific for the differentiation between benign and malignant causes of compression fractures than bone scintigraphy.

Figure 16.10. Old traumatic compression fracture. Sagittal T1-weighted SE image of the lumbar spine shows loss of height of L2, a linear area of decreased signal intensity representing a fracture line *(curved arrow),* and areas of increased signal intensity at the posterior parts of L1 and L2 *(straight arrows),* representing fatty degeneration. Normal signal intensity of residual bone marrow and areas of increased signal intensity on the T1-weighted image are pathognomonic for the diagnosis of an old compression fracture.

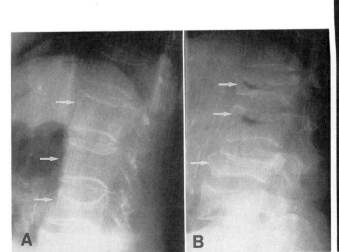

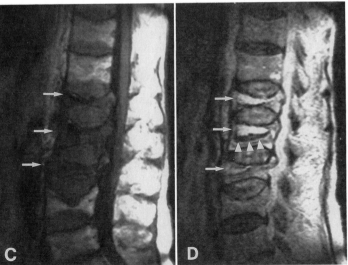

Figure 16.11. Recently collapsed vertebrae, showing signal intensity abnormalities that cannot always be distinguished from metastases. Plain X-ray **(A)** shows osteoporotic lumbar spine in a female patient treated with steroids (L1, L2, and L3 vertebrae marked by arrows). Patient developed compression fractures **(B)** of L1–L3 *(arrows).* Vac- uum phenomenon in L1 and L2 vertebral bodies is a result of avascular necrosis. MR imaging shows low signal intensity on the T1-weighted image **(C)** and high signal intensity on the T2-weighted **(D)** SE images *(arrows).* Gas in L1 and L2 vertebral bodies is seen as linear streaks of low signal intensity on the T2-weighted image *(arrowheads).*

Figure 16.12. MR imaging and bone scintigraphy findings in a group of 71 patients known to have cancer and skeletal metastases. In a prospective study, MR imaging of the spine was compared with bone scintigraphy (BS). Graph shows total number of abnormal vertebrae *(horizontal axis)* identified by MR imaging and bone scintigraphy and the level of the vertebral column *(vertical axis).* Note that MR imaging detects more abnormal vertebrae suspect for metastases than bone scintigraphy at each level of the vertebral column, and that the relative distribution of the metastases detected by MR imaging parallels the relative distribution as detected by bone scintigraphy. (Reprinted with permission from Algra PR, Bloem JL, Verboom LJ, et al. Radiographics 1991;11:219–232.)

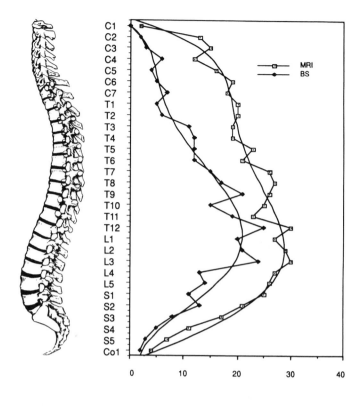

bone response to myelomatous lesions (34). Magnetic resonance imaging of the bone marrow in patients in multiple myeloma will show more lesions than bone scintigraphy (108–110).

Bone marrow-specific radionuclide pharmaceuticals have been recently developed. One of these is ^{99m}Tc-labeled nano colloid (111–116). This radiopharmaceutical accumulates in the intact reticuloendothelial system. Bone marrow replacement is demonstrated as photopenic areas or areas of irregular expansion in such lesions as multiple myeloma. Experience with these newly developed bone marrow agents is as yet found only in European clinical centers (117). In a limited series of patients (118), it was concluded that bone marrow scintigraphy reveals more lesions in patients with multiple myeloma than bone scintigraphy.

Compressive Myelopathy

Magnetic resonance imaging is the modality of choice for diagnosis of compressive myelopathy, as it allows imaging of the tissues surrounding the cord (51, 98, 118–123). Magnetic resonance imaging will accurately indicate the level of compression of the cord (Fig. 16.14). Especially in the case of cord compression at multiple levels, MR imaging will obviate the need for multiples myelographic punctures. Compressive myelopathy can be seen on T2- and T2*-weighted sagittal images of the spine, which show a high signal intensity lesion compressing the cord.

Magnetic resonance imaging is superior to myelography, CT, and plain films in detecting bone and epidural involvement by tumor, and is valuable in clinical decision making (43, 52, 120–123). One early study is more conservative in the estimation of the value of MR imaging in

the diagnostic work-up of patients with suspected malignant spinal disease as compared with myelography (124). However, more recent reports supply us with increasing evidence that surface coil MR images are superior to myelography and CT in the diagnosis of vertebral metastatic disease with or without epidural involvement of tumor (43, 98, 118–125).

Magnetic resonance imaging is recommended as the initial study in patients with suspected metastatic disease within the spinal canal (43). Myelography should be reserved for patients who for some reason cannot undergo a technically adequate or expeditious MR imaging procedure (51).

Also in the case of multiple myeloma, MR imaging can easily demonstrate the upper and lower limits of extradural compression and multiple levels of cord compression without recourse to intrathecal contrast agents (126).

Effects of Radiation Therapy on Magnetic Resonance Signal Intensity

Ionizing radiation exerts a depressive effect on the hematopoietic cells in the irradiated volume and yields eventually a corresponding increase in fat content (65), resulting in an increase of signal intensity on T1-weighted images (51, 64, 127–130). Magnetic resonance imaging shows characteristic areas of increased signal intensity of bone marrow corresponding with radiation ports (Fig. 16.15).

During the first weeks, edema may cause an increase of signal intensity on STIR sequences (44). After 3 to 6 weeks, a heterogeneous increase of signal intensity consistent with yellow marrow regeneration can be observed on T1-weighted SE images. The fatty replacement often starts

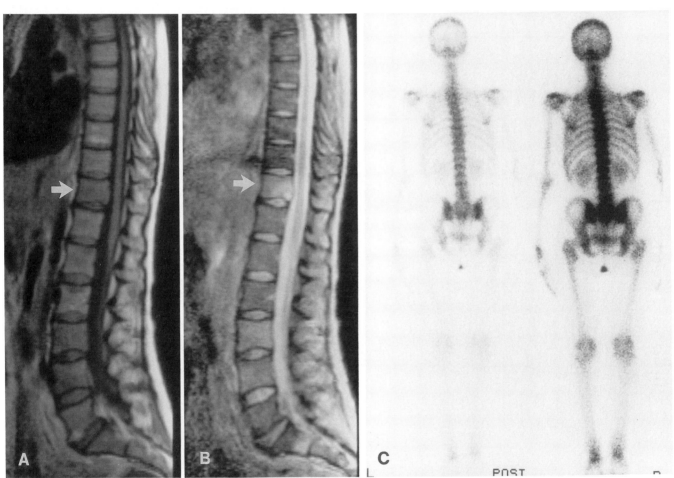

Figure 16.13. False negative bone scan and positive MR imaging. Sagittal T1-weighted SE image **(A)** shows diffuse pattern with decreased signal intensity and diffuse pattern of high signal intensity on T2-weighted SE image **(B)**. Subtle loss of height of vertebral body T11 *(arrow)* has occurred with no signs of compressive myelopathy. At this time the bone scan was negative **(C)**. Follow-up studies showed progressive metastatic disease. Diagnosis was biopsy proven in this patient. (Reprinted with permission from Algra PR, Bloem JL, Verboom LJ, et al. Radiographics 1991;11:219–232.)

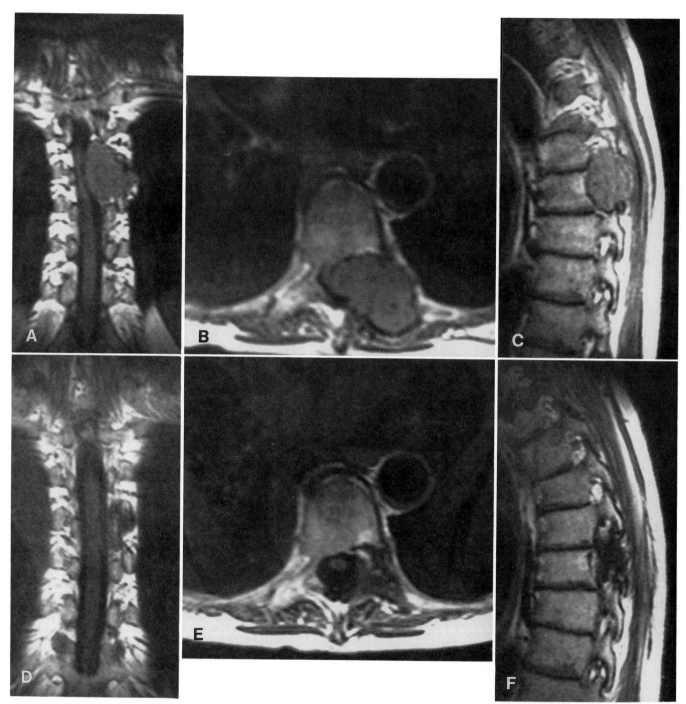

Figure 16.14. Compressive myelopathy. Frontal **(A** and **D)**, transverse **(B** and **E)**, and sagittal **(C** and **F)** T1-weighted SE images. Metastatic lesion at midthoracic level, with prominent epidural involvement, displaces the thoracic cord **(A, B** and **C)**. Marked reduction in diameter of the spinal cord at the site of maximal involvement is best seen in the transverse plane **(C)**. After chemotherapy, a marked reduction in size of the soft-tissue extension is noted **(D, E,** and **F)**. This case illustrates the advantage of multiplanar MR imaging.

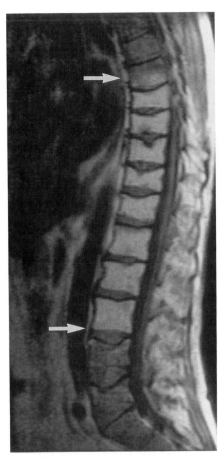

Figure 16.15. Irradiated bone marrow. Sagittal T1-weighted image of the thoracic and lumbar spine in a patient who received radiation therapy. The area of the spine with abnormally high signal intensity on T1-weighted image corresponds to the radiation ports *(arrows).*

ate signal intensity surrounding a central zone of bright signal intensity can be seen on T1-weighted images. The pattern may represent regeneration of hematopoietic marrow (44).

Therapeutic response in patients with osseous metastases is difficult to evaluate. The role of MR imaging in the assessment of therapeutic response in this situation remains to be established. In the case of soft-tissue extension of metastatic disease, MR imaging is very helpful (Fig. 16.9). At present, plain radiography remains the method of choice in follow-up of patients after therapy of metastases confined to the bone and bone marrow (131–134).

Clinical Relevance of Magnetic Resonance Imaging

Currently, bone scintigraphy remains the preferred screening procedure. However, the role of MR imaging is rapidly increasing. Magnetic resonance imaging can visualize metastatic disease when bone scintigraphy is falsely negative. Thus, MR imaging is indicated when a strong clinical suspicion is combined with a negative bone scan.

Magnetic resonance imaging may also elucidate the true nature of hot spots in the vertebral column in cancer patients, as it can often assist in making a distinction between malignant and benign disease. Magnetic resonance imaging can also be helpful in the guidance of vertebral biopsies.

In order to rule out compressive myelopathy or to establish soft-tissue extension of tumor tissue (Fig. 16.16), multiplanar MR imaging offers unique imaging features. The use of CT or MR imaging is helpful in determining the local extent of metastatic disease when planning palliative surgery or radiation therapy (135).

Acknowledgment

The assistance of J. W. Arndt, M.D., is gratefully acknowledged.

around the basivertebral vein (44). This progresses characteristically into homogeneous high signal on T1-weighted SE images of fatty marrow. Especially in young individuals, a band pattern with a peripheral region of intermedi-

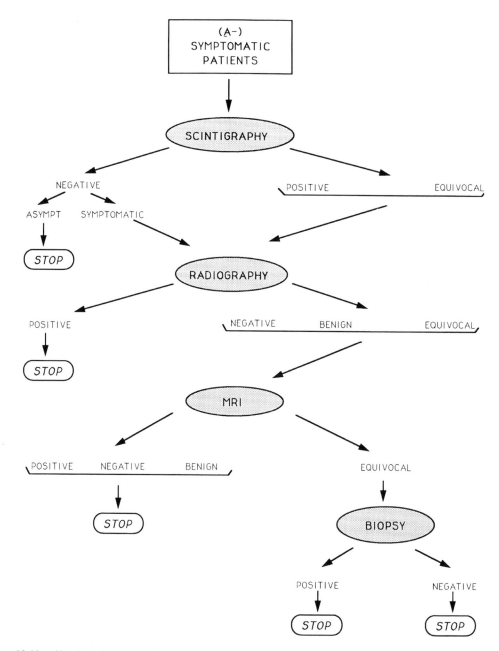

Figure 16.16. Algorithm for diagnostic work-up for patients with suspected metastatic disease (see text).

References

1. Silverberg E, Borin CC, Squires TS. Cancer statistics. Cancer 1990;40:9–26.
2. Moser RP, Madewell JE. An approach to primary bone tumors. Radiol Clin N Am 1987;26:1049–1093.
3. McNeill B. Value of bone scanning in neoplastic disease. Semin Nucl Med 1984;14:277–286.
4. Paget S. The distribution of the secondary growths in cancer of the breast. Lancet 1889;1:571–573.
5. Ewing J. Metastasis. In: Neoplastic diseases. A textbook on tumors. 3rd ed. Philadelphia: W. B. Saunders, 1928:77–89.
6. Johnston AD. Pathology of metastatic tumors in bone. Clin Orthoped 1970;73:8–32.
7. Nicolson GL. Organ specificity of tumor metastasis; role of preferential adhesion, invasion and growth of malignant cells at specific secondary sites.. Cancer Metastasis Rev 1988;7:143–188.
8. de Bruyn PPH. Structural substrates of bone marrow function. Semin Hematol 1981;18:179–193.
9. Brookes M. Blood vessels in the bone marrow. In: The blood supply of bone. An approach to bone pathology. London: Butterworths, 1971:67–91.
10. Coman DR, De Long RP. Role of the vertebral venous system in the metastasis of cancer to the spinal column. Experiments with tumor cell suspensions in rats and rabbits. Cancer 1951;4:610–618.
11. Dodds PR, Caride VJ, Lytton B. The role of vertebral veins in the dissemination of prostatic carcinoma. J Urol 1981;126:753–755.
12. Vider M, Maruyama R, Narvaez R. Significance of the vertebral venous (Batson's) plexus in metastatic spread in colorectal carcinoma. Cancer 1977;40:67–71.
13. Batson OV. The function of the vertebral veins and their role in the spread of metastases. Ann Surg 1940;112:138–149.

14. Batson OV. The role of the vertebral veins in metastatic processes. Ann Intern Med 1942;16:38–45.
15. Batson OV. The vertebral vein system. AJR 1957;78:195–212.
16. Batson OV. The Valsalva manoeuver and the vertebral vein system. Angiology 1960;11:443–447.
17. Messmer B, Sinner W. Der vertebral Metastasierungstyp. Dtsch Med Wochenschr 1966;91:2061–2066.
18. Gowin W. Die bedeutung des Wirbelsäulenvenensystems bei der Metastasenbildung. Strahlentherapie 1983;159:682–689.
19. Johnstone AS. Experimental study of the vertebral venous system. Proc Roy Soc Med 538–540.
20. Galasko CSB. Incidence and distribution of skeletal metastases. In: Skeletal metastases. London: Butterworths, 1986:14–21.
21. Kricun ME. Red-yellow marrow conversion: its effect on the location of some solitary bone lesions. Skel Radiol 1985;14:10–19.
22. Ludwig M, Kumpan W, Sinzinger H. Radiography and bone scintigraphy in multiple myeloma: a comparative analysis. Br J Radiol 1982;55:173–181.
23. Fornasier VL, Horne JG. Metastases to the vertebral column. Cancer 1975;36:590–595.
24. Edelstyn GA, Gillespie PJ, Grebell FS. The radiological demonstration of osseous metastases. Exp Obs Clin Radiol 1967;18:158–162.
25. Ardran GM. Bone destruction not demonstrable by radiography. Br J Radiol 1951;24:107–109.
26. Shackman R, Harrison CV. Occult bone metastases. Br J Surg 1948;35:385–389.
27. Sartoris DJ, Clopton P, Nemcek A, Dowd C, Resnick D. Vertebral body collapse in focal and diffuse disease: patterns of pathologic processes. Radiology 1986;160:479–483.
28. Braunstein EM, Kuhns LR. Computed tomographic demonstration of spinal metastases. Spine 1983;8:912–915.
29. Jacobson HG, Poppel MH, Shapiro JH, Grossberger S. The vertebral pedicle sign. A Roentgen finding to differentiate metastatic carcinoma from multiple myeloma. AJR 1958;80:817–821.
30. Asdourian PL, Weidenbaum M, Dewald RL, Hammerberg KW, Ramsey RG. The pattern of vertebral involvement in metastatic vertebral breast cancer. Clin Orthoped 1990;250:164–170.
31. Arguello F, Baggs RB, Duerst RE, Johnstone L, McQueen K, Frantz CN. Pathogenesis of vertebral metastasis and epidural and epidural spinal cord compression. Cancer 1990;65:98–106.
32. Manabe S, Tanaka H, Higo Y, Park P, Ohno T, Tateishi A. Experimental analysis of the spinal cord compressed by spinal metastasis. Spine 1989;14:1308–1315.
33. Schreiman JS, McLeod RA, Kyle RA, Beabout JW. Multiple myeloma: evaluation by CT. Radiology 1985;154:483–486.
34. Evans CE, Galasko CSB, Ward C. Does myeloma secrete an osteoblast inhibiting factor? J Bone Joint Surg (Br) 1989;71:288–290.
35. Wahner MD, Kyle RA, Beabout JW. Scintigraphic evaluation of the skeleton in multiple myeloma. Mayo Clin Proc 1980;55:739–746.
36. Woolfenden JM, Pitt MJ, Durie BGM, Moon TE. Comparison of bone scintigraphy and radiography in multiple myeloma. Radiology 1980;134:723–728.
37. Jack CR, Berquist TH, Miller GM, et al. Field strength in neuro-MR imaging: a comparison of 0.5T and 0.5T. J Comput Assist Tomogr 1990;14:505–513.
38. Abrahams JJ, Lange RC. Coil holder and marker system for MR imaging of the total spine. Radiology 1989;172:869–871.
39. Requardt H, Offermann J, Erhard P. Switched array coils. Magn Reson Med 1990;13:385–397.
40. Roemer PB, Edelstein WA, Hayes CE, Souza SP, Mueller OM. The NMR phased array. Magn Reson Imag 1990;16:192–225.
41. Moffit B, Reicher M, Lufkin R, Bentson J. Comparison of T1 and T2 weighted images of the lumbar spine. Comput Med Imag Graph 1988;12:271–276.
42. Avrahami E, Tachmor R, Dally O, Hadar H. Early MR demonstration of spinal metastases in patients with normal radiographs, CT and radionuclide bone scans. J Comput Assist Tomogr 1989;13:598–602.
43. Godersky JC, Smoker WRK, Knutzon R. Use of magnetic resonance imaging in the evaluation of metastatic spinal disease. Neurosurgery 1987;21:676–680.
44. Stevens SK, Moore SG, Kaplan ID. Early and late bone marrow changes after irradiation; MR evaluation. AJR 1990;154:745–750.
45. Colman LK, Porter BA, Redmond J, et al. Early diagnosis of spinal metastases by CT and MR studies. J Comput Assist Tomogr 1988;12:423–426.
46. Dwyer AJ, Frank JA, Sank VJ, Reinig JW, Hickey AM, Doppman JL. Short-Tl inversion-recovery pulse sequence: analysis and initial clinical experience in cancer imaging. Radiology 1988;168:827–836.
47. Vogler JB, Murphy WA. Bone marrow imaging. State of the art. Radiology 1988;168:679–693.
48. Baker LL, Goodman SB, Perkash I, Lane B, Enzman DR. Benign versus pathologic compression fractures of vertebral bodies: assessment with conventional spin echo, chemical shift and STIR MR imaging. Radiology 1990;174:495–502.
49. Stimac GK, Porter BA, Olson DO, Gerlach R, Genton M. Gadolinium-DTPA enhanced MR imaging of spinal neoplasms. Preliminary investigation and comparison with unenhanced spin-echo and STIR sequence. AJR 1988;151:1185–1192.
50. Sebag GH, Moore SG. Effect of trabecular bone on the appearance of marrow in gradient echo images of the appendicular skeleton. Radiology 1990;174:855–859.
51. Carmody RF, Yang PJ, Seeley GW, Seeger JF, Unger EC, Johnson JE. Spinal cord compression due to metastatic disease: diagnosis with MR imaging versus myelography. Radiology 1989;173:225–229.
52. Van Dijke C, Ross JS, Tkach J, Masaryk TJ, Modic MT. Gradient-echo MR imaging of the cervical spine: evaluation of extradural disease. Am J Neuroradiology 1989;10:627–632.
53. Dixon WT. Simple proton spectroscopic imaging. Radiology 1984;153:189–194.
54. Wismer GL, Rosen BR, Buxton R, Stark DD, Brady TJ. Chemical shift imaging of bone marrow: preliminary experience. AJR 1985;145:1031–1037.
55. Greco A, Palmer N. MR assessment of spinal metastases using the late 26 sequence. Magn Reson Imag 1989;7:351–356.
56. Rubin J. Basic principles of CSF flow. In: Magnetic resonance of the spine. Enzmann DR, DeLaPaz, JB, Rubin J, eds. St. Louis: C. V. Mosby, 1990.
57. Breger RK, Williams AL, Daniels DL, et al. Contrast enhancement in spinal MR imaging. AJR 1989;153:387–391.
58. Dooms G, Mathurin P, Cornelis G, Malghem J, Maldague B. MR differential diagnosis between benign and malignant vertebral collapses [Abstract]. Proceedings of the 8th annual meeting of the Society of Magnetic Resonance in Medicine, Amsterdam, 1989:16.
59. Krol G, Sze G, Malkin M, Walker R. MR of cranial and spinal meningeal carcinomatosis: comparison with CT and myelography. Am J Neuroradiology 1988;9:709–714.
60. Sze G, Krol G, Zimmerman RD, Deck MDK. Malignant extradural spinal tumors: MR imaging with Gd-DTPA. Radiology 1988;167:217–223.
61. Sze G. MRI of tumor metastases to the leptomeninges. Magn Reson Imag Decis 1988;1:2–8.
62. Sze G. Gadolinium-DTPA in spinal disease. Radiol Clin Am 1988;26:1009–1024.
63. Sze G, Stimac GK, Bartlett C, et al. Multicenter study of gadopentetate dimeglumine as a contrast agent: evaluation in patients with spinal tumors. Am J Neuroradiology 1990;1:967–974.
64. Nyman R, Rehn S, Glimelius B, et al. Magnetic resonance imaging in diffuse malignant bone marrow diseases. Acta Radiol 1987;28:199–205.
65. Sykes MP, Chu FC, Wilkerson WG. Local bone marrow changes secondary to therapeutic irradiation. Radiology 1960;75:919.
66. Custer RP, Ahlfeldt FE. Studies on the structure and function of bone marrow. Variations in cellularity in various bones with advancing years of life and their relative response to stimuli. J Lab Clin Med 1932;17:960.
67. Dooms GC, Fisher MR, Hricak H, Richardson M, Crooks LE, Genant HK. Bone marrow imaging: MR studies related to age and sex. Radiology 1985;155:429–432.
68. Richards MA, Webb JAW, Jewell SE, Gregory WM, Reznek RH. In vivo measurement of spin lattice relaxation time of bone marrow in healthy volunteers: the effect of age and sex. Br J Radiol 1988;61:30–33.

69. Smith SR, Williams CE, Davies JM, Edwards RHT. Bone marrow disorders: characterization with quantitative MR imaging. Radiology 1989;172:805–810.

70. Ricci R, Cova M, Kang YS, et al. Normal age-related patterns of cellular and fatty bone marrow distribution in the axial skeleton: MR imaging study. Radiology 1990;177:83–88.

71. Le Blanc AD, Schonfeld E, Schneider VS, Evans HJ, Taber KH. The spine: change in T2-relaxation times from disuse. Radiology 1988;169:105–107.

72. Algra PR, Bloem JL, Verboom L, Arndt JW, Vogel HJPh, Tissing H. Detection of vertebral metastases; MRI versus bone scintigraphy. Preliminary results. In: Higer, Bielke, eds. Proceedings tissue characterization in MR imaging. Berlin: Springer-Verlag, 1990.

73. Algra PR, Bloem JL. Loss of visibility of the basovertebral vein; a new MR sign for diffuse bone marrow disease. Am J Neuroradiology 1991 (in press).

74. Sugimura K, Yamasaki K, Kitagaki K, Tanaka Y, Kono M. Bone marrow diseases of the spine: differentiation with T1 and T2 relaxation times in MR imaging. Radiology 1987;165:541–544.

75. Li KC, Poon PY. Sensitivity and specificity of MRI in detecting malignant spinal cord compression and in distinguishing malignant from benign compression fractures of vertebrae. Magn Reson Imag 1988;6:547–556.

76. Algra PR, Bloem JL, Verboom LJ, et al. MRI vs scintigraphy in the detection of vertebral metastases. Radiographics 1991;11:219–232.

77. Modic MT, Masaryk TJ, Ross JS, Carter JR. Imaging of degenerative disk disease. State of the art. Radiology 1988;168:177–186.

78. Hajek PC, Baker LL, Goobar JE, et al. Focal fat deposition in axial bone marrow: MR characteristics. Radiology 1987;162:245–249.

79. Bloem JL. Radiological staging of primary malignant musculoskeletal tumors [Thesis]. University of Leiden, Leiden, The Netherlands, 1988.

80. Hayes CW, Jensen ME, Conway WF. Non-neoplastic lesions of the vertebral bodies: findings in magnetic resonance imaging. Radiographics 1989;9:883–903.

81. Grenier N, Grossman RI, Schiebler ML, Yeager BA, Goldberg HI, Kressel HY. Degenerative lumbar disk disease: pitfalls and usefulness of MR imaging in detection of the vacuum phenomenon. Radiology 1987;164:861–865.

82. Goldberg AL, Rothfus WE, Deeb ZL, Khoury MB, Daffner RH. Thoracic disc herniation versus spinal metastases: optimizing diagnosis with magnetic resonance imaging. Skel Radiol 1988;17:423–426.

83. Modic MT, Steinberg PM, Ross JS, Masaryk TJ, Carter JR. Degenerative disk disease: assessment of changes in vertebral body marrow with MR imaging. Radiology 1988;166:193–199.

84. Beltran J, Noto AM, Chakeres DW, Christoforidis AJ. Tumors of the osseous spine: staging with MR imaging versus CT. Radiology 1987;162:565–569.

85. Weigert F, Reiser M, Pfander K. Imaging of neoplastic changes in the spine using MR tomography. ROFO 1987;146:123–130.

86. Von Allgayer B, Flierdt E vd, Gumppenberg S v. Die Kernspintomographie im Vergleich zur Skelettszintigraphie nach traumatischen Wirbelkorperfrakturen. Fortschr Röntgenstr 1990;152:677–681.

87. Castillo M, Malko JA, Hoffman JC. The bright intervertebral disk, an indirect sign of abnormal spinal bone marrow on T1-weighted MR images. Am J Neuroradiology 1990;11:23–26.

88. Sobel DF, Zyroff J, Thorne RP. Diskogenic vertebral sclerosis: MR imaging. J Comput Assist Tomogr 1987;11:855–858.

89. Yuh WTC, Zachar CK, Barloon TJ, Sato Y, Sickles WJ, Hawes DR. Vertebral compression fractures: distinction between benign and malignant causes with MR imaging. Radiology 1989;172:215–218.

90. Horowitz S, Fine M, Azar-Kia B. Differentiation of benign versus metastatic vertebral body compression fractures. Radiology 1990;177(P):370.

91. Frager D, Elkin C, Swerdlow M, Bloch S. Subacute osteoporotic compression fracture: misleading magnetic resonance appearance. Skel Radiol 1988;17:123–126.

92. Naul LG, Peet GJ, Maupin WB. Avascular necrosis of the vertebral body: MR imaging. Radiology 1989;172:219–222.

93. Subramaniam G, McAfee JG. A new complex of ^{99m}Tc polyphosphate for skeletal imaging. Radiology 1971;99:192–196.

94. Galasko CSB. The pathologic basis for skeletal scintigraphy. J Bone Joint Surg 1975;57B:353–359.

95. Galasko CSB. Mechanisms of bone destruction in the development of skeletal metastases. Nature (London) 1976;263:507–508.

96. Israel O, Front D, Frenkel A, Kleinhaus U. 24 hour/4 hour ratio of technetium-99m methylene diphosphonate uptake in patients with bone metastases and degenerative bone changes. J Nucl Med 1985;26:237–240.

97. Bahk YW, Kim OH, Chung SK. Pinhole collimator scintigraphy in differential diagnosis of metastasis, fracture, and infections of the spine. J Nucl Med 1987;28:447–451.

98. Sarpel S, Sarpel G, Yu E, et al. Early diagnosis of spinal-epidural metastasis by magnetic resonance imaging. Cancer 1987;59:1112–1116.

99. Mehta RC, Wilson MA, Perlman SB. False-negative bone scan in extensive metastatic disease: CT and MR findings [case report]. J Comput Assist Tomogr 1989;13:717–719.

100. Delbeke D, Powers TA, Sandler MP. Correlative radionuclide and MR imaging in the evaluation of the spine. Clin Nucl Med 1989;14:742–749.

101. Kattapuram SV, Khurana JS, Scott JA, El-Khoury GY. Negative scintigraphy with positive magnetic resonance imaging in bone metastases. Skel Radiol 1990;19:113–116.

102. Jones AL, Williams MP, Powles TJ, et al. Magnetic resonance imaging in the detection of skeletal metastases in patients with breast cancer. Br J Cancer 1990;62:296–298.

103. Andreasson I, Pentren-Mallmin M, Nilsson S, Nyman R, Hemingsson A. Diagnostic methods in planning palliation of spinal metastases. Anticancer Res 1990;10:731–733.

104. Beatrous TE, Choyke PL, Frank JA. Diagnostic evaluation of cancer patients with pelvic pain: comparison of scintigraphy, CT, and MR imaging. AJR 1990;155:85–88.

105. Daffner RH, Lupeti AR, Dash N, Deeb ZL, Sefczek RJ, Schapiro RL. MRI in the detection of malignant infiltration of bone marrow. AJR 1986;146:353–358.

106. Frank JA, Ling A, Patronas NJ, Carrasquillo JA, Horvath K, Hickey AM, Dwyer AJ. Detection of malignant bone tumors: MR imaging vs scintigraphy. AJR 1990;155:1043–1048.

107. Algra PR, Bloem JL, Tissing H. Sensitivity of MR imaging in detecting vertebral metastases: comparison with bone scintigraphy. Radiology 1989;173(P):142.

108. Ludwig H, Tscholakoff, Neuhold A, Früwald, Rasoul S, Fritz E. Magnetic resonance imaging of the spine in multiple myeloma. Lancet 1987; Aug 15:364–366.

109. Fruehwald FXJ, Tscholakoff D, Schwaighofer B, Wicke L, Neuhold A, Ludwig H, Hajek PC. Magnetic resonance imaging of the lower vertebral column in patients with multiple myeloma. Invest Radiol 1988;23:193–199.

110. De Schrijver M, Streule K, Senekowitsch R, Fridrich R. Scintigraphy of inflammation with nanometer-sized colloidal tracers. Nucl Med Commun 1987;8:895–908.

111. Schmitz R, Hotze A, Bougers H, Joseph K. Knochenmarkszintigraphie: Erste Ergebnisse mit einem neuen Mikrokolloid. Nucl Compact 1983;14:17–23.

112. Ranner G, Fueger GF, Hörmann M. Ergebnisse der Knochenmarkszintigraphie bei hämatologischen Erkrankungen. Fortschr Röntgenstr 1987;146:300–305.

113. Munz DL. The scintigraphic bone marrow status in adult man: a new classification. In: Schmidt HAE, Adam WE, eds. Nuklearmedizin: Darstellung von Metabolismen und Organ-Funktionen. Stuttgart and New York: Schattauer, 1984:640–644.

114. Hotze A, Löw A, Mahlstedt J, Wolf F. Kombinierte Knochenmark-und Skelettszintigraphie bei ossären und myelogenen Erkrankungen. Fortschr Röntgenstr 1984;140:717–723.

115. Munz DL. Knochenmarkszintigraphie: Grundlagen und klinische Ergebnisse. Nuklearmediziner 1984;7:251–268.

116. McAfee JG. Update on radiopharmaceuticals for medical imaging. Radiology 1989;171:593–601.

117. Keßel F, Gamm H, Hahn K. Knochenmarkszintigraphie bei hämatologischen Systemerkrankungen. TumorDiagnostik Therapie 1987;8:5–10.

118. Baker HL, Berquist TH, Kispert DB, et al. MR imaging in a routine clinical setting. Mayo Clin Proc 1985;60:75–90.

119. Barloon TJ, Yuh WTC, Yang CJC, Schultz DH. Spinal subarachnoid tumor seeding from intracranial metastasis: MR findings. J Comput Assist Tomogr 1987;11:242–244.

120. Lien HH, Blomlie V, Heimdal K. Magnetic resonance imaging of malignant extradural tumors with acute spinal cord compression. Acta Radiol 1990;31:187–190.

121. Williams MP, Cherryman GR, Husband JE. Magnetic resonance imaging in suspected metastatic spinal cord compression. Clin Radiol 1989;40:286–290.

122. Wimmer B, Friedburg W, Henning J, Kauffman GW. Möglichkeiten der diagnostischen Bildgebung durch Kernspintomographie. Radiologe 1986;26:137–143.

123. Smoker WRK, Godersky JC, Knutsson RK, Keyes WD, Norman D, Bergmann W. The role of MR imaging in evaluating metastatic spinal disease. AJR 1987;149:1241–1248.

124. Hagenau C, Grosh W, Currie M, Wiley RG. Comparison of spinal magnetic resonance imaging and myelography in cancer patients. J Clin Oncol 1987;5:1663–1669.

125. Karnaze MG, Gado MH, Sartor KJ, Hodges FJ. Comparison of MR and CT myelography in imaging the cervical and thoracic spine. AJR 1988;150:397–403.

126. Joffe J, Williams MP, Cherryman GR, Gore M, McElwain TJ, Selby P. Magnetic resonance imaging in myeloma. Lancet 1988; May 21:1163–1164.

127. Ramsey RG, Zacharias CE. MR imaging of the spine after radiation therapy: easily recognizable effects. AJR 1985;144:1131–1135.

128. Remedios PA, Coletti PM, Raval JK, et al. MRI of bone after radiation. Magn Reson Imag 1988;6:301–304.

129. Rosenthal DI, Hayes CW, Rosen B, Mayo-Smith W, Goodsitt MM. Fatty replacement of spinal bone marrow due to radiation: demonstration by dual energy quantitative CT and MR imaging. J Comput Assist Tomogr 1989;13:463–465.

130. Martin DS, Awwad EE, Sundaram M. The magnetic resonance appearance of spinal radiation: the converse of spinal metastases. Orthopedics 1990;373:378–379.

131. Blomquist C, Elommaa I, Virkunnen P, et al. The response evaluation of bone metastases in mammary carcinoma. The value of radiology, scintigraphy and biochemical markers of bone metabolism. Cancer 1987;60:2907–2912.

132. Scheid V, Buzdar AU, Smith TL, Hortobagyi GN. Clinical course of breast cancer patients with osseous metastasis treated with combination chemotherapy. Cancer 1986;58:2589–2593.

133. Libshitz HI, Hortobagyi GN. Radiographic evaluation of therapeutic response in bony metastases of breast cancer. Skel Radiol 1981;7:159–165.

134. Hayward JL, Carbone PP, Henson JC, Kumaoka S, Segaloff A, Rubens RD. Assessment of response to therapy in advanced breast cancer. Eur J Cancer 1977;13:89–94; and in Br J Cancer 1977;35:292–298.

135. Gold RI, Seeger LL, Bassett LW, Steckel RJ. An integrated approach to the evaluation of metastatic bone disease. Radiol Clin N Am 1990;28:471–483.

CHAPTER 17

Magnetic Resonance Imaging of the Temporomandibular Joint

J. Mark Fulmer and Steven E. Harms

Introduction
Technique

Pathology
Conclusion

Introduction

The temporomandibular joint (TMJ) is a source of significant symptomatology. Headache, facial pain, jaw pain, and ear pain have all been ascribed to disordered TMJ function or structure (1–3). Diagnostic imaging plays a central role in management of TMJ disorders. Magnetic resonance (MR) imaging offers exquisite noninvasive images of the TMJ, and is now considered the primary imaging modality for evaluation of TMJ soft tissues (4–10).

Technique

While techniques for imaging the temporomandibular joint vary, most protocols include sagittal and coronal images, with small field-of-view (FOV) T1- and T2-weighted acquisitions (11–13).

Emphasis has been placed recently on using high-field (1.5 to 2.0 T) imaging systems for TMJ evaluations (14, 15), although early TMJ imaging was performed on midfield imagers (0.5 to 1.0 T). If midfield units are used, signal:noise ratios (SNRs) can best be improved by increasing the number of acquisitions in each view. This will increase the time required to complete an examination. However, increased FOV or increased slice thickness results in suboptimal spatial resolution. High-field imagers are less constrained by signal:noise considerations.

The use of surface coils is mandatory for adequate TMJ imaging. Flat, 3-inch, receive-only coils are most commonly employed. Larger coils are more cumbersome, and may be more difficult to position. In addition, aliasing artifacts will be more severe with larger coils. The coil is centered anterior to the external auditory canal, and the joint is palpated to document position. Stabilization is crucial to avoid motion degradation. Simultaneous bilateral imaging allows assessment of joint symmetry. This technique utilizes dual-surface coils. Both coils operate with a

combiner box, which is connected to the surface coil port of the imager system, allowing the coils to receive signal simultaneously. A commercially available holder is used to position the dual coils (General Electric, Milwaukee, Wisc.) (Fig. 17.1). Switching techniques have been described to alternately isolate each coil. Other workers have used totally separate receivers for both coils (16, 17). However, switching techniques and separate coil systems are not widely available.

Pulse sequences for TMJ imaging include spin-echo and gradient-echo techniques. Spin-echo techniques are ideally suited for imaging TMJ structures. The short T2 values of muscle, fibrocartilage, and fat make these structures well displayed in sequences with short echo times (TE) (TE < 20 ms). By combining this with a short repetition time (TR), a T1-weighted image is achieved (TR < 1000 ms). Signal averaging improves the SNR in these T1-weighted images. Doubling the number of acquisitions improves the SNR by the square root of 2.

A T2-weighted sequence can provide information regarding abnormal joint fluid, bone marrow edema, or soft-

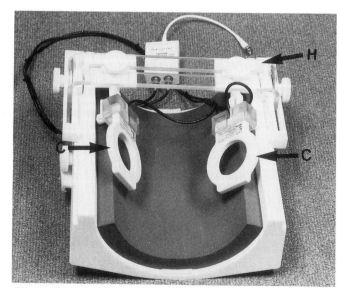

Figure 17.1. Dual-surface coil apparatus. Bilateral 3-inch surface coils, *C*, mounted in adjustable surface coil holder, *H*, allow bilateral simultaneous imaging. (Reprinted from Topics in Magnetic Resonance Imaging, Vol. 1, No. 3, pp. 75–84, with permission of Aspen Publishers, Inc., © 1989.)

Abbreviations (see also glossary): TMJ, temporomandibular joint; FOV, field of view; SNR, signal:noise ratio.

tissue inflammation. This requires lengthening both the TR and TE. The total scan time is therefore increased significantly, and anatomic delineation is less crisp than in T1-weighted images.

Gradient-echo sequences [fast low-angle shot (FLASH), fast imaging with steady state precession (FISP), and gradient-recalled acquisition in the steady state (GRASS)] have all been utilized in TMJ imaging (18–21). The main advantage of these sequences is the rapid scan time they provide. The articular disk is well displayed with these pulse parameters. However, the remaining structures of the TMJ are not visualized as well as in spin-echo sequences.

Functional evaluation is best performed with a cine loop sequence. This is acquired in incremental steps during mandibular opening. Each step in the cycle represents a single image. Many authors advocate gradient-echo techniques for cine loop imaging. The spin-echo technique can also be utilized without significant time penalty. A TR of 300 ms with a TE of 15 ms results in T1-weighted images. Both joints are imaged simultaneously with two acquisitions. The total scan time is approximately 80 s/acquisition (40 s/image). The complete cine series can be completed for both joints in approximately 15 min, and provides high-contrast T1-weighted images. Incremental opening is provided with the Burnett TMJ device (MedRad, Philadelphia, Penn.) (Fig. 17.2). The device is calibrated in 1-mm graduations. Each image in the sequence represents 3 mm of progressive opening, and 6 to 12 images are typically obtained. The result is a sequence of images that demonstrates joint position through passive opening (Fig. 17.3). Static open, midcycle, and closed views can also be used to evaluate condylar motion, although this provides less complete functional evaluation of the joint.

Presaturation pulses applied outside the imaging volume help reduce motion artifacts from nearby vasculature. Presaturation pulses also reduce aliasing artifacts. Gradient moment nulling (GMN) can also be utilized to reduce motion artifacts. This technique requires a lengthening of the TE, however, thus reducing the SNR (11).

Imaging planes are chosen to accommodate the normal obliquity of the joint (Fig. 17.4). Angled coronal and sagittal views performed in the closed mouth position define anatomic features of both joints. The cine loop sequence is usually performed in the true sagittal projection, as the condylar head translates in the sagittal plane.

A FOV of 8 to 10 cm provides excellent spatial resolution when combined with a 192 × 256 imaging matrix. When using a 3-inch surface coil, an 8-cm FOV is unencumbered by aliasing artifacts. If a larger FOV is required, the matrix can be increased to 256 × 256. This will maintain high spatial resolution, but will be less time efficient. Aliasing can be addressed with phase domain oversampling (no phase wrap), frequency domain oversampling (no frequency wrap), and frequency filters if necessary.

A slice thickness of 3 to 5 mm provides adequate signal when using a two-dimensional (2D) acquisition method. Interslice gaps of 1 to 2 mm reduce slice cross-talk and maintain adequate anatomic continuity. Thinner slices can be performed, but this requires increased gradient strength or narrow bandwidth radio-frequency (RF) pulses for 2D acquisition. These changes require longer TE values with resultant signal loss. Three-dimensional (3D) acquisition methods have been described, and can provide very thin slices (22–24). The major disadvantage of these techniques is increased susceptibility to motion-induced artifacts.

A typical TMJ study thus consists of 2D multislice acquisitions in the oblique sagittal and oblique coronal planes.

Figure 17.2. Incremental opening device. The Burnett TMJ device is held by the patient with a mouthpiece, *M*, placed between the incisors. (Reprinted from Topics in Magnetic Resonance Imaging, Vol. 1, No. 3, pp. 75–84, with permission of Aspen Publishers, Inc., © 1989.)

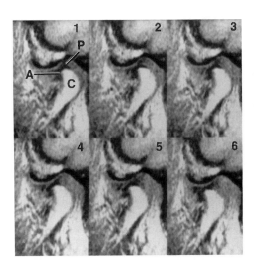

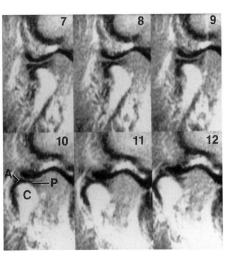

Figure 17.3. Spin-echo (300/12) cine loop sequence. Multiple images demonstrate condylar rotation and translation of the disk-condyle complex. Posterior band, *P*, anterior band, *A*, and condyle, *C*, are labeled. The left side of the image is anterior.

A cine loop sequence in the true sagittal plane is also acquired, and all images are evaluated for functional derangements, structural deformities, and surrounding abnormalities.

Pathology

Internal derangements constitute the most common anatomic abnormality. This spectrum of disorders results from malposition of the articular disk. Anterior displacement of the disk is most common, with sideways, rotational, and posterior displacement being more rare (Fig. 17.5) (25–27).

Mild internal derangement is characterized by anterior displacement of the disk in the closed mouth position. During the opening cycle, the disk reduces to a normal position (Fig. 17.6). This may result in a palpable or audible click at the time of reduction. The shape of the disk is usually unaltered with mild internal derangement. The retromandibular laminae are stretched but not disrupted, and condylar integrity is maintained (Fig. 17.7).

Moderate internal derangement shows anterior disk displacement without reduction (Fig. 17.8). Translation is commonly reduced due to impaction of the codyle into the displaced disk. Patients often experience a decreased range of motion. The disk may be distorted, but there is usually no evidence of degeneration or adhesion (Fig. 17.9).

Severe internal derangement is seen as disk displacement without reduction in association with disk deformity and degeneration. Retrodiskal perforation, disk thickening, and adhesions characterize severe derangements (Fig. 17.10).

Arthroscopic evaluation of the TMJ has placed emphasis on intraarticular adhesions as a manifestation of severe in-

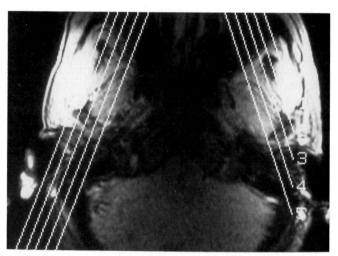

Figure 17.4. Oblique sagittal images. Oblique sagittal views are prescribed for both joints, with the use of a gradient-echo scout view obtained at the level of the external auditory canals. (Reprinted from Topics in Magnetic Resonance Imaging, Vol. 1, No., 3, pp. 75–84, with permission of Aspen Publishers, Inc., © 1989.)

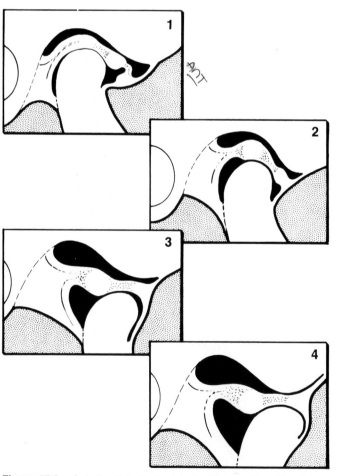

Figure 17.6. Anterior disk displacement with reduction. The disk is displaced anteriorly in the closed mouth position, but reduces to normal position during jaw opening. A physical examination may demonstrate an opening click at the time of reduction. The right side of the image is anterior.

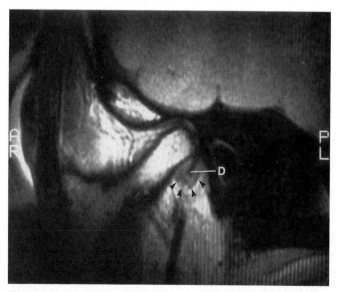

Figure 17.5. Posterior disk displacement. The articular disk, *D*, is displaced posteriorly into the retrodiskal space. This is much less common than anterior disk displacement. The inferior joint space is distended.

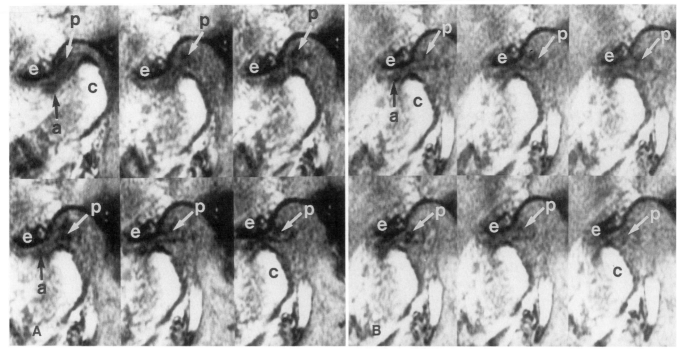

Figure 17.7. Anterior disk displacement with reduction. **(A** and **B)** Cine loop sequence demonstrates anterior disk displacement with early reduction on the second step of the sequence. Anterior band, *A,* poste- rior band, *P,* condyle, *C,* and eminence, *E,* are labeled. (Reprinted from Topics in Magnetic Resonance Imaging, Vol. 1, No. 3, pp. 75–84, with permission of Aspen Publishers, Inc., © 1989.)

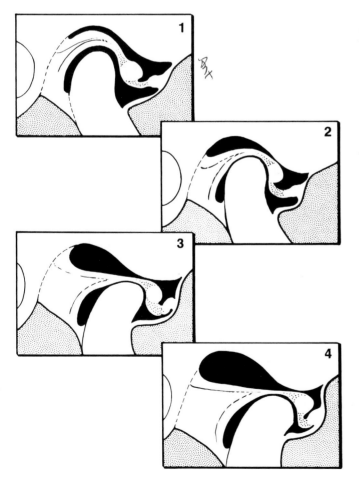

Figure 17.8. Anterior disk displacement without reduction. Disk is displaced anteriorly on closed mouth view. Translation results in deformity of the disk with stretching of the retrodiskal laminae. Patients often experience a decreased range of motion. The right side of the image is anterior.

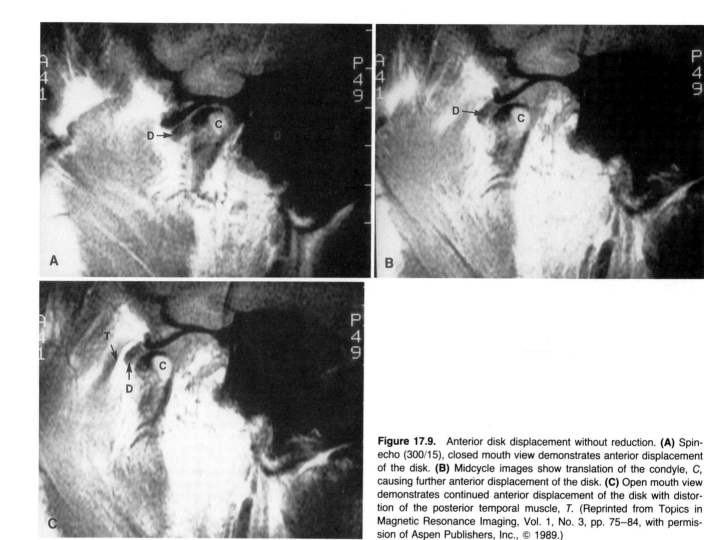

Figure 17.9. Anterior disk displacement without reduction. **(A)** Spin-echo (300/15), closed mouth view demonstrates anterior displacement of the disk. **(B)** Midcycle images show translation of the condyle, *C*, causing further anterior displacement of the disk. **(C)** Open mouth view demonstrates continued anterior displacement of the disk with distortion of the posterior temporal muscle, *T*. (Reprinted from Topics in Magnetic Resonance Imaging, Vol. 1, No. 3, pp. 75–84, with permission of Aspen Publishers, Inc., © 1989.)

ternal derangement. This is best demonstrated on the cine loop sequence, and is seen as reduced motion of the disk, direct visualization of fibrous bands, or evidence of marked limitation of mandibular motion (Fig. 17.11) (28). Arthroscopic treatment consists of lysis of adhesions. This therapy does not typically alter disk position, but often does reduce symptoms associated with the deranged joint (29).

True diskal perforation is less common than perforation of the retrodiskal tissues, and is visualized as a distinct separation between anterior and posterior bands (30, 31). This results in direct contact between the mandible and the eminence.

Certain congenital variations probably predispose to internal derangement. A steep articular eminence and hypermobility of the joint are two such conditions (32). A steep articular eminence appears to predispose to anterior disk displacement. Hypermobility can result in impaction of the disk-condyle complex into the posterior aspect of the temporalis muscle. This may cause symptoms of temporal headaches.

Trauma often results in TMJ disorders (33–35). The most commonly reported trauma is iatrogenic. Extraction of the third molars, orthodontic procedures, and craniofacial plastic surgery have been cited as predisposing causes of internal derangement. One form of iatrogenic intervention has received particular attention. Alloplastic disk implants have been shown to degenerate, resulting in an aggressive intraarticular foreign body reaction (36–38). This is shown as moderate signal, poorly defined soft tissue surrounding the mandibular condyle and filling the retrodiskal space (Fig. 17.12). Severe mandibular and temporal erosion often result in joint distortion and marked decreased motion. This giant cell foreign body reaction requires wide surgical debridement, often followed by total joint reconstruction. Direct facial trauma may result in mandibular fracture, avascular necrosis, or neuromuscular disorders (39). Each of these can have adverse affects on the structure of function of the TMJ.

Osteoarthritis of the joint is a final common pathway of internal derangements, whether precipitated by joint morphology or trauma (40–42). Fibrous adhesions may proceed to fibrous ankylosis. Mandibular erosion, osteophyte

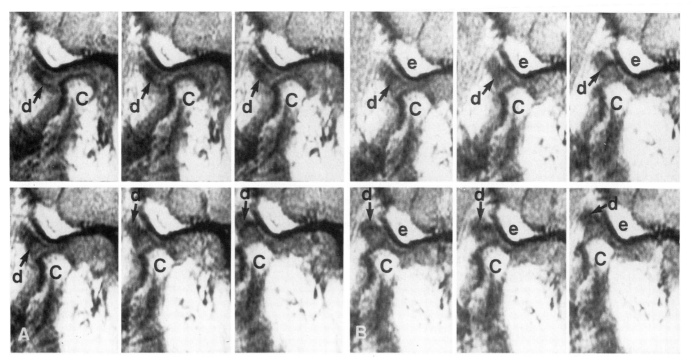

Figure 17.10. Hypermobile joint. **(A** and **B)** The disk, *D,* is displaced anteriorly with thickened and distorted morphologic features. The mandibular condyle, *C,* is shown to translate to a position forward of the articular eminence. (Reprinted from Topics in Magnetic Resonance Imaging, Vol. 1, No. 3, pp. 75–84, with permission of Aspen Publishers, Inc., © 1989.)

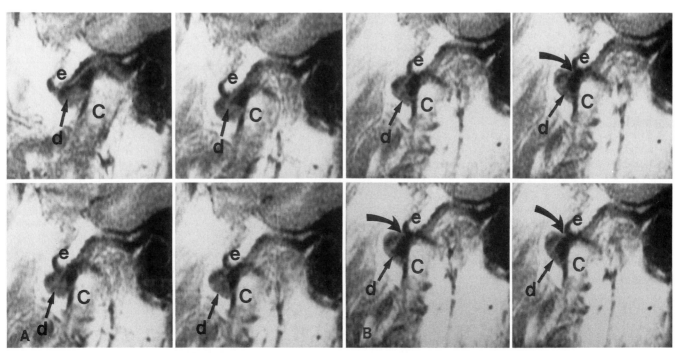

Figure 17.11. Steep articular eminence. **(A** and **B)** The distorted, thickened disk is displaced anteriorly. Separate anterior and posterior bands cannot be defined. Adhesion *(curved arrow)* to the articular eminence is shown on cine images. Articular disk, *d,* articular eminence, *e,* and mandibular condyle, *C,* are labeled. (Reprinted from Topics in Magnetic Resonance Imaging, Vol. 1, No. 3, pp. 75–84, with permission of Aspen Publishers, Inc., © 1989.)

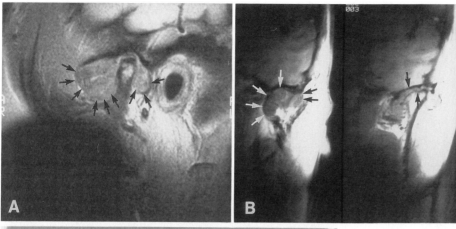

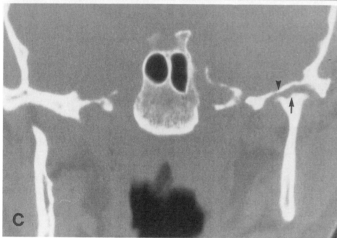

Figure 17.12. Giant cell foreign body reaction. **(A)** T1-weighted closed mouth image (300/15) shows marked distension of joint capsule *(small arrows)* by extensive foreign body reaction. The patient had received an alloplastic disk implant. **(B)** Coronal closed mouth view (300/12) demonstrates marked irregularity and erosion of the mandibular condyle *(arrow)*. The fossa demonstrates marked thinning *(arrowhead)*, and capsular distension is again noted *(small arrows)*. **(C)** A CT scan confirms osseous changes noted on the MR image.

formation, and sclerosis result. Ultimately, the most severely affected joint may undergo osseous ankylosis (41).

Most TMJ sequences will provide partial views of the temporal lobe, mastoids, and surrounding musculature. These areas may harbor unsuspected pathology (42–44) (Fig. 17.13).

Conclusion

Magnetic resonance imaging of the TMJ offers exquisite evaluation of joint anatomy. Joint motion can be reliably assessed using either cine loop or sequential static images. With its multiplanar capabilities, unequaled soft-tissue contrast, and lack of ionizing radiation, MR imaging has become the preferred modality for comprehensive imaging of TMJ structure and function.

References

1. Guarlnick W, Kaban LB, Merril RG. Temporomandibular joint afflictions. N Engl J Med 1978;229:123–129.
2. Shellhas KP, Wilkes CH, Baker CC. Facial pain, headache, and temporomandibular joint inflammation. Headache 1989;29:229–232.
3. Laskin DM, Ryan WA, Green CS, et al. Incident of temporomandibular symptoms in patients with major skeletal malocclusions: a survey of oral and maxillofacial surgery training programs. Oral Surg 1986;61:537.
4. Harms SE, Wilk RM, Wolford LM, et al. The temporomandibular joint: magnetic resonance imaging using surface coils. Radiology 1985;57:133–136.
5. Katzberg RW, Bessette RW, Tallents RH, et al. Normal and abnormal temporomandibular joint: MR imaging with surface coil. Radiology 1986;158:183–189.
6. Helms CA, Gillespy T III, Sims RE, et al. Magnetic resonance imaging of internal derangement of the temporomandibular joint, Radiol Clin N Am 1986;24:189–192.
7. Westesson P-L, Katzberg RW, Tallents RH, et al. CT and MR of the temporomandibular joint: comparison with autopsy specimens. AJR 1987;148:1165–1171.
8. Westesson P-L, Katzberg RW, Tallents RH, et al. Temporomandibular joint: comparison of MR images with cryosectional anatomy. Radiology 1987;164:59–64.
9. Donlon WC, Moon KL. Comparison of magnetic resonance imaging arthrotomography and clinical and surgical findings in temporomandibular joint internal derangements. Oral Surg Oral Med Oral Path 1987;64:2–5.
10. Cirbus MT, Smilack MS, Beltran J, et al. Magnetic resonance imaging in confirming internal derangement of the temporomandibular joint. J Prosthet Dent 1987;57:488–494.
11. Fulmer JM, Harms SE. The temporomandibular joint. Top Magn Reson Imag 1989;1:75–84.
12. Katzberg RW. Temporomandibular joint imaging. Radiology 1989;170:297–307.
13. Harms SE, Wilk RW. Magnetic resonance imaging of the temporomandibular joint. Radiographics 1987;7:521–538.
14. Hansson L-G, Westesson P-L, Katzberg RW, et al. MR imaging of the temporomandibular joint: comparison of images of autopsy specimens made at 0.3T and 1.5T with anatomic cryosections. AJR 1989;152:1241–1244.
15. Schellas KP, Wilkes CH, Fritts HM, et al. Temporomandibular

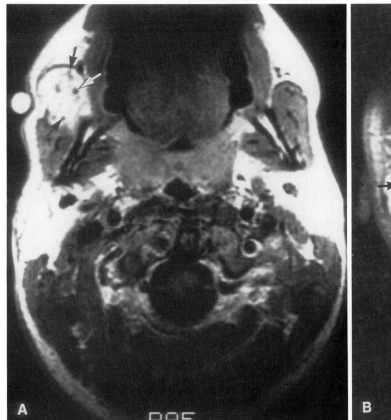

Figure 17.13. Masseter muscle arteriovenous malformation. **(A)** Axial gadolinium-enhanced T1-weighted image (800/20) demonstrates circumscribed enhancing mass in anterior right masseter. Flow void is noted in lesion *(arrow)*. **(B)** T2-weighted (2000/80) coronal image also demonstrates flow void *(arrows)*, as well as marked increased signal in the masseter mass. Surgery confirmed an arteriovenous malformation.

joint: MR imaging of internal derangements and postoperative changes. AJR 1988;150:381–389.

16. Hardy CJ, Katzberg RW, Frey RL, et al. Switched surface coil system for bilateral MR imaging. Radiology 1988;167:835–838.

17. Hyde JS, Jesmanowicz A, Froncisz W, et al. Parallel image acquisition from noninteracting local coils. J Magn Reson 1986;70:512–517.

18. Conway WF, Hayes CW, Campbell RL. Dynamic magnetic resonance imaging of the temporomandibular joint using FLASH sequences. J Oral Maxillofac Surg 1988;46:930–938.

19. Burnett KR, Davis CL, Read J. Dynamic display of the temporomandibular joint meniscus using "fast scan" MR imaging. AJR 1987;149:959–962.

20. Schellhas KP, Fritts HM, Heithoff KB, et al. Temporomandibular joint: MR fast scanning. J Craniomandib Pract 1988;6:209–216.

21. Conway WF, Hayes CW, Campbell RL, et al. Temporomandibular joint motion: efficacy of fast low-angle shot MR imaging. Radiology 1989;172:821–826.

22. Harms SE, Wilk RM. Thin slice, high resolution MRI of the temporomandibular joint with surgical and pathologic correlation. Paper presented at the program of the 5th annual meeting of the Society of Magnetic Resonance in Medicine, Montreal, Canada, 1986.

23. Kramer DM, Compton RA, Yeung HN. A volume (3D) analogue of 2D multislice of "multislab" MR imaging. Paper presented at the program of the 4th annual meeting of the Society of Magnetic Resonance in Medicine, London, August 19–23, 1985.

24. Wilk RM, Harms SE. Temporomandibular joint: multislab, three-dimensional fourier transformation MR imaging. Radiology 1988;167:861–863.

25. Katsberg RW, Westesson P-L, Tallents RH, et al. Temporomandibular joint: MR assessment of rotational and sideways disk displacements. Radiology 1988;169:741–748.

26. Khoury MB, Dolan E. Sideways dislocation of the temporomandibular joint meniscus; the edge sign. Am J Neuroradiology 1986;7:869–872.

27. Thompson JR, Christiansen E, Hasso A, et al. Temporomandibular joint: high-resolution computed tomography. Radiology 1984;150:105–110.

28. Montgomery MT, VanSickels JE, Harms SE, et al. Arthroscopic TMJ surgery: effects on signs, symptoms, and disc position. J Oral Maxillofac Surg 1989;47:1263–1271.

29. Moses JJ, Sartoris D, Glass R, et al. The effect of arthroscopic surgical lysis and lavage of the superior joint space on TMJ disc position and mobility. J Oral Maxillofac Surg 1989;47:674–678.

30. Walter E, Huls A, Schmelzle R, et al. CT and MR imaging of the temporomandibular joint. Radiographics 1988;8:327–348.

31. Shellhas KP, Wilkes CH, Omlie MR, et al. The diagnosis of temporomandibular joint disease: two compartment arthrography and MR. AJR 1988;150:341–350.

32. Harinstein D, Buckingham RB, Braun T, et al. Systemic joint laxity (the hypermobile joint syndrome) is associated with temporomandibular joint dysfunction. Arthr Rheum 1988;31:1259–1264.

33. Norman JE. Post-traumatic disorders of the jaw joint. Ann R Coll Surg Engl 1982;64:27–36.

34. Bessette RW, Katzberg RW, Natiella JR, et al. Diagnosis and reconstruction of the human temporomandibular joint after trauma or internal derangement. Plast Reconst Surg 1985;75:192–205.

35. Weinberg S, La Pointe H. Cervical extension-flexion injury (whiplash) and internal derangement of the temporomandibular joint. J Oral Maxillofac Surg 1987;45:653–656.

36. Timmis DP, Aragon SB, Van Sickels JE, et al. Comparative study

of alloplastic materials for temporomandibular joint disc replacement in rabbits. J Oral Maxillofac Surg 1986;44:541–554.

37. Schellhas KP, Wilkes CH, Eldeeb M, et al. Permanent Proplast temporomandibular joint implants: MR imaging of destructive complications. AJR 1988;151:731–735.

38. Kaplan PA, Ruskin JD, Tu HK, et al. Erosive arthritis of the temporomandibular joint caused by Teflon-Proplast implants: plain film features. AJR 1988;151:337–339.

39. Schellhas KP. Temporomandibular joint injuries. Radiology 1989;173:211–216.

40. Wilkes CH. Internal derangements of the temporomandibular joint. Pathological variations. Arch Otolaryngol Head Neck Surg 1989;115:469–477.

41. Aggarwal S, Mukhopadhyay S, Berry M. Bony ankylosis of the temporomandibular joint: a computed tomography study. Oral Surg Oral Med Oral Pathol 1990;69:128–132.

42. Dolan EA, Voller JB, Angeliud JC. Synovial chondromatosis of the temporomandibular joint diagnosed by magnetic resonance imaging: report of a case. J Oral Maxillofac Surg 1989;47:411–413.

43. Belkin BA, Parageorge MB, Fakitsas J, et al. A comparative study of magnetic resonance imaging versus computed tomography for the evaluation of maxillary and mandibular tumors. J Oral Maxillofac Surg 1988;46:1039–1047.

44. Chisin R, Fasian R, Weber AL, et al. MR imaging of a lymphangioma involving the masseter muscle. J Comput Assist Tomogr 1988;12:690–692.

CHAPTER 18

Computed Tomography of the Temporomandibular Joint

Edwin L. Christiansen

Introduction

Some critics/observers dismiss temporomandibular joint (TMJ) disorders as an affliction of modern woman (85% of patients) and modern man (15% of patients); a consequence of societal stress and distress. However, anthropological and historical medical publications provide evidence to the contrary. In 1823, Sir Astley Cooper, writing in the second London edition of his *Treatise of Dislocations and Fractures of the Joints* (1), stated that the jaw . . . "appears to occasionally quit the interarticular cartilage of the temporal cavity," and that . . . "young women are generally the subjects of this sensation."

In middle and late nineteenth century Europe, TMJ surgeries were devised in an effort to mobilize joints ankylosed by trauma and temporal bone infections. Some of these surgeries were surprisingly innovative and did not differ markedly from modern temporomandibular arthroplastic procedures, for example, autogenous tissue grafts (2, 3). In the United States, animal membranes were surgically substituted for the articular disk during the first two decades of this century (4).

Early on, medical radiologists were responsible for pioneering TMJ plain film and tomographic techniques between 1920 and 1939 (5, 6) while dentists and dental radiologists, slower to become involved in TMJ radiography, refined existing techniques and applied their findings to questions unique to dentistry (7–9). In the 1940s and 1950s, with the important exception of arthrography (10) and tomography (11), interest in most aspects of TMJ imaging waned until approximately two decades later, when plain film arthrography and tomoarthrography were rediscovered and revived (12–14).

Pluridirectional tomography remained the first choice of most radiologists for body section planigraphy until the late 1970s and early 1980s. The strengths and weaknesses of TMJ pluridirectional tomography were thoroughly investigated (15) and fared well in comparisons made with conventional radiography (16). Comparison of conventional tomography with advanced imaging systems was a logical step in the development of TMJ imaging (17).

In the United States, temporomandibular joint computed tomograms were first reported in 1980 by Suarez and coworkers in a remarkable correlative imaging and anatomic study (18) that set the stage for approximately 5 years of intense investigation, which resulted in the publication of numerous papers from many parts of the globe (19–48).

The majority of TMJ computed tomography (CT)-specific publications focused on a single aspect of joint disorders: internal derangement, or displacement of the articular disk. In the historic orthopedic sense, internal derangement refers to a generalized joint disorder (49), but the term has gradually taken on more restricted usage, referring to malpositioning of the fibrocartilagenous articular disk that is normally interposed between the mandibular condyle and the osseous temporal joint component (Fig. 18.1*A* and *B*).

A disordered TMJ involves more than malpositioning of the articular disk and to think otherwise is to adopt a simplistic view that may dilute one's appreciation of the seriousness of the continuum of diseases affecting the TMJ.

Strengths of Computed Tomography

Computed tomography is well suited for a variety of TMJ disorders and broad disorders of the stomatognathic system. Soft-tissue detail of the masticatory muscles (Fig. 18.2) is displayed with remarkable clarity. The articular disk, depending on its shape and location, may frequently, but not always, be readily seen. As a rule, misshapen and thickened disks are more readily visualized with CT than are thin or normally shaped disks.

Osseous joint details are exquisitely displayed with CT (Fig. 18.3). Linear measurement of osseous TMJ components with CT software for clinical and research purposes does not significantly differ from zero except where marked to severe degenerative changes are present (50).

In addition to the selection of a tissue algorithm (soft tissue vs. bone), one of the chief strengths of CT lays in the software capabilities to vertically reformat the display

Abbreviations (see also glossary): TMJ, temporomandibular joint; RFP, relative frequency percent; HU, Hounsfield unit.

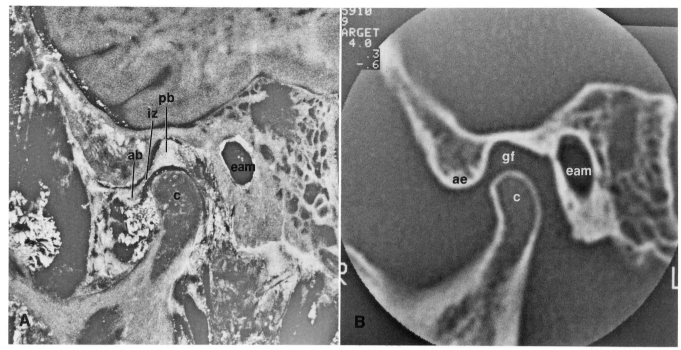

Figure 18.1. **(A)** Sagittal cryosection showing relationship of the articular disk posterior band *(pb)*, intermediate zone *(iz)*, and the anterior band *(ab)* to the mandibular condyle *(c)*. **(B)** Direct sagittal CT section of the same autopsy specimen as above, showing external auditory meatus *(eam)* and the relationship of mandibular condyle to glenoid fossa *(gf)* and articular eminence *(ae)*. Scanning and cryosection-ing were done with the jaws closed and the teeth secured in centric occlusion. (Reprinted with permission from Christiansen EL, Thompson JR, Hasso AN, Hinshaw OB Jr. Connective thin section temporomandibular joint anatomy and computed tomography. Radiographics 1986;6:703–723.)

Figure 18.2. Axial soft-tissue detail of patient scan showing the mandibular condyles *(c)*, the lateral pterygoid muscle *(lpm)*, and, on the patient's left side *(arrows)*, obscuring of the medial margin of the lateral pterygoid secondary to a metastatic squamous cell carcinoma. (Reprinted with permission from Hasso AN, Christiansen EL, Alder M: The Temporomandibular Joint. In: Som PM and Shapiro MD, guest edrs. MRI of the head and neck. Philadelphia: WB Saunders, 1989;27(2):301–314.)

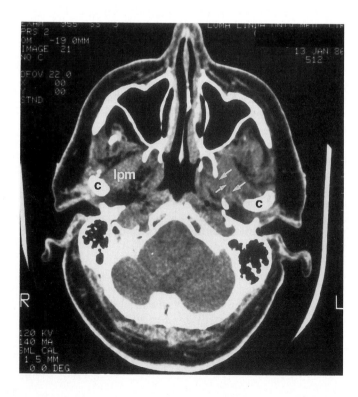

data matrix in anatomic planes *other than* the scanning plane (Fig. 18.4*A* and *B*). Vertical reformatting permits examination of the soft- and hard-tissue joint elements from a variety of perspectives in sections as thin as 0.5 mm. To maximize reformatting efficiency and flexibility, it is best to collect either contiguous or overlapping thin sections (1.5 mm) in the axial plane. Vertical reformatting from sagittally acquired scan sections results in awkwardly displayed images.

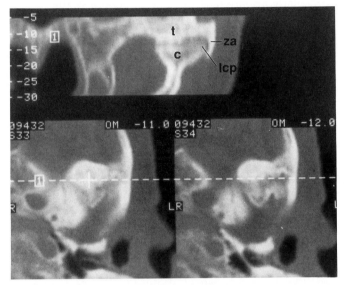

Figure 18.3. Vertically reformatted coronal bone detail image of TMJ ankylosis subsequent to motor vehicle accident, showing altered condylar *(c)* and temporal *(t)* joint components. Laterally, the zygomatic arch *(za)* essentially encapsulates the lateral condylar pole *(lcp)*.

Unilateral, sagittal TMJ scanning precludes the possibility of making (what may prove to be) important comparisons of the size and shape of intra- and extraarticular soft- and hard-tissue joint elements. Sagittal plane scanning (21, 30, 35, 40, 43, 44) became popular because of the emphasis on diagnosing malpositioned articular disks, which were clinically presumed to lie anterior to the mandibular condyle.

Software programs that permit gradient shading of the display data matrix produce three-dimensional TMJ projections that can be rotated about *x*, *y*, or *z* axes (51–53) (Fig. 18.5).

Weaknesses of Computed Tomography

Earlier it was said that the disk was frequently, but not always, seen with CT, depending on the position and shape of the errant disk. Two factors that influence the diagnosis of disk displacement with CT are *(a)* the edge response function and *(b)* the CT blinker or "blink mode" function.

Edge response refers to the inherent inability of the CT system to display abrupt changes in tissue density such as occurs in the relatively short distance from the mandibular condyle across the articular disk to the temporal bone. Rather than displaying the actual abrupt changes in the tissue density differences, there occurs a system-generated gradient that requires several pixels (picture elements) to display at each bone-disk interface. As a result the disk may not be seen as a discrete mass but becomes *lost* in the gradient (Fig. 18.6). If the articular disk is normal in size, shape, and position it will most probably *not* be seen either in vertically reformatted axial sections or direct sagittal soft-tissue detail sections (54). If the disk is

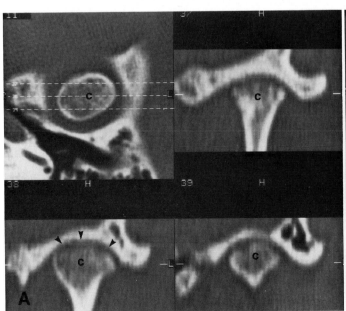

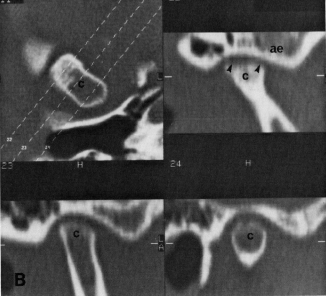

Figure 18.4. **(A)** Cursor lines on axial reference image *(upper left)* of right mandibular condyle *(c)* illustrate the location of vertically reformatted coronal bone detail sections through the anterior *(upper right)*, central *(lower left)*, and posterior *(lower right)* condylar region. Note degenerative changes, including loss of joint space superior to condyle *(arrowheads, lower left)* and cysts or pseudocysts *(arrowheads, upper right)*. **(B)** Corrected sagittal bone detail images show flattening of the articular eminence *(ae)*, sclerosis, and flattening of the mandibular condyle *(c)* *(arrowheads, upper right)*.

thinned it is also unlikely to be seen with CT. However, if the disk is thickened, it will usually be displayed because it will lie outside the gradient of the edge response (Fig. 18.7A and B).

The blinker function (CT number highlighting) is said to be useful for diagnosing displaced TM articular disks. However, the radiodensity of the articular disk is essentially identical to the tendinous attachment of the inferior belly of the lateral pterygoid muscle as it inserts into the pterygoid fovea of the mandibular condyle. Thus the two (tendon vs. disk) may be confused when relying solely on the sensitivity of CT number highlighting (Fig. 18.8A to C).

In Hounsfield units (HUs), we learned that the mean CT numbers of tendon (88 HU) and articular disk (100 HU) were not significantly different ($P > 0.05$). The accuracy of CT for diagnosing disk displacement may be unreliable based solely on CT number-highlighted sections. This weakness in the reliability of CT number highlighting is particularly important if the disk is anterior to the medial one-third to one-half of the condylar pterygoid fovea, where many (actually most) displaced disks reside (55, 56).

If CT is one's modality of choice for TMJ imaging, rather than rely solely on sections windowed for the blinker function (window, 150; level, 120 HU; General Electric 9800 CT/T quick scanner) (General Electric, New Berlin, Wisc.), it is best to use multiplanar soft tissue and highlighted sections. Even with this time-consuming protocol, CT will not adequately image every errant disk.

The question that lingered with most early studies (claiming success rates for CT in excess of 90%) (33, 35, 44) is that they were done on patients with displaced disks rather than normal and symptomatic patients. Thus, the presence of false positive CT results could not be accounted for.

Thickened disks that are lateral, anterolateral, or distinctly anterior to the mandibular condyle are usually readily imaged with CT (54). Disks displaced anteromedial to the mandibular condyle can be confused with the tendon of the lateral pterygoid muscle if viewed solely in the sagittal plane, particularly if one relies on the blinker function.

Applications of Computed Tomography

Articular Disk Displacement

The CT-assisted arthrogram (57) combines the strengths of CT and arthrography. Some aspects of diskal disorders are still best addressed with arthrography, particularly as regards diskal shape and integrity (perforations, etc.) (58).

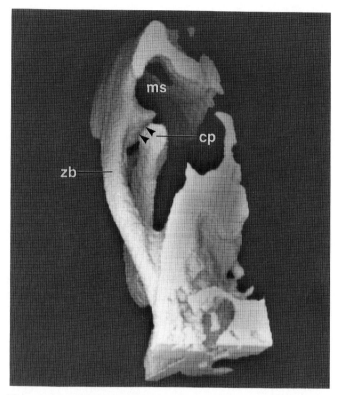

Figure 18.5. Three-dimensional reconstruction of axially acquired images of the left TMJ, showing an enlarged coronoid process *(cp)* which, when it encountered *(arrowheads)* the confluence of the temporal process, zygomatic bone *(zb)*, and the posterior wall of the maxillary sinus *(ms)*, prevented normal and adequate mandibular excursions.

Figure 18.6. Vertically reformatted simultaneous coronal *(above)* and sagittal *(below)* highlighted images. The edge response function prevents visualization of disk between condylar *(c)* and temporal *(t)* joint components in the coronal section. Note also how the edge response encompasses and highlights virtually all osseous surfaces *(arrowheads)*. The highlighted area below the articular eminence *(ae, arrow)* is only partly due to disk displacement and distinction cannot be made between disk and tendon.

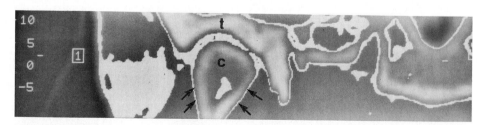

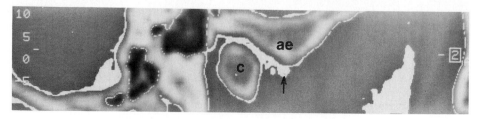

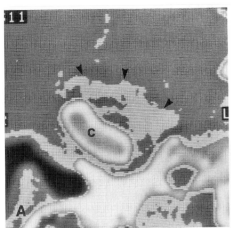

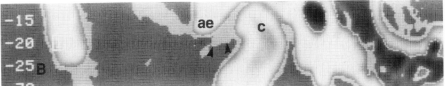

Figure 18.7. **(A)** As illustrated in this axial section, which focuses on the right TMJ, thickened articular disks lying more anterior or anterolateral to the mandibular condyle *(c)* are more readily seen because of the characteristic distribution of the highlighting over the condylar breadth *(arrowheads)*. **(B)** The highlighted mass *(arrowheads)*, inferior to the articular eminence *(ae)* and anterior to the mandibular condyle *(c)* in its lateral aspect, represents a thickened and misshapen articular disk as confirmed with cryosection.

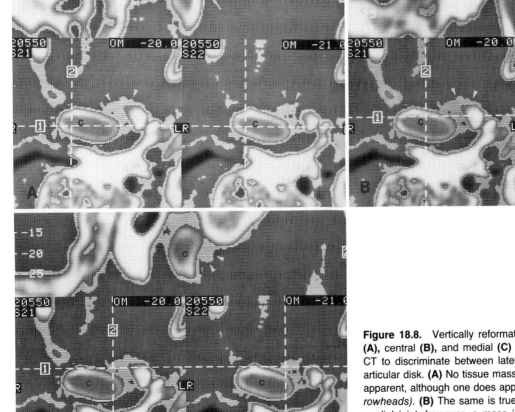

Figure 18.8. Vertically reformatted sagittal sections from the lateral **(A)**, central **(B)**, and medial **(C)** joint aspects illustrate the inability of CT to discriminate between lateral pterygoid muscle tendon and the articular disk. **(A)** No tissue mass anterior *(arrow)* to the condyle *(c)* is apparent, although one does appear in the axial reference section *(arrowheads)*. **(B)** The same is true in the central joint region. **(C)** In the medial joint, however, a mass is evident *(arrowheads)* although, as was learned by cryosection, it was not the articular disk. This is the same autopsy specimen that appears in Figure 18.1*A*. (Reprinted with permission from the Radiological Society of North America.)

Axially acquired thin sections with multiplanar vertical reformatting in conjunction with an arthrogram enhance the diagnostic potential of both modalities because the ultrathin, vertically reformatted sections (0.5 mm) reveal changes that frequently occur in the location and shape of the disk across the breadth of the condyle (Fig. 18.9).

With multiplanar advanced imaging prior to disk-specific therapy, such as attempted recapture or repositioning, clearer indications may emerge as to whether or not certain diskal configurations and/or relationships are more likely to respond favorably to treatment, both noninvasive and invasive.

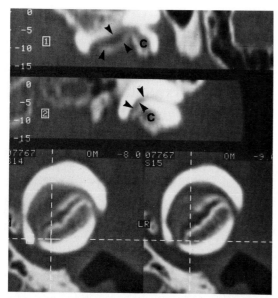

Figure 18.9. CT-assisted arthrography combines the strengths of both modalities by permitting serial thin sections (0.5 mm) vertically reconstructed in any desired plane, enabling evaluation of the position, shape, and condition of the articular disk. In sagittal section *(above)*, the disk is outlined by the envelope of contrast *(arrowheads)* and is anterior to the condyle *(c)*. In the coronal section *(middle)*, the disk is likewise seen within the contrast superior to the condyle *(c)*. (Reprinted with permission from Year Book Medical Publishers, Chicago. from Christiansen EL, and Thompson JR. Temporomandibular joint imaging. Chicago: Year Book Medical, 1990.)

Degenerative Joint Disease

Temporomandibular degenerative joint disease consists of a constellation of symptoms that, in addition to the classical signs (59) of osteosclerosis, osteophytosis, cysts or pseudocysts, and joint space narrowing (Fig. 18.10), also includes increased condylar angulation, diminished transverse condylar dimension, lateral reciprocal reshaping of the condylar pole and lateral wall of the glenoid fossa, loss of condylar height, reshaping of the glenoid fossa, flattening of the slope of the articular eminence (extracapsularly) and increase in the height of the coronoid process (60).

Degenerative changes are detected more readily with CT than with conventional radiographic modalities because CT is not as subject to image blurring and distortion near the steeply curved condylar poles as are linear or pluridirectional tomography.

In one study, 46 patients were scanned for their clinically evident TMJ disorders. On average, three-fourths of the 11 findings described above were seen on each patient. Both joints were involved in slightly more than one-half the patients. Osteosclerosis was the most frequent finding (89 relative frequency percent (RFP), followed by decreased joint space (87 RFP)), changes in the shape of the glenoid fossa (70 RFP), osteophytosis (65 RFP), reciprocal joint reshaping (59 RFP), and changes in condylar shape (50 RFP) (60). The severity of degenerative changes was graded as mild in 30% of cases, moderate in 13%, marked in 41%, and severe in 15%.

Figure 18.10. The degenerative changes in these axial and vertically reformatted sagittal sections are as follow (clockwise from upper left): sclerosis *(arrowheads)* of the lateral three-fifths of the condyle *(c)*; loss of normal contour of the articular eminence *(ae)* and condyle *(c)*; narrowing of joint space *(arrowheads)*.

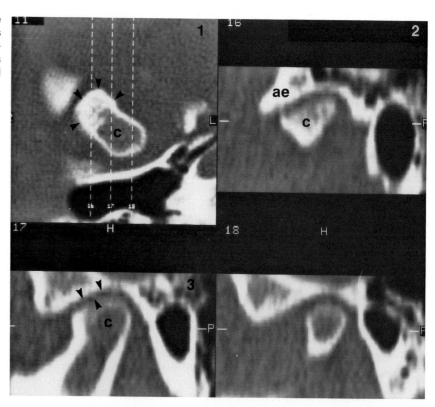

Most osteosclerotic and osteophytic degenerative changes (approximately 50%) are found in the lateral one-third of the joint and the remainder are equally distributed over the middle (25%) and medial thirds (25%).

Variations in the joint space around the mandibular condyle appear to reflect the position of the articular disk in some instances. When measurements were made with CT software across the breadth of the condyle in joints with and without symptoms of disk displacement, it was discovered that the anterosuperior and superior joint spaces were the most significant with respect to disk position (61) (Fig. 18.11).

When the disk was normally positioned the anterosuperior joint space was consistent across the joint (1.5 to 2.0 mm). If the disk was anterior to the condyle the anterosuperior joint space was widened (2.4 to 3.8 mm). When the disk was medial the superior joint space was wider (3.0 to 3.8 mm) than normal (2.0 to 2.4 mm). In those cases where the disk was anteromedial to the condyle the superior joint space was markedly reduced (< 0.2 mm) except in the medial quarter.

Trauma

The value of CT for patients with facial fractures is well recognized (62) and is most probably superior to all other X-ray modalities for detecting TMJ fractures. On average, CT studies of craniomandibular trauma patients reported four findings whereas plain film studies of the same patients reported only one finding (63).

In three studies that investigated the coincidence of trauma and TMJ symptomology the relative frequency ranged from 29 to 38% (64–66). Fractured mandibular condyles often go on to become hypomobile or even ankylosed, depending on the type and extent of injury (see Fig. 18.3).

Whenever facial trauma occurs and skull base injuries are suspected, the temporomandibular joints should be included in the CT examination because early detection of subtle fractures and trauma-induced disk displacements may permit treatment options that may otherwise not be exercised.

Malignant Diseases

Occasionally, an insidious and potentially fatal disease process may either mimic or overlie TMJ pain and/or mandibular dysfunction (see Fig. 18.2) (67). In those cases where TMJ and/or masticatory muscle pain is not influenced by function (i.e., the pain is *not* altered (worsened) by chewing, etc.), and where conservative treatment methods are not soon effective, advanced imaging (CT and/or MR) should be strongly considered (68).

Dosimetry

A comparative dosimetric study was made of the absorbed dose to specific sites of varying radiosensitivity that may be routinely exposed during a TMJ CT scan (69). These sites included the marrow of the mandibular condyle, the pituitary gland, the lens of the eye, brain tissue, and the thyroid gland.

The absorbed soft-tissue radiation doses (expressed in milligrays) for axial and direct sagittal TMJ CT scans (expressed in milligrays) were determined for the General Electric 8800 and 9800 series scanners. One milligray equals 100 mrad. Condylar marrow received doses of 64 and 52 mGy, respectively, for the 9800 and 8800 high-resolution scans; 21 and 17 mGy, respectively, for our standard dynamically sequenced reduced technique scans; and 26 mGy for the GE 9800 direct sagittal CT sections. Direct sagittal scanning is only slightly higher than axial plane scanning (26 vs. 21 mGy), using the same scanner. Of course, the sagittal scan must be repeated for each joint, essentially doubling the dose when both joints are scanned.

The absorbed dose measured for high-resolution and dynamic sequence series are 20 to 30% higher for the General Electric 9800 scanner than for the 8800 scanner, but neither poses significant cataractogenic risk (70, 71).

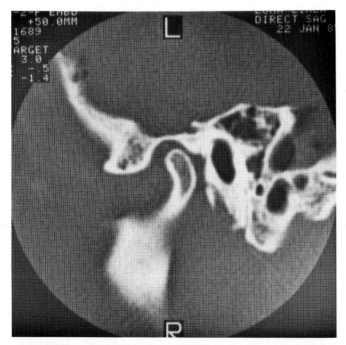

Figure 18.11. Direct sagittal bone detail section illustrates alternate widening of the joint space anterosuperior to the condyle (c) and narrowing of the posterosuperior joint space. Such findings strongly suggest disk displacement and provide indirect support of clinical and CT soft-tissue and highlighted sections.

References

1. Cooper A. A treatise on dislocations and on fractures of the joints. 2nd ed. London: Longman, Hurst, Rees, Orme, and Browne, 1823:386–394.
2. Humphrey GM. Excision of the condyle of the lower jaw. Br Assoc Med J 1856;60:61.
3. Verneuil AAS. De l'ecoulement sanguin dans certaines operations pratiquees sur la face et des moyens propres a en attenuer les inconvenients. Arch Gen Med Paris, 1870, ii.
4. Baer WS. Arthroplasty with the aid of animal membrane. Am J Orthoped Surg 1918;16:1–29.
5. Bishop PA. Roentgen consideration of the temporomandibular joint. AJR 1929;21:556–563.

6. Petrilli A, Gurley JF. Tomography of the temporomandibular joint. J Am Dent Assoc 1939;26:218–224.
7. Lindblom G. Technique for Roentgen-photographic registration of the different condyle positions in the temporomandibular joint. Dent Cosmos 1936;78:1227–1235.
8. Higley LB. Practical application of a new and scientific method for producing temporomandibular Roentgenograms. J Am Dent Assoc 1937;24:222.
9. Maves TW. Radiology of the temporomandibular articulation with correct registration of vertical dimension for reconstruction. J Am Dent Assoc 1938;25:585–594.
10. Norgaard F. Arthrography of the mandibular joint. Acta Radiol Diagn (Stockholm) 1944;23:740.
11. Ricketts RM. Variations of the temporomandibular joint as revealed by cephalometric laminography. Am J Orthodont 1950;36:877.
12. Wilkes CH. Arthrography of the temporomandibular joint in patients with the TMJ pain-dysfunction syndrome. Minn Med 1978;61:645–652.
13. Farrar WB, McCarty WL Jr. Inferior joint space arthrography and characteristics of condylar paths in internal derangements of the TMJ. J Prosthet Dent 1979;41:548.
14. Katzberg RW, Dolwick MF, Bales DJ, Helms CA. Arthrotomography of the temporomandibular joint: new technique and preliminary observations. AJR 1979;132:949–955.
15. Eckerdal O. Tomography of the temporomandibular joint. Acta Radiol Suppl (Stockholm) 1973;329.
16. Petersson A. Radiography of the temporomandibular joint. A comparison of information obtained from different radiographic techniques [Dissertation]. Malmö, Sweden: University of Lund, 1976.
17. Littleton JL, Shaffer KA, Callahan WP, Durzich ML. Temporal bone: comparison of pluridirectional tomography and high resolution computed tomography. AJR 1981;137:835–845.
18. Suarez FR, Bhussry, Huang KH, et al. A preliminary study of computerized tomographs of the temporomandibular joint. Compend Contin Educ Dent 1980;1(3):217–222.
19. Kaban LB, Bertolami CN. The role of CT scan in diagnosis of TMJ ankylosis: report of case. J Oral Surg 1981:39:370–372.
20. Helms CA, Moorish RB, Kircos LT, Katzberg RW, Dolwick MF. Computed tomography of the meniscus of the temporomandibular joint: preliminary observations. Radiology 1982;145:719–722.
21. Manzione JV, Seltzer SE, Katzberg RW, Hammerschlag SB, Chiango BF. Direct sagittal computed tomography of the temporomandibular joint. Am J Neuroradiology 1982;3:677–679.
22. DelBalso AM, Pyatt RS, Busch RF, Hirokawa R, Fink CS. Synovial cell sarcoma of the temporomandibular joint: computed tomographic findings. Arch Otolaryngol 1982;108:520–522.
23. Martignoni M, Bozzao L. The CAT scanning in TMJ examination. Part 1. Centric position. J Gnathol 1982;1(1):37–46.
24. Hüls A, Schulte K, Voigt K, Erlich-Treuenstatt. Computed tomography of the temporomandibular joint; new diagnostic possibilities and initial clinical results. Electromedica 1983;51:14–19.
25. Helms CA, Katzberg RW, Moorish R, Dolwick MF. Computerized tomography of the temporomandibular joint. J Oral Maxillofac Surg 1983,41;8:512–517.
26. Wilkinson T, Maryniuk G. The correlation between sagittal anatomic sections and computerized tomography of the TMJ. J Craniomand Pract 1983;1:37–46.
27. Blankestun J, Boering G, Thun CJP. Arthrography, arthrotomography and computed tomography in the differential diagnosis of temporomandibular joint dysfunction. J Oral Rehabil 1983;10:449.
28. Salon JM, Ross RJ. Computerized tomography with multiplanar reconstruction for examining the TMJ. J Craniomandib Pract 1983;1:27.
29. Hassan TA, Rahman HA, El Din MS. Computed tomography in early diagnosis of temporomandibular joint ankylosis. Egyptian Dent J 1984;30(1):29–38.
30. Sartoris DJ, Neumann CH, Riley RW. The temporomandibular joint: true sagittal computed tomography with meniscus visualization. Radiology 1984;150:250–254.
31. Thompson JR, Christiansen EL, Hasso AN, Hinshaw DB Jr. Dislocation of the temporomandibular joint disk demonstrated by CT. Am J Neuroradiology 1984;5:115–116.
32. Thompson JR, Christiansen EL, Hasso AN, Hinshaw DB Jr. Temporomandibular joints: high resolution computed tomographic evaluation. Radiology 1984;150:105–110.
33. Thompson JR, Christiansen EL, Sauser DD, Hasso AN, Hinshaw DB Jr. Contrast arthrography versus computed tomography for the diagnosis of dislocation of the temporomandibular joint meniscus. Am J Neuroradiology 1984;5:747–750.
34. Christiansen EL, Thompson JR. Anteriorly displaced temporomandibular joint articular disc. A case diagnosed by computed tomography. Oral Surg 1984;58(3):355–357.
35. Manzione JV, Katzberg RW, Brodsky GL, Seltzer SE, Mellins HZ. Internal derangements of the temporomandibular joint: diagnosis by direct sagittal computed tomography. Radiology 1984;150:111–115.
36. Avrahami E, Horowitz, Cohn DF. Computed tomography of the temporomandibular joint. Comput Radiol 1984;8(4):211–216.
37. Larheim TA, Kolbenstvedt A. High-resolution computed tomography of the osseous temporomandibular joint: some normal and abnormal appearances. Acta Radiol Diagn 1984;25(6):465–469.
38. Fjellströem C-A, Olofsson O. Computed tomography of the temporomandibular joint meniscus. A report of preliminary tests. J Maxillofac Surg 1985;13:24–27.
39. Ma Xu-Chen et al. CT diagnoses in TMJ disturbance syndrome. J Beijing Med Univ 1984;16(4):299.
40. Conover GL. Direct sagittal CT scanning of the temporomandibular joint. Calif Dent Serv Rev 1985;6:28–31.
41. Helms CA, Richardson ML, Vogler JB III, Hoddick WL. Computed tomography for diagnosing temporomandibular joint disk displacement. J Craniomandib Pract 3;1:23–26.
42. Swartz JD, Vanderslice R, Hendler BH, Abaza NA, Lansman NA, Popky GL. High-resolution computed tomography. Part 4. Evaluation of the temporomandibular joint. Head Neck Surg 1985;7:468–478.
43. Simon DC, Hess ML, Smilak MS, Beltran J. Direct sagittal CT of the temporomandibular joint. Radiology 1985;2:157.
44. Manco LG, Messing SG, Busino LJ, Fasulo CP, Sordill WC. Internal derangements of the temporomandibular joint evaluated with direct sagittal CT: a prospective study. Radiology 1985;157:407–412.
45. Cohen H, Ross S, Gordon R. Computerized tomography as a guide in the diagnosis of temporomandibular joint disease. J Am Dent Assoc 1985;110:57–60.
46. Schatz SL, Cohen HR, Ryvicker MJ, Deutsch AM, Manzione JV. Overview of computed tomography of the temporomandibular joint. J Comput Tomogr 1985;9:351–358.
47. Ross SR, Cohen HR, Rubenstein HS. Indications for computerized tomography in the assessment and therapy of commonly misdiagnosed internal derangements of the temporomandibular joint. J Prosthet Dent 1987;58(3):360–366.
48. Christiansen EL. Temporomandibular joint x-ray computed tomography [Dissertation]. Stockholm: Karolinska Institutet, 1988.
49. Hey W. Practical observations in surgery, illustrated with cases. 3rd ed. London: Cadell and Davies, 1814.
50. Christiansen EL, Thompson JR, Kopp S. Intra- and interobserver variability and accuracy in the determination of linear and angular measurements in computed tomography; an in-vitro and in-situ study of human mandibles. Acta Odontol Scand 1986;44:221–229.
51. Roberts D, Pettigrew J, Udupa J, Ram C. Three-dimensional imaging and display of the temporomandibular joint. Oral Surg Oral Med Oral Pathol 1984;58:461–474.
52. Hemmy DC, Tessier PL. CT of dry skulls with craniofacial deformities: accuracy of three-dimensional reconstructions. Radiology 1985;157:113.
53. Pettigrew J, Roberts D, Riddle R, Udupa J, Collier D, Ram C. Identification of an anteriorly displaced meniscus in vitro by means of three-dimensional image reconstructions. Oral Surg Oral Med Oral Pathol 1985;59:535–542.
54. Christiansen EL, Thompson JR, Kopp S, Hasso AN, Hinshaw Db Jr. CT number characteristics of malpositioned TMJ menisci; diagnosis with CT number highlighting (Blinkmode). Invest Radiol 1987;22:315–321.
55. Anderson QA, Katzberg RW, Helms CA. Radiographic imaging of the temporomandibular joint. Refresher course #106 given at

the 71st scientific assembly and annual meeting of the Radiological Society of North America, Chicago, November 17–22, 1985.

56. Thompson JR, Christiansen EL, Hasso AN, Hinshaw DB Jr. Normal and abnormal positional relationships of the TMJ disc demonstrated with CT, arthrography and tissue photographs. Paper presented at the 33rd annual meeting of the Association of University Radiologists, Nashville, May 12–17, 1985.

57. Katsberg RW, Dolwick MF, Keith DA, Helms CA, Guralnick WC. New observations with routine and CT-assisted arthrography in suspected internal derangements of the temporomandibular joint. Oral Surg 1981;51:569–574.

58. Tanimoto K, Hansson LG, Petersson A, Rohlin M, Johansen CC. Computed tomography versus single-contrast arthrotomography in evaluation of the temporomandibular joint disc. Int J Oral Maxillofac Surg 1990;18:354–358.

59. Greenfield GB. Radiology of Bone Diseases. 3rd ed. Philadelphia: Lippincott, 1980:774.

60. Christiansen EL, Thompson JR, Dopp S, Hasso AN, Hinshaw DB Jr. Radiographic signs of temporomandibular joint disease; an investigation utilizing x-ray computed tomography. Dentomaxillofac Radiol 1985;14:83–92.

61. Christiansen EL, Thompson JR, Zimmerman G, Roberts D, Hasso AN, Hinshaw DB Jr, Kopp S. Computed tomography correlation of condylar and articular disc positions within the temporomandibular joint. Oral Surg Oral Med Oral Pathol 1987;64:757–767.

62. Baker SR, Gaylord GM, Lantos G, Tabaddor K, Gallagher JE.

Emergency skull radiography: the effect of restrictive criteria on skull radiography and CT use. Radiology 1985;156:409–413.

63. Christiansen EL, Thompson JR, Hasso AN. CT evaluation of trauma to the temporomandibular joint. J Oral Maxillofac Surg 1987;45:920–923.

64. Truelove E, Burgess J, Dworkin S, Lawton L, Sommers E, Schubert M. Incidence of trauma associated with temporomandibular disorders [Abstract]. J Dent Res 1985;64:339.

65. Kaye LB, Moran JH, Fritz ME. Statistical analysis of an urban population of 236 patients with head and neck pain. J Periodontol 1979;50:55–58.

66. Bakland LK, Christiansen EL, Strutz JM. Frequency of dental and traumatic events in the etiology of temporomandibular disorders. Endod Dent Traumatol 1988;4:182–185.

67. Christiansen EL, Thompson JR, Appleton SA. Temporomandibular joint pain/dysfunction overlying more insidious diseases; report of two cases. J Oral Maxillofac Surg 1987;4:335–337.

68. Bavits JB, Chewning LC. Malignant disease as temporomandibular joint dysfunction; review of the literature and report of case. J Am Dent Assoc 1990;120:163–166.

69. Christiansen EL, Moore RJ, Thompson JR, Hasso AN, Hinshaw DB Jr. Radiation dose in radiography, CT and arthrography of the temporomandibular joint. AJR 1987;148:107–109.

70. Gregg EC. Radiation risks with diagnostic x-rays. Radiology 1977;123:447–453.

71. Hall EJ. Radiation cataractogenesis. In: Radiobiology for the radiologist. New York: Harper & Row, 1978:349–356.

CHAPTER 19

Computed Tomography of Craniofacial Fractures

Mini N. Pathria

Introduction

Facial fractures account for 2% of all hospital admissions in the United States and are a significant source of patient morbidity (1). The ever-increasing incidence of assaults and motor vehicle accidents (MVAs), the two most frequent etiologies for facial fractures, has resulted in a rising incidence of facial injury in the past decade. Over 70% of victims of MVAs sustain some cranial and facial trauma, with facial fractures occurring in 10% of such patients (2). Multisystem injury is present in up to 50% of patients and often limits the type and quality of the radiographic examination (1). Inebriated and combative patients are particularly difficult to examine in the acute period. Definitive facial examination can usually be safely delayed until an adequate study is attainable (3, 4). Conventional radiography and computed tomography (CT) are the most frequently performed examinations for assessment of facial trauma. In this chapter, the radiographic and CT characteristics of common fractures will be reviewed. Accurate and timely diagnosis of facial injury is necessary because delayed complications such as persistent diplopia, enophthalmos, cosmetic disfigurement, and infection are difficult to treat (5).

Technique

Conventional Radiography

Facial evaluation should not be performed until cervical spine, cardiopulmonary, vascular, and cranial stability is assured (3, 4). The initial screening examination for suspected facial injury consists of conventional radiographs in the Waters (occipitomental), Caldwell (occipitofrontal), and lateral projections (6). Upright positioning is preferable, since air-fluid levels can be detected, but supine imaging may be necessary in the severely injured patient. The Waters view is obtained posteroanteriorly (PA) with the X-ray beam angled 37° caudad to the canthomeatal line. In this position, the maxillary alveolar ridge is projected above the petrous bone and the entire maxillary sinus is clearly visible. This projection enables optimal evaluation of the midface and allows good visualization of the maxillary bone and sinus, zygoma, orbital rim, and orbital floor. The Caldwell view is also obtained PA but the beam is angled only 15° caudad to the canthomeatal line. This view is useful for evaluation of the frontal bone and sinuses, superior orbital rim, and the orbital floor, which should project just superior to the petrous bone. The greater wing of the sphenoid is seen as an oblique line at the lateral orbital margin of the orbit, whereas the lesser wing produces a horizontal line in the upper portion of the orbit. The lateral view is obtained, by convention, in the left lateral position with the X-ray beam centered at the lateral canthus. Additional radiographs are frequently necessary, depending on the suspected area of injury. Supplemental views of the nose and mandible are routinely utilized for suspected fractures of these areas (6, 7). The base (submentrovertex) view is useful for evaluating the medial and lateral walls of the maxillary sinus, lateral wall of the orbit, sphenoidal margin of the temporal bone, and the zygomatic arches (8). Cervical spine injury should be excluded prior to this examination, as the neck needs to be hyperextended. Numerous other supplemental projections, including facial zonography and stereoradiography, have been utilized for the assessment of complex facial trauma. While these additional views can be helpful in selected cases, most problematic cases are now assessed utilizing conventional tomography or CT. Magnetic resonance (MR) imaging is not indicated for osseous facial trauma but may be helpful for assessment of soft-tissue structures such as the globe, optic nerve, and extraocular muscles. Undisplaced fractures cannot be reliably identified using MR imaging (9).

Conventional and Computed Tomography

Conventional tomography is now infrequently utilized for evaluation of facial fractures, having been supplanted by CT. If CT is not available, conventional tomography can be helpful for characterizing osseous injury (Fig. 19.1). The critical organs irradiated with both techniques include the

---

Abbreviations (see also glossary): MVA, motor vehicle accident; 3D CT, three-dimensional CT; TMJ, temporomandibular joint.

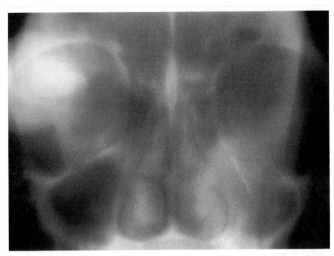

Figure 19.1. Conventional anteroposterior (AP) tomogram illustrates a severely depressed fracture of the medial portion of the left orbital floor. Opacification of the ipsilateral maxillary sinus is present. Since the tomogram is obtained in the supine position, an air-fluid level will not be visualized. Note the blurring produced by the contralateral lens shield.

salivary glands, thyroid gland, and the lens of the eye. Shielding of the thyroid and lens during tomography may result in the loss of valuable information (10). Shielding is not necessary with CT since only a small number of slices are obtained through these sensitive areas. The estimated dose to the lens has been determined to be approximately 6 rads with coronal CT and 4 rads with axial CT (11). This dose is a significant improvement relative to conventional tomography, where doses of 30 rads are not uncommon (5). Tomography of facial fractures is optimally performed with units capable of pluridirectional tube movement (12–14). The anteroposterior (AP) tomograms may be performed in either the true AP or the Caldwell projection, but should not be performed in the Waters position, since the orbital floor is not perpendicular to the tomographic plane in this projection (6, 13). Lateral tomography supplements the AP study and is useful for evaluation of the orbital floor and the anterior and posterior walls of the frontal and maxillary sinuses (15). Each tomographic plane typically results in 10 images through the orbits at 5-mm intervals. While there is evidence that tomography is more accurate than conventional radiographs, it offers no significant advantages compared to CT (13–16). The major advantages of CT compared to conventional tomography are diminished radiation dose, improved soft-tissue contrast, and the ability to perform multiplanar reconstruction (3, 5). In addition to osseous injury, damage to soft-tissue structures such as the globe, optic nerve, and extraocular muscles can be readily ascertained (5, 17).

Computed tomography has become the most frequently utilized technique for evaluation of suspected facial injury when information provided by plain films is inadequate (3, 14, 18, 19). Indications for facial CT include visual or extraocular muscle impairment, fractures that may extend intracranially, unexplained neurological deficit, and massive craniofacial injury (3). Contiguous 4-mm slices are usually adequate, although thinner slices are preferable

when reformations are necessary (3, 5, 20). Both bone and soft-tissue windows should be examined for evidence of injury.

Ideally, both axial and direct coronal images of the facial structures are obtained. The axial projection optimally visualizes the medial and lateral walls of the maxillary sinus and orbit. This projection is preferable in the severely injured patient with concomitant cranial or spinal injury (3). However, fractures of the orbital floor are frequently not identified on the axial images and usually require coronal imaging for detection (11, 15, 21). Direct coronal imaging requires patient motion and may be impossible in the elderly, severely injured, or uncooperative patient (3). Patient movement is not necessary for axial imaging and concomitant evaluation of the cranium is rapidly obtained.

Gentry et al. have reviewed normal bony and soft-tissue facial anatomy with high-resolution CT and have defined horizontal, coronal, and sagittal osseous struts that contribute to facial stability (18). These authors have also reviewed the most common types of facial trauma and categorized them according to the pattern of disruption of the major facial struts (18, 19). In their work they have outlined the optimal planes for examination of a variety of facial injuries. In general, scans should be obtained perpendicular to the structures being evaluated, particularly when thin bony plates need to be visualized (3, 18, 19). An excellent review of the normal sutures, foramina, fissures, and thin bony septa that may be mistaken for fractures can be found in the articles by these authors (18, 19).

Three-dimensional CT (3D CT) can be valuable in cases of severe trauma, providing a clear demonstration of fracture extent and fragment displacement (20). Fracture detection on the 3D CT images is inferior to that obtained on the regular CT study, so this technique complements rather than replaces the conventional examination (20) (Fig. 19.2). Exclusion artifacts from partial volume averaging produce pseudoforamina in areas of papyraceous bone, which can be confused for fractures (20, 22). Errors of inclusion and exclusion can be minimized by using very thin contiguous sections, usually in the range of 1.5 to 3 mm. Overlapping of slices is not necessary since contiguous sections enable satisfactory reconstruction at a lower radiation dose (22).

Nasofrontal Fractures

Frontal Sinus Fractures

Comminuted frontal calvarial fractures may readily extend to involve the thin-walled frontal sinus. Concurrent injury of the nasal and ethmoid regions is frequently present when the frontal sinus is involved (Fig. 19.3). If the sinus is unusually well aerated, these fractures may also extend to involve the superior orbital rim (6). Isolated fractures of the anterior wall of the frontal sinus are common and may be linear or severely comminuted. The latter generally reproduce the shape of the striking object and are stellate and depressed (2). The frontal sinuses are also frequently involved in the more complicated smash injuries in which the major facial buttresses are extensively disrupted. Frontal sinus fractures are easily diagnosed on plain films, as the fracture lines are usually well seen and associated opacification of the frontal sinuses is typically present.

Isolated disruption of the posterior wall is rare but the posterior wall is often fractured in association with comminuted anterior fractures. Disruption of the posterior wall of the sinus allows communication between the sinus and intracranial contents but is frequently difficult to identify on the lateral view. Pneumocephalus or cerebrospinal (CSF) rhinorrhea, in the presence of an opacified sinus, are both indicative of a posterior wall fracture. Axial CT is frequently necessary for evaluating the walls of the frontal sinus when a posterior fracture is suspected. Approximately 8% of fractures involving the paranasal sinuses are associated with pneumocephalus, although this finding is most commonly seen following depressed frontal sinus fractures. Cerebrospinal fluid rhinorrhea is noted in approximately 50% of patients with pneumocephalus and brain abscess or meningitis develops in up to 25% of affected patients (23). While CSF rhinorrhea may develop following a frontal fracture, it is more commonly seen in the presence of an ethmoidal injury (23). Determination of the site of dural tear in those cases that do not resolve spontaneously may require intrathecal contrast or radionuclide examination.

Nasal Fractures

The thin, paired nasal bones are attached to the frontal bone superiorly at the nasofrontal suture and to the maxillary bones laterally at the nasomaxillary sutures. Because of their prominent position, the nasal bones are the most frequently fractured portions of the facial skeleton (4). Suspected nasal injury requires a supplemental underpenetrated coned-down lateral radiograph, including the nasal bones and anterior maxillary spine. Transverse fractures isolated to the nasal tip are best seen on this view (6). Longitudinal fractures of one of the nasal bones are less common and must be differentiated from the normal grooves present in this area. These normal grooves, formed by the paired nasomaxillary sutures and the nasociliary nerve, parallel the dorsum of the nose and are more irregular and less lucent than a fracture line. On the Waters view, longitudinal fractures result in unilateral disruption or depression of the nasal arch. The clinical examination is more reliable than radiography in evaluating the presence and degree of depression of the distal fragment (7). Elevation of the nasal bone may be necessary in severely

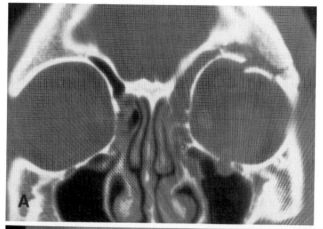

Figure 19.2. A depressed and rotated fracture of the superior orbital rim is seen on coronal CT **(A)** and a 3D CT reconstruction **(B)**. While both techniques show the fracture, the type of displacement is easier to visualize on the reconstructed image. The soft-tissue components of this injury, such as the sinus opacification and soft-tissue prolapse through a concomitant blow-out fracture of the orbital floor, cannot be appreciated on the reconstruction since the soft-tissue elements are excluded with this technique.

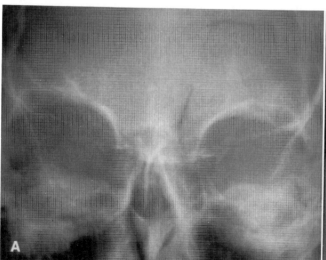

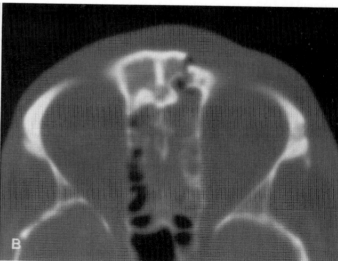

Figure 19.3. The AP view **(A)** in a patient with a calvarial fracture of the left frontal bone demonstrates extension of the fracture into the frontal and ethmoid sinuses. Axial CT **(B)** of the frontoethmoid complex reveals soft-tissue swelling, bony disruption and displacement, sinus opacification, and gas in the soft tissues anterior to the fracture.

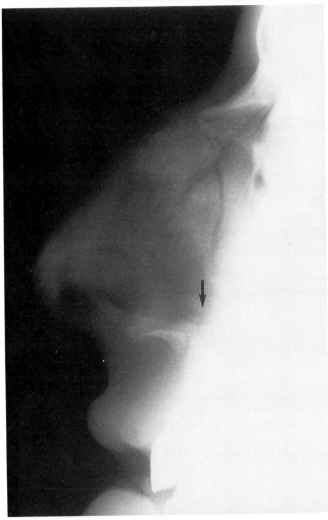

Figure 19.4. The underpenetrated nasal view demonstrates a nasal fracture that is perpendicular to the normal nasal grooves. No depression is evident. An associated avulsion fracture of the anterior maxillary spine is evident *(arrow)*.

depressed fractures to maintain an acceptable appearance of the nasal bridge.

Evaluation of the adjacent osseous structures is essential, as nasal fractures may be associated with more significant injuries involving the bony nasal septum, frontal sinus, or maxillary region. Avulsion of the anterior maxillary spine must be identified since this injury is associated with disruption of the cartilaginous anterior nasal septum (23) (Fig. 19.4). Deviation of the bony nasal septum can be identified on the conventional radiographs but injury of the cartilaginous septum can be directly visualized only by CT.

Fractures of the Orbit

The orbital contents are protected by the peripheral thick orbital rim formed by the frontal bone superiorly, the frontal process of the zygomatic bone laterally, the maxillary process of the zygoma and maxillary bone inferiorly, and the frontal process of the maxillary bone medially (24). The orbital cavity expands immediately posterior to the rim and is cone shaped, with its apex angulated posteromedially. The roof of the orbit is formed by the frontal bone. The sphenoid bone contributes to the lateral orbital wall; the ethmoid and lacrimal bones form portions of the medial wall. The floor of the orbit, formed from the roof of the maxillary sinus, is the thinnest wall of the orbit, consisting of an anterolateral plane region and a superiorly convex posteromedial region (25). The medial wall and floor are the only walls that are not supported by soft tissue, since they abut the ethmoid and maxillary sinuses, respectively.

Fractures of the orbit may occur in isolation or may be components of more complex injuries such as tripod, Le Fort II or Le Fort III fractures. Isolated fractures of the orbit may be divided into those that involve the orbital rim and those that involve the orbital walls. Isolated fractures of the orbital rim most frequently involve its inferolateral portion and result from a direct blow to this area. Clinically, a palpable defect may be present in the orbital margin. The Waters view is optimal for visualization of the superior, lateral, and inferior rim, whereas the Caldwell projection is recommended for evaluation of the superior and medial rim (23). On a properly positioned Waters view, the inferior rim lies above and is slightly thicker than the orbital floor. Fractures of the rim may be associated with rotation of the detached fragment, producing the "disappearing rim" sign (6). The rim fracture may extend to involve a portion of the anterior orbital floor. Fractures of the inferior rim may simulate a blow-out fracture on conventional radiographs and additional views, tomography, or CT may be necessary to define the exact location of the fracture.

Orbital wall fractures may occur anywhere in the orbit, but most frequently involve the posterior orbital floor medial to the infraorbital groove, where the floor is thinnest, or the lamina papyracea of the ethmoid bone (24). The blow-out fracture, by definition, involves the floor or medial wall with sparing of the orbital rim. The classic study by Smith and Regen in 1957 established the etiological mechanism to be a sudden increase in intraocular pressure (26). The fracture is produced when force applied to the globe displaces it and the remaining orbital tissues posteriorly. Blows by objects smaller than the globe tend to rupture this structure, whereas objects larger than the globe dissipate some of their force at the orbital rim. Marked force applied to the orbital rim produces fractures of the rim and floor in combination, but not the true blow-out fracture. When the force is not great enough to fracture the rim, it results in a blow-out fracture (26).

The clinical findings of an orbital floor fracture consist of ecchymosis of the eyelid, chemosis, subconjunctival hemorrhage, and periorbital edema. Displacement of large bony fragments, typically involving over 2.5 cm^2 of the floor, may result in enophthalmos or inferior displacement of the globe (27). Diplopia may result from muscle edema or hemorrhage, nerve damage, and entrapment of posterior orbital fat or the inferior rectus muscle (11, 27, 28). Hammerschlag et al. noted that transient diplopia was present at the time of presentation in 80% of patients with orbital floor fractures. This symptom resolved spontaneously in the majority of these patients and the authors

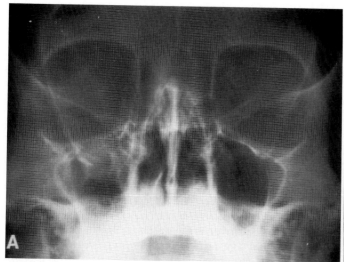

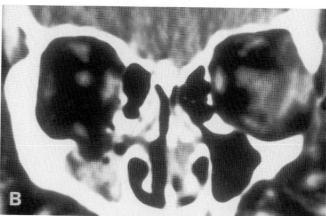

Figure 19.5. Conventional radiograph **(A)** and coronal CT **(B)** of a right orbital floor fracture illustrate the value of CT in evaluating soft-tissue prolapse through the bony defect. The plain film shows disrup-

tion and depression of fracture fragments, but the nature of the soft-tissue prolapse cannot be characterized. The coronal CT demonstrates herniation of orbital fat and displacement of the inferior rectus muscle.

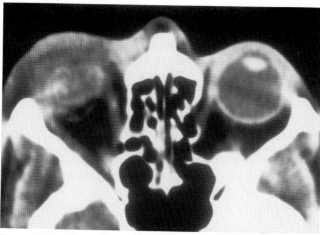

Figure 19.6. Traumatic rupture of the right globe is shown on this axial CT filmed with soft-tissue technique. No fracture or foreign body was associated. The globe is totally disorganized, with loss of definition of the sclera, lens, and vitreous.

suggested that true muscle entrapment is an infrequent etiology for diplopia (27). The delayed complications of orbital floor fractures consist of persistent diplopia and delayed enophthalmos secondary to fibrosis of the orbital fat (29). The incidence of delayed enophthalmos is 8 to 54% within weeks to months after injury (11).

The direct radiographic signs of an orbital floor fracture consist of bone fragmentation, depression of bony fragments, and soft-tissue prolapse through the orbital floor (6, 12, 24, 25). Associated findings include fractures of the orbital rim and medial orbital wall, sinus opacification, and orbital emphysema (24). The depressed bony fragments are best seen on the Waters view but may not be apparent if significantly rotated. Visualization of the fracture fragments is extremely difficult if the sinuses are significantly opacified. Hammerschlag et al. state that virtually all orbital floor fractures are detectable on careful review of con-

ventional radiographs. These authors identified 29 of 30 orbital floor fractures detected by tomography, but they did not review the radiographs prospectively (28). The accuracy of conventional radiography for detection of orbital floor fractures varies widely in different studies, ranging from 5 to 97% (13, 16, 25, 28).

In complex cases of orbital floor fractures where further information is required, CT scanning is more advantageous than tomography because of its ability to display the soft-tissue components of the injury. Soft-tissue prolapse through the orbital floor may consist of hematoma, orbital fat, herniated muscle, or unrelated sinus disease (5, 25, 27). While these entities all have a similar appearance on tomograms, they are easily differentiated by CT (Fig. 19.5). Computed tomography also enables simultaneous assessment of the globe, optic nerve, and extraocular muscles. Rupture of the globe, lens dislocation, and penetrating foreign bodies can be detected (Fig. 19.6). Magnetic resonance imaging can distinguish between choroidal effusion and choroidal hematoma, both of which may be due to ocular trauma (30). Rupture, edema, and entrapment of the optic nerve can also be evaluated.

Most fractures of the orbit and soft-tissue injuries in this region can be readily detected on the axial images. Computed tomography of orbital floor and orbital roof fractures is optimally performed by obtaining both axial and direct coronal images, since these fractures are difficult to recognize and characterize on axial scans alone (11, 14, 17, 19). Difficulties associated with coronal imaging include the need for patient motion and cooperation, artifact from dental fillings, and a higher radiation dose to the lens (11). If direct coronal images perpendicular to the floor cannot be obtained, coronal reconstructions may be extremely helpful in evaluating the orbital floor. Direct oblique sagittal CT or sagittal reformations along the axis of the inferior rectus muscle can be helpful for assessing muscle entrapment (11, 27). Entrapment is suggested when the muscle is acutely displaced into the adjacent sinus and is

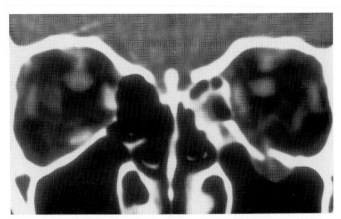

Figure 19.7. This coronal CT image illustrates elongation and entrapment of the inferior rectus muscle in a patient with fractures of the right orbital floor and medial wall. The inferior rectus muscle is impinged on by the fracture fragments and is rotated. Herniation of the orbital fat through the bony defect is also evident.

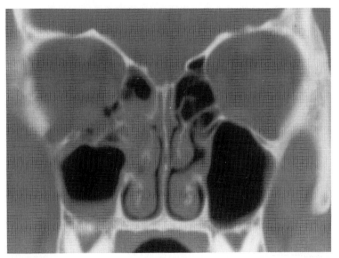

Figure 19.8. Fractures of the floor and medial wall of the right orbit are associated with opacification of the right ethmoid sinus, an air-fluid level in the right maxillary sinus, and depression of the medial floor fragment.

impinged on by surrounding bone fragments (Fig. 19.7). The entrapped muscle is commonly elongated and demonstrates an abrupt kink, unlike the smooth displacement that may occur in association with fat prolapse (14, 27).

While there is controversy regarding the optimal therapy for these injuries, most surgeons agree that a blowout fracture is not a surgical emergency and delay surgery for 10 to 14 days to evaluate the persistence of initial diplopia and enophthalmos (11, 27). Other authors advocate early exploration of the orbital floor, stating that the presence of adhesions and fibrosis following delayed therapy results in a poor outcome (4, 26). Currently, most physicians recommend delayed definitive treatment based on careful examination following resolution of hematoma and edema in the orbital region.

Fractures of the medial orbital wall may occur in isolation but are more frequently found in association with blow-

out fractures of the floor, occurring in 20 to 40% of such patients (12, 31) (Fig. 19.8). Medial wall fractures produce ipsilateral ethmoid sinus opacification and are the most frequent cause of orbital emphysema. These indirect signs are valuable because the bony disruption itself can be extremely difficult to identify, even with tomography or CT (Fig. 19.9). Computed tomography is particularly helpful if entrapment of the medial rectus muscle, an uncommon sequela of this injury, is suspected (31).

Computed tomography is essential for adequate evaluation of the lateral orbital wall since this structure is oriented obliquely on both AP and lateral radiographs and is therefore poorly seen (15). Fractures of the lateral orbital wall are typically associated with ipsilateral zygomatic fractures, whereas fractures of the superior orbital wall are almost invariably associated with fractures of the superior rim (6, 32). Injuries of the lateral orbital wall may be associated with fractures of the orbital apex. The orbital apex is the most posterior portion of the orbital cone and contains the optic foramen and the posterior part of the superior orbital fissure (33). Fractures in this region may produce optic nerve damage, resulting in traumatic blindness or diminished visual acuity, or may produce the superior orbital fissure syndrome, consisting of ophthalmoplegia, ptosis, proptosis, pupillary dilatation, pain, and optic neuralgia (33). Injury to the optic nerve occurs most frequently in its intracanalicular portion (34). Fractures of the orbital apex are difficult to evaluate by conventional radiography and require CT for accurate delineation (33, 34) (Fig. 19.10). Axial CT scanning can identify fractures in this area, determine whether there is transection or impingement of the optic nerve, and may be helpful for planning surgical approach (34).

Fractures of the Zygoma

The paired zygomatic bones are attached to the frontal bone at the zygomaticofrontal sutures, to the temporal bone at the zygomaticotemporal suture in the posterior portion of the zygomatic arches, and to the maxilla along the inferior orbital margin (4, 18, 35). Fractures of the zygoma may involve its prominent malar portion, resulting in the tripod injury, or be isolated to the zygomatic arch. Nakamura and Gross described a series of 323 patients admitted for facial fractures and noted that 79 had a tripod fracture and an additional 28 had an isolated arch fracture, pointing out the frequency of injuries to the zygomatic complex (1).

Zygomatic Arch Fractures

Isolated fractures of the zygomatic arch are produced by a direct blow to the side of the face. Pain, edema, and overlying hematoma are present in all patients but asymmetrical facial flattening and a palpable defect are identified only when the fracture is depressed (35). The most common pattern of injury produces three fracture lines, with depression of the central two fragments and outward displacement of the zygomatic and temporal ends of the arch (6, 12, 35). Isolated zygomatic arch fractures are easily recognized on the routine facial series and infrequently re-

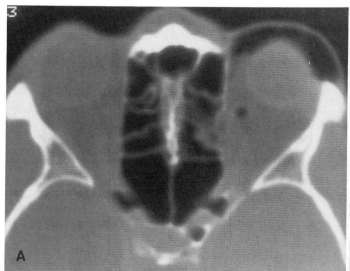

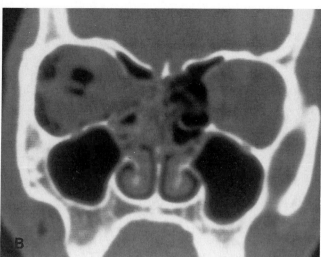

Figure 19.9. CT examinations of two patients with isolated medial wall fractures are shown. The axial image **(A)** shows a left lamina papyracea fractures with ethmoid opacification and orbital emphysema. A coronal CT **(B)** demonstrates similar findings associated with a right orbital fracture. The normal papyraceous bone is often difficult to visualize in this region.

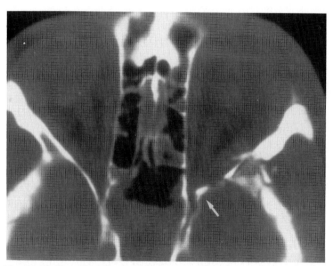

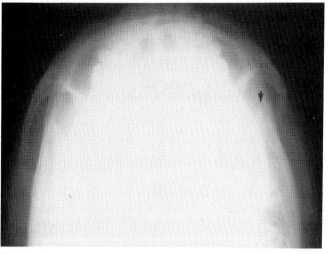

Figure 19.10. This fracture of the lateral wall of the left orbit extends posteriorly to involve the greater wing of the sphenoid and the orbital apex *(arrow)*. An air-fluid level is present within the sphenoid sinus. This patient did not experience any visual deficit.

Figure 19.11. The underpenetrated base view provides excellent visualization of the zygomatic arches. This man has a fracture of the left arch with depression of the central fragments. The relationship between the fracture and the tip of the coronoid process *(arrow)* can be appreciated.

quire additional evaluation. Disruption of the zygomatic arch can usually be identified on the Waters view by noting interruption or overlap of bony fragments. We prefer an underpenetrated base view for assessment of the arch and for determining the degree of depression of the central fragments (7) (Fig. 19.11). The normal zygomaticotemporal suture should not be confused with a fracture. Limited mandibular excursion resulting from bony impingement on the coronoid process by depressed fragment is noted in up to 10% of patients. Computed tomography is very helpful in evaluating the position of the fragments relative to the coronoid process in these patients. Limitation of mouth opening may also be a delayed complication of

zygomatic injury because of fibrosis and ankylosis in the region (1, 35).

Tripod Fracture

The tripod (zygomaticomaxillary) fracture separates the malar eminence of the zygoma from its frontal, temporal, and maxillary attachments. Separation from the frontal bone is typically due to diastasis of the zygomaticofrontal suture but, less commonly, may result from a fracture of the frontal process of the zygoma. This fracture may extend into the lateral orbital wall and into either the greater or lesser wings of the sphenoid (3). Separation from the temporal bone occurs via a simple or comminuted zygomatic arch fracture or, less commonly, by separation of the zygomaticotemporal suture. The typical pattern of separation from the maxilla is a fracture of the inferior orbital rim immediately adjacent to the zygomaticomaxillary suture. Extension into the anterior orbital floor and lateral maxillary wall is commonly seen with this injury (6, 35). Approximately 50% of patients with tripod fractures have significant comminution and displacement of the orbital floor (1).

The patient presents with local pain, hematoma, and edema overlying the malar eminence. The fractures of the zygomatic arch, and orbital rim and orbital floor, are associated with the same complications associated with isolated fractures involving these regions. Significantly displaced tripod fractures are frequently associated with asymmetric facial flattening and diplopia due to loss of the normal orbital contour or bony impingement (35). Injury to the lacrimal canal produces epiphora, whereas separation of the medial canthal ligament results in cosmetic deformity of the medial eyelid (12, 19).

Separation of the zygomaticofrontal suture is best evaluated on the Caldwell or AP projection. The zygomatic arch fracture is best evaluated on the Waters view or with an underpenetrated base view. These features of the tripod injury are well seen on axial CT. The orbital rim, ad-

jacent orbital floor, and lateral maxillary wall fractures are best seen on the Waters view or coronal CT. Rotation and depression of the malar eminence should be evaluated utilizing all available information, as the Waters view alone is inadequate (35, 36). Daffner et al. have described the usefulness of the lateral projection in evaluating the position and symmetry of the paired malar struts. On the lateral view, each malar strut is seen as a triangular structure consisting of the orbital floor superiorly, the anterolateral maxillary wall anteriorly, and the zygomaticomaxillary region posteriorly (36). Significant depression, rotation, or displacement of the malar strut is present in up to 41% of patients with tripod fractures (36) (Fig. 19.12). Depression and displacement may be further evaluated by assessing the relative positions of the frontal processes of the zygoma, inferior orbital rims, and anterior margins of the maxillary walls.

Both tomography and CT scanning have been used for further assessment of tripod complex injuries. Tomography in both the AP and the lateral plane is essential for adequate evaluation of the position of the malar strut. Computed tomography provides similar information, and in a small series of patients was noted to be superior to tomography for accurate evaluation (15) (Fig. 19.13). Computed tomography is highly accurate for determining displacement in the plane of the section but rotation and dis-

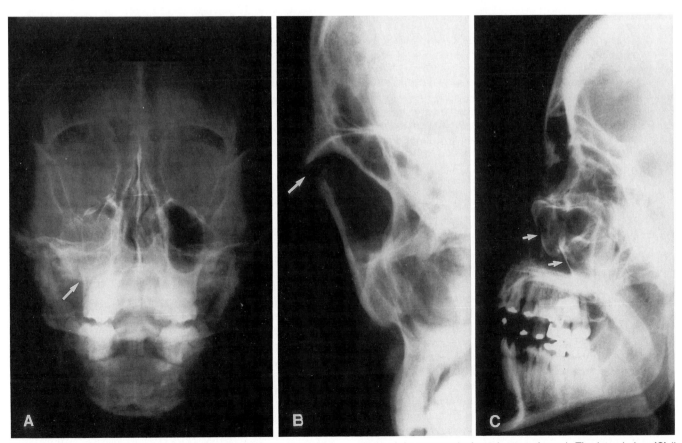

Figure 19.12. The posteroanterior (PA) view **(A)** of a right tripod fracture demonstrates fractures of the orbital rim, orbital floor, and lateral wall of the maxillary sinus *(arrow)*. The Lame view **(B)**, obtained PA with the injured side rotated 37° away from the cassette, shows diastasis of the zygomaticofrontal suture *(arrow)*. The lateral view **(C)** illustrates asymmetry of the triangular malar struts *(arrows)* with displacement and depression of the injured zygoma.

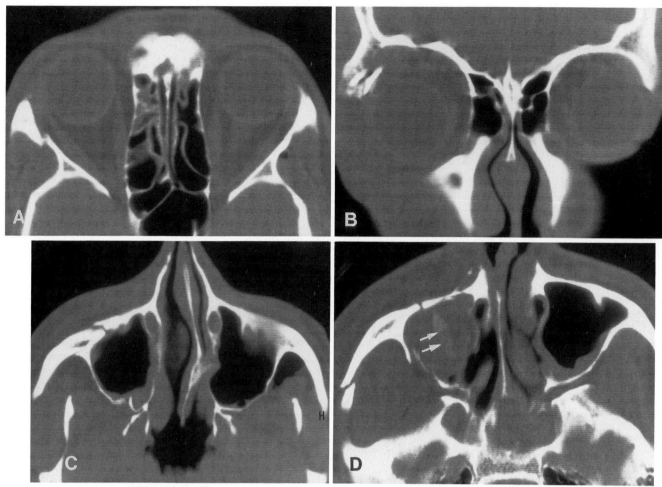

Figure 19.13. CT images from four patients with right tripod fractures demonstrate a variety of abnormalities. The axial CT **(A)** shows simple diastasis of the zygomaticofrontal suture. A coronal CT **(B)** shows that suture diastasis is accompanied by extensive fragmentation of the lateral orbital wall and rim in some patients. An axial CT **(C)** shows fractures of the anterior and lateral maxillary walls, mild malar depression, and mucosal hemorrhage. The final CT **(D)** demonstrates fractures of the zygomatic arch and maxillary walls. The malar eminence is depressed. Note that the right inferior rectus *(arrows)* is visualized secondary to a depressed orbital floor fracture.

placement perpendicular to the section plane are often difficult to recognize without reformations (3). The plain film and CT examinations are complementary in evaluation of the pattern of displacement of separated fragments. All of the components of the tripod injury can be seen on the axial images, including posterior displacement of the zygoma (21). Coronal CT is helpful for full evaluation of the orbital floor when it is disrupted with this injury.

Tripod fractures are typically treated by open reduction and internal interosseous fixation at two of the three fracture sites (4). Typically the zygomaticofrontal suture is wired and the orbital rim and floor are reconstructed. If the fracture is not adequately reduced, significant flattening of the malar eminence may result. Elevation of the zygomatic arch is performed following reduction of the malar strut to prevent coronoid impingement and to provide a satisfactory cosmetic result (4).

Maxillary Fractures

The paired maxillary bones contain the maxillary sinuses and provide bony support for the central face, forming the lateral walls of the nasal cavity, the anterior hard palate and maxillary alveolus, the orbital floor, and the medial portion of the infraorbital rim. Fractures involving the maxillary bones have already been discussed under the headings of orbital and tripod fractures. Isolated fractures of the maxilla are uncommon, constituting less than 5% of midface fractures. All the Le Fort fractures also involve the maxilla and are therefore discussed in this section.

Isolated Maxillary Fractures

The most common isolated fracture of the maxilla results from a direct blow and involves the anterolateral wall of the maxillary sinus (2). Opacification of the maxillary sinus

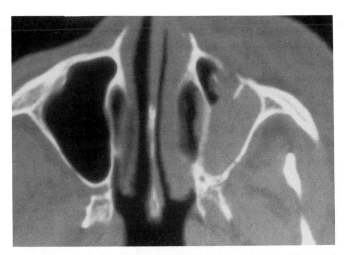

Figure 19.14. An axial CT image through the midplane of the maxillary sinuses demonstrates opacification of the left sinus, a fracture of its anterior wall, and overlying soft-tissue swelling. The fracture could not be visualized on the initial radiographs.

in the absence of a visible fracture is suggestive of this injury, but is more commonly secondary to contusion and tearing of the maxillary mucosa (2). Both CT and MR imaging can distinguish paranasal sinus hemorrhage from mucosal thickening and sinus effusion (9). Subcutaneous emphysema of the cheek or facial paresthesia due to involvement of the infraorbital foramen may be present and are indications for further evaluation (2). This fracture is extremely difficult to visualize on conventional radiographs but an abnormal linear density, produced by a fragment of displaced bone seen on edge, may be seen (37). The anterior and lateral maxillary walls each contain a bony canal through which branches of the maxillary nerve and artery pass. These normal defects within the walls can be differentiated from fractures by the presence of a thin cortical margin completely surrounding the canal (38). Fractures isolated to the anterior maxillary wall are best seen on axial CT or lateral tomography (15, 21) (Fig. 19.14).

Fractures of the maxillary alveolar process typically involve only small portions of the bony ridge and are best demonstrated by malalignment and displacement of the teeth on the Waters view. Occlusal and panoramic radiographs are also valuable for evaluation of the alveolus. The mandible should be carefully scrutinized in the presence of this injury because of the high incidence of associated mandibular fractures (12).

Le Fort Fractures

In a classic series of cadaver experiments in 1901, Le Fort described three planes of weakness in the facial skeleton and the fractures that occur through them. He noted that the alveolar process of the maxilla, the malar eminence of the zygoma, and the nasofrontal process of the maxilla constituted areas of strength in the facial skeleton and that facial fractures tended to avoid these areas. All three Le Fort injuries produce an unstable maxillary fragment that is displaced from the remainder of the craniofacial skele-

ton. They invariably extend posteriorly to involve the pterygoid processes of the sphenoid bone, separated from the maxillary sinus by the pterygopalative fossa (6, 18, 21). Much has been written about Le Fort fractures, although they are relatively uncommon and rarely follow the classic planes described by Le Fort. In a large series of 323 patients with facial fractures, only 9 patients had typical Le Fort fractures; an additional 8 patients had atypical Le Fort fracture complexes (1).

The Le Fort I fracture is a horizontal fracture through the maxilla superior to the alveolus that traverses both the medial and lateral maxillary walls (6, 19). This fracture results in separation of the hard palate and maxillary alveolus from the remainder of the craniofacial skeleton. The Le Fort II fracture extends across the central face in a pyramidal fashion, with fracture lines within the ethmoid complex extending bilaterally across the medial maxillary surface of the orbit, and then through the lateral maxillary walls (12, 19). This injury results in separation of a large fragment containing the upper jaw, the medial walls of the maxillary sinuses, and portions of the nasal complex. Posterior displacement of the hard palate is frequently present in both Le Fort I and Le Fort II fractures and can be identified on lateral radiographs (12). This displacement is more difficult to detect with axial CT, but overlap or distraction of bone fragments is suggestive (3) (Fig. 19.15). The Le Fort III fracture separates the entire face from the base of the skull. Fracture lines extend across the ethmoid to the region of the inferior orbital fissure, and then laterally through the zygomatic processes of the orbit and zygomatic arches. Inferior displacement of the entire face is frequently noted and results in vertical elongation of the orbital cavity (12). Cerebrospinal fluid leaks are common following this injury because of associated involvement of posterior wall of the frontal sinuses or the cribriform plate in some patients (12, 19, 23). Both conventional tomography and CT accurately demonstrate the level of facial injury and the plane of separation.

A common variant is the presence of a unilateral Le Fort II fracture with a contralateral Le Fort III or tripod fracture. When the injuring force is excessive, extensive comminution of the facial structures occurs and no recognizable pattern can be identified. These complex, severely comminuted fractures are frequently referred to as "facial smash" fractures (Fig. 19.16).

Mandible and Temporomandibular Joint

Mandibular fractures are extremely common in all age groups, exceeded only by nasal and zygomatic fractures in the adult, and by nasal fractures in children. The usual mechanism is a blow to the midline or side of the mandible during an assault. Both ends of the mandible are firmly attached to the skull at the temporomandibular joint (TMJ) and it therefore behaves as a complete bony ring. In most series 50% of mandibular fractures are associated with a contralateral fracture or with disruption of the TMJ (39). A blow to the midline symphysis typically produces bilateral body or angle fractures, or bilateral condylar fractures (4, 23). A blow to the side produces a subcondylar fracture

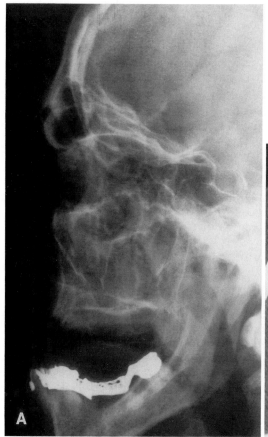

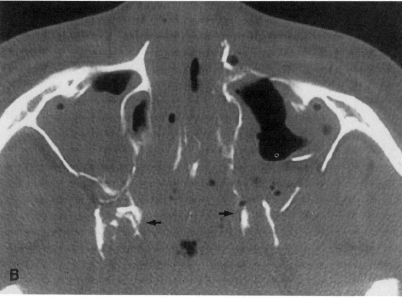

Figure 19.15. The lateral radiograph **(A)** in a case of a Le Fort II fracture shows posterior displacement of the hard palate. The axial CT **(B)** through the fractured pterygoid plates of the sphenoid *(arrows)* also shows disruption of the medial and lateral walls of both maxillary sinuses.

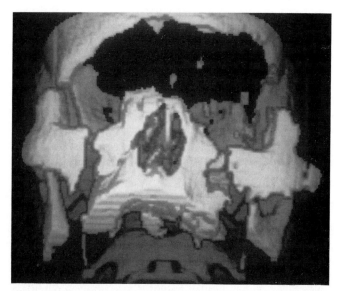

Figure 19.16. The 3D CT of a facial smash injury shows features of the tripod, Le Fort II, and Le Fort III injuries. In this case, the sphenoid was disrupted with a mobile midface and numerous other fractures were present. The configuration of the fracture fragments and their displacement is easier to appreciate on the 3D CT, but more fractures could be seen on the conventional CT.

on the side of impact and a contralateral angle or body fracture (4, 35). The extent of separation and displacement of the bone fragments must be evaluated. It is also important to determine if the fracture line extends to a tooth, since such communication converts a mandibular fracture into an open injury.

Conventional radiography is usually adequate for the assessment of mandibular injury. Mandibular fractures are evaluated utilizing a PA radiograph centered on the mandible, the Towne's projection, and bilateral oblique views (35). The central portion of the mandible overlaps the cervical spine and is extremely difficult to visualize. Panoramic or dental occlusal radiographs may be necessary for adequate visualization of this area (40). While mandibular fractures and TMJ dislocation are typically readily visible on the radiographs, the indirect sign of malocclusion or displacement of the teeth is suggestive. The mandible is difficult to image with CT because of its complex, curving surfaces and the presence of artifact by dental fillings (21, 41). Osborn et al. have reviewed the normal CT appearance of the mandible and note that CT is of little value in examining uncomplicated trauma (41). We reserve CT for assessment of condylar head or neck fractures since these injuries are difficult to delineate with radiography (Fig. 19.17). We have also found CT to be helpful for evaluating

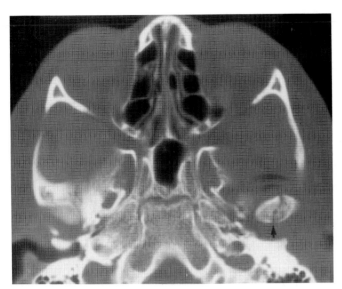

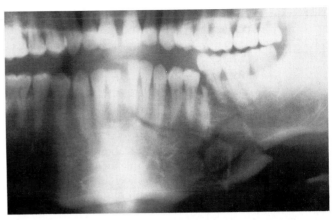

Figure 19.19. A panorex of the mandible illustrates a severely comminuted fracture of the body with displacement of a large portion of the alveolus. The fracture extends to numerous teeth and is severely displaced. The root of the first molar is fractured and the teeth are maloccluded.

Figure 19.17. This 12-year-old girl sustained a mandibular injury and was unable to open her mouth. The initial radiographs, including a panorex, showed opacification of the left mastoid air cells. Axial CT shows vertical fractures of the left condylar head *(arrow)*. Additional images showed a temporal bone fracture responsible for the air in the infratemporal fossa.

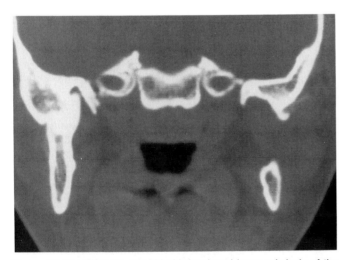

Figure 19.18. This 12-year-old girl developed bony ankylosis of the right TMJ after sustaining a bilateral angulated condylar neck fracture 6 years previously. The left condylar neck is not seen in this section, but was deformed and severely hypoplastic.

complications of condylar fractures such as angular malunion of the neck, avascular necrosis of the mandibular head, and ankylosis of the TMJ (7). Computed tomography is the method of choice for assessment of suspected temporal bone fractures, which may occur concomitantly via force transmission through the TMJ (3).

The mandible is divided into several regions. Between the canine teeth lies the symphysis, which includes the central alveolar ridge. Fractures of the symphysis account for 10 to 20% of all mandibular fractures. These fractures have a tendency to undergo malunion or nonunion due to overriding with entrapment of adjacent muscle (40). The

mandibular body and angle are the most commonly injured portions of the mandible and are involved in 50 to 70% of all mandibular fractures (23) (Fig. 19.18). Fractures angulated posterosuperior to anteroinferior are impacted and stabilized by the masseter and internal pterygoid muscles, whereas fractures angulated in the opposite direction are unstable since they are separated by the action of these muscles (4). The ramus, which extends from the mandibular angle to the bases of the condylar and coronoid processes, is the site of fracture in 3 to 9% of patients. The thin coronoid process is rarely fractured due to overlying support by the pterygoid muscles and the zygomatic arch.

Fractures of the condylar neck comprise 15 to 20% of all mandibular fractures but represent a large proportion of mandibular fractures in children (42). Fractures of the condyle may be missed clinically because dental occlusion is often normal following this injury (40) (Fig. 19.19). The condylar head is typically displaced medially by the lateral pterygoid muscle, which inserts high on its medial aspect. Significant complications following these fractures include angular malunion, mandibular growth arrest due to injury of the condylar growth center, avascular necrosis, and bony ankylosis of the TMJ (40). These complications are noted in 2% of children with mandibular fractures (42).

Closed reduction using arch bars or splints is utilized in most cases of mandibular injury. A variety of intermaxillary fixation devices are utilized and all render satisfactory results if normal occlusion is achieved (39, 40). Open reduction may be required for multiple fractures, overlying laceration, unfavorable obliquity of a fracture line at the mandibular angle, or in severely displaced condylar fractures.

Dislocation of the TMJ can be seen with radiography or CT but MR imaging is the optimal method for evaluating internal derangement of this articulation. Trauma may be a major antecedent of TMJ dysfunction. Nonosseous traumatic alterations of the TMJ include hemarthrosis, internal derangement of the meniscus and ligament, and joint effusion (43). Bony changes other than fractures producing

TMJ dysfunction include osteochondritis dissecans, avascular necrosis, and posttraumatic osteoarthritis (43). The advent of MR imaging has increased our appreciation of the spectrum of posttraumatic abnormalities of the TMJ.

Conclusion

The radiologist plays a critical role in the detection and evaluation of osseous and soft-tissue facial injury. The advent of CT has greatly enhanced the ability to characterize posttraumatic alterations of the facial skeleton and the surrounding soft-tissue structures. Thorough knowledge of normal facial anatomy, patterns of bony injury, and their associated soft-tissue sequalae is necessary for accurate interpretation of this complex area. It is hoped that this review of common facial fractures will aid the radiologist in interpreting studies of this region.

Acknowledgment

Special thanks to Usha Pillai for secretarial assistance.

References

1. Nakamura T, Gross CW. Facial fractures. Arch Otolaryngol 1973;97:288–290.
2. Valvassori GE, Hord GE. Traumatic sinus disease. Semin Roentgenol 1968;3:160–171.
3. Gentry LR, Smoker WRK. Computed tomography of facial trauma. Semin Ultrasound, Comput Tomogr, and Magn Reson 1985;6:129–145.
4. Schultz RC, Oldham RJ. An overview of facial injuries. Surg Clin Am 1977;57:987–1010.
5. Brant-Zawadzki MN, Minagi H, Federle MP, et al. High resolution CT with image reformation in maxillofacial pathology. AJR 1982;138:477–483.
6. Dolan K, Jacoby C, Smoker W. The radiology of facial fractures. Radiographics 1984;4:577–663.
7. Pathria MN, Blaser SI. Diagnostic imaging of craniofacial fractures. Radiol Clin N Am 1983;27:839–853.
8. Whalen JP, Berne AS. The Roentgen anatomy of the lateral walls of the orbit (orbital line) and the maxillary antrum (antral line) in the submentovertical view. AJR 1964;91:1009–1011.
9. Zimmerman RA, Bilaniuk LT, Hackney DB, et al. Paranasal sinus hemorrhage: evaluation with MR imaging. Radiology 1987;162:499–503.
10. Julin P, Kraepelien T. Reduction of absorbed doses in radiography of the facial skeleton. AJR 1984;143:1113–1116.
11. Ball JB. Direct oblique sagittal CT of orbital wall fractures. AJR 1987;148:601–608.
12. Dolan KD, Jacoby CG. Facial fractures. Semin Roentgenol 1978;13:37–51.
13. Haverling M. Diagnosis of blow-out fractures of the orbit by tomography. Acta Radiol 1972;12:347–352.
14. Zilkha A. Computed tomography in facial trauma. Radiology 1982;144:545–548.
15. Kreipke DL, Moss JJ, Franco JM, et al. Computed tomography and thin-section tomography in facial trauma. AJR 1984;142:1041–1045.
16. Gould HR, Titus CO. Internal orbital fractures. The value of laminagraphy in diagnosis. AJR 1966;97:618–623.
17. Grove AS. Orbital trauma evaluation by computed tomography. Comput Tomogr 1979;3:267–278.
18. Gentry LR, Manor WF, Turski PA, et al. High-resolution CT analysis of facial struts in trauma. 1. Normal anatomy. AJR 1983;140:523–532.
19. Gentry LR, Manor WF, Turski PA, et al. High-resolution CT analysis of facial struts in trauma. 2. Osseous and soft-tissue complications. AJR 1983;140:533–541.
20. Gillespie JE, Isherwood I, Baker GR, et al. Three-dimensional reformations of computed tomography in the assessment of facial trauma. Clin Radiol 1987;38:523–526.
21. Johnson DH. CT of maxillofacial trauma. Radiol Clin N Am 1984;22:131–144.
22. Hemmy DC, Tessier PL. CT of dry skulls with craniofacial deformities: accuracy of three-dimensional reconstruction. Radiology 1985;157:113–116.
23. Rogers LF. Radiology of skeletal trauma. Vol. 1. New York: Churchill Livingston, 1982.
24. Zizmor J, Smith B, Fasano C, et al. Roentgen diagnosis of blow-out fractures of the orbit. AJR 1962;87:1009–1018.
25. Fueger GF, Milauskas AT, Britton W. The roentgenologic evaluation of orbital blow-out injuries. AJR 1966;97:614–617.
26. Smith B, Regen WF Jr. Blow-out fracture of the orbit. Am J Ophthalmol 1957;44:733–739.
27. Hammerschlag SB, Hughes S, O'Reilly GV, et al. Blow-out fractures of the orbit: a comparison of computed tomography and conventional radiography with anatomical correlation. Radiology 1982;143:487–492.
28. Hammerschlag SB, Hughes S, O'Reilly GV, et al. Another look at blow-out fractures of the orbit. Am J Neuroradiology 1982;3:331–335.
29. Lewin JR, Rhodes DH Jr, Pavsek EJ. Roentgenologic manifestations of fracture of the orbital floor (blow-out fracture). AJR 1960;83:628–632.
30. MaFee MF, Linder B, Peyman GA, et al. Choroidal hematoma and effusion: evaluation with MR imaging. Radiology 1988;168:781–786.
31. Zilkha A. Computed tomography of blow-out fracture of the medial orbital wall. AJR 1981;137:963–965.
32. Curtin HD, Wolfe P, Schramm V. Orbital roof blow-out fractures. AJR 1982;139:969–972.
33. Unger JM. Orbital apex fractures: the contribution of computed tomography. Radiology 1984;150:713–717.
34. Guyon JJ, Brant-Zawadski M, Seiff SR. CT demonstration of optic canal fractures. AJR 1984;143:1031–1034.
35. Freimanis AK. Fractures of the facial bones. Radiol Clin N Am 1966;4:341–363.
36. Daffner RH, Apple JS, Gehweiler JA. Lateral view of facial fractures: new observations. AJR 1983;141:587–591.
37. Merrell RA Jr, Yanagisawa E, Smith HW. Abnormal linear density. Arch Otolaryngol 1969;90:140–147.
38. Dolan KD, Hayden J Jr. Maxillary "pseudofracture" lines. Radiology 1973;107:321–326.
39. Bernstein L, McClurg FL Jr. Mandibular fractures: a review of 156 consecutive cases. Laryngoscope 1976;87:857–961.
40. Henny FA. Fractures of the jaws. Semin Roentgenol 1971;6:397–402.
41. Osborn AG, Hanafee WH, Mancuso AA. Normal and pathologic CT anatomy of the mandible. AJR 1982;139:555–559.
42. Lehman JA Jr, Saddawi ND. Fractures of the mandible in children. J Trauma 1976;16:773–777.
43. Schellhas K. Temporomandibular joint injuries. Radiology 1989;173:211–216.

CHAPTER 20

Magnetic Resonance Imaging of the Rotator Cuff

Jacques A. van Oostayen, Johan L. Bloem, Milan E. J. Pijl, and Pieter M. Rozing

Introduction

The shoulder impingement syndrome is generally referred to as the condition in which part of the rotator cuff and subacromial bursa are compressed between the humeral head and the coracoacromial arch (1). The latter consists of the anterior third of the acromion, the acromioclavicular joint, and the coracoacromial ligament.

Over the last four years the advent of magnetic resonance (MR) imaging has initiated a renewed interest in shoulder impingement syndrome and rotator cuff tear (2–11). Magnetic resonance imaging offers excellent soft-tissue contrast and is able to depict pathologic changes of the rotator cuff well before a complete rupture has occurred (2–4, 7–9).

In this chapter we discuss the technique used to obtain MR images of the diseased rotator cuff and accompanying structures and the underlying range of pathology as it is known today.

Examination Technique and Normal Anatomy

The use of a surface coil is imperative because the shoulder is located in the periphery (which means relatively insensitive area) of the magnetic field. A Helmholtz configuration will allow visualization of fine detail and a homogeneous signal-to-noise ratio over a larger field of view (12). The MR imaging study is tailored to the problem at hand. Slice thickness varies between 1.5 and 10 mm, but is usually 3 or 5 mm. For T1 weighting we use a spin-echo (SE) sequence with TR 550 or 600 ms and TE 20 or 30 ms (3, 13). This pulse sequence is used for anatomical detail and subtle changes in tendons. Proton density and T2 weighting is emphasized with a TR of 2000 ms and dual-echo technique: TE 20 or 50 and 80 or 100 ms. This pulse sequence will highlight fluid in the joint, cuff, or subacromial-subdeltoid bursa (Fig. 20.1*A* and *B*). The information looked for on these two SE sequences can usually also be obtained on a time-effective combined spin-density T2-weighted sequence (TR 2000, TE 20 and 80 ms). T2*-weighted gradient-recalled echo images can be added to evaluate possible intraarticular pathology. The choice of optimal gradient-recalled echo images depends on the system used. Our T2*-weighted gradient-recalled echo images are made with a small pulse angle of 20 or 45°, a TR of 400 ms, and a TE of 30 ms. To emphasize T1 weighting we use a pulse angle of 90° with a TR of 300 ms and a TE of 10 ms. When a joint effusion is present the T2*-weighted sequence will clearly display the labrum as a structure of low signal intensity within the high signal intensity of fluid (9, 10).

Images are made in our institution with a Philips S15 (Shelton, Conn.) operating at 0.5 or 1.5 T, but in the literature adequate images have been obtained using field strengths as low as 0.3 T (2, 14). The images displayed in this chapter are made at 0.5 T.

Bone marrow, with its relatively short T1 and T2 relaxation times, has a high signal intensity on T1 and T2-weighted images. Cortical bone on the other hand exhibits a low signal intensity line (independent of the pulse sequence used) as a result of low spin density and short T2 relaxation time of compact bone. The surfaces of the glenoid cavity and the humeral head are covered with hyaline cartilage, which has an intermediate to relatively high signal intensity on both T1- and T2-weighted images. Fibrocartilage of the glenoid labrum and the articular surfaces of the acromioclavicular joint are demonstrated by moderately low signal intensity on T1- and T2-weighted images. The rotator cuff tendons have a low signal intensity that cannot be distinguished from that of the cortical bone of the greater tuberosity and humeral head. Subcutaneous and inter- and intramuscular fat has a high signal intensity (12–15).

The selection of imaging planes is important (4, 10, 13–15). The rotator cuff (of which the supraspinatus muscle and tendon are most important) and the inferior part of the labrum are best shown in a coronal oblique plane perpendicular to the glenoid cavity (parallel to the supraspinatus muscle). This plane also offers a view of the articular surfaces of the glenoid and humerus. Redundancy of

---

Abbreviations (see also glossary): SE, spin echo; US, ultrasound; DTPA, diethylenetriamine pentaacetic acid; RA, rheumatoid arthritis; AC, acromioclavicular.

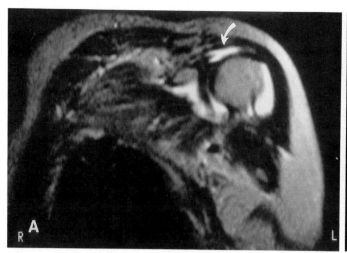 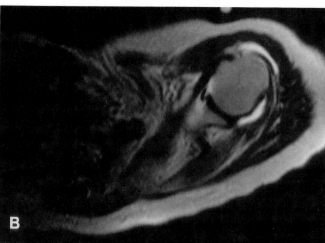

Figure 20.1. **(A)** A 58-year-old woman with breast carcinoma and osseous metastasis. Pain in shoulder region suggestive of metastasis prompted MR imaging. Coronal oblique T2-weighted (TR/TE 2000/100 ms) image shows high-intensity area *(arrow)* in rotator cuff compatible with cuff tear. Note joint fluid and fluid in subdeltoid bursa. No metastasis was found in the shoulder region. **(B)** Axial T2-weighted (TR/TE 2000/100 ms) image shows high-intensity signal of fluid in joint cavity and subdeltoid bursa indicative of rotator cuff tear.

the axillary fold of the capsule, however, may obscure the inferior part of the labrum when no joint effusion is present. The axial plane displays subscapularis muscle and tendon, the biceps tendon, glenohumeral joint, articular surfaces, and the anterior and posterior part of the glenoid labrum. The soft tissues in the axilla, including the neurovascular bundle, are also exhibited well in this plane. A third plane depicting the anterior third of the acromion and its relation to the supraspinatus and infraspinatus tendon is the sagittal oblique plane. This plane parallels the glenoid fossa and is mostly used in equivocal cases. The two oblique planes thus optimally display obliquely orientated structures and are easily reproduced.

The position of the arm is important and should be neutral (3, 6, 15) or in external rotation (10). When the arm is internally rotated (2, 14), the supraspinatus tendon curves anteriorly and leaves the oblique coronal plane. This hampers differentation between supraspinatus tendon and surrounding fat. Fat between biceps tendon and supraspinatus anteriorly, or infraspinatus tendon and supraspinatus posteriorly, may mimic an area of increased signal intensity within the supraspinatus tendon.

Our routine imaging protocol at 1.5 T for evaluation of the rotator cuff consists of 4- to 5-mm thick axial and coronal oblique T1-weighted (600/20) SE sequences and 4- to 5-mm thick coronal and sagittal oblique spin-density and T2-weighted sequences (2000, 20, and 80).

Impingement Syndrome

The impingement syndrome is an important cause of shoulder dysfunction (1). The underlying pathophysiologic mechanism is entrapment of the soft-tissue structures under the coracoacromial arch formed by the anterior third of the acromion, the acromioclavicular joint, and the coracoacromial ligament. When the supraspinatus muscle becomes insufficient, the deltoid muscle has no fulcrum to abduct the arm. This leads to instability and upward pull of the humeral head against the acromion, resulting in late sequelae of cuff arthropathy.

It has been shown that shape and low level of the acromion, relative to the distal clavicle, are of importance (1, 16). As the main position of the arm and hand in general use is in front of the shoulder it will be supraspinatus muscle and tendon (especially the anterior leading edge) that are most susceptible to wear and tear against the coracoacromial arch (1). Impingement is increasingly seen in young athletes because of overuse and excessive subacromial loading. Edema and hemorrhage may result from this excessive overhead use in sports (competitive swimming, tennis, baseball) or work. This is stage 1 of the Neer (1) classification and is usually observed in patients younger than 25 years of age. With time, edema and hemorrhage may progress to rotator cuff tendinitis and fibrosis (Neer stage 2). This is observed in older patients. Finally, incomplete or complete rupture can occur (Neer stage 3). This stage is usually found in patients over 40 years of age. Tears most often occur in the critical zone, which is located anteriorly in the distal aspect of the supraspinatus tendon. This critical zone correlates well with a relative avascularity in the supraspinatus tendon 1 cm from its insertion (4, 10). According to Neer, 95% of cuff tears are associated with chronic impingement syndrome (1).

Rotator cuff impingement syndrome can often be diagnosed on the basis of clinical findings. Patients complain of a dull ache around the shoulder following sports or other overhead activities, sometimes interfering with sleep. Physical examination shows tenderness over the greater tuberosity and the anterior edge of the acromion. A painful arc (40 to 120°) is present. Although in the early stages radiographs often do not correlate with the clinical signs, several radiographic findings associated with this syndrome can be found (11, 17, 18). These include low position of acromion, subacromial proliferation of bone, spur-

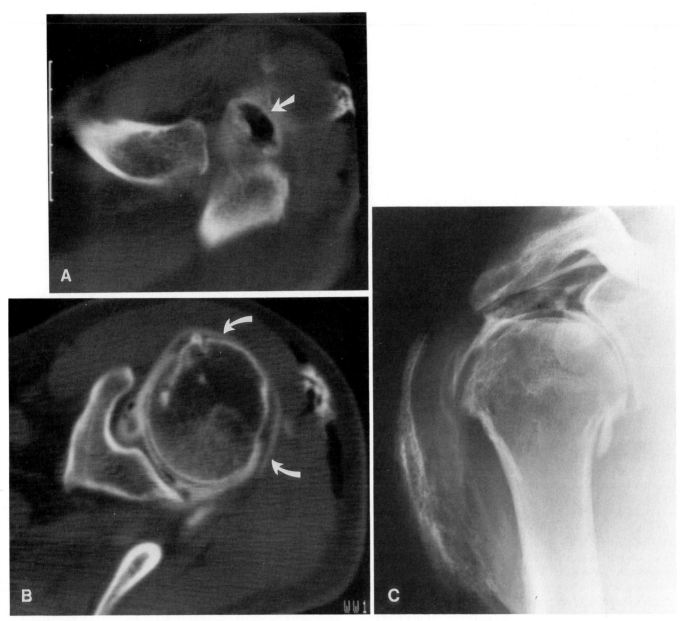

Figure 20.2. **(A)** CT-arthrography performed following traumatic anterior dislocation: 3-mm axial CT image at the level of the supraspinatus tendon. A cuff tear is outlined by air and contrast *(arrow)*. Also note air and contrast in soft tissues ruptured through deltoid muscle plane. **(B)** Air and contrast agent is seen within joint, in subdeltoid bursa *(curved arrows)*, and lateral to the deltoid muscle. This is consistent with rotator cuff tear. **(C)** Radiograph made during arthrography. Air and contrast in joint cavity but also adjoining the acromion are indicative of cuff tear with bursal filling. Note air and contrast in soft tissues.

ring at the inferior aspect of the acromioclavicular joint, and degenerative changes (cystic degeneration, sclerosis, and flattening of the head) in the humeral tuberosities at the insertion of the rotator cuff. Narrowing of the acromiohumeral distance is associated with more advanced degenerative changes and with rotator cuff tear.

Until recently, radiologic evaluation of the rotator cuff has consisted primarily of conventional radiographs, shoulder arthrography, and ultrasound (5). The role of computed tomography (CT)-arthrography, well appreciated when studying the traumatized or unstable shoulder (4), is of limited importance (see also Chapter 21). Full thickness cuff tears will be depicted because of the appearance of air and contrast medium within the subdeltoid bursa. When using thin slices the size of the actual tear can be appreciated (Fig. 20.2). A potential mistake that is easily made here is when the humerus is in low position, for instance in patients with instability. Air and contrast in the joint can be mistaken for a cuff tear. The absence of air and contrast in the bursa, however, will lead to the correct diagnosis (Fig. 20.3). Computed tomography-arthrography is not helpful in the evaluation of early impingement.

Several reports have appeared in the literature claiming

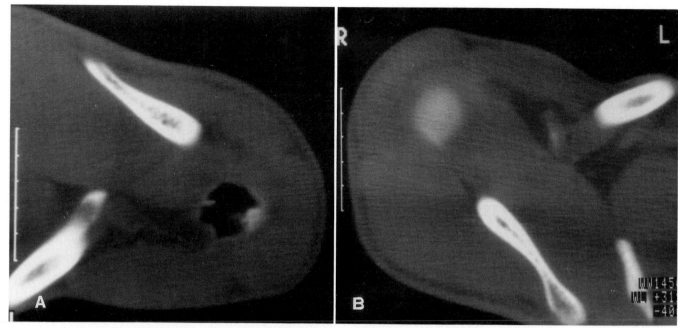

Figure 20.3. **(A)** A 20-year-old patient with dislocatable shoulder without known trauma. Prone position, 3-mm axial CT image following arthrography shows air in joint cavity (humerus in low dislocated position) that could be mistaken for a cuff tear (see also Fig. 20.2A). **(B)** Supine position. Normal position of humerus and a normal supraspinatus muscle are identified. Note that no bursal filling is present.

good results of MR imaging of the impingement syndrome. At an early age (Neer I and II), when arthrography does not show any abnormality, MR imaging exhibits an area of abnormal increased signal intensity located in the tendinous portion of the supraspinatus muscle. This area of abnormal signal intensity is more easily appreciated on T1-weighted and proton density images than on T2-weighted images (Fig. 20.4). Occasionally some fluid is seen in the subacromial bursa. Bursal fluid is not encountered in asymptomatic patients. However, some degree of increased signal, relative to the signal void of normal tendon, may be seen in 57 to 85% of normal volunteers on T1- or proton density weighted images (19, 20). Usually these subtle areas of increased signal intensity can be recognized as normal when no other abnormalities are present. The T2-weighted images do not show an increase of signal intensity. In equivocal cases, sagittal images may be needed to demonstrate a normal tendon.

Surgical and histologic analysis of abnormal signal intensity areas indicated that these changes are due to myxoid degeneration and inflammation (3). In this stage of the disease atrophy of the rotator cuff, seen as fatty replacement and volume loss, may also be observed on MR images.

Even with the arm in neutral position, static impingement of the coracoacromial arch on the rotator cuff may occasionally be found with MR imaging. The sign to look for is an abrupt ending of the fat plane below the distal clavicle at the level of the acromioclavicular joint (Fig. 20.5). Hypertrophy of the rotator cuff, such as occurs in competitive swimmers, may also obliterate this fat plane considerably (Fig. 20.6). If cuff tendinitis progresses to a rotator cuff tear an increase of signal intensity on T2-weighted images, representing fluid, discontinuity of supraspinatus

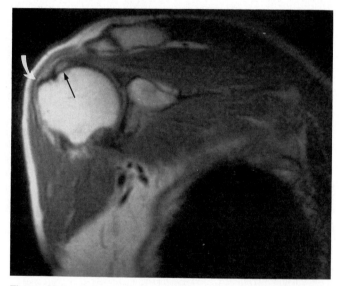

Figure 20.4. Patient with long-standing rotator cuff symptoms: coronal oblique T1-weighted (TR/TE 600/30 ms) image. High signal intensity area in distal supraspinatus tendon *(arrow)* is indicative of tendinitis. Note intact subdeltoid fat plane *(curved arrow)*. At surgery no tear was found. Patient was treated with Neer acromioplasty.

muscle, and retraction of torn edges may be observed (Fig. 20.7). The size of the tear is accurately exhibited on the MR images. A cuff tear is nearly always found in an already diseased muscle. A traumatic event, often related with aggravation of symptoms, only completes or enlarges preexistent partial tears.

Differentation between severe tendinitis and partial tears is very difficult or even impossible. In addition to the dis-

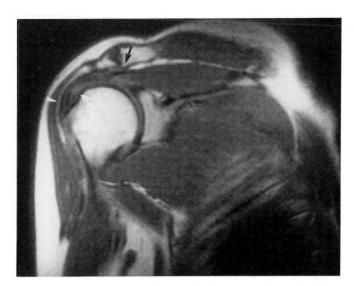

Figure 20.5. Patient with impingement syndrome. Coronal oblique T1-weighted (TR/TE 600/30 ms) image shows acromioclavicular joint osteoarthritis with spurring *(black arrow)*. Note abrupt ending of subacromial fat plane at this level. High signal intensity area in thickened supraspinatus tendon *(curved arrow)* is indicative of tendinitis. Intact subdeltoid fat plane is appreciated *(arrowhead)*.

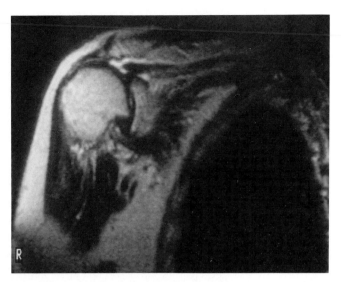

Figure 20.7. Patient with long-standing complaints compatible with impingement syndrome. Coronal oblique T2-weighted (TR/TE 2000/100 ms) image shows area of high intensity within cuff, edges of which are outlined by high signal intensity fluid. Note medial retraction of cuff and fluid in subacromial bursa.

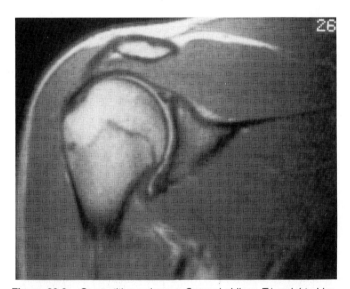

Figure 20.6. Competitive swimmer. Coronal oblique T1-weighted image demonstrates obliteration of subacromial fat plane due to hypertrophy of the supraspinatus muscle. The supraspinatus tendon had a normal signal intensity. (Courtesy of S. Davis, Sir Charles Gairdner Hospital, Perth, West Australia.)

cussed primary signs, secondary signs are helpful in differentiating tendinitis (no secondary signs) from (in)complete cuff tears, which usually display secondary signs (9). The secondary signs relate to the subdeltoid bursa and its accompanying fat plane. The subacromial-subdeltoid bursa is a potential space that, when filled with fluid, will have a high signal intensity on T2-weighted images. A high intensity band on T1-weighted images caused by fat surrounding the bursa can be seen in most, but not all, asymptomatic volunteers (14). Obliteration of this fat plane and especially the appearance of fluid within the bursa are

the important secondary signs in the evaluation of rotator cuff disease (4, 8–10).

Only the last stage of impingement syndrome, a full rotator cuff tear, is easily diagnosed with conventional arthrography (Fig. 20.8A). However, the amount of atrophy and quality of the tendon can only be evaluated with MR imaging (Fig. 20.8B and C), CT-arthrography or ultrasound (US). Furthermore, correlative imaging studies have proven that MR imaging is superior to arthrography in demonstrating both partial and complete tears (6, 8). For incomplete and complete tears the sensitivity of MR imaging is reported to be between 69 and 91% with a specificity between 88 and 94% (5, 6, 8). The role of US in patients with rotator cuff impingement syndrome is controversial (21–24). Whereas MR imaging has markedly diminished arthrographic examinations for the detection of cuff pathology, this has not been the case with the use of US in clinical practice so far (23).

The role of MR arthrography needs further evaluation. Initial results suggest that detection of partial tears at the undersurface of the rotator cuff will be facilitated with intraarticular Gd-labeled diethylenetriamine pentaacetic acid (DTPA) present (Fig. 20.9A and B) (25).

Other abnormalities encountered in patients with impingement syndrome include low position of acromion, osteophytes from the acromion and acromioclavicular joint, concave depression on the inferior surface of the acromion and superior migration of the humeral head, and thickening of the coracoacromial ligament and acromioclavicular joint capsule (2, 3, 9). Bony spurs can present either as a signal void (cortical bone) or as a high-intensity (bone marrow) structure impinging on the rotator cuff. Although calcific deposits are easily missed on MR images they may be seen occasionally as signal void areas, increasing the volume of the supraspinatus tendon.

Treatment depends on age and activity of the patient,

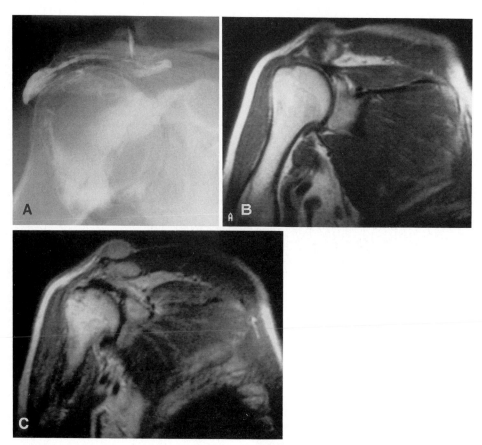

Figure 20.8. (A) Patient with clinical suspicion of cuff tear. Radiograph made during arthrography. Air and contrast are present within the joint cavity and also under the acromion, indicating cuff tear with bursal filling. Extension into acromioclavicular joint is clearly depicted. **(B)** Coronal oblique T1-weighted (TR/TE 600/30 ms) image exhibits atrophy and retraction of supraspinatus muscle, widening of acromioclavicular joint at this level, and extension of fluid into subcutaneous soft tissue. **(C)** Coronal oblique T2-weighted (TR/TE 2000/50 ms) image shows fluid in subcutaneous nodule, cuff tear, supraspinatus atrophy, and retraction to advantage.

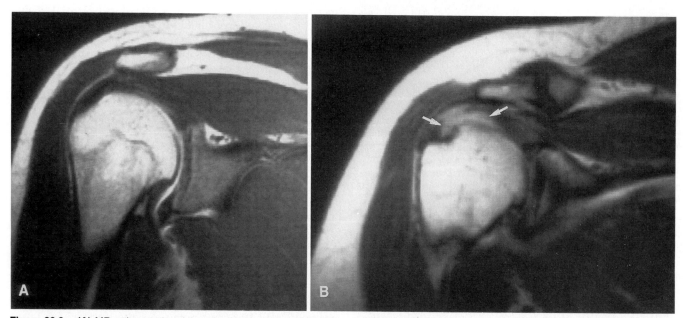

Figure 20.9. (A) MR arthrography of the glenohumeral joint. Coronal oblique T1-weighted SE image. The joint cavity is clearly visualized due to high intensity Gd-DTPA. No filling of subacromial bursa is present. **(B)** Tear of the supraspinatus tendon. Conventional arthrography showed a narrow channel of contrast extravasation with protracted filling of the subacromial bursa. MR arthrography reveals a broad area of high signal intensity *(arrows)*, which represents diffusion of Gd-DTPA into degenerative portions of the supraspinatus tendon and muscle. (Reprinted with permission from Bloem JL, Reiser MF, Vanel D. Magn Reson Q 1990;6(2):136–163.)

associated abnormalities, chronicity and severity of the disorder, and resulting disability. The majority of patients with stage 1 disease respond to conservative treatment although they often have to alter their sporting activities to avoid aggravation. Patients with stage 2 disease can get relief with conservative measures, such as antiinflammatory agents, avoidance of the painful range of motion and physical therapy, but occasionally require decompression after failure to manage the pain by nonoperative treatment. Patients with stage 3 disease are often elderly, who require an anterior decompression or acromioplasty with bursectomy, rotator cuff repair (young patients with good quality of the cuff), or osteophyte resection of the acromioclavicular joint (26). Although surgical intervention may be successful in the sense that it reliefs pain and restores satisfactory function, a return to the former competitive status in sports is not always feasible for the athlete.

Magnetic resonance imaging may thus assist in the diagnosis of impingement syndrome, but more importantly, MR imaging may discriminate early tendinitis from severe tendinitis with a rotator cuff tear or tendinitis with a high risk of tear. Magnetic resonance imaging may therefore be of value in choosing between conservative or surgical therapy and may also be of value in monitoring response to therapy.

Rheumatoid Arthritis

Magnetic resonance imaging has a potential role in the management of patients with crystal-induced synovitis (Milwaukee shoulder), infection (see Chapter 13), and rheumatoid arthritis (RA). Magnetic resonance imaging can exhibit early soft-tissue disease such as pannus (synovial proliferation), as well as early osseous erosions. Destruction of cartilage, bone, ligaments, and tendons (for instance, biceps and supraspinatus rupture) and the development of synovial cysts can be shown by virtue of high-contrast resolution combined with tomographic display of all structures in and around the shoulder. Axillary lymph nodes are frequently observed. Bony erosions (occurring within the first 2 years of the disease) can be shown to better advantage and in an earlier stage than is possible with plain radiographs (Fig. 20.10*A* and *B*) (27). Also, hyaline articular cartilage can be visualized directly and diffuse as well as focal destruction will be shown. Detection of bony and soft-tissue shoulder changes at an early age, before irreversible pathologic changes have occurred, can be useful when planning conservative treatment (Fig. 20.11*A* and *B*).

On the other hand, when considering shoulder replacement, MR imaging may help in choosing between total

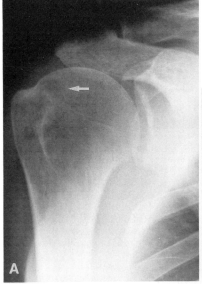
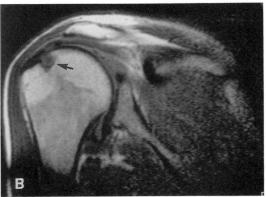

Figure 20.10. **(A)** Radiograph of patient with RA. Right shoulder with early erosion in the humeral head *(arrow)*. **(B)** Coronal oblique T1-weighted (TR/TE 600/30 ms) image exhibits erosion in humeral head *(arrow)* much more clearly than the radiograph.

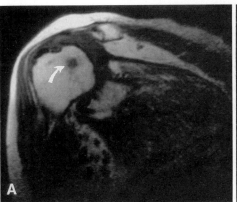
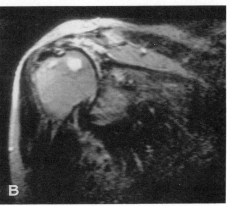

Figure 20.11. **(A)** Patient with RA. Coronal oblique T1-weighted (TR/TE 600/30 ms) image shows erosion in humeral head *(white arrow)* and low to intermediate signal intensity pannus within joint cavity. **(B)** Coronal oblique T2-weighted (TR/TE 2000/100 ms) image exhibits high signal intensity of erosion in humeral head and high intensity pannus. Note the high signal intensity pannus at the level of the destroyed supraspinatus muscle and tendon. The humeral head is cranially displaced.

Figure 20.12. (A) Crystal-induced arthropathy (Milwaukee shoulder). Axial CT image (3 mm) following arthrography shows synovial thickening *(arrow)*. Contrast in subdeltoid bursa *(curved arrow)* indicates presence of cuff tear. Irregularity of humerus represents accompanying cuff arthropathy. **(B)** Level 1 cm cranial to *(A)*: Thickened synovium *(black arrow)* and cuff arthropathy *(white arrow)* shown to advantage.

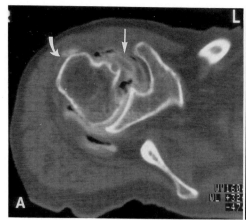
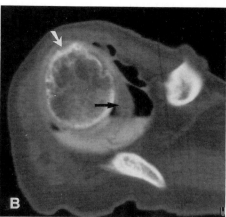

joint or just head replacement. Valuable information can be obtained with MR imaging and also CT-arthrography about bone stock, pannus extension, cartilage and capsular status, and quality of the rotator cuff. A cuff tear may pass undetected on conventional arthrography because of synovial proliferation (Fig. 20.12*A* and *B*).

Pannus formation (adjacent to bony erosions) appears as areas of low signal intensity on T1-weighted images and as areas of either low or high signal intensity on T2-weighted images (28–30). The low signal intensity can be ascribed to hemosiderin content and fibrosis (28). The actively inflamed parts will display a high signal intensity (30). The similarity between intracapsular fluid and pannus can be confusing when displaying high signal intensities on T2-weighted images. When extreme T2 weighting fluid will display a higher signal intensity than pannus (27).

Gd-DTPA-enhanced MR imaging may be used to indicate the activity of inflammatory tissue, as active pannus is expected to enhance more than inactive fibrosis and hemosiderin deposits (29). Gd-DTPA also facilitates differentiation between synovium and joint fluid (29). This evaluation requires imaging shortly after injection, since diffusion of Gd-DTPA into the joint is observed on delayed images (31).

References

1. Neer CS. Impingement lesions. Clin Orthoped 1983;173:70–77.
2. Seeger LL, Gold RH, Bassett LW, Ellman H. Shoulder impingement syndrome: MR findings in 53 shoulders. AJR 1988;150:343–347.
3. Kieft GJ, Bloem JL, Rozing PM, Obermann WR. Rotator cuff impingement syndrome: MR imaging. Radiology 1988;166:211–214.
4. Zlatkin MB, Dalinka MK, Kressel HY. Magnetic resonance imaging of the shoulder. Magn Reson Q 1989;5:3–22.
5. Kneeland JB, Middleton WD, Carrera GF, et al. MR imaging of the shoulder: diagnosis of rotator cuff tears. AJR 1987;149:333–337.
6. Evancho AM, Stiles RG, Fajman WA, et al. MR imaging diagnosis of rotator cuff tears. AJR 1988;151:751–754.
7. Kieft GJ, Sartoris DJ, Bloem JL, et al. Magnetic resonance imaging of glenohumeral joint diseases. Skel Radiol 1987;16:285–290.
8. Zlatkin MB, Iannotti JP, Roberts MC, et al. Rotator cuff tears: performance of MR imaging. Radiology 1989;172:223–229.
9. Tsai JC, Zlatkin MB. Magnetic resonance imaging of the shoulder. Orthopedics 1990;2:279–291.
10. Holt RG, Helms CA, Steinbach L. MRI of the shoulder: rationale and current applications. Skel Radiol 1990;19:5–14.
11. Hardy DC, Vogler JB III, White RH. The shoulder impingement syndrome: prevalance of radiographic findings and correlation with response to therapy. AJR 1986; 147:557–561.
12. Huber DJ, Sauter R, Muller E, Requardt H, Weber H. MR imaging of the normal shoulder. Radiology 1986;158:405–408.
13. Kieft GJ, Bloem JL, Obermann WR, Verbout AJ, Rozing PM, Doornbos J. Normal shoulder: MR imaging. Radiology 1986; 159:741–745.
14. Seeger LL, Ruszkowski JT, Bassett LW, Kay SP, Kahmann RD, Ellman H. MR imaging of the normal shoulder: anatomic correlation. AJR 1987;148:83–91.
15. Middleton WD, Kneeland JB, Carrera GF, et al. High resolution MR imaging of the normal rotator cuff. AJR 1987;148:559–564.
16. Morrison DS, Biglianu LU. The clinical significance of variations in acromial morphology. Orthoped Trans 1986;11:234.
17. Kilcoyne RF, Reddy PK, Lyons F, Rockwood CA Jr. Optimal plain film imaging of the shoulder impingement syndrome. AJR 1989;153:795–797.
18. Burk DL Jr, Karasick D, Mitchell DG, Rifkin MD. MR imaging of the shoulder: correlation with plain radiography. AJR 1990;154:549–553.
19. Holt RG, Neumann CH, Petersen S, et al. MR imaging of the shoulder: rotator cuff and related structures in asymptomatic individuals. Radiology 1990;177(P):294.
20. Hueck A, Appel M, Kaiser E, Luttke G, Lucas P. MR imaging of the shoulder: potential overinterpretation of normal findings in the rotator cuff. Radiology 1990;177(P):103.
21. Brandt TD, Cardone BW, Grant TH, Post M, Weiss CA. Rotator cuff sonography: a reassessment. Radiology 1989;173:323–327.
22. Soble MG, Kaye AD, Guay RC. Rotator cuff tear: clinical experience with sonographic detection. Radiology 1989;173:319–321.
23. Hall FM. Sonography of the shoulder. Radiology 1989;113:310.
24. Middleton WD. Status of rotator cuff sonography. Radiology 1989;173:307–309.
25. Flannigan B, Kursunoglu-Brahme S, Snyder S, et al. MR arthrography of the shoulder: comparison with conventional MR imaging. AJR 1990;155:829–832.
26. Cofield RH. Rotator cuff disease of the shoulder. J Bone Joint Surg 1985;67a:974–979.
27. Kieft GJ, Dijkmans BAC, Bloem JL, Kroon HL. Magnetic resonance imaging of the shoulder in patients with rheumatoid arthritis. Ann Rheum Dis 1990;49:7–11.
28. Beltran J, Caudill JL, Herman LA, et al. Rheumatoid arthritis: MR imaging manifestations. Radiology 1987;165:153–157.
29. Reiser MF, Bongartz GP, Erlemann R, et al. Gadolinium-DTPA in rheumatoid arthritis and related diseases: first results with dynamic magnetic resonance imaging. Skel Radiol 1989;18:591–597.
30. Yulisch BS, Lieberman JM, Newman AJ, Bryan PJ, Mulopulos GP, Modic MT. Juvenile rheumatoid arthritis: assessment with MR imaging. Radiology 1987;165:149–152.
31. Kursunoglu-Brahme S, Riccio T, Weisman MH, et al. Rheumatoid knee: role of gadopentate-enhanced MR imaging. Radiology 1990;1976:831–835.

Magnetic Resonance Imaging and Computed Tomography Arthrography of Shoulder Trauma and Instability

Johan L. Bloem, Anton M. J. Burgers, and Wim R. Obermann

Introduction

The mobile shoulder is prone to dislocations. The relationship between the ensuing injury to the joint and shoulder instability is still controversial. Accurate diagnosis of damage is important in clarifying this relationship and in improving therapy. The potential of magnetic resonance (MR) imaging and computed tomography (CT) in diagnosing potential "at risk" factors for recurrent dislocation is emphasized in this chapter.

Compromised stability, presenting as pain and clicking, is often related to damage of intraarticular structures such as the labrum and hyaline cartilage. This may be caused, for instance, by minor repetitive traumatic events in the elderly or by high-velocity motion in throwing athletes and swimmers (1).

Voluntary multidirectional instability, not related to previous trauma, is a rare but important entity that will also be discussed. Finally, the role of CT-arthrography in the evaluation of the surgically treated shoulder is addressed. The impact of CT on management of humeral head fractures is discussed in Chapter 22.

Examination Technique

When the joint is distended by synovial (hemorrhagic) effusion after dislocation, or by intraarticular injection of fluid or Gd-labeled diethylene triamine pentaacetic acid (DTPA), intra- as well as extraarticular structures can be evaluated with MR imaging. The labrum is best evaluated in the axial and longitudinal coronal oblique plane perpendicular to the plane of the glenoid. T2-weighted spin-echo (SE) images or T2*-weighted gradient-recalled images display the normal labrum as a low signal intensity structure surrounded by high signal intensity fluid. The anterior capsule attachment is also best displayed with these two pulse sequences in the axial plane. Damage to the supraspinatus tendon is evaluated on the coronal oblique plane perpendicular to the glenoid with a combination of T1- and T2-weighted, or spin-density and T2-weighted, SE techniques (2). Axial T1-weighted SE images are used to inspect the humeral head.

We perform an MR examination within 1 week after the traumatic event on a 0.5- or 1.5-T system (Philips S15, Shelton, Conn.). The patient is supine with the arm in neutral position. A surface coil in a Helmholtz configuration is always used. Routinely, we obtain axial T1 (TR 600 ms, TE 20 to 30 ms, two excitations) and long TR multiecho (TR 2000 ms, TE 20 and 80 ms, one excitation) or T2*-weighted fast field echo (FEE) TR 400 ms, TE 30 ms, flip angle 20 to 45°) images. Subsequently a sequence combining spin density and T2 contrast (TR 2000, TE 20 and 80 ms) is made in the coronal oblique plane perpendicular to the glenoid (parallel to the supraspinatus muscle). As an alternative to the axial T2*-weighted gradient-echo technique that we prefer, a sequence with a larger flip angle may also be used (flip angle 70°, TR 600, TE 15 ms). Instead of scout views the axial T1-weighted images are used to plan the oblique longitudinal T1- and T2-weighted SE images. Field of view is 12 to 20 cm, slice thickness varies between 3 and 5 mm for T1-weighted sequences and 4 and 7 mm for T2-weighted sequences. Total acquisition time is less than 30 min.

Computed tomography-arthrography is performed to evaluate patients with unstable glenohumeral articulations. Few conventional radiographs are necessary between contrast material injection and CT. This decreases cost, radiation dose, and examination time (3). For double-contrast arthrography, 2 ml of an iodinated water-soluble contrast agent and 10 to 15 ml of air are injected in the glenohumeral joint. For CT-arthrography the presence of enough air is crucial. The presence of the iodinated contrast agent is less important, a surplus of the positive contrast agent will even seriously degrade the CT images. Computed tomography is capable of demonstrating the soft tissues of the axilla, labrum, cartilage, and capsule, when high-resolution images are made with high milliampere, thin 3-mm contiguous slices, and zoom reconstruction (4–6). Synovial folds and the glenohumeral ligaments may be seen within the normal joint. The patient is usually examined in the supine position with the arm in neutral position. When the posterior part of the labrum is not well exhibited on the standard sequence, additional slices in the prone position with the arm in external rotation will produce images with air surrounding the posterior part of the labrum.

Nonenhanced contiguous CT images with 3- to 6-mm collimation are used to evaluate fractures and are used in patients suspected of having a posterior dislocation with equivocal clinical and radiographic findings.

Glenohumeral Dislocation and Instability

The shoulder is the most frequently dislocated major joint due to its great range of motion. The shallow glenoid cavity allows this large range of motion, but offers little static stability in consequence. Stability is mainly provided by capsule, ligaments, and muscle. Glenohumeral dislocations can be classified as anterior (over 95%), superior, inferior, and posterior (7). Most anterior dislocations occur during abduction and external rotation and are subcoracoid in nature. Less frequent anterior dislocations are subglenoid, subclavicular, and intrathoracic ones.

We focus on anterior glenohumeral dislocations because of its high incidence and high prevalence of resulting glenohumeral instability. Epidemiologic studies reveal that age is the most important factor in developing an unstable glenohumeral articulation after dislocation (8, 9). Under 31 years of age the reported recurrence rate is 55 to 95%. In patients over 40 years of age the recurrence rate is only 14% (9, 10). The influence of sporting activity, immobilization, gender, humeral torsion, bone, and soft-tissue derangements are controversial (8, 9, 11, 12).

Anterior displacement of the humeral head may cause disruption of capsule, stripping of capsule and periosteum from glenoid, labrum avulsion, glenoid rim fracture, subscapularis detachment, fracture of greater tuberosity, rotator cuff tear, and fracture of the posterior aspect of the humeral head. Bankart emphasized the importance of separation of the anterior labrum from the glenoid rim as a cause of recurrent dislocation (13, 14). Others added the stripping of capsule and periosteum from the anterior rim of the glenoid as an important factor in recurrent dislocation (15, 16). Recurrence is reported to be higher in young patients with labral tears (cartilaginous Bankart lesions) and intact capsule than in the elderly, who do not have Bankart lesions but do have capsular tears (5, 12, 17–20). Finally, the role of the subscapularis muscle in the prevention of recurrent dislocation was acknowledged (21, 22). Thus consensus exists about the importance of the anterior capsular mechanism, consisting of the anterior capsule (including glenohumeral ligaments) and anterior labrum, and the subscapularis muscle and tendon in restraining anterior displacement of the humeral head (5, 16, 21). The three glenohumeral ligaments (superior, middle, and inferior) are capsular reinforcements. Posttraumatic insufficiency of the anterior capsule mechanism predisposes to an unstable glenohumeral head and recurrent dislocations.

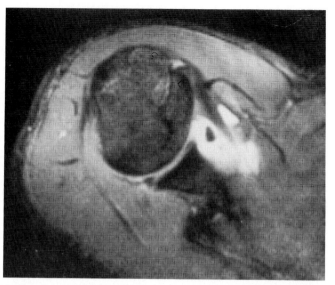

Figure 21.1. Axial T2*-weighted (20°, TR 400, TE 20) image obtained 1 day after first dislocation. The normal posterior labrum is rounded, the anterior labrum is avulsed. The triangular low signal intensity area surrounded by high signal intensity of large joint effusion represents the avulsed labrum. Because of periosteal and capsular stripping, fluid is seen anterior to the scapular neck, and the capsule inserts on the medial third of the scapular neck. The subscapular tendon inserts on the lesser tubercle. The fluid anterior to the subscapular muscle/tendon is located within the subscapular bursa.

Joint effusion, which is always present in the first week after the initial dislocation, allows evaluation of intraarticular structures with MR imaging. Therefore MR imaging is preferred over CT-arthrography in the acute stage. In our patients, who have a MR imaging examination within 1 week after the initial traumatic dislocation, we almost always (over 90%) find labral lesions, irrespective of age. Labral lesions, caused by anterior dislocation, are located anteriorly and inferiorly. The traumatized labrum may be attenuated, truncated, fragmented, or even avulsed completely (Figs. 21.1 and 21.2) (23). These lesions are seen by virtue of the high signal intensity of fluid within the low signal intensity of the labrum on T2- or T2*-weighted images. Complete labral avulsions, seen as complete absence of the labrum, are encountered in almost 70% of our patients after dislocation, whereas glenoid rim fractures (osseous Bankart lesion) are only rarely found (13%) (Figs. 21.3 and 21.4). Globular shaped labral fragments may be seen superior or inferior to the glenoid. Diagnosis of labral lesions is reminiscent of meniscal tear diagnosis in the knee. Attention must be paid to the normal variants (4–6, 24). In this respect the normal recess at the interface of labrum and hyaline cartilage is important (Fig. 21.5). On MR images, a normal high signal intensity line is often seen at the base of the labrum. The normal anterior labrum is triangular (50%), rounded (20%), cleaved (14%), notched (8%), or flat (7%), and the posterior labrum is rounded or triangular (Fig. 21.4) (24, 25). The incidence of labral lesions, after the first traumatic dislocation, detected with MR imaging is much higher than that detected with arthrography (28 to 67%) (17, 18), but similar to that detected with arthroscopy (85%) and CT-arthrography (87%) (19, 26).

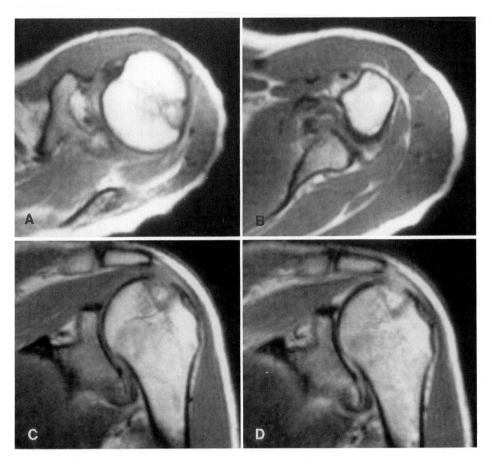

Figure 21.2 MR images obtained 2 days after first anterior shoulder dislocation. **(A)** Axial T1-weighted (550/30) image made at the level of the coracoid process. The Hill-Sachs compression fracture is easily appreciated at the lateral margin of the humerus. Little bone marrow edema is seen as well. **(B)** Axial T1-weighted (550/30) image made at the level of the inferior margin of the glenoid. No labrum is seen at the anterior inferior margin, indicating avulsion of the labrum. Some decreased signal intensity anterior within the bone marrow is consistent with osseous Bankart lesion. **(C)** Oblique coronal T1-weighted (550/30) image. The triangular-shaped Hill-Sachs fracture is again noted. Superior margin of the glenoid has a normal shape and signal intensity. The inferior margin is not visualized, indicating avulsion. The supraspinatus tendon-muscle junction is thickened and displays an increased signal intensity. **(D)** The signal intensity in the supraspinatus tendon/muscle junction increases further on the T2-weighted image (2000/100). These findings are consistent with traumatized muscle/tendon (edema/hemorrhage/partial tears).

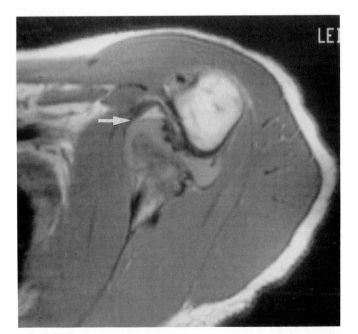

Figure 21.3. Hemorrhagic joint effusion in this patient imaged 1 day after first dislocation is identified by virtue of its intermediate signal intensity on axial T1-weighted (600/20) image. A fluid-fluid level is present *(arrow)*. The high signal intensity of the nondependent part is caused by fat within the joint, indicating presence of intraarticular fracture. The fracture of glenoid and secondary bone marrow edema are appreciated. No anterior labrum is visualized.

Computed tomography-arthrography is preferred in the evaluation of the chronic unstable shoulder and in recurrent dislocation, because a joint effusion is usually absent in these patients. Computed tomography-arthrography is superior to MR imaging in demonstrating osseous, cartilaginous, and capsular damage when no joint effusion is present. The sensitivity of CT-arthrography in detecting labral tears is reported to be 100% with a specificity of 97% (3, 27). The abnormal labrum displayed on CT-arthrography is absent, deformed, irregular, fragmented, or imbedded with contrast agent (Figs. 21.6 and 21.7). A short labrum is not necessarily a result of trauma, since this is not infrequently found as a normal variant in patients with stable articulations.

Labral tears can be found in patients without dislocation. These tears may also be located superiorly, and result mainly from degeneration in the elderly and high-velocity motion in athletes (1).

Posttraumatic stripping of the capsule and periosteum from the anterior rim of the glenoid is difficult to assess because of the large variety in anterior capsule insertion. The place of insertion itself is easily determined on T2- or T2*-weighted images in the axial plane when the joint is distended with fluid or air (Figs. 21.1, 21.7, and 21.8). This insertion has been classified into three types (3, 5). In type 1, the capsule inserts directly into the labrum. In type 2, the insertion is at the middle third of the scapular neck. In type 3, the insertion is at the medial third of the sca-

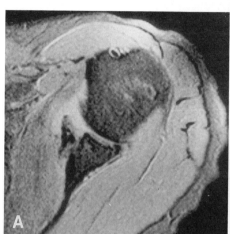
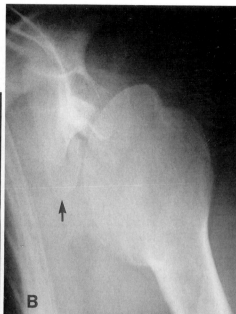

Figure 21.4. **(A)** Axial T2*-weighted (flip angle 20°) image obtained on day of first dislocation. Dislocated glenoid rim fracture is identified. Note normal rounded appearance of posterior labrum. Large Hill-Sachs fracture was seen on more cephalic images. **(B)** Forty-five degree angulated (Didiee) view demonstrates Hill-Sachs fracture and avulsed fragment of glenoid *(arrow)*.

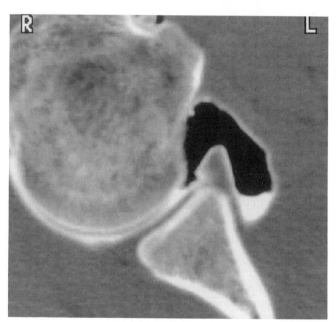

Figure 21.5. CT-arthrography demonstrates normal anterior labrum. The normal recess between the labrum and hyaline cartilage is visualized.

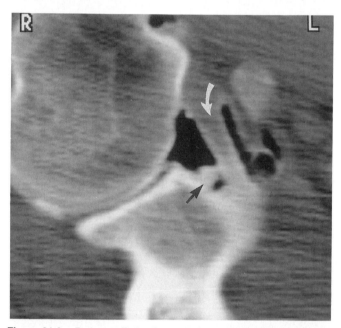

Figure 21.6. Patient suffering from recurrent anterior dislocations. Air within the glenohumeral joint and subscapularis bursa nicely delineates the subscapularis muscle *(curved arrow)* on CT-arthrography. Part of the coracoid process is seen. Anterior aspect of labrum *(arrow)* is avulsed and lies between glenoid and subscapularis muscle.

pular neck. Although it is not entirely clear whether type 3 is a result of posttraumatic stripping or a developmental variant, this medial attachment predisposes to anterior dislocation. Types 1 and 2 may occur as normal variants; the prevalence of glenohumeral dislocations drops substantially in these groups (3). A prominent subscapular bursa is frequently encountered when a type 3 capsular insertion is present (Figs. 21.1, 21.7, and 21.8). This bursa originates through an anterior opening of the capsule between the glenohumeral ligaments. It is located posterior to the subscapularis tendon, but it may extend over the tendon, resulting in a fluid- (on MR images) or air-filled

(on CT-arthrography) compartment seen anterior to the tendon. We encounter posttraumatic capsule stripping, after the first dislocation, in approximately 75% of our MR studies made within 1 week after dislocation, when we define posttraumatic stripping as type 2 or 3 insertions in combination with the above described posttraumatic labral lesions. Detachment of the subscapularis tendon from the lesser tuberosity with or without a bony fragment is very rare, but may be seen with MR imaging.

The fourth important posttraumatic lesion is the Hill-

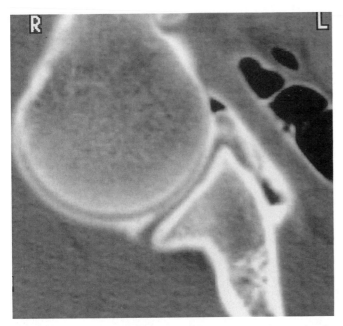

Figure 21.7. CT-arthrography of a patient with recurrent anterior dislocations. Contrast imbibition in the anterior aspect of the glenoid indicates tear. Note the distended subscapularis bursa. The capsule inserts medially on the scapular neck.

Sachs fracture. Despite its often used eponym, it was described earlier and in detail by Eve (28). This compression fracture is appreciated most accurately on a combination of axial (CT or MR imaging) and oblique longitudinal MR images (Figs. 21.2, 21.4, and 21.9). On MR images, the signal intensity is low on T1-weighted images and low to high on T2-weighted images, depending on the amount of intraosseous edema present. It is important to evaluate the humeral head on the axial MR or CT images at or above the level of the coracoid process, since the normal concave margin of the major tuberosity may mimic a Hill-Sachs fracture on the lower axial images. A Hill-Sachs fracture after the initial dislocation is found with MR imaging in 81% of our patients. This prevalence is higher than that detected with standard and special radiographs (32 to 55%) (8, 9, 29), CT-arthrography (67%) (26), or with arthroscopy (47 to 77%) (19, 20). The Hill-Sachs fracture can be arbitrarily classified on axial images as small when the circumference involved is less than 30° (15%), intermediate when the circumference is between 30 and 60° (35%), and large when more than 60° (50%) is involved. When no Hill-Sachs fracture is present, a fracture of the major tuberosity is nearly always found. Typically, the fractured tubercle becomes reduced when the dislocated humeral head is reduced (Fig. 21.10) (12). A concomitant or isolated high sig-

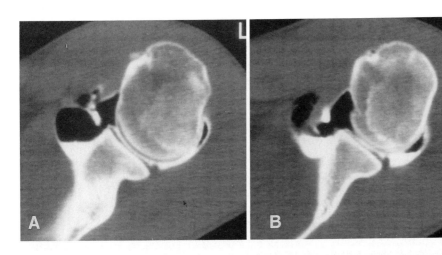

Figure 21.8. CT-arthrography in patient who was treated with arthroscopic repair of anterior capsule mechanism because of recurrent dislocations. Despite this procedure, marked distention of the glenohumeral joint is still possible **(A)**. The anterior aspect of the labrum is avulsed, posterior aspect is normal. The glenohumeral joint remained unstable. The subscapularis bursa is also filled with air **(B)**. Medial insertion of capsule on scapular neck is a result of periosteal capsular stripping **(A** and **B)**. (See also Figs. 21.1, 21.7, 21.14, and 21.15.)

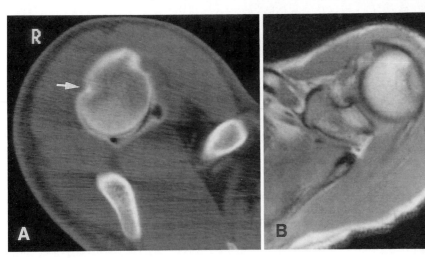

Figure 21.9. (A) Small Hill-Sachs fracture *(arrow)* demonstrated on CT-arthrography. Slice is made cephalic to level of coracoid process. **(B)** Lenticular-shaped decreased signal intensity area of Hill-Sachs fracture is appreciated on T1-weighted (600/20) image made at level of coracoid process.

Figure 21.10. **(A)** Radiograph made after first dislocation shows dislocated shoulder with impaction of glenoid on humeral head resulting in Hill-Sachs fracture. The major tubercle is avulsed. **(B)** T1-weighted (550/30) image made after reduction. The fractured major tubercle is seen in an almost anatomical position.

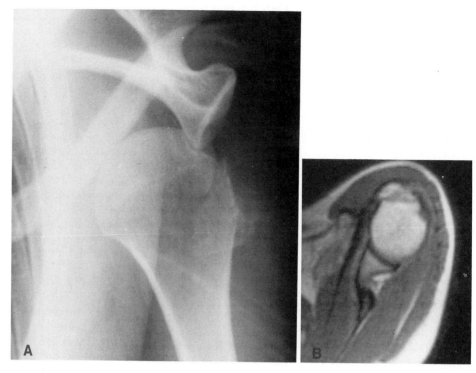

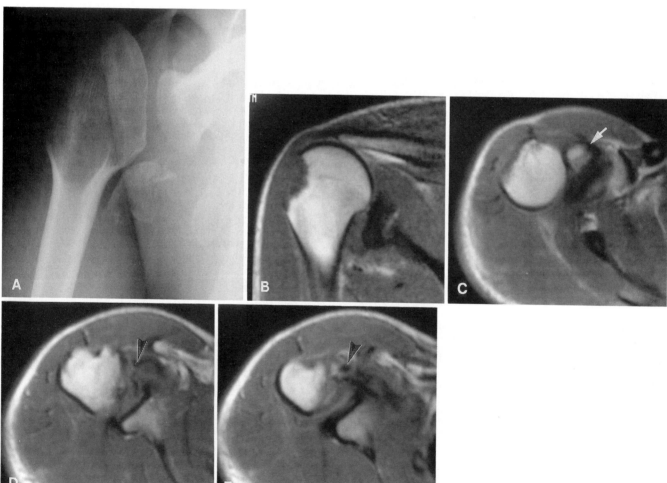

Figure 21.11. **(A)** Radiograph made with 45° angulation (Didiee view) following first anterior dislocation and attempted reduction. The fractured major tubercle is dislocated. The humeral head is not in its normal position. **(B)** A large defect at the side of the major tubercle is seen on the T1-weighted (600/20) oblique coronal image. **(C–E)** Three axial T1-weighted (600/20) images from cranial **(C)** to caudal **(E)** demonstrate the structures obstructing the entrance to the glenoid cavity. The defect at the side of the major tubercle is seen. The major tubercle *(arrow)* is found medial to the humeral head **(C)**. The biceps tendon *(arrowheads)* is medially displaced **(D and E)**. Its intraarticular portion is surrounded by fluid **(D)**. At surgery the entrance to the glenoid cavity was cleared by reduction of both major tubercle and biceps tendon.

nal intensity on T2-weighted images indicative of edema, hemorrhage in the supraspinatus tendon, or even rupture is occasionally observed (Fig. 21.2). Magnetic resonance studies performed in the acute phase demonstrated edema in 25% and an old rotator cuff tear (identified by virtue of marked atrophy with absence of tendon, or high signal intensity in tendon) in 12% of patients. An acute rotator cuff rupture or persistent displacement of a fractured tubercle normally needs surgical intervention (Fig. 21.11) (12).

The relationship between humeral torsion and anterior dislocation and its recurrence rate has not yet been established (12, 30). Similar to anteversion measurements of the femur, torsion of the humerus can be measured when axial CT or MR slices are made at the level of the head and the distal condyles. Preliminary results of an ongoing study in our hospital indicate that the variance of torsion in patients (−14 to 28°) and normal controls (5 to 53°) is rather large. Further studies are needed to determine the importance of an increased external torsion.

As already described by Hippocrates, contraction of the wide space into which the dislocated humerus escapes is the goal of surgical treatment in patients with anterior instability due to dislocation (31). He cauterized the deep tissues in front of the shoulder and immobilized the arm day and night for a long time. Considering the high recurrence rate after one traumatic dislocation in young adults, immediate surgical or athroscopic repair of the anterior capsule mechanism, which is severely damaged in a high incidence of patients, may become the therapy of choice. The anterior capsule may be tightened and sutured to the glenoid rim arthroscopically. The effect of arthroscopic repair awaits further evaluation. Many surgical procedures are used in patients with recurrent anterior dislocations. Reinforcement of the anterior capsule mechanism is the aim of many of these procedures. The most frequently used procedures are as follow: Putti-Platt (shortening of anterior capsule and subscapularis muscle), typically in combination with the Bankart repair (surgical reattachment of capsule to anterior glenoid with staples or sutures), Magnuson-Stack (transfer of subscapularis from lesser to greater tuberosity), Bristow-Helfet (transfer of coracoid process with conjoined tendon to anterior glenoid), and Eden-Hybbinette (anterior glenoid bone block). It is obvious that these procedures can be successful only when an accurate diagnosis of damaged anterior capsular mechanism has been made.

Although posterior glenohumeral dislocation is relatively rare (2 to 4% of all shoulder dislocations), it requires special attention because over 50% of posterior dislocations are not recognized as such at the time of presentation (7, 32). An erroneous diagnosis of frozen shoulder is not infrequently made. Severe osteoarthritis and poor function may ensue and are unnecessary sequelae because radiographs are diagnostic. The internally rotated, posteriorly dislocated humeral head can be recognized on anteroposterior as well as transscapular or axial radiographs. In equivocal cases CT is able to display the dislocation in the acute phase, as well as late changes in patients with missed diagnosis of posterior dislocation (Fig. 21.12). The role of MR imaging and CT-arthrography in the evaluation of the posterior dislocated shoulder is similar to that of the anterior dislocated shoulder when instability ensues. Thus

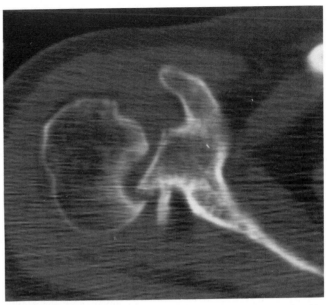

Figure 21.12. Patient complained of pain and impaired shoulder function following trauma 2 years before radiologic consultation. Sequelae of missed posterior dislocation are displayed on CT. Persistent posterior dislocation with large medial compression fracture is seen. Avulsed fragment originates from posterior glenoid.

labral tears, reversed Hill-Sachs fractures (anteromedial aspect of the humeral head), posterior glenoid rim fractures, and increased posterior tilt of glenoid are features that are easily diagnosed with MR imaging or CT-arthrography.

Multidirectional Instability

Multidirectional instability is often related to joint laxity. These patients are typically young males or females, with hypermobility in other joints in approximately 50%. Multidirectional instability may also be caused by repetitive trauma. These patients normally show a variety of the above-described posttraumatic changes both on the anterior and posterior side. Accurate diagnosis is of paramount importance because surgical procedures aimed at tightening or reinforcing the anterior capsule mechanism may result in marked posterior instability. Computed tomography-arthrography is the method of choice in the evaluation of these patients.

Voluntary multidirectional instability is not related to trauma. Patients are young and a psychological factor may be present. It is sometimes associated with systemic laxity, for instance in Ehlers-Danlos or Marfan syndrome. Diagnosis is based on clinical presentation and fluoroscopy (Fig. 21.13). When history and physical examination are confusing, CT-arthrography may be needed to exclude posttraumatic changes. In rare instances an underdeveloped glenoid is found. Identification of these patients is important because surgical intervention is as a rule not indicated.

The Surgically Treated Shoulder

Failed surgical treatment of an unstable shoulder may be due to inaccurate preoperative diagnosis (for instance

Figure 21.13. Nineteen-year-old male patient with voluntary; multidirectional instability. **(A)** At fluoroscopy, following intraarticular injection of air (10 ml) and positive contrast agent (2 ml). The humeral head could be easily subluxated into anterior, posterior, and inferior directions. Note the cephalic collection of air outlining the biceps tendon. **(B)** During CT-arthrography in the supine position the humeral head is posteriorly subluxated. The pliable anterior aspect of the labrum is consequently deformed in this position. Other than the enlarged joint capacity no intraarticular morphologic abnormalities were found.

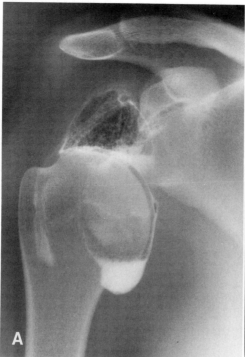
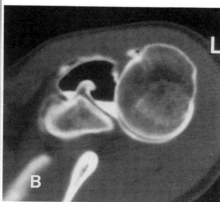

Figure 21.14. CT-arthrography of arthroscopically treated patient suffering from recurrent anterior dislocations. At arthroscopy the anterior aspect of the capsule was sutured to the glenoid rim, thus reducing the space anteriorly. Marked reduction of the anterior joint compartment has been obtained. The level of fixation can be seen (*arrow*, **A**). The drilling hole in which the sutures pass the glenoid is also depicted (*arrow*, **B**). No recurrent dislocations occurred in the first year after repair.

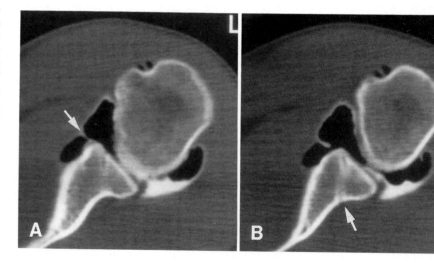

anterior capsule mechanism repair in a multidirectional unstable shoulder, resulting in posterior instability), persistance of large subscapularis bursa, persistence of original abnormality, compromised rotator cuff, and reinjury (Fig. 21.8) (33). We find CT-arthrography the single most useful technique in the evaluation of these patients. The examination technique is the same as in preoperative assessment. Sutures, screws, and drilling holes in the scapula may indicate the nature of the surgical procedure (Fig. 21.14). In arthroscopic or surgical repair of the anterior capsule mechanism (usually a combination of Bankart and Putti-Platt procedures) the subscapularis bursa is obliterated and the capsule is attached to the anterior rim of the glenoid. As a result very little contrast agent and air is

present anterior to the glenohumeral joint space after a successful procedure (Fig. 21.15). The clinical equivalent of this new order within the joint is reduced external rotation. Excessive tightness may even result in a fixed posterior (sub)luxation or posterior instability. This reduced volume of the joint adversely affects the accuracy of MR imaging. The MR images are also degraded by susceptibility artifacts caused by surgical instrumentation and implanted hardware.

Depending on the initial diagnosis and surgical procedure a second intervention can be planned based on clinical and CT-arthrography findings. The most frequent finding in postsurgical redislocation after a combined Putti-Platt and Bankart repair is a large subscapularis bursa with

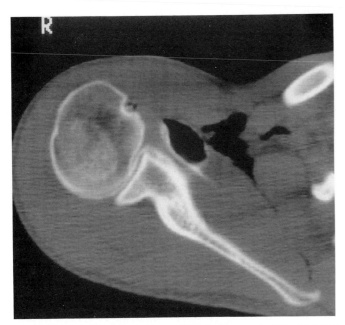

Figure 21.15. CT-arthrography obtained after surgical Bankart and Putti-Platt repair for recurrent anterior dislocations. The anterior joint compartment is almost completely obliterated; only a small amount of contrast material is seen anterior to the glenoid. The intraarticularly introduced air escaped into the subscapularis bursa and axillary fat. The patient did not experience recurrent dislocations and had good function with minor limitation of exorotation.

medial insertion of capsule to the glenoid (3, 12). Kessel advocates anterior block (Eden, Hybbinette) or Bristow procedures in these patients (12).

Miscellaneous Disorders

Rupture of the biceps tendon usually occurs in the elderly, with or without associated rotator cuff tears, and in young athletes. Although it can be diagnosed with MR imaging due to retraction, hemorrhage, edema, and fluid, it is a clinical diagnosis (see also Chapter 14). Synovial fluid surrounding the biceps tendon is normal, but is increased after trauma and in tenosynovitis. An empty intertubercular groove, associated with reactive changes, indicates biceps tendon dislocation (tendon dislocates medially) or rupture. This can be demonstrated with ultrasound, and it may be encountered as incidental finding on MR imaging or CT-arthrography (Fig. 21.11).

Findings of osteonecrosis are similar to those described in Chapter 26 on osteonecrosis of the femoral head.

Conclusion

The choice of therapy of the unstable or recently traumatized shoulder is controversial. Radiologic imaging techniques and arthroscopy, able to exhibit the damaged structures in great detail, have only recently become available. Currently, the indications for and timing of imaging studies depend on the philosophy of the referring physician. Current research will clarify the role of imaging studies in relation to the various therapeutic possibilities.

When imaging studies are considered necessary, MR imaging is preferred in the acute stage because of the high intraarticular contrast caused by joint effusion. Also, extraarticular damage is easily evaluated. When no joint effusion is present, for instance in patients with multidirectional voluntary dislocation or subluxation, or in patients with recurrent dislocations following initial traumatic dislocation, CT-arthrography is preferred. Computed tomography-arthrography is also preferred in the assessment of the surgically treated shoulder.

References

1. Kohn D. The clinical relevance of glenoid labrum lesions. Arthroscopy 1987;3:223–230.
2. Kieft GJ, Bloem JL, Obermann WR, et al. Normal shoulder MR imaging. Radiology 1986;159:741–745.
3. Wilson AJ, Totty WG, Murphey WA, Hardy DC. Shoulder joint: arthrographic CT and long-term follow-up, with surgical correlation. Radiology 1989;173:329–333.
4. Deutsch AL, Resnick D, Mink JH, et al. Computed and conventional arthrotomography of the glenohumeral joint: normal anatomy and clinical experience. Radiology 1984;153:603–609.
5. Zlatkin MB, Bjorkengren AG, Gylys-Morin V, Resnick D, Sartoris DJ. Cross-sectional imaging of the capsular mechanism of the glenohumeral joint. AJR 1988;150:151–158.
6. Resnik CS, Deutsch AL, Resnick D, et al. Arthrotomography of the shoulder. Radiographics 1984;4(6):963–966.
7. Resnick D, Goergen TG, Niwayama G. Physical injury. In: Resnick D, Niwayama G, eds. Diagnosis of bone and joint disorders. Philadelphia: W. B. Saunders, 1988:2824–2844.
8. Hovelius L, Eriksson GK, Fredin H, et al. Recurrences after initial dislocation of the shoulder. Results of a prospective study of treatment. J Bone Joint Surg (Am) 1983;65A:343–349.
9. Rowe RC. Prognosis in dislocation of the shoulder. J Bone Joint Surg (Am) 1956;38A:957.
10. Rowe CR, Sakellarides HT. Factors related to the recurrence of anterior dislocation of the shoulder. Clinical Orthoped 1961;20:40–48.
11. Watson-Jones R. Recurrent dislocation of the shoulder [editorial]. J Bone Joint Surg (Br) 1948;30-B:6.
12. Kessel L. Anterior dislocation of the shoulder. In: Kessel L, ed. Clinical disorders of the shoulder. Edinburgh: Churchill Livingstone, 1982:137–149.
13. Bankart ASB. Recurrent or habitual dislocation of the shoulder joint. Br Med J 1923;2:1132–1133.
14. Bankart ASB. The pathology and treatment of recurrent dislocation of the shoulder joint. Br J Surg 1938;26:23–29.
15. Townley CO. The capsular mechanism in recurrent dislocation of the shoulder. J Bone Joint Surg (Am) 1950;32-A:370–380.
16. Mosley HF, Overgaard B. The anterior capsular mechanism in recurrent anterior dislocation of the shoulder. J Bone Joint Surg (Br) 1962;44:913–928.
17. Kuriyama S, Fujimaki E, Katagiri T, Uemara T. Anterior dislocation of the shoulder joint sustained through skiing. Am J Sports Med 1984;12:339–346.
18. Beck E. Die Arthrographie zur erfassung der Prognose der erstmal Shulterluxation. Unfallchirurgie 1988;14:22–25.
19. Baker CL, Uribe JW, Whitman C. Arthroscopic evaluation of the acute initial shoulder dislocations. Am J Sports Med 1990;18:25–28.
20. Calandra JJ, Baker CL, Uribe J. The incidence of Hill-Sachs lesions in initial anterior shoulder dislocations. Arthroscopy 1989;5:254–257.
21. DePalma AF, Cooke AJ, Prabhakar M. The role of the subscapularis in recurrent anterior dislocations of the shoulder. Clin Orthoped 1967;54:35–49.
22. Magnusson PB. Treatment of recurrent dislocation of the shoulder. Surg Clin N Am 1945;25:14–20.
23. Kieft GJ, Bloem JL, Rozing PM, Obermann WR. MR imaging of recurrent anterior dislocation of the shoulder: comparison with CT arthrography. AJR 1988;150:1083–1087.

24. McNiesh LM, Callaghan JJ. CT arthrography of the shoulder: variations of the glenoid labrum. AJR 1987;149:963–966.
25. Neumann CH, Jahnke AH, Peterson SA, Morgan FW. MR imaging of the glenohumeral joint: vagaries of the normal labral-capsular anatomy. Radiology 1990;177(p):104.
26. Ribbons WJ, Mitchell R, Taylor GJ. Computerized arthrotomography of primary anterior dislocation of the shoulder. J Bone Joint Surg (Br) 1990;72-B:181–185.
27. Shuman WP, Kilcoyne RF, Matsen FA, Rogers JV, Mack LA. Double-contrast computed tomography of the glenoid labrum. AJR 1983;141:581–584.
28. Eve FS. A case of sub-coracoid dislocation of the humerus, with formation of an indentation on the posterior surface of the head, the joint being unopened. Proc Roy Soc Med 1880;8:511.

29. Rozing PM, de Bakker HM, Obermann WR. Radiographic views in recurrent anterior shoulder dislocation. Acta Orthoped Scand 1986;57:328–330.
30. Debevoise NT, Hyatt GW, Townsend GB. Humeral torsion in recurrent shoulder dislocations. Clin Orthoped Rel Res 1971;76:87–93.
31. Hippocrates. The genuine works of Hippocrates, vol. 2 [trans. Adams F]. London: Sydenham Society, 1849:553–654.
32. Rowe CR, Yee L. Recurrent posterior dislocation of the shoulder. J Bone Joint Surg (Am) 1944;26A:582.
33. Singson RD, Feldman F, Bibliani LU, Rosenberg ZS. Recurrent shoulder dislocation after surgical repair: double contrast CT arthrography. Radiology 1987;164:425–428.

CHAPTER 22

Computed Tomography of Shoulder Trauma

Ray F. Kilcoyne

Introduction

As the most movable joint in the body, the shoulder is subjected to many different injuries and fractures. The ability to image the injured shoulder is limited by the anatomy of the region, which prevents obtaining a conventional lateral radiograph without significant superimposition of other structures, and by the fact that the injured shoulder may be too painful to move for specialized radiographic projections.

For the above reasons, computed tomography (CT) is ideally suited for examining the injured shoulder. The patient may lie on his or her back and not have to move the injured limb. The transverse plane of CT imaging is very useful for seeing the alignment of the glenohumeral joint and supplements the information that may be obtained from an axillary radiograph.

Technique of the Computed Tomography Examination

Some earlier model scanners may have difficulty in imaging the shoulder region of large patients because of "streaking" caused by regions of the body outside the computed scan path. The use of "bolus bags" (intravenous fluid bottles laid across the anterior chest) may help to even out the density across the scanning area.

For acute fractures the image should be centered over the shoulder in question. In these cases, it is not usually necessary to obtain comparison images of the uninjured shoulder. Comparison with the opposite side is useful in chronic conditions: the angulations of the glenohumeral

joints can be compared, as can the contour of the injured glenoid and humeral head with the normal side. In principle, it is better to center on the shoulder of interest. The alternative process of scanning both shoulders and magnifying the area of interest retrospectively is usually not as effective, since there will be a loss in spatial resolution.

The anteroposterior (AP) scanning radiograph is used to plan the level of axial slices. If the proximal humerus or the glenoid fossa is the area of interest, the scans need only be obtained from the superior margin of the humeral head down to the inferior lip of the glenoid. Restricting the slices to only the area of interest will be a great time saver, especially helpful in the severely injured patient. The slice thickness should be set at 2 to 4 mm in order to obtain the maximum amount of information. Because thin slices mean less radiation per slice (photon depletion) the scans should be done at a high milliampere setting or with longer scan slice time (if the patient can hold still).

By using thin slices with the best spatial resolution it is possible to go back later and obtain reformatted images in the coronal or sagittal planes if this should become necessary. In most cases, axial CT images will suffice when interpreted with the AP radiograph.

Fractures of the Proximal Humerus

Orthopedic surgeons depend on a classification system for proximal humeral fractures described by Codman in 1934 (1). This work was amplified by Neer in 1970 (2). According to Neer, treatment of proximal shoulder fractures depends on the location of the fractures and whether or not displacement of the fracture fragments has occurred. If there is significant displacement of the fracture fragments, it is unlikely that closed reduction will be successful in producing a clinically acceptable result. Displaced fragments have important muscular attachments, such as the rotator cuff. Abduction or internal-external rotation will be severely limited if the displaced pieces are not reduced. In addition, if there is so much comminution that the humeral head is stripped of its blood supply, osteonecrosis is a likely sequela. Displaced fractures are treated surgically for reduction of the fragments; severely comminuted fractures or fracture-dislocations have a humeral head prosthesis inserted. Before a fracture of the surgical neck can be considered as a "two-part" fracture the two parts

---

Abbreviations (also see glossary): AC, acromioclavicular; CC, coracoclavicular.

A   **B**   **C**

Figure 22.1. Cross-sectional drawings of the humeral head, showing the location of the greater and lesser tuberosities adjacent to the bicipital groove. This plane of section is similar to what is seen with axial CT. **(A)** Intact tuberosities. **(B)** Two-part fracture of the greater tuberosity with posterior displacement of the fragment due to the pull of the infraspinatus tendon. **(C)** Two-part fracture of the lesser tuberosity with medial displacement of the fragment due to the pull of the subscapularis tendon. (Reprinted with permission of the Williams & Wilkins Co., from Kilcoyne RF, Shuman WP, Matsen III FA, Morris M, Rockwood CA. AJR 154:1029–1033, 1990.)

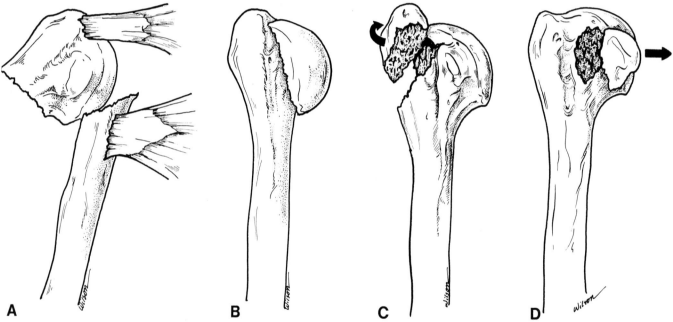

A   **B**   **C**   **D**

Figure 22.2. Anatomical drawings of two-part fractures. **(A)** Surgical neck (subtuberous): the supraspinatus tendon is attached to the greater tuberosity. The pectoralis major pulls the humeral shaft medially. **(B)** Anatomical neck (articular segment): the deltoid pulls the humeral shaft superiorly. **(C)** Greater tuberosity: the external rotators (infraspinatus and teres minor) pull the fractured greater tuberosity posteriorly. **(D)** Lesser tuberosity: the subscapularis pulls the fractured tuberosity medially. (Reprinted with permission of the Williams & Wilkins Co., from Kilcoyne RF, Shuman WP, Matsen III FA, Morris M, Rockwood CA. AJR 1990;154:1029–1033.)

must be displaced by a centimeter or more, or angulated 45° or more. Sometimes it is difficult to see the exact position of the fracture fragments, or the patient may be difficult to position. Kristiansen et al. found wide interobserver variation in classifying proximal humeral fractures using only plain radiographs (3). On the other hand, CT is very helpful in classifying proximal humeral fractures. The more serious the injury, the more likely CT is to be useful (Fig. 22.1) (4, 5).

CT is especially helpful in the following cases: displaced fractures of the greater and lesser tuberosities, impaction fractures of the humeral head, posterior dislocation of the humeral head, head-splitting fractures, fracture of the anterior or posterior glenoid rim, loose bodies in the shoulder joint, osteonecrosis of the humeral head, and displaced malunion of healed fractures.

The Neer Classification of Proximal Humeral Fractures

The term *part* refers to a displaced fracture fragment. One part is always the humeral shaft, so the simplest displaced fracture is called a two-part fracture. The figures depict two-part fractures (Fig. 22.2), three part fractures (Fig. 22.3), and four-part fractures and head-splitting fractures (Fig. 22.4).

Radiographic examples of various displaced proximal humeral fractures are shown in Figures 22.5 to 22.11.

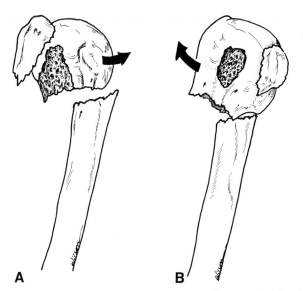

A   **B**

Figure 22.3. Anatomical drawings of three-part fractures. **(A)** Surgical neck and greater tuberosity fractures: the intact subscapularis attachment pulls the humeral head into internal rotation. **(B)** Surgical neck and lesser tuberosity fractures: the intact infraspinatus and teres minor pull the humeral head into external rotation. (Reprinted with permission of the Williams & Wilkins Co., from Kilcoyne RF, Shuman WP, Matsen III FA, Morris M, Rockwood CA. AJR 1990;154:1029–1033.)

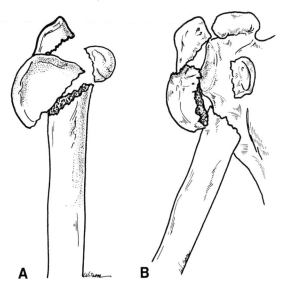

A   **B**

Figure 22.4. Anatomical drawings of four-part fractures and fracture-dislocations. **(A)** Surgical neck and greater and lesser tuberosities fractures: the humeral head is free of its muscular (and vascular) attachments and tends to rotate superiorly. **(B)** Head-splitting fracture: one piece of the humeral head tends to move superior to the glenoid and another inferior. (Reprinted with permission of the Williams & Wilkins Co., from Kilcoyne RF, Shuman WP, Matsen III FA, Morris M, Rockwood CA. AJR 1990;154:1029–1033.)

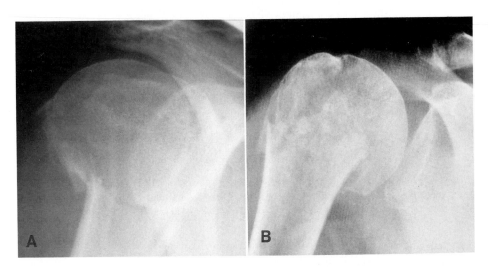

A   **B**

Figure 22.5. **(A)** Anteroposterior radiograph shows two-part fracture of the surgical neck. **(B)** Anteroposterior radiograph shows two-part fracture of the anatomical neck. (The fracture of the greater tuberosity is not displaced.) CT is usually not needed with these two-part fractures. (Reprinted with permission of the Williams & Wilkins Co., from Kilcoyne RF, Shuman WP, Matsen III FA, Morris M, Rockwood CA. AJR 1990;154:1029–1033.)

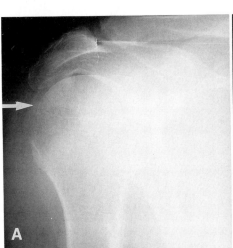
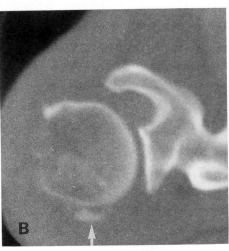

A   **B**

Figure 22.6. **(A)** Anteroposterior radiograph shows a two-part fracture of the greater tuberosity *(arrow).* **(B)** C tomography shows how far posteriorly the greater tuberosity fragment is pulled *(arrow),* and the fact that failure to reduce the displacement will cause severe limitation of external rotation of the humerus. (Reprinted with permission of the Williams & Wilkins Co., from Kilcoyne RF, Shuman WP, Matsen III FA, Morris M, Rockwood CA. AJR 1990;154:1029–1033.)

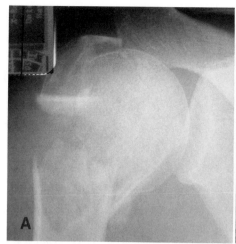

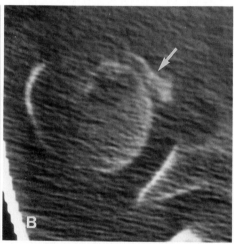

Figure 22.7. **(A)** Three-part surgical neck and greater tuberosity fractures on AP radiograph. **(B)** The greater tuberosity displacement is seen well on the CT image *(arrow)*. (Reprinted with permission of the Williams & Wilkins Co., from Kilcoyne RF, Shuman WP, Matsen III FA, Morris M, Rockwood CA. AJR 1990;154:1029–1033.)

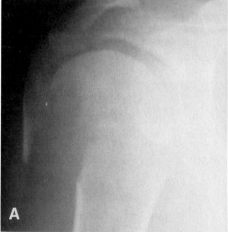

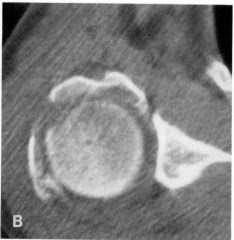

Figure 22.8. **(A)** Three-part surgical neck and lesser tuberosity fractures on AP radiograph. **(B)** The lesser tuberosity displacement is seen well on the CT image *(arrow)*. (The greater tuberosity is fractured but not displaced.) (Reprinted with permission of the Williams & Wilkins Co., from Kilcoyne RF, Shuman WP, Matsen III FA, Morris M, Rockwood CA. AJR 1990;154:1029–1033.)

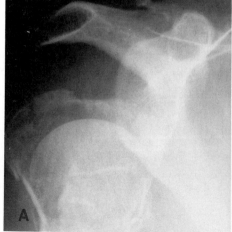

Figure 22.9. **(A)** Four-part surgical neck greater and lesser tuberosity fracture. **(B)** The displaced greater and lesser tuberosities are shown by CT. (Reprinted with permission of the Williams & Wilkins Co., from Kilcoyne RF, Shuman WP, Matsen III FA, Morris M, Rockwood CA. AJR 1990;154:1029–1033.)

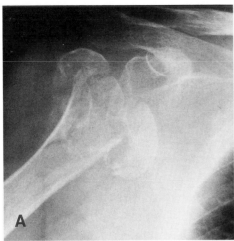

Figure 22.10. **(A)** Head-splitting fracture with dislocation of the articular fragments. **(B)** Computed tomography shows the portion of the head that is dislocated anteriorly. In addition displacement of the lesser tuberosity fragment is seen. (Reprinted with permission of the Williams & Wilkins Co., from Kilcoyne RF, Shuman WP, Matsen III FA, Morris M, Rockwood CA. AJR 1990;154:1029–1033.)

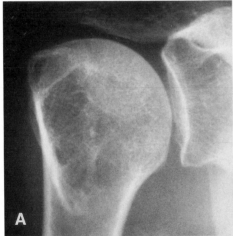

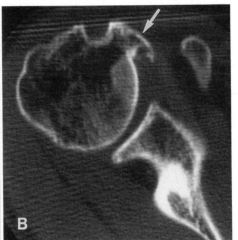

Figure 22.11. **(A)** The proximal humeral fracture has healed with only slight deformity. **(B)** Computed tomography shows that the lesser tuberosity is displaced, severely limiting internal rotation *(arrow)*. (Reprinted with permission of the Williams & Wilkins Co., from Kilcoyne RF, Shuman WP, Matsen III FA, Morris M, Rockwood CA. AJR 1990;154:1029–1033.)

Fractures of the Scapula

Fractures of the scapula can be divided into the following types (6): neck, acromial process, spinous process or base of the acromion, glenoid fracture (intra- or extraarticular) (Figs. 22.12 to 22.14), coracoid process (Figs. 22.15 to 22.16), and body (sagittal or transverse).

Most fractures of the *body* of the scapula do not require any specific orthopedic treatment. The scapula is encased in muscles both anteriorly and posteriorly, which allows very little movement of the fractured fragments. However, fractures through the glenoid articular surface associated with dislocation may require operative intervention. In those cases CT may be very useful to look for incongruity of the glenoid fossa or for the presence of loose fragments within the shoulder joint. In cases of displaced fractures of the coracoid or the acromion process, CT may help to show the degree of displacement.

Fractures of the Clavicle

Fractures of the clavicle usually involve the middle third. These common injuries do not require CT for diagnosis. The plain radiographs are quite adequate in estimating the degree of displacement or angulation present. The less common injuries that involve the medial and lateral ends of the clavicle may be observed with benefit by CT, which can show the amount of displacement and the degree of articular involvement. In medial injuries disruption of the sternoclavicular joint can be very difficult to detect on conventional radiographs. Not only is a *lateral* view difficult to obtain, the usual AP or oblique AP radiograph is frequently difficult to interpret because of overlap with the spine, mediastinum, etc. It is difficult to distinguish between sternoclavicular dislocation and fracture into the joint. Computed tomography is needed in almost all cases and is better than a conventional tomography at showing

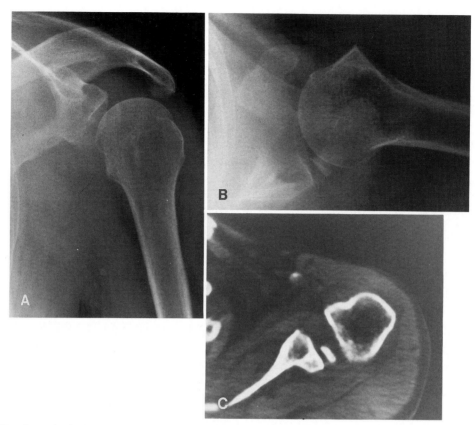

Figure 22.12. **(A)** AP radiograph of a fracture of the scapula involving the glenoid articular surface. **(B)** Axillary view shows an osteochondral fragment in the shoulder joint. **(C)** CT shows that the fracture fragment came from the anterior rim of the glenoid and the fracture fragment is displaced posteriorly.

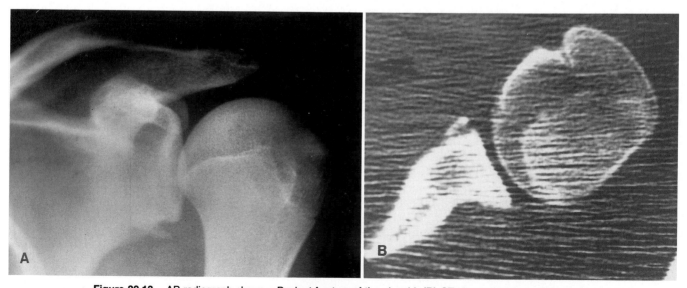

Figure 22.13. AP radiograph shows a Bankart fracture of the glenoid. **(B)** CT shows that the anterior lip is fractured but not displaced.

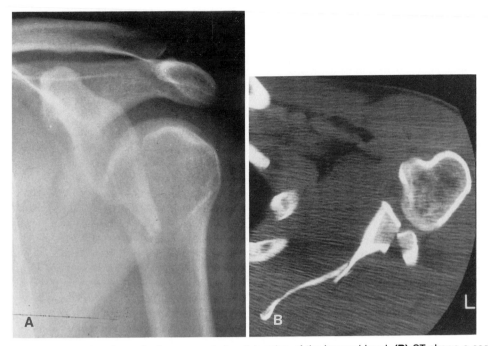

Figure 22.14. **(A)** AP radiograph shows a posterior dislocation of the humeral head. **(B)** CT shows a sagittal fracture of the lateral scapula that involves the posterior glenoid rim.

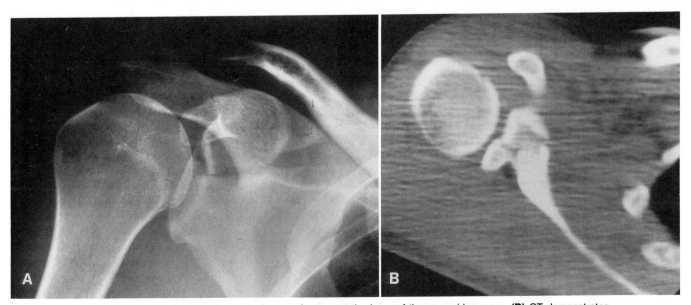

Figure 22.15. **(A)** AP radiograph shows a fracture at the base of the coracoid process. **(B)** CT demonstrates the intraarticular extension of the fracture.

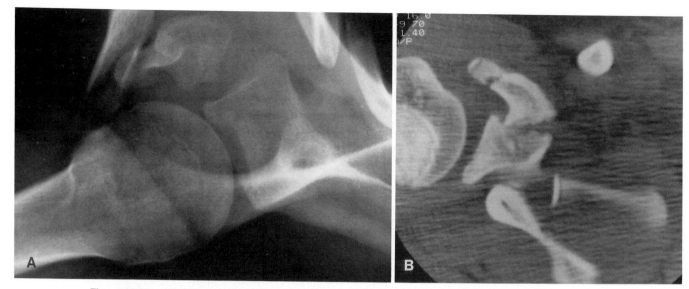

Figure 22.16. **(A)** Axillary radiograph shows residuals of a 1-year-old coracoid fracture. The patient continues to have pain. **(B)** Computed tomography taken 6 months earlier shows that the coracoid is fractured at two levels.

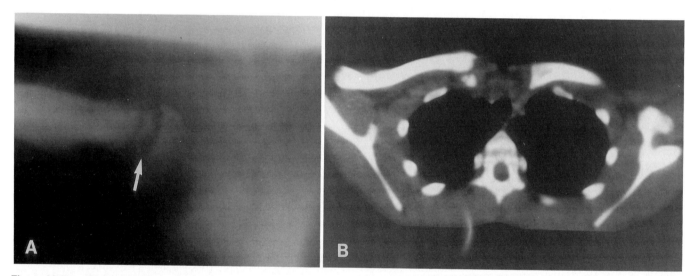

Figure 22.17. **(A)** Computed tomography shows a fracture of the medial end of the right clavicle *(arrow)*. **(B)** The medial end of the clavi- cular shaft is displaced anteriorly, with a small fragment of the medial clavicle remaining in its normal position.

displacement (Fig. 22.17). This information is very valuable if there is suspicion that there may have been compression of the subclavian artery. The CT examination showing posterior displacement of the clavicle may be used as evidence that a subclavian arteriogram needs to be performed.

Disruption of the acromioclavicular joint with or without fracture of the distal clavicle may require CT for adequate diagnosis. Injury to the acromioclavicular joint is classified according to the severity of the injury. The structures at risk are the acromioclavicular and the coracoclavicular ligaments. There are six injury patterns (7):

1. Sprain of the acromioclavicular (AC) ligaments
2. Complete disruption of the ligaments of the acromioclavicular joint

3. AC separation plus disruption of the coracoclavicular (CC) ligaments (or coracoid fracture)
4. Type 3 plus superior and posterior displacement of the clavicle
5. Type 3 plus marked superior displacement of the clavicle
6. AC ligament disruption with clavicle displaced inferiorly below the acromion or the coracoid. The CC ligaments may be intact

Displacement of the lateral clavicle in type 4, 5, or 6 injuries may be demonstrated better with CT (Fig. 22.18).

Fractures of the outer clavicle similarly are divided into three categories (8):

1. Nondisplaced or minimally displaced fracture through the intact coracoclavicular ligaments

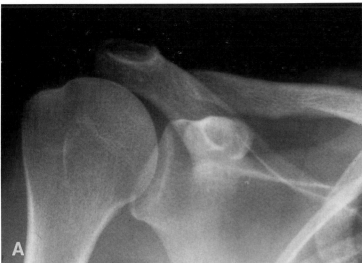
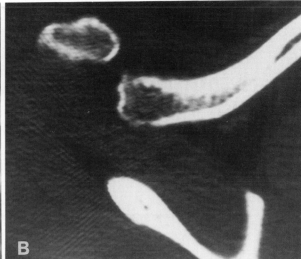

Figure 22.18. **(A)** AP radiograph shows a separation of the right acromioclavicular joint. The distal clavicle is displaced inferiorly. **(B)** Computed tomography shows that this is a type 4 injury with posterior displacement of the distal clavicle through the trapezius muscle.

2. Fracture medial to the coracoclavicular ligaments or a fracture in the interligamentous area with detachment of the coracoclavicular ligaments from the proximal fragment
3. Fracture of the articular surface of the distal clavicle

Conclusion

Computed tomography of the injured shoulder is useful whenever the plain X-rays are inadequate for whatever reason. Displaced fragments, loose bodies in the joint, or dislocation-subluxations may be seen more easily. Computed tomography has replaced routine tomography in most shoulder injuries, and it aids the surgeon in deciding whether an operation needs to be performed or not.

Acknowledgment

Acknowledgment is given to Dinah Wilson for the anatomical drawings in the figures.

References

1. Codman EA. The shoulder. Rupture of the supraspinatus tendon and other lesions in or about the subacromial bursa. Brooklyn: G. Miller, 1934:262–287.
2. Neer CS. Displaced proximal humeral fractures. Part I. Classification and evaluation. J Bone Joint Surg 1970;52-A:1077–1089.
3. Kristiansen B, Andersen LS, Olsen CA, Vasmarken JE. The Neer classification of fractures of the proximal humerus: an assessment of interobserver variation. Skel Radiol 1988;17:420–422.
4. Kilcoyne RF, Shuman WP, Matsen FA III, Morris M, Rockwood CA. The Neer classification of displaced proximal humeral fractures—the spectrum of findings on plain radiographs and CT. AJR 1990;154:1029–1033.
5. Castagno AA, Shuman WP, Kilcoyne RF, Haynor DR, Morris ME, Matsen FA. Complex fractures of the proximal humerus: role of CT in treatment. Radiology 1987;165:759–762.
6. Rockwood CA, Green DP. Fractures in adults. Philadelphia: Lippincott, 1984:713–715.
7. Williams GR, Nguyen VD, Rockwood CA Jr. Classification and radiographic analysis of acromioclavicular dislocations. Appl Radiol 1989;18:29–33.
8. Rockwood CA, Green DP. Fractures in adults. Philadelphia: Lippincott, 1984:707–708.

CHAPTER 23

Magnetic Resonance Imaging of the Elbow

Charles P. Ho

Introduction
Technique

Normal Anatomy
Pathology

Introduction

Imaging of the elbow may draw on multiple techniques in the radiology armamentarium. Conventional radiography remains the basic screening modality, while such imaging techniques as conventional tomography, computed tomography, radionuclide bone scans, arthrography, and computed arthrotomography all may have great utility in specific clinical settings. While magnetic resonance (MR) imaging has not yet been proven the primary modality of choice in most clinical settings, it has enormous potential and already can surely be a useful adjunct, given its ability to image in multiple planes and to delineate soft-tissue structures with relatively high contrast. Examples of MR imaging demonstrations of impressive detail of both normal anatomy and various specific pathologic processes involving the elbow have been described (1–4).

Technique

Imaging Parameters

As with other peripheral small joints and other body parts, optimal imaging of the elbow is realized with a dedicated surface or extremity coil that allows small fields of view (FOV) for high resolution while maintaining good signal-to-noise ratio. Combining a 10- to 12-cm FOV with 256 × 256 matrix size and thin 5-mm slice thickness yields small pixel size and good resolution. One or two signal averages (number of excitations, NEX) may be needed for acceptable signal-to-noise ratio, depending on the type of pulse sequence and the pixel size, with more averages needed for smaller pixel size.

T1-weighted images, achieved with a relatively short repetition time (TR) of about 600 ms and a short echo time (TE) of about 20 ms, generally show excellent anatomic detail of joints, soft tissues, and bone marrow, with rea-

sonably short acquisition times because of the short TR. T2-weighted images require long TR and long TE times, about 2000 ms and 60 to 80 ms, respectively. The benefit of T2-weighted images lies in the T2 prolongation of most pathologic processes, which therefore appear relatively brighter on T2-weighted compared to T1-weighted images and stand out from normal tissues. This unfortunately is balanced by the longer scan times required by the long TR, and the often decreased signal-to-noise ratio due to long TE.

Gradient-echo techniques can achieve an effective T2 or T2*-weighting with short TR and TE times of about 400 ms and 20 to 30 ms, respectively, and a low flip angle of 25°, giving the benefit of T2* weighting with short scan times. Also, flow effects such as in vascular structures are highlighted with increased signal. In general, however, gradient-echo images have decreased soft-tissue contrast and reduced signal-to-noise ratio (5).

Finally, fat-suppression techniques such as short τ inversion recovery (STIR) or HYBRID (6) can increase tremendously the sensitivity for subtle areas of signal change that may signify lesions by removing the often overwhelming background fat signal. This has great promise for bone marrow imaging, for example. However, this benefit is weighed against the decreased signal to noise ratio and the greater complexity of the pulse sequences and software needed.

In most instances, then, a satisfactory screening examination can be achieved with a T1-weighted sequence for overall anatomic detail combined with a T2-weighted sequence, generally in an orthogonal plane, for increased sensitivity for detection of pathologic processes. Additional or oblique planes and gradient-echo or specialized pulse sequences such as STIR or HYBRID are then used as needed to supplement the examination.

Imaging Planes

The sagittal (Fig. 23.1) and coronal imaging planes generally are very useful for analyzing the elbow joint, including capsule, articulations, and articular hyaline cartilage. Marrow characteristics and the overall extent of marrow and soft-tissue processes similarly are well demonstrated in these planes. Axial imaging (Figs. 23.2 and 23.3), on the other hand, identifies specific structures and demonstrates anatomic relationships, including involvement of particu-

Abbreviations (see also glossary): FOV, field of view; NEX, number of excitations; STIR, short τ inversion recovery.

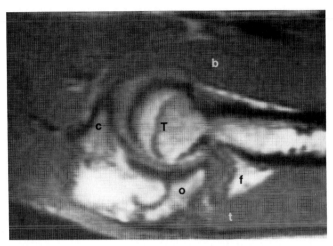

Figure 23.1. Normal parasagittal anatomy at coronoid/trochlea articulation on T1-weighted image (TR 800 ms, TE 20 ms): coronoid *(c)*, trochlea *(T)*, olecranon *(o)*, posterior fat pad *(f)*, brachialis *(b)*, triceps *(t)*.

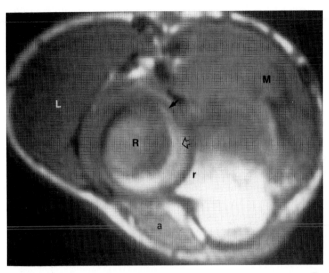

Figure 23.3. Normal axial anatomy at proximal radioulnar joint on T1-weighted image (TR 800 ms, TE 20 ms): radial head *(R)*, radial notch *(r)* of ulna, lateral muscle group *(L)*, medial muscle group *(M)*, anconeus *(a)*, annular ligament *(black arrow)*. Articular hyaline cartilage shows intermediate signal intensity *(open arrow)* on T1 weighting.

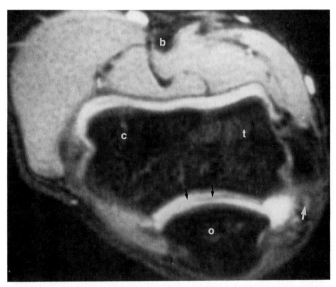

Figure 23.2. Normal axial anatomy of elbow on T2*-weighted gradient echo image (TR 500 ms, TE 17 ms, flip angle 25°): biceps tendon *(b)*, capitellum *(c)*, olecranon *(o)*, trochlea *(t)*, ulnar nerve *(white arrow)*. Articular hyaline cartilage demonstrates bright signal intensity *(small black arrows)* on T2* weighting.

lar muscle groups, crossing of fascial planes, or involvement of neurovascular structures, particularly well.

The choice of imaging planes will vary with the suspected location and orientation of the pathologic process. Sequences in at least two orthogonal planes are helpful to convey a sense of the three-dimensional nature and extent of the lesion. Thus, generally either a sagittal or coronal T1-weighted sequence is chosen, depending on the plane that better separates the lesion from the underlying bone or other adjacent structures of interest, to demonstrate the overall size of the lesion. A dual-echo intermediate and T2-weighted axial sequence then are used to increase sensitivity for detection of lesions through the typical

brightening of most pathology on T2-weighted images, to assess the specific anatomic margins and to complete the evaluation of the signal characteristics for possible clues to tissue characterization of the pathologic process. Additional planes or specialized sequences may then be helpful to complement the examination.

Normal Anatomy

The elbow joint proper and the superior radial ulnar joint are bounded by a common capsule and may be considered together, with a total of three functioning articulations. The trochlea and capitellum of the distal humerus articulate with the ulna and radial head, respectively, acting as a hinge joint for flexion/extension, while the radial head further articulates with the radial notch of the proximal ulna, allowing rotation in supination/pronation of the forearm.

Thickenings of the common capsule medially and laterally constitute the supporting collateral ligaments. The ulnar collateral ligament extends from the medial epicondyle to the trochlear notch of the proximal ulna, while the radial collateral ligament runs from the lateral epicondyle to the annular ligament surrounding the radial head.

Further support of the elbow is provided by surrounding muscles, conveniently and functionally considered as anterior, posterior, medial, and lateral muscle groups (1–3). The anterior group of elbow flexors includes the biceps brachii, extending from the shoulder to insert distally at the radial tuberosity with, additionally, a superficial biceps aponeurosis fanning medially over the medial muscle group surface, and the brachialis, coursing from distal humerus to the ulnar tuberosity.

In the posterior group of elbow extensors are the triceps and the small anconeus muscle. The triceps attaches between shoulder and the posterior superior aspect of the olecranon, while the anconeus extends from the posterior lateral epicondyle to the posterior lateral portion of the olecranon.

Medially grouped are the forearm pronator and the wrist and finger flexors. The pronator teres, which is the most medial and anterior/superficial of this group, originates in two parts, from the distal humerus above the medial epicondyle and also from the proximal ulna, and then sweeps obliquely laterally to the forearm. A common flexor tendon from the medial epicondyle gives rise, from anterior to posterior, to the flexor carpi radialis, palmaris longus, flexor digitorum superficialis, flexor carpi ulnaris, and flexor digitorum profundus. Arising from the common flexor tendon, these muscles are difficult to distinguish as separate muscles at the elbow level.

The lateral group is composed of brachioradialis, supinator, and wrist/finger extensors. The brachioradialis extends from the lateral aspect of the distal humerus to the lateral portion of the radial styloid at the wrist. Originating just distal to the brachioradialis, also from the distal humerus, is the extensor carpi radialis longus. A common extensor tendon from the lateral epicondyle gives rise to the extensor carpi radialis brevis, extensor digitorum, extensor carpi ulnaris, and extensor digiti minimi, which are difficult to differentiate as distinct muscles at the level of the elbow. Finally, the deep muscle supinator crosses from the lateral epicondyle and posterolateral proximal ulna to attach at the lateral proximal radius.

Of the major vascular structures, the brachial artery travels medial to the biceps along the anterior superficial surface of the brachialis, and divides into radial and ulnar arteries, usually just distal to the elbow joint. The major veins are anterior and superficial to the muscles, the cephalic vein laterally and the basilic medially located.

The ulnar nerve extends just posterior to the medial epicondyle and is quite superficial; it is easily located on axial views at this level (Fig. 23.2) and can be followed distally on sequential axial views. The median nerve parallels medially the brachial artery along the anterior surface of the brachialis. The radial nerve courses laterally between the brachialis and brachioradialis.

Pathology

In many instances of patients with elbow symptoms, the clinical presentation, physical examination, and conventional radiographs (if needed) are sufficient to establish the diagnosis, and no further imaging is required. In other cases, however, such as when the clinical presentation is not characteristic, physical examination is difficult, or radiographs are negative or nonspecific, additional diagnostic imaging may be of great value. Magnetic resonance imaging should be considered in particular when articular processes, marrow abnormalities, or soft-tissue (including ligament, tendon, muscle, and nerve) lesions are suspected.

Joint Effusions

In suspected elbow pathology, the joint effusion has been an important finding on radiographs to support the presence of articular involvement, with displacement of the

posterior and anterior fat pads in the olecranon and coronoid fossae, respectively, of the distal humerus by joint fluid indicating indirectly on the lateral radiograph a large effusion. The radiographic fat pad sign may not be sensitive for smaller effusions, however.

Magnetic resonance imaging is able to image directly even small amounts of joint fluid in the elbow, as in other joints. Effusions image with low signal intensity on T1 weighting, and brighten on intermediate through T2 weighting. When seen in the sagittal plane, fluid can be seen to displace the fat pads in the MR imaging equivalent of the radiographic fat pad sign.

While quite sensitive for fluid, however, MR imaging may be relatively nonspecific concerning the nature of the effusion, as with many other MR imaging findings. Traumatic effusions (Fig. 23.4) may appear identical to infected fluid, and similar to other inflammatory and noninflammatory effusions. Of course, hemarthroses or lipohemarthroses might be more specifically imaged, with bright signal on T1-weighted images from hemorrhagic or fat components and fluid-fluid levels from layering of the different components.

Intraarticular Bodies, Osteochondral Lesions, and Other Trauma

Intraarticular fragments may be readily identified by MR imaging, particularly when significant joint fluid is also present to outline the margins of the fragments (Fig. 23.5).

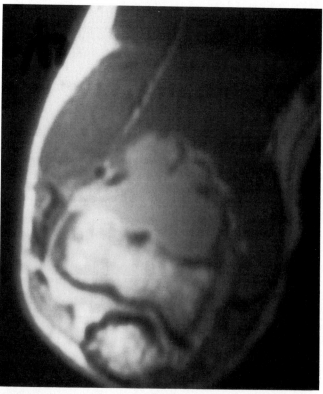

Figure 23.4. Traumatic elbow effusion, images with intermediate signal intensity on proton density axial image (TR 2000 ms, TE 20 ms), extending from olecranon posteriorly to a distended capsule anteriorly.

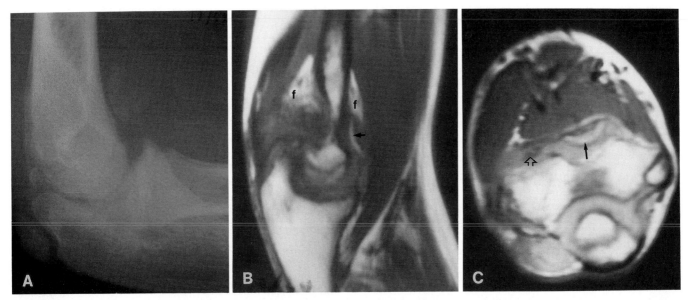

Figure 23.5. Intraarticular bodies: **(A)** Lateral radiograph shows multiple osseous fragments, the largest located anterior to the distal humerus. **(B)** Parasagittal T1-weighted image (TR 800 ms, TE 20 ms) also shows the large osseous fragment *(black arrow)* with fatty marrow signal; joint effusion is displacing the anterior and posterior fat pads *(f)*. **(C)** Transaxial intermediate image (TR 2000 ms, TE 20 ms) demonstrates the large fragment *(black arrow)*, and also shows a smaller, purely calcific or cortical bone fragment *(open arrow)* with very low signal outlined by intermediate signal joint effusion.

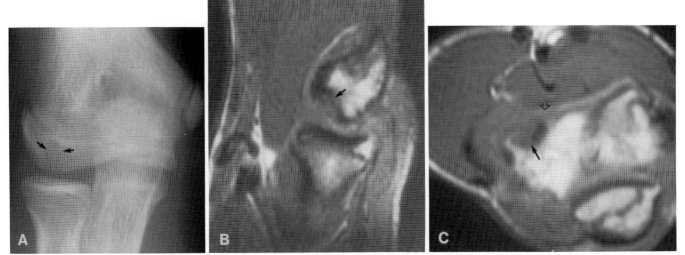

Figure 23.6. Osteochondritis dissecans (Panner's disease) of elbow. **(A)** Oblique frontal radiograph shows vague lucency of capitellum *(black arrows)*. **(B)** Parasagittal image and **(C)** axial T1-weighted image (TR 800 ms, TE 20 ms) of capitellum/radial head articulation show focal concave defect of subchondral bone of capitellum with sclerotic margins *(black arrow)*. Axial image also shows intermediate signal articular hyaline cartilage *(open arrow)* disrupted over the focal bone defect.

In the absence of joint fluid, however, MR imaging may not be sensitive for small fragments, and even larger bodies could be missed. Small calcific or osseous fragments will show very low signal and be difficult to detect. Purely cartilaginous fragments demonstrate intermediate signal similar to intact articular cartilage and other soft tissue about the joint and may be difficult to distinguish. Conceivably, even large osseous bodies, with fatty marrow signal similar to the fat pads and other periarticular fat planes, may be rather subtle and a challenge to identify, while being readily apparent on radiographs. Correlation with radiographs is very helpful. Air-contrast computed arthrotomography is likely more sensitive than MR imaging for detection of intraarticular bodies, particularly for cartilage fragments (7).

Similarly, the articular cartilage can be assessed directly by MR imaging for signal inhomogeneity, thinning, surface irregularity, or focal defects that have been staged for chondromalacia in other joints (8), particularly when joint fluid is present to outline the cartilage surface. When little or no fluid is found, however, MR imaging may not be sensitive for surface irregularity or defects. Computed arthrotomography is likely more sensitive for examination of the cartilage surface. Magnetic resonance arthrography may be even more efficacious, enabling assessment of both the cartilage substance and surface (9).

Osteochondral injuries may be suspected on conventional radiographs by lucency abutting the subchondral bone. However, the lucency may be subtle, particularly when not tangential to the X-ray beam angle in positioning for the radiographs. Magnetic resonance imaging is likely more sensitive for and facilitates more complete examination of osteochondritis dissecans in the elbow (Fig. 23.6), as in other joints such as the knee, where MR imaging has been reported to allow assessment of stability of the fragment and the integrity of the overlying hyaline cartilage (10).

In trauma cases, as always, conventional radiographs are the appropriate screening modality for possible fractures. Magnetic resonance imaging may be a useful adjunct in selected cases when radiographically occult fractures or bone contusions/edema are suspected. By increasing T1 and T2 relaxation time, marrow edema will image as relatively low signal, in sharp contrast to the normal bright fatty marrow signal, on T1-weighted images, with brightening on T2-weighted images. Magnetic resonance imaging may be particularly helpful when cartilage or surrounding soft-tissue injury is suspected. Muscle contusions or hematomas may be identified by the brightening of edema on T2-weighted images and the high signal of subacute hemorrhage on T1-weighted images. The supporting ligaments and tendons about the elbow can be assessed for possible tears or tenosynovitis, searching for the tell-tale brightening of edema on T2-weighted images as well as evaluating the continuity of the ligament and tendon fibers. The neurovascular structures may be examined for injury as well. For example, the ulnar nerve is quite superficial, relatively exposed, and susceptible to trauma or entrapment as it courses just posterior to the medial epicondyle at the elbow. When ulnar nerve injury is suspected, it is readily identified at this level on axial images, and may be followed distally or proximally on sequential axial images.

Arthritis and Other Inflammatory Processes

Traditionally, arthritides and other inflammatory processes involving the elbow, as in other joints, have been imaged with conventional radiographs for findings such as erosions, joint space narrowing, and effusions. However, such radiographic findings may in fact reflect relatively advanced stages of these processes. Magnetic resonance imaging, with its high soft-tissue contrast, is able to demonstrate directly the articular cartilage, synovium and capsule, joint fluid, and the surrounding tendons and ligaments, as well as the subchondral and periarticular bone,

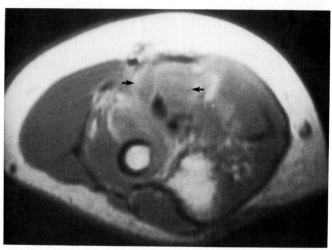

Figure 23.7. Ruptured synovial cyst in rheumatoid arthritis patient who noted growing anterior elbow mass that suddenly "disappeared." Axial proton density image (TR 2000 ms, TE 20 ms) at the proximal radioulnar joint shows lobulated anterior intermediate signal intensity cyst *(black arrows)* with neck extending posteriorly toward the joint. Poorly defined intermediate signal soft-tissue edema is seen surrounding the cyst.

Figure 23.8. Osteochondroma of elbow. **(A)** Parasagittal image and **(B)** axial T1-weighted image (TR 800 ms, TE 20 ms). Parasagittal image shows direct continuity of cortex and marrow of osteochondroma with underlying humerus, and mass effect of osteochondroma displacing the brachial artery *(black arrow)*. Axial view shows thin cartilage cap with low signal *(white arrow)* on T1 weighting.

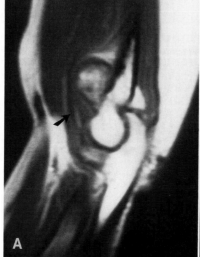

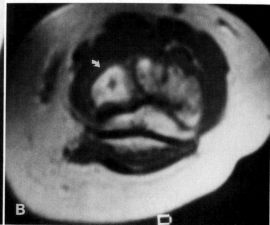

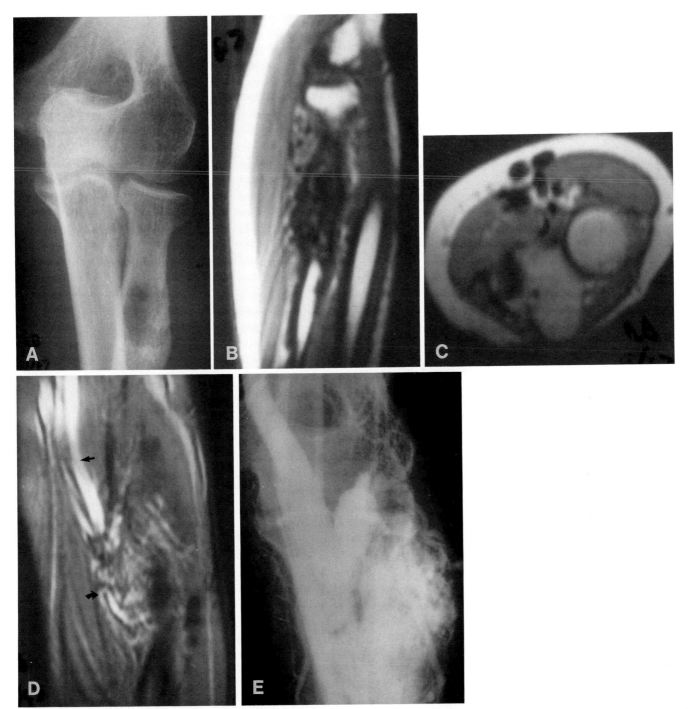

Figure 23.9. Arteriovenous malformation of elbow. **(A)** Frontal oblique radiograph shows lytic lesion of proximal radius disrupting the lateral cortex. **(B)** Parasagittal image and **(C)** axial T1-weighted image (TR 800 ms, TE 20 ms) show irregular tubular structures with signal void involving radius and surrounding soft tissues. **(D)** Coronal gradient-echo image shows angiographic effect of bright flow in large feeding and draining vessels *(black arrow)* extending to tangled mass of abnormal vessels of malformation *(curved black arrow).* **(E)** Frontal view of angiogram confirms MR imaging findings.

and may be useful and more sensitive for assessment of earlier as well as relatively advanced stages of disease, and for following response to therapeutic interventions. Cartilaginous as well as osseous erosions may be demonstrated, in addition to gross subluxations and deformity. Joint effusions and synovial cysts (Fig. 23.7) are easily identified with their pronounced brightening on T2 weighting. Surrounding soft-tissue edema, also with relatively low signal on T1-weighted images and brightening on T2-weighted images, can be demonstrated accurately. Integrity or disruption of surrounding tendons and ligaments as well as possible tenosynovitis, as in trauma patients, can be assessed by direct imaging of these structures.

In addition, the thickened, irregular, and inflamed synovium can be detected and potentially followed for activity and response to therapy. Enhancement with gadolinium intravenous contrast, in particular, as has been demonstrated in the knee (11), may also increase sensitivity and accuracy of evaluation of pannus and other synovial inflammatory processes.

Neoplasms

When neoplasms (see also Chapter 15) or other masses are suspected about the elbow, again conventional radiographs are the appropriate screening modality, as radiographic findings may be diagnostic or relatively specific. Magnetic resonance imaging should be considered as an adjunct, for assessment of the extent of soft-tissue or marrow involvement, crossing of fascial planes, possible neurovascular compromise, or possible intraarticular extension. Some degree of tissue characterization is possible, such as identification of fatty or hemorrhagic components of the neoplasm.

With this detail of soft-tissue and marrow involvement possible, MR imaging is very helpful for staging and preoperative planning as well as for selecting a site and approach for biopsy. Scanning should generally be performed prior to biopsy such that postbiopsy changes such as hemorrhage or edema will not confuse tumor margins and signal characteristics. If nonoperative therapy is used, subsequent scans may be used to follow therapeutic response.

In general, both benign/indolent and malignant/aggressive neoplasms will prolong both T1 and T2 time constants, resulting in low to intermediate signal on T1-weighted images, with brightening on T2-weighted images. Extremely fibrous lesions, however, may show persistent low signal on both T1 and T2 weighting, as will large areas of tumor matrix calcification or ossification. Gradient-echo scans tend to accentuate the apparent size of the areas of calcification due to the sensitivity of this pulse sequence to rapid changes of magnetic susceptibility, creating an edge enhancement effect. Thus, gradient-echo imaging may be useful to find smaller areas of matrix calcification, although CT remains superior for assessment of tumor bone or calcification.

Reactive marrow or soft-tissue edema surrounding the lesion is not uncommon, and also brightens on T2-weighted images. Differentiation of tumor margin from surrounding edema remains a difficulty.

In addition to its general efficacy for assessment of most neoplasms, specific features of MR imaging may be helpful and advantageous for particular lesions or lesion categories.

Cartilage tumors will have a cartilage matrix, often with imaging characteristics similar to that of the articular cartilage in the elbow. Matrix calcification is commonly seen, with imaging approach as discussed above. Magnetic resonance imaging is particularly helpful and should be considered for symptomatic osteochondromas (Fig. 23.8). The direct continuity of cortex and marrow space of the osteochondroma with that of the underlying bone is readily demonstrated with the multiplanar capability of MR imaging, while the overlying cartilage cap can be directly imaged. Symptomatic size effect and impingement on adjacent muscles, tendons, or neurovascular structures can be assessed readily. When possible malignant transformation of the osteochondroma is feared, the thickness and margins of the cartilage cap can be directly assessed, as thicknesses greater than 1 to 2 cm and surface irregularity and indistinctness with adjacent soft-tissue change may suggest malignant transformation (12). Relatively decreased signal on T1-weighted images and increased brightening on T2-weighted images also may be supportive of transformation.

Vascular malformations may involve the bone and soft

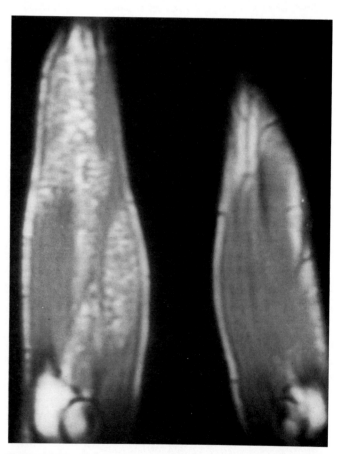

Figure 23.10. Hemangioma of elbow and forearm. Coronal oblique T1-weighted image (TR 800 ms, TE 20 ms) shows extensive hemangioma involving muscles showing mottled intermediate and high signal intensity of abnormal vessels/mass (including fat) of hemangioma.

tissue, and cross the joint (Fig. 23.9). The complex mass of abnormal vessels tends to image with signal void because of rapid flow. The tubular shape of the abnormal structures should suggest their vascular origin. Gradient-echo images demonstrating bright flow signal produce an angiographic effect that can also be helpful to identify the abnormal vessels.

Hemangiomas, however, may demonstrate areas of intermediate signal due to slow flow, with brightening on T2 weighting. Depending on the size of the abnormal vessels, they may or may not be apparent within the hemangioma mass on MR imaging (Fig. 23.10).

Acknowledgment

The assistance of Janet Schuster in preparation of this manuscript is gratefully acknowledged.

References

1. Middleton WD, Macrander S, Kneeland JB, et al. MR imaging of the normal elbow: anatomic correlation. AJR 1987;149:543–547.
2. Bunnell DH, Fisher DA, Bassett LW, et al. Elbow joint: normal anatomy on MR images. Radiology 1987;165:527–531.
3. Macrander SJ. The elbow. In: Middleton WD, Lawson TL, eds. Anatomy and MRI of the joints. New York: Raven Press, 1989:49–86.
4. Stoller DW. The hand, wrist, and elbow. In: Stoller DW, ed. Magnetic resonance imaging in orthopaedics and rheumatology. Philadelphia: Lippincott, 1989:284–321.
5. Crues JV III, Morgan FW. Physical and technical considerations in magnetic resonance imaging. In: Stoller DW, ed. Magnetic resonance imaging in orthopaedics and rheumatology. Philadelphia: Lippincott, 1989:1–22.
6. Szumowski J., Eisen JK, Vinitski S, Haake PW, Plewes PB. Hybrid methods of chemical-shift imaging. Magn Reson Med 1989;9:379–388.
7. Resnick D. Arthrography, tenography, and bursography. In: Resnick D, Niwayama G, eds. Diagnosis of bone and joint disorders. 2nd ed. Philadelphia: W. B. Saunders, 1988:336–341.
8. Yulish BS, Montanez J, Goodfellow DB, Bryan PJ, Mulopulos GP, Modic MT. Chondromalacia patellae: assessment with MR imaging. Radiology 1987;164:763–766.
9. Gylys-Morin VM, Hajek PC, Sartores DJ, Resnick D. Articular cartilage defects: detectability in cadaver knees with MR. AJR 1987;148:1153–1157.
10. DeSmet AA, Fisher DR, Graf BK, Lange RH. Osteochondritis dissecans of the knee: value of MR imaging in determining lesion stability and the presence of articular cartilage defects. AJR 1990;155:549–553.
11. Kursonoglu-Brahme S, Riccio T, Weisman MH, et al. Rheumatoid knee: role of gadopentetate-enhanced MR imaging. Radiology 1990;176:831–835.
12. Resnick D, Kyriakos M, Greenway GD. Tumors and tumor-like lesions of bone: imaging and pathology of specific lesions. In: Resnick D, Niwayama G, eds. Diagnosis of bone and joint disorders. 2nd ed. Philadelphia: W. B. Saunders, 1988:3713–3714.

CHAPTER 24

Magnetic Resonance Imaging of the Wrist and Hand

Cooper R. Gundry, Mayur M. Patel, and Sevil Kursunoglu-Brahme

Introduction

Magnetic resonance (MR) imaging can make significant contributions to the work-up and management of hand and wrist abnormalities. The use of dedicated surface coils permits imaging of problems related to wrist instability and ligamentous abnormalities. The high contrast resolution inherent in MR imaging allows accurate assessment of cartilage, bone marrow, nerves, tendons, and soft tissues. (1–8). This chapter will review imaging techniques, pertinent anatomy, and specific wrist abnormalities.

Imaging Techniques

High-resolution imaging of the wrist is best performed through the use of a high field strength magnet in conjunction with dedicated surface coils. We use a Signa 1.5-T magnet (General Electric, Milwaukee, Wisc.), with a 4-inch transmit/receive wrist coil (Medical Advances, Milwaukee, Wisc.) Proper patient positioning is mandatory in order to maximize patient comfort so that any motion will be minimized. Patients may be placed prone with their arm extended over their head, or they may be placed supine with their arm at their side. Supine positioning is best but requires the use of an off-axis field of view, and may not be possible in larger patients. Surface coil options include small wraparound extremity coils, single circular or rectangular surface coils, and dual-surface coils in a Helmholtz configuration (3, 8, 9).

The choice of pulse sequences and imaging planes depends on the clinical situation. As in any imaging procedure, images obtained in more than one plane are optimal. In general, both T1- and T2-weighted images should be acquired in the plane that best demonstrates the anatomy pertinent to the clinical question at hand. For example, if the clinical question relates to wrist instability, then T1 and T2 images should be obtained in the coronal plane in order to allow assessment of the intercarpal ligaments and triangular fibrocartilage complex (TFCC). Sagittal T1 scans should be added to allow assessment of carpal alignment as well as the dorsal and volar radiocarpal ligaments. Alternatively, if the clinical question is regarding the carpal tunnel, T1 and T2 images should be obtained in the axial plane (4, 5). The key is to tailor each examination so that the images obtained will be appropriate to answer the clinical question that is posed.

T1-weighted images may be obtained with a repetition time (TR) of 600 ms and an echo time (TE) of 20 to 25 ms. Spin-echo proton density and T2-weighted images may be obtained with a TR of 2000 to 2500 ms and TE of 20 and 60 to 80 ms. Spin-echo images are best obtained using thin 3-mm sections, two signal excitations (NEX), a 256 × 128 matrix, and a small field of view (FOV) ranging from 8 to 10 cm. Gradient-echo (GRE) images may be helpful in the evaluation of articular cartilage and are most often obtained in the coronal plane. Zlatkin and coworkers found that a GRE sequence using a TR of 200 ms and a TE of 15 ms with a 70° flip angle permitted the best evaluation of articular cartilage (8). When GRE imaging sequences are employed, four excitations (NEX) are preferable.

Anatomy

The various components of wrist anatomy display characteristic signal intensities on MR imaging. Cortical bone and tendons are devoid of signal due to lack of mobile protons and appear black on both T1- and T2-weighted images, whereas ligaments display low signal intensity on both T1- and T2-weighted images. Owing to less water content than hyaline cartilage, fibrocartilage demonstrates relatively low signal intensity on most imaging sequences.

Anatomy of the wrist important to MR imaging can best be considered in terms of the major imaging planes.

Axial Plane

The axial plane allows assessment of the carpal tunnel. The flexor retinaculum is of low signal intensity and courses from the hook of the hamate and pisiform medially to the

Abbreviations (see also glossary): TFCC, triangular fibrocartilage complex; CTS, carpal tunnel syndrome; DRUJ, distal radial ulnar joint; AVN, avascular necrosis: TS, tenosynovitis; DISI, dorsal intercalated segmental instability; VISI, volar intercalated segmental instability.

tubercle of the trapezium and scaphoid laterally and forms the volar border of the carpal tunnel. Carpal bones form the dorsal aspect of the carpal tunnel. The contents of the carpal tunnel include the flexor digitorum superficialis, and the profundus and flexor pollicis longus tendons. The median nerve is shown as a structure of intermediate signal intensity in the superficial radial aspect of the carpal tunnel.

Coronal Plane

The coronal plane allows direct assessment of the relationship between the carpal bones, intercarpal ligaments, and the TFCC. The TFCC is seen as a low-signal band arising from the ulnar aspect of the distal radius and extending to the base of the ulnar styloid process. The ulnar aspect of the TFCC may demonstrate slightly increased signal intensity, perhaps related to the meniscus homolog. The scapholunate ligament is identified as a thin band of low signal intensity joining the proximal aspects of the scaphoid and lunate. The lunatotriquetral ligament has a similar appearance, but it is more difficult to consistently identify. The use of T2-weighted images may aid in the identification of intercarpal ligaments by accentuating joint fluid and thereby outlining the lower signal intercarpal ligaments. The radial and ulnar collateral ligaments are easily identified as low-signal structures on coronal images. Thin-section coronal images may allow visualization of the radiocarpal ligaments, but these are more consistently visualized on sagittal images.

Sagittal Plane

Sagittal images facilitate assessment of the capsular anatomy, including the dorsal and volar radiocarpal ligaments. Carpal alignment is also easily assessed in this plane. The radiocarpal ligaments are seen as low-intensity structures coursing along the volar and dorsal aspects of the wrist. Intraarticular fluid on T2 weighted sequences may aid in visualization of these ligaments due to bright fluid signal outlining their intraarticular borders. Precise identification of these ligaments may be difficult because of their oblique course. The sagittal plane also allows visualization of the longitudinal aspects of the flexor and extensor tendons.

Pathology

Carpal Tunnel Syndrome

The carpal tunnel syndrome (CTS) is a median nerve neuropathy that results in nocturnal discomfort, paresthesias, and thenar muscle weakness (4). Carpal tunnel syndrome results when the median nerve is compressed or damaged in its course through the carpal tunnel. The carpal tunnel consists of a space along the volar aspect of the wrist that is bordered dorsally by the carpal bones and volarly by the flexor retinaculum. The contents of the carpal tunnel are tendons of the flexor muscles of the hand (flexor digitorum profundus, flexor digitorum superficialis, and flexor pollicus longus) and the median nerve. Any abnormality affecting the median nerve as it courses in the carpal tunnel may result in CTS, such as space-occupying processes

(excessive fat or muscle, local tenosynovitis, neoplasms, synovial cysts, edema associated with pregnancy, and amyloid deposition) or processes that narrow the carpal tunnel (carpal fracture deformities, callous formation, or hypertrophic changes) (4).

The diagnosis of CTS has depended on clinical examination and nerve conduction studies and radiographic studies have had a very limited role, except in the occasional cases caused by bony abnormalities. MR imaging is appealing because it may determine the exact cause of CTS in some patients. Regardless of the cause, Mesgarzadeh and colleagues have described four MR findings seen in patients with CTS (4). These findings include swelling of the median nerve at the level of the pisiform, flattening of the median nerve at the level of the hook of the hamate, palmar bowing of the flexor retinaculum, and increased signal intensity of the median nerve on T2-weighted images (Fig. 24.1). The subjective assessment of median nerve swelling and flattening may be difficult, and objective swelling and flattening ratios that evaluate the size of the median nerve at various levels in the carpal tunnel have been proposed (10). Similarly, objective assessment of palmar bowing of the flexor retinaculum may be accomplished through the use of a bowing ratio that has been described by Mesgarzadeh and coworkers (10).

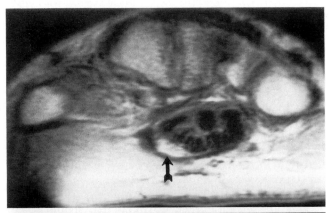

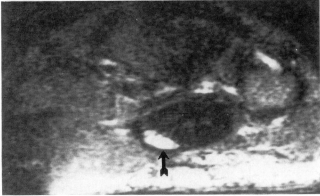

Figure 24.1. Carpal tunnel syndrome. Axial proton density (top) and T2-weighted (bottom) images (TR = 2000, TE = 20/80) demonstrate enlargement of the median nerve, which images with increased signal intensity on both proton density and T2-weighted sequences *(arrows)*. (Reprinted with permission from Mosure JC, Belran J. The Hand and wrist. In: MRI musculoskeletal system, Beltran J, ed. Philadelphia: Lippincott; New York: Gower Medical Publishing, 1990.)

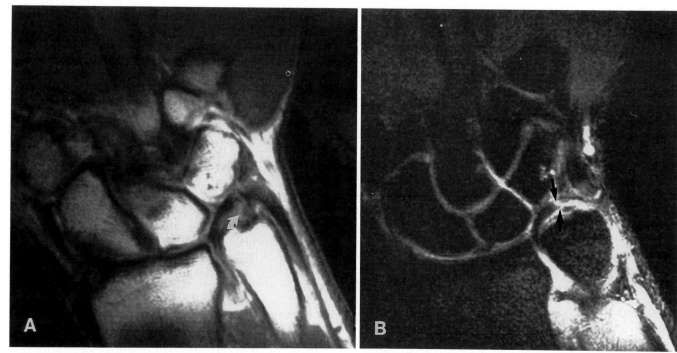

Figure 24.2. Triangular fibrocartilage tears. **(A)** Coronal T1-weighted image (TR = 600, TE = 29) showing abnormal zone of intermediate signal intensity *(arrow)* interrupting the low signal intensity band representing the fibrocartilage. **(B)** Gradient-recalled echo image (TR = 600, TE = 30, flip angle of 65°) revealing disruption of the triangular fibrocar- tilage with high signal fluid interposed between the torn triangular fibro- cartilage margins *(arrows).* (Reprinted with permission from Golimbu CN, Firooznia H, Melone CP, Rafii M, Weinreb J, Leber C. Tears of the triangular fibrocartilage of the wrist: MR imaging. Radiology 1989;173:731–733.)

Specific causes of CTS that may be identified on MR imaging include tenosynovitis, ganglia, rheumatoid arthritis, excess fat or muscle occupying the carpal tunnel, and neoplasms. Magnetic resonance imaging has a potential role in the postoperative assessment of recurrent CTS, since it may detect specific causes such as incomplete flexor retinaculum incision, the presence of a persistent median artery, chronic median nerve neuritis, or postoperative incisional neuromas (4).

There are several normal variations that should be remembered when assessing the carpal tunnel (5). If the study is performed with partial flexion of the fingers, the lumbrical muscles attached to the flexor tendons may migrate into the carpal tunnel and present as space-occupying masses. Anatomical variations that may contribute to CTS include a persistent median artery and anomalous lumbricals, which may be detected in the carpal tunnel in the absence of flexion of the fingers (5).

Ligamentous Abnormalities

Persistent wrist pain in posttraumatic patients remains a difficult diagnostic problem. In the past the diagnosis of ligamentous abnormalities of the wrist has relied on the use of wrist arthrography. Magnetic resonance imaging permits direct noninvasive assessment of the triangular fibrocartilage as well as the intrinsic and extrinsic ligaments of the wrist. Recent studies have shown that MR imaging is comparable to arthrography for the assessment of the triangular fibrocartilage as well as intercarpal ligaments,

especially the scapholunate ligament (3, 11). The normal TFCC appears as a biconcave band of low signal originating on the medial aspect of the distal radius and inserting on the base of the ulnar styloid. Areas of increased signal intensity may normally be seen at both the radial and ulnar attachments of the triangular fibrocartilage. Traumatic disruptions of the triangular fibrocartilage most frequently occur near the ulnar insertion and are to be distinguished from the degenerative perforations that tend to occur in the central portion of the triangular fibrocartilage (3). On T1-weighted images, disruptions and perforations appear as focal linear areas of intermediate signal that bridge both articular surfaces of the triangular fibrocartilage. T2-weighted imaging will often demonstrate fluid interposed in these linear areas of signal change (Fig. 24.2). Golimbu and colleagues evaluated 20 patients with surgical correlation and found that MR had an accuracy of 95% in the detection of triangular fibrocartilage tears (3). In a larger study, Zlatkin and coworkers found accuracies of 95 and 90% when MR imaging was compared to arthrography and surgery, respectively (8).

The high spatial resolution of MR imaging has also allowed assessment of individual intercarpal ligaments, the most important being the scapholunate and the lunatotriquetral ligaments. Normal intercarpal ligaments appear as thin bands of dark signal that join the proximal portions of the scaphoid and lunate as well as the proximal portions of the lunate and triquetral bones, and when intact they are seen on at least two contiguous images (Fig. 24.3) (8). Due to its smaller size, consistent visualization of the lunatotriquetral ligament may not be possible. Disruptions of intercarpal ligaments appear as areas of frank discontinuity. T2-weighted images may demonstrate fluid bridging the space normally occupied by these intercarpal ligaments (Fig. 24.4). For the evaluation of the scapholunate ligament Zlatkin et al. found accuracies of 90 and 95% when MR imaging was compared to arthrography and surgery, respectively. Lunatotriquetral ligament tears were made more difficult to diagnose and accuracies ranged from 80 to 90%, depending on the standard of comparison (8).

Wrist Instability

Magnetic resonance imaging of the wrist permits assessment of carpal instability patterns and has advantages over plain films in that direct visualization of important stabilizers is possible. These stabilizers include the TFCC, the intercarpal ligaments, and the dorsal and volar radiocarpal ligaments. Carpal alignment abnormalities resulting in instability patterns are easily assessed on sagittal and coronal images. The most common instability pattern is a dorsal intercalated segmental instability pattern (DISI), most often caused by scapholunate dissociation, which requires disruption of the scapholunate intercarpal ligament as well as the volar radioscapholunate ligament (Fig. 24.5) (12). This pattern is easily recognized on sagittal images by visualizing abnormal alignment between the lunate and capitate and noting an increased scapholunate angle (13). Volar intercalated segmental instability (VISI) is also easy to recognize by observing abnormal alignment of the lunate and capitate as well as a narrowed scapholunate angle on sagittal images. Disruption of the lunatotriquetral ligament with resultant triquetrolunate dissociation often results in a VISI pattern (14). Other static instability patterns that may be encountered include dorsal subluxation, volar subluxation, and ulnar translocation (14). Dorsal subluxation typically results from a malunited Colles' fracture, and ulnar translocation is common in patients with rheumatoid arthritis.

Current investigations are assessing the use of dynamic and stress MR imaging in the evaluation of carpal instability, the advantage being that MR allows direct visualization of intrinsic and extrinsic ligament stabilizers (8). It is hoped that the direct visualization of these wrist stabilizers may aid the orthopedic surgeon in more accurate diagnosis and better treatment planning.

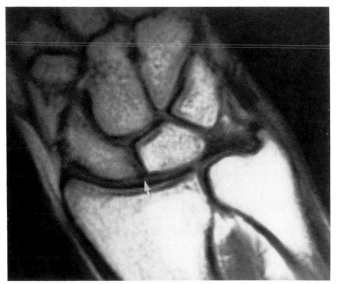

Figure 24.3. Normal scapholunate ligament. Coronal T1-weighted image (TR = 600, TE = 20) of an intact scapholunate ligament, which appears as a thin continuous band of low signal bridging the proximal portions of the scaphoid and lunate *(arrow)*.

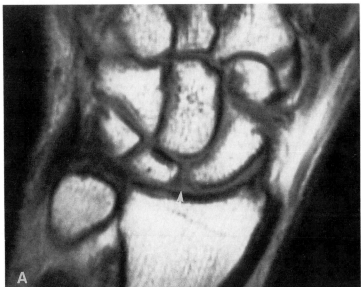
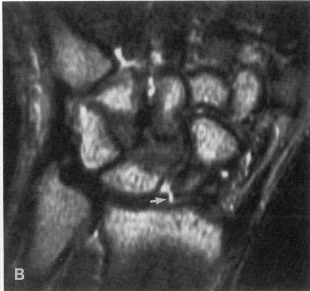

Figure 24.4. Scapholunate tear. **(A)** T1-weighted image (TR = 600, TE = 20) demonstrating abnormal intermediate linear signal interrupting the low signal scapholunate ligament *(arrow)*. **(B)** T2-weighted image (TR = 2000, TE = 80) showing high signal intensity synovial fluid bridging this gap in the scapholunate ligament *(arrow)*.

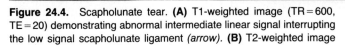

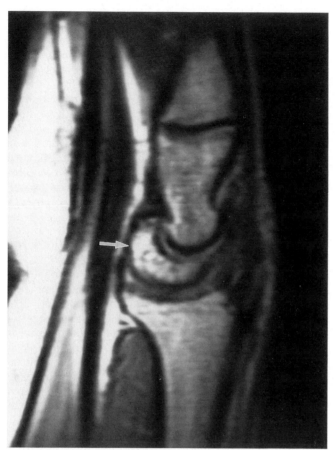

Figure 24.5. DISI instability pattern. Sagittal proton density image (TR = 2000, TE = 20) revealing abnormal dorsal tilt of lunate *(arrow)*. An adjacent sagittal image revealed abnormal volar tilting of scaphoid. This patient was subsequently found to have a scapholunate ligament disruption.

Distal Radioulnar Joint Congruity

Axial MR imaging enables assessment of distal radioulnar joint congruity (DRUJ). Radiographic and clinical assessment of DRUJ congruity may be impossible without the use of transaxial imaging because of the difficulty in obtaining a true lateral view. Mino and associates were the first to describe criteria for the assessment of DRUJ congruity using axial CT images obtained at the level of the sigmoid notch (15). They described a line through the dorsal border of the radius and a second line through the palmar border of the distal radius. Congruity was present if the distal ulna was located between these lines, and dorsal and volar subluxation resulted in the distal ulna crossing each respective line (Fig. 24.6). Wechsler and coworkers described two additional sets of criteria useful in assessment of DRUJ congruity using transaxial imaging (16). Unlike the Mino criteria, which work best with wrists imaged in a neutral position, these latter criteria allow the assessment of joint congruity in any forearm position (16). The same criteria described for CT imaging may be applied to axial MR images. An advantage of magnetic resonance is the ability to visualize associated soft-tissue abnormalities, such as disruptions of the TFCC.

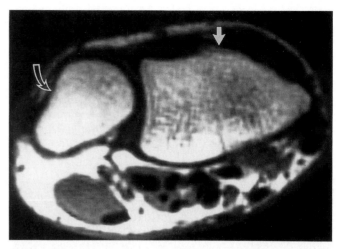

Figure 24.6. Normal inferior radioulnar joint congruity. Axial T1-weighted (TR = 600, TE = 20) image at level of radial sigmoid notch, demonstrating a normal relationship between the distal radius *(solid arrow)* and the adjacent distal ulna *(curved open arrow)*.

Trauma and Ischemic Necrosis

The scaphoid bone is the most frequently injured carpal bone due to its unique position in linking the carpal rows. This, coupled with lack of direct blood supply to the proximal pole, accounts for the high incidence of delayed healing and ischemic necrosis in fractures involving the proximal pole and waist of the scaphoid (17). Prompt diagnosis and early treatment of acute injury are essential to minimize complications. Kienböck's disease (lunatomalacia) represents osteonecrosis of lunate bone; various etiologies have been proposed, including inadequate or marginal blood supply and trauma. The etiology of Kienböck's disease is most likely multifactorial, and there is an increased incidence of Kienböck's in patients with an ulna minus variance, which presumably places increased forces on the lunate and compromises a preexisting tenuous blood supply. Prompt treatment of ischemic necrosis is important, and ideally the diagnosis should be made in early stages, before radiographs become abnormal. Radionuclide scintigraphy, particularly with pinhole collimation, has high sensitivity but lacks specificity and the degree of spatial resolution offered by MR imaging. The role of scintigraphy for AVN is especially limited in the presence of fracture, which itself causes increased radiopharmaceutical uptake. Owing to these limitations of scintigraphy, MR imaging is the procedure of choice for the early diagnosis of ischemic necrosis (Figs. 24.7 and 24.8). Thin-section (3 mm) coronal T1- and T2-weighted images are adequate for the assessment of ischemic necrosis. Reinus et al. demonstrated a sensitivity of 87.5% and specificity of 54.5% using decreased marrow signal on T1-weighted images as the sole criterion, and when data from T2-weighted images were also utilized, the specificity improved to 100% (6). In both T1- and T2-weighted images, there was diminished signal intensity of the marrow.

Inflammation of Tendon and Tendon Sleath

Tenosynovium is composed of a visceral and parietal layer of synovium that lines, lubricates, and nourishes the tendon. When fluid accumulates within this potential space, it results in tenosynovitis (TS) (18). The etiology of TS includes various forms of arthritides, particularly rheumatoid, as well as trauma and infection. In rheumatoid arthritis, TS is seen as part of a spectrum that includes radiocarpal synovitis, radioulnar synovitis, and capsular synovitis. Purulent TS requires early diagnosis and appropriate treatment because of the potential of extensive damage to the delicate structures of the flexor sheath and spread to the deep palmar surface (18). A common site of tenosynovitis is the first extensor compartment involving the abductor pollicis longus and extensor pollicis brevis ten-

dons at the level of the radial styloid known as De Quervain's tenosynovitis. In the setting of TS, MR images demonstrate swelling of the tendon with an abnormal accumulation of fluid surrounding the inflamed tendon sheath (Fig. 24.9). Peritendinous edema is manifested with

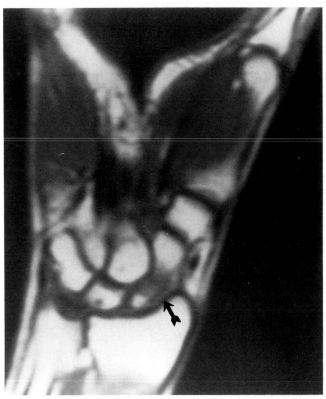

Figure 24.8. Posttraumatic ischemic necrosis of the lunate. T1-weighted coronal image (TR = 600, TE = 20) in a patient with prior scaphoid wrist fracture demonstrating abnormally diminished signal in the proximal pole of the scaphoid, consistent with posttraumatic ischemic necrosis *(arrow)*. (Courtesy of David Sartoris, M.D.)

Figure 24.7. Kienböck's disease. Coronal proton density-weighted image (TR = 2000, TE = 20) showing abnormal diminished signal intensity of the lunate *(arrow)*, indicative of ischemic necrosis.

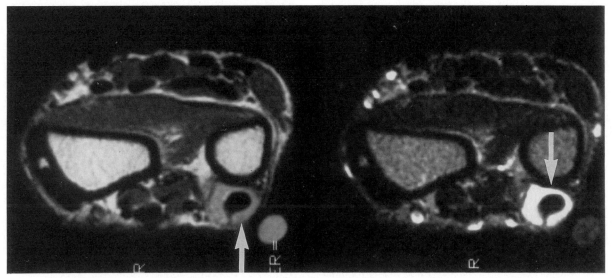

Figure 24.9. Tenosynovitis. Proton density *(left)* and T2-weighted *(right)* axial images of the wrist, demonstrating abnormal fluid signal intensity surrounding the extensor carpi ulnaris tendon *(arrows)*.

low signal on T1- and proton-weighted images and as high signal on T2-weighted images.

Injury of Tendon

The superficial course of the flexor and extensor tendons of the hand render them vulnerable to injury. Clinical diagnosis of tendon injuries may be difficult, particularly in cases of partial tendon lacerations. Tendon ruptures may occur in the absence of trauma, particularly in patients with rheumatoid arthritis who may have pathological degeneration of the tendon. Magnetic resonance imaging is a useful adjunct in the evaluation of tendon injury because of exquisite anatomic detail and a high degree of inherent soft-tissue contrast (19). Due to their low water content, tendons have low signal on both T1- and T2-weighted images. Tendon tears are manifested as foci of higher signal intensity, and complete tears may result in discontinuity of the tendon. Incomplete tears result in abnormal signal foci within the tendons, but there is no frank discontinuity (Fig. 24.10). Unlike the tendon injuries about the ankle, the literature is still lacking in hand tendon injuries. Magnetic resonance imaging may be useful in the assessment of postoperative healing, especially to determine continuity of repaired tendon and assess peritendinous scarring.

Soft-Tissue Masses

Magnetic resonance imaging is the best modality for the work-up of soft-tissue masses about the hand and wrist because of its high soft-tissue contrast resolution combined with multiplanar imaging capability. Osseous, ligamentous, and muscle involvement is easily assessed and this enables more accurate preoperative planning.

Dorsal wrist ganglia arise from the capsule adjacent to the scapholunate joint and are the most common masses of the hand (20). While the exact etiology of these ganglia is unknown, they are commonly felt to represent protrusion of a pseudocapsule that results from remodeling of the fibrous capsular tissue of the joint. The ganglion cavity is usually single and filled with mucinous material, which is transilluminant with a pen light—a classic diagnostic test (7). Magnetic resonance imaging demonstrates a well-demarcated mass with low signal intensity on T1-weighted images and high signal intensity on T2-weighted images (Figs. 24.11 and 24.12). Occasionally, a ganglion appears loculated and trabeculated due to fibrous septations, but on sectioning, a single communicating cavity is demonstrable. Ganglia are occasionally also found along the ulnar aspect of the wrist joint or the trapezioscaphoid joint and may present near the radial artery, causing compression or distortion. Rarely they may be located within the carpal tunnel or in Guyon's canal and thereby result in compression of the deep branch of the ulnar nerve (20, 21). Magnetic resonance imaging is useful in evaluating distal ulnar neuropathy when a mass is clinically not palpable. Binkovitz et al. (21) demonstrated three cases of

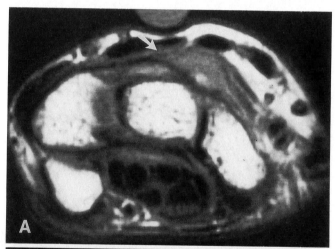

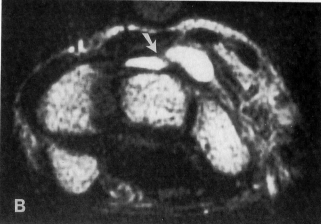

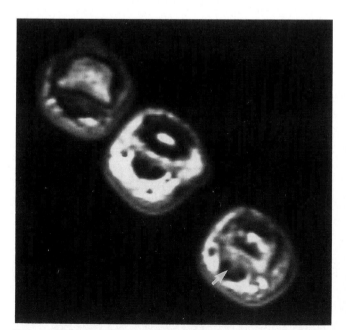

Figure 24.10.   Partial tear of flexor digitorum profundus tendon. Proton density-weighted image (TR = 2000, TE = 20) through distal index finger demonstrating abnormal intermediate signal intensity in the substance of the flexor tendon, indicating a partial tear (arrow).

Figure 24.11.   Dorsal ganglion. **(A)** Proton density-weighted image (TR = 2000, TE = 20) revealing intermediate signal intensity structure dorsal to the midcarpal row (arrow). **(B)** T2-weighted image demonstrating characteristic bright signal of the dorsal ganglion (arrow).

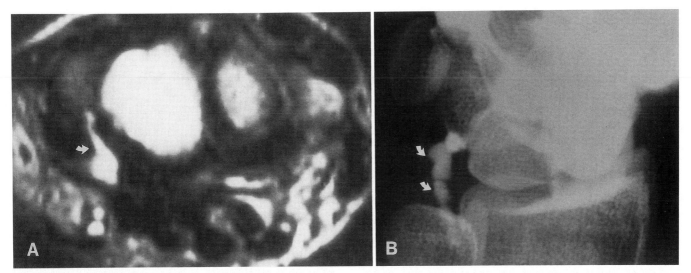

Figure 24.12. Wrist ganglion. **(A)** Axial T2-weighted image showing irregular area of increased signal intensity projecting volar to the lunatotriquetral junction, suggestive of a ganglion *(arrow)*. **(B)** Midcarpal compartment arthrogram demonstrating abnormal serpiginous collec-

tion of contrast adjacent to the lunatotriquetral interface, consistent with a communicating wrist ganglion *(curved arrows)*. (Reprinted with permission from Kursunoglu-Brahme S, Gundry C, Resnick D. Advanced imaging of the wrist. Radiol Clin Am 1990;28:307–320.)

Figure 24.13. Intramuscular lipoma. **(A)** T1-weighted coronal image (TR = 600, TE = 200) revealing large, high signal intensity structure with fat signal intensity interposed between the distal ulna and radius *(arrows)*. **(B)** Proton density *(left)* and T2-weighted *(right)* axial images (TR = 2000, TE = 20, 60) demonstrating lobulated soft-tissue mass that images with fat signal intensity. *u,* Ulna; *r,* radius. (Reprinted with permission from Kursunoglu-Brahme S, Gundry C, Resnick D. Advanced imaging of the wrist. Radiol Clin Am 1990;28:307–320.)

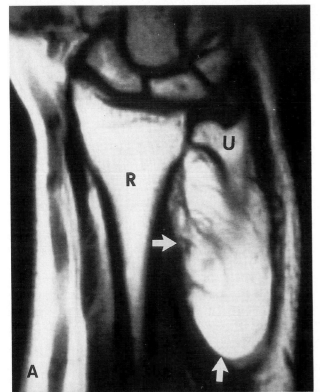

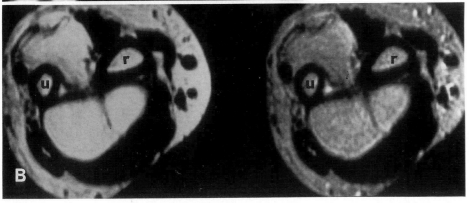

nonpalpable ganglionic cysts at the level of the Guyon canal that manifested as distal ulnar neuropathy. When a ganglion dissects and grows to reach a subcutaneous location, MR imaging is particularly helpful in defining its origination for planning resection and differentiating from neoplastic masses.

Excluding ganglia, tumors of the wrist and hand are rare. Most of the lesions that occur are benign, although malignant masses may rarely be seen and include primary bone neoplasms as well as a variety of sarcomas and, rarely, metastatic disease. Binkovitz et al. (21) have reported the largest series to date of MR imaging findings in benign and malignant masses involving the hand and wrist.

Certain benign and malignant masses have signal characteristics that may aid in specific MR imaging diagnosis. In general, benign masses are well marginated and do not invade or encase neurovascular or tendinous structures. Except for giant cell tumors of tendon sheaths, which have areas of low or absent signal on T1- and T2-weighted images secondary to hemosiderin deposit, benign masses usually have a homogeneous signal intensity. In a series of 112 soft-tissue masses, Kransdorf et al. found "no reliable criteria to distinguish the MR images of malignant from benign soft-tissue masses" (22). More recently Binkovitz et al., using parameters of signal homogeneity and intensity, margination, soft-tissue invasion, and knowledge of clinical findings, were able to correctly distinguish benign and malignant masses in a study of 22 mass lesions of the hand and wrist (21). Malignant masses were noted to have inhomogeneous signal intensity that increased with T2 weighting. Margins of malignant masses were less well demarcated than benign lesions and, also, malignant masses cause distortion of tissue planes and encasement of tendons and neurovascular structures. Clearly, the distinction between benign and malignant lesions is not always possible, but it is useful to use these criteria in order to predict the relative aggressiveness of a particular lesion.

Lipomas are most commonly seen on the palmar aspect of the hand, where fat is found in relatively large quantities, and on MR imaging are well-demarcated masses with high signal intensity on T1-weighted images and intermediate signal intensity on T2-weighted images (Fig. 24.13). The presence of fibrous elements in fibrolipomas is demonstrated as low signal intensity bands within a high signal intensity mass on T1-weighted images. Nerve sheath neoplasms (neurofibromas) produce a fusiform mass along the course of the affected nerve and also demonstrate low to intermediate signal intensity on T1-weighted images and high signal intensity on T2-weighted images (Fig. 24.14). Arteriovenous malformations are recognized as a serpiginous mass with signal void resulting from flowing blood. Magnetic resonance imaging might replace the more invasive angiographic studies that are presently performed. Giant cell tumor of the tendon sheath represents a proliferative disorder of the synovium that is considered by many to be a localized form of pigmented villonodular synovitis. The presence of hemosiderin may result in paramagnetic effects that image with focal areas of low signal intensity on T1- and T2-weighted images. Diffuse pigmented villonodular synovitis manifests as a periarticular mass with areas of low or absent signal of both T1- and T2-weighted

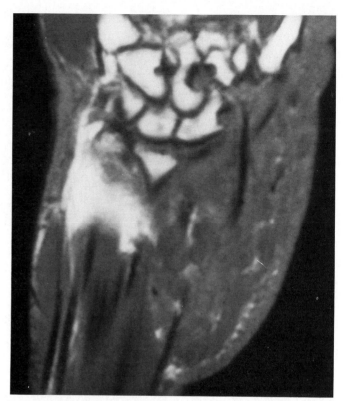

Figure 24.14. Neurofibromatosis. Coronal T1-weighted image (TR = 600, TE = 20) revealing large, somewhat irregular intermediate signal intensity mass surrounding the radial aspect of the wrist. Biopsy revealed a large plexiform neurofibroma. (Courtesy of David Sartoris, M.D.)

images, due to the paramagnetic effect of hemosiderin deposits. Several reports suggest this to be characteristic enough to make an accurate preoperative diagnosis; however, any cause of hemosiderin deposition such as rheumatoid arthritis and hemophilia may image similarly (14, 23).

Soft-tissue infection of the hand and wrist may present as a soft-tissue mass. Magnetic resonance is useful to determine the precise extent of the process so that the surgeon may plan any drainage procedure if necessary (Fig. 24.15).

Arthritis

Magnetic resonance applications in the evaluation of arthritis have not been fully determined. Due to superior inherent contrast sensitivity, MR imaging is useful in evaluating arthritis in both early and advanced stages. Subtle early erosions may be evident prior to positive findings on plain radiography. The inflammatory pannus found in rheumatoid arthritis is demonstrated as thickened, irregular synovium with intermediate signal on T1-weighted images. Acutely inflamed synovium usually demonstrates marked increased signal intensity on T2-weighted images due to the presence of edema. Coexistent joint effusions present with high signal intensity on T2-weighted images (Fig. 24.16). In advanced stages, findings of subluxation and erosions are well demonstrated. Pannus in chronic in-

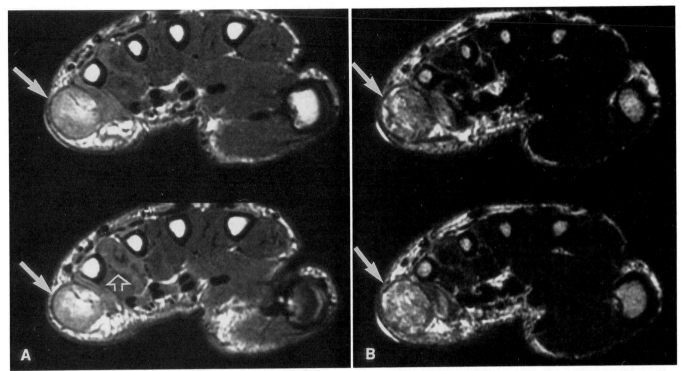

Figure 24.15. Cellulitis. **(A)** Axial proton density image (TR = 2000, TE = 20) revealing abnormal intermediate and increased signal intensity in the soft tissues adjacent to the fifth metacarpal shaft *(arrows)*. The patient had experienced a puncture wound 3 weeks prior to the MR scan. Notice the abnormal intermediate signal intensity infiltrating the lumbrical muscles between the fourth and fifth metacarpals *(open arrow)*. **(B)** Axial T2-weighted image showing inhomogeneous increased signal intensity throughout the focal masslike lesion. Surgery revealed cellulitis.

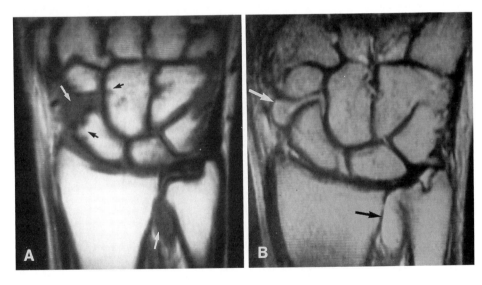

Figure 24.16. Rheumatoid arthritis. **(A)** Coronal T1-weighted image (TR = 600, TE = 20). Small erosions are seen involving the radial aspect of the capitate and the distal scaphoid *(black arrows)*. Pannus is identified as areas of intermediate signal in the midcarpal compartment and inferior radioulnar joint *(white arrows)*. **(B)** Coronal T2-weighted image (TR = 2000, TE = 60) revealing diffuse brightening of pannus in midcarpal compartment and inferior radioulnar joint *(arrows)*. (Reprinted with permission from Kursunoglu-Brahme S, Gundry C, Resnick D. Advanced imaging of the wrist. Radiol Clin Am 1990;28:307–320.)

flammation may appear as low to intermediate signal intensity on T2-weighted images, well distinguished from adjacent bright-appearing joint effusions. Utility of MR imaging in assessing activity of disease and response to therapy needs to be further investigated.

Conclusion

Magnetic resonance imaging of the wrist is now a useful tool in the evaluation of many wrist disorders. No other single modality allows direct visualization of both osseous and soft-tissue components of the wrist. The ability of MR to image cartilage, bone marrow, tendons, and ligaments enables it to be useful in the assessment of a whole host of wrist disorders. Cost considerations remain a concern to many clinicians, however, and this probably represents the single largest deterrent to wider use of MR imaging of the wrist. In some abnormalities such as ischemic necrosis, MR is superior to other imaging techniques. In other disorders, such as the evaluation of ligamentous abnormalities of the wrist, it remains to be seen whether MR visualization of these structures is clearly superior to the proven track record of multicompartment arthrography (24).

Acknowledgments

The authors thank Howard J. Mindell, M.D., for reviewing the manuscript and Carol McClure for help in manuscript preparation.

References

1. Beltran J, Noto AM, Herman LJ, Lubbers LM. Tendons: high-field-strength surface coil MR imaging. Radiology 1987;162:735–740.
2. Ehman LR, Berquist TH, McLeod RA. MR imaging of the musculoskeletal system: a 5 year appraisal. Radiology 1988;166:313–320.
3. Golimbu CN, Firooznia H, Melone CP, Rafii M, Weinreb J, Leber C. Tears of the triangular fibrocartilage of the wrist: MR imaging. Radiology 1989;173:731–733.
4. Mesgarzadeh M, Schneck CD, Bonakdarpour A, Mitra A, Conaway D. Carpal tunnel: MR imaging. Part II. Carpal tunnel syndrome. Radiology 1989;171:749–754.
5. Middleton WD, Kneeland JB, Kellman GM, et al. MR imaging of the carpal tunnel: normal anatomy and preliminary findings in the carpal tunnel syndrome. AJR 1987;148:307–316.
6. Reinus WR, Conway WF, Totly WG, et al. Carpal avascular necrosis: MR imaging. Radiology 1986;160:689–693.
7. Weiss KL, Beltran J, Lubbers LM. High-field MR surface coil imaging of the hand and wrist. Part II. Pathologic correlation and clinical relevance. Radiology 1986;160:147–152.
8. Zlatkin MB, Chao PC, Osterman AL, Schnall MD, Dalinka MK, Kressel HY. Chronic wrist pain: evaluation with high-resolution MR imaging. Radiology 1989;173:723–729.
9. Weiss KL, Beltram J, Shamam OM, Stilla RF, Levey BA. High-field MR surface-coil imaging of the hand and wrist. Part I. Normal anatomy. Radiology 1986;160:143–146.
10. Mesgarzadeh M, Schneck CD, Bonakdarpour A. Carpal tunnel: MR imaging. Part I. Normal anatomy. Radiology 1989;171:743–748.
11. Jelinek JS, Kransdorf MJ, Utz JA, et al. Imaging of pigmented villonodular synovitis with emphasis on MR imaging. AJR 1989;152:337–342.
12. Blatt G. Scapholunate instability. In: Lichtman DM, ed. The wrist and its disorders. Philadelphia: W. B. Saunders, 1988:251–273.
13. Gilula LA, Destout JM, Weeks PM, et al. Roentgenographic diagnosis of the painful wrist. Clinical Orthoped 1984;187:52–64.
14. Talesnik J. Current concepts review carpal instability. J Bone Joint Surg 1988;70-A:1262–1268.
15. Mino DE, Palmar AK, Levinsohn EM. The role of radiology and computerized tomography in the diagnosis of subluxation and dislocation of the distal radio-ulnar joint. J Hand Surg (Am) 1983;8:23.
16. Wechsler RJ, Wehbe MA, Rifkin MD et al. Computed tomography of distal radioulnar subluxation. Skel Radiol 1987;16:1.
17. O'Brien ET. Acute fractures and dislocations of the carpus. In: Lichtman DM, ed. The wrist and its disorders. Philadelphia: W. B. Saunders, 1988:129–132.
18. Carter PR. Hand infections. In: Common hand injuries and infections a practical approach to early treatment. Philadelphia: W. B. Saunders, 1983.
19. Erickson SJ, Kneeland JB, Middleton WD, et al. MR imaging of the finger: correlation with normal anatomic sections. AJR 1989;152:1013–1019.
20. Bogumill GP. Tumors of wrist. In: Lichtman DM, ed. The wrist and its disorders. Philadelphia: W. B. Saunders, 1988:373–384.
21. Binkovitz LA, Berquist TH, McLeod RA. Masses of the hand and wrist: detection and characterization with MR imaging. AJR 1990;154:323–326.
22. Kransdorf MJ, Jelinek JS, Moser RP, et al. Soft tissue masses: diagnosis using MR imaging. AJR 1989;153:541–547.
23. Zeiss J, Skie M, Ebraheim N, Jackson NT. Anatomic relations between the median nerve and flexor tendons in the carpal tunnel: MR evaluation in normal volunteers. AJR 1989;153:533–536.
24. Zinberg EM, Pulman AK, Coren AB, et al. The triple injection wrist arthrogram. J Hand Surg (Am) 1988;13:803.

CHAPTER 25

Computed Tomography of the Wrist

Brian A. Howard and Joel D. Rubenstein

Introduction

The wrist is a complex anatomic region containing multiple bones with numerous articular surfaces, ligaments, tendons, and neurovascular structures (1). Wrist pain, with or without dysfunction, is a common diagnostic problem, but the clinical signs are often subtle, and conventional imaging (plain film radiography, scintigraphy, tomography, and arthrography) frequently proves inadequate or equivocal in determining the source of the pain (2, 3). High-resolution computed tomography (CT), with its enhanced soft-tissue contrast and multiplanar imaging capabilities, is therefore used with increasing frequency for the assessment of wrist disease (2). Recent reports describe the superiority of CT in demonstrating occult carpal fractures (3), fracture healing, surgical arthrodesis solidity, graft incorporation (4), foreign bodies (5), and soft-tissue masses (6).

This chapter describes the relevant cross-sectional anatomy of the wrist and the technical considerations for performing wrist CT, followed by a review of established and potential applications for CT.

Anatomy

The detailed anatomy of the hand and wrist is discussed in Chapter 4. Certain anatomic features, however, deserve emphasis to permit an understanding of the normal appearance and spatial orientation of the bones in the standard imaging planes. The bones of the wrist are uniquely arranged into transverse and longitudinal carpal arches (7). The cross-sectional appearance of the individual bones and the articular alignment of the carpometacarpal arches depend on the plane of imaging and the posture of the wrist (8, 9). It is also important to recognize that change in wrist posture alters the orientation of the multiple articular compartments of the wrist; these changes

in alignment are the result of normal inherent rotational and translational motion of the carpal bones (10).

Coronal Cross-Sectional Landmarks

The appearance of a midplanar coronal section of the wrist closely resembles that seen on a frontal radiograph (Fig. 25.1) (9). The articular margins of the distal radioulnar, radiocarpal, midcarpal, and carpometacarpal joints are seen. The rigid, chevron-shaped articulation of the second carpometacarpal joint is readily appreciated. Reliable assessment of ulnar variance can be made only when the ulnar styloid is in a medial position (11). The intermetacarpal articular surfaces of the ulnar four metacarpals demonstrate subtle bony ridges for the attachment of joint capsule and interosseous ligaments (1).

The dorsal and volar sections of the carpus pass, respectively, through the convexity and concavity of the car-

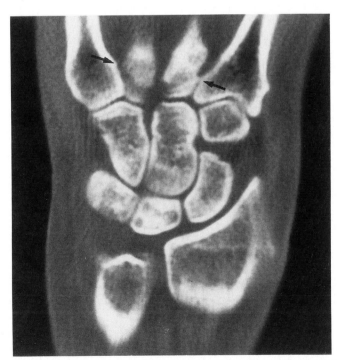

Figure 25.1. Coronal image of the left wrist, demonstrating normal midplanar bony anatomy. Radial styloid, physeal scar, and the chevron-shaped articular surface of the proximal second metacarpal are readily identified. Arrows indicate the subtle proximal metacarpal articular ridges.

Abbreviations (see also glossary): DRUJ, distal radioulnar joint; TFCC, triangular fibrocartilage complex; CCM, common carpometacarpal.

313

pal arches and appear distinct from the midplanar section (9, 12). The dorsal section emphasizes the prominent rectangular-shaped carpometacarpal joint of the middle ray, and the dorsal articular lip of the distal radius with Lister's tubercle. The volar image accentuates the scalloped volar articular lip of the distal radius, the thenar and hypothenar muscles, and the flexor tendons between the trapezial ridge radially and the hook of the hamate ulnarly. The fat-encased contents of the ulnar canal are seen between the hook of the hamate and the pisiform.

Axial Cross-Sectional Landmarks

The small bony prominences on the dorsal and volar aspects of the carpus are best shown on images obtained in the transverse plane. The cross-sectional anatomy in the axial plane varies at the levels of the distal radioulnar joint (DRUJ), the pisotriquetral joint, the hook of the hamate, and the base of the metacarpals (8).

The following structures must be identified at the level of the DRUJ (Fig. 25.2): the flat, well-defined volar surface of the distal radius, the dorsal convexity with Lister's tubercle, and medially the sigmoid notch with the congruent ulnar head. Of the six separate extensor tendon compartments on the dorsal aspect of the carpal arch, the extensor carpi ulnaris is the most mobile and lies in the dorsal groove of the ulnar styloid. The extensor pollicis longus tendon is readily identified on the ulnar aspect of Lister's tubercle. On the volar aspect of the radius, the midline soft tissues, from deep to superficial, consist of the pronator quadratus muscle, Parona's space and the common flexor tendons. The flexor carpi ulnaris, which inserts into the proximal pole of the pisiform, is a superficial volar soft-tissue landmark on the ulnar aspect of the wrist; the ulnar nerve is interposed between this structure and the digital flexor tendon group.

At the level of the pisotriquetral joint, the horizontal articular plane of this joint is readily differentiated from the other midcarpal joints. The canal of Guyon, containing the ulnar neurovascular bundle encased in fat, is situated on the radial aspect of the pisotriquetral joint.

At the level of the hamulus, the carpal tunnel lies between the thenar and hypothenar muscles, where its axial dimensions are smallest (Fig. 25.3). The transverse carpal ligament is prominent at this level and stretches from the volar trapezial ridge to the hamulus; this ligament stabilizes the transverse carpal arch and confines the contents of the carpal tunnel (7,13). The ulnar artery and nerve, surrounded by fat, are readily identified at the tip of the hamulus.

Distally, the cuboidal metacarpal bases are arranged transversely in the form of a Roman arch that encloses the long flexor tendons centrally with the thenar and hypothenar muscles peripherally (14).

Sagittal Cross-Sectional Landmarks

The anatomy in the sagittal and parasagittal sections closely resembles that of a lateral radiograph with the hand in neutral position. This imaging plane provides optimal depiction of the longitudinal carpal arch of the wrist, the

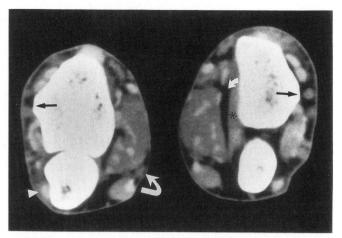

Figure 25.2. Axial image at the level of the DRUJ demonstrates Lister's tubercle *(black arrows)*, the extensor carpi ulnaris *(arrowhead)*, the pronator quadratus muscle *(asterisk)*, Parona's space *(short curved arrow)*, and the ulnar nerve *(solid bent white arrow)* superficially interposed between the flexor carpi ulnaris and the central digital flexor tendons.

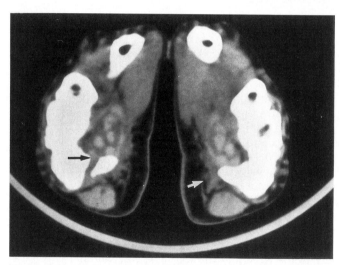

Figure 25.3. Axial image (3 mm thick) through the hook of the hamate. On the right the normal hamate hook, flexor tendons in the carpal tunnel, transverse carpal ligament are demonstrated. The ulnar nerve *(white arrow)* is readily identified on ulnar aspect of the tip of the hamulus. Demonstrated on the left *(black arrow)* is a displaced hamulus fracture with interposition of a common flexor tendon.

sagittal alignment, and the articular congruency of the carpometacarpal, intercarpal, and radiocarpal joints. These bones are functionally arranged into a central force-bearing column plus mobile radial and ulnar columns (Fig. 25.4) (1). The longitudinal course of the flexor and extensor tendons can be best assessed in this plane.

Technique

Optimal information from high-resolution CT of the wrist is obtained by tailoring each study. All pertinent clinical information (e.g., previous trauma, surgery, functional disability, and palpable mass) and the results of conven-

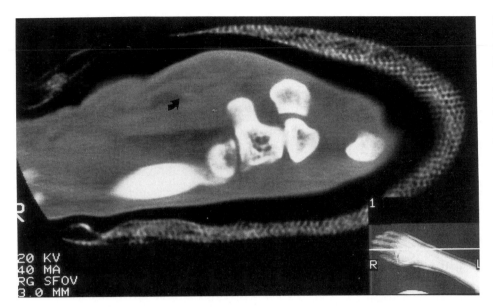

Figure 25.4. Sagittal image (3 mm) through the mobile ulnar column of the wrist. Medial ulnar styloid, pisotriquetral joint, and hamulus are readily identified. Portions of the proximal shafts of the fourth and fifth metacarpals are also seen. The ulnar neurovascular bundle is noted at the volar tip of the hamulus encased in fat *(curved arrow)*. (Top of image is volar side.)

tional imaging are required to plan the examination. These factors dictate patient positioning, the use of skin markers, the use of intravenous contrast, and the choice of single or multiplanar imaging with the wrist positioned in one or more orientations.

The majority of imaging is performed with the patient positioned comfortably to minimize motion-induced artifact. The patient can be positioned on the cradle in the supine, prone, or semiprone position with appropriate support of the limbs and chest. Alternatively, the patient may be able to stand or sit behind the gantry of some CT scanners with the hand and wrist adequately supported and stabilized. Radiolucent Plexiglas splints are not essential, but help to stabilize the wrist and standardize wrist positioning (9).

Initial axial images of both wrists are usually obtained with the wrists in neutral position, the palms symmetrically placed at right angles to the gantry, and the elbows together and flexed. The wrists, separated by a foam sponge, are taped together. Beam-hardening artifacts are decreased by maintaining the thumbs in an abducted position.

Subsequently, coronal or sagittal images of the abnormal wrist are obtained for optimal display of the abnormal anatomy in an orthogonal plane; for example, imaging planes perpendicular to fracture clefts best demonstrate the fractures (3). With the patient supine, the forearm is placed above the head—for coronal imaging, the palm and distal forearm are oriented parallel to the gantry, while for sagittal imaging, the palm and distal forearm are placed perpendicular to the gantry (2). The hand and wrist must be adequately supported on radiolucent sponges to prevent motion and excessive limb fatigue, with the proximal forearm oriented obliquely to the plane of section to decrease longitudinal beam-hardening artifacts. Plaster or fiberglass casts do not degrade the CT image (15); however, the presence of aluminum splints, external fixators, and other metallic objects causes objectionable image degradation due to the starburst streak artifacts (16).

Multipositional studies have been used to demonstrate instabilities of the DRUJ (17, 18), and other techniques (e.g., forced radioulnar deviation, clenched fist, and distraction with Chinese finger traps) may be used to elicit dynamic effects on the wrist during CT; however, the role of these techniques for detecting carpal instabilities is controversial.

Intravenous contrast enhancement is a useful adjunct in assessing soft-tissue masses and for identifying the neurovascular bundles (19). The role of intraarticular contrast (air, iodinated contrast, or both) has not been completely defined. At least one report suggests there is no role for CT arthrographic assessment of the triangular fibrocartilage complex (20); however, another report advocates CT arthrotomography to determine the course and extent of dissecting synovial cysts, especially in the volar soft tissues (21).

Computed tomography scanners vary in their technical hardware and software. The majority of our images were obtained on General Electric (Milwaukee, Wis.) 9800 or 9800 quick scanners. Digital scout images (80 kV, 40 mA) are normally obtained in the coronal plane and photographed with a magnification factor of 1.2, which allows accurate slice level correlation. If patient positioning is limited, gantry angulation of up to 20° can be used to obtain true coronal and sagittal imaging, but gantry angulation markedly prolongs subsequent reformatting time.

Although long exposure times are suggested to decrease sampling errors, this must be balanced against the potential for motion artifacts. Contiguous 3.0-mm slices are usually obtained in the coronal plane, and 5.0-mm slices in the axial and sagittal planes; if necessary, these may be supplemented by 1.5-mm slices. A small calibration file is used if the wrists are positioned in the head holder and a large calibration file is usually used when the wrists are positioned on the cradle. Often the wrists are positioned eccentrically within the gantry; therefore, a large scan field of view is suggested initially to obtain the appropriate coordinates.

Both a soft-tissue and bone (edge-enhancing) algorithm are used to reconstruct the images that should be displayed on a 512 × 512 matrix. A small display field of view optimizes resolution, with pixel sizes varying from 0.18 to 0.34 mm. Direct measurement of areas, volumes, distances, and articular arcs can be performed at an independent console, if indicated. Hard copies are usually photographed on soft-tissue mode (narrow settings window 500/level 50) and bone mode (wide settings window 2000/level 250) but this can be individualized. In the presence of metallic implants, we use a bone reconstruction algorithm and display images on an extended window to decrease the effect of metal artifact. Starburst streak artifacts of metal may be reduced by the use of special commercially available reconstruction algorithms or by image reformation into other planes (16). Two-dimensional sagittal and coronal reformatted images are limited in resolution compared to images obtained directly in the sagittal and coronal plane.

Three-dimensional reformatting (3D CT), an established visual aid, provides the clinician with a global picture of the altered bony morphology (22–25). The role of 3D CT in preoperative planning and custom prosthetic design for the wrist has not been established, but the three-dimensional display of fractures may aid the surgeon's decision on which method of fixation to use. The use of 3D CT for the detection of dynamic carpal instability is currently under investigation (26).

Radiation dose is an important consideration because of increased utilization and increased demand for thinner tomographic cuts (27). Radiation exposure is nonuniform within the slice and along the axis of the limb, with the skin and the center of the body part (isocenter kernel) representing the regions of highest exposure (28). The radiation dose is dependent on the milliamperage and kilovoltage.

Pathology

Computed tomography is ideally suited to evaluate the complex anatomy of the wrist because it can provide multiplanar sections of the bone and the soft tissues. Historically, wrist CT has been limited to the evaluation of traumatic instability of the distal radioulnar joint (DRUJ), fractures of the hook of the hamate, carpal tunnel compromise, and neoplasms of bone and soft tissue. In recent years, however, there has been increasing use of CT in the evaluation of articular and osseous diseases of the wrist, although the specific indications are still evolving (2–6). Acute and chronic wrist pain, weakness, and palpable masses are common clinical problems initially evaluated with routine radiography. In spite of its nonspecificity and poor spatial resolution, three-phase bone scanning is a sensitive screen for osseous disease in the persistently painful wrist that appears normal with routine radiography (3). In those wrists with a positive bone scan, CT has been able to identify subtle fractures, osteoid osteomas, and intraosseous ganglia (Fig. 25.5) (3, 29). On the other hand, for the persistently painful wrist with a negative radiograph and suspected soft-tissue injury, magnetic reso-

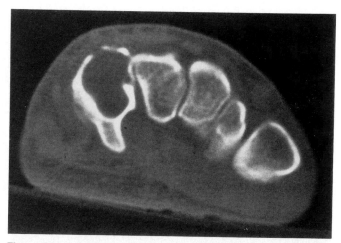

Figure 25.5. Axial image at the level of the hamate, in a 45-year-old male manual worker with long history of wrist pain and increasing weakness. Note the large, expansile, well-corticated lytic lesion replacing hamate body, with thinning of cortex dorsally and radially without soft-tissue mass. CT appearance and clinical features are compatible with expansile intraosseous ganglion. Patient refused surgical intervention.

nance MR imaging is probably the imaging modality of choice.

Trauma

It must be understood that current knowledge of wrist injuries and carpal instabilities is incomplete. Occupational and sports-related wrist trauma is common, with solitary or repetitive insults leading to disability and chronic wrist pain. Treatment of the injured wrist depends on the nature and extent of the injury and accurate assessment of the injury often requires a combination of imaging modalities. Routine radiographs and conventional tomography are difficult to interpret because of overlapping cortices, superimposition of cast material, and osteopenia (Fig. 25.6A). Specialized radiographic views are often not obtainable because pain and limb splints prevent appropriate positioning.

Multiplanar CT overcomes these limitations and accurately demonstrates articular and bony injuries (Fig. 25.6B and C). A global overview of complex osteoarticular disruptions can be provided by three-dimensional reformatting of the cross-sectional images and at least one report has demonstrated the fidelity of 3D CT reformatted images in displaying fracture gaps (30); however, surface-rendering computer software programs, due to inadequate thresholding techniques, may produce spurious fracture clefts and abnormal osseous fusions (23).

The detection of tendon and neurovascular structure entrapment or disruption is aided by comparison with the corresponding images of the uninjured wrist (Fig. 25.3) (29). Three-dimensional evaluation of dynamic carpal instabilities is still experimental (26). Magnetic resonance imaging and arthrography are superior to CT in evaluating the integrity of the intra- and extracapsular ligamentous restraints (2).

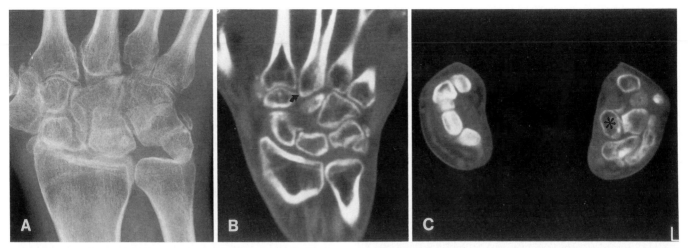

Figure 25.6. **(A)** Frontal radiograph of the wrist in a patient with an old industrial crush injury. Loss of carpal height, incongruent carpal arcs, and secondary degenerative changes are suggestive of a chronic midcarpal dislocation. **(B)** Midplane coronal image through the left wrist. Note absence of capitate, proximally subluxated third carpometacarpal joint *(short curved arrow)*, and moderate degenerative change involving the scapholunate and posterior aspect of the triquetrum with the proximally displaced pisiform. **(C)** Normal appearing axial slice through the level of the pisotriquetral joint on the right. On the left, the carpal tunnel contents have been effaced by the volar dislocated capitate *(asterisk)*. Degenerative sclerosis is demonstrated in the radioscaphoid, lunatotriquetral joints, and between the interposed articular surfaces of capitate and lunate. The pisiform is displaced ulnarly and proximally.

Figure 25.7. Axial image of the distal radial ulnar joint with the wrist in pronation, demonstrating dorsal subluxation of the right ulnar head.

Joint Injuries

Instability of the DRUJ is a clinically recognized source of disabling wrist pain. Motion of the radius relative to the ulna is controlled by the interosseous membrane, the triangular fibrocartilage complex (TFCC), and the extensor carpi ulnaris tendon (18, 31). In supination, the DRUJ is stabilized by the stronger palmar ligaments, the pronator quadratus muscle, and the interosseous membranous ligament. Soft-tissue disruptions resulting in DRUJ instability, therefore, occur more often with hyperpronation injuries (32). Conventionally, volar and dorsal subluxation of the ulnar head is described, although it is the distal radius that exhibits the aberrant motion. Instability of the DRUJ may be isolated, or associated with fractures or fracture-dislocation of the forearm. Inflammatory disorders such as rheumatoid arthritis, and congenital disorders such as Madelung's deformity, may also be associated with instability of the DRUJ and chronic wrist pain. Distal radioulnar joint instability leads to painful snapping of the extensor carpi ulnaris tendon, chondromalacia of the ulnar head, and degenerative arthrosis of the DRUJ with limited painful supination and pronation (33–35).

Fixed or dynamic articular incongruity of the DRUJ can be difficult to evaluate with routine radiography; it is best evaluated with CT using multiple contiguous 3-mm axial slices through the DRUJ of both wrists sequentially obtained in neutral, fully supinated and fully pronated positions. In the neutral position, the ulnar head articulates concentrically in the sigmoid notch of the distal radius. In supination, the ulnar head is in an eccentric dorsal distal position (positive variance), and with pronation moves to a volar proximal position (negative variance) by a combined rotatory and axial translational movement. Volar and dorsal dislocation are readily recognized (Figs. 25.7 and 25.8); however, subluxation is deceptively difficult to assess despite the use of Mino's tangential lines, or Wechsler arcs of congruency (17, 34).

In the author's opinion, there is a limited role for CT evaluation of acute complex intraarticular fractures and fracture-dislocations of the wrist. In the polytraumatized patient, initial CT assessment may not be feasible; but CT evaluation after initial stabilization and treatment may detect occult articular surface disruption, interposed intraarticular fracture fragments, and entrapped soft tissues that may be associated with persistent symptoms and routine radiographic malalignment (Figs. 25.3 and 25.9) (36–38). Preoperative CT of old crush injuries accurately identifies joint incongruities and bony mechanical blocks that need to be addressed in reconstructive surgery (Fig. 25.6*B* and *C*). Postoperative CT allows assessment of graft position, osseous alignment, and the position of surgical implants (4).

Common carpometacarpal (CCM) joint injuries are difficult to detect and assess clinically and radiographically. Unrecognized articular marginal avulsion or impaction fractures associated with dorsal diastasis of the CCM joint can lead to chronic painful instability requiring limited arthrodesis (39, 40). Computed tomography in the coronal and sagittal planes depicts the degree of osteoarticular malalignment and disruption (Fig. 25.10).

Osseous Injuries

Impaction and avulsion fractures involving the volar and dorsal processes, ridges, and bodies of the carpus are often difficult to detect with conventional radiography. Imaging of the traumatized wrist in at least two planes is essential.

Imaging planes perpendicular to the fracture cleft are recommended for optimal CT demonstration of fractures. The scaphoid is the most commonly fractured carpal bone; thus, coronal and sagittal scans through the long axis of the scaphoid are preferred (41–43). Malunion, posttraumatic cyst formation, and proximal pole osteonecrosis are significant posttraumatic sequelae (Fig. 25.11) (44, 45). Scaphoid malunion, with its characteristic humpback deformity, can be quantified by measuring the interpolar angle on direct sagittal CT images (46).

Associated soft-tissue entrapment or fracture fragment displacement can cause compromise of adjacent volar neurovascular bundles within the central or ulnar carpal canals (Fig. 25.6C). Axial imaging best demonstrates compromise of the flexor tendons, median nerve, and ulnar nerve associated with displaced fractures of the hamulus (Fig. 25.3) (47). Postoperative CT may also confirm the cause of residual pain and ulnar nerve dysthesias in patients with retained fracture fragments, perineural fibrosis of the canal of Guyon (Fig. 25.12), and ulnar artery thrombosis or aneurysms (48, 49).

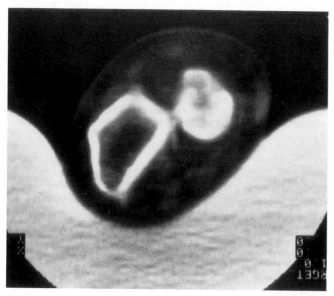

Figure 25.8. Axial image of the distal radial ulnar joint with the wrist in supination, demonstrating volar dislocation of the right ulnar head with impaction fracture and bony impingement.

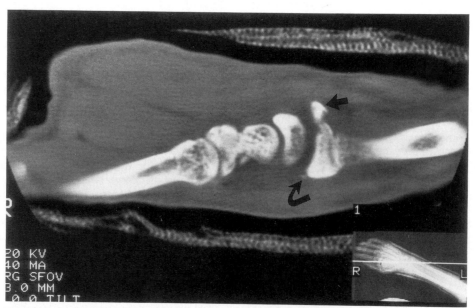

Figure 25.9. Parasagittal CT of wrist in plaster, demonstrating larger volar osteochondral fracture fragment off the distal radius *(solid arrow)* associated with volar subluxation of the carpus. Note the small avulsion fracture fragment off the dorsal radius *(curved arrow)* due to ligamentous disruption. (Top of image is volar side.)

Posttraumatic ischemic necrosis of the proximal pole of the scaphoid, lunate, and rarely the proximal pole of the capitate may lead to fragmentation and carpal collapse, resulting in a degenerative, painful arthrosis (50). Computed tomography demonstration of the degree of fragmentation, carpal collapse, and articular surface remodeling may change the surgical management from a limited to an extended arthrodesis. Postoperative CT of limited (triscaphe, lunatocapitate, triquetrohamate) or extensive carpal arthrodesis is recommended when routine radiography is inadequate in determining solidity (4, 51–53).

Accurate assessment of the degree of osseous union may help in deciding when to begin active rehabilitation (4).

Soft-Tissue Injuries

Injuries of the intra- and extraarticular carpal soft tissues may be caused by a direct blow, overuse, or chronic impingement. This may result in attenuation, disruption, or entrapment of the volar and dorsal tendons, with resultant tendinous edema, hematoma, scarring, and reparative changes that are better appreciated with MR imaging (see Chapter 24). Quinn et al. have reported on the supplemental role of postarthrography CT of soft-tissue injuries to demonstrate TFCC thinning, tears, and ulnar capsuloligamentous disruptions (20). Traumatic nerve injuries

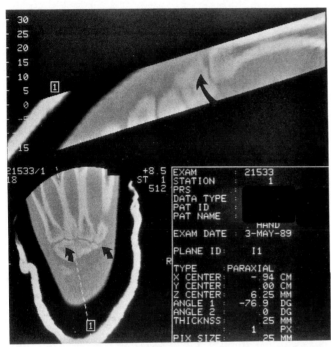

Figure 25.10. Postreduction dorsal coronal CT image of the left hand in plaster demonstrates osteochondral fracture fragments, widened fourth and fifth carpometacarpal joint spaces, and suspicion of impaction of the dorsal articular surface of the hamate *(small curved arrows)*. Dorsal impaction fracture through the hamate confirmed on reformatted sagittal images *(large curved arrow)*.

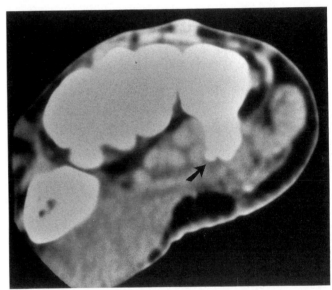

Figure 25.12. Axial images of the wrist at the level of the hamate, demonstrating postsurgical scarring in the palmar soft tissues overlying the surgically amputated hook of hamate. Residual ossific fragment *(arrow)* and loss of tissue planes around ulnar neurovascular bundle suggest perineural fibrosis involving the canal of Guyon, explaining patient's persistent painful parathesias in the distribution of the ulnar nerve.

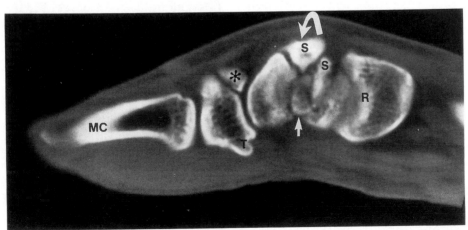

Figure 25.11. Parasagittal view through the scaphotrapeziometacarpal column. The volar trapezial ridge *(T)*, the base of the trapezoid *(asterisk)*, and the proximal metacarpal shaft *(MC)* are normal. The scaphoid demonstrates humpbacked deformity *(bent arrow)* with nonunion, sclerosis, and fragmentation of the avascular proximal pole of scaphoid *(S)*, and nonincorporation of interposed graft *(straight arrow)*.

or chronic entrapment may lead to neuroma formation and focal nodular degeneration, either mucoid or pseudocystic (54, 55). Computed tomography has also been used to localize intraarticular loose bodies and foreign bodies within the soft tissues (56, 4).

Neoplasms and Pseudotumors

Palpable wrist masses may or may not be painful. The clinical history, physical examination, and routine radiography often suggest a specific diagnosis or a limited differential. Computed tomography evaluation of bony tumors delineates the anatomic extent of the lesion (Fig. 25.13) and may narrow the differential diagnosis by more clearly defining the growth characteristics and features such as matrix calcification (57). Serial CT is of particular importance for the radiotherapist or surgeon trying to determine tumor response, tumor recurrence, and healing of interposed grafts in limb salvage procedures; however, MR imaging utilizing dedicated wrist coils is proving superior in staging tumor extent and determining treatment response.

Except for ganglia, soft-tissue tumors of the wrist are uncommon. Small, sharply demarcated lytic lesions of the carpal bones are common and probably represent degenerative cysts or intraosseous ganglia. Ganglia may present as progressively painful enlarging cystic lesions of bone and soft tissue, which must be differentiated from focal amyloid deposition and the subchondral pseudotumors of rheumatoid arthritis (58).

In the distal radius and the metacarpals, the typical lytic lesions of giant cell tumor or aneurysmal bone cyst can be readily differentiated from the permeative pattern of destruction seen with round cell tumors (i.e., Ewing's sarcoma, lymphoma, and leukemia).

Of the benign osteoblastic lesions, osteoid osteoma is not uncommon in the carpus and patients have often been symptomatic for over a year. Although three-phase bone scanning is the most sensitive screening modality, CT has been shown to be the most specific method of identifying the nidus (Fig. 25.14) (57, 59, 60). Primary malignant tumors and distal metastases are uncommon.

The most common soft-tissue tumor of the wrist is the ganglion. The etiology of the wrist ganglion is uncertain, but most likely represents a secondary manifestation of periscaphoid and intercarpal ligamentous injury (61). Ganglia commonly arise in the capsular soft tissues and tendon sheaths. Two-thirds of patients are women between the ages of 30 and 40. Ganglia may also occur in subperiosteal and intraosseous sites (62), and may extend into the carpal or ulnar tunnels to cause neurovascular compromise.

Superficial or deep lipomas of the wrist are common. Computed tomography provides easy, accurate assessment of the extent of fat-containing tissues as well as pseudoencapsulation or infiltration (63). Intraneural lipomas and lipofibromas have been reported to cause focal enlargement of the radial and median nerve (64, 65). Focal enlargement from a single traumatic neuroma should be differentiated from the diffuse changes in plexiform neurofibromatosis (66).

Nodular fasciitis and other forms of fibromatosis may produce focal or diffuse infiltrating fibrous masses that may

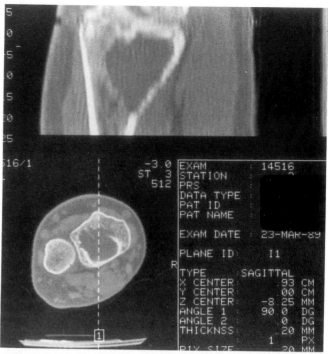

Figure 25.13. Axial image with sagittal reformatted image demonstrates a large eccentric lytic lesion involving the distal radial metaphysis. The lesion abuts the subchondral plate of the distal radial articular surface. The cortices are intact and no soft-tissue extension is apparent. Appearance is compatible with the pathological diagnosis of giant cell tumor of the distal radial metaphysis.

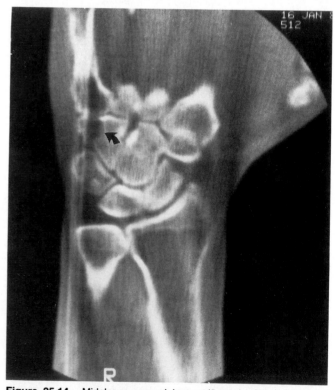

Figure 25.14. Midplanar coronal image (3 mm) of the right wrist, demonstrating beam-hardening artifact obscuring the ulnar aspect of the carpus. Subtle lytic lesion with well-defined margins is demonstrated in the ulnar aspect of the hamate *(small curved arrow)*. This was confirmed on the axial images not shown. Diagnosis of osteoblastoma was made at surgery, undertaken for unremitting wrist pain.

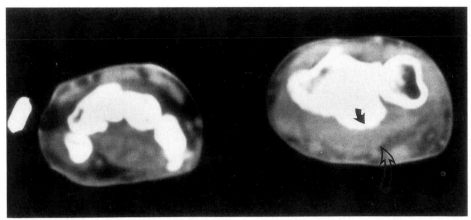

Figure 25.15. Axial images at the level of the distal radioulnar joint demonstrate marked volar periosteal new bone formation and reactive sclerosis *(small curved arrow)*. Soft tissues demonstrate loss of definition of tissue fat planes, with obliteration of Parona's space, and marked increase in density that extends into the common flexor tendinous compartment *(open arrow)*. Biopsy revealed nodular fascitis without evidence of granulomatous infection.

elicit adjacent bony reaction (Fig. 25.15) (67). Focal or diffuse benign capillary or cavernous hemangiomatous malformations frequently involve the wrist (68). Ectopic and anomalous muscle bundles are more common than muscle tumors (69). Reparative giant cell tumors of the tendon sheaths and synoviomas are uncommon in the wrist.

Arthropathies

Although CT has been used, MR imaging is the imaging modality of choice for the assessment of the carpal tunnel and its contents, especially for patients with carpal tunnel syndrome (13). Surgical decompression of the carpal tunnel, with sectioning of the transverse carpal ligament, may be associated with painful aberrant motion of the pisotriquetral joint, secondary osteoarthritis, and mechanical compromise of the ulnar canal contents that may be depicted on axial CT (70, 71).

Chronic pain on the ulnar aspect of the wrist may be due to osteoarthritis involving the pisotriquetral joint (72), osteoarthritis in an incomplete congenital coalition of the lunatotriquetral joint (73), and osteoarthritis of the DRUJ due to traumatic disruption of the TFCC (74). Computed tomography may be indicated in those cases with focal osteoarthritic processes in which lesions are atypical and routine radiography is inconclusive.

In severe rheumatoid arthritis (75), silicone-reactive granulomatous arthropathy (76), pigmented villonodular synovitis, and synovial chondrometaplasia (56), CT may aid the surgeon's preoperative planning and postoperative assessment by demonstrating the extent of osteoarticular involvement, adjacent soft-tissue disease, and residual bony stock.

Infection

Infection may involve the carpal bones, joints, tendon sheaths, and the potential closed spaces of the hand and forearm. Hematogenous osteomyelitis and septic arthritis of the wrist are uncommon; more commonly, infection involves the extraarticular portions of the distal radial or ulnar metaphysis. Accidental or occupation-related penetrating injuries may infect the soft tissues and articular compartments of the wrist. Chronic granulomatous synovial infections due to atypical mycobacteria are now recognized with increasing frequency (77). Preoperative CT of the infected wrist can identify deep sequestra, loculated abscesses, proximal or distal compartmental extension, and foreign bodies.

Conclusion

Wrist CT is one of several sophisticated imaging techniques that is selectively indicated when conventional imaging studies are inconclusive. Careful consultation with the referring physician is necessary to tailor the examination to the clinical problem.

Computed tomography has proven most useful in defining osteoarticular abnormalities, particularly following trauma. Computed tomography is also of value in identifying subtle osseous disease in the presence of an abnormal bone scan and normal radiographs. Soft-tissue processes, in contrast, are better delineated by MR imaging.

As in other areas of the body, attention to technical detail and an understanding of the normal anatomy will maximize the successful use of CT in the wrist.

References

1. Fisk GR. The wrist. J Bone Joint Surg 1984;66(b):396–407.
2. Quinn SF, Belsole RJ, Greene TL, Rayhack JM. Advanced imaging of the wrist. Radiographics 1989;9:229–246.
3. Hindman BW, Kulik WJ, Lee G, Avolio RE. Occult fractures of the carpals and metacarpals: demonstration by CT. AJR 1989;153:529–532.
4. Quinn SF, Murray W, Watkin, T, Kloss J. CT for determining the results of treatment of fractures of the wrist. AJR 1987;149:109–111.
5. Rhoades CE, Soye I, Levene E, Reckling FW. Detection of wooden foreign body in the hand using computed tomography; case report. J Hand Surg (Am) 1982;7:306–307.
6. Heiken JP, Lee JKT, Smathers RL, Totty WG, Murphy WA. CT of benign soft tissue masses of the extremities. AJR 1984;142:575–580.
7. Garcia-Elias M, An K-N, Cooney WP, Linscheid RL, Chao EYS. Stability of the transverse carpal arch: an experimental study. J Hand Surg (Am) 1989;14:277–282.

8. Cone RO, Szabo R, Resnick D, Gelberman R, Taleisnik J, Gilula L. Computed tomography of the normal soft tissue of the wrist. Invest Radiol 1983;18:546–551.

9. Biondetti PR, Vannier MW, Gilula LA, Knapp P. Wrist: coronal and transaxial CT scanning. Radiology 1987;163:149–151.

10. Kauer JMG. Functional anatomy of the wrist. Clin Orthoped 1980;149:9–20.

11. Epner RA, Bowers WH, Guildford WB. Ulnar variance—the effect of wrist positioning and Roentgen filming technique. J Hand Surg (Am) 1982;7:298–305.

12. Paley D, Axelrod TS, Martin C, Rubenstein J, McMurtry RY. Radiographic definition of the dorsal and palmar edges of the distal radius. J Hand Surg (Am) 1989;14:272–276.

13. Zucker-Pinchoff B, Hermann G, Srinivasan R. Computed tomography of the carpal tunnel: a radioanatomical study. J Comput Assist Tomogr 1981;5:525–528.

14. Rawles JG. Dislocations and fracture-dislocations at the carpometacarpal joints of the fingers. Hand Clin 1988;4:103–112.

15. Patel RB. Evaluation of complex carpal trauma; thin-section direct longitudinal computed tomography scanning through plaster cast. J Comput Assist Tomogr 1985;9:107–109.

16. Robertson DD, Weiss PJ, Fishman EK, Magid D, Walker PS. Evaluation of CT techniques for reducing artifacts in the presence of metallic orthopedic implants. J Comput Assist Tomogr 1988;12:236–241.

17. Wechsler RJ, Wehbe MA, Rifkin MD, Edeiken J, Branch HM. Computed tomography diagnosis of distal radioulnar subluxation. Skel Radiol 1987;16:1–5.

18. King GJ, McMurtry RY, Rubenstein JD, Gertzbein SD. Kinematics of the distal radioulnar joint. J Hand Surg (Am) 1986;11:798–804.

19. Young SW, Noon MA, Marincek B. Dynamic computed tomography time density study of normal human tissue after intravenous contrast administration. Invest Radiol 1981;16:36–39.

20. Quinn SF, Belsole RS, Greene TL, Rayhack JM. Work in progress: post arthrography computed tomography of the wrist: evaluation of the triangular fibrocartilage complex. Skel Radiol 1989;17:565–569.

21. Kuschner SH, Gelberman RH, Jennings C. Ulnar nerve compression at the wrist. J Hand Surg (Am) 1988;13:577–580.

22. Weeks PM, Vannier MW, Stevens WG, Gayou D, Gilula LA. Three-dimensional imaging of the wrist. J Hand Surg (Am) 1985;10:32–39.

23. Vannier MW, Totty WG, Stevens WG, et al. Musculoskeletal applications of three dimensional surface reconstructions. Orthoped Clin N Am 1985;16:543–555.

24. Sutherland CJ. Practical application of computer-generated three-dimensional reconstructions in orthopedic surgery. Orthoped Clin N Am 1986;17:651–656.

25. Ney DR, Fishman EK, Magid D, Kuhlman JE. Interactive real-time multiplanar CT imaging. Radiology 1989;170:275–276.

26. Belsole RJ, Hilbelink D, Llewellyn JA, Dale M, Stenzler S, Rayhack JM. Scaphoid orientation and location from computed, three dimensional carpal models. Orthoped Clin N Am 1986;17:505–510.

27. Huda W, Sandison GA, Lee TY. Patient doses from computed tomography in Manitoba from 1977 to 1987. Br J Radiol 1989;62:138–144.

28. Mostrom U, Ytterbergh C. Spatial distribution of dose in computed tomography with special reference to thin slice techniques. Acta Radiol 1987;28:771–777.

29. Quinn SF, Belsole RJ, Greene TL, Rayhack JM. Ct of the wrist for the evaluation of traumatic injuries. Crit Rev Diagn Imag 1989;29:357–380.

30. Drebin RA, Magid D, Robertson DD, Fishman EK. Fidelity of three dimensional CT imaging for detecting fracture gaps. J Comput Assist Tomogr 1989;13:487–489.

31. Hotchkiss RN, An K-N, Sowa DT, Basta S, Weiland AJ. An anatomic and mechanical study of the interosseous membrane of the forearm: pathomechanics of proximal migration of the radius. J Hand Surg (Am) 1989;14:256–261.

32. Drewniany JJ, Palmer AK. Injuries to the distal radioulnar joint. Orthoped Clin N Am 1986;17:451–459.

33. Essex-Lopresti P. Fractures of the radial head with distal radialulnar dislocations. J Bone Joint Surg 1951;33B:244–247.

34. Mino DE, Palmer AK, Levinsohn EM. The role of radiography and computerised tomography in the diagnosis of subluxation and dislocation of the distal radioulnar joint. J Hand Surg (Am) 1983;8:23–31.

35. Scheffler R, Armstrong D, Hutton I. Computed tomographic diagnosis of distal radioulnar joint disruption. J Can Assoc Radiol 1984;35:212–213.

36. Green DP, O'Brien ET. Classification and management of carpal dislocations. Clin Orthoped 1980;49:55–72.

37. Garcia-Elias M, Dobyns J, Cooney WP, Linscheid RL. Traumatic axial dislocations of the carpus. J Hand Surg (Am) 1989;14:446–457.

38. Bellinghausen HW, Gilula LA, Young LV, Weeks PM. Post traumatic palmar carpal subluxation. J Bone Joint Surg 1983;65(A):998–1006.

39. Joseph RB, Linscheid RL, Dobyns JH, Bryan RS. Chronic sprains of the carpometacarpal joints. J Hand Surg (Am) 1981;6:172–180.

40. Carroll RE, Carlson E. Diagnosis and treatment of injury to the second and third carpometacarpal joints. J Hand Surg (Am) 1989;14:102–107.

41. Bush CH, Gillespy T III, Dell PC. High resolution CT of the wrist: initial experience with scaphoid disorders and surgical fusions. AJR 1987;49:757–760.

42. Friedman L, Yong-Hing K, Johnston GH. Forty degree angled coronal CT scanning of scaphoid fractures through plaster and fiberglass casts. J Comp Assist Tomogr 1989;13:1101–1104.

43. Pennes DR, Jonsson K, Buckwalter KA. Direct coronal CT of the scaphoid bone. Radiology 1989;171:870–871.

44. Amadio PC, Berquist TH, Smith DK, Ilstrup DM, Cooney WP, Linscheid RL. Scaphoid malunion. J Hand Surg (Am) 1989;14:679–687.

45. Ferlic DC, Morin P. Idiopathic avascular necrosis of the scaphoid: Preiser's disease. J Hand Surg (Am) 1989;14:13–16.

46. Sanders WE. Evaluation of the humpback scaphoid by computed tomography in the longitudinal axial plane of the scaphoid. J Hand Surg (Am) 1988;13:182–187.

47. Egawa M, Asai T. Fracture of the hook of the hamate; report of six cases and suitability of computerised tomography. J Hand Surg (Am) 1983;8:393–398.

48. Foster EJ, Palmer AK, Levinsohn EM. Hamate erosion: an unusual result of ulnar artery constriction. J Hand Surg (Am) 1979;4:536–539.

49. Watson HK, Rogers WD. Nonunion of the hook of the hamate: an argument for bone grafting the nonunion. J Hand Surg (Am) 1989;14:486–490.

50. Viegas SF, Calhoun JH, Eng M. Osteochondritis dissecans of the lunate. J Hand Surg (Am) 1987;12:130–133.

51. Eckenrode JF, Louis DS, Greene TL. Scaphoid-trapezium-trapezoid fusion in the treatment of chronic scapholunate instability. J Hand Surg (Am) 1986;11:497–502.

52. Kleinman WB. Long term study of chronic scapho-lunate instability treated by scapho-trapezio-trapezoid arthrodesis. J Hand Surg (Am) 1989;14:429–445.

53. Rogers W, Watson KW. Radial styloid impingement after triscaphe arthrodesis. J Hand Surg (Am) 1989;14:297–301.

54. Allieu PY, Cenac PE. Peripheral nerve mucoid degeneration of the upper extremity. J Hand Surg (Am) 1989;14:189–194.

55. Shea JD, McClain EJ. Ulnar nerve compression syndrome at and below the wrist. J Bone Joint Surg 1969;51A:1095–1103.

56. Tehranzadeh J, Gabriele OF. Intra-articular calcified bodies: detection by computed arthrotomography. South Med J 1984;77:703–710.

57. Stoker DJ. Management of bone tumours—the radiologist role. Clin Radiol 1989;40:233–239.

58. Rennell C, Mainzer F, Multz CV, Genant HK. Subchondral pseudocysts in rheumatoid arthritis. AJR 1977:129;1069–1072.

59. Shaw JA. Osteoid osteoma of the lunate. J Hand Surg (Am) 1987;12:128–130.

60. Marck KW, Dhar BK, Spauwen PHM. A cryptic cause of monoarthritis in the hand: the juxta-articular osteoid osteoma. J Hand Surg 1988;13B:221–223.

61. Watson HK, Rogers WD, Ashmead D. Re-evaluation of the cause of the wrist ganglion. J Hand Surg (Am) 1989;14:812–817.

62. Schajowicz F, Clavel Sainz M, Slullitel JA. Juxta-articular bone

cysts (intra-osseous ganglia): a clinicopathological study of eighty-eight cases. J Bone Joint Surg 1979;61B:107–116.

63. Oster LH, Blair W, Steyers CM. Large lipomas in the deep palmar space. J Hand Surg (Am) 1989;14:700–704.

64. Jacob RA, Buchino JJ. Lipofibroma of the superficial branch of the radial nerve. J Hand Surg (Am) 1989;14:704–706.

65. Houpt P, Storm van Leeuwen JB, van den Bergen HA. Intraneural lipofibroma of the median nerve. J Hand Surg (Am) 1989;14:706–709.

66. Case report: peripheral neurofibromatosis. J Comput Assist Tomogr 1983;7(2):374–375.

67. Hajdu SI. Benign soft tissue tumours: classification and natural history. CA 1987;37:66–76.

68. Rauch RF, Silverman PM, Korobkin M, et al. Computed tomography of benign angiomatous lesions of the extremities. J Comput Assist Tomogr 1984;8(6):143–146.

69. Still JM, Kleinert HE. Anomalous muscles and nerve entrapment in the wrist and hand. Plast Reconstr Surg 1973;52:394–400.

70. Richman JA, Gelberman RH, Rydevik BL, et al. Carpal tunnel syndrome: morphologic changes after release of the transverse carpal ligament. J Hand Surg (Am) 1989;14:852–857.

71. Seradge H, Seradge E. Piso-triquetral pain syndrome after carpal tunnel release. J Hand Surg (Am) 1989;14:858–862.

72. Paley D, McMurty RY, Cruickshank B. Pathologic conditions of the pisiform and pisotriquetral joint. J Hand Surg (Am) 1987;12:110–119.

73. Gross SC, Watson HK, Strickland JW, Palmer AK, Brenner LH. Triquetral-lunate arthritis secondary to synostosis. J Hand Surg (Am) 1989;14A:95–102.

74. Palmer AK. Triangular fibrocartilage complex lesions: a classification. J Hand Surg (Am) 1989;14:594–606.

75. Cope R. The surgery of the rheumatoid wrist: postoperative appearances and complications of the more common procedures. Skel Radiol 1989;17:576–582.

76. Smith RJ, Atkinson RE, Jupiter JB. Silicone synovitis of the wrist. J Hand Surg (Am) 1985;10:47–60.

77. Dillon J, Millson C, Morris I. *Mycobacterium kansasii* infection in the wrist and hand. Br J Rheumatol 1990;29:150–153.

CHAPTER 26

Magnetic Resonance Imaging Evaluation of Avascular Necrosis of the Hip

Cheryl L. Kirby, Susan J. F. Meyer, and Murray K. Dalinka

Introduction

Avascular necrosis (AVN), also known as osteonecrosis, ischemic necrosis, or aseptic necrosis, is a condition resulting from insufficient blood supply to subchondral bone. By convention, when bone necrosis occurs in the metaphyseal or diaphyseal region, the disease process is referred to as bone infarction. The histology of AVN and bone infarction is identical; the only difference is the location of the ischemia.

Trauma is the most common cause of AVN (1). The ischemia follows femoral neck fractures or hip dislocations that interrupt the nutrient, retinacular, and periosteal vessels. The incidence of AVN is related to the degree of fracture displacement or the delay in reduction of the dislocation (1).

Nontraumatic AVN is usually a bilateral disease that most commonly affects the proximal and distal femora and the proximal humeri. There are many predispositions for AVN (2); causes of AVN of the femoral head are as follows:

1. Traumatic
   a. Following fracture
   b. Following reduction
   c. Barotrauma
   d. Postradiation
2. Idiopathic
   a. Legg-Perthes disease
   b. Idiopathic aseptic necrosis in adult
3. Systematic diseases
   a. Hemoglobinopathies
   b. Storage diseases
   c. Alcoholism
   d. Collagen vascular disease
4. Steroids

Predispositions most commonly encountered are high-dosage exogenous steroids, sickle cell disease, and alcoholism (3, 4). Although an association with collagen-vascular disease has been reported, many of these patients have also been treated with corticosteroids (1, 5).

An idiopathic form of AVN is most commonly found in men (6). After separating patients with alcoholism and other predisposing conditions, one group arrived at an incidence of 11.5% for the idiopathic form (7). Legg-Perthes disease represents the idiopathic variety in childhood (8).

This chapter will discuss nontraumatic AVN of the femoral head and the use of magnetic resonance (MR) imaging in the diagnosis, staging, and treatment of this entity.

Etiology and Pathogenesis

The insufficient blood supply associated with the development of AVN may be secondary to decreased or obstructed blood flow. The decreased blood flow may be arterial, venous, or sinusoidal and obstructed blood flow may be secondary to an intravascular or extravascular process (6). Intravascular obstruction may result from vessel wall disease or a thrombotic or embolic event. Extravascular sinusoidal occlusion leading to AVN may cause osteonecrosis in patients with Gaucher's disease (9). In this disorder glucocerebrosides are deposited in the reticuloendothelial system and bone marrow, occluding the extravascular sinusoids (1). In trauma, transection of the vessel is the obvious cause of vascular interruption.

The etiologic basis of nontraumatic AVN is unknown although many theories have been proposed. Jones (10) believes that AVN results from fatty emboli that may arise from a fatty liver, destabilization and coalescence of plasma lipoproteins, and/or disruption of adipose tissue depots such as fatty bone marrow. Increased deposition of fat in the liver and other tissues has been seen with hypercortisolism and alcohol abuse (11, 12). Fat emboli have been identified histologically in patients with AVN and 13 predisposing conditions (10). Jacobs (4) found fatty emboli in 89% of 269 patients with nontraumatic AVN. The presence of fatty emboli does not, however, indicate a causal relationship.

Abbreviations (see also glossary): AVN, avascular necrosis; SPECT, single-photon emission computed tomography; BMS, bone marrow scanning; BS, bone scan; MPR, multiplanar reformation; ROC, receiver operating characteristic.

Solomon (1, 11) studied the marrow changes in patients with unilateral AVN secondary to steroid therapy or associated with alcoholism. Intraosseous pressures and core biopsy specimens were obtained from both the affected hip and the X-ray-negative hip. Increased intraosseous pressure, a paucity of vascular sinusoids, and increased size and deposition of fat cells were found in both hips. Solomon postulated that the increase in size and number of fat cells led to sinusoidal compression, resulting in venous stasis and eventually bone necrosis.

Many authors (6, 13–15) have demonstrated increased bone marrow pressure within the femoral necks of patients with AVN; this suggests that the ischemia may be secondary to a "compartment syndrome," with the femoral neck acting as an unyielding bony envelope. Hungerford and Lennox (6) believe that ischemia within this compartment produces marrow edema. This edema increases bone marrow pressure and creates further edema, perpetuating a cycle leading to worsening ischemia and eventually to necrosis.

Arlot et al. (16) believe subclinical metabolic bone disease is important in the development of osteonecrosis. They obtained transiliac bone biopsies from 77 patients with AVN of the hip. Sixty-two percent of their patients were alcoholics or on steroids. Seventy-four of 77 patients had abnormal bone histology; 9 patients had osteomalacia and 65 had osteoporosis. They believed that the decrease in bone formation was secondary to a direct depressive effect on osteoblasts and/or secondary hyperparathyroidism caused by steroids or alcohol.

Saito and colleagues (17) believe that hemorrhage is a major factor in the pathogenesis of AVN. Core biopsy specimens were obtained from 16 asymptomatic patients with positive bone scans and negative X-rays (the "silent stage of AVN"). They found multifocal and multiphasic areas of hemorrhage associated with bone necrosis. Although hemorrhagic episodes may occur elsewhere in the bone marrow, the combination of hemorrhage with in-creased mechanical stress may explain the increased incidence of AVN in the femoral head.

Kenzora and Glimcher (18) feel that AVN is multifactorial in etiology and results from abnormal underlying bone and accumulated cell stresses. The abnormal underlying bone may be produced by steroids, alcohol, drugs, neoplasm, or systemic diseases such as systemic lupus erythematosis, renal disease, or hemoglobinopathies. The underlying abnormality may be intensified by increased stress, which may lead to mechanical failure.

Early Diagnosis

Plain film radiography detects changes in bone mineral content. Dead bone has the same radiographic appearance as live bone. Plain film abnormalities in AVN are a result of the osteoclastic and osteoblastic activity associated with healing, and hence appear as late rather than early findings (5). Plain film radiography is not a sensitive modality for early detection of AVN (15).

Radionuclide scanning, although more sensitive than plain films (19–21), is less sensitive than MR imaging in the early detection of AVN (Fig. 26.1) (22–24). Planar scintigraphy uses technetium-99m (^{99m}Tc)-tagged phosphates as a scanning agent. Although the precise mechanism of action of Tc phosphates is unknown, factors affecting the accumulation of tracer in bone include blood supply, chemisorption, and bone remodeling (21, 25). In very early AVN (just after the vascular insult) the bone scan may show decreased activity (cold spot) (Fig. 26.2). Diagnosis at this stage is unusual because the patient is frequently asymptomatic and the defect may be small and difficult to detect (21). Increased scintigraphic activity occurs with repair (Fig. 26.3) and can be depicted in rabbits within 3 weeks of the insult, well in advance of radiographic abnormalities. The resolution of bone scanning is limited; hence, comparative images of both hips and pinhole col-

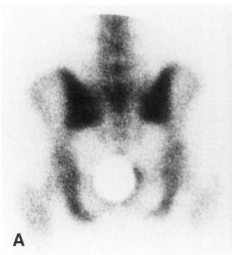 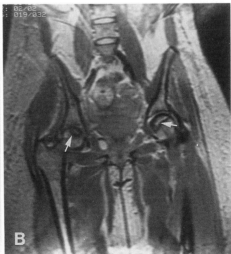

Figure 26.1. Bilateral AVN with normal bone scan and positive MR imaging in a 23-year-old female with systemic lupus erythematosis. **(A)** Normal bone scan. **(B)** Coronal TR 2500/TE 40 image demonstrating classic MR imaging findings of AVN in both hips *(arrows)*, despite the poor quality of the image.

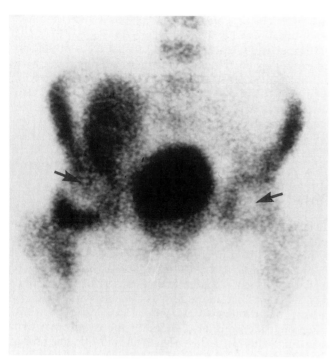

Figure 26.2. Anterior bone scan in a patient with bilateral AVN after renal transplantation. Photopenic defects of AVN *(arrows)* are present in both femoral heads, with increased uptake on the right probably representing reparative phase. Incidental note is made of a right renal transplant.

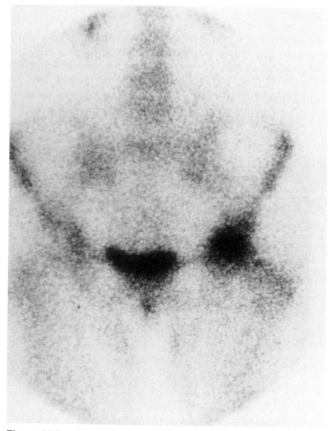

Figure 26.3. Anterior bone scan depicting increased tracer activity in the left hip secondary to AVN.

limation are crucial for interpretation (26, 27). False negative results may occur in bilateral disease.

Although planar scintigraphy is more sensitive than radiography, it is relatively nonspecific. Increased tracer uptake may be seen with multiple other disorders, including osteomyelitis, fractures, arthritis, and neoplasm (20).

Single-photon emission computed tomography (SPECT) is more sensitive than planar scintigraphy (28, 29) and less sensitive than MR imaging (29, 30). The reported sensitivity of SPECT varies from 58% (30) to 87% (29). The radioisotope used is ^{99m}Tc-labeled medronate. The diagnosis of osteonecrosis is determined by the presence of a radionuclide-deficient area (photopenic defect) with or without a surrounding area of increased uptake. The increased sensitivity of SPECT in comparison to planar scintigraphy is related to the elimination of the uptake of superimposed structures by the use of cross-sectional tomography. Miller et al. believe that SPECT is more sensitive in diagnosing early AVN; advanced disease is often associated with nondiagnostic areas of increased uptake (30). Disadvantages of SPECT include the need for bladder catheterization and the intravenous injection of the radionuclide. The SPECT examination may require 4 to 5 hr to complete while MR imaging of the hip may be performed in 20 min (30).

Bone marrow scanning (BMS) utilizing ^{99m}Tc-labeled sulfur colloid (SC) has been utilized in differentiating osteonecrosis from osteomyelitis in patients with sickle cell disease. Normally fat marrow replaces red marrow in the appendicular skeleton, beginning in childhood. In patients with sickle cell disease there is increased marrow production with expansion, or lack of regression, of the hematopoietic marrow (31).

Rao and colleagues (32) performed bone scans (BS) and BMS on 42 children with sickle cell disease and bone pain. Bone scans depicted increased uptake in most patients with either AVN or osteomyelitis. Of 16 patients with AVN, 15 had decreased uptake on BMS and all five patients with osteomyelitis had normal BMS. Decreased uptake on BMS was not seen in osteomyelitis; however, the studies were performed less than 10 days after the onset of symptoms.

Alavi (33) and Alavi and Heyman (34) studied induced osteomyelitis in a rabbit model and reported normal tracer uptake on BMS for 5 to 6 days and decreased uptake after 1 week.

Kim et al. (35) believed that osteonecrosis could be differentiated from osteomyelitis in patients with sickle cell disease by utilizing a combination of BMS and BS within 3 days of the onset of bone pain. Patients with infarction and less than 3 days of bone pain showed decreased uptake on BMS and decreased or normal uptake on BS. In patients with bone pain of more than 7 days' duration, decreased uptake on BMS and increased uptake on BS were seen in most cases of infarction or infection. Hence, late imaging findings of infarction may not differ from those seen with osteomyelitis. The increased sensitivity and specificity of MR imaging probably obviates the use of marrow scanning in sickle cell patients with suspected osteonecrosis.

Computed tomography (CT) is more sensitive than radiography (36) and less sensitive than MR imaging in the evaluation of AVN of the hip (22). As the result of in-

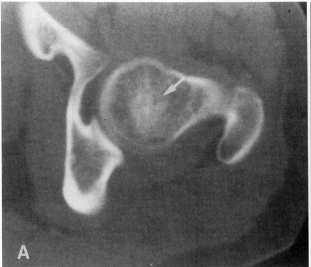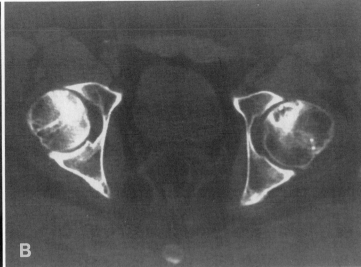

Figure 26.4. **(A)** Axial CT image of left femoral head, illustrating the normal weight-bearing trabeculae resembling an asterisk *(arrow).* **(B)** Axial CT of both hips, demonstrating bilateral ischemic necrosis. The right hip has a sclerotic area of osteonecrosis anteriorly with small areas of lucency within it. A core decompression tract enters the sclerotic area from the posterior aspect of the femor. On the left is a cystic area of osteonecrosis.

creased resolution and axial tomography, the bony architecture of the hip can be seen without overlapping structures and with greater detail than seen on radiographs or planar scintigraphy (36, 37). Dihlmann (38) described the appearance of the physiologic weight-bearing trabeculae of the femoral head as an asterisk of increased central density radiating to the periphery (Fig. 26.4*A*). With AVN there is a change in the appearance of this asterisk secondary to new bone formation and fractures. In the early stages of AVN clumping and fusion of the peripheral raylike structures occur. In advanced AVN, the asterisk is lost and fractures are seen in association with femoral head collapse (Fig. 26.4*B*).

Casteleyn et al. (37) used CT and disulphine blue dye to evaluate bone destruction and viability of the femoral head in osteonecrosis. They believed both techniques were helpful in diagnosis and treatment planning.

Magid and coworkers (36) utilized computed tomography with multiplanar reformation (CT/MPR) in the evaluation of AVN of the hip. With CT/MPR, they were able to obtain sagittal and coronal images to evaluate the weight-bearing portions of the joint. The use of these techniques resulted in an increase in the stage of the disease in 30% (14/45) of cases and a diagnosis of AVN in the contralateral hip in 22% (7/32) of patients. These authors concluded that CT/MPR was useful in the diagnosis of early disease, staging, and treatment planning of AVN.

Sartoris et al. (39) evaluated CT scanning with multiplanar and three-dimensional computerized tomography (3D CT) reformation in patients with advanced AVN of the hip. These techniques did not add new diagnostic information to that achieved by routine CT; however, the information attained may aid in preoperative planning with respect to the extent of the lesion and the degree of subchondral collapse. The information generated by 3D CT

can also be used in manufacturing polyethylene models or designing custom prostheses.

In a controlled study comparing MR imaging, CT, and planar scintigraphy, MR imaging was more sensitive than CT, which was more sensitive than nuclear medicine in diagnosing early AVN (22). Mitchell et al. (40) showed that CT was more accurate than MR imaging in detecting fractures. Thus the information derived from CT may alter the staging of the disease and affect treatment (22, 40).

Magnetic resonance imaging is the most sensitive imaging modality in the early detection of ischemic necrosis of the hip (22, 23, 41–44). Glickstein et al. (45) utilized receiver operating characteristic (ROC) curves to determine the usefulness of MRI in differentiating AVN from other hip diseases (Figs. 26.5 and 26.6). Magnetic resonance imaging was 97% sensitive in differentiating AVN from normal hips, and 85% sensitive in differentiating AVN from other diseases.

The high resolution of MR imaging is dependent on intrinsic tissue relaxation times (T1 and T2) as well as proton density. Tissues have different signal characteristics that may change with different pulse sequences. Fat has a short T1 relaxation time and produces high signal intensity on short TR/TE (T1-weighted) images. The femoral head is composed primarily of fatty marrow. Replacement of the normal fatty marrow lengthens the T1 relaxation time and decreases the MR imaging signal intensity.

Ficat, Hungerford, and others have utilized more invasive techniques in the early diagnosis of AVN (13, 14). These techniques consist of bone marrow pressure measurements followed by intramedullary venography and core biopsy of the femoral head. They have termed this study functional exploration of bone and perform it under local or general anesthesia.

The normal baseline bone marrow pressure in the inter-

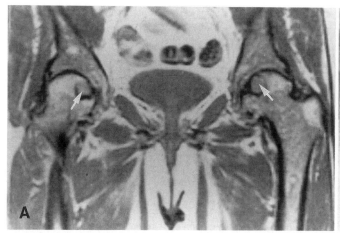

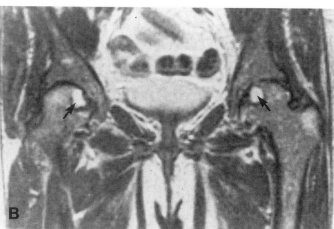

Figure 26.5. Degenerative cysts in osteoarthritis. Coronal **(A)** TR 2000/ TE 30 and **(B)** TR 2000/TE 80 images of a 67-year-old female, depicting degenerative cysts in both femoral heads *(arrows)*. Note the uniform intensity of the cysts on both imaging sequences and the degenerative spurs about the acetabula. The position of the defects is more medial than the defects of AVN.

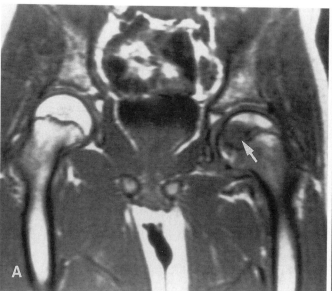

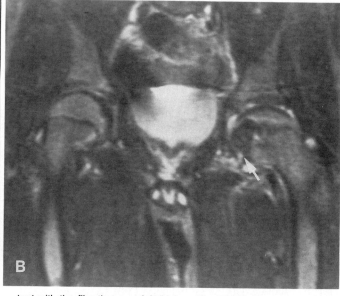

Figure 26.6. Chondroblastoma. Coronal images of hips in a 15-year-old male with left hip pain. **(A)** TR 600/TE 20 and **(B)** TR 2500/TE 80 images show a low-intensity lesion in the femoral head *(arrow)* that is associated with a joint effusion. The low-signal lesion could be consistent with the fibrotic type of AVN (see Fig. 26.13); however, the lesion is located more inferomedially than typical AVN. Additionally, despite the large size of the lesion, the femoral head contour is maintained.

trochanteric region is less than 30 mm Hg. If this is normal, a saline stress test is performed. This consists of injecting 5 ml of isotonic saline into the bone and recording the change in baseline pressure. A pressure change of 10 mm Hg or less is normal. This is followed by intramedullary venography, which consists of injecting 10 ml of contrast into the trochanteric area. Normally there is rapid clearing of the contrast by the efferent vessels. In patients with AVN, the injection is often painful and difficult, with diaphyseal reflux and stasis of contrast material 15 min after injection. If these are normal, the trocar is placed in the femoral head and the bone marrow pressure measurements are repeated. A core biopsy of the femoral head is then obtained for histologic evaluation.

In a series by Beltran et al. (46), bone marrow pressure measurements were more sensitive than MR imaging (92% vs. 89%); however, they were nonspecific with a false positive rate of 57%. Increased bone marrow pressure may also be seen in reflex sympathetic dystrophy, degenerative arthritis, and venous disease (46).

Histopathology of Avascular Necrosis

Osteonecrosis can be subdivided into four histopathologic zones: a central area of cell death surrounded by successive zones of ischemia, hyperemia, and normal tissue.

A reactive interface representing repair encompasses the hyperemic zone and part of the ischemic zone (5).

The cells found within bone marrow vary in their response to ischemia. The hematopoietic cells are the most sensitive and usually die after 6 to 12 hr of anoxia. Osteocytes, osteoclasts, and osteoblasts are more resistant to anoxia and survive for 12 to 48 hr. Marrow fat cells are the most resistant and can survive anoxia for 2 to 5 days (5).

Although cell death occurs within hours to days, dead bone has the same radiographic appearance as living bone and hence plain film findings are absent early in the disease. Complete cell autolysis may take days to months (5). Once necrosis occurs, the cellular breakdown products provoke a hyperemic response at the margin of the ischemic tissue, forming the hyperemic zone. This leads to fibrous tissue deposition and osteoclastic resorption of bone. Plain radiographs may demonstrate focal osteopenia surrounding the unresorbed central zone (dead bone), which appears relatively more sclerotic. As healing continues, bone resorption stimulates compensatory bone production at the outer margin of the reactive interface, which results in a true sclerotic margin with a central "cystic" lesion. With time, the reactive interface encroaches on the central zone. Further bone resorption weakens the weight-bearing trabeculae and creates microfractures ("crescent" sign) (Fig. 26.7) that eventually lead to flattening and collapse of the femoral head (Fig. 26.8).

Staging

The primary classification system used in AVN is that described by Arlet and Ficat (13, 47). In this classification stage 0 represents biopsy-proven AVN in asymptomatic patients with normal X-rays. This constellation of findings is referred to as the "silent hip." Hungerford confirmed the existence of AVN in the silent hip by measuring bone marrow pressures in the asymptomatic, X-ray-negative hips of 27 patients with documented osteonecrosis of the contralateral hip. Seventeen of these patients had increased bone marrow pressures and 11 patients developed biopsy-proved AVN in 1 to 5 years (15).

In stage I AVN, signs and symptoms of hip disease are present, particularly pain and decreased range of motion. The pain is usually sudden in onset and may radiate to the thigh. The limitation of motion may be multidirectional or primarily in abduction or medial rotation. In this stage, radiographs are normal or questionably abnormal and depict a subtle loss of cortical or trabecular clarity or slight osteoporosis.

In stage II disease, the clinical symptoms persist or worsen and the radiographs show mixed sclerotic and osteoporotic changes in the femoral head. The sclerotic

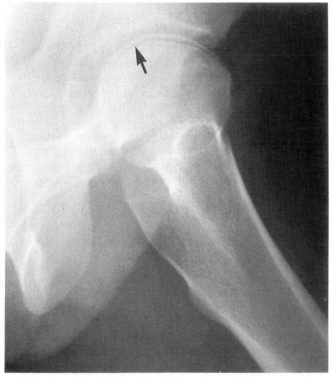

Figure 26.7. Frog lateral view of left hip demonstrates the "crescent" sign *(arrow)* of AVN.

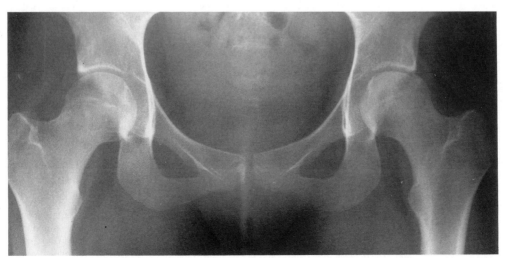

Figure 26.8. Anteriorposterior view of both hips depicts AVN with subchondral collapse of the left hip.

changes may be diffuse, localized, or arcuate and the osteoporotic areas may appear cystlike.

In stage III, there is collapse of the femoral head with maintenance of the joint space. The appearance of a subchondral fracture without visible cortical collapse, also known as the crescent sign (Fig. 26.7), represents a transition between stages II and III.

In stage IV disease, there is marked collapse of the femoral head with loss of the joint space, indicating the development of degenerative joint disease.

This classification has been modified by Steinberg et al. (48, 49). Subclassifications were established by planimetric measurements in an attempt to quantify the area involved and relate it to the results of treatment (48, 49). They eliminated clinical findings and incorporated the new imaging modalities such as bone scanning (48, 49) and, more recently, MR imaging (50).

Magnetic Resonance Imaging Technique

The early diagnosis of AVN can be easily established with virtually any currently available MR imaging unit. Initially, we utilized a 0.13-T resistive magnet, which was later replaced by a 1.5-T superconducting unit (General Electric, Milwaukee, Wis.). Our standard protocol consists of coronal T1-weighted spin-echo images with a TR of 400 to 600 ms and a TE of 11 to 20 ms, utilizing a body coil. The slice thickness is 5 mm with a 2.5-mm interslice gap. Coronal proton density and T2-weighted images are then obtained with a long TR of 2000 to 2500 ms and multiecho TE of 40 and 80 msec with a 5-mm slice thickness and a 1-mm interslice gap. The field of view is 32 to 48 cm and

the images are reconstructed using a 128×256 data matrix acquired with two excitations.

The diagnosis of AVN is easily made with standard short TR/TE (T1-weighted) coronal images (51). The long TR/multiecho TE images are used to increase the diagnostic specificity and may aid in the depiction of other abnormalities about the hip. We feel that other imaging planes or sequences, including gradient-echo techniques, chemical shift imaging, or high-resolution surface coil images, are unnecessary in the vast majority of patients with AVN. These techniques may occasionally be used to answer specific questions although they are of limited value in diagnosis.

Magnetic Resonance Imaging: The Normal Hip

With increasing age, there is a normal physiologic conversion of bone marrow from hematopoeitic to fatty, proceeding distally to proximally in the appendicular skeleton (52). The factors responsible for this are unclear; however, temperature and vascularity are thought to play a role (52, 53). The physiologic conversion from red to fatty marrow is associated with a decrease in intramedullary blood flow and a change from a sinusoidal blood supply to a more sparse capillary system (52, 54). A causal or temporal relationship between the red to fatty marrow conversion and the above-mentioned physiologic and histologic changes is unclear (55). By age 25, an adult distribution pattern is usually reached and remaining hematopoietic marrow is found primarily in the vertebrae, sternum, ribs, pelvis, skull, and proximal shafts of the femora and humeri (52).

The epiphyses and apophyses contain fatty marrow even

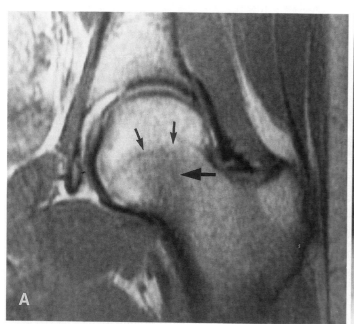 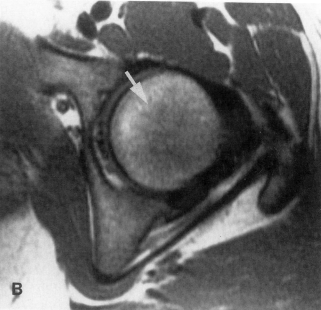

Figure 26.9. Normal MR imaging of hip in a 25-year-old patient. **(A)** Coronal TR 800/TE 25 image depicts a narrow, low-intensity line *(small arrows)* that represents the closed physis. The broad, vertically oriented low-intensity band *(thick arrow)* represents the primary weight-bearing trabeculae. High-intensity fatty marrow is seen in the femoral head and greater trochanter. Moderate-intensity hematopoietic marrow is present in the femoral neck and intertrochanteric region. **(B)** Axial TR 800/TE 25 image depicts a low-intensity area in the central portion of the femoral head *(arrow)*, representing the normal weight-bearing trabeculae and corresponding to the asterisk seen on CT.

in early childhood (Fig. 26.9*A*). The normal femoral head consists of fatty marrow traversed by linear weight-bearing trabeculae. The femoral head is separated from the femoral neck by the low-intensity physeal line, which is seen in all normal patients regardless of age (Fig. 26.9*A*). The weight-bearing trabeculae appear as oblique low-intensity bands extending anterolaterally to posteromedially through the high-signal femoral head and neck (Fig. 26.9*A*) (23, 56). On axial images these trabeculae correspond to the asterisk sign seen on CT (Fig. 26.9*B* and 26.4) (38).

The femoral neck and intertrochanteric region are composed of hematopoietic marrow, which decreases with increasing age (55) and consists of predominantly fatty marrow in 88% of normal patients over the age of 50 (Fig. 26.10) (53).

Magnetic Resonance Imaging Appearance of Avascular Necrosis

The MR imaging findings of AVN reflect the histopathological changes (56A). On short TR/TE (T1-weighted) images the reactive interface and its sclerotic border are depicted as a low-intensity line or rim in the anterosuperior portion of the femoral head (Fig. 26.11*A*). This low-signal margin is present in over 90% of patients with AVN (57). The area within this central zone may be isointense, hypointense, or hyperintense in comparison with the normal fatty marrow on short TR/TE images (57).

The long TR/TE (T2-weighted) images add specificity to the diagnosis (45, 57). Commonly, a high-intensity line is seen within the low-signal margin. This appearance has

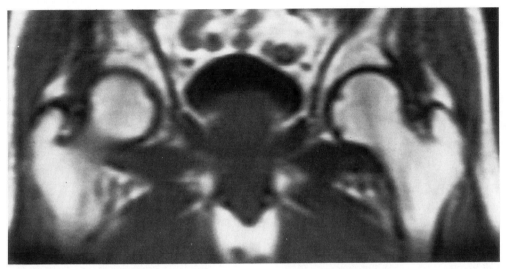

Figure 26.10. Magnetic resonance image of normal hip in a 67-year-old male. Coronal TR 600/TE 20 image demonstrates high-intensity fatty marrow throughout the proximal femurs. A large prostate gland is also depicted.

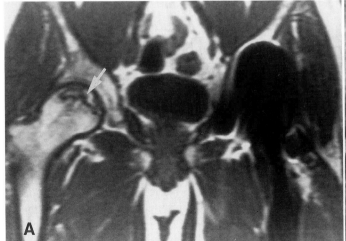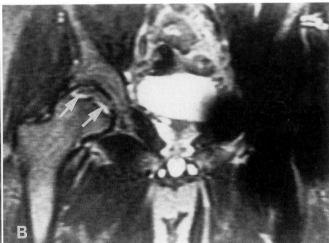

Figure 26.11. Double line sign of AVN. Coronal images of hips in a 40-year-old male. **(A)** TR 400/TE 25 image depicts a lesion of the right femoral head with low-intensity margin. The area within the low-intensity margin is isointense with fat *(arrow)*. Note the high-intensity marrow in the femoral neck and intertrochanteric region, representing premature conversion to fatty marrow. Signal void in the left hip area is secondary to metallic prosthesis. **(B)** TR 2500/TE 80 image illustrates high signal intensity line inside the low-intensity margin of the right hip, the "double line" sign *(arrow)*. This represents the reactive interface of granulation tissue. The area above the interface has signal characteristics of fat.

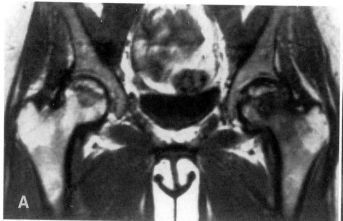
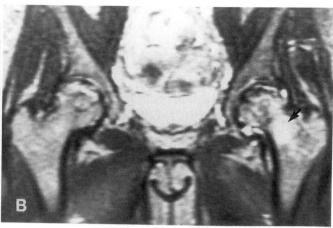

Figure 26.12. Coronal images of hips in a 30-year-old female with chronic active hepatitis and AVN. **(A)** TR 600/TE 20 image illustrating low-intensity margins in both femoral heads secondary to AVN. The left femoral neck has decreased signal intensity. **(B)** TR 2500/TE 80 image shows marked increased intensity in the left femoral neck *(arrow)* consistent with marrow edema. Small left effusion is seen.

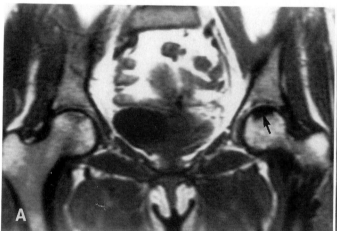
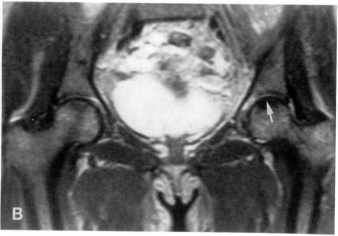

Figure 26.13. Fibrotic appearance of AVN. Coronal **(A)** TR 600/TE 25 and **(B)** TR 2500/TE 80 images illustrating a low-intensity area in the left femoral head *(arrows)*, representing fibrotic appearance of AVN. Low-intensity border is also seen in the superior aspect of the right femoral head representing AVN.

been called the "double line" sign (Fig. 26.11*B*) and was present in 80% of patients in one series (57). The high-intensity rim is presumably related to the large extracellular water content in the granulation tissue of the reactive interface. In some cases, chemical shift artifact contributes to the double line sign by displacing the granulation tissue in respect to the frequency-encoding gradient, leaving behind a signal void. In other instances, sclerosis and the rim of granulation tissue account for the appearance (58). This double line sign is highly specific for AVN and is less common in patients with early or X-ray-negative disease (24). This can be explained by the fact that the MR imaging diagnosis can be made prior to the formation of detectable granulation tissue (24).

The signal characteristics of AVN can be subclassified secondarily to the appearance of the central zone (area within the low-intensity rim) on short and long TR/TE images (57). The lesions may simulate fat (Fig. 26.11), blood, edema (Fig. 26.12), or fibrotic tissue (Fig. 26.13) (57). The

use of selective chemical shift imaging techniques does not yield a direct correlation between tissue and signal type (55). Histologic evaluation of the fatlike central zone revealed granulation tissue and chondroid metaplasia. Presumably this contains large amounts of protein that decrease T1 and T2 relaxation times and produce an MR imaging signal similar to fat (55).

The MR imaging signal characteristics of AVN of the hip correlated roughly with radiographic staging (57). Fatlike lesions (high signal on short TR/TE images and intermediate intensity on long TR/TE images) were frequently seen in X-ray-negative or early disease. Fibrotic-like areas (low signal on both sequences) were seen mostly in advanced disease. Edema and bloodlike lesions were often heterogeneous and occurred less frequently (57).

In patients with an edema-like pattern (low signal on short TR/TE (T1-weighted) and high signal on long TR/TE (T2-weighted) images), the signal abnormalities often extend into the femoral neck. This pattern is not specific for

AVN unless it is accompanied by a typical lesion with a low-signal margin. Marrow edema can also be seen in transient osteoporosis (59, 60), osteomyelitis (61), and secondarily to occult fractures (62–64).

Transient osteoporosis of the hip is a poorly understood entity. This condition is most commonly seen in young to middle-aged adults, particularly men. In women, it is often discovered in the third trimester of pregnancy. Either hip may be involved in men but in women the left hip is almost exclusively affected. Hip pain may be of rapid or gradual onset and often leads to an antalgic gait within weeks. Radiographs reveal marked osteopenia of the femoral head and often the femoral neck with a normal joint (65). Bone scans depict focal increased uptake in the region of the joint (59). The clinical symptoms and edema-like pattern of the femoral head and neck on MR imaging regress spontaneously (59, 60, 66) unlike osteonecrosis, which is a progressive disorder (14, 24, 48, 67).

The relationship between transient osteoporosis and AVN is unclear. Turner et al. (66) reported six cases with proven AVN presenting with signal characteristics of marrow edema without detectable areas of AVN. The diagnosis of AVN was later confirmed by core biopsy in three patients and by the development of the characteristic focal MR imaging appearance in the other three hips. Hence, the edema pattern in the hip is nonspecific and may revert to normal

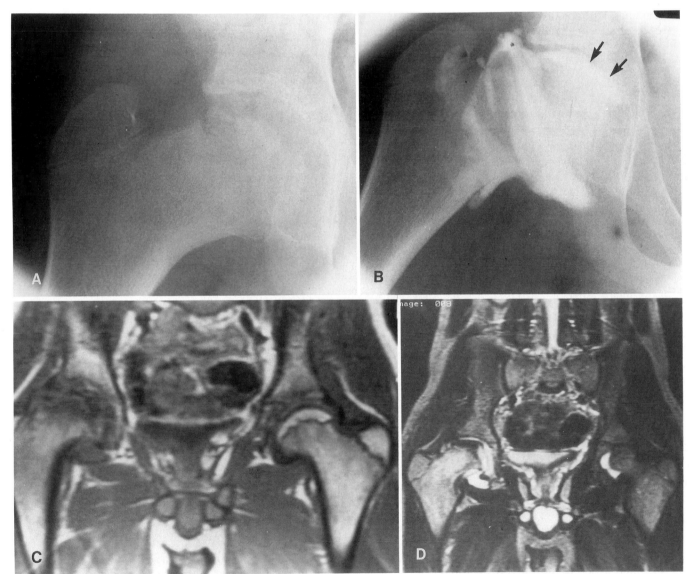

Figure 26.14. A 14-year-old male with Legg-Perthes disease. **(A)** AP radiograph of right hip depicts classic changes of Legg-Perthes disease with irregularity, collapse, and sclerosis of the femoral head. **(B)** Arthrogram reveals collapse of the femoral head and increased contrast between the head and medial aspect of the acetabulum *(arrows)*. The head is contained within the joint on this abduction film. **(C)** Coronal TR 600/TE 20 image illustrating collapsed low-signal femoral head margin with lateral displacement. Note the deformed head on the left with normal fatty signal intensity. **(D)** Coronal TR 2500/TE 80 image depicts flattened right femoral head with low signal intensity, large effusion, and lateral displacement of the head with respect to the acetabulum. A large effusion is present on the left, but the deformed, remodeled head has signal characteristics of normal fatty marrow.

or progress to AVN. Transient osteoporosis of the hip may represent transient ischemia similar to that reported in the knee (68).

Premature conversion of hematopoietic to fatty marrow is often seen in patients with AVN (Fig. 26.11). Mitchell et al. (53) studied 141 hips of patients below the age of 50 and compared controls to patients with AVN of the hip. In this study, premature conversion from red to fatty marrow was seen in at least 62% of patients with AVN. Mitchell et al. postulated that this premature conversion may be a reflection of decreased blood flow. The increased fatty marrow content is also consistent with Solomon's theory (1, 11) that AVN is secondary to increased fat deposition in the marrow and Jones's theory (10) that fatty emboli are responsible for ischemic necrosis.

A small amount of fluid can be seen in normal hips by MR imaging (69). In patients with X-ray-negative disease, a large joint effusion correlates with pain (24). Why some patients develop symptoms early in the course of the disease and others remain asymptomatic is unknown.

Treatment

Treatment options for AVN depend on many factors, including stage of the disease, degree of marrow involvement, location of the necrotic area, and patient age (70–74). The options include conservative nonweight-bearing techniques (41, 71), core decompression with or without bone grafting and electrical stimulation (13–15, 48, 49, 75, 76), rotational osteotomy (73, 74), vascularized grafts, arthrodesis (71), and hip replacement (12, 71). Although the utility of hip-preserving operations is controversial, it is generally agreed that the outcome of these procedures is more successful if performed in the early stages of AVN (70–74). Results of the operative procedures vary. In many cases the major etiologic factors such as steroids and alcoholism persist after treatment, making the results difficult to interpret. Tooke et al. (70) evaluated the influence of steroids following a core biopsy procedure and found that 50% of the hips progressed when steroids were continued compared with a 22% progression when steroids were discontinued.

In children, MR imaging of the hip can be used for diagnosis and follow-up in patients with Legg-Perthes disease. The use of MR imaging obviates the need for arthrography in evaluating the extent of cartilagenous destruction in these patients (Fig. 26.14) (77).

In a recent study (78) evaluating the MR imaging appearance of AVN following core biopsy and bone grafting, the signal characteristics of most lesions were unchanged after surgery. Exceptions included four of nine large lesions that progressed to femoral head collapse, two cases of marrow edema that resolved after surgery, and one patient in whom fatty marrow reconverted to red marrow. Size appeared to represent the most important prognostic factor. Small lesions usually remained stable and the clinical course either improved or was nonprogressive.

Conclusion

Nontraumatic AVN of the hip is a disorder mainly affecting patients between the ages of 20 and 50 years. Although many predispositions exist, its etiology is still unknown. Bilateral involvement is seen in 50 to 88% of patients (48, 79). Of these patients, 15% will also have involvement of one shoulder (48).

Magnetic resonance imaging is the most sensitive and specific noninvasive imaging modality available for the early diagnosis of AVN. The primary indication for MR imaging is the detection of early or occult disease in patients at increased risk for AVN. They include patients with predispositions to AVN (see "Introduction"), hip pain and normal X-rays, or those with known unilateral disease.

References

1. Solomon L. Mechanisms of idiopathic osteonecrosis. Orthoped Clin N Am 1985;16:655–667.
2. Dalinka MK, Alavi A, Forsted DH. Aseptic (ischemic) necrosis of the femoral head. JAMA 1977;238:1059–1061.
3. Jones JP Jr. Osteonecrosis. In: McCarthy DJ, ed. Arthritis and allied conditions. 10th ed. Philadelphia: Lea & Febiger, 1985:1356–1373.
4. Jacobs B. Epidemiology of traumatic and nontraumatic osteonecrosis. Clinical Orthopaedics and Related Research 1978;130:51–67.
5. Sweet DE, Madewell JE. Pathogenesis of osteonecrosis. In: Resnick DK, Niwayama G, eds. Diagnosis of bone and joint disorders. 2d ed. Philadelphia: W. B. Saunders, 1988:3187–3237.
6. Hungerford DS, Lennox DW. The importance of increased intraosseous pressure in the development of osteonecrosis of the femoral head: implications for treatment. Orthoped Clin N Am 1985;16:635–654.
7. Antti-Poika I, Karaharju E, Vankka E, et al. Alcohol-associated femoral head necrosis. Ann Chir Gynaecol 1987;76:318–322.
8. Pinot MR, Peterson HA, Berquist TH. Magnetic resonance imaging in early diagnosis of Legg-Calve-Perthes disease. J Pediatr Orthoped 1989;9:19–22.
9. Edeiken J, Dalinka M, Karasick D. Bone ischemia and osteochondroses. In: Edeiken J, Dalinka M, Karasick D, eds. Roentgen diagnosis of diseases of bone. 4th ed. Baltimore: Williams & Wilkins, 1990:909–972.
10. Jones JP Jr. Fat embolism and osteonecrosis. Orthoped Clin N Am 1985;16:595–633.
11. Solomon L. Idiopathic necrosis of the femoral head: pathogenesis and treatment. Can J Surg 1981;24:573–578.
12. Jones JP Jr, Engleman EP. Osseous avascular necrosis associated with systemic abnormalities. Arthr Rheum 1986;9:728–735.
13. Ficat RP. Idiopathic bone necrosis of the femoral head: early diagnosis and treatment. J Bone Joint Surg (Br) 1985;67B:3–9.
14. Hungerford DS. Pathogenetic considerations in ischemic necrosis of bone. Can J Surg 1981;24:583–590.
15. Hungerford DS. Bone marrow pressure, venography and core decompression in ischemic necrosis of the femoral head. In: The hip: proceedings of the 7th open scientific meeting of the Hip Society. St. Louis: C. V. Mosby, 1979:218–237.
16. Arlot ME, Bonjean M, Chavassieux PM, Meunier PJ. Bone histology in adults with aseptic necrosis. J Bone Joint Surg (Am) 1983;65A:1319–1327.
17. Saito S, Inove A, Ono K. Intramedullary haemorrhage as a possible cause of avascular necrosis of the femoral head: the histology of 16 femoral heads at the silent stage. J Bone Joint Surg (Br) 1987;69-B:346–351.
18. Kenzora JE, Glimcher MJ. Accumulative cell stress: the multifactorial etiology of idiopathic osteonecrosis. Orthoped Clin N Am 1975;16:669–679.
19. Gregg PJ, Walder DN. Scintigraphy versus radiography in the early diagnosis of experimental bone necrosis with special reference to Caisson disease of bone. J Bone Joint Surg (Br) 1980;62-B:214–220.
20. Alavi A, McCloskey JR, Steinberg ME. Early detection of avascular necrosis of the femoral head by 99m technetium diphosphonate bone scan: a preliminary report. Clin Orthoped 1977;127:137–141.

21. Bonnarens F, Hernandez A, D'Ambrosia R. Bone scintigraphic changes in osteonecrosis of the femoral head. Orthoped Clin N Am 1985;16:697–703.

22. Mitchell MD, Kundel HL, Steinberg ME, et al. Avascular necrosis of the hip: comparison of MR, CT, and scintigraphy. AJR 1986;147:67–71.

23. Totty WG, Murphy WA, Ganz WI, et al. Magnetic resonance imaging of the normal and ischemic femoral head. AJR 1984;143:1273–1280.

24. Coleman BG, Kressel HY, Dalinka MK, et al. Radiographically negative avascular necrosis: detection with MR imaging. Radiology 1988;168:525–528.

25. Mettler FA Jr, Guiberteau MJ. Bone scanning. In: Essentials of nuclear medicine. 2d ed. Orlando: Grune & Stratton, 1986:247–286.

26. Thickman D, Axel L, Kressel HY, et al. Magnetic resonance imaging of avascular necrosis of the femoral head. Skel Radiol 1986;15:133–140.

27. Murray IP, Dixon J. The role of single photon emission computed tomography in bone scintigraphy. Skel Radiol 1989;18:493–505.

28. Collier BD, Carrera GF, Johnson RP, et al. Detection of femoral head avascular necrosis in adults by SPECT. J Nucl Med 1985;26:979–987.

29. Hawkins RA, Flynn R, Bassett LW, et al. SPECT and MRI for evaluation of aseptic necrosis of the femoral heads. J Nucl Med 1987;28:564.

30. Miller IL, Savory CG, Polly DW Jr, et al. Femoral head osteonecrosis: detection by magnetic resonance imaging versus single-photon emission computed tomography. Clin Orthoped 1989;247:152–162.

31. Lutzker LG, Alavi A. Bone and marrow imaging in sickle cell disease: diagnosis of infarction. Semin Nucl Med 1976;6:83–93.

32. Rao S, Solomon N, Miller S, et al. Scintigraphic differentiation of bone infarction from osteomyelitis in children with sickle cell disease. J Pediatr 1985;107:685–688.

33. Alavi A. Scintigraphic detection of bone and bone marrow infarction in sickle cell disorders. In: Bohrer SP, ed. Bone ischemia and infarction in sickle cell disease. St. Louis: Warren H. Green, 1981:274–304.

34. Alavi A, Heyman S. Bone marrow imaging. In: Gottschalk A, Hoffer PB, Potchen EJ, eds. Diagnostic nuclear medicine. Vol II. Baltimore: Williams & Wilkins, 1988:707–724.

35. Kim HC, Alavi A, Russel MO, et al. Differentiation of bone and bone marrow infarcts from osteomyelitis in sickle cell disorders. Clin Nucl Med 1989;14:249–254.

36. Magid D, Fishman EK, Scott WW Jr, et al. Femoral head avascular necrosis: CT assessment with multiplanar reconstruction. Radiology 1985;157:751–756.

37. Casteleyn PP, DeBoeck H, Handelberg F, et al. Computed axial tomography and disulphine blue in the evaluation of osteonecrosis of the femoral head. Int Orthoped (SICOT) 1983;7:149–152.

38. Dihlmann W. CT analysis of the upper end of the femur: the asterisk sign and ischaemic bone necrosis of the femoral head. Skel Radiol 1982;8:251–258.

39. Sartoris DJ, Resnick D, Gershuni D, et al. Computed tomography with multiplanar reformation and 3-dimensional image analysis in the preoperative evaluation of ischemic necrosis of the femoral head. J Rheumatol 1986;13:153–163.

40. Mitchell DG, Kressel HY, Arger PH, et al. Avascular necrosis of the femoral head: morphologic assessment by MR imaging, with CT correlation. Radiology 1986;161:739–742.

41. Jergesen HE, Heller M, Genant HK. Magnetic resonance imaging in osteonecrosis of the femoral head. Orthoped Clin N Am 1985;16:705–716.

42. Markisz JA, Knowles RJ, Altchek DW et al. Segmental patterns of avascular necrosis of the femoral heads: early detection with MR imaging. Radiology 1987;162:717–720.

43. Mikhael MA, Paige ML, Widen AL. Magnetic resonance imaging and the diagnosis of avascular necrosis of the femoral head. Comput Radiol 1987;11:157–163.

44. Mitchell DG, Steinberg ME, Dalinka MK, et al. Magnetic resonance imaging of ischemic hip: alterations within the osteonecrotic, viable and reactive zones. Clin Orthoped 1989;244:60–77.

45. Glickstein MF, Burke DL Jr, Schiebler ML, et al. Avascular necrosis versus other diseases of the hip: sensitivity of MR imaging. Radiology 1988;169:213–215.

46. Beltran J, Herman LJ, Burk JM, et al. Femoral head avascular necrosis: MR imaging with clinical pathologic and radionuclide correlation. Radiology 1988;166:215–220.

47. Arlet J, Ficat P. Diagnostic de l'osteonecrose femoroccipitale primitive au stade I (stade preradiologic). Rev Chir Orthop 1968;54:637; as cited by Warner JJ, Philip JH, Brodsky GL, Thornhill TS. Studies of nontraumatic osteonecrosis: the role of core decompression in the treatment of nontraumatic osteonecrosis of the femoral head. Clin Orthoped 1987;225:104–126.

48. Steinberg ME, Brighton CT, Steinberg DR, et al. Treatment of avascular necrosis of the femoral head by a combination of bone grafting, decompression, and electrical stimulation. Clin Orthoped 1984;186:137–153.

49. Steinberg ME, Brighton CT, Hayken GD, et al. Early results in the treatment of avascular necrosis of the femoral head with electrical stimulation. Orthoped Clin N Am 1984;15:163–175.

50. Steinberg ME, Brighton CT, Corces A, et al. Osteonecrosis of the femoral head: results of core decompression and grafting with and without electrical stimulation. Clin Orthoped 1989;249:199–208.

51. Shuman WP, Castagno AA, Baron RL, et al. MR imaging of avascular necrosis of the femoral head: value of small field-of-view sagittal surface-coil images. AJR 1988;150:1073–1078.

52. Kricun ME. Red-yellow marrow conversion: its effect on the location of some solitary bone lesions. Skel Radiol 1985;14:10–19.

53. Mitchell DG, Rao VM, Dalinka MK, et al. Hematopoietic and fatty bone marrow distribution in the normal and ischemic hip: new observations with 1.5T MR imaging. Radiology 1986;161:199–202.

54. Trueta J, Harrison MHM. The normal vascular anatomy of the femoral head in adult man. J Bone Joint Surg 1983;65:442–461.

55. Mitchell DG, Joseph PM, Fallon M, et al. Chemical-shift MR imaging of the femoral head: an in vitro study of normal hips and hips with avascular necrosis. AJR 1987;148:1159–1164.

56. Littrup PJ, Aisen AM, Braunstein EM, et al. Magnetic resonance imaging of femoral head development in roentgenographically normal subjects. Skel Radiol 1985;14:159–163.

56A. Lang P, Jergesen HE, Moseley ME, et al. Avascular necrosis of the femoral head: high-field strength MR imaging with histologic correlation. Radiology 1988;169:517–524.

57. Mitchell DG, Rao VM, Dalinka MK, et al. Femoral head avascular necrosis: correlation of MR imaging, radiographic staging, radionuclide imaging and clinical findings. Radiology 1987;162:709–715.

58. Mitchell, D. Bone marrow imaging. Paper presented at the magnetic resonance musculoskeletal imaging symposium. Milwaukee, October 21, 1989.

59. Wilson AJ, Murphy WA, Hardy DC, Totty WG. Transient osteoporosis: transient bone marrow edema? Radiology 1988;167:757–760.

60. Bloem JL. Transient osteonecrosis of the hip: MR imaging. Radiology 1988;167:753–755.

61. Unger E, Moldofsky P, Gatenby R, et al. Diagnosis of osteomyelitis by MR imaging. AJR 1988;150:605–610.

62. Yao L, Lee JK. Occult intraosseous fracture: detection with MR imaging. Radiology 1988;167:749–751.

63. Stafford SA, Rosenthal DI, Gebhardt MC, et al. MRI in stress fracture. AJR 1986;147:553–556.

64. Lee JK, Yao L. Stress fractures: MR imaging. Radiology 1988;169:217–220.

65. Resnick D, Niwayama G. Osteoporosis. In: Resnick DK, Niwayama G, eds. Diagnosis of bone and joint disorders. 2d ed. Philadelphia: W. B. Saunders, 1988:2043–2054.

66. Turner DA, Templeton AC, Selzer PM, et al. Femoral capital osteonecrosis: MR findings of diffuse marrow abnormalities without focal lesions. Radiology 1989;171:135–140.

67. Lee CK, Hansen HT, Weiss AB. The "silent hip" of idiopathic ischemic necrosis of the femoral head in adults. J Bone Joint Surg (Am) 1980;62-A:795–800.

68. Schneider R, Goldman AB, Vigorita V. Localized transient osteoporosis of the knee. Paper presented at the 86th annual

meeting of the American Roentgen Ray Society, April 15, 1986, Washington, D. C.

69. Mitchell DG, Rao V, Dalinka M, et al. MRI of joint fluid in the normal and ischemic hip. AJR 1986;146:1215–1218.

70. Tooke SM, Nugent PJ, Bassett LW, et al. Results of core decompression for femoral head osteonecrosis. Clin Orthoped 1986;226:99–104.

71. Meyers MH. Osteonecrosis of the femoral head: pathogenesis and long-term results of treatment. Clin Orthoped 1988;231:51–61.

72. Warner JJ, Philip JH, Brodsky GL, Thornhill TS. Studies of nontraumatic osteonecrosis: the role of core decompression in the treatment of nontraumatic osteonecrosis of the femoral head. Clin Orthoped 1987;225:104–126.

73. Maistrelli G, Fusco U, Avai A, et al. Osteonecrosis of the hip treated by intertrochanteric osteotomy. J Bone Joint Surg (Br) 1988;70-B:761–766.

74. Saito S, Ohzono K, Ono K. Joint-preserving operations for idiopathic avascular necrosis of the femoral head: results of core decompression, grafting, and osteotomy. J Bone Joint Surg (Br) 1988;70-B:78–84.

75. Bonfiglio M. Technique of core biopsy and tibial bone grafting (Phemister procedure) for the treatment of aseptic necrosis of the femoral head. Iowa Orthoped J 1982;2:57.

76. Camp JF, Colwell CW. Core decompression of the femoral head for osteonecrosis. J Bone Joint Surg (Am) 1986;68A:1313–1319.

77. Scoles PV, Yoon YS, Makley JT, et al. Nuclear magnetic resonance imaging in Legg-Calve-Perthes disease. J Bone Joint Surg (Am) 1984;66A:1357–1363.

78. Chan T, Dalinka MK, Steinberg ME, et al. Aseptic necrosis of the femoral head: progression of MRI appearance following core decompression and grafting. Skeletal Radiology 1991;20:103–107.

79. Hauzeur JP, Pasteels JL, Orloff S. Bilateral non-traumatic aseptic osteonecrosis in the femoral head. J Bone Joint Surg (Am) 1987;69-A:1221–1225.

CHAPTER 27

Computed Tomography of Pelvic Trauma

Donna Magid and Elliot K. Fishman

Introduction

Trauma is the leading cause of death in nearly every age group below 40 years of age. Pelvic and acetabular traumas, with their associated soft-tissue, vascular, and visceral injuries, account for approximately 10% of the morbidity (1). Accurate and rapid assessment of pelvic injury is therefore critical for patient triage.

Prior to the introduction of computed tomography (CT) in the late 1970s, the traditional mode of pelvic evaluation in the traumatized patient was standard radiographs followed by tomography if indicated (2, 3). A series of pelvic views, including pelvic inlet and outlet views and 45° Judet oblique views, was often obtained in an attempt to maximize the information available in a traumatized patient. Even so, up to 29% of acetabular fractures were missed on the initial plain film examination (4).

Although plain films remain the initial screening study obtained in the traumatized patient, computed tomography has come to play an increasing role in the evaluation of pelvic and acetabular trauma. Early articles demonstrated the increased accuracy of CT in determining the presence and extent of injury, and in assessing associated complications (5–11). One of the obvious advantages of CT is the visualization of the extent of soft-tissue, vascular, and visceral injuries in a single examination (12, 13) (Fig. 27.1).

The improvements in scanner technology have increased the speed with which such an examination can be done. Initial acquisition times of 1 min/slice have now dropped to under a second, making the rapid evaluation of the traumatized patient a reality. This chapter will pre-

sent the current status of the use of CT in the evaluation of pelvic and acetabular trauma and will discuss the newer imaging displays that are increasing the clinical efficacy of CT.

Computed Tomography vs. Plain Radiography

Standard transaxial computed tomography has several significant imaging advantages over standard radiographs. Numerous articles in the literature (14–16) have shown that in up to 40% of cases the obtained views of the traumatized pelvis (anteroposterior (AP), frog leg, Judet oblique, inlet, outlet) will not be adequate for interpretation. This may be due to a number of factors, including poor patient positioning, inadequate exposure, or overlying foreign matter. Computed tomography therefore has a significant advantage. The patient is positioned comfortably in the scanning gantry, with minimal positioning maneuvers, and can be left on a trauma board if significant spinal injury is considered. Newer trauma CT tables even eliminate the need for patient transfer, with patients both transported and studied on a single examination table. The CT examination takes only a few minutes, especially if dynamic scanning is used. Since computed tomography of the abdomen and pelvis is now recognized as the baseline of standard trauma evaluation, several additional scans through the pelvis and acetabulum add little time to the actual CT examination (17–19). Once the CT scans are obtained the patient can be transferred promptly to surgery, intensive care, or back to the emergency room.

A series of articles in the early 1980s addressed the use of CT in the patient with pelvic trauma. Sauser et al. (6) found that the information from transaxial CT significantly influenced treatment in 4 of 13 patients. Shirkhoda et al. (5) presented five cases of acetabular injuries where CT was very helpful. Vas et al. (17) compared plain radiographs with CT in 24 trauma patients and found that in 13 patients, CT provided more information as to the extent of fracture and the number, size, and position of bone fragments. Harley et al. (7) did a similar study in 26 adult patients and found CT to be particularly helpful in detecting fractures of the acetabular roof and posterior acetabular lip and intraarticular loose bodies. Mack et al. (8) used the transaxial CTs to help think of the "interrelations of

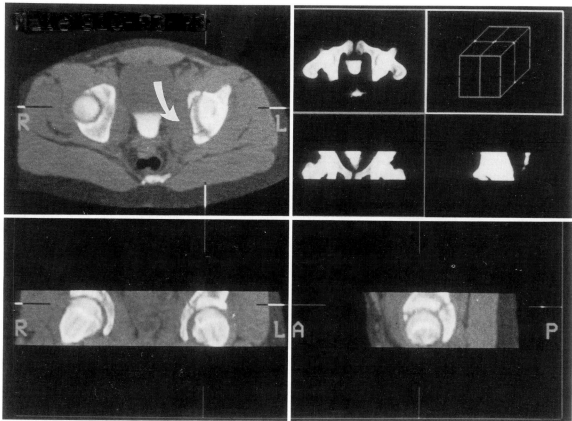

Figure 27.1. Left acetabular fracture with associated pelvic hematoma *(arrows).* One screen showing axial *(top left),* coronal *(lower left),* sagittal *(lower right),* and 3D reference images *(top right).* Smaller right pelvic hematoma with associated right acetabular fracture is also seen. Hematoma involves both obturator internus muscles.

fragments at multiple levels, [so that] a three-dimensional image of the total lesion may be created."

Multiplanar Imaging of the Pelvis

Although transaxial CT scans are excellent in detecting the subtleties of fractures and in giving a proper orientation of fracture extent, they may not overcome specific limitations, particularly in the more complicated pelvic injuries. The hip and pelvis are three-dimensional structures; the orthopaedic surgeon needs to consider the extent of injury in three dimensions in order to properly manage the patient. The display of a series of between 20 and 50 transaxial CT slices through an injured pelvis and acetabulum often presents a confusing array of information that can be difficult to assimilate and use. Reformation of these data into coronal, sagittal, and oblique displays may be used to provide a more optimal display of pelvic injuries.

Initial multiplanar displays provided reformation of single user-selected planes. Several companies, such as Siemens Medical Systems (Iselin, N.J.), took this one step further. Automated software programs now create a series of contiguous coronal, sagittal, or oblique images on the CT satellite console. This package, termed MPR/D (multiplanar reconstruction and display), provides an easy alternative to the acquisition of individual user-selected coronal, sagittal, or oblique images. We have previously

documented pitfalls in arbitrary user-selected orthogonal views; unexpected abnormalities may be missed or underestimated. The ability to display a full contiguous data set increases understanding of extent of injury and avoids the potential errors that may be made if only a few selected images are reformatted (18, 19) (Fig. 27.2).

One of the early limitations of the multiplanar package was the limited display format (127 × 127, or one-quarter of the CT image size). This meant that in cases of trauma the contralateral extremity could not be displayed simultaneously for comparison. Similarly, this meant that if the patient had bilateral injuries, two separate reconstructions were required. Although the reconstructions took only 6 to 12 min, the reconstruction and filming of two sets became time consuming. Also, inability to compare both sets of images on a single screen could be frustrating and left the physician with an unwieldy, large series of films for comparison.

The introduction of faster, less expensive computers during the 1980s provided the impetus for increasing sophistication in medical image displays. We are currently using a Sun Microsystems workstation (3/160 or 4/280) (Sun Microsystems, Inc., Mountain View, Calif.) and the Pixar Image computer (Pixar, Inc., San Rafael, Calif.) as the platform for our two- and three-dimensional reconstructions. Using the OrthoTool (HipGraphics, Baltimore, Md.) software and a mousedriven series of commands, we can

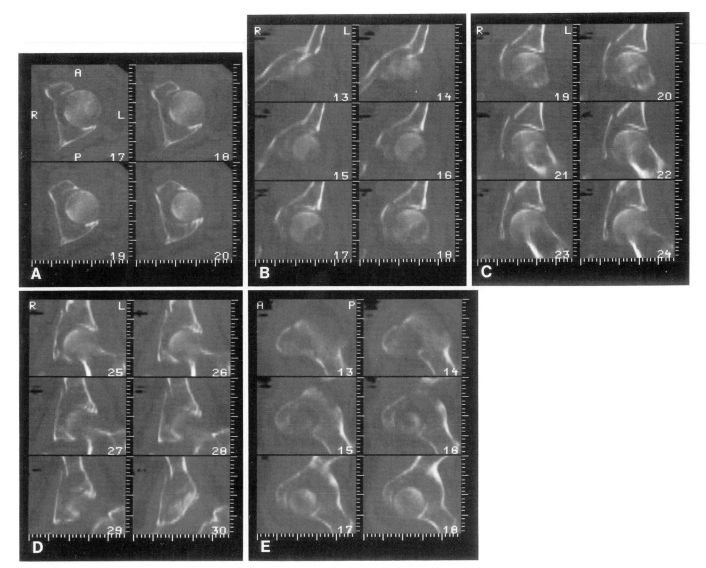

Figure 27.2 Left acetabular fracture following trauma in a patient with a history of multiple myeloma. **(A)** Transaxial views demonstrate a complex fracture involving both anterior and posterior acetabular walls. **(B–D)** A series of coronal images better defines the full extent of the fracture with displacement of medial acetabular wall. Extension of the fracture through the posterior acetabular wall is clearly defined. **(E)** Sagittal view demonstrates disruption in weight-bearing surface.

display and interact with transaxial, coronal, and sagittal images (20). The entire 512 × 512 CT image of up to 192 contiguous slices is included, allowing for careful comparison of contralateral structures. The program also provides the ability to obtain arbitrary oblique planes from any of the image displays (Fig. 27.3). The increased speed of the computer systems, using a more sophisticated reconstruction algorithm, allows the use of bicubic interpolation rather than the standard linear reconstruction used on CT scanner-based multiplanar displays. This allows for a higher quality reconstruction with better detail of both bone and soft-tissue structures. One of the major advantages of a full image display with freely selectable oblique planes is the ability to construct specific views that may be particularly valuable in the traumatized patient. For example, if one places an oblique line paralleling the posterior sacral cortex on sagittal views (approximately 45° oblique in most

patients), one creates a standard oblique series that displays the entire pelvic ring (Fig. 27.4) and most of the sacrum and sacroiliac (SI) joints *en face*. The displacement of the symphysis pubis, sacroiliac joints, or medical walls of the acetabulum can be evaluated simultaneously. This is a particularly valuable view in patients with complex pelvic ring fracture and/or dislocation. The *en face* sacral views enhance evaluation of the sacrum, SI joints, and sacral foramina.

Multiplanar reconstruction of transaxial CT images into coronal, sagittal, and oblique planes provides both the radiologist and orthopaedic surgeon with several specific imaging advantages. These include the following.

1. Better evaluation of the articular surface. Partial averaging limits evaluation of the joint space on routine transaxial images. Reformatting of images overcomes this

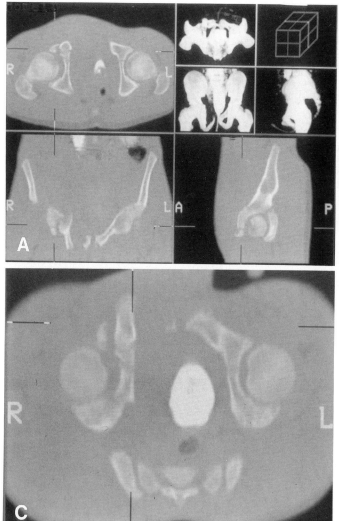

Figure 27.3. Bilateral acetabular fractures following an automobile accident. Axial *(top left)*, coronal *(lower left)*, sagittal *(lower right)*, and 3D reference images are displayed. **(A)** A complex right acetabular fracture is seen involving anterior and posterior acetabular walls. Disruption of symphysis pubis is seen with associated pelvic hematoma. The bladder is compressed. **(B)** Scan inferior to image **(A)** shows the full extent of pelvic fractures. A 45° line is drawn through the sagittal image *(lower right)*. **(C)** Rotation of **(B)** presents the image in pelvic axis view. Note disruption of the anterior portion of the ring.

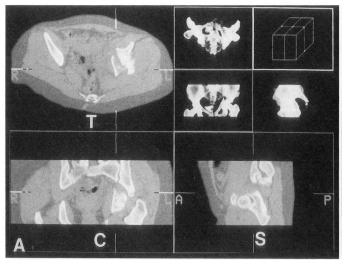

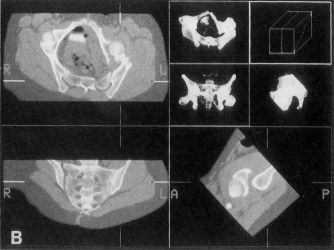

Figure 27.4. Comminuted left acetabular fracture. **(A)** Standard sequence of transaxial *(T)* coronal *(C)* and sagittal *(S)* images demonstrates fracture through the acetabular roof. **(B)** Rotation of dataset along a 45° axis off the sagittal plane now provides a pelvic ring view *(upper left)*, which shows disruption of the pelvic ring *(upper left)*. Diastasis of right SI joint is also seen *(lower left)*.

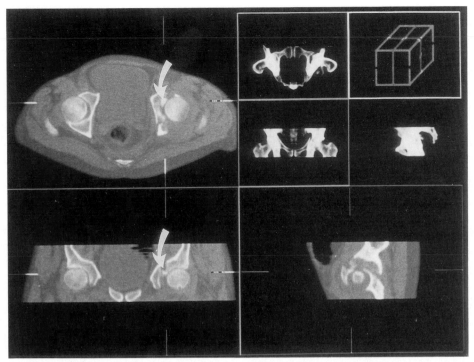

Figure 27.5. Fracture of left acetabulum following an auto accident. The fracture could not be reduced by traction and a CT scan was done. Axial *(top left)*, coronal *(lower left)*, sagittal *(lower right)*, and 3D refer- ence images are displayed. Sequence of orthogonal images demon- strates intraarticular fragment in joint space precluding reduction *(ar- rows)*.

problem and provides visualization of the joint in all planes (Figs. 27.3 and 27.4).

2. "Cross-haired" localization of intraarticular fragments, for arthroscopic removal (Fig. 27.5).
3. Better definition of subtle fractures, which may extend along several transaxial slices.
4. Improved detection of muscle or soft-tissue injury or joint effusions.
5. Accurate measurement of fracture displacement, used as a template for surgical planning (Fig. 27.6).
6. Spatially accurate data for surgical planning, measure- ment, and, if desired, custom hardware design.

We reviewed a series of 34 patients with pelvic trauma and roentgenographically definite (24 cases) or suspected (10 cases) fractures. Computed tomography with MPR de- tected four fractures missed on plain radiographs (11). The findings on the CT/MPR led to an alteration in patient management in seven cases. Of these, four required more invasive management (open surgical reduction) and three more conservative management (traction and closed re- duction) than had been originally anticipated.

Three-Dimensional Imaging

Three-dimensional (3D) imaging was introduced during the late 1970s as a means of providing a more complete understanding of complex skeletal anatomy. Three-dimen- sional imaging is an ideal approach to the pelvis and ace- tabulum due to the inherent three-dimensional nature of the anatomy to be imaged (21–23).

Regardless of the system used for generating 3D recon-

structions a successful study requires careful attention to scanning detail. The techniques used in multiplanar im- aging (4-mm thick sections, 3-mm spacing with scan pa- rameters of 125-kVpeak (kVp), 230 mA, 3s or 125 kVp, 250 mA, 1 s) can be used for 3D studies. Although these pa- rameters are for a Siemens Somatom scanner (DRH or PLUS), this technique is readily modified for any commer- cially available scanner. The use of a 1- or 2-mm slice over- lap is particularly valuable in creating accurate 2D or 3D images with a minimum of associated artifact and aliasing.

Three-dimensional reconstruction has evolved rapidly over the last few years in terms of image quality, speed of reconstruction, and interactive flexibility. Imaging systems may be either scanner based or free standing in nature. In general, the free-standing workstations provide better qual- ity reconstructions but are more expensive. Since they are dedicated to a single task, their software is usually more user friendly and flexible. The specific advantages and dis- advantages of the different scanning systems and of the different types of 3D reconstruction algorithms used are beyond the scope of this chapter. Rather, we will briefly review the techniques we find useful and list several ref- erences from which further information can be obtained.

The technique we use is volumetric rendering; unlike standard binary 3D reconstruction techniques, it is a per- centage classification. Volume rendering, unlike surface rendering, preserves all of the information from the origi- nal CT scans (24–26). Therefore, object depth can be pre- served in the simulated 3D projections. In volumetric ren- dering the CT scans are conceptually stacked up as a volume in the computer and the gray-scale intensity of the pixels in the volume is replaced with simulated gels of

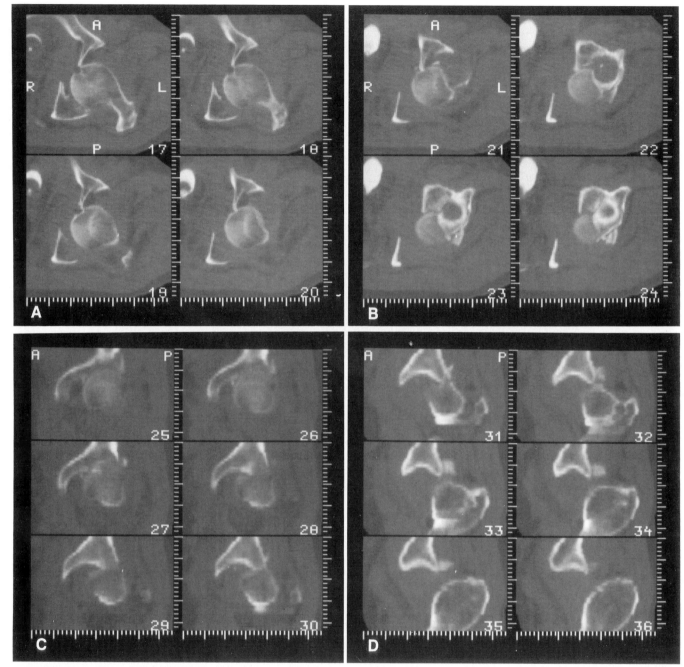

Figure 27.6. Fracture of left acetabulum with dislocation of femur. (**A** and **B**) Transaxial images define acetabular fractures with posterior dislocation. (**C** and **D**) Sagittal reconstructions better define the extent of fracture as well as defining in greater detail the extent of posterior dislocation.

varying color and transparency. Color or black-and-white views of the data set can be computed from arbitrary angles with varying degrees of object transparency. Shading can also be varied to help show texture and interfaces in the volume and to provide monocular depth cues.

All 3D examinations are done with at least three different reconstructions: a pelvic or side-to-side rotation, a somersaulting rotation, and an edited view. The edited view, usually bisecting the pelvis vertically, provides an unimpeded view of the acetabular medial wall (Fig. 27.7). Depending on the extent of injury and individual physi-

cian preference, we can disarticulate the hip to view the acetabular articular surface, which has been helpful when acetabular reconstruction is needed.

Although the majority of our pelvic and acetabular fractures have been in adults we have scanned a number of patients under the age of 15. The examination poses no additional problem except for the usual compliance difficulties in examining pediatric patients (Fig. 27.8). We have found that although sedation may be needed, the majority of patients can be done without any premedication. In order to lower the radiation dose to these patients we have

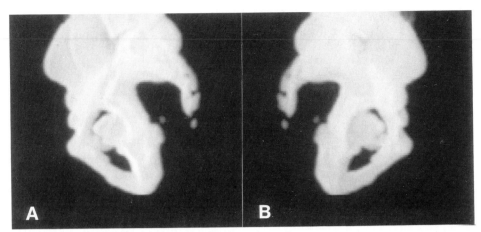

Figure 27.7. Acetabular fracture with 3D edited views. **(A** and **B)** Edited views with contralateral hemipelvis removed and femur disarticulated optimally define the fracture of the medial acetabular wall. These views are especially valuable in surgical planning.

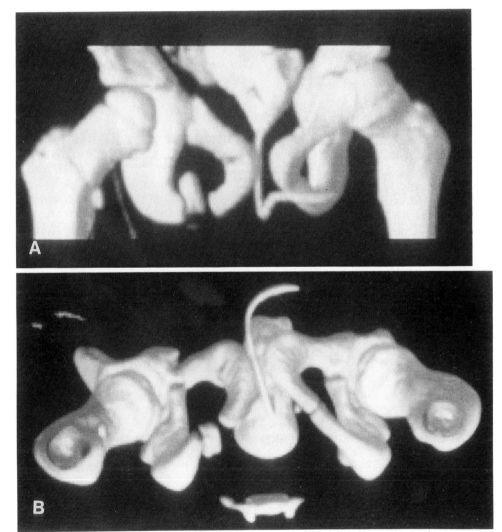

Figure 27.8. Comminuted right acetabular fracture as well as proximal femoral fracture. Note the Foley catheter and contrast in the bladder. **(A)** 3D pelvic view demonstrates fracture of superior and inferior right pubic rami with displacement of a fragment of inferior column. **(B)** Somersaulting view optimally demonstrates the extent of pelvic fracture, as well as defining the degree of displacement of fragments.

developed a low-dose scanning protocol (3 s, 140 mA, 125 kVp) that lowers our standard dose by approximately half (T. J. Beck, Johns Hopkins Hospital), (personal communication).

Clinical Applications

Three-Dimensional Imaging

The reconstruction of routine transaxial images into a 3D display is helpful to the radiologist but is more valuable to the orthopaedic surgeon. Pelvic and acetabular fractures must be dealt with clinically in three dimensions. An understanding of the extent of injury, the fracture components, their orientation, and degree of displacement must all be understood to reach a consensus on optimal therapeutic approach. We believe that augmenting initially available information should provide better short-term and long-term results. In a series of 21 patients with acetabular fractures, 3D imaging has been shown (27) to change either the timing of surgery, surgical approach, and/or hardware used in nearly 50% of patients.

Burk et al. (21) reviewed a series of 20 patients with acetabular fractures and 3D reconstructions. The authors found the 3D images provided an increased understanding of complex injuries. Since in some cases surgical reduction and fixation will be necessary, the 3D images can play a major role in patient triage.

At our institution, 3D reconstructions are not routinely ordered on every patient with a pelvic or acetabular fracture. Those with nondisplaced or minimally displaced fractures can often be managed with plain radiographs alone. If the plain films are suboptimal or if more information is needed, a CT scan is ordered. It is our experience that if the injury is significant enough for a CT scan then a 2D/3D reconstruction should be done. The minimal additional scanning time and radiation dose involved is offset by the important additional diagnostic and therapeutic information generated. The detailed information provided by 2D/3D CT frequently blurs distinction between various categories used to assign a fracture to one pattern or another.

Scott et al. (28) reviewed a series of 19 cases of acetabular fractures in which 3D volumetric reconstructions and plain radiographs were available. Overlying bowel content or foreign matter obscured detail in 35 of 45 plain radiographic views, while positioning or radiographic technique were suboptimal in 21. The 3D reconstructions eliminated these problems by creating accurate simulated oblique, inlet, and tangential views in all cases. In 10 of 19 cases a nonconventional oblique view, and in 13 of 19 a cephalocaudal angulation other than the conventional inlet, tangential, or anteroposterior view, best demonstrated the fracture. It was therefore concluded that 3D reconstructions of pelvic injuries had the following advantages:

1. Evaluation of all possible obliquities to obtain optimal projections to demonstrate the extent of fracture.
2. Elimination of overlying densities such as bowel contents or foreign matter.
3. Elimination of improperly exposed radiographs.
4. Elimination of potentially painful and dangerous positioning of the patient.
5. Decrease in radiation dose to the patient.

Surgical Planning and Decision Making

The surgical approach to the fractured pelvis or acetabulum varies from institution to institution and from surgeon to surgeon (2, 3, 29). Although certain basic principles of reconstruction remain constant, individual experience and preference results in a variety of approaches and of selected hardware. In the past, the surgeon's final hardware selection often was made at surgery, following fracture exposure and exploration. This had several potential problems, including potential lack of availability of desired shelf or custom hardware. Three-dimensional imaging preoperatively provides the orthopaedic surgeon with all of the necessary views of the traumatized pelvis, allowing for a completely noninvasive presurgical evaluation. In our experience, this allows selection of the correct hardware in advance, thereby facilitating the surgery. Additionally, we have had numerous cases where the extent of injury, once fully understood, resulted either in a deferment (often for 6 to 8 weeks) or indefinite postponement of surgery. This is not surprising when one considers the potential information gap between standard radiographic studies and 2D/3D CT in the evaluation of complex joint structures like the hip.

Another potential advantage of the acquisition of a set of digital data such as 3D CT study is the ability to design custom orthopedic hardware if indicated. Using CAD/CAM (computer-assisted design/computer-assisted manufacturing) technology, the orthopedic surgeon can design or have designed components varying from custom hardware pins and plates to arthroplasty components. Although this is more expensive than standard off-the-shelf hardware, in the more complex cases it may prove critical in providing the corrective surgery necessary.

In time, using artificial intelligence (AI), the computer may be able to suggest a surgical approach and/or the type of hardware needed for the procedure. Such an analysis may be used in the future, to supplement the judgment of the orthopedic surgeon, providing a computer-generated second opinion.

Surgical Simulation

The advanced computer workstation is capable of far more sophisticated tasks than the simple display of 2D and 3D CT reconstruction. Manipulation of the CT data to simulate potential surgical procedures is now possible and is in fact currently being used at several centers across the United States. The impetus for such application has been driven by developments in industries as diverse as the automotive (i.e., General Motors) and aerospace (i.e., McDonnell Douglas) industries, who have for years relied on computer simulations for the design and testing of new products. Computer simulations provide a rapid, low-cost method of testing various designs, relationships, and results without having to create the actual product.

Similar efforts are now being applied to medicine. Plastic surgeons are using computers to simulate reconstructive surgery in difficult and complex craniofacial deformities. Different combinations of incisions and modifications of bone can be simulated to illustrate potential surgical results long before the operation ever begins. This capability allows unprecedented precision in preoperative planning. In addition, the apprehensive patient can be shown the potential technique and results, often alleviating some of the preoperative fear (30).

Although these capabilities are available only on the more expensive workstations at present, design packages that would work on PCs and Macintoshes are under development or just becoming available.

The Postoperative Patient

Computed tomography has been shown to be useful in the postoperative hip and pelvis (31). Complications related to prior trauma repair, inappropriate healing (i.e., avascular necrosis) or internal fixation (i.e., pins extending into the joint, broken hardware), or to the initial injury (i.e., osteoarthritis), can be evaluated with CT with multiplanar reformation. These reformatted images are especially helpful in those cases in which the transaxial CT scan quality is degenerated by beam-hardening artifact from implanted hardware.

This examination technique is identical to that described earlier in this chapter. Although metal artifact reduction techniques (32–34) have been described in the literature, no manufacturer has implemented these techniques on current version scanner software releases. In our experience, the reformatted images minimize the transaxial metal artifacts and may salvage the clinical utility of the study (31, 32).

The "starburst" artifacts are due to the severe X-ray attenuation (missing data) caused by the metal in certain views. The missing data (hollow projections) cause starburst artifact to streak the cross-sectional images. Left uncorrected, these artifacts make it nearly impossible to obtain useful information anywhere near the metal implant. These starburst artifacts can be markedly reduced by reformation of the axial CT images into orthogonal or oblique plane images. Data reformation into other planes will weight the true signal over the randomly distributed artifact when integrating two adjacent axial images. In this way the artifact seen in the axial images is averaged out in the multiplanar reformations, and the resulting reformatted image may appear less degraded than the axial from which it is generated.

Other theoretical methods to reduce artifact include increased effective X-ray energy (limited by acceptable dose), use of smaller appliances and/or metals with lower attenuation coefficients (i.e., titanium rather than stainless steel), and more advanced metal correction algorithms. At this time, reformatting is the simplest and most readily accessible mode of artifact reduction in daily practice.

Osteoarthritis

Regardless of the surgical repair attempted, the excellence of the surgeon, or the cooperation of the patient, posttraumatic osteoarthritis develops in a significant percentage of patients. Narrowing of the joint space or deformity of articular surfaces can be a result of the initial injury or a result of abnormal biomechanics following the initial injury. Computed tomography with two-dimensional reconstruction is an excellent method of evaluating these patients. Involvement of the joint space can best be defined on vertical planar images, especially the sagittal projection. Decisions for patient management are often made based on these images.

Conclusion

Computed tomography supplemented by 2D and 3D image reconstructions provides the maximum information available from a specific data set. This allows a more complete understanding of inherently 3D normal anatomy and pathology. This additional understanding can then be transferred to patient management decisions, leading to better patient care.

References

1. Melton LJ III, Sampson JM, Morrey BF, Ilstrup DM. Epidemiologic freatures of pelvic fractures. Clin Orthoped 1981;155:43–47.
2. Judet R, Judet J, Letournel E. Classification and surgical approaches for open reduction. J Bone Joint Surg (Am) 1964;46:1615–1646.
3. Letournel E. Acetabulum fractures: classification and management. Clin Orthoped 1980;151:81–106.
4. Pearson JR, Hargadon EJ. Fractures of the pelvis involving the floor of the acetabulum. J Bone Joint Surg (Br) 1962;44:550–561.
5. Shirkhoda A, Brashear HR, Stabb EV. Computed tomography of acetabular fractures. Radiology 1980;134:683–688.
6. Sauser DD, Billimoria PE, Rouse GA, Mudge K. CT evaluation of hip trauma. AJR 1980;135:269–274.
7. Harley JD, Mack LA, Winquist RA. CT of acetabular fractures: comparison with conventional radiography. AJR 1982;138:413–417.
8. Mack LA, Harley JD, Winquist RA. CT of acetabular fractures: analysis of fracture patterns. AJR 1982;138:407–412.
9. Blaquiere RM. Computed tomography of acetabular trauma. Clin Radiol 1985;36:5–11.
10. Tile M. Fractures of the acetabulum. Orthoped Clin N Am 1980;11(3):481–506.
11. Magid D, Fishman EK, Brooker AF Jr, Mandelbaum BR, Siegelman SS. Multiplanar computed tomography of acetabular fractures. J Comput Assist Tomogr 1986;10(5):773–778.
12. Federle MP. CT of upper abdominal trauma. Semin Roentgenol 1984;19:269–280.
13. Federle MP. Computed tomography of blunt abdominal trauma. Radiol Clin N Am 1983;21:461–475.
14. Griffiths HJ, Nordenstam CG, Burke J, Lamont B, Kimmel J. Computed tomography in the management of acetabular fractures. Skel Radiol 1984;11:22–31.
15. Walker RH, Burton DS. Computerized tomography in assessment of acetabular fractures. J Trauma 1982;22:227–231.
16. Hubbard LF, McDermott JH, Garrett G. Computed axial tomography in musculoskeletal trauma. J Comput Assist Tomogr 1982;22(5):388–394.
17. Vas WG, Wolverson MK, Sundaram M, et al. The role of computed tomography in pelvic fractures. J Comput Assist Tomogr 1982;6(4):796–807.
18. Fishman EK, Magid D, Mandelbaum BR, et al. Multiplanar (MPR) imaging of the hip. Radiographics 1986;6(1):7–53.

19. Magid D, Fishman EK, Sponseller PD, Griffin PP. 2D and 3D computed tomography of the pediatric hip. Radiographics 1988;8:901–933.

20. Ney DR, Fishman EK, Magid D, Kuhlman JE. Interactive real-time multiplanar imaging: The 2D/3D orthotool. Radiology 1989;170:275–276.

21. Burk DL Jr, Mears DC, Kennedy WH, et al. Three-dimensional computed tomography of actabular fractures. Radiology 1985; 155:183–186.

22. Pate D, Resnick D, Andre M, et al. Perspective: three-dimensional imaging of the musculoskeletal system. AJR 1986;147:545–551.

23. Totty WG, Vannier MW. Complex musculoskeletal anatomy: analysis using three-dimensional surface reconstruction. Radiology 1984;150:173–177.

24. Fishman EK, Magid D, Drebin RA, Brooker AF Jr, Scott WW Jr, Lee RH Jr. Advanced three-dimensional evaluation of acetabular trauma: volumetric image processing. J Trauma 1989;29(2):214–218.

25. Fishman EK, Drebin B, Magid D, et al. Volumetric rendering technique: applications for 3-dimensional imaging of the hip. Radiology 1987;163:737–738.

26. Ney DR, Drebin RA, Fishman EK, Magid D. Volumetric rendering computed tomography data: principles and techniques. Comput Graph Appl 1990;10(2):24–32.

27. Scott WW Jr, Magid D, Fishman EK, et al. 3-D evaluation of acetabular trauma. Contemp Orthoped 1987;15:17–24.

28. Scott WW Jr, Fishman EK, Magid D. Acetabular fractures: optimal imaging. Radiology 1987;165:537–539.

29. Pennal GF, Tile M, Waddell JP, Garside H. Pelvis disruption: assessment and classification. Clin Orthoped 1980;151:12–21.

30. Gerber J, Ney DR, Magid D, Fishman EK. Simulated femoral repositioning with three-dimensional CT. J Comput Assist Tomogr 1991;15(1):121–125.

31. Baird RA, Schobert WE, Pais MJ, et al. Radiographic identification of loose bodies in the traumatized hip joint. Radiology 1982;145:661–665.

32. Weese JL, Rosenthal MS, Gould H. Avoidance of artifacts on computerized tomograms by selection of appropriate surgical clips. Am J Surg 1984;147:684–687.

33. Glover GH, Pelc NJ. An algorithm for the reduction of metal clip artifacts in CT reconstructions. Med Phys 1981;8:799–807.

34. Fishman EK, Magid D, Robertson DD, et al. Metallic hip implants: CT/MPR imaging. Radiology 1986;160:675–681.

Magnetic Resonance Imaging and Computed Tomography of the Sacroiliac Joint

Mark D. Murphey, John M. Bramble, and Louis H. Wetzel

Introduction

Numerous pathologic conditions involve the sacroiliac joint. Sacroiliiitis frequently occurs as a component of the spondyloarthropathies, including ankylosing spondylitis, psoriatic arthritis, and Reiter's syndrome. In addition, other arthritides such as gout, rheumatoid arthritis, and the enteropathic arthropathies cause sacroiliitis (1). The sacroiliac joint is not uncommonly involved in pyogenic arthritis and trauma may disrupt the articulation as well. Adjacent neoplasms can also secondarily extend into the sacroiliac joint. Symptoms from these various abnormalities are frequently indistinguishable from mechanical causes of low back pain.

Multiple modalities have been used to examine the sacroiliac joint. These include conventional radiography and tomography, scintigraphy, computed tomography (CT), and recently magnetic resonance (MR) imaging. The complex anatomy of the sacroiliac joint often makes interpretation of conventional radiographs difficult and unrewarding. Overlying bowel adds to the frustration in visualizing the articulation. Computed tomography and MR imaging alleviate these difficulties and are ideally suited to evaluation of the sacroiliac joint. Both techniques allow accurate assessment of pathologic abnormalities involving the sacroiliac joint, their severity, and effects on adjacent structures.

Technique

Computed tomography technique is important to optimize visualization of the sacroiliac joint, particularly for evaluation of sacroiliitis. Initially a lateral computed radiograph (scout view) should be obtained to establish the correct gantry angle and anatomic level (Fig. 28.1A) The CT gantry is titled such that image orientation is as parallel as possible to the long axis of the upper sacrum and sacroiliac joint. The gantry angle is typically 20 to 30°, although some CT scanners have a limitation on the amount of tilt allowed. Images obtained are thus paraaxial, or coronal relative to the upper sacrum. Contiguous 4- to 5-mm thick sections are reconstructed such that the sacroiliac joint and adjacent structures occupy the majority of the pixels in the image. This optimizes spatial resolution as opposed to zooming after image reconstruction. Eight to 10 scans are usually required to evaluate the entire sacroiliac joint. Bone algorithm is utilized to improve trabecular and cortical spatial resolution (30 to 50% improvement in spatial resolution), particularly in evaluation of sacroiliitis (2). This may require an increase in milliamperage to reduce noise. Use of the bone algorithm, however, does increase noise, which degrades evaluation of the adjacent soft tissues. Therefore if analysis of the surrounding soft-tissue anatomy is more important, such as in suspected neoplasm, the soft-tissue algorithm for reconstruction is more appropriate. Similarly, routine axial imaging best evaluates both osseous and soft-tissue structures in this clinical situation, with coronal images being ancillary. Traumatized patients are frequently scanned in the axial plane as well. Scans are displayed for both bone (1500- to 2000-HU (Hounsfield unit) window width; 300- to 700-HU window level) and soft tissue (400- to 500-HU window width; 50- to 100-HU window level) assessment. Higher window width and level is helpful in younger patients with relative increase in thickness of cortical margins.

The incident radiation dose with an entire CT of the sacroiliac joints is approximately 1.5 to 2.0 rads (15 to 20 mGy) (3). Testicular dose is negligible owing to the tightly collimated beam. Ovarian dose is difficult to determine; with variable position of the ovaries, however, it has been estimated at 30 mGy. The CT radiation dose is significantly improved from that in conventional tomography of approximately 6.12 mGy per image (four to five images per examination) (4).

Magnetic resonance imaging for the identification of intrinsic sacroiliac joint abnormality requires use of a surface

---

Abbreviations (see also glossary): HU, Hounsfield unit; DISH, diffuse idiopathic skeletal hyperostosis; WBC, white blood cell; MFH, malignant fibrous histiocytoma.

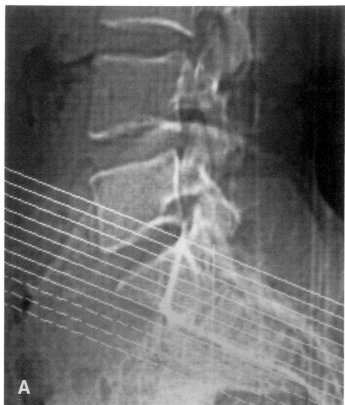

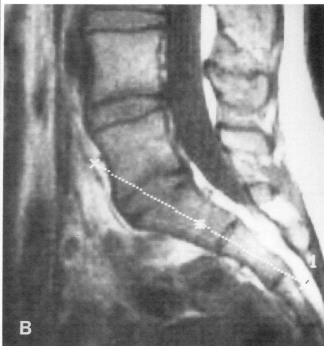

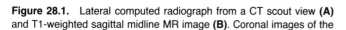

Figure 28.1. Lateral computed radiograph from a CT scout view **(A)** and T1-weighted sagittal midline MR image **(B)**. Coronal images of the sacroiliac joints are obtained parallel to the long axis of the upper sacrum *(lines)*.

coil for adequate spatial resolution. Patients are examined supine with their knees flexed for comfort. At our institution, we utilize an elliptical surface coil that is also used in imaging other areas of the spine. Initially sagittal short acquisition scout views are obtained. Images in a true coronal plane relative to the upper sacrum are then obtained by prescribing an oblique plane from the midline sagittal scout view (Fig. 28.1*B*). Unlike CT, there is no limitation on the plane of imaging.

Both T1- and T2-weighted images are necessary to evaluate the sacroiliac joint. For all pulse sequences, 4-mm sections with 1-mm intersection gaps are suggested. As in other musculoskeletal techniques, T1-weighted images are best for anatomic detail while T2-weighted images provide for improved contrast between different tissues. Investigation of possible sacroiliitis places high demands on spatial resolution in order to delineate initial cartilage involvement. T1-weighted images are obtained with four or eight acquisitions and a 256×256 matrix. In order to limit examination time, T2-weighted images are performed with two or four acquisitions and a 256×256 or 256×128 matrix, respectively, in evaluation of sacroiliitis. Imaging time is approximately 45 min with these parameters at our institution. Application of fast imaging sequences (gradient echo) and the use of other surface coil technology may also have a role in evaluation of the sacroiliac joint (5). In MR imaging of other suspected abnormalities, such as neoplastic involvement, distinction of sacroiliac cartilage is less important. In these cases, two acquisitions are suffi-

cient for both T1- and T2-weighted images and the body coil is typically utilized. The coronal imaging plane is useful with investigation of neoplastic processes to evaluate extension into the sacroiliac joint. The axial projection, however, is usually the primary imaging plane (T1- and T2-weighted images) because of the ability to best demonstrate both osseous and soft-tissue involvement. Sagittal T1-weighted imaging is an additional imaging plane that can be particularly important for evaluation and surgical planning of sacral neoplasms.

Normal Anatomy and Development

The precartilaginous ilium can be identified in the first embryonic weeks, and by the 8th week it attaches to the sacrum. Ossification begins much earlier in the ilium (9th week) than in the sacrum (3rd month) (6). Precursors of the sacroiliac joint appear in the 10th week, with cavities forming at the previous site of the iliac and sacral union. It is not until the second trimester that a true sacroiliac joint is present. The sacrum and ilium develop at different rates after birth. In the 1st year, the sacroiliac articulation is parallel; with further growth, they converge caudally as seen in the adult.

The sacroiliac joint is morphologically a diarthrosis; however, since movement is limited it is more correctly considered an amphiarthrosis (1). The articulation is formed by the auricular surfaces of the ilium and sacrum (first, second, and often the third segments). The sacroiliac joint consists of discrete synovial and ligamentous components (Fig. 28.2) (7). The articulation is widest in the ligamentous compartment, which represents the cranial and posterior portion. The interosseous ligaments, associated loose connective tissue, and adipose tissue fill the space between the superior sacrum and ilium. The interosseous ligaments are short bands attached to opposing osseous surfaces in variable directions and provide internal stability. Accessory sacroiliac joints are common, being reported to occur in 10 to 50% of the population and are more frequent in Caucasians (8). The sacral accessory articular facets arise from the posterior surface adjacent and lateral to the second sacral foramen. The iliac accessory articular facets originate from the medial surface of the posterior superior iliac spine and/or at the tuberosity of the ilium. There is typically only one accessory articulation, although rarely two joints are present on each side.

The surrounding ligaments about the sacroiliac joint are well developed (Fig. 28.3). The joint capsule blends with the ventral sacroiliac ligament, which is relatively weak. Superficial to the interosseous ligament lies the strong dorsal sacroiliac ligament, which runs medially and inferiorly. A component of these fibers extends from the posterior superior iliac spine to the third and fourth sacral segments and may unite with the sacrotuberous ligament. The vertical sacrotuberous (posterior anterior iliac spine to the third, fourth, and fifth sacral segments) and oblique sacrospinous ligaments (ischial spine to the lower sacral and upper coccygeal segments) are posterior to the dorsal sacroiliac ligament (9). The extrinsic sacroiliac ligaments provide further stability to the articulation.

The ventral and inferior one-half to two-thirds of the sacroiliac joint is the synovial compartment. In the synovial compartment the sacral auricular surface is covered by thicker (up to 4 mm) hyaline cartilage while on the iliac side there is thinner fibrocartilage (up to 1 mm) (9). The iliac cartilage may also contain normal areas of splitting, which progress with age, as well as degenerative clefts (10). There is a fibrous capsule closely attached to the associated osseous structures. In younger patients, a joint cavity is present. The joint space narrows with age and the joint cavity may become obliterated with fibrous adhesions. Motion is restricted to small amounts by the irregularities of the articular surfaces, intrinsic articular ligaments, as well as the extrinsic ligaments. The gluteus maximus muscle also limits motion and with contraction decreases the

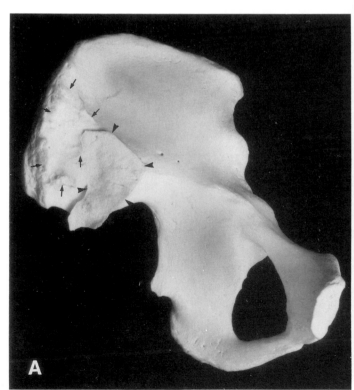

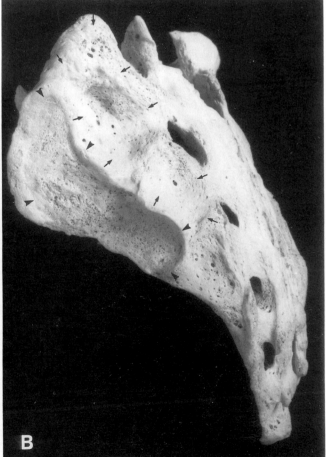

Figure 28.2. Photographs of disarticulated cadaveric ilium (posterior aspect, **A**) and sacrum (lateral aspect, **B**). The auricular surfaces *(arrowheads)* that form the synovial compartment are more anterior and in a different plane than the more posterior ligamentous portion *(arrows)*.

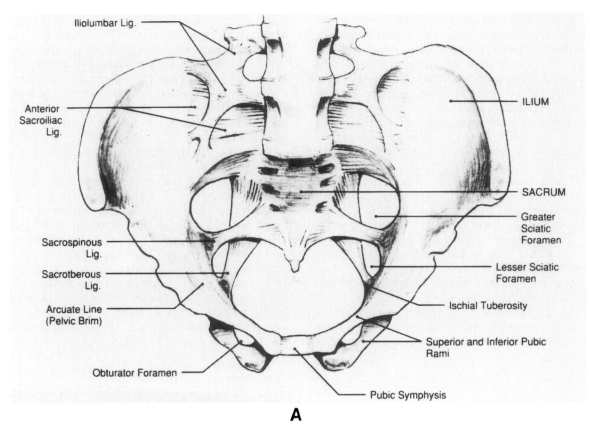

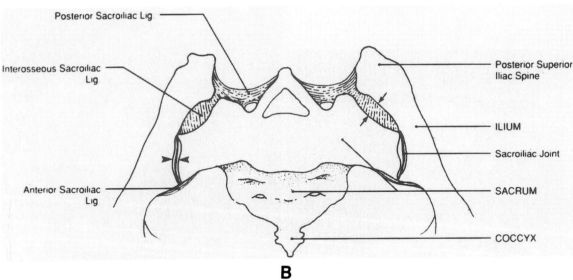

Figure 28.3. Line drawing depicting the ligamentous structures stabilizing the sacroiliac joint both from the anterior aspect **(A)** and from above **(B)**. The different planes of the ligamentous compartment *(ar-rows)* and synovial compartment *(arrowheads)* are clearly delineated. (Courtesy of Fred Reckling, M.D. with modification.)

width of the sacroiliac joint. In pregnancy, range of motion increases due to hormonal effects producing ligamentous laxity.

Roentgenographic detection of sacroiliac disease requires a knowledge of the normal appearance. Computed tomography identifies the two compartments of the normal sacroiliac joint by differences in orientation (Fig. 28.4). The ligamentous compartment, located posteriorly and

cranially, is oblique while the more anterior synovial compartment is vertically oriented (11–18). The sacroiliac joints are symmetric in younger patients (under age 30). Older individuals, however, may show a significant degree of normal asymmetry, seen in 77% of people over the age of 30 and in 87% over the age of 40 by CT (7). The synovial compartment varies in width from 2 to 5 mm with frequent focal joint space narrowing, particularly anteriorly,

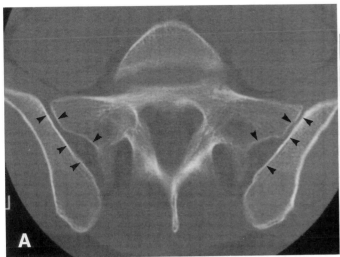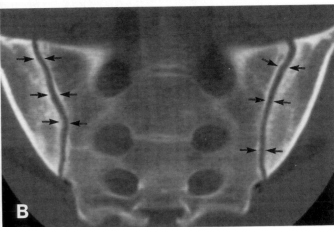

Figure 28.4. Coronal CT of the normal sacroiliac joints. The posterior ligamentous compartment **(A)** is obliquely oriented *(arrowheads)* while the anterior synovial compartment **(B)** is vertically oriented *(arrows)*.

in individuals over the age of 40. Subchondral sclerosis (eburnation) is normally symmetric and uniform (less than 3 mm) on the sacral side of the articulation. Iliac sclerosis is usually symmetric as well, but is often nonuniform (83% of normal patients) with focal areas of increased eburnation (greater than 5 mm) most prominent anteriorly. Cortical margins and the adjacent subchondral bone are usually well defined. Focal areas of ill definition may be present with a predilection for the iliac side (67% of normal subjects) of the articulation (7). These areas are most frequent at the transition zone between the ligamentous and synovial compartments, inferiorly, and should not be mistaken for erosions. Similar changes occur at the joint margins as well as in undulating segments and are often symmetric. They may be caused by a combination of partial volume averaging and sites of insertion for the interosseous ligaments. These insertional "pits" at times are quite prominent in the ligamentous compartment of the sacroiliac joint (3).

Magnetic resonance imaging clearly distinguishes the synovial and ligamentous compartments of the normal sacroiliac joint, both by orientation and appearance (Fig. 28.5). On T1-weighted images high signal intensity adipose tissue is seen in the oblique posterior ligamentous portion of the joint. Focal areas of low intensity within this region represent loose connective tissue and the interosseous sacroiliac ligaments. Insertional "pits" are seen on MR imaging as sacral bone marrow defects and irregularities. Magnetic resonance imaging directly identifies cartilage in the more anterior and vertically oriented synovial compartment. On T1-weighted images a thin zone of intermediate intensity represents the cartilage. The cartilage has a maximum thickness of 4 to 5 mm posteriorly, being somewhat thinner anteriorly and inferiorly (2 to 3 mm). Cartilage thickness is greater in younger patients. In approximately 85% of patients the sacral and iliac components of the cartilage are focally separable, by an intervening low-intensity linear zone (T1-weighted images), superiorly within the joint. The signal intensity suggests hyaline cartilage, although it is likely that there is volume

averaging. The iliac and sacral cortices are seen as thin, linear, low-intensity zones on all pulse sequences on either side of the cartilage. The bone marrow adjacent to the sacroiliac joint is sharply defined without areas of marginal irregularity. T1-weighted images are best for visualizing the cartilage and joint anatomy. Cartilage signal cannot usually be defined on T2-weighted images; however, no areas of increased intensity should be seen, normally, on longer TR images.

Sacroiliitis

Clinical diagnosis of sacroiliitis is often difficult, particularly early in the disease process (19). Physical findings may be obscured by the overlying soft tissues and symptoms are similar to mechanical causes of low back pain. Consequently, radiologic assessment plays a major role in the diagnosis of sacroiliitis. Conventional radiography remains the most widely accepted and available initial screening method. While in the majority of patients (70 to 80%) plain radiographs will confirm or refute the diagnosis of sacroiliitis, findings are at times equivocal. Despite multiple projections, there is still significant inter- and intraobserver variation in interpretation. In one series, 20% of readings were incorrect (14, 20). The adolescent sacroiliac joint is even more difficult to evaluate with normal ill-defined subchondral bone and widened joint space simulating sacroiliitis. Because of these difficulties CT and MR imaging are useful in evaluation of patients with suspected sacroiliitis.

Computed tomography manifestations of sacroiliitis are similar to those in conventional radiography and predominantly involve the synovial compartment. These radiographic changes include joint space widening or narrowing, juxtaarticular demineralization, erosion of the cortical surface and subchondral bone, paraarticular sclerosis, and intraarticular bone ankylosis (14–16). The earliest finding of juxtaarticular osteopenia, a result of inflammation and hyperemia, is difficult to assess if no additional alterations are present. Progressive articular cartilage destruction will

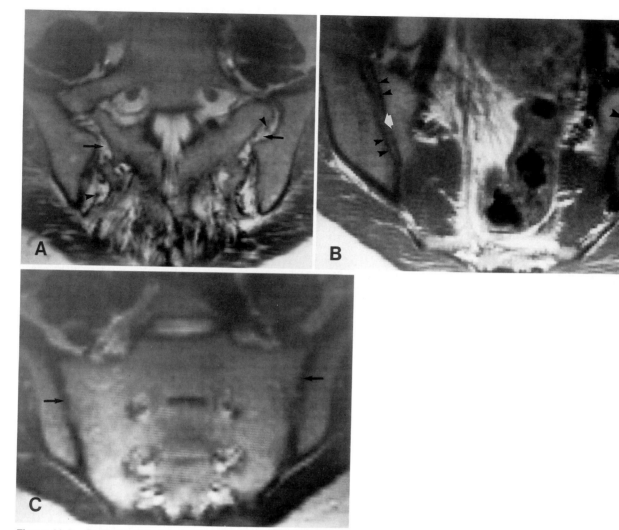

Figure 28.5. Coronal MR images of the normal sacroiliac joint. **(A)** T1-weighted image (TR = 600 ms, TE = 25 ms) of the ligamentous compartment shows high signal intensity adipose tissue *(arrows)*. Areas of low intensity represent loose connective tissue and interosseous sacroiliac ligaments *(arrowheads)*. **(B)** T1-weighted image (TR = 600 ms, TE = 25 ms) of the synovial compartment shows thin zone of intermediate intensity representing cartilage *(white arrows)* with adjacent low-intensity iliac and sacral cortex *(small arrowheads)*. Note the sharp definition of the subjacent marrow margins *(large arrowheads)*. **(C)** T2-weighted image (TR = 2100 ms, TE = 90 ms) shows loss of cartilage definition with low intensity within the sacroiliac joint *(arrows)*; no areas of increased intensity are noted. (From Murphey MD, Wetzel LH, Bramble JM, Levine E, Simpson KM, Lindsley HB. Sacroiliitis: MR imaging findings. Radiology 1991;180:239–244. Reprinted with permission of *Radiology*.)

result in uniform joint space narrowing to less than 2 mm in width. Erosions involving the articulation on one or both sides are seen as focal irregular areas of destruction of the articular cortex and underlying subchondral bone, which may result in widening of the sacroiliac joint (Figs. 28.6*A* and 28.7*A*). Erosions are usually more extensive on the iliac side of the joint where cartilage is normally thinner and subchondral bone thus less protected. This is analogous to marginal erosions in peripheral joints in the inflammatory arthritides. An additional factor that may allow earlier access of pannus to cortical bone on the iliac side of the sacroiliac joint is the possible presence of normal cartilage clefts (21). Increased subchondral sclerosis is seen as areas of higher density and is usually related to erosions. These areas are probably on a reparative basis with new bone formation, a finding often associated with the spondyloarthropathies (1). While subchondral scle-

rosis is frequent in sacroiliitis, it is seldom an isolated finding. Iliac sclerosis usually predominates paralleling the erosive disease; however, sacral eburnation is more distinctive of inflammatory disease in the sacroiliac joint (7). Iliac sclerosis is common normally and osteitis condensans ilii also causes extensive iliac sclerosis. Intraarticular bone ankylosis, either focal or diffuse, is a relatively late manifestation in sacroiliitis (Fig. 28.8*A*). Initially, it may be seen as sclerotic foci within the articulation. With maturation, trabeculae with associated bone marrow may traverse the sacroiliac joint. At this stage, sclerosis if present is limited. Initially, osseous ankylosis involves the synovial compartment. However, it is one of the few alterations that frequently extends into the ligamentous compartment, particularly in ankylosing spondylitis.

Multiple studies have shown the improved sensitivity in identifying changes of sacroiliitis with CT vs. conven-

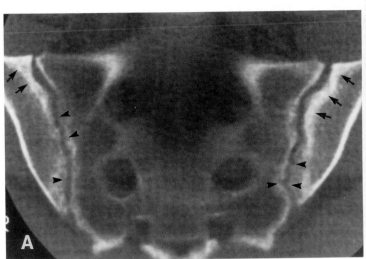 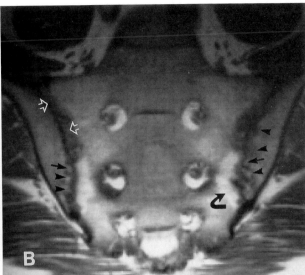

Figure 28.6. Coronal CT and MR images of a 41 year old with ankylosing spondylitis. **(A)** CT demonstrates bilateral sacroiliac joint erosions with irregular cortical margins *(arrowheads)* more prominent on the iliac side of the synovial compartment. Subchondral sclerosis *(arrows)* also predominates on the ilial aspect. **(B)** T1-weighted image (TR = 600 ms, TE = 25 ms) shows loss of cartilage with replacement by inhomogenous mixed signal intensity tissue. Cortex has areas of increased intensity *(small arrows)* with irregularity of the marrow margin representing erosions *(arrowheads)*. Remaining cartilage superiorly appears thickened *(open white arrows)* and hyperintense reactive bone marrow is seen *(curved arrow)*.

tional radiography (14). In investigations by Carrera and colleagues, approximately 40 to 60% of patients with normal or equivocal findings on plain films had CT evidence of sacroiliitis (15). In addition, CT identified all patients in which conventional radiographs were also positive. For the inexperienced observer, particularly, CT changes of sacroiliitis are more distinctive and easier to detect. The best indicators of sacroiliac disease by CT are erosions, osseous ankylosis, sacral subchondral sclerosis, and uniform joint space narrowing (less than 2 mm) (7).

The most characteristic and prominent finding on MR imaging in patients with sacroiliitis is the loss of the normal thin band of intermediate intensity representing cartilage on T1-weighted images (Figs. 28.6*B* and 28.7*B*). In our experience this is seen in over 90% of cases. Cartilage is replaced by areas of inhomogeneous mixed signal intensity tissue. Pathologically, this probably corresponds to cartilage destruction by pannus. Focal areas of residual cartilage are frequent. Cartilage can also appear thickened on T1 weighting, which may represent earlier stages of involvement with partial volume averaging of synovial proliferation and residual cartilage.

Erosions are seen as increased signal intensity in the sacral and iliac cortex about the sacroiliac joint with adjacent marginal irregularity and deeper defects in the subjacent marrow on T1-weighted images (Figs. 28.6*B* and 28.7*B*). These changes cause an apparent widening of the sacroiliac joint and are also seen in more than 90% of patients with sacroiliitis. Erosions are more prominent anteroinferiorly and on the iliac side of the joint, again owing to less subchondral bone protection by the thinner cartilage. Erosions are almost invariably associated with cartilage replacement (22). Higher intensity in bone marrow adjacent to erosions is frequent and suggests focal reactive

atrophy with increased marrow fat. This is similar to findings seen in the lumbar spine adjacent to chronic degenerative disk disease (23). Subchondral sclerosis is more difficult to detect on MR imaging than by CT. On all pulse sequences areas of eburnation demonstrate decreased signal intensity in the bone marrow adjacent to the joint. This appearance may accentuate apparent widening of the sacroiliac joint. Long-standing osseous ankylosis shows marrow signal bridging the sacroiliac joint (Fig. 28.8*B*). The MR imaging findings are almost exclusively localized to the synovial compartment except bone ankylosis.

Areas of increased intensity may be seen on T2-weighted images both within the sacroiliac joint and less commonly in erosions (Fig. 28.7*C*). These regions may be linear or focal and suggest a more acute inflammatory and edematous tissue. Linear intraarticular components may correspond to fluid within the joint. The increased intensity on T2-weighted images is less than is seen in a pure inflammatory process. Fibrous proliferation within pannus, seen pathologically in the spondyloarthropathies, may lower the signal intensity (1).

Computed tomography or MR imaging findings in sacroiliitis do not allow differentiation of the specific etiology. Certain characteristics may, however, narrow the differential diagnosis. Unilateral or significantly asymmetric disease essentially excludes ankylosing spondylitis or enterocolitic arthropathy (1, 6). Reiter's syndrome or psoriatic sacroiliitis would be likely considerations. Extensive subchondral sclerosis and osseous ankylosis would be an unusual manifestation of rheumatoid sacroiliitis or involvement by gout. However, this is commonly seen with the rheumatoid variants (ankylosing spondylitis, Reiter's syndrome, and psoriatic arthritis) and the enteropathic arthropathies.

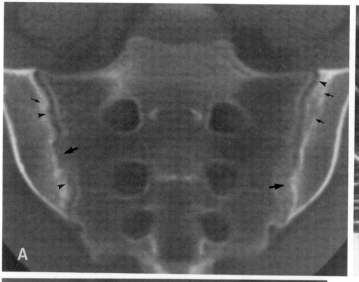
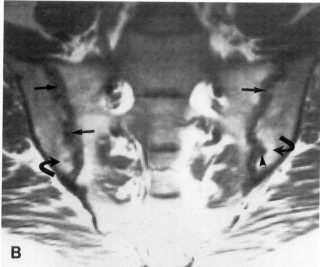
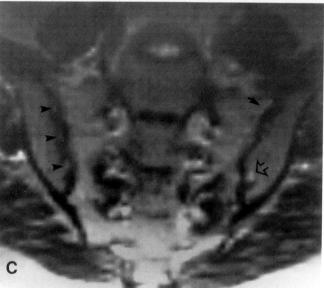

Figure 28.7. Coronal CT and MR images of a 31 year old with psoriatic sacroiliitis. **(A)** CT reveals erosions *(small arrowheads)*, and subchondral sclerosis *(small arrows)*, and possible areas of osseous fusion *(large arrows)*. **(B)** MR image (analogous level to CT), T1 weighted (TR = 600 ms, TE = 25 ms), shows cartilage replacement bilaterally. Adjacent irregularity of marrow margin and increased intensity in cortex representing erosions *(arrows)* results in sacroiliac joint space widen-

ing. Deeper erosion *(arrowhead)* is noted on the left, as is hyperintensive reactive bone marrow *(curved arrows)*. **(C)** MR image, T2 weighted (TR = 2100 ms, TE = 90 ms), shows both focal *(arrowheads)* and linear *(arrow)* increased intensity in sacroiliac joints and in the most inferior erosion on the left *(open arrow)*. MR images disprove areas of possible fusion seen on CT.

Advantages of CT in evaluating sacroiliitis include its general availability and our familiarity as radiologists with the osseous abnormalities. In addition, the cost and length of time imaging (15 vs. 45 min) are less than in MR imaging. There is also no limitation on which patients can be examined, as there is in MR imaging, where claustrophobia, surgical clips, and cardiac pacemakers place restrictions on scanning. Advantages of MR imaging include lack of ionizing radiation, direct visualization of cartilage alterations, and unrestricted "true" coronal imaging. Imaging without radiation exposure may be particularly important as many patients with sacroiliitis are young and in their reproductive years. In addition, conventional radiographs in adolescents often simulate sacroiliitis, making differen-

tiation between normal and abnormal difficult. In our experience, the ability of MR imaging to obtain a "true" coronal plane can improve visualization of subtle erosions. We have also seen several examples of suspected focal areas of osseous ankylosis on CT disproved by MR imaging (Fig. 28.7). Finally, MR imaging has a unique ability to directly evaluate cartilage abnormalities. This may allow detection of early synovitis prior to secondary osseous changes visualized by other modalities (Fig. 28.9).

Osteoarthritis and Other Joint Disorders

Degenerative disease in the sacroiliac joint is common. Radiographic changes are frequently present after the age

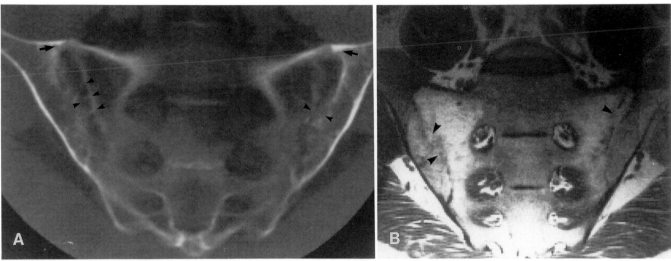

Figure 28.8. Coronal CT and MR images in a 51 year old with ankylosing spondylitis. **(A)** CT reveals trabecular bridging within the sacroiliac joint *(arrowheads)* and at the articular margins superiorly *(ar-*

rows). **(B)** T1-weighted MR image (TR = 600 ms, TE = 25 ms) shows marrow signal traversing the sacroiliac articulation *(arrowheads)* due to osseous ankylosis.

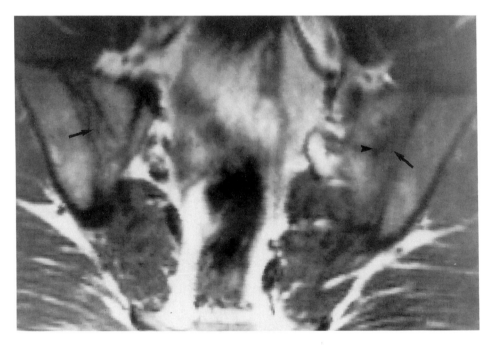

Figure 28.9. Coronal T1-weighted MR image (TR = 600 ms, TE = 25 ms) in an 18 year old with sacroiliitis from Reiter's syndrome shows poorly defined, thickened cartilage signal intensity in anterior portion of the sacroiliac joints *(arrows)*. Compare this with normal thin cartilage at a similar level in Fig. 28.5*B*. Corresponding CT (not shown) was normal because osseous manifestations were not present.

of 40 years. Osteoarthritis can be unilateral, particularly when altered stress is restricted to one side of the pelvis (contralateral hip disease or chondral atrophy associated with ipsilateral paralysis), or bilateral. Osteoarthritic involvement of the accessory sacroiliac joints has been implicated as a cause of low back pain (8). Initially, joint space narrowing is seen, which is frequently focal and localized to the inferior and anterior aspect of the synovial compartment. Pathologically, this is caused by superficial cartilage fibrillation with cleft formation and necrosis (9, 10, 24). Cartilage fusion and intraarticular fibrous bands also develop. Subchondral sclerosis, either linear or focal, is common, particularly on the iliac side, both superiorly and inferiorly. Osteophyte formation is most prominent at the

anterosuperior aspect of the synovial compartment and extension may cause periarticular osseous ankylosis (Fig. 28.10). True intraarticular bone ankylosis is exceedingly unusual in osteoarthritis of the sacroiliac joints, unlike the inflammatory arthritides (9, 24). Computed tomography better demonstrates these two manifestations of bone production as areas of increased density (subchondral sclerosis) and spurlike excrescences (osteophytes). Subchondral sclerosis may be subtle on MR imaging, as previously noted, and causes mild irregularity of the marrow margin. Osteophytes on MR imaging appear as low-intensity spurlike projections, which with maturation may contain bone marrow. Subchondral cyst formation is unusual in degenerative disease of the sacroiliac joint and on CT appears as

Figure 28.10. Axial CT in this 65-year-old man reveals changes of osteoarthritis with large anterior-superior bridging osteophytes *(arrows)*.

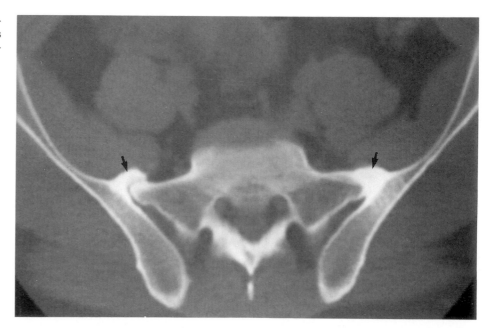

Figure 28.11. Pneumatocyst *(large arrow)* is demonstrated on this axial CT near the junction of the synovial *(small arrows)* and ligamentous compartments *(small arrowheads)*. Normal nonuniform iliac subchondral sclerosis and area of ill-defined cortex at the transition zone between the ligamentous and synovial compartments *(open arrow)* are seen. Deep insertional pits *(curved arrow)* are also present, as is an incidental right iliac bone island.

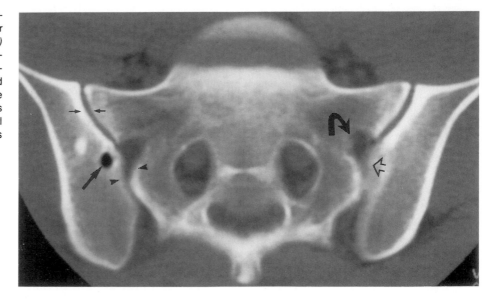

fluid-density areas with intact overlying cortex. On MR imaging subchondral cysts have low intensity on T1-weighted images and marked increased intensity on T2-weighted images, typical of fluid. Cortex and bone marrow may extend circumferentially about the cyst, helping differentiation from an erosion. A vacuum phenomenon may occur within the joint and can extend into paraarticular "pneumatocysts" on the basis of osteoarthritis or other benign etiology (Fig. 28.11) (25, 26). Computed tomography demonstrates low density (−600 to −1000 HU) corresponding to gas (nitrogen) while on MR imaging low intensity is present on all pulse sequences. Superficial erosions occur rarely (6%) in pathologic investigations and may be partially responsible for cortical irregularities on CT (10). They are localized on the iliac side of the joint without the deeper defects that are characteristic of true sacroiliitis. With the overlap in CT findings of sacroiliitis

and osteoarthritis of the sacroiliac joint, the distinction is occasionally difficult. In our experience, MR imaging has been helpful in clarifying these cases. Magnetic resonance imaging has the ability to demonstrate intact cartilage, although thinning may be present, typical of osteoarthritis vs. cartilage replacement characteristic of sacroiliitis.

Diffuse idiopathic skeletal hyperostosis (DISH) is a bone-forming diathesis characterized by enthesophytes (ossification at ligamentous and tendinous attachments). Computed tomography findings in DISH include osseous ankylosis of the ligamentous joint compartment, enthesophytes and osteophytes anterosuperiorly with or without bridging, subchondral sclerosis, and vacuum phenomenon (Fig. 28.12) (27, 28). These changes as well as the absence of erosions and intraarticular fusion of the synovial compartment should aid in distinction from sacroiliitis or osteoarthritis. Magnetic resonance imaging should demonstrate intact cartilage, although we are not aware of any such reported cases.

Osteitis condensans ilii is a condition of unknown etiology that causes a well-defined triangular area of sclerosis on the iliac side of the sacroiliac joint, inferiorly. Theories concerning the cause include urinary tract infections, mechanical stress with increased vascularity often related to pregnancy, and an inflammatory process similar to the spondyloarthropathies (1, 29, 30). The condition is usually symmetric and women are far more frequently affected. Patients are often asymptomatic or have only mild complaints. Conventional radiographs are usually diagnostic. Computed tomography and MR imaging will also show the bilateral iliac sclerosis, preserved joint space, and cartilage, without erosions or intraarticular osseous ankylosis.

Subchondral resorption at the sacroiliac joints as a manifestation of primary or secondary hyperparathyroidism closely stimulates the radiographic findings of erosive sacroiliitis. The increased bone resorption is caused by excess parathyroid hormone. Both conventional radiographs and CT reveal marginal erosion resulting in "pseudowidening" of the sacroiliac joint space and adjacent sclerosis (Fig.

28.13). Erosion is more severe on the iliac side of the joint and changes are symmetric. The marked irregularity of the osseous margins and lack of joint narrowing should raise suspicion of hyperparathyroidism (31, 32). Other characteristic changes and sites of involvement as well as appropriate clinical history are invariably present.

Infection

Infection of the sacroiliac joint occurs most frequently by hematogenous contamination. Other routes of spread include seeding from contiguous infection, direct "implantation," and a postsurgical basis. Pyogenic sacroiliitis is associated with trauma, pregnancy, endocarditis, primary infections of the genitourinary system, bone, or skin, bowel disease, and intravenous drug abusers (33). Intravenous drug abusers have a particular predilection for involvement of the sacroiliac joint (34).

The ilium is the most frequently infected flat bone. The slow circulation in this region resembles long bone metaphyses and favors hematogenous implantation. The association of sacroiliac joint infection with pelvic infection and surgery may be explained by a hematogenous route in-

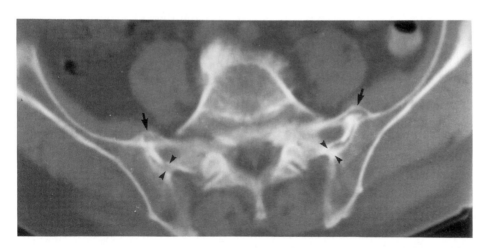

Figure 28.12. Axial CT with involvement of the sacroiliac joints (ligamentous compartment) by DISH. Bridging anterosuperior enthesophytes *(arrows)* and areas of intraarticular osseous ankylosis *(arrowheads)* without erosions allow distinction from osteoarthritis. Spinal manifestations of DISH were also present (not shown).

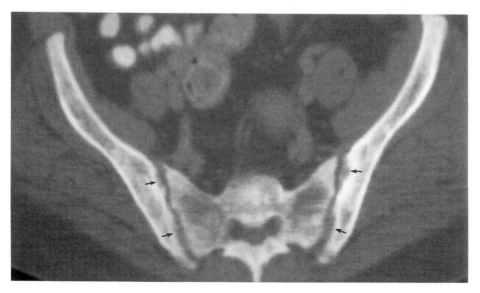

Figure 28.13. Axial CT of the synovial compartment in a chronic renal failure patient with secondary hyperparathyroidism. Bone resorption predominates on the iliac side of the sacroiliac joint, resulting in "pseudowidening" *(arrows)*. Diffuse osseous sclerosis is also present from renal osteodystrophy.

volving Batson's paravertebral venous plexus. Direct extension from adjacent pelvic infection may occur by disruption of the anterior joint capsule or the periosteum of the sacrum or ilium. Trauma may also predispose to sacroiliac infection by this mechanism as well as visceral and other soft-tissue injury. *Staphylococcus aureus* is the most commonly isolated organism (35). Streptococci and pneumococci are not infrequent causes. In intravenous drug abusers gram-negative bacteria, especially *Pseudomonas,* are frequently the offending agent. Unusual organisms including *Brucella,* myobacteria, and fungi may also involve the sacroiliac joint.

The diagnosis of pyogenic sacroiliitis is difficult because it is not common, with only 96 cases described in the English language literature in a recent report (36). There are poorly localizing symptoms and a lack of early findings by conventional radiography. The most common symptoms are a limp or pain in the buttock and gluteal region (sciatic distribution). Physical findings may incorrectly suggest abdominal, pelvic, hip, or lumbar disk abnormality as the causative factor. Delay of accurate diagnosis may result and in one study ranged from 12 to 200 days (37). This may allow progression and contamination of extraarticular sites with abscess or sequestra formation. Infection may extend in multiple directions, including along the iliac fossa or iliopsoas tendon, into the gluteal muscles and lumbar region, or penetrating the pelvic floor.

Pyogenic sacroiliitis is almost invariably a unilateral process with only rare reports of bilateral disease. Thus, an important radiographic tenet is that unilateral sacroiliitis represents infection until proven otherwise. Unfortunately, conventional radiographic findings of blurring of the subchondral osseous line and joint space widening with erosion are not apparent for 2 to 3 weeks. Technetium bone scanning may demonstrate abnormal activity (or photon deficiency) earlier, within 1 week of symptoms; however, it is nonspecific. Other radionuclide examinations such as gallium- and indium-labeled white blood cell (WBC) scanning may be beneficial and add specificity (38, 39). Computed tomography and MR imaging, however, are invaluable, not only in detection of disease earlier than conventional radiography but more importantly in delineating the pattern of soft-tissue involvement (Fig. 28.14).

Computed tomography will reveal focal areas of osseous destruction with erosion more prominent on the ilial side of the joint, again owing to its thinner cartilage (40, 41). It has been our experience that, because of the more rapid cartilage destruction and perhaps the delay in diagnosis, sacral destruction is more prominent than might be expected from other causes of erosive sacroiliitis. In addition, involvement may extend to the ligamentous compartment of the joint early in the disease, which is also atypical for other causes of sacroiliitis. The changes are most extensive in the anteroinferior aspect of the joint (synovial compartment). Soft-tissue density is increased within the joint on CT, which usually results in widening although narrowing may also occur. On T1-weighted MR imaging unilateral osseous and joint alterations are similar to other causes of erosive sacroiliitis. This includes replacement of the thin cartilage band by intermediate mixed intensity tissue, with extension of destruction to focally

erode adjacent cortex and underlying bone marrow. T2-weighted MR images reveal increased intensity owing to the inflammatory edematous process (42). It is often difficult to determine if the bone or the joint was the initial site of infection, because in the majority of cases findings coexist in both regions. Occasionally CT or MR imaging may demonstrate only an increase in soft-tissue density or cartilage replacement, respectively, within the sacroiliac joint, suggesting pyarthrosis as the infectious origin. Sclerosis of adjacent bone is best demonstrated by CT and is influenced by the degree of healing and virulence of the organism. Detection of sequestra as possible residual foci of infection is important as surgical debridement may be required. Computed tomography detects these dense sclerotic bone fragments within or related to the joint to best advantage. Magnetic resonance imaging will show sequestra as avascular fragments with low intensity on T1-weighted images. Increased intensity may be present with T2-weighted images (43). Soft-tissue extension of the inflammatory process frequently causes thickening of surrounding musculature on CT and MR imaging, particularly the iliacus, psoas, piriformis, gluteal, and erector spinae muscles. This is optimally demonstrated by axial CT and MR imaging. Other causes of sacroiliitis do not involve the adjacent soft tissues and this is an important distinguishing feature. Focal fluid collections on CT and MR imaging suggest abscess formation and may require surgical or percutaneous drainage (33). Intraosseous gas, or air within the soft tissues or abscess, may also be present. With resolution of the pyogenic arthritis, after treatment intraarticular osseous ankylosis may result. Both CT and MR imaging are used for evaluation of recurrent infection. Soft-tissue changes should resolve, without new areas of bone destruction. The T2-weighted MR images should show improvement in areas of increased intensity with response to antibiotic therapy.

Identifying the causative organism is frequently difficult in patients with negative blood cultures. Aspiration of the sacroiliac joint may be necessary in an attempt to isolate the offending organism (Fig. 28.15). This procedure should be performed under CT guidance. The patient is scanned in the prone position to identify an appropriate plane for needle entry into the inferior aspect of the synovial compartment of the sacroiliac joint from a posterior approach. A 20- or 22-guage, 3.5-inch spinal needle is then inserted into the joint with intermittent CT scans to follow the needle path. Aspiration is performed when the needle is within the joint. Lavage with several cubic centimeters of nonbacteriostatic saline and respiration may be necessary. The fluid is then sent for culture, staining, and sensitivity testing. Intraarticular location can be confirmed after successful aspiration with an injection of a small amount (1 to 2 cm^3) of water-soluble contrast (33, 43, 44).

Trauma

Pelvic fractures constitute only a small percentage (2 to 3%) of total skeletal injuries, with reported mortality ranging from 8 to 50% (45, 46). They are often, however, major components in the multisystem trauma patient owing to high speed motor vehicle, industrial, or falling accidents.

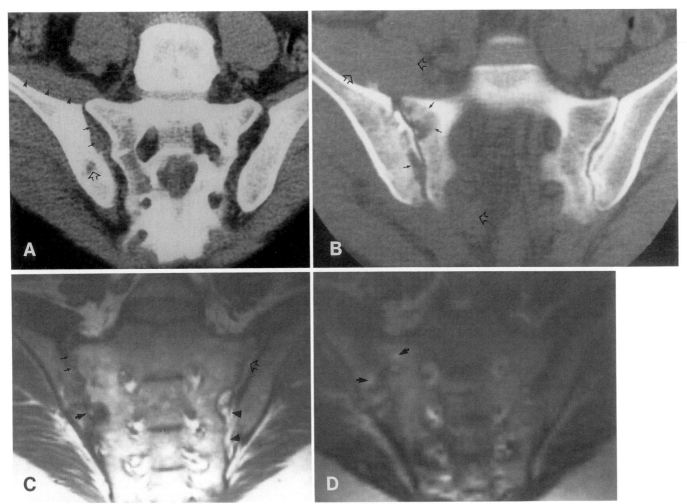

Figure 28.14. Staphylococcal septic sacroiliitis and osteomyelitis. **(A)** Coronal CT of the ligamentous compartment shows increased density within the right saroiliac joint *(arrows)* and thickening of the iliacus muscle *(arrowheads)* owing to inflammatory involvement. Focus of osteomyelitis is present in the ilium *(open arrow)*. **(B)** Coronal CT in the synovial compartment reveals more extensive osseous destruction, particularly on the sacral side *(arrows)*. Iliacus and piriformis muscle enlargement is seen *(open arrows)*. **(C)** T1-weighted coronal MR image (TR = 600 ms; TE = 25 ms) several months after antibiotic therapy shows resolution of adjacent soft-tissue inflammation. Destruction involving both the synovial (cartilage replacement, *small arrows*) and ligamentous compartments persists *(large arrow)*. Normal fat in the ligamentous compartment on the left inferiorly *(arrowheads)* and intact cartilage superiorly *(open arrow)* are present. **(D)** Coronal T2-weighted image (TR = 2100 ms; TE = 90 ms) shows areas of hyperintensity suggesting residual inflammatory tissue *(arrows)* in the sacrum and ilium.

The pelvis is a ring structure composed of the sacrum, coccyx, and innominate bones with strong ligamentous support. Disruption of any portion of the pelvic ring should be associated with another disruption in the ring (47). The sacroiliac joint and adjacent ligaments act as the major posterior stabilizing structures to the pelvis. Diastasis and fractures about the joint are significant components in pelvic injuries, with the anterior counterpart represented by injuries to the symphysis and pubic rami. Conventional radiography, however, has been estimated to allow 30 to 70% of posterior injuries to remain undetected (48). Computed tomography is ideally suited to evaluate the traumatized pelvis and is the modality of choice for assessment of both osseous (see more extensive discussion in Chapter 27) and soft-tissue abnormalities (49–54). The use of MR imaging in acute trauma to the pelvis appears limited at this time.

Posterior pelvic injury on CT may be seen as sacroiliac diastasis or manifested by vertically oriented fractures that often parallel the articulation either on the sacral or ilial side. Fractures may enter the joint with injuries being unilateral, bilateral, or contralateral to the anterior component. Younger patients are predisposed to sacroiliac joint involvement (53). In the immature skeleton sequelae of injury may include undergrowth of the hemipelvis with sacroiliac fusion and limb length discrepancy (55). In CT evaluation of sacroiliac joint diastasis it is important to determine which aspect of the joint is involved and particularly the presence of subluxation and its severity. The alignment of the posterior segment (ligamentous compartment) of the articulation dictates the presence of stability because of the associated supporting ligaments (posterior sacroiliac, sacrospinous, and sacrotuberous). Any CT finding that suggests disruption of the posterior ligaments, such

Figure 28.15. Biopsy needle with its tip *(arrowhead)* in place in the midsacroiliac joint from a posterior approach (patient prone). Changes of osteomyelitis are identified more anteriorly *(arrows)*.

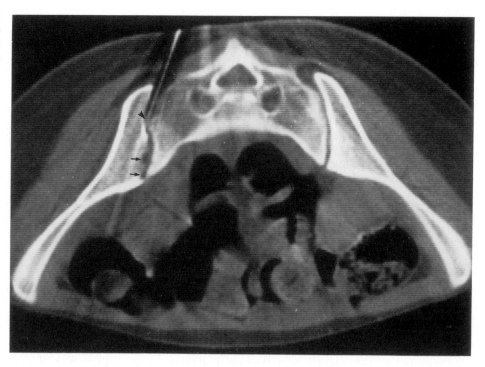

as posterior joint diastasis as well as vertical or anteroposterior subluxation of the articulation, indicates instability is present (54). In an analogous fashion, if fractures instead of diastasis are seen, their alignment posteriorly determines stability.

This can best be understood by utilizing the classification scheme of Pennal et al., based on the mechanism of injury (Fig. 28.16) (56). In anteroposterior compression injuries diastasis of the symphysis is caused by forces on the anterosuperior iliac spines. Progressive force and symphysis diastasis of greater than 2.5 cm will widen the anterior aspect (synovial compartment) of the sacroiliac joint but the posterior (ligamentous compartment) articulation remains intact, preserving stability in all but rare instances. This is the open-book or sprung pelvis injury. Lateral compression is the most common mechanism of injury and causes impaction forces in the anterior aspect of the sacroiliac complex. The extent of force determines whether posterior distraction and instability exist. This is evaluated on CT by the presence or absence of an increased gap posteriorly either at the sacroiliac joint or fracture site. Vertical shearing injuries (Malgaigne type) extensively disrupt the posterior structures and are unstable (Fig. 28.17) (57). The forces displace the hemipelvis superiorly, medially, and posteriorly. Depending on the series reviewed, between 40 and 60% of injuries are sacroiliac joint disruptions rather than fractures of the adjacent sacrum or ilium (58). The outcome with sacroiliac joint disruption is also less favorable as opposed to the presence of fractures. Trauma to the pelvis is frequently due to multidirectional forces, with injuries sustained representing a mixed pattern. Coronal and three-dimensional (3D) reconstructions (particularly views from above) are helpful in determination of stability and severity of subluxation (51). These are important factors in orthopedic surgical planning for either internal or external fixation.

Sacral stress fractures are not uncommon but are often difficult to detect with conventional radiography (59, 60). They are usually insufficiency-type fractures, on the basis of osteoporosis, with fatigue fractures being only rarely reported (61). Previous radiation, rheumatoid arthritis, and steroid administration predispose to these fractures. Technetium bone scanning is frequently the initial abnormal imaging modality and shows linear increased activity paralleling the sacroiliac joint. Fractures are often bilateral, with a horizontal component forming an H-like appearance on bone scanning with the superior margins more lateral ("Honda" sign) (62, 63). Computed tomography confirms the nonspecific radionuclide findings and in our experience sclerosis adjacent to the sacroiliac joint is the most frequent alteration (Fig. 28.18) (64, 65). This represents endosteal callus formation. A low-density fracture line with surrounding sclerosis has also been described. Computed tomography identification may require coronal 5-mm tilted gantry images in earlier stages. Magnetic resonance imaging will reveal linear or more diffuse marrow replacement on T1-weighted images. On T2-weighted images the fracture zone (low intensity) may be surrounded by higher intensity marrow edema and form a "pseudo" sacroiliac joint appearance medial to the true joint (Fig. 28.19).

Neoplasm

Neoplastic involvement of the sacrum and ilium is not infrequent. Common benign lesions include giant cell tumors, aneurysmal bone cysts, unicameral bone cysts, and neurofibromas. Primary malignant neoplasms also affect the ilium and sacrum, and include chordoma, chondrosarcoma, Ewing's sarcoma, osteosarcoma, malignant fibrous histiocytomas (MFH), and myeloma or plasmocytoma (1, 66). Secondary tumors of metastatic origin (breast, renal,

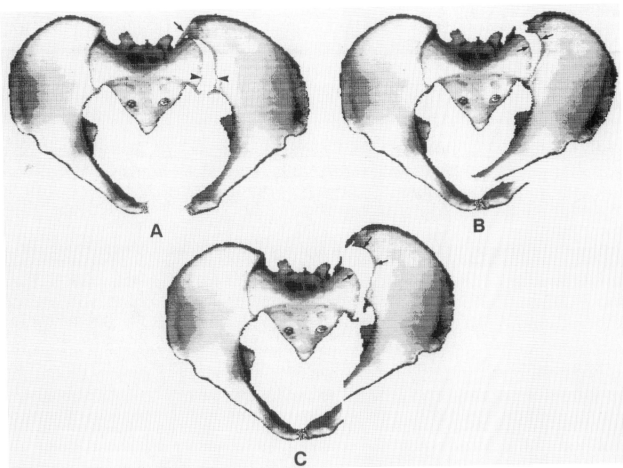

Figure 28.16. Schematic representation of injuries involving the sacroiliac joint. Anterior compression injuries **(A)** can disrupt the anterior sacroiliac joint *(arrowheads)* when symphysis diastasis exceeds 2.5 cm; however, the posterior sacroiliac ligaments remain intact as evidence by lack of posterior disruption *(arrow)*. Severe lateral compression **(B)** disrupts the posterior sacroiliac ligaments; widening *(arrows)* and impaction fractures of the anterior sacrum and ilium may be present. Vertical shearing injuries **(C)** extensively disrupt the sacroiliac joint *(arrows)*, with superior, posterior, and medial displacement of the hemipelvis.

and colon) and malignant degeneration of Paget's disease involve this region as well. These widely diverse lesions may occasionally extend into the sacroiliac joint. In our experience the tumors that commonly involve the sacroiliac joint are expansile in nature, arise from the sacrum, and include giant cell tumor, chordoma, chondrosarcoma, and metastatic disease (Fig. 28.20). In the study by Smith and associates, 5 of 13 giant cell tumors extended into the adjacent articulation (67). This predilection on the part of sacral neoplasms to involve the sacroiliac joint may be related to the sacrum being a more confined space, limiting expansion in other directions.

Magnetic resonance imaging and CT are well suited to best demonstrate both osseous and soft-tissue components (68, 69). The neoplasm is seen as bone destruction on CT, with T1-weighted MR images showing marrow replacement with increased intensity on T2-weighted images (Fig. 28.21). As in other neoplasms, qualitative assessment of tumor intensities is not reliable to differentiate among the various neoplasms. Sacroiliac joint involvement is present when the tumor extends through the cortical margin and may involve the adjacent osseous structure, either ilium or sacrum. Magnetic resonance imaging will no longer differentiate the band of cartilage in the synovial compartment, because it is engulfed by tumor. Ligamentous compartment extension of tumor is detected by soft tissue replacing the normal adipose tissue within this area. Oblique coronal imaging may be more helpful to evaluate early joint and sacral foraminal extension. Advantages of CT are detection and evaluation of matrix formation, calcification, and sometimes early cortical penetration (see also Chapter 15). As in other musculoskeletal applications, MR imaging advantages include improved soft-tissue contrast to evaluate areas of invasion and the multiple available imaging planes. Sagittal imaging is very helpful in assessment of sacral lesions to define tumor extent. Coronal and sagittal MR imaging is extremely important in determination of surgical respectability as well as in preoperative and radiation therapy planning. Computed tomography and MR imaging are useful in patient follow-up, both for response to therapy and evaluating tumor recurrence.

Figure 28.17. Vertical shearing injury. **(A)** Conventional radiograph shows displacement of the right hemipelvis superiorly and medially with offset at the inferior sacroiliac joint *(arrow)* and left pubic rami fractures. **(B)** Axial CT shows the disrupted posterior ligamentous portion of the sacroiliac joint *(arrows)* with several fracture fragments *(arrowheads)*. The remainder of the hemipelvis is displaced posteriorly and medially.

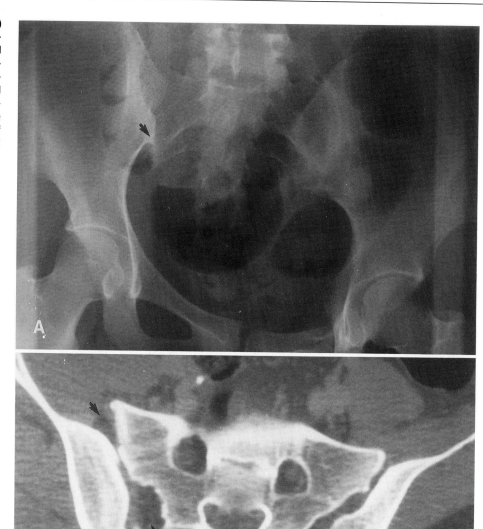

Figure 28.18. Coronal CT of an osteoporotic patient with a sacral stress fracture as evidenced by sclerosis representing callus paralleling the right sacroiliac joint *(arrowheads)*.

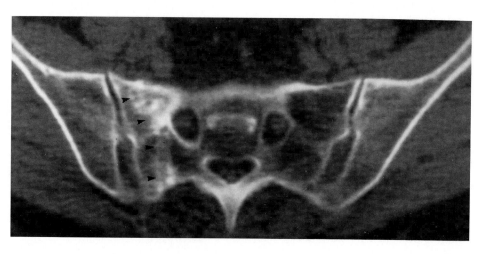

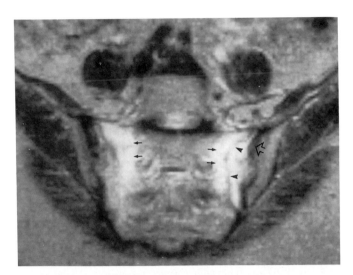

Figure 28.19. Coronal T2-weighted MR image (TR = 2100 ms, TE = 90 ms) shows bilateral sacral insufficiency fractures in an osteoporotic patient who had received radiation for colon carcinoma. Increased intensity in the marrow corresponds to edema *(arrows)* with low intensity on the left *(arrowheads)* forming a "pseudo" sacroiliac joint appearance medial to the articulation *(open arrow)*. T1-weighted images (not shown) demonstrated marrow replacement.

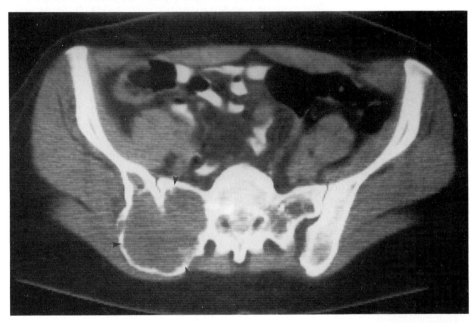

Figure 28.20. Axial CT showing an expansile giant cell tumor involving both the iliac bone and sacrum bridging the sacroiliac joint *(arrowheads)*.

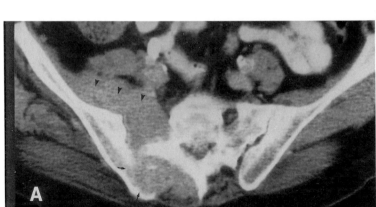

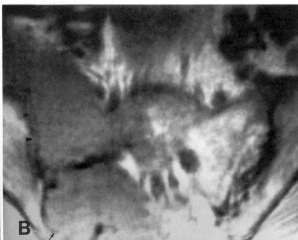

Figure 28.21. Malignant fibrous histiocytoma extending across the sacroiliac joint. **(A)** Axial CT reveals a destructive sacral lesion involving the adjacent ilium *(arrows)*. The iliacus muscle is infiltrated *(arrowheads)*. **(B)** T1-weighted axial MR (TR = 600 ms, TE = 25 ms) also demonstrates the neoplastic marrow replacement with articular extension posteriorly *(arrows)*; however, intact cartilage is present anteriorly *(arrowheads)*.

Conclusion

Clinical assessment of patients with symptoms related to the sacroiliac joint and evaluation of this articulation by conventional radiography is difficult. Physical and roentgenographic findings referable to the sacroiliac joint and surrounding structures are frequently obscured by the complex anatomy and overlying soft tissues. As we have seen, multiple pathologic conditions involve or extend into the sacroiliac joint, including arthritis, pyarthrosis, trauma, and neoplasm. Magnetic resonance imaging and CT are excellent noninvasive techniques that allow optimal evaluation of the wide variety of abnormalities. They are the modalities of choice in investigation of this complex articulation.

References

1. Resnick D, Niwayama G, eds. Diagnosis of bone and joint disorders. 2nd ed. Philadelphia: W. B. Saunders, 1988.
2. Anonymous. General Electric CT 9800 scan manual. Milwaukee: GE Medical Systems, 1988:5–14.
3. Tiggelen RV, Breunsbach J. Sacroiliitis: difficulties in the radiographic diagnosis. Advantage of CT? J Belgel Belgisch Tijdschriff Radiologie 1987;70:1–4.
4. De Smet AA, Gardner JD, Lindsley HB, Goin JE, Fritz SL. Tomography for evaluation of sacroiliitis. AJR 1982;139:577–581.
5. Konig H, Sauter R, Deimling M, Vogt M. Cartilage disorders: comparison of spin echo CHESS, and FLASH sequence MR images. Radiology 1987;164:753–758.
6. Dihlmann W. Diagnostic radiology of the sacroiliac joints. Chicago: Year Book Medical Publishers, 1980.
7. Vogler JB, Brown WH, Helms CA, Genant HK. The normal sacroiliac joint in CT study of asymptomatic patients. Radiology 1984;151:433–437.
8. Ehara S, El-Khoury GY, Bergman RA. The accessory sacroiliac joint: a common anatomic variant. AJR 1988;150:857–859.
9. Sashin D. A critical analysis of the anatomy and the pathologic changes in the sacroiliac joints. J Bone Joint Surg 1930;12:891–910.
10. Cohen AS, McNeil JM, Calkins E, Sharp JT, Schubart A. The "normal" sacroiliac joint: analysis of 88 sacroiliac roentgenograms. AJR 1967;100:559–563.
11. Murphey MD, Wetzel LH, Bramble JM, Levine E, Simpson KM, Lindlsey HB. Sacroiliitis: MR imaging findings. Radiology 1991;180:239–244.
12. Borlaza GS, Seigel R, Kuhns LR, Good AE, Rapp R, Martel W. Computed tomography in the evaluation of sacroiliac arthritis. Radiology 1981;139:437–440.
13. Whelan MA, Gold RP. Computed tomography of the sacrum. 1. Normal anatomy. AJR 1982;139:1183–1190.
14. Ryan LM, Carrera, GF, Lightfoot RW Jr, Hoffman RG, Kozin F. The radiographic diagnosis of sacroiliitis: a comparison of different views with computed tomograms of the sacroiliac joint. Arthr Rheum 1983;26:6–10.
15. Carrera GF, Foley WD, Kozin F, Ryan L, Lawson TL. CT of sacroiliitis. AJR 1981;136:41–46.
16. Lawson TL, Foley WD, Carrera GF, Berland LL. The sacroiliac joints: anatomic roentgenographic and computed tomographic analysis. J Comput Assist Tomogr 1982;6:307–314.
17. Wilkinson M, Meikle JAK. Tomography of the sacroiliac joints. Ann Rheum Dis 1966;25:433–440.
18. De Somer F, De Queker J, Baert AL. Comparison of computed tomography and conventional tomograms in evaluation of sacroiliitis. J Belgel Belgisch Tijdschrift Radiologie 1985;63:351–354.
19. Forrester DM, Hollingsworth PN, Dawkins RL. Difficulties in the radiographic diagnosis of sacroiliitis. Clin Rheum Dis 1983;9(2):323–332.
20. Yazici H, Turunc M, Ozdogan H, Yurdakul S, Akinci A, Barnes CG. Observer variation in grading sacroiliac radiographs might be a cause of "sacroiliitis" reported in certain disease states. Ann Rheum Dis 1987;46:139–145.
21. Resnick D, Niwayama G, Goergen TG. Degenerative disease of the sacroiliac joint. Invest Radiol 1975;10:608–621.
22. Pasion EG, Goodfellow JW. Pre-ankylosing spondylitis. Histopathological report. Ann Rheum Dis 1975;34:92–97.
23. Modic MT, Steinberg PM, Ross JS, Masaryk TJ, Carter JR. Degenerative disk disease: assessment of changes in vertebral body marrow with MR imaging. Radiology 1988;166:193–199.
24. Resnick D, Niwayama G, Goergen TG. Degenerative disease of the sacroiliac joint. Invest Radiol 1975;10:608–621.
25. Ramirez H Jr, Blatt ES, Cable HF, McComb BL, Zornoza J, Hibri N. Intraosseous pneumatocysts of the ilium. Radiology 1984;150:503–505.
26. Feldman F, Johnston A. Intraosseous ganglion. AJR 1973;118:328–343.
27. Durback MA, Edelstein G, Schumacher R Jr. Abnormalities of the sacroiliac joints in diffuse idiopathic skeletal hyperostosis: demonstration by computed tomography. J Rheumatol 1988;15:1506–1511.
28. Resnick D, Shapiro RF, Weisner KB, Niwayama G, Utsinger KB, Shaul SR. Diffuse idiopathic skeletal hyperostosis (DISH) (ankylosing hyperostosis of Forestier and Rotes-Querol). Semin Arthr Rheum 1978;7:153–187.
29. Numaguchi Y. Osteitis condensans ilii, including its resolution. Radiology 1971;98:1–8.
30. Wells J. Osteitis condensans ilii. AJR 1956;76:1141–1143.
31. Resnick D, Dowsh IL, Niwayama G. Sacroiliac joint in renal osteodystrophy: roentgenographic-pathologic correlation. J Rheumatol 1985;2:287.
32. Hooge WA, Li D. CT of sacroiliac joints in secondary hyperparathyroidism. J Can Assoc Radiol 1981;31:42–45.
33. Kerr R. Pyogenic sacroiliitis. Orthopedics 1985;8:1030–1034.
34. Gordon G, Kabin SA. Pyogenic sacroiliitis. JAMA 1980;69:50–56.
35. Oka M, Möttönen T. Septic sacroiliitis. J Rheumatol 1983;10:475–478.
36. Pollack M, Gurman A, Salis JG. Pyogenic infection of the sacroiliac joint. Orthoped Grand Rounds 1986;3:2–9.
37. Coy JT, Wolf CR, Brower TD. Pyogenic arthritis of the sacroiliac joint. J Bone Joint Surg 1976;58A:845–849.
38. Horgan JG, Walker M, Newman JH, Wate I. Scintigraphy in the diagnosis and management of septic sacroiliitis. Clin Radiol 1983;34:337–346.
39. Ailsby RL, Staheli LT. Pyogenic infections of the sacroiliac joint in children. Radioisotope bone scanning as a diagnostic tool. Clin Orthop 1974;100:96–100.
40. Guyot DR, Manoli AII, Kling GA. Pyogenic sacroiliitis in IV drug abusers. AJR 1987;149:1209–1211.
41. Rosenberg D, Baskies AM, Deckers PJ, Leiter BE, Ordia JI, Yablon IG. Pyogenic sacroiliitis: an absolute indication for computerized tomographic scanning. Clin Orthop 1984;184:128–132.
42. Wilbur AC, Langer BG, Spigos DG. Diagnosis of sacroiliac joint infection in pregnancy by magnetic resonance imaging. Magn Reson Imag 1988;6:341–343.
43. Hendrix RW, Lin PJP, Kane WJ. Simplified aspiration or injection technique for the sacroiliac joint. J Bone Joint Surg 1982;64A:1249–1252.
44. Mieskew DB, Block RA, Witt PF. Aspiration of infected sacroiliac joints. J Bone Joint Surg 1979;61A:1071–1072.
45. Rogers LF, ed. Radiology of skeletal trauma. New York: Churchill Livingstone, 1982.
46. McMurty R, Walton D, Dickinson D, Kellam J, Tile M. Pelvic disruption in the polytraumatized patient: a management protocol. Clin Orthop 1980;151:22–30.
47. Rockwood CA Jr, Green DP, eds. Fractures in adults. 2nd ed. Philadelphia: Lippincott, 1984.
48. Montana MA, Richardson ML, Kilcoyne RF, Harley JD, Shuman WI, Mack LA. CT of sacral injury. Radiology 1986;161:499–503.
49. Dunn EL, Berry PH, Connally JD. Computed tomography of the pelvis in patients with multiple injuries. J Trauma 1983;23:378–383.
50. Raffii M, Firooznia H, Golimbu C, Waugh T, Naidich D. The impact of CT in clinical management of pelvic and acetabular fractures. Clin Orthop 1983;178:228–235.

51. Fishman EK, Magid D, Brooker AF, Siegelman SS. Fractures of the sacrum and sacroiliac joint: evaluation by computed tomography with multiplanar reconstruction. South Med J 1988;81:171–177.

52. Vas WG, Wolverson MK, Sundaram M, et al. The role of computed tomography in pelvic fractures. J Comput Assist Tomogr 1982;6(4):796–801.

53. Donoghue V, Daneman A, Krajbich I, Smith CR. CT appearance of sacroiliac joint trauma in children. J Comput Assist Tomogr 1985;9(2):352–356.

54. Heare MM, Heare TC, Gillespy III T. Diagnostic imaging of pelvic and chest wall trauma. Radiol Clin N Am 1989;27(5):873–889.

55. Heeg M, Visser JD, Oostovogel HJM. Injuries of the acetabular triradiate cartilage and sacroiliac joint. J Bone Joint Surg (Br) 1988;1:34–37.

56. Pennal GF, Tile M, Waddell JP, Garside H. Pelvic disruption: assessment and classification. Clin Orthop 1980;151:12–21.

57. Malgaigne JF. The classic. Double vertical fractures of the pelvis. Clin Orthop 1980;151:8–11.

58. Simpson LA, Waddell JP, Leighton RK, Kellam JF, Tile M. Anterior approach and stabilization of the disrupted sacroiliac joint. J Trauma 1987;27:1332–1339.

59. Lourie H. Spontaneous osteoporotic fractures of the sacrum. An unrecognized syndrome in the elderly. JAMA 1982;248:715–717.

60. Laasonen E. Missed sacral fractures. Ann Clin Res 1977;9:84–87.

61. Hoang TA, Nguyen TH, Daffner RH, Lupetin AR, Deeb ZL. Case report 491. Skel Radiol 1988;17:364–367.

62. Schneider R, Yacovone J, Ghelman B. Unsuspected sacral fractures: detection by radionuclide bone scanning. AJR 1985;144:337–341.

63. Ries T. Detection of osteoporotic sacral fractures with radionuclides. AJR 1983;146:783–785.

64. De Smet AA, Neff JR. Pelvic and sacral insufficiency fractures: clinical course and radiologic findings. AJR 1985;145:601–606.

65. Gacetta DJ, Yandow DR. Computed tomography of spontaneous osteoporotic sacral fractures. J Comput Assist Tomogr 1984;8(6):1190–1191.

66. Wilner D, ed. Radiology of bone tumors and allied disorders. Philadelphia: W. B. Saunders, 1983.

67. Smith J, Wixon D, Watson RC. Giant cell tumor of the sacrum: clinical and radiologic features in 13 patients. J Can Assoc Radiol 1979;30:34–38.

68. Wetzel LH, Levine E. MR imaging of sacral and presacral lesions. AJR 1990;154:771–775.

69. Levine E, Batnitzky S. Computed tomography of sacral and perisacral lesions. CRC Crit Rev Diagn Imag 1984;21:307–374.

CHAPTER 29

Magnetic Resonance Imaging of the Pediatric Hip

Cees F. A. Bos and Johan L. Bloem

Introduction

The pediatric hip is subject to a larger number of mal-developments than any other joint in the body. As any hip abnormality early in life may have far-reaching consequences later in life, numerous studies of the intrauterine development and anatomy of the hip have been performed for a better understanding of the pathogenesis of some hip disorders.

The hip of a neonate and infant is composed mainly of nonradiopaque structures. These are most important for the diagnosis of maldevelopments and acquired disorders of the hip. Diagnostic imaging of these structures is important. Traditional imaging techniques are conventional radiography and arthrography. More recently computed tomography (CT) and ultrasound were introduced.

Magnetic resonance (MR) imaging is a noninvasive modality that allows high-contrast, multiplanar tomographic imaging without the use of ionizing radiation. Reports on the pediatric orthopedic application of MR imaging have been relatively sparse (1, 2, 3). It gives us insight in the developmental anatomy and pathologic conditions unique to children. A major drawback of MR imaging is that the very young child must lie still in the magnet for at least 20 min.

Magnetic resonance imaging can play an important role in the diagnosis of pediatric hip disorders, as it provides diagnostic information that otherwise can be only partially obtained with arthrography. Thus MR imaging may have a major impact on patient management.

Examination Technique

Children between 1 and 5 years of age should be sedated before MR imaging, whereas older children can usually lie still for the required imaging time. Feeding the babies prior to the study facilitates the examination, because it induces sleep. General anesthesia is sometimes necessary when the routine procedure is not successful.

Magnetic resonance imaging is performed on a 0.5- or 1.5-T imager. Most images demonstrated in this chapter were made with a 0.5-T Philips scanner (Philips, Shelton, Conn.). A flexible surface coil is used when one hip needs visualization and a head coil is used when both hips must be evaluated. A short repetition time/echo time (TR/TE) spin-echo (SE) sequence is used for T1 weighting (TR less than 600 ms, TE 20 ms). T2 weighting is emphasized by using SE sequences with a TR of 1500 to 2000 ms and a dual-echo technique with a TE of 50 to 100 ms. Slice thickness is 3 to 5 mm, and the slice gap is 0.1 mm. T2*-weighted gradient-echo techniques can also be used instead of T2-weighted SE sequences to visualize intraarticular structures such as the labrum.

Congenital hip dysplasia and dislocation are imaged in coronal, sagittal, and axial planes with short T1-weighted sequences. Generally both hips are visualized. When the hip is dislocated T2- or T2*-weighted coronal images are obtained to assess the position of the labrum. In Legg-Perthes disease coronal T1- and T2-weighted, as well as sagittal T1-weighted, images are obtained of the symptomatic hip. When symptoms are bilateral, or if radiographs show bilateral abnormality, both hips are examined.

Congenital Dysplasia

It has been generally accepted that patients with congenital dysplasia of the hip should be treated before learning to walk. This is because persistent displacement of the femoral head causes a variety of changes in the soft tissues. The continuing growth of the child aggravates deformation of bone and soft tissues in and around the hip. This may lead to contracture of soft-tissue structures.

Concentric position of the femoral head in the acetabular socket is the aim of treatment in children with dysplasia. Closed reduction is the preferred treatment when the hip is dislocated. This is usually possible before the child is 3 years old (4, 5), but rarely thereafter.

The main soft-tissue changes in congenital dysplasia of the hip that occur when early treatment is not installed are inversion of the labrum, hypertrophy of the pulvinar, stretching of the ligamentum teres and transverse ligament, and adherence of the capsule to the ilium. When

---

Abbreviations (see also glossary): AP, anteroposterior; SE, spin echo.

these soft-tissue structures undergo irreversible changes, an attempted closed reduction will be unsuccessful and harmful. The importance of these soft-tissue changes in congenital dislocation and dysplasia of the hip is reflected in the classification system of Dunn (6) and Ogden (7). In grade I, the hip can be subluxated and is positionally unstable. In grade II, the hip is dislocatable and there is progressive eversion of the labrum. In grade III, the femoral head is completely displaced posterosuperiorly and a false acetabulum and an infolded labrocapsular fold is present.

Until now arthrography has been the most accurate way to demonstrate obstacles to closed reduction. The disadvantages of hip arthrography are its invasiveness, exposure to ionizing radiation, and indirect visualization of soft-tissue structures.

In MR imaging the important soft-tissue structures are directly visualized with thin tomographic sections. The signal intensities of normal structures in the pediatric hip are relatively constant with different pulse sequences (8). The pulvinar (fat pad) has the highest signal intensity and its signal intensity is identical to that of subcutaneous fat and yellow bone marrow. Cortical bone and ligaments emit no signal and are thus seen as black lines on both T1- and T2-weighted images. The fibrous (outer) part of the labrum has a low signal intensity identical to that of cortical bone. Hyaline cartilage has an intermediate signal intensity.

Magnetic resonance images reveal anatomic features that are not seen on plain radiographs, arthrograms, or computed tomograms (9, 10). Magnetic resonance images, which demonstrate a redundant capsule, a sharply angled insertion of the capsule to the ilium, and a laterally displaced femoral head with an everted labrum (grade I or II), represent hips that are amenable for closed reduction (Fig. 29.1).

Magnetic resonance images that demonstrate a folded capsule, closely fitted to the ilium with an inverted labrum, represent hips with irreversible soft-tissue changes. The proximal part of the capsule resembles a hairpin. One leg of the hairpin is formed by the periosteum of the ilium. The other leg is formed by capsule adhesion to the ilium, while the connecting part is formed by the capsular fold (Fig. 29.2). These hips can be reduced only by surgical procedure.

Magnetic resonance imaging can also be used, even when a cast has been applied, to evaluate persistent joint incongruity after closed or open reduction (Fig. 29.3). It is well known that the future of these hips when not treated will be unsatisfactory (11). Magnetic resonance imaging is superior to anthrography in the evaluation of these patients. With MR imaging it is feasible to differentiate between incongruity caused by thickening of structures such as cartilage and pulvinar, and incongruity due to interposition of soft-tissue structures such as labrum and psoas tendon. The "rose thorn" sign of arthrography, which indicates a normally positioned labrum, is often absent in these patients even when the labrum has a normal position. This is probably related to the lateral position of the femoral head and/or adhesions.

In difficult cases, such as redislocations following open reduction, MR imaging provides the orthopedic surgeon

with more clarifying information about changes occurring in and around the dislocated hip.

Magnetic resonance imaging has great potential in assisting the orthopedic surgeon to select the appropriate form of therapy in patients with congenital dysplasia of the hip. The main indications for MR imaging are late-detected congenital dysplasia of the hip, difficult cases such as redislocations, and serious doubts about concentric reduction. Magnetic resonance imaging is not indicated in uncomplicated hip dysplasia of the newborn.

Acetabular Residual Dysplasia

Unsatisfactory response of acetabular development to closed or open reduction is commonly designated as acetabular residual dysplasia. As a rule acetabular residual dysplasia is identified radiographically as persistent deficiency of the acetabulum, which may be characterized by a shallow acetabulum or an inadequately covered femoral head. The degree of lateral coverage of the femoral head is conventionally assessed by the center edge (CE) angle of Wiberg (12). This measurement, however, is inaccurate in children younger than 6 years of age. Radiographic evaluation obviously is limited to bony margins and does not offer a separate evaluation of the anterior, lateral, or posterior cartilaginous coverage of the femoral head.

The cartilaginous part of the acetabular roof in children is formed by articular and growth cartilage. Delayed ossification is a major cause for insufficient osseous coverage of the femoral head as displayed on plain radiographs. Therefore, visualization of cartilage with MR imaging, in coronal, sagittal, and axial planes, is important in the assessment of the cartilaginous congruity within the joint (Figs. 29.4 and 29.5). Measurements of the acetabular index and CE angles, including the chondroepiphyseal cartilage, can be taken from MR images (13). These measurements are a reliable guide in determining whether an acetabular reconstruction is needed or not. Surgery is not needed when the radiographically determined acetabular dysplasia reflects a delay in endochondral ossification. In these instances sufficient acetabular cartilaginous coverage is identified on MR images (Fig. 29.4) (10). When surgery is needed the choice of the most appropriate form of acetabular reconstruction is facilitated, as information about acetabular coverage anteriorly, laterally, and posteriorly is available (Fig. 29.5).

Legg-Perthes Disease

In Legg-Perthes disease an ischemic process is responsible for avascular necrosis of the capital femoral epiphysis. The extent of involvement, which later on in the disease may involve the metaphysis, is prognostic for hip deformity and early disability. This may result in growth disturbance of the femoral head.

The earliest radiographic signs are a small epiphysis, an increased density of the epiphysis, and a subchondral fracture line. The radiographic classification of Salter and Thompson (14) is based on the extent of the subchondral fracture line. The extent of this subchondral fracture is also prognostic for the extent of the ischemic zone. However,

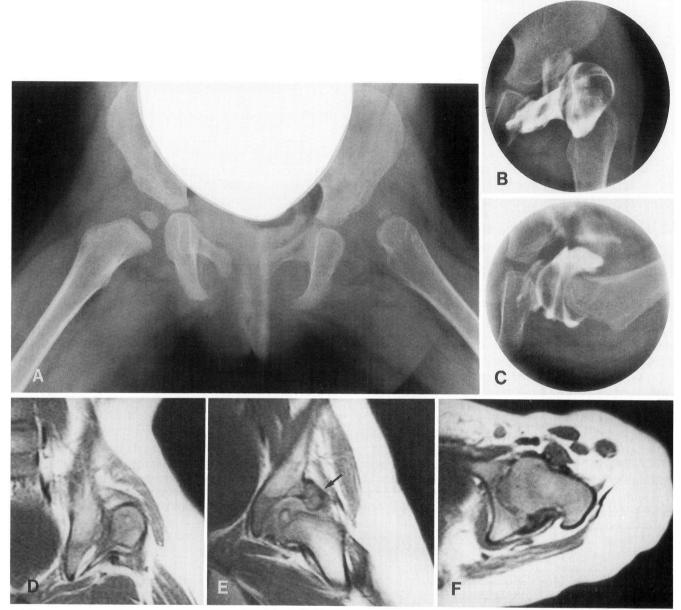

Figure 29.1. (A) Bilateral dysplasia and left-sided dislocation were demonstrated on admission radiograph in this 23-month-old girl. **(B)** Arthrogram made before reduction. **(C)** A concentrically reduced femoral head can be appreciated on arthrogram made after reduction. Contrast medium is squeezed into the superior aspect of the capsule, giving an appearance of increased capacity. **(D)** T1-weighted coronal image made prior to reduction demonstrates obtuse capsular insertion on pelvis. The femoral head is surrounded by a wide capsule and stands away from the ilium. No impeding structures are identified. **(E)** T1-weighted coronal image made after reduction shows proximal capsular pouch *(arrow)*. The capsule inserts proximally to the cartilaginous rim of the acetabulum on the ilium. The low signal intensity labrum is well extended over the femoral head. Hyaline cartilage is thickened. The femoral head is concentrically reduced. **(F)** T1-weighted axial MR image after reduction. The capsule has been stripped from the posterior acetabular rim, giving the appearance of a redundant capsule.

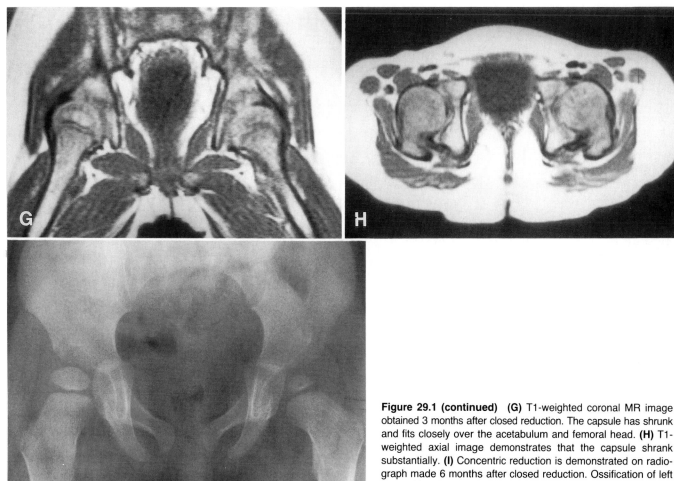

Figure 29.1 (continued) **(G)** T1-weighted coronal MR image obtained 3 months after closed reduction. The capsule has shrunk and fits closely over the acetabulum and femoral head. **(H)** T1-weighted axial image demonstrates that the capsule shrank substantially. **(I)** Concentric reduction is demonstrated on radiograph made 6 months after closed reduction. Ossification of left femoral head is retarded. (Reprinted with permission from Bos CFA, Bloem JL. J Bone Joint Surg 1989;71A:1523–1530.)

the subchondral fracture line is, unfortunately, a transient radiographic phenomenon and is seen only in the early stages of Legg-Perthes disease. The radiographic classification of Catterall (15) is determined by the extent of epiphyseal involvement. The major disadvantage of this classification is that the precise degree of involvement cannot be established before the process of resorption is completed. A limitation of conventional radiography in determining the extent of necrosis is the considerable loss of trabecular bone required before abnormalities can be appreciated.

Magnetic resonance imaging has been demonstrated to depict detailed marrow changes in the vascular compromised femoral head (16, 17). Diagnosis of epiphyseal abnormalities is facilitated by the fact that, even in very young children, the proximal femoral epiphysis contains yellow and not red bone marrow. Consecutive MR imaging manifestations in Legg-Perthes disease show a constant pattern (18). In the initial phase of the disease a subchondral low-intensity line is visible both on T1- and T2-weighted images. (Figs. 29.6C–E, 29.7A,B, and 29.8C). The normal signal intensity of the underlying capital epiphyseal marrow will change into an intermediate to low signal intensity on T1- and T2-weighted images. However, in the very early stage of the disease necrosis may display normal signal intensities (19). In these patients the subchondral fracture line is the only osseous abnormality.

The repair tissue interface separates the intact trabeculae from the necrotic area situated proximal to the repair tissue interface. The double line sign of this interface is similar to that seen in the adult with avascular necrosis (Figs. 29.6H and 29.8E) (16, 20). This zone corresponds to thickened trabecular bone and mesenchymal and fibrous repair tissue. The inner portion of the repair tissue interface, characterized by a high signal intensity on T2-weighted images, correlates with radiolucent areas on radiographs

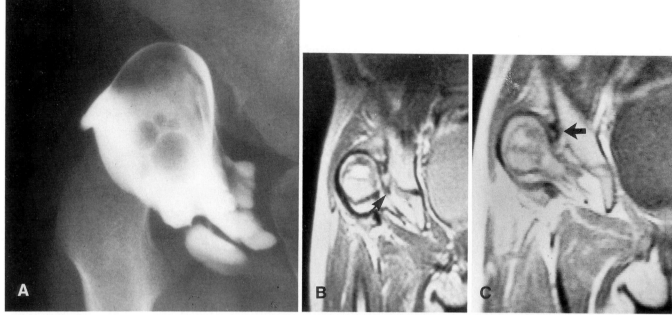

Figure 29.2. **(A)** Arthrogram of a girl, age 2 years and 10 months, with grade III dysplasia of right hip shows a narrowed entrance to the acetabular socket. **(B)** Coronal T2-weighted image displays inverted labrum *(arrow)*. The labrum, transverse ligament, capsule, and cortex have a low signal intensity. The ligamentum teres is stretched. The high signal intensity surrounding the femoral head and inverted labrum represents joint effusion. **(C)** A more ventral T1-weighted coronal image demonstrates folded capsule (low signal intensity) closely fitted to the ilium *(arrow)*. This explains the narrow entrance to the acetabular socket, seen on the arthrogram. The area of high signal intensity between the ligamentum teres and acetabular wall represents thickened pulvinar. (Reprinted with permission from Bos CFA, Bloem JL. J Bone Joint Surg 1989;71A:1523–1530.)

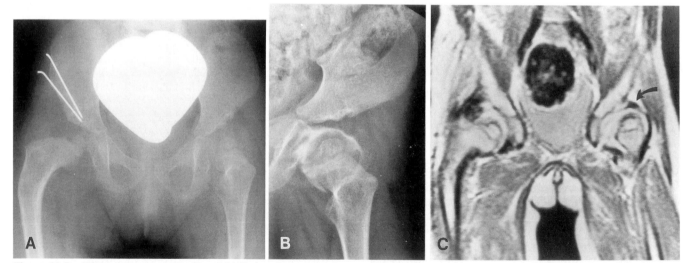

Figure 29.3. Bilateral hip dysplasia. The right hip has been treated surgically, whereas the left hip had been treated conservatively. **(A)** The left hip shows severe residual dysplasia. **(B)** Arthrography of the left hip fails to show a medially directed "rose thorn" sign of contrast medium. This suggests the presence of an inverted labrum. Medial pooling of contrast medium is noted. **(C)** Coronal T2-weighted MR image clearly demonstrates incongruity with a subluxated femoral head. The capsule is thickened. Labrum is everted *(arrow)*. Absence of the rose thorn sign does not correlate with the presence of an inverted labrum.

(Fig. 29.8). The outer zone of the repair tissue interface on MR images (low signal intensity on T1- and T2-weighted images) is similar to the sclerosis at the periphery of the lesion as seen on radiographs. This is defined as new bone formation, also known as creeping substitution (21).

Trabecular bone, either intact or fractured, that has not been invaded by the repair tissue typically remains visible on T1- and T2-weighted images as a low signal intensity area in the dome of the epiphysis. This uninvaded necrotic bone, which has been designated a sequestrum, is absorbed in a later stage.

A metaphyseal area of decreased signal intensity on T1-

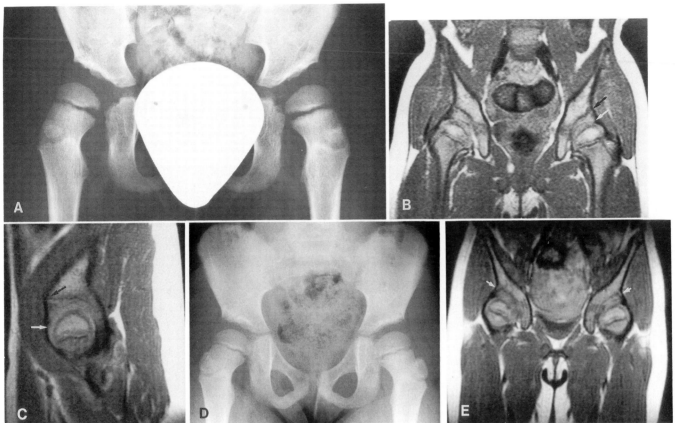

Figure 29.4. **(A)** Marked bilateral dysplasia is noted on this radiograph of a 5-year-old girl. **(B)** Coronal T1-weighted MR image made at the level of the posterior aspect of the acetabulum reveals the concentric position of both femoral heads. They are covered by a thick layer of unossified acetabular cartilage. The labrum is closely fitted over the femoral heads on both sides. The cartilaginous *(white arrow)* coverage was considered to be sufficient in both hips. The black arrow points to the bony margin. **(C)** T1-weighted sagittal image of the right hip shows limited anterior osseous *(black arrow)* and sufficient cartilaginous *(white arrow)* coverage. **(D)** Radiograph made 18 months later demonstrates marked improvement. **(E)** Coronal T1-weighted MR image, made at this time, demonstrates bone marrow, indicative for ossification, in the lateral outer edge of acetabular cartilage *(arrows)*. (Reprinted with permission from Bos CFA, Bloem JL, Verbout AJ. Clin Orthop 1991; 265:207–217.)

weighted images, exhibiting a high signal intensity on T2-weighted images (Fig. 29.8F), may be found in hips with more than 50% of necrosis. This probably represents edema. Joint effusion is present in all affected hips. Fatty marrow conversion may be encountered after, or during, long-standing disease.

The repair tissue interface can indicate whether or not and to which extent the infarcted bone includes the growth plate. The extent of the infarcted zone below the subchondral fracture line is precisely delineated on MR images within 6 months after onset of the symptoms. This means that MR images demonstrate the true extent of necrosis prior to radiographs.

Magnetic resonance imaging displays four consistent courses of signal patterns, similar to the classification of Catterall (15), depending on the extent of ischemic necrosis.

Group I

The initial phase is characterized on MR images by a low signal intensity area in the superior part of the epiphysis on T1- and T2-weighted images. The signal intensity of the underlying epiphyseal bone does not change during the course of the disease and remains isointense with subcutaneous fat. The size of the infarcted area in the dome of the epiphysis decreases and finally shows normal high signal intensity on T1-weighted images, consistent with the presence of yellow bone marrow (Fig. 29.6, left hip). No metaphyseal changes are depicted on MR images in this group.

Group II

On coronal MR images a subchondral low-intensity line, covering the lateral half of the epiphysis except for a small part of the utmost lateral part, is seen. On sagittal MR images the subchondral low-intensity line covers the complete anterior half of the epiphyseal dome. Although the underlying bone marrow may initially have a normal signal intensity, an intermediate or low signal intensity develops within 6 months.

The first sign of repair is represented by the characteristic semilunar shaped repair tissue interface in the epiphysis on T1-weighted coronal MR images, demarcating normal signal intensities of viable epiphyseal bone medially

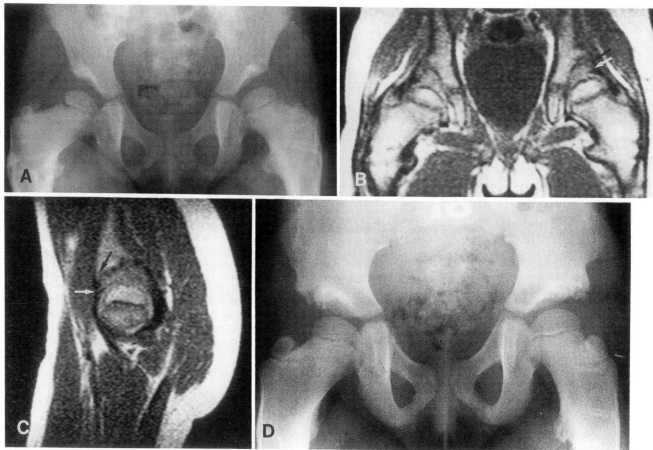

Figure 29.5. **(A)** Bilateral acetabular dysplasia is appreciated on this anteroposterior (AP) radiograph of the pelvis in a 5-year-old girl. **(B)** Coronal T1-weighted image, made at the level of the anterior aspect of the acetabulum, displays insufficient acetabular osseous *(black arrow),* and cartilaginous *(white arrow)* coverage of both hips. **(C)** T1-weighted sagittal MR image of the right hip shows sufficient anterior osseous and cartilaginous coverage. Thus MR imaging indicates the need for a surgical procedure aimed at improving lateral coverage. **(D)** Follow-up radiograph demonstrates the result of bilateral innominate osteotomy. (Reprinted with permission from Bos CFA, Bloem JL, Verbout AJ. Clin Orthop 1991;265:207–217.)

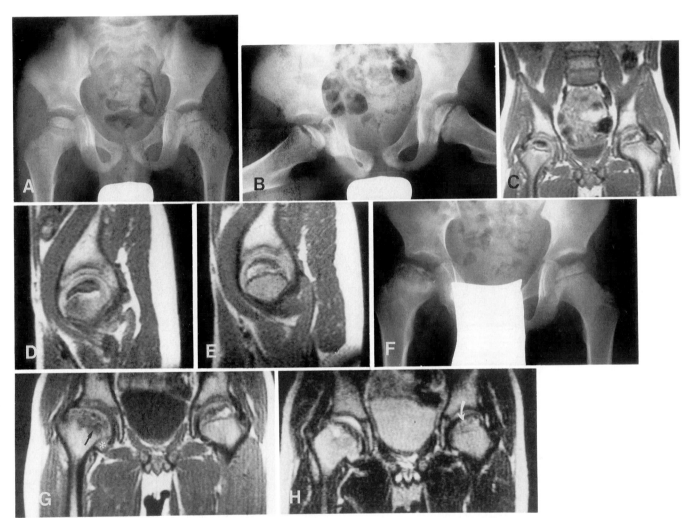

Figure 29.6. **(A)** AP and **(B)** frog-leg (Lauenstein) view demonstrate the initial features of Legg-Perthes disease in both hips. A lytic defect is visible in the dome of the epiphysis of the left hip. The corresponding coronal **(C)** and sagittal **(D)** T1-weighted MR images of the right hip demonstrate a subchondral fracture line covering the whole epiphysis. Necrotic bone underneath the fracture has a decreased (intermediate) signal intensity. The signal intensity of the epiphyseal marrow below the fracture in the left hip is relatively normal **(E)**. Therefore prognosis of the left hip (group I) is better than that of the right hip (group IV). **(F)** Radiograph made 1 year after onset of Legg-Perthes disease shows collapse, lateral uncovering, completed resorption, and new bone formation in the epiphysis of the right hip. The left hip has almost completely healed. **(G)** Corresponding coronal T1-weighted image demonstrates subchondral fracture in the right hip as an interrupted low-intensity line. The repair tissue interface is composed of a curved, low-intensity line *(arrow)* and an inner zone of intermediate signal intensity. A joint effusion *(asterisk)* is present. The signal intensity of the left epiphysis is, with the exception of the epiphyseal dome, normal. **(H)** The large area of high signal intensity in the right hip on the corresponding T2-weighted image represents viable repair tissue. The increase of signal intensity in the dome of the left hip also indicates the presence of viable repair tissue *(arrow)*. (Reprinted with permission from Bos CFA, Bloem JL, Bloem RM. J Bone Joint Surg (Br) 1991;73-B:219–224.)

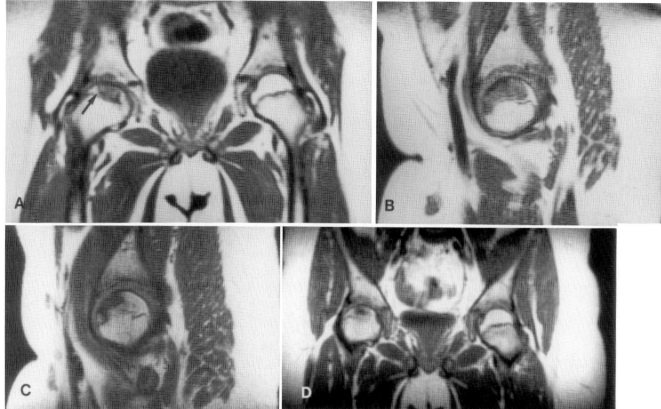

Figure 29.7. **(A)** Coronal T1-weighted MR image, made 6 months after first symptoms, reveals low intensity repair tissue interface *(arrow)* surrounding an intermediate signal intensity of necrosis. The medial and lateral aspects of the epiphysis are uninvolved. **(B)** Corresponding sagittal T1-weighted image of right hip. The subchondral fracture remains visible as a low intensity line. The necrotic epiphysis has an intermediate signal intensity. Only a small part of the posterior aspect of the epiphysis is normal. **(C)** One year later the sagittal T1-weighted MR image shows further healing. This is reflected by extension of normal bone marrow in the anterior direction. **(D)** The corresponding T1-weighted coronal image demonstrates an increase of normal signal intensity area in the medial and lateral aspect of the epiphysis. (Reprinted with permission from Bos CFA, Bloem JL, Bloem RM. J Bone Joint Surg (Br) 1991;73-B:219–224.)

and laterally from centrally located intermediate signal intensities of necrotic bone. Sagittal T1-weighted images reveal an uninvolved posterior part of the epiphysis (Fig. 29.7).

Further healing is seen as an increase of normal yellow marrow and reduction of the extent of necrosis (low-intensity area) in the anterior aspect of the epiphysis (Fig. 29.7). Uninvaded necrotic bone immediately under the articular cartilage remains visible as a core-shaped signal-void structure on T1- and T2-weighted images. This is consistent with the presence of nonresorbed necrotic bone as seen on radiographs.

When present, cystlike changes in the methaphysis display an intermediate signal intensity on T1-weighted images and a high signal intensity on T2-weighted images. The sclerotic margin has a low signal intensity on all pulse sequences. Articular cartilage is thickened on the medial, dorsal, and lateral aspect of the femoral head.

Group III

The subchondral fracture line extends from the anterior surface to the posterior surface. Only a small part of the lateral and posterior aspect and a larger part of the medial aspect of epiphysis are not involved. The femoral capital epiphysis has a decreased signal intensity. Subsequently, MR images exhibit the typical repair tissue interface on T1- and T2-weighted images. A centrally located core of low signal intensity on T1- and T2-weighted images represents a sequestrum (Fig. 29.8). The metaphyseal lesions described in group II may also be encountered.

Flattening of the femoral head may be apparent. Further healing is symbolized on MR images by a decreased size of the sequestrum and cephalic movement of the repair tissue interface. Progression of femoral head flattening may occur. Articular cartilage is thickened, particularly in the medial and posterior aspects of the femoral head.

Group IV

In the earliest phase of this group T1- and T2-weighted MR images show a subchondral low-intensity fracture line, covering the complete epiphysis on coronal and sagittal images (Fig. 29.6). The underlying bone of the epiphysis is characterized by an intermediate signal intensity on T1- and T2-weighted images. The repair tissue interface extends in the metaphysis beyond the growth plate. The epiphysis shows low-intensity fragments and lateral un-

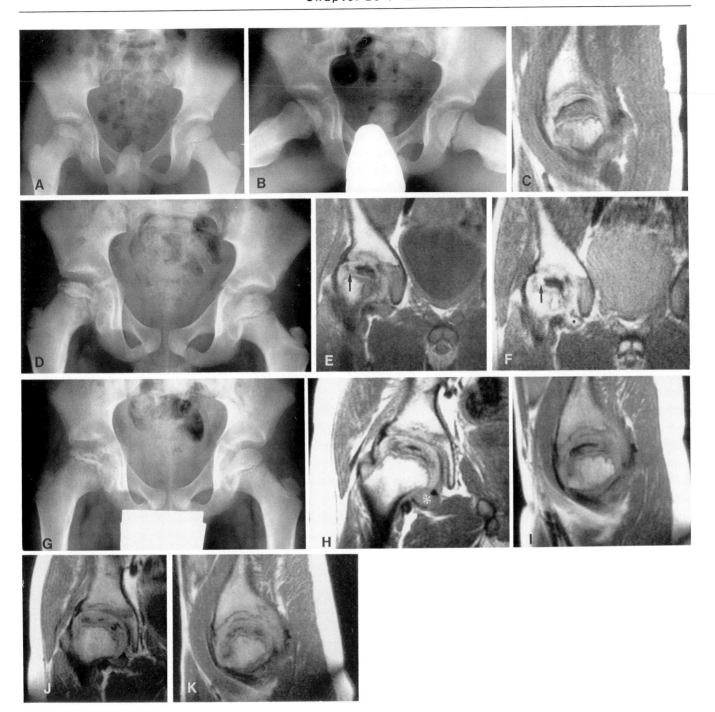

Figure 29.8. (A) AP radiograph of the pelvis demonstrates an increased density of small right epiphysis. (B) The frog-leg view demonstrates a subchondral fracture of the right hip. (C) The sagittal T1-weighted MR image demonstrates a low signal intensity subchondral fracture line in the anterior and middle aspects of the femoral epiphysis. The underlying necrotic bone marrow has an intermediate signal intensity. (D) Six months later the AP radiograph shows loss of epiphyseal height, a dense area in the dome of the epiphysis, a medial and lateral fragment, and a metaphyseal lesion. (E) The corresponding T1-weighted coronal MR image shows the repair tissue interface, which is subdivided in an outer cup-shaped low-intensity band *(arrow)* and an inner zone of intermediate signal intensity. The sequestrum is visible as a low signal intensity structure in the dome of the epiphysis. The metaphyseal lesion has an intermediate signal intensity and a low-intensity rim. (F) Corresponding T2-weighted image demonstrates an in-

crease of signal intensity of the inner area of the repair tissue interface *(arrow)*. Size and signal intensity of the sequestrum do not change with increased T2 weighting. Joint effusion is noted *(asterisk)*. The signal intensity of the central area of the metaphyseal lesion is high. This is consistent with cartilage. (G) One year later the radiographs reveal further loss of epiphyseal height, a large sequestrum, progressive fragmentation, and an ill-defined metaphyseal lesion. (H) Progressive epiphyseal flattening of the right hip is noted on a corresponding T1-weighted coronal MR image (compare to E). Effusion is still present. (I) The corresponding sagittal T1-weighted MR image of the right hip demonstrates flattening of the sequestrum that extends from the anterior to midsuperior aspect of the epiphysis. (J) Coronal T1-weighted MR image and (K) corresponding sagittal T1-weighted image of the right hip another 6 months later demonstrate progressive fragmentation of the sequestrum, indicating progressive healing.

covering. Diffuse metaphyseal involvement is characterized by intraosseous edema.

Progression of the repair tissue interface in a cephalic direction is identical to repair seen in group III.

Magnetic resonance imaging allows diagnosis of Legg-Perthes disease at an earlier stage than when radiographs are used. The extent of the infarcted bone can be determined more precisely, and at an earlier stage, than with radiographs. Magnetic resonance imaging improves classification and can give an indication for different treatment at an earlier stage in the seriously involved hips (groups III and IV).

References

1. Scoles PV, Yoon YS, Makley JT, Kalamchi A. Nuclear magnetic resonance imaging in Legg-Calvé-Perthes disease. J Bone Joint Surg 1980;62B:300–306.
2. Toby EB, Koman LA, Bechtold RE. Magnetic resonance imaging of pediatric hip disease. J Pediatr Orthop 1985;5:665–671.
3. Hall Rush B, Bramson RT, Ogden JA. Legg-Calvé-Perthes disease: detection of cartilaginous and synovial changes with MR imaging. Radiology 1988:153;138–145.
4. Mardam-Bey TH, MacEwen GD. Congenital hip dislocation after walking age. J Pediatr Orthop 1982;2:478–486.
5. MacEwen GD. Treatment of congenital dislocation of the hip in older children. Clin Orthop 1987;225:86–92.
6. Dunn PM. Perinatal observations on the etiology of congenital dislocation of the hip. Clin Orthop 1976;119:11–22.
7. Ogden JA. Dynamic pathobiology of congenital hip dysplasia. In: Tachdjian MO, ed. Congenital dislocation of the hip. New York: Churchill Livingstone, 1982:93–144.
8. Bos CFA, Verbout AJ, Bloem JL. A correlative study of MR images and cryo-sections of the neonatal hip. Surg Radiol Anat 1990;12:43–51.
9. Bos CFA, Bloem JL, Obermann WR, Rozing PM. Magnetic resonance imaging in congenital dislocation of the hip. J Bone Joint Surg 1988;70B:174–178.
10. Bos CFA, Bloem JL. Treatment of dislocation of the hip, detected in early childhood, based on magnetic resonance imaging. J Bone Joint Surg 1989;71A:1523–1530.
11. Renshaw TS. Inadequate reduction of congenital dislocation of the hip. J Bone Joint Surg 1981;63A:1114–1121.
12. Wiberg G. Studies in dysplastic acetabular and congenital subluxation of the hip joint: with special reference to the complication of osteoarthritis. Acta Chir Scand 1939;83(suppl):58.
13. Bos CFA, Bloem JL, Verbout AJ. MR imaging in acetabular residual dysplasia. Clin Orthop 1991;265:207–217.
14. Salter RB, Thompson GH. Legg-Calvé-Perthes disease. J Bone Joint Surg 1984;66A:479–489.
15. Catterall A. The natural history of Perthes disease. J Bone Joint Surg 1971;53B:37–53.
16. Lang P, Jergesen HE, Moseley ME, Block JE, Chafetz NI, Genant HK. Avascular necrosis of the femoral head: High-field strength MR imaging with histologic correlation. Radiology 1988;169(2):517–524.
17. Mitchell DG, Steinberg ME, Dalinka MK, Rao VM, Fallon M, Kressel HY. Magnetic resonance imaging of the ischaemic hip. Alterations within the osteonecrotic viable and reactive zones. Clin Orthop 1989;244:60–77.
18. Bos CFA, Bloem JL, Bloem RM. Sequential magnetic resonance imaging in Perthes' disease. J Bone Joint Surg (Br) 1991;73-B:219–224.
19. Genez BM, Wilson MR, Houk RW, Weiland FL, Unger HR, Shields NN, Rugh KS. Early osteonecrosis of the femoral head: detection with MR imaging. Radiology 1988;168:521–524.
20. Vogler JB, Murphy WA. Bone marrow imaging. Radiology 1988;168:679–693.
21. Phemister DB. Repair of bone in the presence of aseptic necrosis resulting from fractures, transplantations, and vascular obstruction. J Bone Joint Surg 1930;12:769–787.

CHAPTER 30

Computed Tomography of the Pediatric Hip

Donna Magid and Elliot K. Fishman

Introduction

For those familiar with the adult pelvis and hip, the pediatric pelvis will be both familiar and foreign. While certain general principles of biomechanics, structure, and function will translate from the adult to the child, there are significant differences in physiology and anatomy between the child and adult. Further consideration must be given to the differences introduced by normal sequential bone development, growth and maturation, and by congenital or acquired deviations from these normal sequences.

The pediatric pelvis is a challenge to diagnostic imaging. The structures are smaller and less well ossified than in the adult, but no less complex. Less body fat contributes to the poorer definition of soft-tissue planes. Each study must be custom designed to minimize radiation dose while adequately addressing the clinical question. The resulting images must make a definite and significant contribution to diagnosis and management to justify the economic cost and medical risk of such an approach (1).

Preservation or restoration of normal anatomy, function, and growth are the goals of pediatric orthopaedics. In any patient, therapeutic decisions reflect the balance of a cost:benefit ratio. At any age and with any disease process, the best outcome usually requires early and accurate diagnosis and appropriate treatment. Computed tomography (CT) is a rapid, reliable, relatively noninvasive way to examine the pediatric pelvis. This chapter will outline the applications of conventional transaxial CT, reformatted coronal and sagittal (two dimensional, 2D) CT, and simulated three-dimensional (3D) CT in pediatric pelvic

trauma, tumor, infection, and congenital or acquired dysplasias.

Methods

All CT studies are performed on a Siemens Somatom PLUS or DRH scanner (Siemens Medical Systems, Iselin, N.J.). Conventional protocol for studies to be reformatted is 125-kVp 280 mAs, 3-s scans, 4-mm collimation, 3-mm interval. Screening or follow-up examinations may be done with wider collimation and interval (e.g., 8 or 4 mm at 8 mm) but must be individually adjusted relative to patient size; an 8-mm slice through a small structure in a small patient introduces too much partial volume effect. In the last 18 months we have tested this protocol with 140 mAs and found no discernible change in transaxial or reformatted image quality. The average number of slices per study varies according to area of interest and patient size; it can be estimated in any patient by measuring the cephalocaudal height in millimeters of the area of interest, and dividing by three. Patients are made as comfortable as possible with small cushions or foam pads, and paper taped across the lower abdomen, knees, and feet to discourage motion. Young or predictably noncompliant patients may be sedated; compliant patients are instructed and reassured. Intravenous contrast, which would add to the risk and discomfort of the study, is not necessary for trauma studies but may be advisable for cases of inflammatory disease or tumor, particularly with soft-tissue components.

A more limited study, or "mini-CT," can be performed in patients in whom the area of interest and clinical question of interest are sharply and narrowly defined (1). This is best exemplified by the infant in plaster following hip reduction, where a question of lost reduction may be answered with a very few thin slices through the acetabulum (Fig. 30.1).

In the studies requiring 2D and 3D reformatting, the transaxial CT data are transferred from the scanner to the stand-alone workstation, which is a Sun workstation (Sun 3/180 or 3/280; Sun Microsystems, Mountain View, Calif.). The workstation is used for operator interaction with the PIXAR Image Computer (PIXAR, San Rafael, Calif.), a high-speed parallel-architecture computer with a 20-megabyte image memory, 1024×768 high-resolution color display, and four parallel bit-section processors that can process up to 40 million instructions per second (2, 3).

Abbreviations (see also glossary): MPR/D, multiplanar reconstruction/display; SI, sacroiliac; CHD, congenital hip dysplasia; SCFE, slipped capital femoral epiphysis.

377

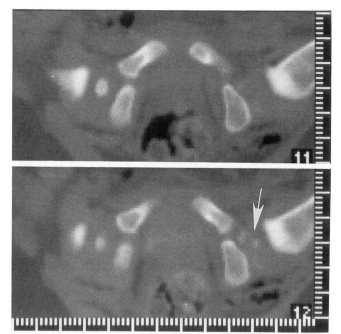

Figure 30.1. Congenital hip subluxation in a 6-month-old infant: status, post open reduction of persisting left hip subluxation. "Mini-CT" was obtained in a spica cast to assess possible loss of reduction. Transaxial images at the level of the normal right acetabulum and ossification center show the smaller left ossification center *(arrow)* to be in an acceptable position relative to the triradiate cartilage and acetabulum. Slight positioning asymmetry requires comparison of two contiguous slices to have comparable levels on the right *(top image)* and left *(bottom image).*

The transferred slices are interpolated in the Z (cephalocaudal) axis, to equalize voxel elements determined by field of view (X and Y dimensions) to that determined by slice interval (Z dimension). Interpolation produces cubic voxels, using a bicubic interpolation algorithm with a Catmull-Rom spline base function (2). The entire transaxial CT data set can be interpolated in under 10 s. Interactive animated coronal and sagittal images are generated and displayed using the OrthoTool (HipGraphics, Baltimore, Md.) software and the interpolated data set. The displayed transaxial, coronal, and sagittal images can be roamed in real time. There are distance and angle measurement options. If desired, a new baseline oblique axis can be chosen and a new set of orthogonal images oriented to that baseline will be generated and displayed in approximately 10 s. Traveling reference axes on each image define the level from which the displayed slices are drawn. These 2D images are similar to the multiplanar reconstruction/display (MPR/D) option on the CT scanner, but offer several advantages. The format is larger and the resolution of the display monitor better. The OrthoTool can reconstruct up to 192, vs. 49, consecutive scans, and preserves the entire dataset—i.e., entire field of view—which allows comparison of the opposite side to the side of interest. The images are animated and interactive, making it easier to integrate a large number of images and a large volume of information.

In most reformatted studies, the final step is 3D reconstruction and animation. This has been previously described (3, 4). Image series rotating in a spinal or side-to-side axis and a head-over-heels or somersaulting axis are generated and displayed. Extraneous or superimposed structures can be edited off for unimpeded viewing of the structure of interest.

Pelvic Fractures

Pelvic Ring: Anatomy

The pelvic ring is formed by the sacrum and paired innominate bones, joined posteriorly by the strongest ligaments in the body and anteriorly by the weaker symphysis pubis and associated ligaments. In the absence of adequate interlocking bony stability, these ligaments provide stabilization and support for the pelvic ring. The concept of the pelvis as a rigid ring dictates the conventional wisdom that pelvic disruptions must occur at two or more sites but understates the complex biomechanics of trauma and of the pediatric pelvis. Even in adults, the anterior ring can undergo minor disruption without compromising the posterior ring (5); the ligaments and cartilages provide a degree of elasticity that keeps the ring from being perfectly rigid. This elasticity is somewhat greater in the child, who may tolerate more deformity without fracturing. The more abundant cartilages and growth centers, however, also leave the pediatric pelvis more vulnerable to "invisible" trauma through these structures (6).

Pelvic Ring Fractures

Classification systems for adult fractures are applicable to children and are based on the mechanism of injury, complexity and location of injury, stability of the injury, and/or associated injuries. While pediatric pelvic fractures tend to be less displaced than their adult counterparts and to require less treatment of the fracture itself for a favorable outcome, they are associated with more significant trauma to the viscera and soft tissues (one series found 90% of pediatric pelvic fractures to be the result of car vs. pediatric pedestrian, with 87% of these patients having severe associated injuries) (6, 7).

While many adult pelvic fractures may be seen as the result of relatively minor and therefore isolated trauma (falls in the osteoporotic and elderly), pediatric pelvic ring fractures are usually not seen until there has been major trauma (the pediatric avulsion type injury is an exception, but does not compromise the ring). With so many of these patients being the pedestrian victims of collisions, associated head, neck, chest, or abdominal injuries often bring the patient to emergency CT shortly after arriving at the hospital. In patients with only minor or no apparent lower abdominal or pelvic injuries, a screening pelvic CT examination may be sufficient but is advised. We routinely screen all major trauma patients coming for upper abdominal studies, extending imaging caudally through the symphysis pubis. Unanticipated soft-tissue or bony injuries may be detected in this manner that were overlooked in the presence of more life-threatening injuries, but which will need attention once the patient is stabilized (Fig. 30.2). Computed tomography also may document more extensive hemor-

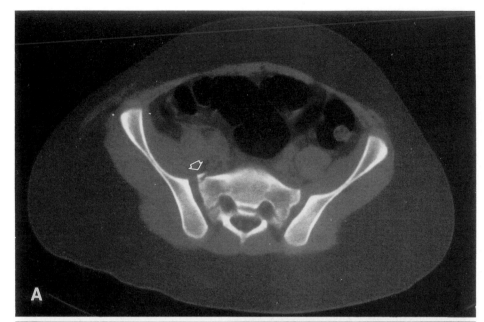

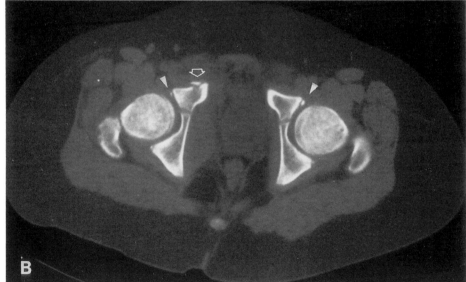

Figure 30.2. Pelvic trauma in a 12-year-old boy involved in a low-velocity motor vehicle accident. Transaxial CT through the upper pelvis **(A)** showed a small right anterior sacral fracture *(arrow)* not seen on plain film and subtle air densities (vacuum cleft phenomena) in the right sacroiliac (SI) joint. The right psoas and iliacus muscles are swollen compared to the more normal left. Comparison to the contralateral side becomes more important in the acetabulum **(B),** where a small anterior column fracture was noted *(arrow).* The contralateral side reaffirms the normal findings of small accessory ossification centers *(arrowheads)* that mimic adult chip fractures, and the symmetric, intact appearance of the triradiate cartilages.

rhage than is clinically suspected, with large amounts of blood pooled in the dependent pelvis.

Computed tomography may confirm or suggest the mechanism of injury (anterior or lateral compression, vertical shear), determination of which is part of the initial treatment and stabilization process (Fig. 30.3). Prompt early reduction and stabilization, involving a component of reversal of the initial deforming force, may be an effective way to acutely control or reduce hemorrhage (8–11).

The sacrum and sacroiliac (SI) joints are the major components of the posterior pelvic ring; the integrity of these structures and their associated ligaments determines pelvic integrity. This is, however, the part of the pelvis most likely to be inadequately seen on routine plain film examination (12). This area is well seen on transaxial CT. In the patient with a high suspicion of injury, multiplanar reformatting is advised. Coronal images allow simultaneous comparison of both SI joints and of the sacral foramina. Sagittal images confirm sacral integrity or fracture and demonstrate the central canal and presacral soft tissues. The OrthoTool oblique option allows selection of a true frontal plane paralleling the long axis of the sacrum, which gives a true frontal or coronal series through the sacrum for better visualization of cortical margins and the paired foramina. This option takes less than 10 s of additional time in studies already being reconstructed on the OrthoTool program (2).

The Acetabulum: Anatomy and Fractures

The triradiate cartilage is formed by the junction of the iliac, ischial, and pubic ossification centers. Expansion of these centers enlarges the acetabulum with growth, while the presence of a normal spherical femoral head induces the correct concavity at each stage in growth. Central ace-

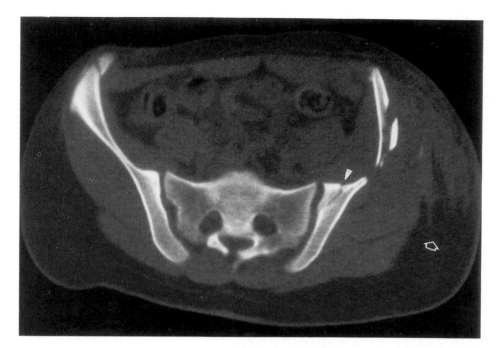

Figure 30.3. Pelvic ring fractures in a 12-year-old female passenger whose father's car was struck by an intoxicated driver. Transaxial CT graphically outlines the lateral compression mechanism of trauma. Marked left lateral soft-tissue contusion and edema *(arrows)* are evident, with "closing book" medial displacement of the fractured left iliac wing. One fracture extends into the left anterior SI joint *(arrowhead)*. There are less extensive fractures of the anterior superior right iliac wing. Other images showed a vacuum cleft in the right anterior SI joint, suggesting it too had been forcibly distracted at time of injury.

tabular fractures through this structure are rare in childhood, but when present may be impossible to detect on plain film. Adult acetabular fractures may be missed or underestimated on plain film, but CT is far more sensitive to subtle fracture lines or fragments (10, 11).

In the child, an acetabular fracture is more likely to involve the triradiate cartilage than bony structures alone. This may be impossible to appreciate unless CT is used to detect subtle asymmetry of the injured acetabulum relative to the normal side, and to look for evidence of diastasis, displacement, or disruption. Transaxial CT in the well-positioned patient gives a built-in normal comparison baseline for each image. Where there is a high index of suspicion, reformatting of both sides (as two sets of images on conventional CT scanner MPR software, or as one set on the OrthoTool) will increase appreciation of subtle asymmetries or abnormalities of contour or density. The availability of the normal comparison also enhances appreciation of the normal appearance of this region at different ages, which, like other ossification centers, may appear to closely simulate a fracture otherwise (Fig. 30.2). As the triradiate cartilage fuses, the acetabulum begins to approximate the adult acetabulum in its response to trauma and appearance on CT.

Developmental Hip Dysplasias

The hip dysplasias are congenital and acquired abnormalities that are expressed along a broad spectrum of severity. It is beyond the scope of this chapter to discuss the many aspects of acquired dysplasias such as are seen in patients with meningomyelocele or cerebral palsy, where differences are introduced by the variable age of onset, degree of weight bearing, and muscle tone. Even in assumed idiopathic congenital dysplasias, there is variability introduced by the severity at presentation, age at diagnosis, and adequacy and timing of treatment. The common

basic clinical principles and the use of CT in the dysplasias will be illustrated by a discussion of its applications in congenital hip dysplasia (CHD).

In the neonate, the diagnosis of hip dysplasia or subluxation is a clinical one. While much has been made of the plain film landmarks and measurements, the persistence of fetal hip flexion and the paucity of bony landmarks combine to introduce a high degree of uncertainty in the earliest films and the subtle cases. With growth, the femoral capital epiphysis begins to ossify and the bony acetabulum becomes more distinct, so that plain films become more reliable as the first year progresses.

The goal in treating CHD is to restore normal congruity and concentricity of the femoral head and acetabulum, to restore normal function, and to induce normal growth and development. The earliest possible treatment is required to reach this goal; it has been estimated that such prompt diagnosis and treatment could allow up to 95% of such infants to have normal hips (13). In these youngest patients, CT does not play a primary role in diagnosis, although it will be more effective than plain films in assessing the relationship of the minimally or nonossified capital epiphysis to the acetabulum. (See also Chapter 29.)

The reduced neonate or infant hip is usually stabilized in spica cast. Plain films of these treated hips are often inadequate to define the anatomy of interest. In these cases, a carefully tailored "mini-CT" may be a far more effective examination (Fig. 30.1). In a carefully positioned patient, four to eight contiguous thin (3 or 4 mm) slices through the acetabulum should confirm maintenance or loss of reduction and provide an idea of femoral head congruity and concentricity. The dose is sharply curtailed from a conventional study and localized only to the area of interest (1). Reduced milliamperage further controls local dose.

With age, the increasing ossification on both sides of the joint allow both plain film and CT to play a larger role in assessment. However, advancing age also decreases the likelihood that entirely normal hip development will be restored. In the older child, closed reduction is often inadequate. Open reduction in these children often requires reconstruction of both the acetabulum and proximal femur. In these patients, where complex structural surgery is required, the increased cost and dose of a conventional (low mAs) CT scan with 2D and 3D reformatting are warranted. In our institution, each child's study is custom designed depending on patient age, proposed surgery, and severity of dysplasia (Figs. 30.4 and 30.5) (1, 14). The coronal and sagittal images of the 2D study give a definition of the deformities of the head and acetabulum difficult to appreciate on transaxial alone, and help document the potential for congruity and concentricity. The 3D study as-

similates these findings, and allows the surgeon to become completely familiar with the individual anatomy prior to surgery. Our pediatric orthopaedic surgeons use extensive review of the animated 2D and 3D images to plan surgery and often bring the tapes into the operating room for further reference as desired. Such extensive review may also be helpful when dealing with relatively rare variations in dysplasia, or when training less experienced personnel. Eventually, such interactive 2D/3D studies will be used for rehearsal surgery and to test the biomechanical outcome of proposed procedures.

Hip dysplasia may be seen in the older child as a sequela of any early childhood disease process altering the morphology and function of the femoral head. in Legg-

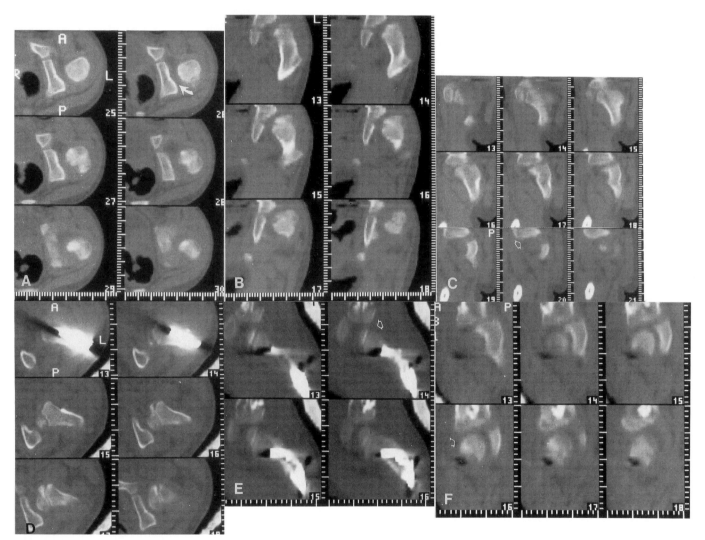

Figure 30.4. Pre- and postoperative CT of congenital hip dysplasia in a 6-year-old boy. Preoperative transaxial images **(A)** demonstrate a posteriorly subluxed head articulating with a shallow pseudoacetabulum *(arrow)* that abuts a nearly flat native acetabulum. The coronal images **(B)** map the extent of lateral and minimal superior subluxation, leaving most of the dysplastic head uncovered, and the nearly vertical, nonconcave dysplastic acetabulum. Serial sagittal images **(C,** anterior is to viewer's left) confirm a flattened acetabular roof, with little development of anterior and superior concavity *(arrow).* Surgical reduction

of the proximal femur would be futile without sufficient acetabular reconstruction to provide containment and congruity. Baseline CT following innominate and proximal femoral osteotomies **(D–F)** show relative anterior translation of the head in the acetabulum on transaxial **(D)** compared to baseline transaxial **(A),** with lateral anterior displacement of the acetabular dome fragment now providing better coverage and containment of that head, best seen on reformatted coronal **(E,** *arrow)* and sagittal **(F,** *arrow)* images.

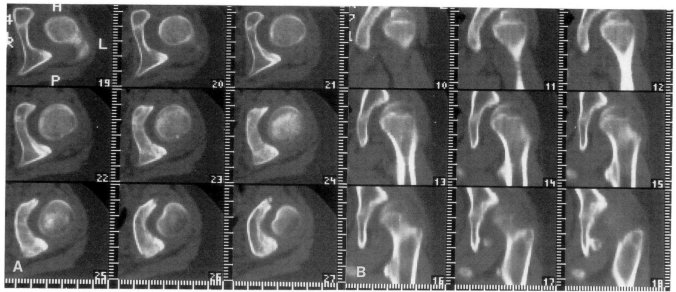

Figure 30.5. Preoperative CT of congenital hip dysplasia in an 11-year-old girl. Transaxial images **(A)** show a dysplastic acetabulum, with the position of the head relative to the triradiate cartilage suggesting superior subluxation of the head. This is easier to quantify on coronal images **(B)**, which also depict the more vertical, shallow acetabulum

and show potential premature cartilage loss as narrowing at the lateral acetabular rim. Adequate surgical reduction of this hip required acetabular as well as proximal femoral repair, to provide a deeper acetabulum covering a reduced proximal femur.

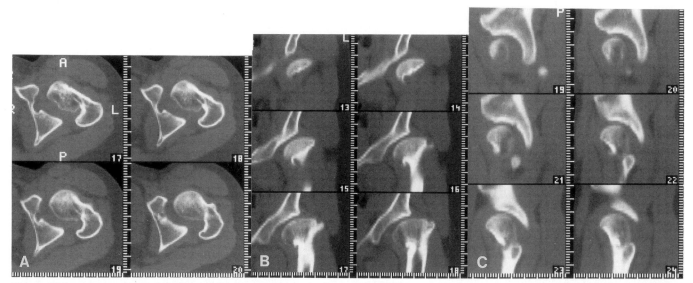

Figure 30.6. An 11-year-old girl with Legg-Perthes disease and subsequent hip dysplasia. Transaxial images **(A)** document femoral head and neck deformities with a dysplastic, anteverted acetabulum. Coronal images **(B)** show profound flattening of the head on a short, broad

neck, left more than 50% uncovered by subluxation from a shallow, inadequate acetabulum. Sagittal images **(C)** confirm a broad, shallow, acetabulum. Adequate repair required reconstruction on both sides of this joint.

Perthes disease, for example, further growth of and altered weight bearing on a misshapen femoral head eventually can produce well-established dysplasias of both the proximal femur and acetabulum (Fig. 30.6). These, like the dysplastic subluxated hip seen in older children with inadequately treated CDH, require surgical reconstruction of both the acetabulum and proximal femur.

Two-dimensional and 3D CT are also the best way to

document and depict the postoperative result, and to follow subsequent changes with growth and remodeling.

Femoral Neck Anteversion

The angle of femoral neck version is defined relative to the transcondylar plane. Anterior angulation of the neck relative to this baseline is anteversion; neck angulation posterior to this baseline is retroversion. Normal infants start out markedly anteverted (25 to 40°) and, with growth, retrovert slowly toward the normal adult range of 8 to 18° anteversion (15, 16). This normal femoral regression is influenced by many factors, including weight bearing with erect posture, muscular forces, growth and remodeling, and genetic influences. Alterations in any of these interacting factors may modify the remodeling process, altering the resultant femoral version. Cerebral palsy, with increased and unbalanced spastic forces acting on the hip, neuromuscular diseases with markedly altered and reduced weight bearing, long-standing hip dislocations, and marked obesity are among the childhood conditions known to modify femoral neck version.

Transaxial CT allows simple, reproducible measurements of femoral neck version with only negligible additional time or dose. The appropriate transaxial images of the neck are obtained as part of the routine hip examination. Additional slices through the distal femoral condyles are necessary when version calculations are desired. In the asymmetrically positioned patient or patient with real or apparent leg length discrepancy (e.g., scoliosis or pelvic tilt), this may take several slices to obtain the appropriate transcondylar slice bilaterally. Otherwise, two additional slices often suffice. We are looking for a slice through the midcondyle, where the posterior notch is deep and well defined; this is usually (but not always) at a midpatellar level in a conventionally positioned patient. The "tabletop" method, which draws a line connecting the posterior condylar cortices, is used for the baseline because it is simple and easily reproducible (17, 18). On the transaxial slice best representing the midfemoral neck, a line bisecting the anterior and posterior cortices provides a longitudinal axis (Fig. 30.7). The angle between the transcondylar plane and the femoral neck axis is the angle of version (positive for anteversion, negative for retroversion). The images can be compared either by double exposing an image, or by translating measurements from one image to the next. The accuracy of either method, however, depends on the compliance of the patient between images, since motion or rotation of the leg between the proximal and distal slice negates accurate measurement. (In those patients undergoing multiplanar reformatting, a nonaliased reconstruction provides further incidental reassurance that motion has not compromised measurements.) Extreme coxa valga also compromises the accuracy of these measurements.

Measurements of femoral neck version may be of interest in the older child with hip dysplasia, in whom reconstruction of both the proximal femur and the acetabulum will be required; it may allow better planning of rotational osteotomies and better match of the angles of the reconstructed femur and acetabulum. We also believe that significant alterations of femoral neck version correlate with other risk factors in the older child at risk for slipped capital femoral epiphysis (SCFE), or with an early, often clinically occult or confusing slip (1). In our population, children with SCFE have about 15° less anteversion (relative retroversion) than normal periadolescent children; we define a hip as "at risk" if there is 5° or less of anteversion (Fig. 30.8). In the first 15 children with unilateral SCFE studied by CT, 5 were found to have bilateral true retro-

version. Relative retroversion relates to the clinical observation in these children of decreased internal rotation in flexion (1, 19).

Tumor

Computed tomography may be used to characterize known or suspected tumors in the pelvis (see also Chapter 15). With luck, a suspected tumor may prove at CT to be a stress fracture (rare in the pediatric pelvis) or healing avulsion injury. When the presence of tumor is confirmed, CT characterizes size, borders and matrix, apparent aggression, and compartmental (medullary, cortical, periosteal, muscle, soft tissue) components (Figs. 30.9 and 30.10). This provides a reasonable working differential diagnosis, dictates the progression of the subsequent workup, and provides the basis of the therapeutic plan. Tumors to be excised, biopsied, or amputated are approached using the CT (or magnetic resonance (MR) imaging) study as a guide, to allow invasive procedures to be as accurate but as tissue sparing as possible (Fig. 30.11). Radiation therapy portals are planned around the CT data. Computed tomography allows accurate documentation of known premalignancies such as pelvic exostoses or previously irradiated lesions, and with subsequent follow-up examinations detects the subtle changes that may herald degeneration (Figs. 30.7 and 30.8). In treated lesions it establishes the baseline against which recurrence can be detected (20).

We study all potentially malignant lesions using the 2D/3D protocol. The reformatted 2D images have proven particularly useful in assessing matrix and borders or interfaces of the mass with normal tissue. Coronal and sagittal 2D images at soft-tissue windows may show muscle planes being displaced, rather than invaded by, tumor mass. This indicates a more indolent, although not necessarily benign, lesion and may help in planning the approach to and extent of resection (Figs. 30.9 and 30.10). Similarly, a sharply defined or sclerotic bony margin may also be evidence of a less aggressive lesion. Such characterization may encourage the surgeons to use an *en bloc* but local excision rather than a more mutilating amputation. The adequacy of such a procedure, of course, must be confirmed by conventional histologic review of surgical margins.

Computed tomography may also be used to follow benign tumors or tumor-like processes such as fibrous dysplasia for complications such as pathologic fracture (Fig. 30.12). Where resection or curettage of such lesions is required because of fracture or biomechanical compromise, CT again provides an excellent preoperative guide and postoperative baseline.

Infection

As has been discussed elsewhere, CT may be the best way to document infection of bone or soft tissue in the pediatric pelvis. Besides confirming and characterizing abnormal findings, CT may be used to guide diagnostic or therapeutic punctures (Fig. 30.13) for the most precise but least invasive procedure. Computed tomography may localize persisting fluid collections, soft-tissue or intrame-

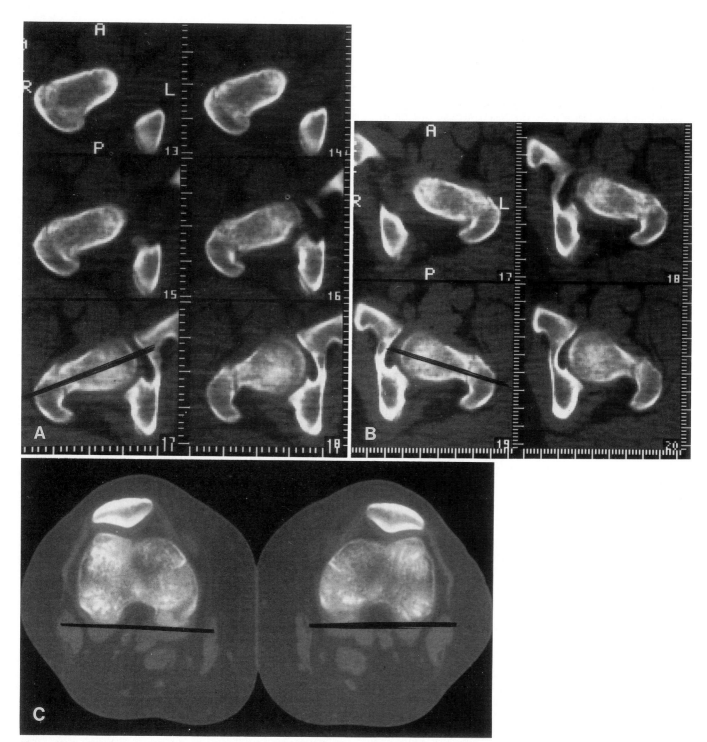

Figure 30.7. Normal anteversion in a 14-year-old black female. Comparison of the longitudinal neck axes of the right **(A)** and left **(B)** hips to the posterior transcondylar planes **(C)** reveals 20° anteversion on the right, 12° anteversion on the left. There is usually less variance between sides, but no definite symptoms or signs were associated with this slight asymmetry.

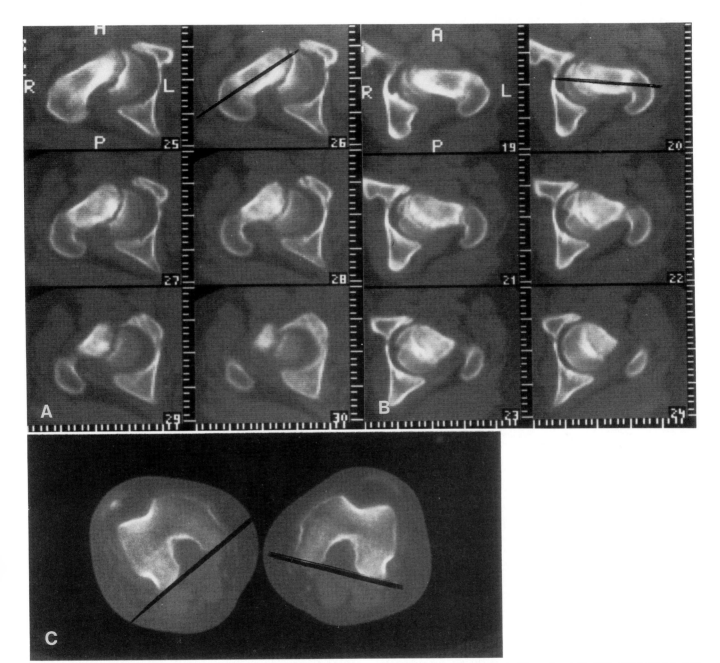

Figure 30.8. Slipped capital femoral epiphysis with retroversion. A 15-year-old black male with acute right hip pain was found to have acute right grade III (marked) slip and asymptomatic chronic grade I (mild) left slip. Serial transaxials of the right **(A)** and left **(B)** proximal femora are used to establish the central longitudinal neck axes *(lines)*. The angle formed by these axes on the posterior transcondylar base-lines **(C)** is 5° retroversion (−5°) on the right, 10° retroversion (−10°) on the left. The right epiphyseal plate is wide and sharply defined; the marked posterior subluxation of the right epiphysis exposes the anterior metaphysis. The right hip is held in external rotation. The left plate is less sharply defined and the left capital epiphysis is less posteriorly displaced.

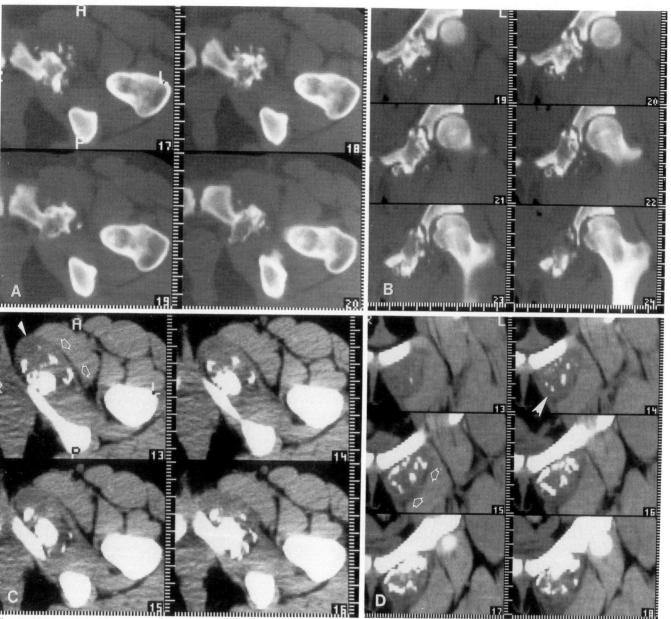

Figure 30.9. Left ischial chondrosarcoma in a 16-year-old female. Preoperative 2D/3D CT was used to characterize the tumor and its margins, in hopes of designing as conservative a resection as possible. Four transaxial images through the ischium (**A**) and six coronal images (**B**, moving posteriorly from image 19) show an exostotic lesion, with a stippled cartilaginous cap. Transaxial (**C**) and coronal (**D**) images at soft-tissue window settings define the noncalcified portion of the mass *(arrowheads)* and demonstrate that it displaces, rather than invades, adjacent pectineus and adductor magnus muscle; the fascial planes are preserved *(arrows)*. It was unclear whether the abutting obturator externus was displaced or invaded. Conservative resection allowed *en bloc* excision with clean margins and confirmed the diagnosis, but preserved the architectural and functional integrity of the pelvis. (Case reproduced with permission from RadioGraphics.)

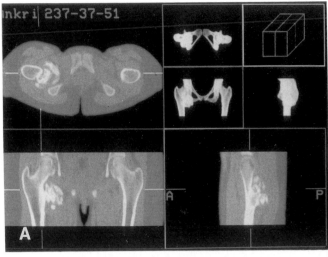
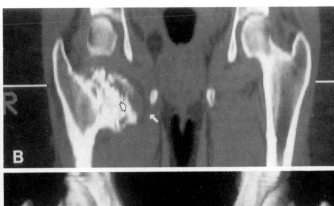
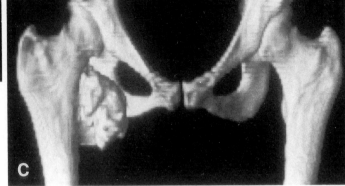

Figure 30.10. Sarcomatous degeneration and exostosis in a 16-year-old girl with newly painful right hip. OrthoTool images **(A)** show a large exophytic lesion of the posterior aspect of the right lesser trochanter, as seen on transaxial *(top left)*, coronal *(lower left)*, and sagittal *(lower right)* images. Zoomed coronal image **(B)** shows a focal area of apparent bony erosion under the less dense cartilaginous cap *(black arrow)*, which would suggest potential malignant degeneration. Balancing this was a more reassuring finding, preservation of a medially displaced fascial plane *(white arrow)*, implying a more indolent mass that has displaced but not yet definitely invaded adjacent muscle. Animated 3D images **(C)** were used to plan the *en bloc* resection, which revealed low-grade sarcoma within an exostosis.

dullary abscesses, or small sequestra in cases where postherapeutic recurrence or chronicity is suspected.

Ischemic Necrosis

The use of CT in the assessment of ischemic necrosis of the femoral head (see also Chapters 26 and 29) has been described in detail previously (21). As in the adult, 2D CT provides an excellent assessment of morphologic changes for staging, follows the progression of disease, and often detects unsuspected, earlier disease in the contralateral hip of patients at risk (from sickle cell anemia, organ transplants, or steroid use, for example) for nontraumatic ischemic necrosis (Figs. 30.14 and 30.15).

Dwarfism

In the various types of dwarfism, it is clinically useful to divide orthopaedic abnormalities into two groups: those morphological alterations that characterize the normal but altered development, and those that are actually complications or disabling consequences of the morphologic variation (22). We do not use CT to document the expected abnormalities that are "normal" for any given type of dwarfism—rather, we reserve CT of the pelvis and hips for those patients in whom morphologic variation has become painful or disabling, and for whom corrective treatment is planned. In diastrophic dwarves, for example, the combination of severe flexion deformities and hip dysplasia can produce secondary osteoarthritis in childhood. To preserve mobility in these patients, and to slow the course of osteoarthritis, early proximal femoral and/or acetabular reconstruction may be desirable. We use 2D/3D CT in such patients to map the extent of femoral head and neck and acetabular dysplasia and arthritis (Fig. 30.16). The 2D coronal and sagittal images best show the extent of osteoarthritis and are useful for following muscle planes and soft tissues for those patients in whom tenotomies or transfers are planned. The surgeon finds that the animated 3D images give an overview of these complicated abnormalities; particularly in those patients exhibiting a unique or idiosyncratic variation of dwarfism or dysplasia.

Other Congenital Anomalies

Congenital anomalies range along a spectrum from common and relatively predictable patterns of developmental error to profound and idiosyncratic multisystemic malformations. In the more pattern-oriented anomalies, such as bladder exstrophy, pelvic CT may be useful in providing pre- and postoperative baselines of interrelated musculoskeletal, genitourinary, and gastrointestinal system anomalies (Fig. 30.17). In more predictable anomalies, reviews of accumulated series have led surgeons in our institution to revise previous treatment protocols. Even with experience in imaging more predictable anomalies, it is

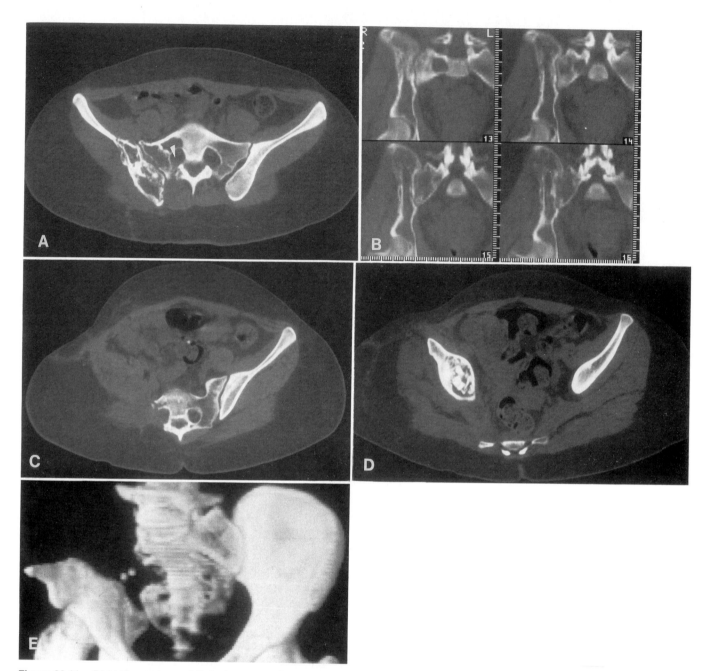

Figure 30.11. Right iliac and sacral Ewing's sarcoma in a 14-year-old girl. The initial preoperative transaxials **(A)** and coronals **(B)** showed far more extensive sacral involvement than had been appreciated from plain films, with the lytic expansile lesion extending deeply into the ala and with soft-tissue density in the right neural foramina *(arrowhead)*. The right iliacus muscle was minimally enlarged. The CT data were used to mill a plastic 3D model for a prosthetic implant. However, it was decided that an *en bloc* resection would be too mutilating; it was hoped that extensive debulking and radiotherapy would control the disease. The postoperative baseline examination **(C–E)** showed partial resection of the right iliac and sacrum **(C** and **E)** with curettage and bone graft packing of the preserved inferior ilium **(D).** The 3D image **(E)** was used to consider prosthetic options while the patient underwent therapy. Subsequent CT studies have followed response to therapy and have been used to monitor for early recurrence, since clean margins could not be obtained.

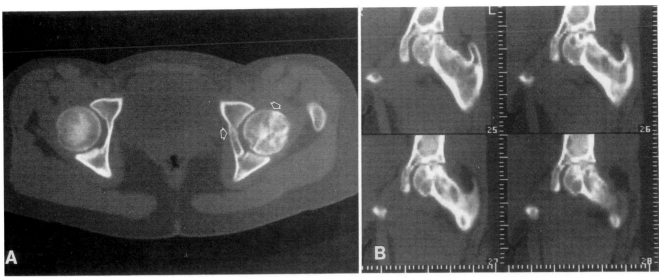

Figure 30.12. CT of a 15-year-old girl with known left hip fibrous dysplasia and increasing hip pain. CT was requested to rule out pathologic fracture. Transaxial CT **(A)** through the head and acetabulum show loculated, minimally expansile lesions on both sides of the joint *(arrows)*. Reformatted coronal images **(B)** precisely characterized the flattening of the superior pole of the femoral head and the loss of joint space cartilage, and defined a volume of involvement exceeding that estimated from plain film. No pathologic fracture was found; pain was attributed to early arthritis. (Case reproduced with permission from RadioGraphics.)

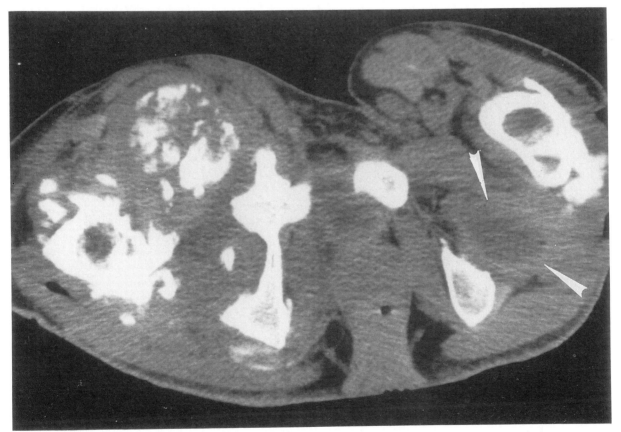

Figure 30.13. Soft-tissue abscess in a 16-year-old male paralyzed 7 years earlier by a thoracic spine gunshot injury. Massive heterotopic bone around the hips and femoral shafts and marked flexion deformity severely limited assessment via physical examination or fluoroscopy, and restricted percutaneous access for joint or tissue aspiration. Transaxial CT was used to look for a source of systemic infection. A low-density fluid collection *(arrowheads)* was localized posterior to the left subtrochanteric femur and medial to a large decubitus ulcer. This study was used to choose the most direct path for drainage that would avoid both the heterotopic bone and the decubitus margin. Ten cubic centimeters of purulent fluid were obtained.

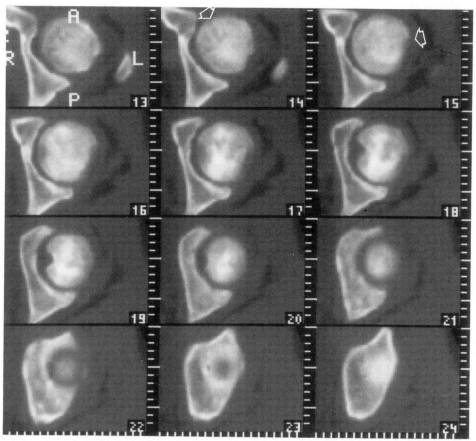

Figure 30.14. Ischemic necrosis and sickle cell disease in a 14-year-old boy homozygous for sickle cell anemia, with left hip pain. A series of transaxial images (ascending from top left) show diffuse geographic sclerotic changes in the femoral head, a radiolucent line with subtle cortical step off in the anterior segment *(arrows),* and early or impending loss of head sphericity. Earlier, clinically and radiographically unsuspected ischemic necrosis of the contralateral head was discovered.

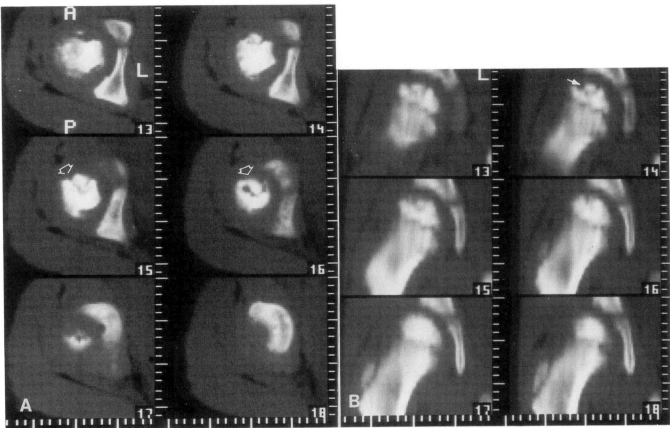

Figure 30.15. CT of a 5-year-old boy with spondyloepiphyseal dysplasia (SED) and ischemic necrosis of the right femoral head. This is a less common complication of SED than retarded ossification or premature osteoarthritis of the femoral head. Serial transaxial images **(A)** through the superior half of the femoral head revealed sclerotic, irregular bone with a fracture and/or collapse of the anterior quadrant *(arrows)*. Coronal images **(B,** moving posteriorly from upper left) confirm the collapse and impaction of the superior anterior pole *(arrow),* with joint cartilage grossly intact at this time.

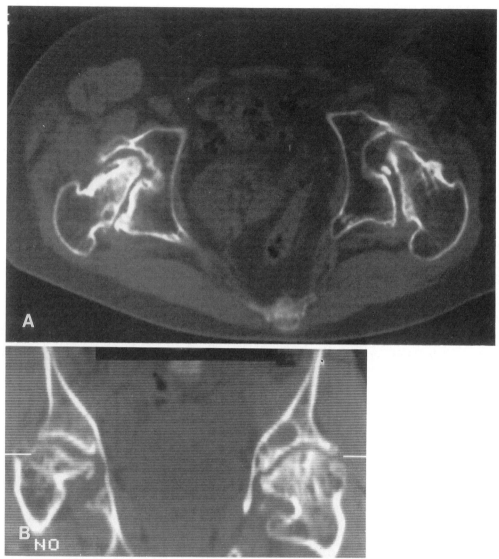

Figure 30.16. Diastrophic dwarfism. Transaxial CT **(A)** defines the marked deformity of both femoral necks, with severe premature degenerative changes on the right greater than left. There are corresponding acetabular dysplastic and degenerative changes, right greater than left, and marked anteversion. Coronal reformatting **(B)** on a second patient shows the characteristic dysplastic changes, with broad necks and flattened, dysplastic heads articulating with shallow, dysplastic and degenerative acetabula. Secondary degenerative cysts are seen in the left acetabulum and proximal femur.

helpful to use the transaxial data for 2D and 3D reformatting, to provide a better overview and easier orientation. In more esoteric one-of-a-kind anomalies such as partial pelvic duplication or incomplete twinning (Fig. 30.18), the anatomy and pathology are even more confusing. In such a case, a series of transaxial images will be informative, but the addition of 2D and 3D images will greatly enhance the accessibility of that information. Three-dimensional images, in particular, provide the best orienting overview for the design of complicated, unique procedures modified to answer the complex challenge of such difficult anomalies.

Conclusion

Transaxial CT can be a valuable tool in assessing the pediatric pelvis. It has the advantage, compared to con-ventional radiographs, of being unimpeded by overlying tissue and bowel and of showing bone, joint, and soft tissue in superior detail. It is more sensitive to subtle differences in contour, size, or density. Computed tomography has a key role to play in the diagnosis, therapeutic planning, and posttherapeutic follow-up of many pediatric problems.

In our institution we strongly advocate the use of a custom-tailored 2D and 3D examination for many problems in this age group. All commercially available CT scanners have some sort of 2D software allowing coronal and sagittal reformatting, making this a potentially ubiquitous approach. We approach each problem individually, to minimize dose and maximize information.

Acknowledgment

I wish to thank Dr. George P. Saba II for his steadfast encouragement and support.

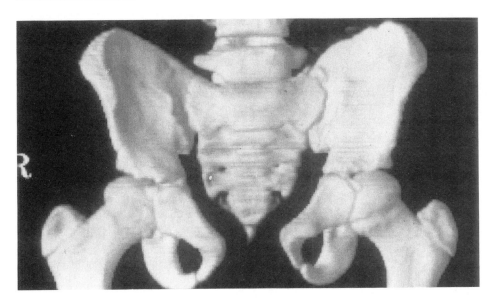

Figure 30.17. Pelvic exstrophy in a 7-year-old child. CT was obtained primarily to assess the genitourinary system and soft tissues prior to repair. However, the orthopaedic surgeons performing the staged closure procedures found that the reconstructed, animated 3D images facilitate planning for the iliac osteotomies, and simplify assessment of secondary hip dysplasia, where present. Postoperative animated 2D/3D studies provide the best confirmation of results and are the baseline for planning subsequent procedures.

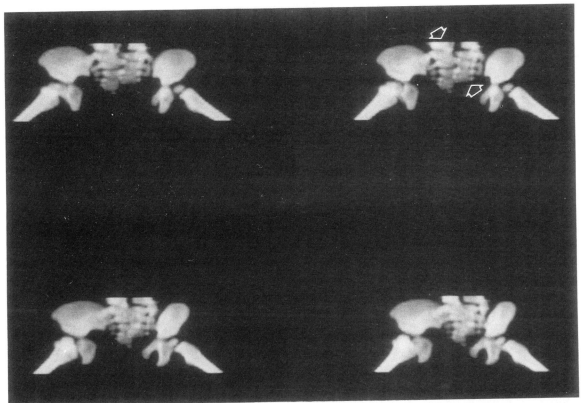

Figure 30.18. Infant with incomplete twinning of the sacrum, with associated bladder exstrophy and diastatic anterior pelvic ring. This patient had two nearly complete but dysplastic and fused sacrums *(arrows).* Preoperative 2D/3D CT was used to help plan and coordinate the neurosurgical, urologic, and orthopaedic procedures that were needed. In a patient with unique and complex multisystemic anomalies, multiplanar and 3D images may be necessary to orient to and integrate the unusual findings.

References

1. Magid D, Fishman EK, Sponseller PD, Griffin PP. 2D and 3D computed tomography of the pediatric hip. Radiographics 1988;8:901–933.
2. Ney DR, Fishman EK, Magid D, Kuhlman JE. Interactive real-time multiplanar CT imaging. Radiology 1989;170:275–276.
3. Fishman EK, Drebin RA, Magid D, et al. Volumetric rendering techniques: applications for three-dimensional imaging of the hip. Radiology 1987;163:737–738.
4. Fishman EK, Magid D, Ney DR, Drebin RA, Kuhlman JE. Three-dimensional imaging and display of musculoskeletal anatomy. J Comput Assist Tomogr 1988;12:465–467.
5. Bucholz RW. The pathological anatomy of Malgaigne fracture dislocations of the pelvis. J Bone Joint Surg 1981;63-A:400–404.
6. Canale ST, King RE. Pelvic and hip fractures. In: Rockwood CA, Wilkins KE, King RE, eds. Fractures in children. Philadelphia: Lippincott, 1984;733–872.
7. Rang M. Children's fractures. Philadelphia: Lippincott, 1983;233–263.
8. Tile M. Pelvic fractures: operative versus nonoperative treatment. Orthoped Clin N Am 1989;11:423–464.
9. Young JWR, Burgess AR, Brumback RJ, Poka A. Pelvic fractures: value of plain radiography in early assessment and management. Radiology 1986;160:445–451.
10. Magid D, Fishman EK, Brooker AF Jr, Mandelbaum BR, Siegelman SS. Multiplanar computed tomography of acetabular fractures. JCAT 1986;10:778–783.
11. Magid D, Fishman EK. Computed tomography of acetabular fractures. Semin Ultrasound Comput Tomogr Magn Reson 1986;7:351–361.
12. Scott WW Jr, Fishman EK, Magid D. Acetabular fractures: optimal imaging. Radiology 1987;165:537–539.
13. Harris NH, Lloyd-Roberts GC, Gallien R. Acetabular development in congenital dislocation of the hip. J Bone Joint Surg 1975;57-B:46–52.
14. Magid D, Fishman EK, Sponseller PD, Thompson JD. The pediatric hip: 2D and 3D CT analysis. Contemp Orthoped 1989;18:53–59.
15. Visser JD, Jonkers A, Hillen B. Hip measurements with computed tomography. J Pediatr Orthoped 1982;2:143–146.
16. Shands AR, Steele M. Torsion of the femur. J Bone Joint Surg 1958;40A:803–816.
17. Galbraith RT, Gelberman RH, Hajek PC, et al. Obesity and decreased femoral anteversion in adolescence. J Orthopaed Res 1987;5:523–528.
18. Weiner DS, Cook AJ, Hogt WA Jr, Oravec CE. Computerized tomography in the measurement of femoral anteversion. Orthopaedics 1978;1:299–306.
19. Magid D, Fishman EK, Sponseller PD, Griffin PP, Thompson JD. Analysis of slipped capital femoral epiphysis: CT with multiplanar reconstructions. Paper presented at the 74th scientific assembly and annual meeting of the Radiologic Society of North America, Chicago, 1988.
20. Hernandez RJ, Pozanski AK. CT evaluation of pediatric hip disorders. Orthoped Clin N Am 1985;16:513–541.
21. Magid D, Fishman EK, Scott WW Jr, et al. Femoral head avascular necrosis: CT assessment with multiplanar reconstruction. Radiology 1985;157:751–756.
22. Kopits, SE. Orthopedic complications of dwarfism. Clin Orthoped Rel Res 1976;114:153–179.

CHAPTER 31

Computed Tomography of Knee Trauma

Roberto Passariello, Carlo Masciocchi, Eva Fascetti, and Antonio Barile

Introduction
Technique
Normal Anatomy

Meniscal Lesions
Capsuloligamentous Lesions

Introduction

Intraarticular blood effusion and painful contracture adversely affect clinical evaluation of the traumatized knee. This creates the necessity for noninvasive diagnostic techniques, which offer a panoramic visualization of all the articular structures and an early and reliable diagnosis, which favors the planning of the successive therapeutic phases. Computed tomography (CT) has been the method used for many years in the study of articular pathologies of the knee joint (1–5). It has been proven, in numerous cases, to be able to reach high levels of diagnostic accuracy. The possibility of visualizing various pathological conditions is obviously founded on its capacity to recognize all articular formations and their anatomical relationships. In order to obtain an accurate imaging, the CT study must be employed with the use of high-resolution technique.

Technique

In studying the knee joint with CT the patient is positioned supine, with only the leg under examination in the gantry tunnel; the contralateral leg is supported by a stirrup attached to the top of the gantry itself. The leg under examination is immobilized on a wooden support, strapped in place at the foot, leg, and thigh levels. This support, which moves with the CT table, has a pedal at the distal end that holds the foot in a physiological plantar position. A plastic triangular wedge placed under the knee holds the limb in a semiflexed position (8 to 10°). Both the wedge and the foot support may be adapted to fit any leg by sliding them longitudinally. Particular attention must be paid to positioning the patient properly and orienting the gantry so that the scanning plane is parallel to the articular surface of the tibia. In fact, menisci and their lesions may be identified correctly only if they are in precise par-

allel alignment with the axial plane. Two digital radiograms (anteroposterior and lateral) (Fig. 31.1A and B) are first obtained to align the patient correctly. On lateral digital radiography the distal scanning level is centered a few millimeters caudal to the head of the tibia with the gantry angled, so that the scanning plane is parallel to the articular surface of the tibia itself. Starting from this first slice at the tibiofibular level, 12 to 14 contiguous scans are taken, moving cranially, until the proximal insertion of the cruciate ligaments is reached. If for clinical reasons the patellofemoral joint must be studied, the number of scans may reach 20 to 25. The scans of the menisci overlap slightly to improve the quality of the images. Two millimeter slice thickness and 256 × 256 reconstruction matrix are used. The scans are made with 100 to 120 kV and 300 to 400 mA; for the entire examination the surface dose, determined by thermoluminescence dosimetry, is 4 ± 0.25 rads. The examination takes 20 to 30 min, most of which is used to place the patient's limb in the proper position. More time is needed to evaluate the images on the diagnostic console (6).

Normal Anatomy

The best visualization of meniscal structures on CT scans is obtained on the planes of the intercondylar eminence and intercondylar spines. They appear as a characteristic C shape for the medial meniscus and as an incomplete O for the lateral meniscus. They both have homogeneous fibrocartilagineous density (70 to 90 Hounsfield units (HU)). They are easily distinguishable from the adjacent structures and visible at their site of capsular ligament connection and bone insertion. The connections to the medial meniscus most relevant to radiologic examination are, at its middle part, the medial collateral ligament and, more posteriorly, the Hughston ligament and posterior joint capsule. The relevant connections for the lateral meniscus are the popliteus muscle tendon and the fibular collateral ligament at its middle and posterior parts and the meniscofemoral ligaments at the posterior horn. The anterior cruciate ligament (ACL) appears with its points of attachment well demarcated in the distal section, with a characteristic "horseshoe" appearance in the middle section, and with a ribbon-like form at the proximal insertion (Fig. 31.2). The average density values range between 50 and 70 HU. The posterior cruciate ligament (PCL) (average density 80 to 100 HU) has an oval shape at the distal insertion and is triangular in the middle section and at the proximal inser-

---

Abbreviations (see also glossary): ACL, anterior cruciate ligament; PCL, posterior cruciate ligament; MCL, medial collateral ligament; LCL, lateral collateral ligament.

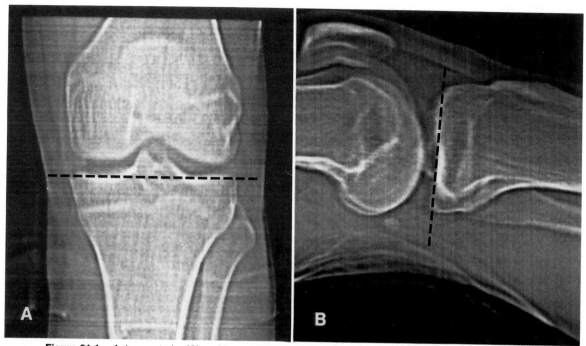

Figure 31.1. Anteroposterior (**A**) and lateral (**B**) digital scouts are used for paralleling tibial plateau slicing.

tion. Furthermore, its accessory bundle can be running either behind (most frequently) or in front of the ligament itself (Wrisberg and Humphry ligaments, respectively). The medial collateral ligament (MCL) has a ribbon-like appearance and is homogeneous with a density between 50 and 70 HU. The fibular or lateral collateral ligament (LCL) has a density similar to the MCL and a cordlike appearance. It has an important anatomical relationship with the biceps tendon in the distal section and with the popliteus muscle tendon in the proximal section. The patellofemoral joint is optimally visualized on the most cranial plane. Computed tomography allows evaluation of articular relationships and surfaces, even in semiextended position. The diagnostic accuracy in recognizing lesions of meniscal or capsuloligamentous formations is very high (7–9). This high accuracy is achieved through careful examination techniques, with immobilization of the limb under examination and a high-resolution imaging program. Diagnostic accuracy for the medial meniscus oscillates between 89.2 and 97.7% and for the lateral meniscus between 96.1 and 100%, while for the ACL and PCL it is 93.3 and 100%, respectively (10). It is convenient to describe the various CT findings, considering first the meniscal tears, then capsuloligamentous lesions, and the various pathological associations, also including the osteochondral lesions.

Meniscal Lesions

Meniscal tears appear as alterations in normal morphology and in homogeneous density of the menisci. They are characterized by hypodense streaks or gaps in a longitu-

dinal direction or, less frequently, in a transverse or oblique direction. The hypodense streaks may be single or multiple and vary in size and irregularity, according to the type of tear. Longitudinal tears of the medial meniscus (Fig. 31.3*A* and *B*) are easily diagnosed using CT. A large, irregular, hypodense streak along the meniscal structures that is present on all scanning planes allows an accurate diagnosis. Peripheral tears at the middle part of the posterior horn commonly cause greater problems in their diagnosis. In fact, there are CT patterns of tears on the posterior horn in which it is possible to have a false-positive diagnosis. Lack of experience may significantly affect the diagnosis in some cases. In other cases, the hypodense appearance at the posterior meniscal horn may be important pathologically when it is seen on several scanning planes. Typically, these hypodense streaks run in a continuous line with clearly visible edges. Overlap of a meniscal flap onto the remnant structures produces a hyperdense lesion (Fig. 31.3*C*).

Transverse tears at the middle part of the medial meniscus appear as an area of marked hypodensity that interrupts the axial outline and is present on several scanning planes. Complex lesions are most frequently identified with meniscal fragment displacements. The clear morphologic alteration and presence of meniscal fragments that appear as hyperdense formations permit an easy diagnosis and a definition of the extent of the tear. The meniscal detachments may or may not be associated with lesions of the capsular ligament structures of the internal compartment. The appearance on the CT scan is characterized by a hypodense longitudinal area. This produces a plane separating the meniscal structures from the capsular ligament,

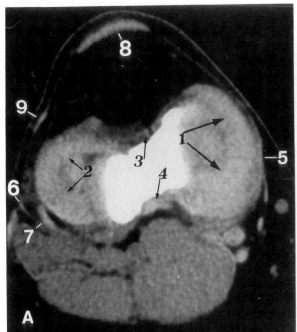 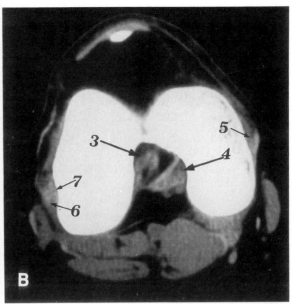

Figure 31.2. Demonstration of principal anatomical structures identified at two different levels: **(A)** intercondylar eminence plane and **(B)** intercondylar fossa plane. *1,* Medial meniscus (MM); *2,* lateral meniscus (LM); *3,* anterior cruciate ligament (ACL); *4,* posterior cruciate ligament (PCL); *5,* medial collateral ligament (MCL); *6,* lateral collateral ligament (LCL); *7,* tendon of popliteous muscle; *8,* patellar tendon; *9,* iliotibial band.

which normally adheres closely to the posterior horn (11). Medial meniscal cysts (Fig. 31.3*D*) are found in rare cases. They are located, normally, at the level of the posterior meniscal horn. They are quite small, monolocular, and may be associated with a complete longitudinal lesion of the meniscus. The CT diagnosis of both the lesion and the cystic degeneration is complete and accurate.

In other cases (Fig. 31.4*A* and *B*) in which a "bucket-handle" tear is present and the inner fragment is capsized and dislocated into the intercondylar fossa, a characteristic hyperdense "knot" is seen at the anterior meniscal horn. The meniscal fragment should not be confused with the anterior cruciate ligament in the intercondylar fossa, especially in those patients in whom this kind of tear is linked to an old and untreated lesion of the same ligament (Fig. 31.4*C* and *D*). The anatomic position, morphology, and density values on the images can help in differential diagnosis. Lateral meniscal tears have the same CT characteristics as medial meniscal tears. The most common clinical findings are complete longitudinal tears (Fig. 31.5*A*), complete transverse tears (Fig. 31.5*B*), and multiple fragmentations. It is possible to show with CT lateral discoid menisci. Not always does a discoid meniscus cause pain, most often it shows itself only through the clinical sign of "snapping-knee syndrome." A painful condition may indicate such complications as lesion or cystic degeneration (12) (Fig. 31.5*C*). Computed tomography can show a discoid meniscus, define the anatomopathological type, and, above all, the existence of an associated lesion. A total dis-

coid meniscus is attached throughout its perimeter, whereas a subtotal discoid meniscus is not inserted in its central part. These two forms represent 75 to 80% of discoid menisci.

The lateral meniscus, more frequently than the medial meniscus, can also be affected by cystic degeneration. This may be localized in the outer edges or it may encompass the entire meniscus, subjecting it to a complete morphologic change. The lateral meniscal cysts can be classified into diffuse and localized (plurilocular and monolocular) types. The diffuse type causes a complete morphological alteration of the meniscus with dishomogeneity and marked hypodensity. The localized type, on the other hand, originates from the meniscus and forms a cleavage plane with the nearby tissues and can be pedunculated. Normally, the localized type of meniscal cyst is monolocular with mucoid content and is usually situated at the level of the middle third and/or anterior third of the meniscus. Sometimes the meniscal cyst tends to exteriorize itself, thus involving the capsule, which appears thickened, and/or the lateral collateral ligament, which can be englobed. The pathogenesis of the meniscal cyst is not yet clearly understood. For this reason various theories have been formulated regarding degenerative, traumatic, or degenerative-traumatic pathogenesis. These theories attribute various degrees of importance to the association between the meniscal cyst and the meniscal lesion. In our opinion, the degenerative-traumatic theory is more dependable.

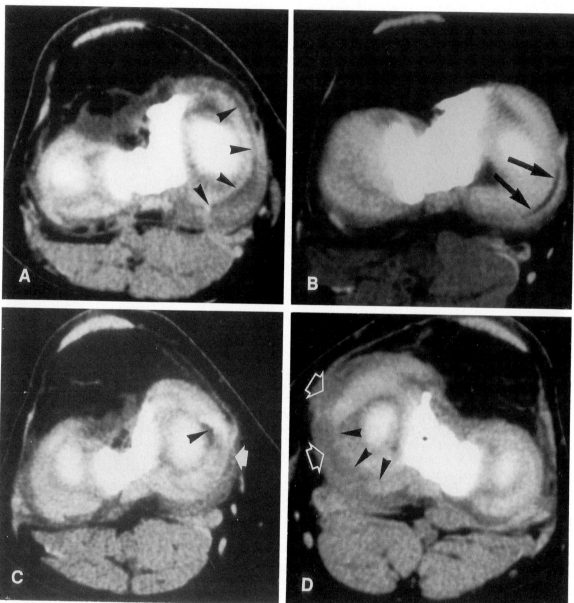

Figure 31.3. Different types of medial meniscus lesion. **(A)** Complete longitudinal lesions detected by an irregular hypodense line *(black arrowheads).* **(B)** Two-centimeter lesion localized at the level of peripheral portion of posterior horn *(black arrows).* **(C)** Small lesion of the body *(white arrow)* with flap formation *(black arrowhead).* **(D)** Lesion of the body and posterior horn of the medial meniscus *(black arrowheads)* with associated large degenerative changes *(open white arrows).*

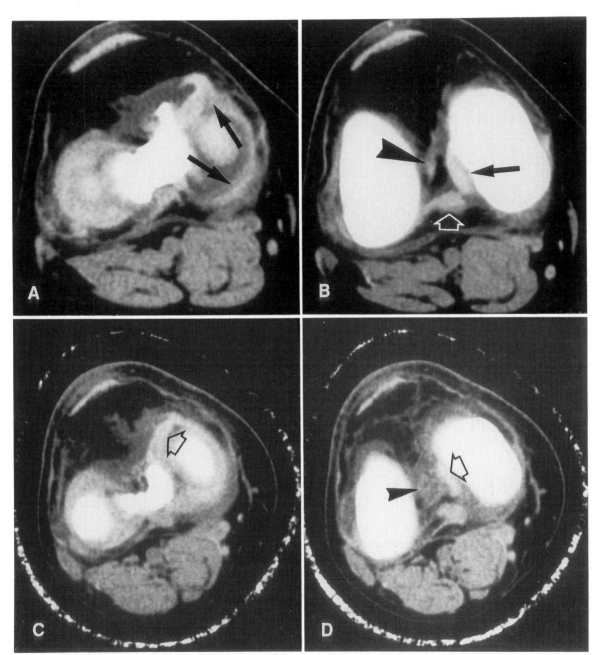

Figure 31.4. Bucket-handle lesion of the medial meniscus with associated ACL lesion in two different patients. (**A** and **B**) Bucket-handle lesion with flap subluxating in intercondilar fossa *(long black arrows)* and partial ACL lesion *(arrowhead);* the PCL is normal *(open white arrow).* (**C** and **D**) Bucket-handle lesion *(open black arrow)* associated with acute complete ACL lesion *(arrowhead).*

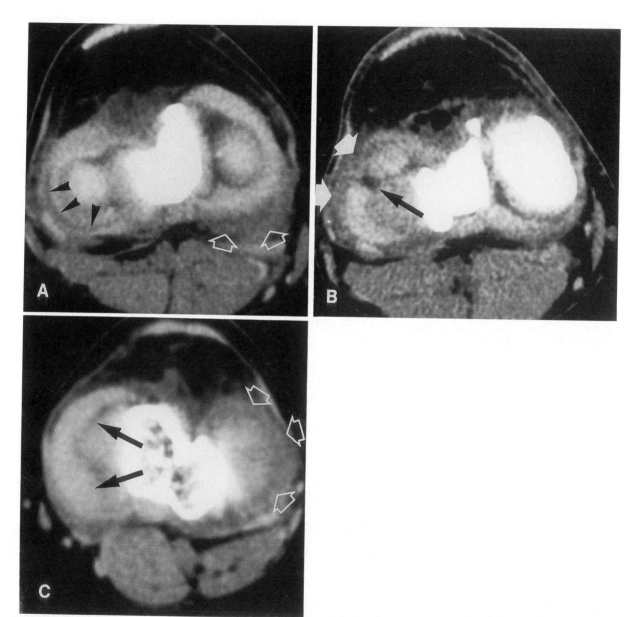

Figure 31.5. Lateral meniscus lesions. **(A)** Longitudinal lesion of the body and posterior horn *(black arrowheads)* with associated detachment of the medial meniscus *(open white arrows)*. **(B)** Transverse lesion localized between the anterior horn and body *(long black arrow)* with degenerative changes of the peripheral border. **(C)** Degenerative changes *(open white arrows)* involving the subtotal discoid meniscus; the medial meniscus is normal *(long black arrows)*.

Capsuloligamentous Lesions

Acute lesions produce similar changes in both the ACL (Fig. 31.6*A* and *B*) and PCL (Fig. 31.7*A* and *B*), which on CT examination appear nonhomogeneous, hypodense, and enlarged (13). On the other hand, when a complete laceration is present, there may be an absence of the ligament on some of the scanning planes or detachment of its insertion with osteochondral fragments. In cruciate ligament injuries CT can define the incomplete injuries. A "black area" indicates a total laceration, whereas a persistent re-

duction in volume of the ligament on all the scanning planes suggests a partial laceration of only one ligament bundle.

An acute lesion of the collateral ligaments (Fig. 31.8*A* and *B*) is characterized by nonhomogeneity, hypodensity, and enlargement of the ligament by CT. An old, healed laceration presents a morphological alteration with thickening, hyperdensity, and, sometimes, calcification. Only axial scans accurately establish the seriousness of acute lesions by demonstrating the degree of hypodensity, nonhomogeneity, and enlargement of the ligament with respect to normal. Information obtained by axial scans

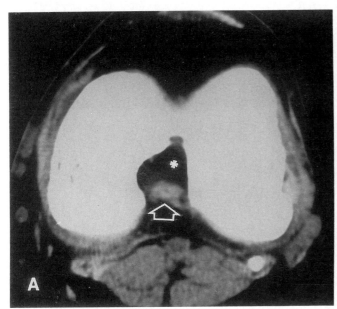

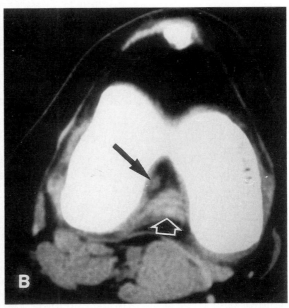

Figure 31.6. **(A)** Complete rupture of the ACL that is no longer recognizable *(asterisk)*; only the PCL is present *(open white arrow)*. **(B)** ACL partial lesion with a small residual band *(long black arrow)*; the PCL is normal *(open white arrow)*.

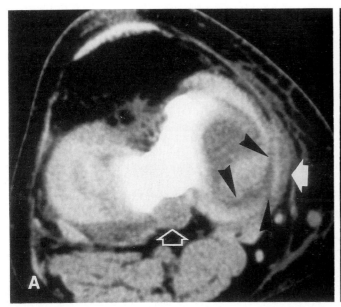

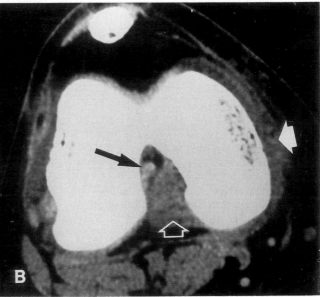

Figure 31.7. **(A and B)** Complete lesion of the PCL *(open white arrow)* associated with a complex tear of the posterior horn of the medial meniscus *(black arrowheads)* and a lesion of the MCL *(solid white arrow)*. Normal ACL *(black arrow)*.

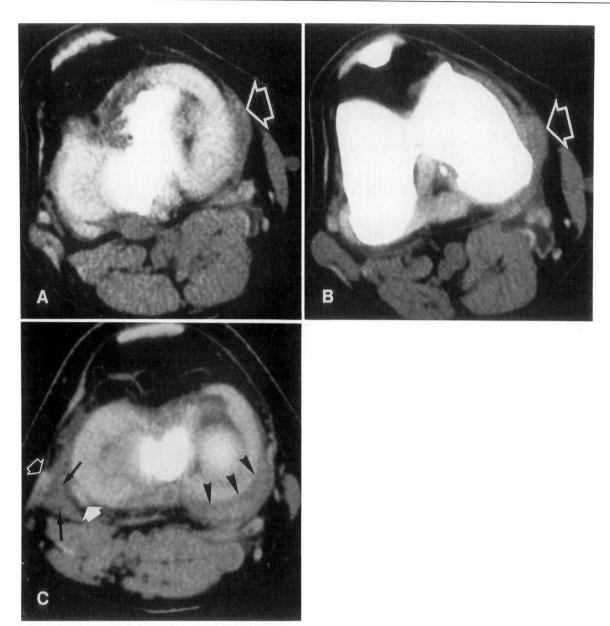

Figure 31.8. Lesions of the collateral ligaments. **(A** and **B)** Acute lesion of the MCL that appears enlarged, thickened, and hypodense at two different levels *(open white arrow)*. **(C)** Lesion of the LCL that is inhomogeneous and hypodense *(long black arrows)*, while the tendons of the popliteous muscle *(solid white arrow)* and of the biceps *(open white arrow)* are normal. A lesion of the posterior horn of the medial meniscus is associated *(black arrowheads)*.

furnishes sufficient and often definitive data to define the site and extent of a lesion, of the collateral ligaments, and of the PCL. Furthermore, with CT, acute lesions are evident at the site of the ligament most damaged by the trauma. Evaluation of damage to the ACL is more problematic because of its anatomical proximity to structures of higher density and the presence of the surrounding synovial sheath. In our CT experience we have a high diagnostic accuracy for the acute capsuloligamentous lesions (93.3% for the ACL, 100% for the PCL and the collaterals). However, to make a correct CT diagnosis, it is indispensable that (*a*) the ligament is completely visible; (*b*) all the CT signs identifying lesions are constantly present, since the presence of only one sign is of dubious value and does not establish the diagnosis with certainty; and (*c*) the clinical impression supports the CT diagnosis.

In 80% of traumatic events associated lesions are present (Fig. 31.9). The most frequent are medial meniscus and ACL lesions, although osteochondral lesions can also be present. In these cases CT can show not only osseous loose bodies but also cartilagenous fragments not evident on plain radiography (14). The articular cartilage has a CT density included between 40 and 60 HU and is well visualized without the use of intraarticular contrast medium. Therefore, in the case of medium- and high-degree chondropathies that are secondary to chronic instability or meniscal injuries, tapers, fraying, and erosions of the cartilage can be demonstrated by CT.

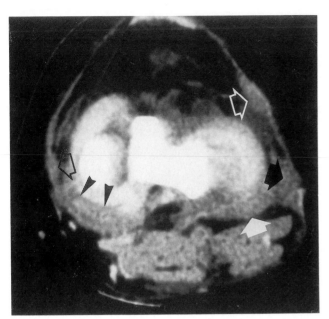

Figure 31.9. Acute trauma with multiple tears. Lesion of the posterior horn of the medial meniscus *(black arrowheads)*, the MCL *(open black arrow)*, the tendon of the popliteous muscle *(solid white arrow)*, the LCL *(solid black arrow)*, and the iliotibial band *(open white arrow)*.

References

1. Manco LG, Kavanaugh JH, Lozman J, Colman ND, Bilfield BS, Fay JJ. Diagnosis of meniscus tears using high resolution computed tomography. J Bone Joint Surg (Am) 1987; 69:498–502.

2. Manco LG, Lozman J, Colman ND, Kavanaugh JH, Bilfield BS, Dougherty J. Noninvasive evaluation of knee meniscal tears: preliminary comparison of MR imaging and CT. Radiology 1987; 163:727–730.

3. Passariello R, Trecco F, de Paulis F, Masciocchi C, Bonanni G, Beomonte Zobel B. Meniscal lesions of the knee joint: CT diagnosis. Radiology 1985; 157:29–34.

4. Pavlov H, Freiberger RH, Deck MF, Marschall JL, Morrissey JK. Computer assisted tomography of the knee. Invest Radiol 1978;13:57–62.

5. Steinbach LS, Helms CA, Sims RE, Gillespy T, Genant HK. High resolution computed tomography of knee menisci. Skel Radiol 1987; 16:11–16.

6. Passariello R, Trecco F, de Paulis F, Bonanni G, Masciocchi C, Beomonte Zobel B. Computed tomography of the knee joint: technique of study and normal anatomy. J Comput Assist Tomogr 1983; 7:1035–1042.

7. Manco LG, Kavanaugh JH, Fay JJ, Bilfield BS. Meniscus tears of the knee: prospective evaluation with CT. Radiology 1986; 159:147–151.

8. Pavlov H, Hirschy JC, Torg JS. Computed tomography of the cruciate ligaments. Radiology 1979; 132:389–393.

9. Reiser M, Rupp N, Karpf PM, Feuerbach ST, Anacher H. Evaluation of the cruciate ligaments by CT. Eur J Radiol 1981;1:9–15.

10. Passariello R, Trecco F, de Paulis F, De Amicis R, Bonanni G, Masciocchi C. Computed tomography of the knee joint: clinical results. J Comput Assist Tomogr 1983;7:1043–1049.

11. Manco LG, Berlow ME, Czajka J, Alfred R. Bucket-handle tears of the meniscus: appearance at CT. Radiology 1988;168:709–712.

12. Passariello R, Masciocchi C, Beomonte Zobel B, de Paulis F, Trecco F. Computed tomography of meniscal cysts of the knee joint. Eur J Radiol 1987;7:83–86.

13. Passariello R, Trecco F, de Paulis F, Masciocchi C, Bonanni G, Beomonte Zobel B. CT demonstration of capsuloligamentous lesions of the knee joint. J Comput Assist Tomogr 1986;10:450–456.

14. Sartoris DJ, Kursunoglu S, Pineda C, Kerr R, Pate D, Resnick D. Detection of intraarticular osteochondral bodies in the knee using computed arthrotomography. Radiology 1985;155:447–450.

CHAPTER 32

Computed Tomography Arthrography of the Knee

Bernard Ghelman

Introduction

The advent of magnetic resonance (MR) imaging resulted in lesser use of arthrography of the knee since this new noninvasive, nonionizing imaging technique gives good demonstration of normal and abnormal anatomical structure. At times, however, computed tomography (CT) scan combined with arthrography of the knee provides not only an alternative to MR imaging, but in certain situations a more precise diagnostic tool for its internal derangements. This chapter is focused on the use of CT arthrography of the knee.

Menisci of the Knee (Normal and Abnormal)

The double-contrast arthrogram of the knee is performed in the usual manner (1–4). In general, the CT images following the arthrogram give better demonstration of anatomy if a large amount of air (at least 40 cm³ in the average patient) and a small amount of positive contrast (at most 4 cm³) (5) are introduced in the knee. The use of a small amount of epinephrine is advisable to delay the absorption of the contrast agent by synovium.

Computed tomography imaging of menisci has already been described in the literature (6). Every effort should be made to have the X-ray beam coming off the CT gantry tangential to the tibial plateau. Therefore, the tilt of the gantry is determined by the position of the tibial plateau as seen on the scout lateral image. The CT images should be of high resolution (1.5 mm thick) and overlap is advisable (1-mm spacing). The set of images extends from the most inferior surface of the femoral condyle to the subchondral surface of the tibial plateau and, in general, consists of 9 to 12 images.

Two sets of images are obtained, one for the medial and one for the lateral meniscus. The patient is in the lateral

decubitus position to take advantage of the double-contrast effect around each meniscus. A CT scan of the left knee to demonstrate the medial meniscus is done, therefore, with the patient in a left lateral decubitus. For the demonstration of the lateral meniscus, the patient is in a right lateral decubitus position. The medial meniscus of the right knee is examined with the patient in a right lateral decubitus and lateral meniscus of the right knee is scanned with the patient in a left lateral decubitus position (Fig. 32.1).

Images obtained with the dynamic mode shorten the study, and in some cases avoid motion on the part of the patient.

The meniscus is demonstrated on an axial plane as seen in the CT images. Often, the entire meniscus is seen in one single image; however, at times different parts of the semilunar cartilage are displayed in separate images.

The medial meniscus is seen as a crescent-shaped structure resembling the letter C (Fig. 32.2). The most anterior portion of the meniscus is small, gradually enlarging toward its posterior horn (Fig. 32.3). The meniscus is completely attached to the adjacent joint capsule. Recesses at the base of the meniscus can be seen on CT images as collections of air and contrast agent, which may be confused with tears (Fig. 32.4).

The most inferior surfaces of the femoral condyles are demonstrated central to the meniscus (Fig. 32.2). The retinaculum of the knee joint is clearly identified as a thin, soft-tissue structure extending bilaterally from the patella and patellar tendon toward the posterior surface (Fig. 32.5).

The lateral meniscus is also a crescent-shaped structure, but more circular than the medial meniscus (Fig. 32.5). The size of the lateral meniscus is more uniform as one compares its anterior horn, body, and posterior horn (Fig. 32.6). The popliteal tendon bursa is readily identified at the base of the junction of the body and posterior horn of the lateral meniscus. The popliteus tendon is seen as a structure that bulges into the posterolateral surface of the bursa. The extent to which the tendon bulges into the bursa is variable; however, it is rare to find it as a free structure inside the bursa, as is the case, for instance, of the long head of the biceps in the shoulder joint.

Meniscal tears are most commonly seen in the posterior half of the medial meniscus. The mechanism of the tear is generally thought to be due to a rotational force while the knee is partially flexed, leading to catching of the meniscus between the femoral condyle and tibia. Meniscal tears

---

Abbreviations (see also glossary): DJD, degenerative joint disease.

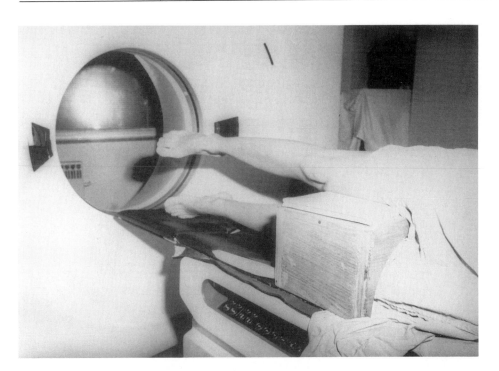

Figure 32.1. Position of patient for CT imaging of the right lateral meniscus. A wood table is placed between the patient's thighs.

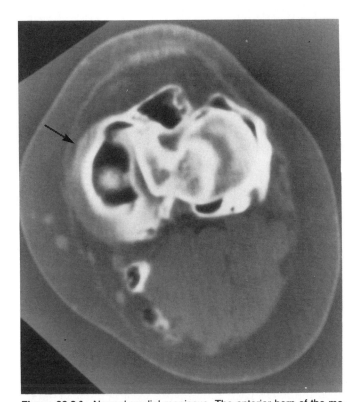

Figure 32.2.[a] Normal medial meniscus. The anterior horn of the medial meniscus *(arrow)* is smaller than its posterior horn. The meniscus is attached to the capsule in its entire periphery. The bottom of the medial femoral condyle is seen in the center of the medial compartment.

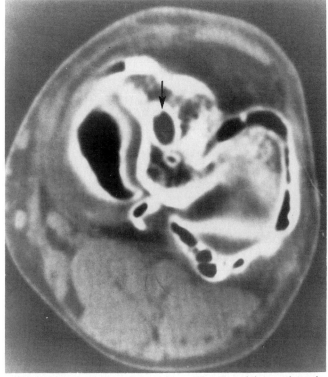

Figure 32.3.[a] Normal medial meniscus. The size of the meniscus increases progressively from the anterior to the posterior horn. The medial meniscus is in direct continuity with the joint capsule. The round lucent defects in the center of the tibial plateau *(arrow)* are due to previous surgical reconstruction of the anterior cruciate ligament.

[a] In Figs. 32.2 to 32.11, 32.13 to 32.17, 32.21, 32.27 to 32.30, 32.32 to 32.34, and 32.37 to 32.40, the image is presented with the anterior surface of the knee at the top of the figure, posterior surface at the bottom, medial aspect of the knee to the left side, and lateral aspect of the knee to the right side.

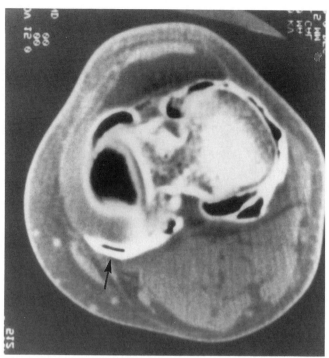

Figure 32.4.ª Normal medial meniscus. The collection of air and contrast agent next to the posterior horn of the medial meniscus *(arrow)* is the result of normal synovial recesses at the meniscal capsular junction. This finding could be confused with peripheral separation of the meniscus if the CT arthrogram is interpreted without the double-contrast arthrogram radiographs.

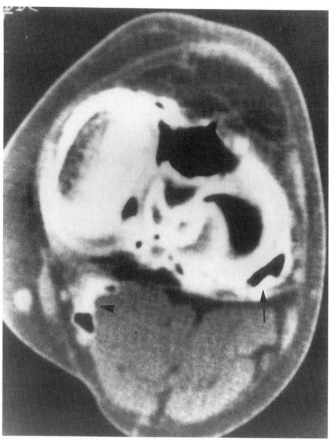

Figure 32.6.ª Normal lateral meniscus. The anterior horn, body, and posterior horn are uniform in size. The popliteus tendon *(arrow)* bulges into the recess at the base of posterior horn of the lateral meniscus. A small popliteal cyst is seen posteromedial to the joint *(arrowhead).*

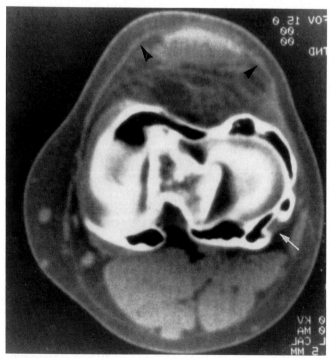

Figure 32.5.ª Normal lateral meniscus. The anterior horn, body, and posterior horn are similar in size. The popliteus tendon *(arrow)* bulges into the recess at the base of the junction of body and posterior horn of the meniscus. The synovial recesses around the lateral meniscus are larger, therefore air and contrast agent are seen at most of the periphery of the cartilage. The patellar retinaculum is outlined *(arrowheads).*

are usually classified as longitudinal (most common), horizontal (most common in older patients), oblique, and radial tears. Tears can be associated with degenerative changes, congenital anomalies such as discoid menisci, and with other joint trauma as, for instance, injury to the anterior cruciate ligament.

Tears of the menisci are seen on CT arthrography images as deformities in the contour of these structures (Figs. 32.7 and 32.9) (7). Radial and concentric tears (Fig. 32.11) are demonstrated. In general, vertical tears have a better chance of being demonstrated in the CT arthrogram images than horizontal tears because of the plane of the X-ray beam in the CT scanner. Loose fragments or torn tags are also readily displayed (Figs. 32.8 and 32.10).

Tears of the posterior horn of the medial meniscus are easily differentiated from popliteal cyst. Tears of the posterior horn of the lateral meniscus (Figs. 32.11 and 32.12) are also differentiated from the adjacent popliteus tendon bursa.

The advantages of CT arthrography of the knee in diagnosis of meniscal tears reside above all in the demonstration of the abnormality in a manner that closely resembles the intraarticular pathology. Arthrography is a reliable test for the diagnosis of meniscal tears; however, the spacial relationship between the torn fragments cannot be properly assessed. The alignment between the meniscus and the torn fragment, the contour of the tear, and the

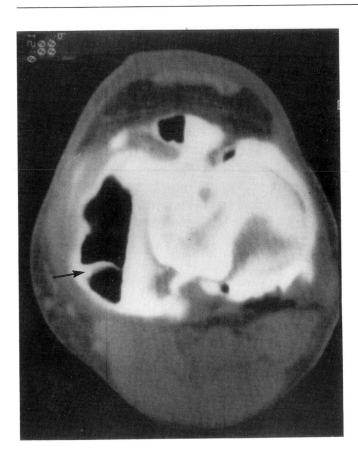

Figure 32.7.[a]   Torn medial meniscus. The inner edge of the meniscus is irregular. A torn fragment extends into the joint from the posterior horn of the cartilage *(arrow)*.

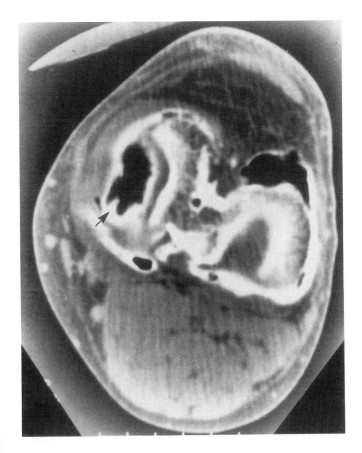

Figure 32.8.[a]   Torn medial meniscus. There is irregularity in the inner edge of the meniscus at its body and posterior horn. A small torn fragment *(arrow)* protrudes into the joint from the posterior horn.

Figure 32.9.[a] Tear and cyst of medial meniscus. The inner edge of the body is irregular. Contrast agent pools in a cyst at the periphery of the body of the meniscus *(arrow)*.

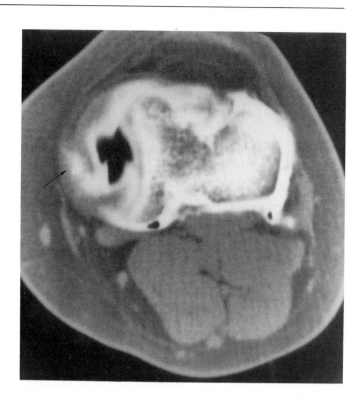

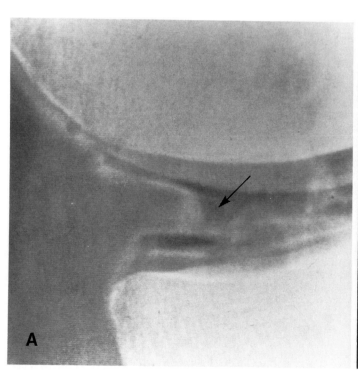

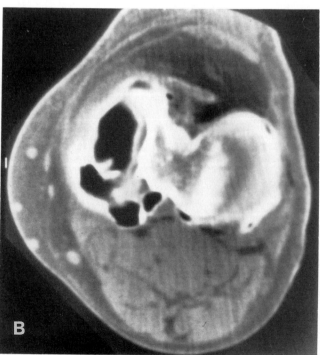

Figure 32.10.[a] Tear of the medial meniscus. **(A)** Arthrogram spot film: the inner edge of the meniscus is blunted and irregular, indicating a tear *(arrow)*. **(B)** Computed tomography arthrogram: the posterior horn is deformed and a torn fragment extends into the joint.

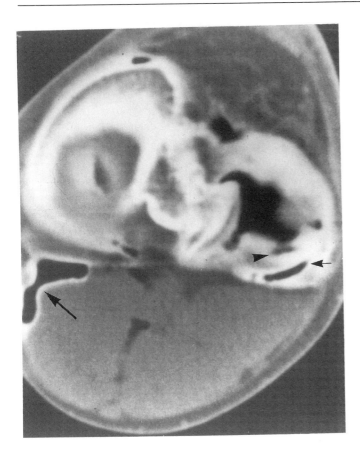

Figure 32.11.[a] Tear of the lateral meniscus. Air collects in a small concentric tear *(arrowhead)* close to the inner edge of the posterior horn of the lateral meniscus. The popliteal tendon sleeve is seen close by *(small arrow)*. A popliteal cyst is present in the medial side of the joint *(large arrow)*.

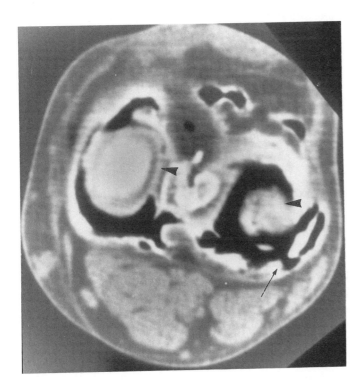

Figure 32.12. A tear of the lateral meniscus. There is marked irregularity in the contour of the posterior horn of the cartilage *(arrow)*. The lowermost surfaces of the condyles project in the center of the medial and lateral compartment *(arrowheads)*.

extent of the tear are all well seen in the CT arthrogram images. The advantages also include the easy separation of the tears of the posterior horn of the medial meniscus from adjacent popliteal cyst, and the easy distinction between abnormalities of the lateral meniscus and the adjacent popliteus tendon bursa.

Disadvantages for CT arthrography in the diagnosis of meniscal tears, however, are significant. Technically, this is a difficult study to perform. Any slight motion of the patient results in images that miss the meniscus altogether, or display them only partially. The recesses seen at the base of both medial and lateral menisci can be easily confused with peripheral tears. In the opinion of this author, the diagnosis of meniscal tear should not be based only on the CT arthrogram images. Tears are properly diagnosed if one correlates the findings in the arthrogram with the images of the CT arthrogram. The information thus provided allows the surgeon to plan for arthroscopic repair or removal of the damaged cartilage.

Meniscal cysts

These are areas of cystic degeneration (Figs. 32.9 and 32.13) usually associated with tears, often seen at the base of the menisci, most commonly in the lateral side (8). The CT arthrogram displays these cysts as round or elongated soft-tissue masses. Tears often extend into these cysts and, therefore, air and/or contrast agent are present in the cystic mass. Obviously if the tear does not communicate with the cyst, these are not opacified (Fig. 32.14). The differential diagnosis should include a soft-tissue ganglion (Fig. 32.15).

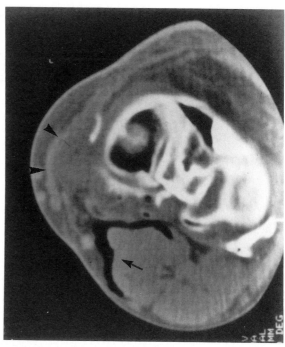

Figure 32.14.[a] Noncommunicating cyst of the medial meniscus. A round, soft-tissue mass is visible at the posteromedial surface of the joint *(arrowheads)*. Neither air nor contrast agent extend into the meniscal cyst. There is air within a popliteal cyst *(arrow)*.

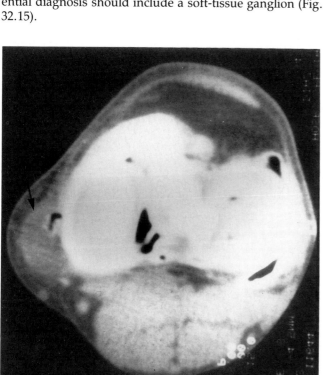

Figure 32.13.[a] Medial meniscus cyst. Soft-tissue mass *(arrow)* projects close to the medial joint line. A small amount of air has collected in the cyst, indicating its communication with the joint space through a tear of the meniscus. The tear is not demonstrated in this image.

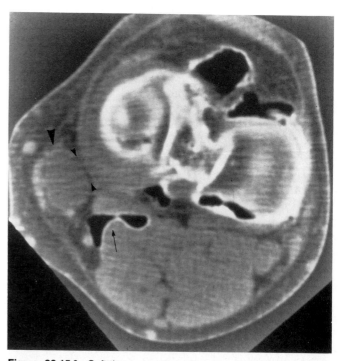

Figure 32.15.[a] Soft-tissue ganglion. Round soft-tissue mass *(large arrowhead)* projects close to the posterior horn of the medial meniscus. Thin, lucent fat plane projects between the ganglion and the adjacent meniscus *(small arrowheads)*. A small popliteal cyst is present *(arrow)*.

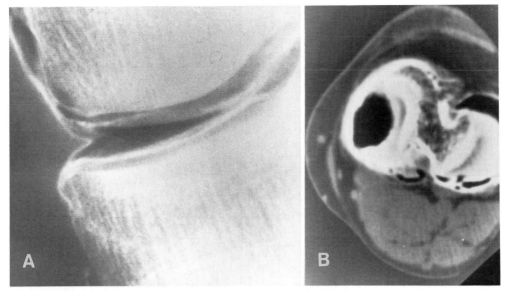

Figure 32.16.[a] Normal postmeniscectomy remnant of medial meniscus. **(A)** Arthrogram shows small triangular shape of remaining cartilage. **(B)** Computed tomography arthrogram shows that most of the meniscus is absent.

Postsurgical Menisci

The evaluation of a surgically repaired meniscus can be difficult (Fig. 32.16) (9). These cases, in which the surgeon reattached the torn meniscus to its capsular insertion (10) or sutured the torn fragments, often give abnormal signal intensity on MR images. Arthrography and CT arthrography allow the demonstration of the surface anatomy of the meniscus and proper diagnosis of whether a new tear is present or not.

Popliteal or Baker's Cyst

These are cysts resulting from abnormal communication of the bursa of the gastrocnemius/semimembranosus tendons with the knee joint space (Figs. 32.11, 32.14, and 32.15). Popliteal cysts always arise behind the medial femoral condyle (11). During arthrography, these cysts can obscure the posterior horn of the medial meniscus, a problem that does not exist in the CT arthrogram images. Popliteal cysts are also easily demonstrated in plain CT images (14) of the knee, MR images, and with ultrasound.

Plain Computed Tomography Scans for the Diagnosis of Meniscal Tears

Passariello and others (12, 13) have proposed the use of plain CT scans in the diagnosis of meniscal tears (see also Chapter 31). These CT images are also of high resolution (1.5 mm thick). The examination is carried out with the patient supine, the knee slightly flexed, and the CT gantry parallel to the tibial plateau. The meniscus is seen as an area of slightly greater absorption of radiation (Fig. 32.17) than the surrounding structures. Tears can be demonstrated as irregular areas or lines of diminished absorption. In general, the tears are not as well seen on plain CT images as compared to MR imaging, arthrography, and CT arthrography.

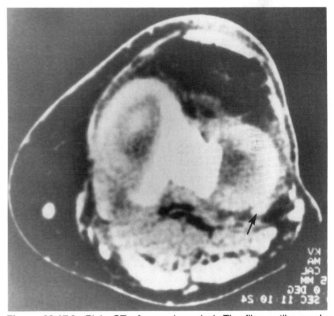

Figure 32.17.[a] Plain CT of normal menisci. The fibrocartilages absorb radiation somewhat more than surrounding structures. The popliteus tendon is identified at the posterior horn of the lateral meniscus *(arrow)*.

Chondromalacia Patellae

Chondromalacia patellae refers to changes in articular cartilage that lead to softening and in some cases can progress to abrasion, ulceration, thinning, and finally lysis. The term should be applied to articular cartilage in any joint; however, it is in the patella where this condition is most often diagnosed and treated.

Chondromalacia is probably due to changes in the cartilage ground substance and collagen fibers. These alterations result in less elastic cartilage, fibrillation, fissuring, fragmentation, crater formation, and finally eburnation of the adjacent bone.

The condition in the patella can be the result of trauma (acute or repetitive) (15), inflammation of the synovium (rheumatoid arthritis, infection, etc.), or degenerative diseases, in which case the disease may have a familiar distribution.

The traumatic type of chondromalacia usually starts with softening of the cartilage, while fibrillation is the usual first stage of the disease if it is of nontraumatic origin. Superficial lesions are supposed to have less ability to heal since this layer of the cartilage is avascular. Active healing can occur in deeper lesions.

Convery and others (16) have shown that defects less than 3 mm in diameter healed in 3 months while defects larger than 9 mm did not heal. Outerbridge (17) classified chondromalacia patellae according to the size of the lesion with grade 1 (measuring less than 0.5 cm), grade 2 (between 0.5 and 1 cm), grade 3 (1 to 2 cm), and grade 4 (lesions larger than 2 cm). Bentley (18) and Bentley and Dowd (19) classified the condition in different types of grades. Grade 1 is a localized softening, swelling, or fibrillation of the articular cartilage, grade 2 is a fragmentation fissuring in an area 1.3 cm or less in diameter; grade 3 is fragmentation and fissuring in an area more than 1.3 cm in diameter; grade 4 is erosion of articular cartilage down to subchondral dome (see also Chapters 34 and 35).

The CT examination of the patella is performed with the patient supine. The articular cartilage of the patella is best demonstrated when a large amount of air (at least 40 cm^3) is injected, and when separation is achieved between the patella and the adjacent femoral condyles (20). Even though the best detail is obtained with double-contrast arthrography, the injection of air alone (21) can provide diostic studies (Figs. 32.22, 32.25, and 32.26).

The CT can be obtained with the knee slightly flexed (not more than 30°) or with the knee fully extended. The patient should be instructed not to tighten the quadriceps tendon, so as to allow separation between the patella and the underlying femur (22, 23).

Normally, the articular cartilage of the patella has smooth surfaces (Figs. 32.18). The three facets of the patella (odd, medial, and lateral) are well demonstrated. The lateral facet is the longest facet. The articular cartilage should be thickest in the medial articular facet. The shape and size of the odd facet is variable. Different types of contour of the patella are readily identified (Fig. 32.19). In some cases, there is discrepancy between the contour of the bone as compared to the contour of the different articular facets of the patella. Measurements of the congruency between patella and adjacent femoral condyles (24) can be readily performed with the information available in the CT arthrogram (25–27).

Chondromalacia of the articular cartilage of the patella is well demonstrated in the CT arthrogram images as areas of fibrillation (Fig. 32.20), imbibition of contrast agent (Fig. 32.21), blister formation (Fig. 32.22), and different degrees of thinning and erosion (Figs. 32.23–32.26). Most often chondromalacia patellae occurs close to the junction of the medial and lateral facets. At times, chondromalacia patellae can be associated with thickened synovial plica and

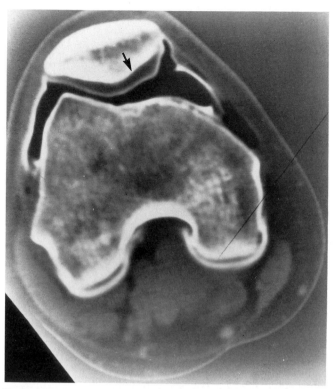

Figure 32.18. Normal articular cartilage of the patella (medial side of the knee is to the reader's right side). Three facets (odd, medial, and lateral) are identified. The articular cartilage is thickest in the medial facet *(arrow)*. The surfaces of the cartilage are smooth. The apex of the articular surfaces is located lateral to the apex of the subchondral bone surfaces.

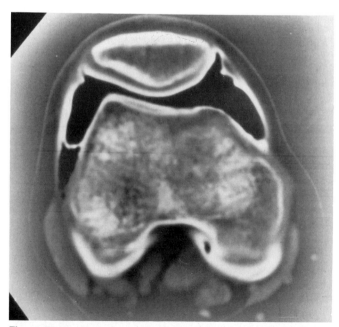

Figure 32.19. Normal articular cartilage of the patella (medial side of the knee is to the reader's right side). The surface of the articular cartilage is smooth. The articular surface has a round contour with less well-differentiated apices between odd, medial, and lateral facets.

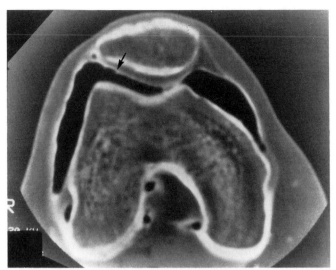

Figure 32.20. Chondromalacia patella: fibrillation (medial side of the knee is to the reader's right side). There is minimal irregularity of the cartilagenous surface of the lateral facet *(arrow)*.

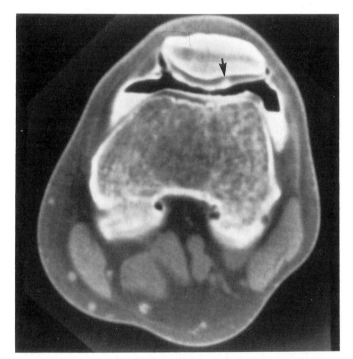

Figure 32.21.[a] Chondromalacia patella: fibrillation and imbibition of contrast agent. There is irregularity and imbibition in the medial facet and contrast opacifies part of the cartilage in the lateral facet *(arrow)*.

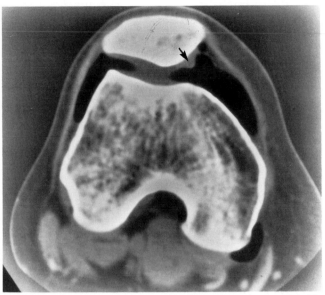

Figure 32.22. Chondromalacia patella: blister formation (medial side of the knee is to the reader's right side) (CT arthrogram performed with air only). Round soft-tissue mass, a blister *(arrow)*, is present in the medial articular facet.

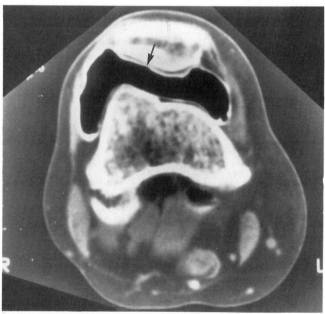

Figure 32.23. Chondromalacia patella: thinning and erosion (medial side of the knee is to the reader's right side). There is marked thinning of the articular cartilage with small erosion *(arrow)* in the lateral facet.

also with erosion in the articular cartilage of the adjacent femoral condyle. Chondromalacia of the femoral condyles can also be demonstrated with CT arthrograms (Figs. 32.27 and 32.28).

Softening of the articular cartilage obviously cannot be demonstrated by CT arthrography. This can be evaluated only by direct inspection (arthroscopy or arthrotomy), which, however, cannot assess the thickness of the cartilage that can be easily measured in the CT arthrogram images.

Review of 124 consecutive knees CT arthrograms performed at the Hospital for Special Surgery demonstrated chondromalacia patellae in close to 50% (60 abnormal cases). The most common findings of chondromalacia were thinning (25 cases), medium sized erosion (19 cases), fibrillation (12 cases), imbibition of contrast agent (11 cases), small erosions (8 cases), large erosions (7 cases), and blister formation (4 cases).

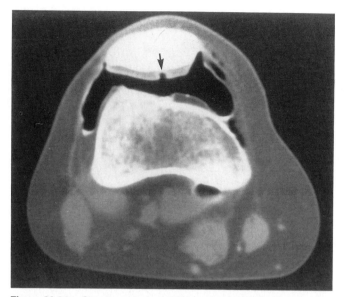

Figure 32.24. Chondromalacia patella: erosion (medial side of the knee is to the reader's right side). Well-defined medium-sized erosion *(arrow)* is visible at the junction of the medial and lateral facets. The thickness of the remaining articular cartilage is normal.

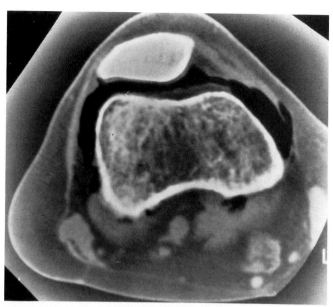

Figure 32.26. Chondromalacia patella: thinning and erosion (medial side of the knee is to the reader's right side) (CT arthrogram performed with air only). Marked thinning of the cartilage is visible in the medial facet. The cartilage of most of the lateral facet is absent.

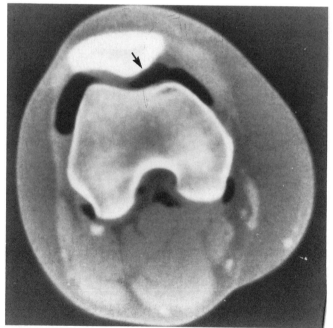

Figure 32.25. Chondromalacia patella: erosion (medial side of the knee is to the reader's right side) (CT arthrogram performed with air only). The cartilage of the medial facet is absent *(arrow)*.

Synovial Plicae

Synovial plicae in the knee represent varying degrees of persistence of embryonic synovial septae. During intrauterine development of the knee joint, the medial and lateral compartments, as well as the suprapatellar pouch, exist as three separate synovial cavities. The septae separating these cavities later involve so that the knee becomes a

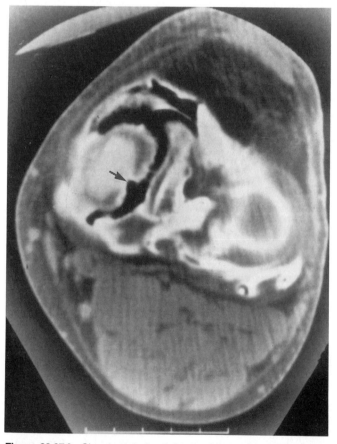

Figure 32.27.ᵃ Chondromalacia of the femoral condyle. Medium-sized erosion *(arrow)* is seen in the intercondylar surface of the medial femoral condyle.

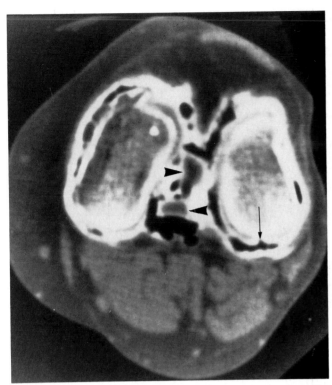

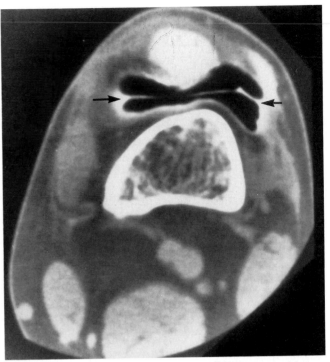

Figure 32.29.ᵃ Normal suprapatellar plica. A thin band of soft tissue *(arrows)* extends from the medial to lateral surfaces of the suprapatellar bursa.

Figure 32.28.ᵃ Chondromalacia of femoral condyle. Small erosion *(arrow)* is visible in the posterior surface of the lateral femoral condyle. The anterior and posterior cruciate ligaments are outlined *(arrowheads)*.

single space. In approximately 20% of the population, however, this process is incomplete, and these synovial plicae can become a source of knee pain in certain patients (28, 29).

The three most common forms of synovial plicae are the suprapatellar plica (medial or lateral) (Fig. 32.29), the infrapatellar plica (ligamentum mucosum, vertical septum) (Fig. 32.30), and the medial patellar plica (shelf plica) (Fig. 32.31).

The infrapatellar plica is fan shaped and lies between the intercondylar notch and the infrapatellar fat pad. This is the most common plica in the knee. Although in the literature, this has not been implicated in any pain syndromes, its presence often makes diagnosis of anterior cruciate ligament insufficiency more difficult by both arthrography and arthroscopy, especially if it is well developed.

The suprapatellar plica is usually seen as a medially or laterally based synovial fold of varying thickness and size. It originates beneath the quadriceps tendon and extends from just proximal to the superior pole of the patella to the medial or lateral wall of the joint. Occasionally, the septum will remain complete except for a central opening of variable diameter called the porta. Irritation and thickening of the suprapatellar plica may cause clinically significant knee pain.

The medial patellar plica (30) lies along the medial wall of the joint, originating near the suprapatellar plica and coursing obliquely downward to attach distally to the syn-

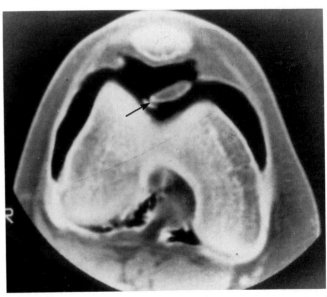

Figure 32.30.ᵃ Infrapatellar plica. There is a moderate thickness of the plica *(arrow)* as it crosses by the anterior surface of the lateral femoral condyle. The patient is asymptomatic.

ovial membrane covering the infrapatellar fat pad. The structure has also been referred to as a "shelf" and has been implicated as a possible cause of chondromalacia of the medial and odd facets of the patella as well as of the medial femoral condyle, due to a bow-stringing effect when the knee is flexed.

Most synovial folds or plicae are asymptomatic. How-

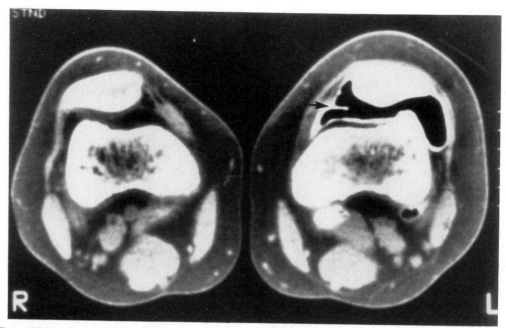

Figure 32.31. Normal medial patellar plica. A thin band of soft tissue *(arrow)* arises from the medial aspect of the joint close to the inner edge of the patella.

ever, following acute trauma, such as a direct blow, or in chronic repetitive trauma seen in competitive athletes, inflammation with associated edema and thickening may render the fold rather inelastic. The plica becomes symptomatic as it traverses the femoral condyle (Figs. 32.32 and 32.33) (31). Over time, fibrous changes can occur in the plicae and conceivably lead to mechanical erosion of the underlying articular cartilage (32, 33). The pain is characteristically intermittent, related to activity, located above the joint line, and often produces a clicking or snapping sensation. Generalized swelling of the knee is not common and, if present, should alert the physician to suspect other causes for the patient's knee problem.

Conservative treatment includes avoidance of further repetitive trauma and antiinflammatory agents (34). Although local steroid injections have been utilized, their use is controversial at this time. Operative treatment for refractory cases includes arthroscopic division of the synovial band, or excision of the band and partial synovectomy by either arthroscopy or arthrotomy. Computed tomography arthrography at the present time is the best available method for demonstrations of these synovial plicae (35). Magnetic resonance imaging is able to demonstrate synovial plica if the joint is distended with synovial fluid.

Following surgery (arthroscopy or arthrotomy) adhesions can be formed in the synovial cavity that are easily confused with synovial plicae (Fig. 32.34).

Osteochondritis Dissecans

Osteochondritis dissecans refers to abnormalities involving the subchondral bone and the adjacent articular cartilage. This lesion is most frequently seen in the intercondylar surface of the medial femoral condyle, but it can

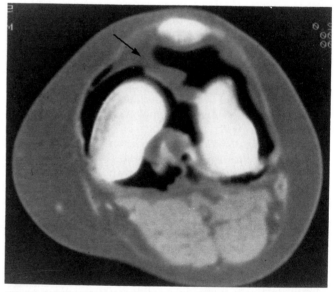

Figure 32.32.ª Plica syndrome: infrapatellar plica (CT arthrogram performed with air only). The plica is thickened and irregular in contour *(arrow)*. The plica projects close to the front of the medial femoral condyle.

be found in other sites such as the lateral femoral condyle, patella, talus, and capitellum of the elbow joint.

The changes of the subchondral bone are those of osteonecrosis while the adjacent cartilage may be thinned, eroded, or normal. The condition can lead to formation of an intraarticular osteochondromatous loose body. The condition is most commonly seen in young individuals, adolescents, and young adults.

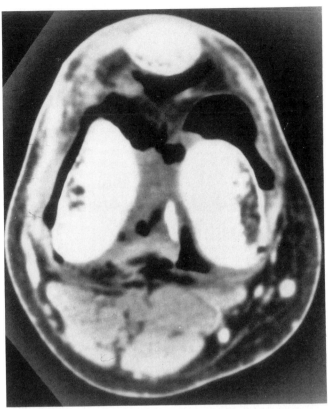

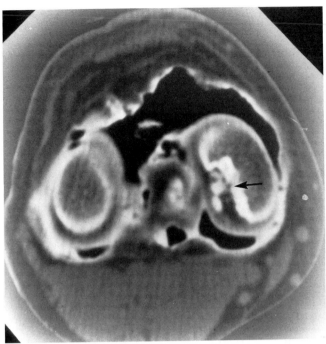

Figure 32.35. Osteochondritis dissecans with intact articular cartilage (medial side of the knee is to the reader's right side). There is fragmentation of subchondral bone *(arrow)* in the posterior surface of the medial femoral condyle. The adjacent articular cartilage is normal.

Figure 32.33.ᵃ Plica syndrome (CT arthrogram performed with air only). The infrapatellar and medial plica are thickened.

Computed tomography arthrography demonstrates the changes of the subchondral bone as well as the abnormalities of the articular cartilage. The subchondral changes are seen as irregular areas of increased and decreased density. The articular cartilage may be normal (Fig. 32.35), thinned, or eroded (Fig. 32.36). Contrast agent may extend into the bed of the osteochondritis dissecans, indicating separation of necrotic bone and underlying cartilage from the surrounding bone.

The assessment of the status of the articular cartilage is fundamental in the decision of how to approach these lesions. Intraarticular approach is indicated only if the articular cartilage that covers the area of osteochondritis is abnormal; when the cartilage is normal, the surgical approach should be extraarticular.

In unusual situations such as intraarticular loose bodies (Fig. 32.37) (36, 37) or pigmented villonodular synovitis (38) (Figs. 32.38 and 32.39), CT arthrography adds a new dimension in the localization of the abnormalities. One unusual case of synovial calcinosis (Fig. 32.40) was examined with CT arthrograms. This is a condition that has some similarity to tumoral calcinosis, but with the soft-tissue deposits of calcium being limited to synovium.

Conclusion

Computed tomography arthrography is a valuable method in the evaluation of internal derangements of the knee and should be used as a complement to the usual arthrograms. Computed tomography arthrography is the ideal test in the diagnosis and evaluation of chondromalacia patella (39) and synovial plicas. Computed tomogra-

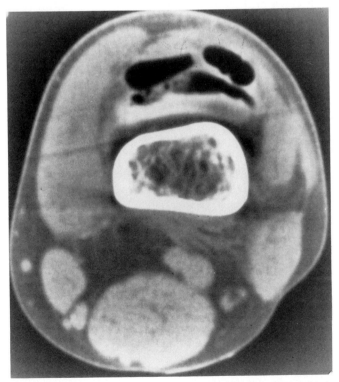

Figure 32.34.ᵃ Postoperative adhesions. Following arthroscopy, irregular adhesions are present in the suprapatellar bursa.

Figure 32.36. Osteochondritis dissecans with abnormal articular cartilage (medial side of the knee is to the reader's right side). Subchondral fragmentation *(arrow)* is present in the posterior surface of the lateral femoral condyle. The adjacent articular cartilage is irregular in contour, thinned out, and eroded.

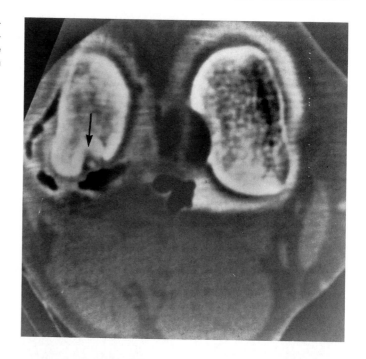

Figure 32.37.[a] Loose body. Small osteocartilagenous loose body is trapped in a synovial recess behind the medial femoral condyle *(arrow)*.

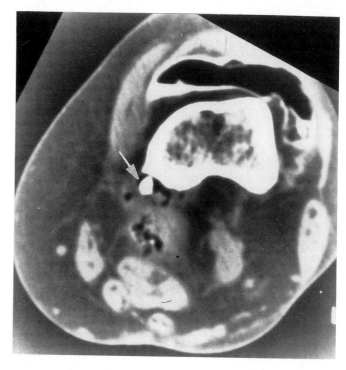

phy arthrography should also be considered in the following situations.

1. **The arthrogram is equivocal:** in cases of inconclusive arthrographic demonstration of a meniscal tear, the CT arthrogram images can be diagnostic. The evaluation of chondromalacia patellae, synovial plicae, and osteochondritis dissecans is more precise in the CT arthrogram images than on the arthrogram films.
2. **The plain CT scan is equivocal:** meniscal tears can be difficult to detect on plain CT images of the knee. Even

though the articular cartilage of the patella is seen on plain CT scans, the detail is never as good as demonstrated on CT arthrogram images. Even in the presence of effusion, synovial plicae are difficult to diagnose on plain CT.
3. **The MR imaging scan of the knee is equivocal:** this is a situation often encountered in the evaluation by MR imaging of chondromalacia patellae, synovial plicae (especially in the absence of intraarticular effusion), and at times, in the assessment of the articular cartilage covering areas of osteochondritis dissecans. Surgical

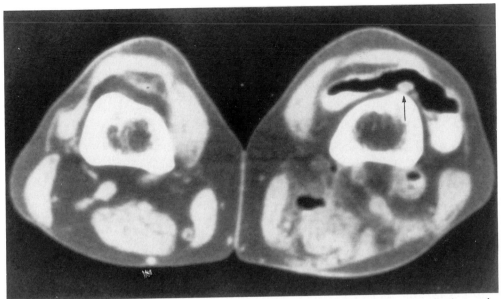

Figure 32.38.[a] Pigmented villonodular synovitis. A small nodule is visible in the suprapatellar bursa of the knee *(arrow)*.

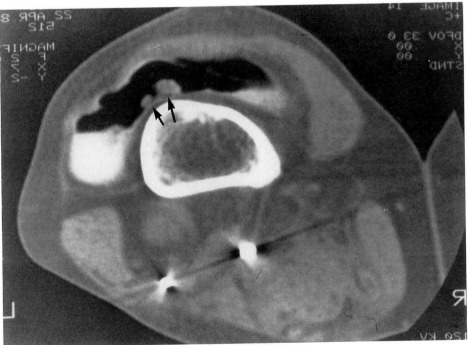

Figure 32.39.[a] Pigmented villonodular synovitis, recurrent. Small nodules *(arrows)* are visible in the suprapatellar bursa. The surfaces of the synovium are irregular. Surgical clips are present behind the femur.

metallic devices, such as those used at times in the reconstruction of the anterior cruciate ligament, produce artifactual shadows on the MR images. These artifacts often can be avoided in the CT arthrogram images by the use of high resolution (1.5- to 3.0-mm slices) and careful positioning of the knee.

4. **Claustrophobia:** this is a problem found at times in MR imaging of anxious individuals. Most of these patients are able to tolerate a CT scan without significant fear.

When used in a careful and logical manner and also under direct supervision of the radiologist, CT scans combined with arthrography can provide an added and useful dimension in the evaluation of the internal derangements of the knee.

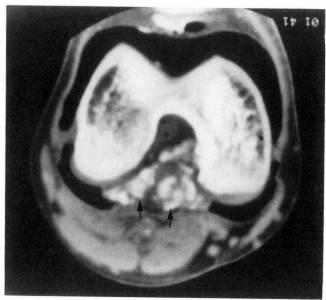

Figure 32.40.^a Synovial calcinosis in an 11-year-old boy with pain in the knee. Multiple irregular collections of calcium *(arrows)* are present immediately adjacent to, but not inside, the joint compartment. The location of the calcifications in the synovium was confirmed by surgical biopsy.

References

1. Freiberger RH, Kaye JJ. Arthrography. New York: Appleton-Century-Crofts, 1975.
2. Ingram C, Stoker DJ. Contrast media in double-contrast arthrography of the knee: a comparison of ioxaglate and iothalamate preparations. Br J Radiol 1986;59(698):143–146.
3. Nielsen F, de Carvalho A, Hjallund Madsen E. Omnipaque and urografin in arthrography of the knee. Acta Radiol Diagn 1989;25(2):151–154.
4. Tallroth K, Vanka E. Comparison of iohexol and meglumine iothalamate in single contrast knee arthrography. A double-blind investigation. Ann Clin Res 1986;18(3):144–147.
5. Apple JS, Martinez S, Khoury MB, Royster R, Allen S. A comparison of hexabrix and renografin-60 in knee arthrography. AJR 1985;145(1):139–142.
6. Ghelman B. Meniscal tears of the knee: evaluation by high-resolution CT combined with arthrography. Radiology 1985;157(1):23–27.
7. Pavlov H, Ghelman B, Vigorita VJ. Atlas of the knee menisci—an arthrographic-pathologic correlation. New York: Appleton-Century-Crofts, 1983.
8. Lee KR, Cox CG, Neff JR, Arnett GR, Murphey MD. Cystic masses of the knee: arthrographic and CT evaluation. AJR 1989;148(20):329–334.
9. Cavina C, Cossi CG, Pagliazzi A, Giorgi B, Grasso A. Arthrography in the investigation of persistent pain after meniscectomy. Ital J Orthoped Trauma 1984;10(3):363–368.
10. Miller DB Jr. Arthroscopic meniscus repair. Am J Sports Med 1988;16(4):315–320.
11. Guercio N, Solini A. Arthrography in the study of extrameniscal pathology of the knee. Ital J Orthoped Trauma 1988;14(20):257–265.
12. Passariello R, Trecco F, de Paulis F, Bonanni G, Masciocchi C, Zobel BB. Computed tomography of the knee joint: technique of study and normal anatomy. J Comput Assist Tomogr 1983;6(7):1035–1042.
13. Passariello R, Trecco F, de Paulis F, Masciocchi C, Bonanni G, Zobel BB. Meniscal lesions of the knee joint: CT diagnosis. Radiology 1985;157(1):29–34.
14. Schwimmer M, Edelstein G, Heiken JP, Gilula LA. Synovial cysts of the knee: CT evaluation. Radiology 1985;154(1):175–177.
15. Abernethy PJ, Townsend PR, Rose RM, Radin EL. Is chondromalacia patellae a separate clinical entity? J Bone Joint Surg 1978;60-B:205–210.
16. Convery FR, Akeson WH, Keown GH. The repair of large osteochondral defects: an experimental study in horses. Clin Orthoped 1972;82:253–258.
17. Outerbridge RE. The etiology of chondromalacia patellae. J Bone Joint Surg 1961;43-B:752–757.
18. Bentley G. Chondromalacia patellae. J Bone Joint Surg 1970;52-A:221–232.
19. Bentley G, Dowd G. Current concepts of etiology and treatment of chondromalacia patellae. Clin Orthoped 1984;189:209–227.
20. Tijn CJ, Hillen B. Arthrography and the medial compartment of the patello-femoral joint. Skel Radiol 1984;11(3):183–190.
21. Totty WG, Murphy WA. Pneumoarthrography: reemphasis of a neglected technique. J Can Assoc Radiol 1984;35(3):264–266.
22. Schutzer SF, Ramsby GR, Fulkerson JP. Computed tomographic classification of patellofemoral pain patients. Orthoped Clin Am 1986;17:235–247.
23. Schutzer SF, Ramsby GR, Fulkerson JP. The evaluation of patello-femoral pain using computerized tomography: a preliminary study. Clin Orthoped 1986;204:286–292.
24. Martinez S, Korobkin M, Fondren FB, Hedlund L, Goldner JL. Diagnosis of patello-femoral malignment by computed tomography. J Comput Assist Tomogr 1983;6:1050–1053.
25. Boven F, Bellemans M, Geurts J, Potvliege R. A comparative study of the patello-femoral joint on axial roentgenogram, axial arthrogram, and computed tomography following arthrography. Skel Radiol 1982;8:179–181.
26. Delgado-Martins H. A study of the position of the patella using computerized tomography. J Bone Joint Surg 1979;61-B:443–444.
27. Imai N, Tomatsu T, Takeuchi H, Noguchi T. Clinical and roentgenological studies on malignment disorders of the patello-femoral alignments using dynamic sky-line view arthrography with special consideration of the mechanism of mal-alignment disorders. J Orthoped Assoc 1987;61(1):1–15.
28. Patel D. Arthroscopy of the plicae-synovial folds and their significance. Am J Sports Med 1978;6(5):217–225.
29. Patel D. Plica as a cause of anterior knee pain. Orthoped Clin N Am 1986;17(2):273–277.
30. Dory MA. Arthrographic recognition of the mediopatellar plica of the knee. Radiology 1984;150(2):608.
31. Broom M, Fulkerson J. The plica syndrome: a new perspective. Orthoped Clin 1980;17(2):279–291.
32. Kinnard P, Levesque RY. The plica syndrome. Clin Orthoped 1984;183:141–143.
33. Mital MA, Hayden J. Pain in the knee in children: the medial plica shelf syndrome. Orthoped Clin Am 1979;10(3):713–722.
34. Munzinger U, Ruckstuhl J, Scherrer H, Gschwend N. Internal derangement of the knee joint due to pathologic synovial folds: the medio-patellar plica syndrome. Clin Orthoped 1981;155:59–64.
35. Aprin H, Shapiro J, Gershwind M. Arthrography (plica views): a non-invasive method for diagnosis and prognosis of plica syndrome. Clin Orthoped 1984;183:90–95.
36. Sartoris DJ, Kursunoglu S, Pineda C, Kerr R, Pate D, Resnick D. Detection of intra-articular bodies in the knee using computed arthrography. Radiology 1985;155(2):447–450.
37. Tehranzadeh J, Gabriele OF. Intra-articular calcified bodies: detection by computed arthrography. South Med J 1984;77(6):703–710.
38. Cavina C, Cossi CG, Pagliazzi A, Giorgi B, Grasso A, Cosenza M. Evaluation of double-contrast arthrography in pigmented villonodular synovitis of the knee. Ital J Orthoped Trauma 1984;10(1):121–126.
39. Ihara H. Double-contrast CT arthrography of the cartilage of the patello femoral. Joint Clin Orthoped 1985;198:50–55.

CHAPTER 33

Magnetic Resonance Imaging of Knee Trauma

Andrew L. Deutsch and Jerrold H. Mink

Introduction

In no musculoskeletal area has the marked growth and expansion of interest in the application of magnetic resonance (MR) imaging been more evident than in the assessment of internal derangement of the knee. The noninvasive nature of MR imaging has facilitated evaluation of even acutely traumatized patients. These attributes have allowed MR to rapidly challenge and indeed replace arthrography in many institutions. This chapter will focus on the application of MR toward the assessment of traumatic abnormalities involving the knee.

Technique

Regardless of the MR system or field strength of the magnet on which the examination is performed, the use of a surface coil is mandatory for evaluation of the knee. Circumferential extremity coils, preferably with both send and receive functions, are utilized to provide uniform signal to noise across the joint (1–8). As the pre-MR diagnosis is not often known or correct, a standard protocol must be utilized that permits rapid but comprehensive and accurate assessment of virtually any potential knee disorder.

The standard examination of the knee in our institution is performed with the patient supine and the affected knee in 10 to 20° of external rotation and full extension, a position that optimizes visualization of the anterior cruciate ligament (ACL) (6–9). Coronal and sagittal scan sequences are performed on each patient. The coronal series is performed first. The repetition time (TR) is 600 to 800 ms, depending on the number of slices needed to cover the knee; the echo time (TE) is 20 ms; and the image sections

are 5 mm, with no interslice gap. A 16-cm field of view (FOV) and a 128×256 matrix acquisition provides spatial resolution of 1.25 mm in the phase-encoded direction and 0.6 mm in the frequency-encoded direction. The sagittal sequence is prescribed from the coronal. Again, a 16-cm FOV and 5-mm section thickness with no interslice gap are selected. A spin-echo multiecho sequence (TR/TE 2000/ 20, 80) provides intermediate and mildly T2-weighted images. For assessment of pediatric patients, the FOV may be reduced to 12 cm and still provide adequate visualization of the quadriceps-patellar mechanism in the sagittal plane, and slice thickness may be reduced to 3 mm for evaluation of meniscal lesions. The above detailed scan protocol can be accomplished in 7 min of imaging time utilizing one excitation (NEX). In patients with acute trauma, a coronal T2-weighted spin-echo sequence is substituted for the initial coronal T1-weighted sequence for optimal evaluation of the collateral ligaments.

Axial scans are routinely obtained and provide the best evaluation of the patellofemoral joint and for assessment of the articular cartilage. We utilize a short tau inversion recovery (STIR) sequence (TR 2200 TE 35 TI 160) for assessment of the articular cartilage (10–16). Several investigators have proposed the use of fast scan gradient-echo (refocused) sequences in place of T2-weighted spin-echo sequences.

These methods utilize flip angles of less than 90° and provide effective T2 or T2* contrast. With three-dimensional Fourier transform (3D FT) techniques, gradient-echo imaging can further reduce both slice thickness and imaging times. These methods can be particularly helpful for assessment of articular cartilage abnormalities. We have not, however, routinely employed gradient-echo techniques for standard imaging because of decreased contrast resolution for soft tissue and medullary bone. Additionally, these sequences are not significantly "faster" than the spin-echo pulse sequences described above.

High-contrast photography is utilized for optimal assessment of intrameniscal signal. Window widths are set extremely narrow, and window level is adjusted so that slight inhomogeneities in the normally signalless meniscus can be seen. These images are then magnified 1.5 times and filmed on a 9:1 "format." Many of the illustrations of meniscal abnormalities utilized in this chapter were filmed utilizing these "meniscal" windows.

Abbreviations (see also glossary): ACL, anterior cruciate ligament; FCL, fibular collateral ligament; FOV, field of view; MCL, medial collateral ligament; NPV, negative predictive value; AMB, anteriomedial bundle; PLB, posterolateral band; PCL, posterior cruciate ligament; TCL, tibial collateral ligament; LCL, lateral capsular ligament; ITB, iliotibial band; OCD, osteochondritis dissecans.

Menisci

Anatomy

The fibrocartilagenous menisci or semilunar cartilages are C-shaped structures that are attached to the tibial condylar surface and function to provide enhanced rotational stability and mechanical load transmission. Both menisci are 3 to 5 mm in height. The average width of the anterior horn of the medial meniscus is 6 mm and that of the posterior horn 12 mm. In addition to its central condylar attachments, the medial meniscus is directly attached to the deep layer of the medial collateral ligament and capsule. On the initial parasagittal section, the medial meniscus appears as a homogeneously dark band, or ''bow tie,'' situated between the femoral and tibial articular cartilages. Typically by the third 5-mm section the free edge of the meniscus has been crossed and separate anterior and posterior horns are evident. The posterior horn of the medial meniscus is larger in both height and breadth than the anterior horn, which may normally be imaged on only one or two sections. This relationship is important to critically evaluate because focal diminution in size of an otherwise normal appearing posterior horn may be the only indication of a displaced tear.

The lateral meniscus forms a more tight C shape and has an approximate width of 12 mm throughout. The central ligamentous attachments of the lateral meniscus may normally contain signal and must be differentiated from the actual fibrocartilaginous menisci to avoid diagnostic error. This appearance is typically most confusing in the anterior horn of the lateral meniscus, where the anterior ligamentous attachment may be imaged on one to two sections. In contrast to the triangular meniscal horn, the anterior ligamentous attachment is rhomboid and points obliquely upward. The transition point between the meniscus and ligamentous attachment is the site of origin of the transverse ligament, which secures the meniscus directly to the tibial intercondylar eminence, ACL, and medial meniscus. Beyond the central attachment, the lateral meniscus appears as two triangles for two to three 5-mm sections. Next the body is encountered as a homogeneous dark or bow-tie structure, which appears more symmetric than on the medial side. In distinction to the medial meniscus, which is firmly attached to the joint capsule throughout, the popliteal tendon violates the meniscocapsular junction of the lateral meniscus, creating two struts or fascicles that function as the peripheral meniscal attachments. The intricate anatomy of the fascicles is not routinely demonstrated on standard T1-weighted sequences, but can be demonstrated to advantage on T2-weighted images, particularly when joint fluid is present (Fig. 33.1). The lateral meniscus is separated from the lateral collateral ligament, which is extracapsular, and has attachments to the posterior cruciate ligament and medial femoral condyle via the ligament of Wrisberg.

The vascular supply to the meniscus is derived from the genicular arteries, which provide an arborizing network of perimeniscal capillary vessels that supplies the peripheral border of the meniscus throughout its attachment to the joint capsule (17). The degree of peripheral vascular penetration ranges between 10 and 30% of the width of the meniscus. The vascular supply of the meniscus is the principal determining factor in providing the ability of this tissue to heal and be repaired.

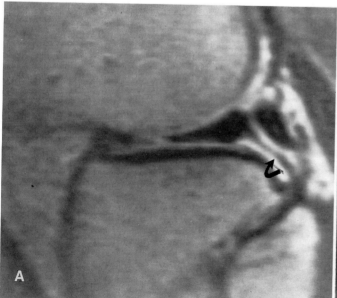 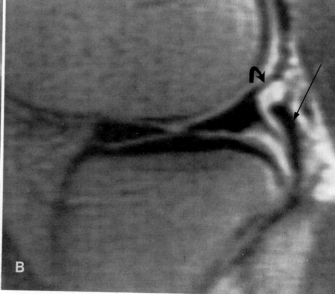

Figure 33.1. Peripheral attachments of the posterior horn lateral meniscus. **(A)** Sagittal image (TR/TE 2000/80). On this extreme sagittal section through the lateral meniscus, the presence of fluid allows depiction of the complex relationships of the posterior horn to the joint capsule. The popliteal tendon violates the meniscocapsular junction.

Only the inferior strut is intact *(curved arrow)*. **(B)** Sagittal image (TR/TE 2000/80). On a more centrally located sagittal section, the popliteal tendon *(long arrow)* has moved caudally, and the superior strut *(curved arrow)* is now the peripheral meniscal attachment.

Meniscal Grading System

Areas of meniscal degeneration and frank meniscal tears both image as areas of increased signal intensity within the normally low signal intensity meniscus. In an attempt to separate minor degenerative abnormalities from arthroscopically detectable and significant tears, a grading system of intrameniscal signal based on its morphology has been developed. The most popular system divides intrameniscal signal into three principal grades (18). Grade 1 signal is a globular focus of increased signal intensity that is entirely contained within the meniscus without evidence of extension to an articular surface (Fig. 33.2). Histologically, menisci in which grade 1 signal is seen demonstrate foci of mucinous, hyaline, or myxoid degeneration in areas of chondrocyte deficiency (8, 10, 19). The findings are frequently seen in asymptomatic individuals and may represent a response to mechanical stress and loading with increased production of mucopolysaccharide ground substance. Grade 2 intrameniscal signal is primarily a linear signal within the meniscus that again does not extend to an articular surface. In nearly all cases, grade 2 signal arises within the substance of the posterior horn of the medial meniscus at the meniscosynovial junction; it is oriented in the midplane of the meniscus without a vertical component (Fig. 33.3). Histologically these foci are characterized by more extensive bands of mucinous degeneration bordering hypocellular regions of the meniscus. Microscopic areas of collagen fragmentation may be observed, although no distinct fibrocartilagenous separation is present. Grade 2 signal abnormality represents a continuum of progressive degeneration. Patients with grade 2 signal may or may not be symptomatic, although the histologic stage has been described as a precursor to frank tears (8, 10, 19).

Grade 3 signal is defined as intrameniscal signal that unequivocally extends to an articular surface. Grade 3 signal has been further subdivided into grade 3A, which is

linear intrameniscal signal abutting an articular margin (Fig. 33.4), and grade 3B, which has irregular morphology. For purposes of this grading system, an articular surface refers to the upper, lower, or free edge of the meniscus (8, 19). Grade 3 signal reflecting a meniscal tear is commonly seen on more than two contiguous 5-mm sections. In small radial tears of the lateral meniscus, however, the abnormal signal may characteristically be seen on only one section along the free edge. Histologically, frank fibrocartilagenous separation is identified in menisci demonstrating grade

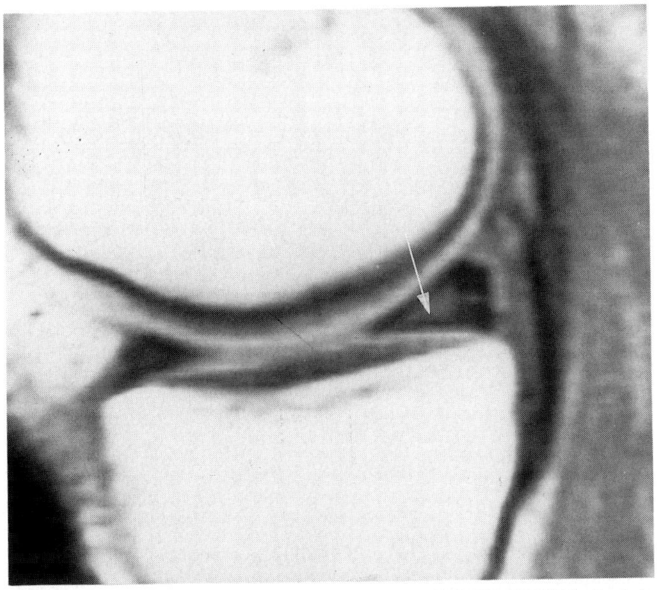

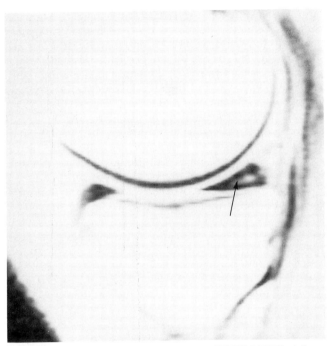

Figure 33.3.   Grade 2 signal. Sagittal image (TR/TE 800/20). A linear focus of increased signal intensity is seen within the posterior horn of the medial meniscus *(arrow)*. The signal does not extend to the tibial articular surface. It is important to evaluate the coronal sequence to further exclude articular surface extension.

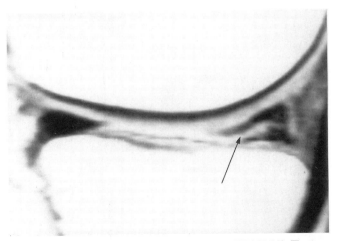

Figure 33.2.   Grade 1 signal. Sagittal image (TR/TE 800/20). A globular focus of increased signal *(arrow)* as seen as entirely contained within the meniscus. No extension to an articular surface is present.

Figure 33.4.   Grade 3 signal. Sagittal image (TR/TE 800/20). The typical appearance of a degenerative cleavage tear within the posterior horn of the medial meniscus is illustrated *(arrow)*. The signal demonstrates unequivocal extension to the meniscal articular surface.

3 signal along with extensive regions of mucinous degeneration and chondrocyte death. Grade 3B lesions have been associated with more extensive degenerative change in the surrounding meniscus than the more linear grade 3A tears. Menisci containing linear grade 3A signal may appear initially near normal at gross inspection. Extensive probing at the time of arthroscopic evaluation may be required to delineate these tears.

Meniscal Tears

Undoubtedly, the most common indication for MR imaging of the knee is for assessment of the menisci. Meniscal tears can generally be sorted into the named categories familiar to the arthroscopist. A common meniscal tear encountered on MR in our experience is the *horizontal cleavage* or *"fish mouth"* tear (2, 7, 8). Horizontal tears roughly parallel the superior and inferior surfaces of the meniscus and separate the meniscus into two halves (Fig. 33.4). Cleavage tears most often occur in older individuals and are considered degenerative in nature, in distinction to vertical tears, which are classically associated with acute trauma (8). The posterior half of the medial meniscus and midsegment of the lateral meniscus are the most common sites of horizontal tears. These tears commonly communicate with the margin of the meniscus at the free edge. At arthroscopy, the meniscal free edge may resemble the tips of a fish's scales, with the tear itself suggesting the appearance of the mouth of a fish, hence the common designation of a fish mouth tear.

As a consequence of age-related changes, the elasticity and resistance of the meniscus is gradually reduced, predisposing the meniscus to injury. During flexion and extension, the femur moves with the superior half of the meniscus, and the tibia moves with the inferior half, creating a mechanical shear stress within the central meniscus. Histologically, myxoid material preferentially accumulates along the "middle perforating bundle," a collagen bundle that roughly divides the meniscus into superior and inferior halves. With increasing degeneration, a rent forms parallel to the middle perforating bundle corresponding to the intrasubstance tear first described by Smille (20). Central extension results in the common horizontal cleavage tear. Many of these tears detected on MR represent confined or closed intrasubstance cleavage tears. Diagnosis of these closed meniscal tears may not be possible by visual inspection alone and frequently requires extensive probing at the time of surgery to allow their detection. It is important to note that a horizontal cleavage lesion is part of the degenerative aging process and that these tears may or may not be symptomatic.

Vertical longitudinal tears are classically associated with acute trauma (21). Displaced longitudinal tears are also referred to as bucket-handle tears because of the resemblance of the centrally displaced fragment to the handle of a bucket with the remaining peripheral section representing the bucket. The presence of a displaced bucket-handle tear should be suspected when focal alteration in meniscal size is encountered regardless of whether there is increased signal intensity (8). With displaced longitudinal tears, the plane of separation and displacement of the "handle" often occur through the region of abnormal intrameniscal signal, resulting in a signalless and often triangular meniscal rim that can appear surprisingly normal.

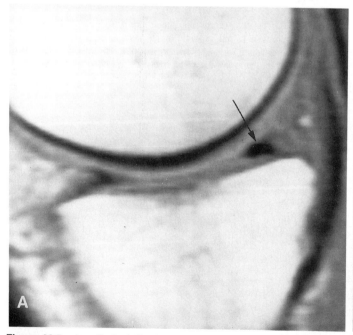

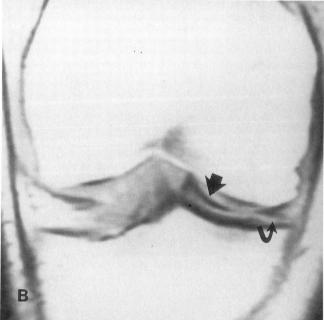

Figure 33.5. Longitudinal (bucket-handled) tear. **(A)** Sagittal image (TR/TE 800/20). A roughly triangular shaped posterior horn of the medial meniscus is identified *(arrow)*. No abnormal intrameniscal signal is seen. The posterior horn is smaller than the anterior horn, an inversion of the normal relationship. This should strongly suggest a displaced tear. **(B)** Coronal image (TR/TE 800/20). The triangular shaped peripheral meniscus *(bucket)* is again identified *(curved arrow)*. A large displaced fragment *(handle)* is seen extending into the intercondylar notch *(large arrow)*.

Detection of these subtle MR findings requires appreciation that the meniscus is "too small" and/or a fragment is displaced into the intercondylar notch. Coronal sequences can be essential to the diagnosis of displaced meniscal tears (Fig. 33.5). *Peripheral tears* occur through the outer one-

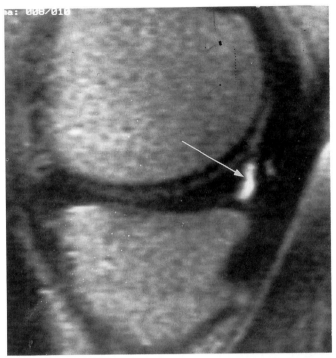

Figure 33.6. Meniscocapsular separation. Sagittal image (TR/TE 2000/80). On this T2-weighted image through the posterior horn of the medial meniscus a continuous line of increased signal is seen separating the meniscus from its capsular attachment *(arrow)*. This appearance must be distinguished from prominent synovial recesses.

third of the meniscus or at the meniscosynovial junction. These tears represent an important subgroup of meniscal injuries as a consequence of their ability to heal or be repaired. These tears may be difficult to detect at the time of arthroscopy, necessitating extensive probing at the time of surgery. The posterior horn of the medial meniscus is a common site secondary to its relative immobility and fixation by the coronary ligaments. Peripheral tears are recognized by either the presence of grade 3 signal within the outer third of the meniscus or by the presence of fluid completely interposed between the meniscus and its capsular attachment (Fig. 33.6).

Radial or *transverse tears* occur across the body of the meniscus and extend from the inner edge to the periphery (8, 21). These tears may be complete or incomplete and are most common in the lateral meniscus, particularly at the junction of the body and posterior horn. The occurrence of radial tears in the lateral meniscus is principally related to its more circular shape and smaller radius of curvature, which places the greatest strain along the concave inner margin during rotational stress. An incomplete radial tear may at times extend in an anterior and posterior direction. This creates both vertical and horizontal components in an oblique plane at the free edge of the meniscus and is commonly referred to as a "parrot beak" tear (22).

Radial tears can be quite small and therefore subtle in their MR appearance. As stated, they typically occur at the junction of the posterior horn and body of the lateral meniscus and may be seen on only one contiguous 5-mm sagittal MR image. Increased signal along the free edge related to partial volume effects must be distinguished from true tears. High-resolution coronal images of the meniscus as well as radial scans may allow better evaluation of the location and extent of these tears than sagittal images alone (Fig. 33.7). Small foci of signal localized to the tip of the

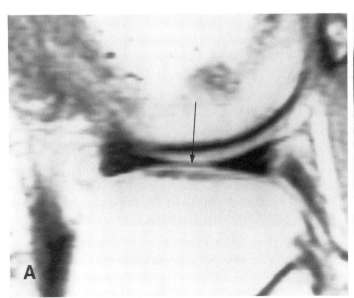 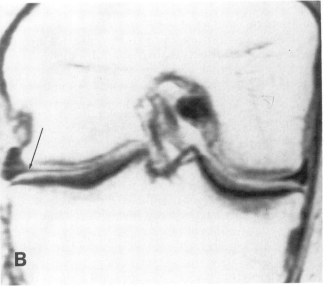

Figure 33.7. Radial tear. **(A)** Sagittal image (TR/TE 800/20). A small focus of increased signal intensity is seen along the free edge of the lateral meniscus *(arrow)*. This is the typical site of a radial tear but must be distinguished from partial volume effect and minimal free edge fibrillation. **(B)** Coronal image (TR/TE 800/20). Coronal image demonstrates subtle blunting of the apex of the meniscus, confirming the presence of a meniscal tear *(arrow)*.

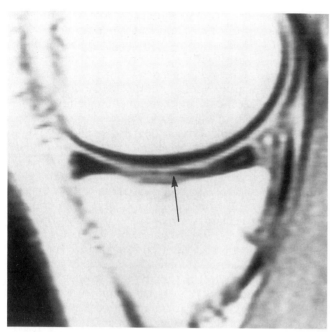

Figure 33.8. Flap tear. Sagittal image (TR/TE 800/20). No abnormal signal is present. A meniscal tear is identified on the basis of the abnormal elongated morphology of the posterior horn, representing a meniscal flap (arrow).

free edge of the lateral meniscus may only be associated with visual fraying of the free edge at arthroscopy and evidence of fibrillation of the margin on histology (8, 10, 19). It may be extremely difficult to differentiate between simple fraying of the free edge and a small parrot break tear by MR.

The *oblique tear* is closely related to the radial tear and represents a full-thickness tear through the body of the meniscus without any evidence of a horizontal cleavage plane (21). The tear is produced by sudden straightening of the meniscus with a consequent strain placed on the thin, concave, unattached inner edge. This strain precipitates the oblique tear, which commonly involves the medial meniscus, in distinction to the radial tear, which involves the body of the lateral meniscus. *Flap tears* are the same as oblique tears, but these occur in a degenerated meniscus that already contains the horizontal cleavage split. The flaps may involve either the superior or inferior half of the meniscus and are referred to as superior flap or inferior flap tears, respectively (Fig. 33.8) (21). Meniscal tears are considered complex when there are several tears, each in different places, such as a flap tear, a horizontal tear, and possibly a coexistent radial or longitudinal tear. These usually occur in the degenerated meniscus that already has a preexisting horizontal cleavage split.

Postoperative Menisci

Patients having undergone prior meniscal surgery represent a heterogeneous group. A basic tenet of present arthroscopic surgery is restraint in meniscal resection (22, 23). The challenge for the orthopedic surgeon is in balancing the need for debridement with the need for preserva-

tion of tissue. In some patients, particularly with degenerative meniscal tears, the surgeon will excise any excessive flaps but may elect to preserve some shaggy meniscal tissue. Furthermore, complete resection of a horizontal cleavage tear is not often considered necessary. This underscores the difficulty in MR morphological analysis of the postoperative meniscus for stability. The postoperative meniscus will demonstrate an altered morphology directly dependent on the degree of surgical intervention. We have found it useful to divide postoperative patients into three groups (24). The first group is composed of patients in whom relatively minimal meniscal resection has been previously performed, with the meniscus appearing "near normal" in overall morphology. In these patients application of the meniscal grading system predicts the presence of a meniscal retear with a degree of accuracy comparable to that established in the virgin meniscus. Minimal blunting of the meniscal tip may be allowable in these patients. The second group is composed of patients in whom a more extensive meniscal resection has been performed but in whom a qualitatively substantial remnant remains present. Evaluation of these menisci is more problematic and the presence of grade 3 signal is a far less accurate predictor of meniscal retear than in the virgin meniscus. The third group of menisci are those in whom extensive prior meniscal resection has been performed and in whom only a small peripheral rim remains present. Diffuse increased signal often extending to and involving an articular surface is not uncommonly identified in these small remnants and has not proven to be an accurate discriminator of meniscal stability. Extensive degeneration of the articular cartilage is a universal finding in these patients. It is arguable how significant the determination of extensive degeneration or retear of these remnants is in patient management. Attention should be directed toward assessment of the articular cartilage and in detection of potential loose osteocartilagenous bodies in these patients. T2-weighed images or gradient-refocused sequences may be of particular value in this regard.

Recently there has been increasing interest in the conservative management or suture repair of tears involving the peripheral or outer third of the meniscus (25–27). The basis for meniscal healing relates to the unique vascular supply to the peripheral third of the meniscus. A perimeniscal capillary plexus derived from the genicular arteries provides an arborizing network of vessels that supplies the peripheral border of the meniscus throughout its attachment to the capsule. Experimental studies in animals have demonstrated healing of complete radial meniscal tears with fibrovascular scar tissue by 10 weeks (28). This scar tissue undergoes progressive transformation into meniscus-like tissue with the histologic appearance of fibrocartilage. This fibrocartilage, however, is histologically distinguishable from normal meniscus. In a recent study, we performed longitudinal follow-up MR scans on patients with documented peripheral meniscal tears that underwent either conservative management or arthroscopically guided suture repair (29). All patients met accepted orthopedic criteria for healed tears and four menisci were evaluated by second-look arthroscopy and noted to be healed. Persistence of grade 3 signal was noted in an overwhelming majority of patients at the site of the previous tear. The significance of this signal with regard to the structural quality of the repair tissue and the degree of potential risk for retear is not known. Certainly, at a minimum, the persistence of grade 3 signal should be recognized by those interpreting MR examinations and should not provide the sole basis for the diagnosis of meniscal retear. Additionally, the persistence of grade 3 signal reflecting prior but healed meniscal tears may provide an important explanation for some cases of MR/arthroscopy disagreement.

Meniscal Cysts

Meniscal cysts are fluid collections that accumulate in the parameniscal region of either meniscus, although they are four times more common laterally than medially. In a detailed study of the histopathology of meniscal cysts, Ferrer-Roca found that cysts were invariably associated with horizontal meniscal tears, which extended into the parameniscal soft tissues (30). On the lateral side of the knee, the cysts tend to remain confined to the joint line. They insinuate themselves between the fibular collateral ligament (FCL) and iliotibial band or occasionally dissect into the Hoffa fat pad. Medially, cysts frequently dissect along soft-tissue planes. If they occur posterior to the medial collateral ligament (MCL), they expand in an unrestricted manner and may ultimately present as a mass at some distance from the joint (9, 31). Rarely, a meniscal cyst may erode bone. Recurrence of a cyst is common after surgery if the relationship to an underlying meniscal tear goes unrecognized. A ganglion represents a fluid-filled mass invariably communicating with a nearby joint but without connection to the underlying meniscus. Meniscal resection is not necessary in treatment of ganglia although the capsular connection must be recognized and resected to avoid recurrence.

Meniscal cysts demonstrate increased signal intensity on long TR/TE sequences (Fig. 33.9). Differential considerations include popliteal cysts, angiomas, and distended bursae (e.g., anserine, semimebranosus-gastrocnemius, infrapatella), ganglia, and venous varices.

Interpretive Pitfalls

A number of areas of interpretive difficulty and diagnostic confusion have become recognized with regard to meniscal evaluation (8, 31–33). As previously discussed, assessment of the meniscal free edge has proven to be one of the most challenging areas of meniscal diagnosis. Fibrillation or fraying of the free edge may be manifest as increased signal confined to the meniscal tip and must be distinguished from a small radial or parrot beak tear. Abnormal signal along the free edge may be manifest on only a single 5-mm sagittal section and must be actively sought. Coronal images are essential for proper assessment of these lesions. Any accompanying morphological alteration (e.g., truncation) provides information invaluable to making the proper distinction between these two entities. Another area of potential diagnostic difficulty is in detection of tears involving the extreme periphery of the meniscus and men-

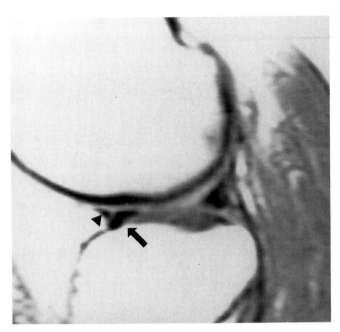

Figure 33.9. Pseudotear. Sagittal image (TR/TE 800/20). A tear of the anterior horn of the lateral meniscus is mimicked by the juxtaposition of the transverse meniscal ligament *(arrowhead)* and anterior horn *(arrow)*. Tracing the transverse ligament on contiguous sections allows differentiation from a meniscal tear.

iscosynovial separations. T2-weighted images may be of particular value in this setting in depiction of fluid interposed between the meniscus and joint capsule. Radial images have also been advocated for depiction of extreme peripheral tears. The transverse meniscal ligament can simulate an oblique tear adjacent to the anterior horn of the lateral meniscus. This ligament connects the anterior horns of both menisci, and in up to 30% of MR examinations a tear is simulated by fat interposed between the low signal intensity ligament and adjacent meniscus (Fig. 33.9). The popliteus tendon sheath can mimic a meniscal tear within the posterior horn of the lateral meniscus. The sheath courses in a characteristic oblique anterosuperior-to-posteroinferior direction, allowing confident differentiation from a true tear (Fig. 33.10). In approximately 30% of MR examinations, a horizontal high signal intensity line is seen apparently within the meniscus on an extreme sagittal section. This line does not represent a meniscal tear but rather is believed to be related to volume averaging of the high signal intensity tissue at the concave meniscal edge.

Accuracy

Multiple studies of varying size have compared the accuracy of MR to arthroscopy (1, 2, 4, 18, 34–37). The largest study performed by Mink and associates compared the preoperative MR reports of 459 virgin menisci to the findings as observed at arthroscopy (7). The overall agreement of MR and arthroscopy in meniscal evaluation was 93%. Thirty-seven different orthopedic surgeons participated in the review, but in a subgroup in which the surgical procedures were performed by a single knee subspecialist arthroscopist, the MR-arthroscopy agreement rate increased

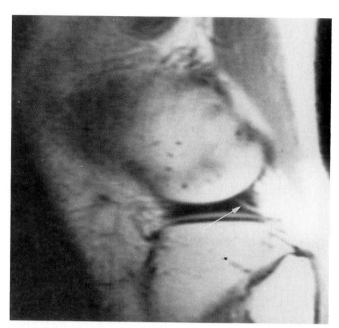

Figure 33.10. Pseudotear of the posterior horn of the lateral meniscus. Sagittal image (TR/TE 800/20). A tear of the posterior horn of the lateral meniscus is mimicked *(arrow)* by the interface between the popliteal tendon and posterior horn of the lateral meniscus. Compare with Figure 33.1.

to 95%. Several other large studies have confirmed this extremely high agreement rate of MR and arthroscopy. Jackson and colleagues examined 87 knees that had both MR and arthroscopy; they reported an MR accuracy of 93% for the medial and 97% for the lateral meniscus (36). Mandelbaum et al. (37) reported an MR arthroscopy agreement of 90%, and Polly et al. (35) reported a value of 94%.

If MR is to be utilized as a screening technique, one of the most important statistics is the negative predictive value (NPV) of the test. The NPV is defined as the percentage of patients with a negative MR who do not have a tear arthroscopically. The four largest published series of MR of the knee have reviewed a total of 1072 menisci that have been examined by both MR and arthroscopy (6). The overall NPV was 94% or, in other words, only 6% of patients with a negative MR report had a tear. Of the false-negative examinations, 5 of the 11 missed tears reported by Mink and two-thirds of the missed tears reported by Jackson et al. (36) were only minimal tears that did not require partial meniscectomy. Therefore, the likelihood of MR to fail to detect a clinically significant meniscal tear should be quite low.

False-positive MR examinations have been a greater source of error than have false-negative studies. A number of factors likely contribute to this situation. Interpretive experience on behalf of the radiologist and widely varying image quality related to magnetic field strength, coil, and protocol design undoubtedly contribute to diagnostic error. In a similar manner, the operator dependence of arthroscopy, the difficulty for arthroscopists to optimally examine the posterior horn of the medial meniscus (the most common site of false-positive MR examinations), and the frequent inability to detect degenerative

cleavage tears without extensive probing, all likely contribute to possible false-negative arthroscopies. Distinguishing areas of free edge fibrillation from meniscal tears, and persistent signal from "healed" tears, remain areas of interpretive difficulty.

Cruciate Ligaments

Anterior Cruciate Ligament

The cruciate ligaments are intraarticular yet extrasynovial structures that are critical to the maintenance of joint stability (38, 39). The ACL originates from the nonarticular area anterior and lateral to the tibial spine 23 mm from the anterior tibial cortex. At this site it is in immediate continuity to the central attachments of the anterior horn of the lateral meniscus. The tibial attachment of the ACL is wider and stronger than the femoral attachment. The ACL passes beneath the transverse meniscal ligament and courses obliquely upward and backward along the lateral aspect of the intercondylar notch to insert with a broad attachment over the posterior medial surface of the lateral femoral condyle. The width of the ACL averages 11.1 mm and its length averages 31 to 38 mm. The ACL consists of two major fiber bundles: *(a)* a wider and stronger anteromedial bundle (AMB) and *(b)* a smaller, posterolateral band (PLB). Each band is composed of a collection of individual fascicles. Although a majority of the fibers of the ACL are longitudinally oriented, some spiral from one attachment site to the other in a slight outward (lateral) direction (21). The spiral orientation is significant in allowing winding and unwinding of the ligament along its long axis, thus providing tension on some portion of the ligament throughout the range of motion. The ACL is taut in full extension and relaxed at about 45° of flexion.

The normal ACL most commonly appears on MR as a single dark continuous band of varying thickness, or as two or three separate fiber bundles extending from the medial aspect of the lateral femoral condyle to the anterior tibial plateau. The identification of "individual bundles" is most often recognized on coronal scans near the femoral attachment and on sagittal scans near the tibial attachment. Visualization of the ACL on MR is optimized when the leg is positioned in mild (10 to 15°) external rotation. This position is both natural and comfortable for the patient and allows visualization of the ligament in over 90% of individuals. In cases in which visualization of the ligament in the sagittal plane is suboptimal a number of alternatives exist. Assessment of the ligament in the coronal plane has proven valuable in our experience, particularly for assessment of the proximal femoral attachment. Oblique sagittal imaging of the ACL utilizing an axial localizer and oblique coronal sections utilizing a sagittal localizer have been advocated by some investigators (Fig. 33.11). Partial volume effects may be minimized utilizing 3-mm or thinner sections (three-dimensional (3D) acquisitions) in selected cases. Utilizing spin-echo techniques, T2-weighted images have been demonstrated to provide higher sensitivity for the detection of ACL injuries and are strongly recommended (1).

The ACL is the most frequently torn knee ligament (21, 40–42). In patients sustaining acute knee trauma sufficient

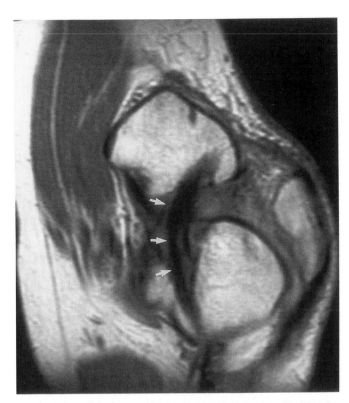

Figure 33.11. ACL reconstruction. Oblique sagittal image (TR/TE 2000/20). This section, angled along the true long axis of the reconstructed ligament, optimizes assessment of the ligament *(arrows)*.

to produce hemarthrosis, an accompanying ACL tear is reported in more than 70% of cases. The majority of ACL tears are initially interstitial. The rupture is more often (90%) within the main substance of the ligament and toward the proximal (femoral) side. Less commonly the ACL is avulsed from bone. Tears with bony or chondral fractures classically demonstrate tense joint effusions with gross fat in the aspirate as compared with interstitial tears. These findings may be demonstrable with MR. Tears with attached bone usually occur in younger individuals. The MR appearance of an ACL tear depends on the age of the injury and the degree of disruption. In an acute tear, the ACL is either clearly discontinuous or demonstrates a serpiginous or grossly concave anterior margin. Acute partial or complete tears may demonstrate a mass of intermediate signal on T1-weighted images, most often at the proximal end of the tendon, with or without an associated identifiable discontinuity of the tendon itself (2, 3, 8, 10). The mass consists of fluid, hemorrhage, and acute synovitis, which often increases in signal on T2-weighted images (Fig. 33.12). The appearance of a torn ACL may be mimicked on T1-weighted images by partial volume effect with the lateral femoral condyle producing a pseudomass (Fig. 33.13) T2-weighted images are of value in allowing the distinction of pseudomass from true tear to be accomplished. With the partial volume pseudomass there is no signal change on T2-weighted images in distinction to the true ACL tear, in which signal may increase secondary to fluid and hemorrhage within and around the torn ligament. Differentia-

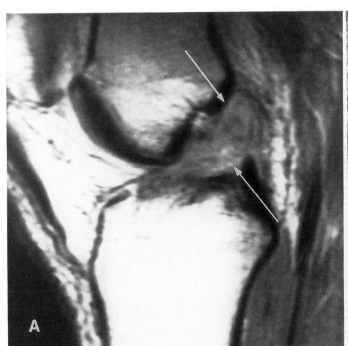

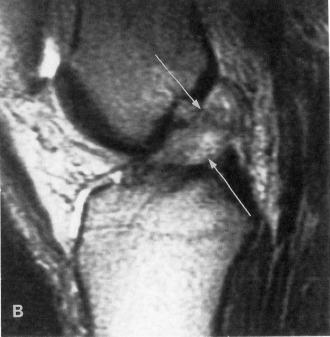

Figure 33.12. Complete ACL tear. **(A)** Sagittal image (TR/TE 2000/20). An amorphous "mass" is seen within the intercondylar notch *(arrows)*, totally replacing the anterior cruciate ligament that should be seen on this section. **(B)** TR/TE 2000/80. On the more heavily T2-weighted image, the mass increases in signal intensity *(arrows)*. No recognizable ACL can be seen. This is the typical appearance of an acute ACL tear.

tion between a complete and partial disruption may be more difficult but generally can be accomplished (Fig. 33.14). Minimal degrees of attenuation, fraying, and chronic partial tears are unreliably detected by MR and, indeed, usually require extensive probing at the time of arthroscopy for detection.

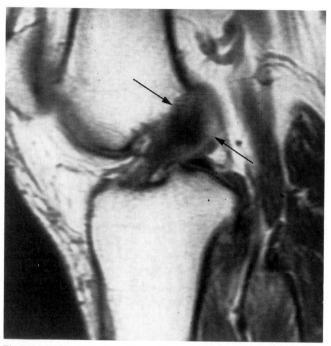

Figure 33.13. ACL pseudotear. Sagittal image (TR/TE 2000/20). On this image a mass is seen within the proximal intercondylar notch, mimicking the appearance of an ACL tear *(arrows)*. Compare with Fig. 33.12*A*. This results from partial volume effect. On a T2-weighted image, this would not increase in signal intensity.

Rapid synovial overgrowth is characteristic of ACL injuries and likely relates to the blood supply of the ligament. Within 2 weeks of an ACL tear, arthroscopic inspection reveals a contracted ligament that is covered by a synovial cap. The natural history of an ACL tear is usually attachment to the PCL as the torn ACL falls on the PCL and gains a secondary source of attachment and blood supply (41). A chronically torn ACL attached to the PCL may be recognized on MR by its low-lying position, its lack of proximal attachment, and abnormal course through the knee (Fig. 33.15). In cases in which there has been an extensive and violent disruption of the ligament, there may be no ACL tissue arthroscopically visible. On MR, a chronically torn ACL often appears completely absent; no structure is identified in the lateral intercondylar notch. Occasionally, a low signal intensity structure, representing the remnant of the ACL, is seen to lie on the tibial plateau (Fig. 33.16).

Treatment of acute ACL injuries is controversial and often centers on the presence of an associated meniscal or collateral ligament injury. Therapy is individualized with regard to the patient's age, level of activity, and extent of associated knee injury. Primary repair of the ACL is most successful in cases of avulsion. In the more common intrasubstance tear, the ligament is usually insufficient to sustain a primary repair. Anterior cruciate ligament reconstruction has been performed utilizing a number of biologic (semitendinosus, patella tendon) and synthetic (Dacron, Gore-Tex) materials.

In patients who have undergone ACL reconstruction, the course of the reconstructed ligaments is best appreciated on sagittal and oblique sagittal MR sequences (8, 43). The normal graft is of low signal intensity on most imaging sequences, as is the normal ACL, and runs in a similar oblique plane. When a tibial tunnel has been created, the

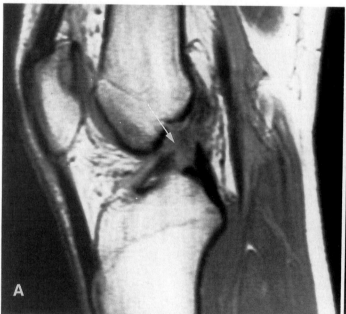

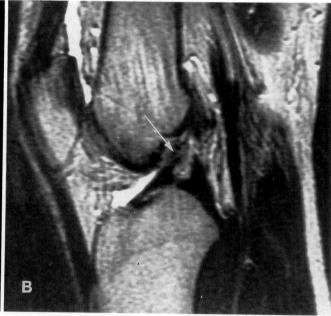

Figure 33.14. Partial ACL tear. **(A)** Sagittal image (TR/TE 2000/20). On this section a smaller mass is seen, partially obscuring the ACL *(arrow)*. The ligament both proximally and distally can be identified. **(B)** The mass, representing synovitis and hemorrhage, increases slightly in signal intensity. This is a typical appearance for a partial ACL tear.

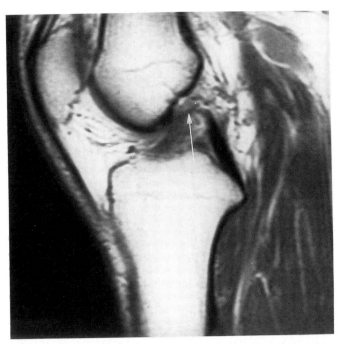

Figure 33.15. Torn ACL attached to the PCL. Sagittal image (TR/TE 2000/20). The distal remnant of an ACL has taken attachment to the midportion of the PCL *(arrow)*. This is a typical appearance following a complete ACL tear.

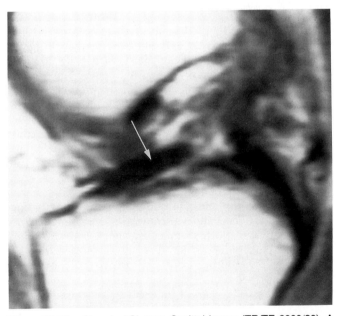

Figure 33.16. Chronic ACL tear. Sagittal image (TR/TE 2000/20). A small remnant of the distal ACL is seen lying freely on the intercondylar notch *(arrow)*. The remainder of the ligament is absent with no identifiable structure in the intercondylar notch.

tunnel itself with its low signal intensity ligament can be followed to the anterior aspect of the tibia. Like the normal ACL, Gore-Tex and other synthetic graft materials have a solid low signal intensity appearance across the intercondylar notch. The attachment site on the lateral femoral condyle can be evaluated for isometric graft placement. Focal collections of fluid associated with a reconstructed

ligament, and/or discontinuity of the graft material, must be regarded as overwhelmingly suggestive of a partial or incomplete graft tear. Chronic synovial effusions have recently been reported in patients with Gore-Tex grafts and may relate to an irritant effect of particles of Gore-Tex on the synovium.

Posterior Cruciate Ligament

The PCL is longer, wider, and stronger than the ACL and is a fundamental stabilizer of the knee in flexion, extension, and internal rotation (21). It originates from a fan-shaped attachment on the posterior aspect of the lateral surface of the medial femoral condyle and courses posteriorly and distally across the joint to attach to the extreme posterior intercondylar (nonarticular) region of the tibia. The PCL is narrowest in its midportion. The fascicles comprising the PCL are divided into an anterior portion, from which the majority of the ligament is composed, and a smaller posterior segment, which courses obliquely across the joint. The fibers of the PCL twist on their longitudinal axis in a clockwise direction during flexion and unwind in extension. The dominant anterolateral bundle tightens on flexion and relaxes in extension.

The normal PCL is generally the structure within the knee most easily depicted utilizing MR. With the knee in extension, the normal PCL demonstrates a smooth, mildly convex posterior curve and is visualized as a uniform low signal intensity band. The meniscofemoral ligament is intimately associated with the PCL. It extends from the posterior horn of the lateral meniscus to the lateral aspect of the medial femoral condyle. The ligament may divide to pass anteriorly and/or posteriorly to the PCL. The anterior division (ligament of Humphrey), when present, can vary considerably in size and may be up to one-third the size of the PCL. The posterior meniscofemoral ligament (ligament of Wrisberg) lies behind the PCL and also varies considerably in size. One or the other ligament is seen in up to 70% of knees examined (39).

The PCL is most typically injured following a posteriorly directed force against the flexed knee (motor vehicle accident), resulting in posterior displacement of the tibia in relation to the femur. Hyperextension injuries are a second major mechanism for PCL tears. Disruption of the PCL can be "isolated" but often is associated with other serious knee injuries, including posterior lateral capsular disruptions and tears of the arcuate ligament complex. Acute and chronic tears of the PCL most commonly appear as foci of increased signal intensity, replacing the normally homogeneously dark ligament on T1-weighted images. The ligament is often focally widened in caliber, but the tear is characteristically less masslike than those involving the ACL. With complete tears, a gap can be identified separating the two ends of the ligament (Fig. 33.17). The ligament may demonstrate a "buckled" or S-shaped appearance. In subacute tears, focal areas of signal relating to areas of hemorrhage may be identified. Chronic tears or fibrous scar may manifest little abnormal signal and may be demonstrable only by subtle contour changes or by demonstration of abnormal tibial shift. Tears of the PCL are most common in its midsubstance. Bony avulsions oc-

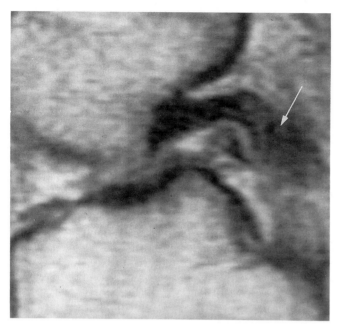

Figure 33.17. PCL tear. Sagittal image (TR/TE 2000/20). A complete intrasubstance tear of the mid-PCL is demonstrated *(arrow)*. These tears are typically less masslike than acute ACL tears.

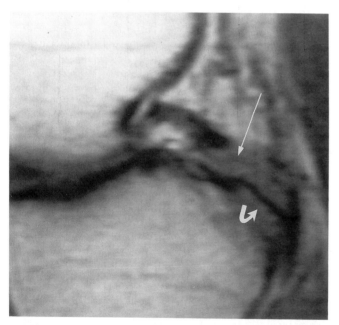

Figure 33.18. Distal PCL avulsion. Sagittal image (TR/TE 2000/20). A distal avulsion of the PCL is present *(arrow)*. This is often accompanied by decreased signal intensity within the subchondral bone of the tibia at the level of PCL insertion *(curved arrow)*.

cur most often at the tibial attachment. A potential problem in diagnosis occurs when a torn or avulsed PCL remains closely apposed to the tibia (6, 44). A clue to correct diagnosis can be demonstrated in the immediate subchondral bone, which will often demonstrate diminished signal intensity secondary to edema (Fig. 33.18).

Collateral Ligaments

Medial Collateral Ligament

The MCL is classically divided into two components: superficial and deep. The superficial component, the tibial collateral ligament (TCL), arises from the medial femoral epicondyle and inserts onto the medial aspect of the tibia approximately 5 cm below the joint line. The TCL is separated from the medial capsular ligament and attached meniscus by a bursa that allows movement between the two. The ligament itself is best depicted on coronal spin-echo images and is seen as a thin, low signal band on both T1- and T2-weighted images. The ligament blends imperceptibly with the low-signal tibial and femoral cortices. The capsular ligament attaches firmly onto the middle third of the meniscus, securing the meniscus to the femur by the meniscofemoral ligament, and to the tibia by the meniscotibial (coronary) ligament. The deep layer of the MCL can occasionally be demonstrated on high-resolution coronal images, as can the intervening bursa. Recent studies suggest that the TCL is the principal medial stabilizer of the knee (45).

As a consequence of the normal valgus position of the knee, the medial ligamentous structures are more vulnerable to injury than those on the lateral side. Most major ligamentous injuries are caused by contact and severe valgus stress often associated with external rotation ("clipping injury"). Ligamentous injuries are commonly subclassified into three clinical grades: disruption of only a few fibers (grade 1), disruption of up to 50% of the ligamentous fibers (grade 2), and complete disruption (grade 3). Tears of the medial ligamentous structures follow a predictable progression of injury. With grade 1 tears, only a limited number of deep capsular fibers are torn, and the ligament appears normal in thickness and contour on MR. Extracapsular soft-tissue edema and hemorrhage are well demonstrated and manifest on MR by decreased signal on T1-weighted images and increased signal on T2-weighted images (8, 10) (Fig. 33.19). In grade 3 tears, both the deep capsular ligament and superficial ligament (TCL) are completely disrupted. Complete TCL rupture is associated with marked thickening, discontinuity, and serpiginous contours of the affected ligament on MR. Increased distance is seen between the high-intensity signal from the medullary and subcutaneous fat. The site of the tear can frequently be localized precisely, particularly on T2-weighted images which we routinely obtain in all cases of acute knee trauma for best depiction of collateral ligament injuries (Fig. 33.20). Grade 2 injuries have features of both grade 1 and grade 3 tears and are less precisely characterized by MR (44). Tears of the MCL may be "isolated" or more commonly associated with a number of related injuries. Among the most commonly encountered associated injuries with high-grade MCL tears are bone bruises (medullary trabecular microfractures) (44, 46). When seen in association with MCL tears, these lesions may involve the tibia, femur, or both sides of the joint. Tears of the ACL are also commonly associated with both medial collateral and lateral compartment bony injuries. Tears of the medial meniscus, in particular longitudinal tears of the posterior horn, and peripheral tears including meniscal capsular separations are also often identified.

Lateral Collateral Ligament

The ligamentous anatomy of the lateral aspect of the knee is complex and often divided into anterior, middle, and

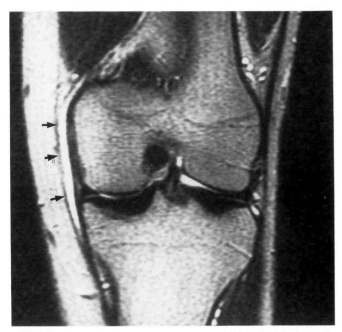

Figure 33.19.   Grade 1 MCL tear. Coronal image (TR/TE 2000/80). A band of increased signal intensity representing fluid overlays an intact tibial collateral ligament *(arrows).*

posterior subdivisions (47). The lateral capsular ligament (LCL) is the deepest layer of the lateral supporting structures and extends from the patella to the PCL. The anterior subdivision consists of that section of the LCL that extends from the patella to the iliotibial band (ITB). The ITB is an extracapsular structure that is superficial to the LCL and inserts proximally on the epicondylar tubercle of the femur and distally on the Gerdy tubercle of the tibia. The middle section of the lateral capsuloligamentous complex consists of the ITB and the LCL deep to it and extends posteriorly to the fibular collateral ligament (FCL). The middle section of the LCL is strong and has attachments both to the femur and tibia.

The posterior third consists of the LCL and extracapsular ligaments, which form a single functional unit called the arcuate complex. The components of the complex are the FCL, arcuate ligament, and tendoaponeurotic unit formed by the popliteus muscle. The FCL is an extracapsular structure that originates from the lateral epicondyle of the femur directly anterior to the insertion of the gastrocnemius and passes directly beneath the lateral knee retinaculum to form a cordlike structure that joins with the biceps femoris tendon to insert into the head of the fibula as a conjoined tendon. The popliteus tendon passes under the FCL to perforate the posterior horn of the lateral meniscus and join its muscle belly on the posterior tibia. The arcuate ligament forms a triangular sheet that diverges upward from the fibular styloid.

The normal ITB, representing the thickened portion of the fascia lata, is identified on coronal MR images as a delicate, longitudinal low signal intensity line coursing longitudinally in the high signal intensity subcutaneous fat

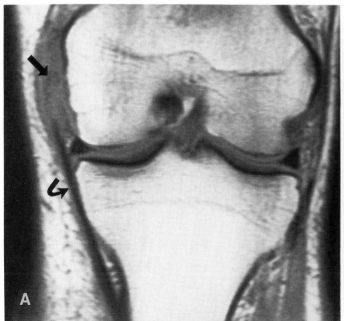

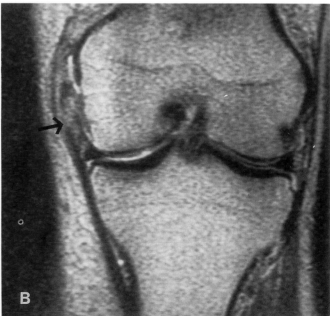

Figure 33.20.   Grade 3 MCL tear. **(A)** Coronal image (TR/TE 2000/20). Amorphous intermediate signal intensity entirely replaces the proximal attachment of the tibial collateral ligament *(arrow).* The distal tibial collateral ligament is intact *(curved arrow).* **(B)** Coronal image (TR/TE 2000/80). Signal intensity is mildly increased within the region of the proximal MCL, representing a combination of edema and hemorrhage *(arrow).* The underlying ligament cannot be defined and the appearance is typical of a high-grade MCL injury.

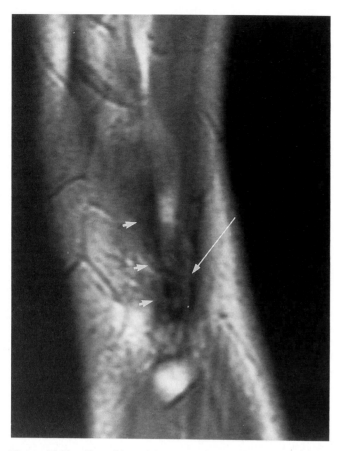

Figure 33.21. Normal lateral ligaments. Sagittal image (SE 800/20). On this extreme lateral section, the FCL *(arrows)* is seen coursing posteroinferiorly to join the biceps femoris tendon *(long arrow)* just above their common insertion onto the fibula.

of the lateral thigh. The FCL and its common insertion with the biceps tendon on the head of the fibula are depicted on coronal images and occasionally on extreme lateral sagittal views (Fig. 33.21). As in other articulations, the normal ligaments are of low signal intensity on all pulse sequences.

Injuries of the lateral compartment are less common than those on the medial side of the joint. Disruption of these ligamentous structures is manifest by their complete absence or interruption of their normal contour. A wavy appearance or localized fluid within or around them is characteristic of lateral compartmental injuries. Medial compartmental bony injuries may be seen secondary to varus stress that disrupts the lateral compartment. Capsular disruptions may be recognized by fluid extravasation into the surrounding soft tissues and is often best depicted tracking from the joint superficial to the popliteus muscle and tendon.

Bone and Cartilage

Bone Bruise

Magnetic resonance imaging is capable of detecting a progressive spectrum of trauma related abnormalities to bone and cartilage (44–46, 48–50). One of the most commonly encountered abnormalities is that of the bone bruise, which was first described in association with major injuries to the contralateral collateral ligament. In the initial description, these lesions, thought to be areas of edema and hemorrhage, were found in five of eight patients who had significant MCL injury. In a more detailed analysis of acute bone fractures, bone bruises were found in five of five severe MCL injuries. Additionally, lateral bone bruises were

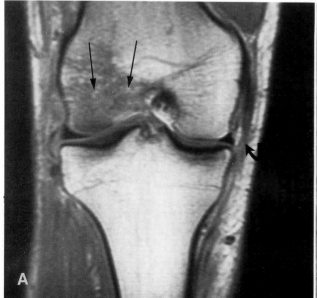
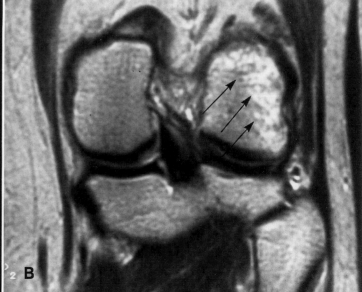

Figure 33.22. Bone bruise. **(A)** Coronal (TR/TE 800/20). Amorphous decreased signal intensity is visible within the subchondral bone of the lateral femoral condyle *(arrows)*. An MCL injury is identified *(curved arrow)*. **(B)** Coronal image (TR/TE 2000/80). A different patient, who also sustained a valgus stress injury. Amorphous increased signal intensity within the lateral femoral condyle reflects an underlying bruise *(arrows)*.

identified in 18 of 25 acute ACL tears without collateral ligament injuries (46).

The MR imaging appearance of bone bruises is thought to represent the suspected basic pathology, that is, trabecular disruption with edema and hemorrhage. On MR imaging, bruises appear as poorly defined areas of decreased signal on T1-weighted images that irregularly increase in signal on T2-weighted sequences (Fig. 33.22). The lesion is primarily epiphyseal, although it frequently extends into the metaphysis. The lesion is by definition confined to the medullary bone. The cortical bone and articular cartilage remain intact.

Bruises are generally considered to be self-limiting and essentially benign abnormalities. Multiple cases have been followed to rapid MR imaging resolution, and no reported cases of progression have been documented (46, 50). There is no arthroscopic correlate to the lesion as would be expected with a purely subchondral injury. Transient symptoms have been associated with these lesions.

Osteochondral Injuries

Multiple mechanisms may contribute to injury of the chondral surface. Within the knee, dislocation-relocation of the patella is the most common mechanism, resulting in injury to the patella or the anterior, nonweight-bearing surface of the lateral femoral condyle (51–54). Rotary motions in which the tibial spine impacts the femur may cause injury to the medial femoral condyle; this mechanism is thought by some to be responsible for the production of osteochondritis dissecans.

Mink and Deutsch classified osteochondral fractures occurring in the knee into two subgroups based on their MR imaging appearance: (a) displaced and (b) impacted (46). The displaced type is seen most commonly involving the inferior pole of the patella and the anterior lateral femoral condyle. The lesions may be purely chondral or involve a small segment of subchondral bone/calcified cartilage (Fig. 33.23). In adults, the fracture occurs at the "tide mark," the junction between calcified and uncalcified cartilage. Purely chondral fractures are virtually impossible to diagnose by conventional radiographs. Chondral fractures may be easy to overlook on MR imaging unless one maintains a high index of suspicion for such a lesion. Even when the lesions are large, there may be little abnormality of subchondral bone. Loose fragments may be displaced into the joint and can be detected with MR imaging. The impaction type of osteochondral fracture may be etiologically related to a bone bruise (46). Whether one or the other is produced depends on the magnitude of the causative compressive force. Within the knee, the most common impaction type of osteochondral injury observed was a localized abnormality directly over the anterior horn of the lateral meniscus. The subchondral injury predominates over any abnormality at the articular surface, and the lesions may be arthroscopically occult. On MR imaging, they are seen as poorly defined zones of decreased signal intensity in the medullary bone of the femur directly overlying the anterior horn of the lateral meniscus (Fig. 33.24). This lesion is commonly associated with an acute ACL tear. In a study of 25 acute complete ACL tears, this particular im-

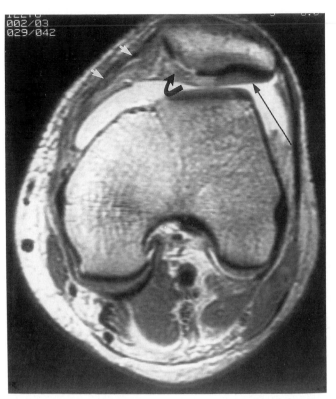

Figure 33.23. Displaced osteochondral fracture. Axial image (TR/TE 2000/80). The normal articular cartilage of the lateral facet of the patella is seen well outlined by high signal intensity joint fluid in the supra patellar bursa *(long arrow).* A gap in the cartilaginous surface is seen along the medial facet at the site of a displaced osteochondral fracture *(curved arrow).* The medial retinaculum has been disrupted *(small arrows)* in this patient with a recent patella dislocation.

paction fracture was found in five cases (46). Eighteen patients in the same group suffered lateral compartment bone bruises. Thus, 23 of 25 knees with ACL injuries sustained lateral compartment bony and cartilaginous injuries.

The term osteochondritis dissecans (OCD) refers to fragmentation and often complete separation of the articular surface of a joint. Most investigators accept that trauma is the etiology of OCD, although a history of a single acute event is not usually available. The fracture in OCD occurs parallel to the joint surface and may involve cartilage alone, or cartilage plus a variable amount of cortical or subchondral bone, or both. The fate of the fracture fragment is variable (9, 55). It may remain in situ and heal in place, undergo resorption, or become detached. Radiographs may be normal with small lesions and with purely chondral lesions. Since young cartilage is quite elastic, the deforming force may be transmitted to the underlying cortical and medullary bone; the cartilage may "rebound" into a normal position. In such cases, arthrography and arthrotomography may be normal, and the subchondral abnormality may be the only manifestation of the injury. Magnetic resonance imaging may be utilized to advantage for detection of radiographically "occult" lesions, assessment of the integrity of the overlying cartilage, detection of loose bodies, and precise determination of the extent and loca-

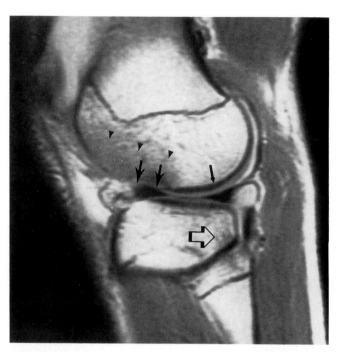

Figure 33.24. Impaction-type osteochondral fracture. Sagittal image (TR/TE 2000/20). A focus of poorly defined low signal intensity is visible within the lateral femoral condyle *(small arrowheads)*. A slight concavity immediately overlays the anterior horn of the lateral meniscus *(arrows)*. The articular cartilage is mildy deformed. An associated bruise is seen within the subchondral bone of the posterior lateral tibial metaphysis *(open arrow)*.

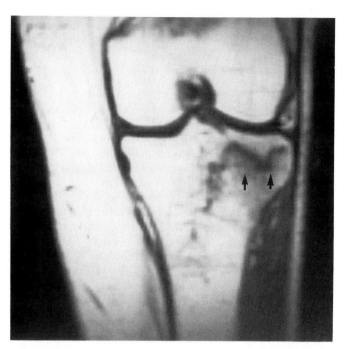

Figure 33.25. Stress fracture. Coronal image (TR/TE 800/20). A serpiginous line of decreased signal intensity is seen coursing perpendicular to the long axis of the tibial metaphysis *(arrows)*. Adjacent low signal intensity represents bone marrow edema.

tion of the lesion. All these factors may be critical to the choice and timing of surgical intervention.

Stress Fracture

Two MR imaging patterns of stress fractures have been described: *(a)* linear and *(b)* amorphous (45, 56–59). The former is characterized by a linear zone of decreased signal on T1-weighted images that is surrounded by a more broad and poorly defined area, which demonstrates less intensity. The linear component remains dark on T2-weighted images, but the surrounding zone (of presumed edema) becomes more intense. The linear areas are short and straight, or long and serpiginous, and course perpendicular to the adjacent cortex (Fig. 33.25). The amorphous pattern is characterized by a geographic area of decreased signal on T1-weighted images in which there are frequently more focal globular foci. Some portions of the lesions increase in signal intensity on T2-weighted images. There is no linear component. Amorphous stress fractures are differentiated from bone bruises by clinical history, because the appearance on MR imaging can be similar.

Plateau Fractures

Magnetic resonance imaging has also been effectively utilized in the assessment of tibial plateau fractures, both for detection of occult lesions as well as for further characterization of known fractures. Tibial plateau fractures result from excessive axial loading of the knee combined with a

valgus strain, during which time the lateral femoral condyle is driven into the plateau (60). Degrees of comminution, depression, and displacement are the factors that determine operative vs. nonoperative therapy. In general, types I, II, and III (nondisplaced, local compression, split compression), characterized by depression of less than 8 mm and without significant displacement, can be treated nonoperatively. Conventional radiography underestimates these critical features; CT and planar tomography have been used adjunctively. The multiplanar capabilities of MR may facilitate three-dimensional perception and evaluation of such fractures and permit assessment of the collateral and cruciate ligaments and the menisci (Fig. 33.26). In addition to assessing known fractures, MR may have a role in the detection of such fractures that are radiographically occult. In one study 13 tibial plateau and distal femoral fractures were found on MR images that were normal (11 cases) or only slightly suspicious (2 cases) for fracture on plain radiographs. Magnetic resonance has also been utilized in the detection of radiographically occult fractures of the hip and was found to be more sensitive than conventional radiographs (58).

Extensor Mechanism

The term *extensor mechanism* refers to a functional unit of muscle, tendon, and bone that serves to extend the knee joint (61). All of the components, with the exception of the articular surface of the patella, are extraarticular, and thus not available to assessment utilizing arthroscopy. Extensor mechanism injuries may occur at any one of multiple sites, including *(a)* muscle-tendon junction, *(b)* intratendinous

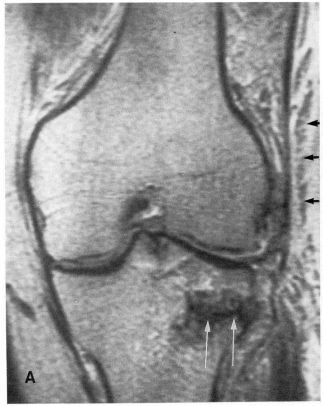
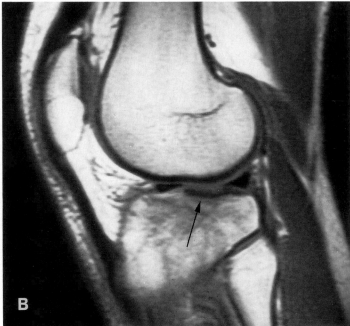

Figure 33.26. Tibial plateau fracture. **(A)** Coronal image (TR/TE 2000/80). A broad-based band of low signal intensity is seen coursing perpendicular to the long axis of the proximal tibia *(long arrows)*. This fracture did not extend to the articular surface and no significant displacement was present. Extensive soft-tissue edema is seen superficial to the ITB *(small arrows)*. **(B)** Sagittal image (TR/TE 2000/20). A central localized depression of the tibial plateau is visible *(arrow)*. The degree of comminution and depression is well depicted by MR.

(quadriceps or patella), *(c)* tendoosseous junction (supra- or infrapatella), and *(d)* tendotubercle insertion. Magnetic resonance imaging is well suited to the depiction of injuries involving all of the constituents of the extensor mechanism.

Extensor Muscle

The quadriceps muscle serves as the principal extensor of the knee, and is composed of the rectus femoris and the three vasti muscles. These four muscles converge to form a common tendon that crosses the knee joint and attaches to the tibial tuberosity via the patella. The quadriceps tendon is composed of three laminae, which can normally be resolved on MR scans. The superficial layer is contributed by the rectus femoris, the middle layer from the tendons of the vastus lateralis and medialis, and the deep layer from the vastus intermedius. Extensor mechanism injuries are seen both in deconditioned (often elderly) individuals and in highly trained athletes. Tears may occur from either indirect or direct trauma. Normal skeletal muscle is characterized by an intermediate to slightly long T1 relaxation time and by a short T2 relaxation time relative to other soft tissues (44, 62–64). Muscle tears and other soft-tissue injuries are most conspicuous on T2-weighted images, which optimize contrast between edema and hemorrhage (processes with prolonged T2 relaxation times) and normal muscle (44, 65). T1-weighted images, while less sensitive to depiction of soft-tissue abnormalities, provide improved anatomic detail and may also be useful in providing specificity in regard to the presence of hemorrhage. Muscle injuries demonstrate a broad spectrum of severity, ranging from simple bruises to complete ruptures and compartment syndromes. Muscle contusions are characterized by intraparenchymal edema or hemorrhage, or both. On MR imaging, this is seen as an infiltrative (feathery) process that is of increased signal intensity on T2-weighted images and typically isointense on T1-weighted images. Increased girth of the affected musculature is commonly observed. Partial muscle tears are manifest most typically as stellate areas of intermediate signal intensity on proton-density images that significantly increase in signal on T2-weighted images (65) (Fig. 33.27). Most tears occur within the muscle belly and constitute first and second degree strains. In complete tears, a true gap in the affected muscle can be demonstrated along with extensive edema and hemorrhage.

Intramuscular hemorrhage and edema (infiltrative pattern) are to be distinguished from hematoma formation. Hematomas are typically well defined and confined to a single muscle or tissue plane (Fig. 33.28). The appearance of hematomas is affected by both the field strength of the system as well as the stage of evolution of the mass (66–69). On partial saturation sequences, almost all early (less

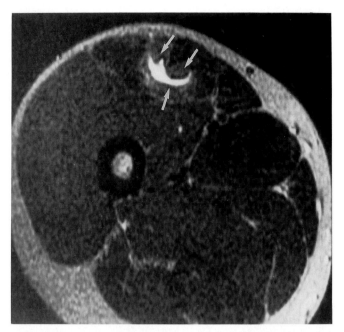

Figure 33.27. Partial muscle tear. Axial image (TR/TE 2000/80). The semilunar focus of increased signal intensity represents a combination of edema and hemorrhage within the rectus femoris muscle *(arrows)*. The appearance is typical of an acute partial tear. No complete gap was seen on sagittal images.

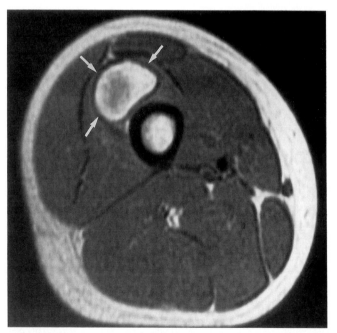

Figure 33.28. Intramuscular hematoma. Axial image (TR/TE 2000/20). A well-circumscribed high signal intensity mass is seen within the vastus intermedius on this spin-density image. A lower signal intensity peripheral rim is also evident *(small arrows)*.

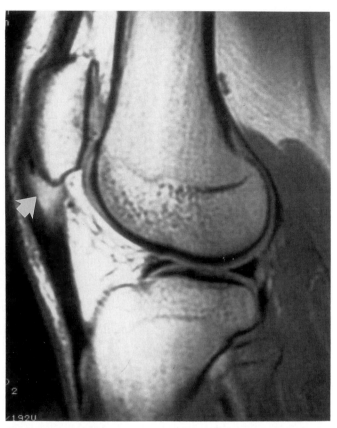

Figure 33.29. Patella tear. Sagittal image (TR/TE 800/20). A globular focus of high signal intensity is seen within a focally widened patellar tendon, reflecting hemorrhage within a partial tear *(arrow)*.

than 30 days) hematomas have higher signal intensity than surrounding muscle at both low and high field strength. As hematomas age, their water content decreases and their protein content increases, leading to decreases in T1 and T2 relaxation times. In older hematomas, a low-intensity margin may be noted at high field strength related to the presence of hemosiderin-laden macrophages.

Quadriceps and Patella Tendons

Tendons are composed of fascicles of very densely arranged nearly parallel collagenous fibers, elongated tendon cells, and only minimal ground substance. As a reflection of this basic composition and low water content, normal tendons demonstrate diminished signal intensity on all MR imaging sequences (9, 44, 70, 71). This affords excellent contrast against surrounding high signal intensity fat, allowing for excellent depiction on MR imaging.

Complete quadriceps rupture is characterized on MR imaging by tendon discontinuity that is readily demonstrable. The tear most commonly involves the tendon approximately 2 cm superior to the upper pole of the patella. There may be associated redundancy of the patella tendon. Hemorrhage and edema are present within and surrounding the tendon. Partial tears are characterized by focal areas of augmented signal intensity, often accompanied by focal tendon thickening. No complete gap is identified, a finding utilized to distinguish partial from complete tears. T2-weighted spin-echo and gradient-refocused pulse sequences are most sensitive for depiction of intratendonous signal abnormality. Patella tendon tears, in distinction to those involving the quadriceps, are more often identified in younger individuals (72, 73) (Fig. 33.29). Avulsion in these younger individuals may occur at either the tendotubercle or infrapatellar insertion. Bone fragments containing high signal intensity marrow may be identified. Tendon laxity with a wavy contour may be seen with both acute and chronic tears.

Patella Dislocation

As a consequence of the inclination of the femur, the quadriceps muscle does not pull in a direct line with the patella (74). The angle formed between the two is valgus and is known clinically as the Q angle. This angle results in a natural tendency toward lateral displacement of the patella. The vastus medialis muscle, and in particular its inferior component, the vastus medialis obliquus, serve to counteract this naturally occurring lateral deviating vector. Additionally both the vastus medialis and vastus lateralis contribute fibrous expansions critical to the maintenance of normal patellar alignment and motion. The expansions, known as the medial and lateral retinacula, are attached to the margins of the patella and patella tendon and extend backward on both sides as far as the corresponding collateral ligament and downward to the condyles of the tibia.

Traumatic dislocation of the patella may result from a direct blow or secondary to severe rotary stress imposed on the weight-bearing knee (60). The clinical presentation may at times be difficult to differentiate from acute ACL disruption. Patellar dislocation is often accompanied by traumatic disruption of the medial retinacular supporting structures, or by avulsion of a small medial fragment of patella at the site of retinacular insertion. The MR appearance of retinacular injuries varies with the extent of disruption. With lateral dislocation, the medial patellar ligament is torn from the intramuscular septum directly beneath the vastus medialis obliquus. Increased signal intensity, representing edema and/or hemorrhage, is seen along the medial aspect of the retinaculum on T2-weighted images. A discontinuous or wavy contour to the retinaculum itself is commonly evident (Fig. 33.30).

Osteochondral injuries are also commonly associated with patella dislocation. The medial facet of the patella and lateral femoral condyle are most commonly involved, and injury occurs at the time of spontaneous reduction. Intraarticular osteocartilagenous fragments may result and are common in the patellofemoral space as well as the lateral gutter adjacent to the popliteal tendon. Axial sections are ideal for assessment of the patellar articular cartilage defects. The curved contour of the lateral femoral condyle is often best evaluated on sagittal sections. Hemarthrosis is common and the presence of lipohemarthrosis is diagnostic of osteochondral fracture with intraarticular extension.

Conclusions

The advent of MR imaging has dramatically changed the diagnostic imaging approach toward traumatic abnormalities of the knee. Increasingly MR is being utilized in many centers as the primary imaging examination for patients with knee complaints. The increased contrast resolution of MR has provided the capability of depicting a wide spec-

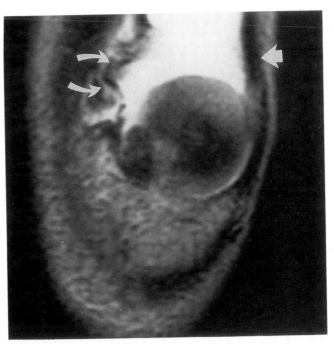

Figure 33.30. Retinacular tear. Coronal image (TR/TE 2000/80). High signal intensity fluid is seen within the suprapatellar bursa. The lateral retinaculum is intact *(arrow)*. The medial retinaculum is wavy and buckled, consistent with disruption *(curved arrows)*. The patient has experienced a patellar dislocation.

trum of trauma-related abnormalities, many of which have likely previously escaped detection. Magnetic resonance imaging has largely replaced diagnostic arthrography and can obviate the need for arthroscopy as a diagnostic test in many situations. As a single technique, MR may obviate the need for multiple other traditional examinations, including scintigraphy and tomography. It is expected that with increasing experience and future technical developments, the role of MR in this area will continue to expand.

References

1. Reicher MA, Hartzman S, Basset LW, et al. Magnetic resonance imaging of the knee joint. Clinical update I: Injuries to menisci, patellar tendon and cruciate ligaments. Radiology 1987;162:547–553.
2. Gallimore GW, Harms SE. Knee injuries: High-resolution MR imaging. Radiology 1986;160:457–461.
3. Burk DL Jr, Kanal E, Brunberg JA. High resolution MR imaging of the knee. In: Kressel HY, ed. Magnetic resonance annual. New York: Raven Press, 1988:1–36.
4. Burk DL Jr, Kanal E, Brunberg JA, Johnstone GF, Swensen HE, Wolf GL. 1.5-T surface-coil MRI of the knee. AJR 1986;147:293–300.
5. Beyer D, Steinbrich W, Friedmann G, Ermers JWLM. Use of surface coils in magnetic resonance imaging of orbit and knee. Diagn Imag Clin Med 1986;55:84–91.
6. Mink JH, Deutsch AL. Magnetic resonance imaging of the knee. Clin Orthoped 1989;244:29–47.
7. Mink JH, Levy T, Crues JV III. Tears of the anterior cruciate ligament and menisci of the knee: MR imaging evaluation. Radiology 1988;167:769–774.
8. Mink JH, Reicher MA, Crues JV III. Magnetic resonance imaging of the knee. New York: Raven Press, 1987.

9. Deutsch AL, Mink JH. Articular disorders of the knee. Top Magn Reson Imag 1989;1(3):43–46.
10. Stoller DW. The knee. In: Stoller DW, ed. Magnetic resonance imaging in orthopedics and rheumatology. Philadelphia: Lippincott, 1988:97–205.
11. Hagger AM, Froelich JW, Hearshen DO, Sadasivan K. Meniscal abnormalities of the knee: 2D ft fast-scan GRASS MR imaging. AJR 1988;150:1341–1344.
12. Reeder JD, Matz SO, Becker L, Andelman SM. MR imaging of the knee in the sagittal projection: comparison of three-dimensional gradient-echo and spin-echo sequences. AJR 1989;153:537–540.
13. Spritzer CE, Vogler JB, Martinez S, et al. MR imaging of the knee: preliminary results with a 3dft GRASS pulse sequence. AJR 1988;150:597–603.
14. Tyrell RI, Gluckert K, Pathria M, Modic MT. Fast three-dimensional MR imaging of the knee: comparison with arthroscopy. Radiology 1988;166:865–872.
15. Konig H, Sauter R, Diemling M, Vogt M. Cartilage disorders: comparison of spin-echo CHESS, and FLASH sequence MR images. Radiology 1987;164:753–758.
16. Reiser MF, Bongartz G, Erlemann R, et al. Magnetic resonance imaging in cartilaginous lesions of the knee joint with three-dimensional gradient-echo imaging. Skel Radiol 1988;17:465–471.
17. Arnoczky SO. Blood supply of the meniscus. In: McGinty JB, ed. Techniques in orthopedics, Vol. 5. Arthroscopic surgery update. Rockville, Md.: Aspen Publications, 1985.
18. Crues JV III, Mink J, Levy TL, Lotysch M, Stoller DW. Meniscal tears of the knee: accuracy of MR imaging. Radiology 1987;164:445–448.
19. Crues JV, Stoller DW. The menisci. In: Mink JH, ed. Magnetic resonance imaging of the knee. New York: Raven Press, 1987.
20. Smille IS. Diseases of the knee joint. New York: Churchill Livingstone, 1980.
21. Shahriaree H, ed. O'Conner's textbook of arthroscopic surgery. Philadelphia: Lippincott, 1984.
22. Casscells SW, ed. Arthroscopy: diagnostic and surgical practice. Philadelphia: Lea & Febiger, 1984.
23. Johnson LL. Diagnostic and surgical arthroscopy. St. Louis: C. V. Mosby, 1981.
24. Deutsch AL, Mink JH. MRI of the post operative meniscus. Radiology 1988;169(P):20.
25. De Haven DE. Rationale for meniscus repair or excision. Clin Sports Med 1985;4:267–273.
26. Miller DB. Arthroscopic meniscus repair. Arthroscopy 1985;1:170–172.
27. Arnoczky SP, Warren RF, Spivak JM. Meniscal repair using an exogenous fibrin clot: an experimental study in dogs. J Bone Surg 1988;70A:1209–1217.
28. Arnoczky SP, Warren RF. The microvasculature of the meniscus and its response to injury. An experimental study in the dog. Am J Sports Med 1983;11:131–141.
29. Deutsch AL, Mink JH, Rothman B: Longitudinal follow-up of patients with conservatively treated and arthroscopically repaired peripheral meniscal tears: MR findings and implications. Radiology 1989;173(P):232.
30. Ferrer-Roca O, Vilalha C. Lesions of the meniscus. Clin Orthoped 1980;146:289–307.
31. Burk DL Jr, Dalinka MK, Kanal E, et al. Meniscal and ganglion cysts of the knee: MR evaluation. AJR 1988;150:331–336.
32. Watanabe AT, Carter BCV, Teitelbaum GP, Bradley WG Jr. Common pitfalls in magnetic resonance imaging of the knee. J Bone Joint Surg (Am) 1989;71:857–862.
33. Herman LJ, Beltran J. Pitfall in MR imaging of the knee. Radiology 1988;167:775–781.
34. Beltran J, Noto AM, Mosure JC, Weiss KL, Zuelzer W, Christoforidis AJ. The knee: surface coil MR imaging at 1.5T. Radiology 1986;159:747–751.
35. Polly CW, Callaghan JJ, Sikes RA, McCabe JM, McMahon K, Savory CG. The accuracy of selective magnetic resonance imaging compared with the findings of arthroscopy of the knee. J Bone Joint Surg (Am) 1988;70:192–198.
36. Jackson DW, Jennings LD, Maywood RM, Berger PE. Magnetic resonance imaging of the knee. Am J Sports Med 1988;16:29–38.
37. Mandelbaum BR, Finerman GA, Reischer MA, Hartzman S, et al. Magnetic resonance imaging as a tool for evaluation of traumatic knee injuries. Am J Sports Med 1986;14:361–368.
38. Feagin JA. Case studies 1–15. In: Feagin JA, ed. The cruciate ligaments. New York: Churchill Livingstone, 1988.
39. Arnoczky SP, Russel RF. Anatomy of the cruciate ligaments. In: Feagin JA, ed. The cruciate ligaments. New York: Churchill Livingstone, 1988.
40. Hughston JC, Andrews JR, Cross MJ, et al. Classification of knee ligament instabilities. Part 1. The medial compartment and cruciate ligaments. J Bone Joint Surg. 1976;58:159–172.
41. Jackson RW. The torn ACL: natural history of untreated lesions and rationale for selective treatment. In: Feagin JA, ed. The cruciate ligaments. New York: Churchill Livingstone, 1988.
42. Steadman RJ, Higgins RW. ACL injuries in the elite skier. In: Feagin JA, ed. The cruciate ligaments. New York: Churchill Livingstone, 1988.
43. Moeser P, Bechtold RE, Clark T, Rovere G, Karstaedt N, Wolfman N. MR imaging of anterior cruciate ligament repair. J Comput Assist Tomogr 1989;13:105–109.
44. Deutsch AL, Mink JH. MRI of musculoskeletal trauma. Radiol Clin N Am 1988;27:983–1002.
45. Indelicato PA. Injury to the medial capsuloligamentous complex. In: Feagin JA, ed. The cruciate ligaments. New York: Churchill Livingstone, 1988.
46. Mink JH, Deutsch AL. Occult osseous and cartilaginous injuries about the knee: MR assessment, detection and classification. Radiology 1989;170:823–829.
47. Dietz GW, Wilcox GM, Montgomery JB. Second tibial condyle fracture; lateral capsular ligament avulsion. Radiology 1986;159:467.
48. Yao J, Lee JK. Occult intraosseus fracture: detection with MR imaging. Radiology 1988;168:749–751.
49. Lynch TC, Crues J, Sheehan W, et al. Stress fractures of the knee; MRI evaluation [Abstract]. Magn Reson Imag 1988;6:10.
50. Lynch TCP, Crues, JV III, Morgan FW, Sheehan WE, Harter LP, Rye R. Bone abnormalities of the knee: prevalence and significance of MR imaging. Radiology 1989;171:761–766.
51. Matthewson MH, Dandy DJ. Osteochondral fractures of the lateral femoral condyle. J Bone Joint Surg 1978;60B(2):199–202.
52. Milgram JW, Rogers LF, Miller JW. Osteochondral fractures: mechanisms of injury and fate of fragments. AJR 1978;1130:651–658.
53. Rosenberg NJ. Osteochondral fractures of the lateral femoral condyle. J Bone Joint Surg 1964;46A(5):1013–1026.
54. Gilley JS, Gelman MI, Edison DM, et al. Chondral fractures of the knee. Radiology 1981;138:51–54.
55. Mesgarzadeh M, Sapega AA, Bonakdarpour A, et al. Osteochondritis dessicans: analysis of mechanical stability with radiography, scintigraphy, and MR imaging. Radiology 1987;165:775–780.
56. Stafford SA, Rosenthal DI, Gebhardt MC, Brady TJ, Scott JA. MRI in stress fracture. AJR 1986;147:553–556.
57. Berger PE, Ofstein RA, Jackson DW, Rossison DS, Silvino N, Amador R. MRI demonstration of radiographically occult fractures: What have we been missing. Radiographics 1989;9:407–436.
58. Deutsch AL, Mink JH, Waxman AD. MR imaging of occult fractures of the proximal femur. Radiology 1989;170:113–116.
59. Lee JK, Yao L. Stress fractures: MRE imaging. Radiology 1988;169:217–220.
60. Hohl M, Larson RI. Fractures and dislocation of the knee. In: Rockwood CA, Green DP, eds. Fractures. Philadelphia: Lippincott, 1975;1131–1286.
61. Cailliet R. Knee pain and disability, 2nd ed. Philadelphia: F. A. Davis, 1983.
62. Fisher MR, Dooms GC, Hricak H, Reinhold C, Higgens CB. Magnetic resonance imaging of the normal and pathologic muscular system. Magn Reson Imag 1986;4:491–496.
63. Kuno S, Katsuta S, Inouye T, Anno I, Matsumoto K, Akisada M. Relationship between MR relaxation time and muscle fiber composition. Radiology 1988;169:567.
64. Murphy WA, Totty WG, Carroll JE. MRI of normal and pathological skeletal muscle. AJR 1986;146:565–574.
65. Fleckenstein JL, Weatherall PT, Parkey RW, Payne JA, Peshock

RM. Sports-related muscle injuries: evaluation with MR imaging. Radiology 1989;172:793–798.

66. Dooms GC, Fisher MR, Hricak H, Higgins CB. MR imaging of intramuscular hemorrhage. J Comput Asst Tomogr 1985;9:908–913.

67. Pakter RI, Fishman EK, Zerhouni EA. Calf hematoma—computed tomographic and magnetic resonance findings. Skel Radiol 1987;16:393–396.

68. Rubin JI, Gomori JM, Grossman RI, Gefter WB, Kressel HY. Highfield MR imaging of extracranial hematomas. AJR 1987;148:813–817.

69. Unger EC, Glazer HS, Lee JKT, Ling D. MRI of extracranial hematomas: preliminary observations. AJR 1986;146:403–407.

70. Beltran D, Noto AM, Herman LJ, Lubbers LM. Tendons: high field strength surface coil MR imaging. Radiology 1987;162:735–740.

71. Daffner RH, Riemer BL, Lupetin AR, Dash N. Magnetic resonance imaging in acute tendon injuries. Skel Radiol 1986;15:619–621.

72. Gould ES, Taylor S, Naidich JB, Furie R, Lane L. MR appearance of bilateral spontaneous patellar tendon rupture in systemic lupus erythematosus. J Comput Assist Tomogr 1987;11:1096–1097.

73. Bondne D, Quinn SF, Murray WE, et al. Magnetic resonance image of chronic patellar tendinitis. Skel Radiol 1988;17:24–28.

74. Fox JM, Sheman OH, Pevsner D. Patellofemoral problems and malalignment. In: McGinty JB, ed. Techniques in orthopedics, Vol. 5. Arthroscopic surgery update. Rockville, Md.: Aspen Publications, 1985.

CHAPTER 34

Magnetic Resonance Imaging of Miscellaneous Knee Joint Disorders

Curtis W. Hayes and William F. Conway

Introduction

In the evaluation of suspected nontraumatic disorders of the knee, the radiologist must take a different approach to the magnetic resonance (MR) imaging examination when compared with the typical examination performed for "internal derangements." A great many disease processes may present with nonspecific symptoms, and thus an informed, tailored approach to the imaging examination is critical in order to arrive at the most accurate assessment in the most efficient manner. In this chapter, miscellaneous nontraumatic disorders of the knee have been divided into the following categories: abnormalities of the patellofemoral joint (including chondromalacia patellae), primary and secondary arthritides, osteonecrosis and related disorders, infections, and tumors and tumor-like conditions. Emphasis is placed both on technical considerations in the evaluation of these disorders and their specific MR imaging appearances. Tumors, infection, and arthritis are briefly covered with respect to their unique appearances about the knee only, since general considerations of these topics are covered elsewhere.

Technique

General Approach

We believe that a tailored or problem-solving approach to musculoskeletal MR imaging should be followed. At our institution, nearly all musculoskeletal examinations are performed with constant physician monitoring. The radiologist's role begins prior to the examination, with review of pertinent history (often quite brief) as well as review of any available radiographic studies. Direct com-

munication with the referring physician is often essential in the appropriate planning of the examination. We find that this is best performed when the study is initially scheduled, since occasionally a more appropriate alternative imaging modality may be suggested in lieu of MR imaging.

Once the decision has been made to perform the MR examination, objectives should be set. In the case of nonspecific symptoms, the objective is to perform an overview study of the entire knee, without sacrificing accuracy in evaluation of the menisci and ligaments of the knee. This is the *overview* examination. A *staging* examination is one in which spatial resolution may be partly sacrificed in order to image a large area—often the entire bone in question. This is particularly important in the evaluation of primary bone tumors and suspected infection. *Follow-up* examinations for tumors should be performed with similar positioning and technique as the baseline examination, with monitoring available for the occasional unexpected result. Finally, the *arthritis* examination must demonstrate fine anatomic detail in multiple planes, utilizing at least one T2- or T2*-weighted sequence to demonstrate the arthrogram effect. In all cases, it is important to consider, prior to the start of the examination, the optimal patient position, type of coil to be used, expected field of view, and pulse sequences for best evaluation. Anticipated use of intravenous Gd is important, since an intravenous line must be started prior to the beginning of the examination in such cases. Patient comfort and the possible use of sedation or monitoring must also be considered. By taking these factors into consideration prior to the beginning of the examination, the best possible diagnostic information may be obtained in the most efficient manner.

Two imaging systems are currently in use at our institutions (Medical College of Virginia Hospitals, Richmond, Va.). All clinical imaging is performed on either a 1.0-T superconducting unit (Siemens Magnetom; Erlangen, Germany) or a 1.5-T Signa system (General Electric, Milwaukee, Wis.). Nearly all of the images in this chapter were derived from these machines.

Coils

A large number of commercial coils are available. In cases of nonspecific pain or suspected arthritis of the knee, we

---

Abbreviations (see also glossary): CMP, chondromalacia patellae; CA, congruence angle; 3D FT, three-dimensional Fourier transform; FISP, fast imaging with steady-state free precession; FLASH, fast low-angle shot; PVNS, pigmented villonodular synovitis.

prefer a commercially available transmit/receive extremity coil, such as the unit shown in Figure 34.1. This coil produces homogeneous signal throughout the field of view, allowing easier comparison of structures throughout the image. Cases of suspected intraosseous tumor require a large field of view; therefore the body coil is generally preferred in order to visualize the entire length of the involved bone. In this case, both legs are included in the field of view, enabling comparison. With infants or small children, both lower extremities often fit within the extremity coil, thus allowing excellent spatial resolution and visualization of the entire bone. With respect to positioning within the coil, if a single knee is to be imaged, the standard 15° to 20° of external rotation with near full extension is preferred (1–3). This allows optimal visualization of the anterior cruciate ligament and will not interfere with other imaging objectives. If both legs are included, symmetric positioning is a necessity. When imaging in the coronal plane, it is often desirable to elevate the patient's heels slightly to allow visualization of the entire shaft of either the femur or tibia on a single coronal view. This is particularly helpful in assessing the intraosseous extent of a tumor process.

Occasionally, the area of interest is specifically in the anterior portion of the knee. In these cases, such as "jumper's knee" and chondromalacia follow-up, we may employ a 3.5-inch receive-only circular surface coil positioned directly over the area of interest. This coil maximizes spatial resolution in the anterior aspect of the knee. Rapid signal drop-off limits the usefulness of the study for the posterior structures, however, and makes both photography and interpretation difficult (Fig. 34.2).

Imaging Sequences

In most nonspecific knee cases, we begin with T1-weighted spin-echo sequences in two planes, followed by a T2- or T2*-weighted sequence in at least one plane. T1-weighted images are sensitive to intraosseous changes and are best suited for defining the extent of intraosseous lesions. In the absence of effusion, or when advanced changes of arthritis are present, T1-weighted images are often adequate for the evaluation of cartilage. However, when evaluating specifically for cartilage abnormalities, a T2- or T2*-weighted gradient-echo sequence is always performed in at least one plane. We find it useful to perform the T1-weighted sequences in two planes first, then decide which plane is preferable for the more time-consuming T2- or T2*-weighted sequence. We currently use both conventional T2-weighted images and high flip angle (75°) 3D FT (three-dimensional Fourier transform) gradient-echo sequences (FISP; fast imaging with steady state free precession) to obtain an "arthrogram effect" for the evaluation of cartilage thickness and defects. The specific gradient-echo pulse sequences

Figure 34.1.   Typical transmit/receive extremity coil (Siemens, Erlangen, Germany).

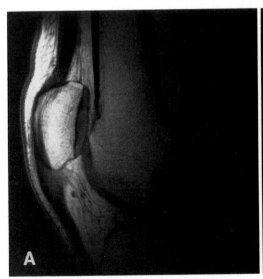 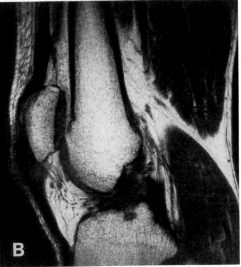

Figure 34.2.   Anteriorly positioned 3.5-inch circular surface coil produces excellent detail of anterior structures **(A)**. Rapid signal drop-off posteriorly results in images that are both difficult to interpret and photograph compared with the homogeneous images produced by a typical transmit-receive extremity coil **(B)**.

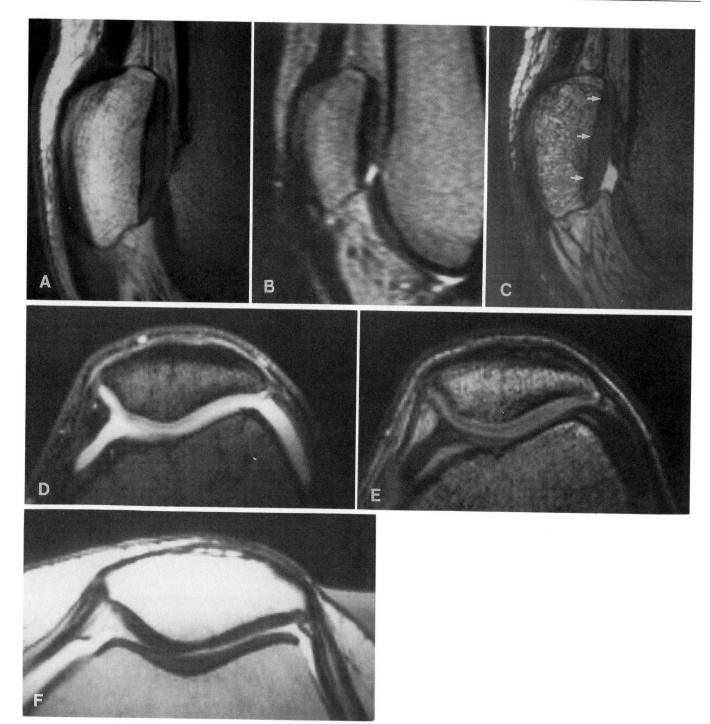

Figure 34.3. Pulse sequences. T1-weighted spin-echo image (TR = 700 ms, TE = 20 ms) **(A)** shows fine anatomic detail. In the absence of effusions, cartilage defects may be seen. T2-weighted spin-echo image (TR = 2500 ms, TE = 90 ms) **(B)** shows excellent contrast between synovial fluid and cartilage, although overall resolution and distinction between subchondral bone and cartilage are sacrificed. 3D FISP 75 (TR = 40, TE = 20) **(C)** demonstrates high-signal fluid, intermediate cartilage, and well-defined cartilage/bone interface *(arrows).* Differing gradient-echo techniques produce varying levels of contrast between fluid and cartilage. **(D)** 2D FLASH 10 (TR = 12, TE = 11); **(E)** 3D FISP 45 (TR = 40, TE = 15); **(F)** T1-weighted spin-echo image following intraarticular Gd injection may allow a combination of increased detectability of cartilage defects in the presence of fluid as well as optimal resolution (**F** reprinted, with permission, from Conway WF, Hayes CW, Loughran T, et al. Cross-sectional imaging of the patellofemoral joint and surrounding structures. RadioGraphics 1991;11:195–217.)

should be chosen to enhance contrast between cartilage, subchondral bone, and synovial fluid.

There is no consensus regarding the optimal sequence choice for the detection of cartilage lesions. Surface irregularities, internal cartilage changes, and underlying bone changes all need to be considered. It is not known whether one single sequence can suffice. For example, Yulish et al. (4) found T1-weighted images were sufficient to accurately show advanced patellar cartilage changes. T2-weighted images were most helpful when joint fluid was present. In an examination of cadaveric knees, Hayes et al. (5) also found T1-weighted images could accurately demonstrate chondral lesions of moderate and advanced degrees. In both reports, early changes were not reliably seen. In examining surgically created cartilage defects, Gylys-Morin et al. (6) found T1-weighted images to be ineffective when saline was present in the joint. T2-weighted images allowed the detection of defects as small as 3 mm. Various gradient-echo (7, 8) and chemical shift techniques (9) have recently been evaluated as well. Finally, in spite of its invasiveness, MR arthrography using intraarticular Gd may prove to be the most accurate method in the evaluation of articular cartilage (10). A comprehensive comparison of sequences will be necessary before this issue can be agreed on (Fig. 34.3).

In cases of suspected infection or tumor we utilize at least one conventional T2-weighted spin-echo sequence. T1-weighted sequences following intravenous Gd are also helpful in the assessment of suspected tumors. We find Gd useful in determining nonenhancing cystic from enhancing solid masses. It is possible that patterns of Gd enhancement may also aid in differentiating viable from necrotic tumor, peritumoral edema from actual mass, and in assessing response to adjuvant therapy (11).

Patterns of signal intensity on T1- and T2-weighted images, and following Gd enhancement, are generally nonspecific in terms of tissue-specific diagnosis and for determination of benign vs. malignant masses (12, 13). Several fairly specific patterns have been described, however, including fat (lipomas, liposarcomas), rim enhancement or lobular, homogeneous patterns with chondroid tumors (12), and fluid-debris levels with aneurysmal bone cysts (14) (see also Chapter 15). It should be stressed, however, that a deliberate analysis of plain films is still most helpful in determining the differential diagnosis of bone lesions. Although still necessary (i.e., for the detection of calcifications), plain radiographs generally are less helpful in the evaluation of soft-tissue tumors.

Artifacts and Normal Variants

The most common technical artifacts to be encountered in the evaluation of the knee are the chemical shift effect, motion artifact, volume averaging, and artifacts created by metallic particles.

The chemical shift artifact may cause over- or underestimation of the thickness of a structure, often cartilage or cortical bone, in certain areas. This well-known phenomenon occurs along the frequency-encoding axis, due to inherent differences in the precessions of fat and water hydrogen protons. This difference results in a misrepresentation of the position of fatty tissue relative to tissue containing high water content, such as muscle, or tissue with little signal, such as cortical bone. Asymmetric thickness of the cortical bone may be observed along the frequency-encoding direction on axial images. This may produce distortion of the margins between an intraosseous lesion and the cortical bone, giving the appearance of endosteal bone erosion (Fig. 34.4). The chemical shift artifact is more prominent on sequences performed using a large field of the view (low zoom) (15). If a chemical shift artifact is suspected, then a repeat of the sequence, following rotation of the frequency- and phase-encoding gradients by 90°, may be necessary.

Motion artifacts along the phase-encoding direction generally involve the popliteal artery and may occasionally cause confusion where the artifact overlies the articular cartilage, giving the false impression of a significant cartilage defect (Fig. 34.5).

Volume averaging may present a problem when rounded surfaces or diagonally oriented structures are being evaluated. In such cases, it is usually necessary to image in two planes. In the case of the patellar cartilage, usually the axial and sagittal planes are required. In the evaluation of the condyles, coronal and sagittal images are generally complementary.

Large metallic hardware may present unavoidable artifacts in postoperative patients. In some cases, however, tiny metallic fragments—virtually undetectable by plain radiographs—may cause significant metallic artifact. These tiny metallic fragments are sometimes seen in patients following arthroscopic or open surgery and should be considered when small sharp areas of signal drop-out are found associated with the joint space (Fig. 34.6).

A number of anatomic variations are present about the knee. The so-called "ring defect" of the patella is quite common and should not be mistaken for an area of osteonecrosis (2) (Fig. 34.7). The bipartite patella is usually an asymptomatic variant. Magnetic resonance imaging may be of value in demonstrating the status of the overlying articular cartilage crossing the ununited segment of the patella (Fig. 34.8). Meniscal anatomy is described elsewhere (see Chapter 33).

It has been observed that patchy areas of residual red marrow are often found in the diaphysis and metaphyseal regions of the femur and tibia. This pattern is most common in obese, middle-aged women and is of uncertain etiology (16) (Fig. 34.9). Finally, patterns of residual marrow and of the unfused growth plate in adolescents may present a confusing appearance when imaged in the axial projection. Attention to this possibility will help to avoid the misinterpretation of such benign collections as metastatic disease or other infiltrative processes.

Patellofemoral Joint and Extensor Mechanism

Introduction

Abnormalities of the patellofemoral joint and extensor mechanism are common, especially in young adults and adolescents (17–19). Patellofemoral abnormalities com-

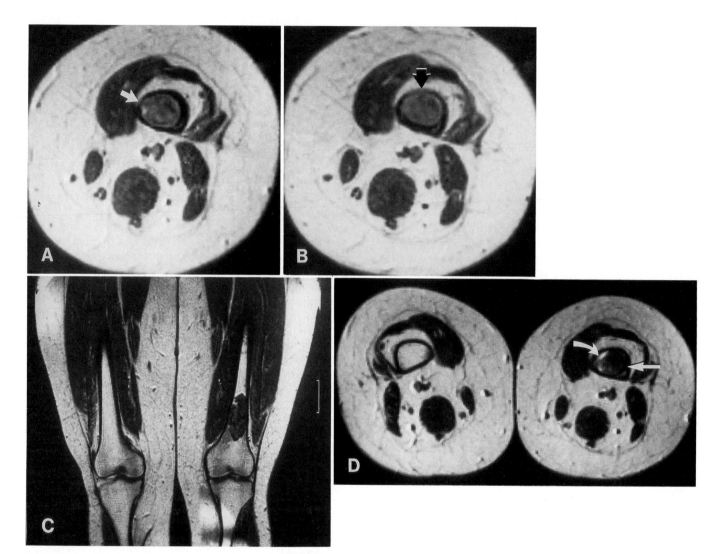

Figure 34.4. Chemical shift artifacts. Transverse, T1-weighted (TR = 700, TE = 20) image through a chondroid lesion in the distal femur shows apparent thinning of the medial femoral cortex (**A,** *arrow*). Following rotation of the frequency-encoding gradient, an image at the identical location demonstrates apparent thinning of the anterior cortex (**B,** *arrow*). These findings are due to the misrepresentation of fat (extraosseous adipose tissue in this case) and soft tissue (intraosseous lesion) along the frequency-encoding direction. Coronal T1-weighted image **(C),** with frequency-encoding gradient in the vertical direction, demonstrates the mass filling the medullary cavity, but causing no endosteal scalloping. A transverse, T1-weighted image at a second location demonstrates typical low-signal (**D,** *straight arrow*) and high-signal *(curved arrow)* rims on either side of the lesion. These are caused by separation and summation of fat and water signals, respectively.

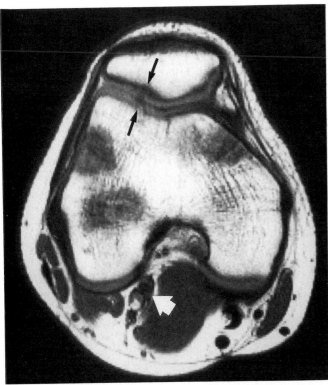

Figure 34.5. Motion artifact. Axial T1-weighted image (TR = 700, TE = 20) shows a vertically oriented motion artifact extending from the popliteal artery *(large arrow)*, through the trochlear and patellar cartilages *(small arrows)*, giving the false impression of cartilage defects. Focal low-signal regions in the marrow are secondary to close proximity to the growth plate in this adolescent patient.

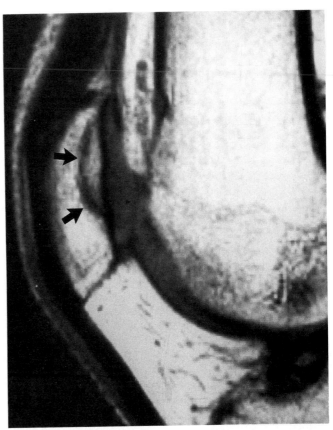

Figure 34.7. The ring defect *(arrows)* is a normal variant appearance of the subchondral bone of the lateral facet of the patella.

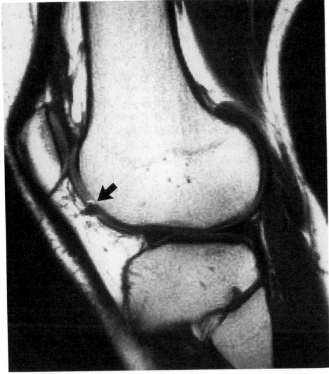

Figure 34.6. Metallic artifact. Tiny metallic fragments, undetected by plain films, may cause small but significant artifacts, as demonstrated on this sagittal T1-weighted image *(arrow)*.

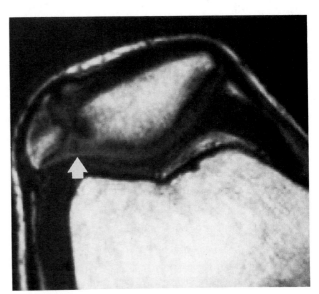

Figure 34.8. Bipartite patella. T1-weighted axial image (TR = 700, TE = 20) demonstrates an ununited bony segment at the lateral aspect of the patella. The cartilage is demonstrated to be intact across the defect *(arrow)*.

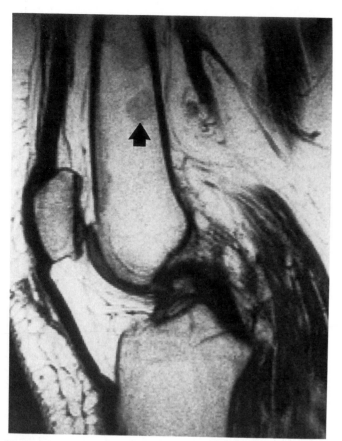

Figure 34.9. Residual hematopoietic marrow. Focal areas of residual red marrow are common in obese, middle-aged women. T1-weighted image (TR = 700, TE = 20) demonstrates focal low-signal areas *(arrow)*. Biopsies have shown mildly hypercellular marrow, but without other significant abnormalities.

monly present with nonspecific anterior knee pain (20, 21). Many other abnormalities may produce similar symptoms and thus it is important to accurately detect these abnormalities in order to ensure prompt, appropriate treatment. Many imaging modalities have been utilized for the evaluation of the patellofemoral joint. Plain radiography (17, 22–26), bone scintigraphy (27), arthrography (28), and computed tomography (29–32) have shown some usefulness in the diagnosis of bony and cartilaginous changes and alignment abnormalities of the patellofemoral joint. In our opinion, MR imaging is the noninvasive modality of choice for the evaluation of nonspecific anterior knee pain and suspected patellofemoral disorders due to its unique ability to display both superior soft-tissue contrast and intraosseous changes in multiple planes.

Anatomy and Biomechanics of the Patellofemoral Joint

The patella is an integral part of the extensor mechanism of the knee, consisting of the vastus lateralis, medialis, and intermedius and rectus femoris muscles, their combined tendons, the patella, and the inferior patellar tendon, which inserts on the tibial tubercle. The patella is a sesamoid bone; it may be first identified as a cluster of cells developing

within the future quadriceps tendons at 8 weeks of gestation. The patellar ossification center appears at approximately 2 to 3 years of age. The adult size is attained around age 15 (33). It is roughly triangular in shape, with its apex pointing inferiorly. Hyaline cartilage covers the posterior surface of the patella except at the inferior pole. The posterior surface is divided by longitudinal ridges into three articular facets: lateral, medial, and medialmost odd facet. The patellar cartilage is the thickest cartilage in the body, with an average depth of 4 to 5 mm on both medial and lateral facets. We have noted cartilage thickness of up to 6 mm in asymptomatic individuals. The lateral facet is generally the largest in area, although a wide variety of patellar shapes exists.

The patella serves to protect the anterior knee joint, but its primary function is to increase the efficiency of the extensor mechanism by increasing the lever arm of the quadriceps muscle. In performing this function the patella is subjected to extreme forces—an estimated 1.5 times body weight for walking, 3.3 times body weight for stair climbing, and approximately eight times body weight in squatting (34, 35). In order to function adequately and without pain, the patella must balance the forces pulling laterally, medially, and superiorly against the fixed patellar tendon and groove-shaped trochlea of the femur, while allowing smooth gliding motion during flexion and extension.

At full extension the patella usually rests against the lateral femoral metaphysis at the superior extent of the trochlear cartilage. Usually only a small portion of the inferior aspect of the lateral facet is in contact with the articular cartilage in extension. In asymptomatic individuals there may be lateral positioning of the patella in both the relaxed and tensed, fully extended knee. Patellar alignment in 0° of flexion (full extension) has been studied by several authors, utilizing computed tomography (CT) and, more recently, MR imaging. Lateralization of the patella in extension was found in from 5% (30) to 87% (36) off normal volunteers. Pathologic subluxation of the patella may occur between 0 and 20° of flexion. Such cases may be impossible to demonstrate by plain radiographic techniques, but are readily shown with axial MR imaging.

With mild flexion, the lateral facets of the patella and trochlea are normally brought into contact, followed by medial facet contact, usually at around 20° (37). Further flexion produces more contact along the medial side of the joint. The odd facet does not come into contact with its opposing cartilage until extreme flexion, at which times it opposes the inner aspect of the medial femoral condyle. This low degree of contact has been implicated as one possible cause of the relative high frequency of chondromalacia involving the odd facet of the patella (21, 34, 38).

Normal Magnetic Resonance Imaging Appearance

In normal adults the marrow of the patella is predominantly yellow marrow and thus shows high signal on T1-weighted spin-echo images (Fig. 34.10). In the subchondral bone of the lateral facet, a focal curvilinear area of decreased signal of variable size is often seen in asymptomatic individuals. This appearance may sometimes be

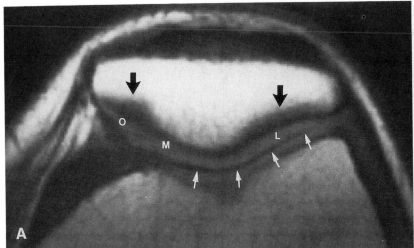

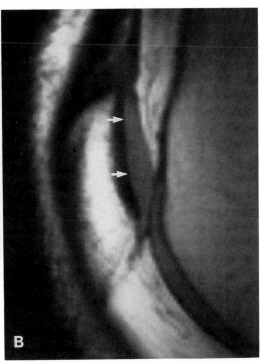

Figure 34.10. Normal anatomy. Axial image (TR = 700, TE = 20) **(A)** demonstrates lateral *(L)*, medial *(M)*, and odd *(O)* facets. In mild flexion (approximately 5°) this volunteer showed smooth cooptation between the lateral facet of the patella and the trochlea, separated by a fine, thin, low-signal line *(arrows)*. Low-signal subchondral regions may be found along the lateral and occasionally medial facets *(large arrows)*. Sagittal view **(B)** demonstrates separation between subchondral bone and cartilage *(arrows)*.

quite pronounced and may occasionally be mistaken for osteonecrosis (Fig. 34.7). A focal area of slightly decreased signal may also be found along the medial facet, although we have found this to be less frequent. Such areas of decreased signal may be distinguished from pathologic subchondral changes by their focal, well-defined appearance. True subchondral abnormalities frequently show increased signal on corresponding T2-weighted or T2*-weighted images. Decreased signal at the patellar apex is, in comparison, most often pathologic in our experience.

The subchondral cortex is most often thin and regular, but may occasionally merge with the previously described area of decreased signal to form a thick, low-signal band adjacent to the articular cartilage. On axial views, the bony surface of the lateral facet is most often slightly concave, while the medial facet is frequently straight or slightly convex. On sagittal views, the cortical margin of the lateral facet is often straight, slightly concave, or occasionally mildly wavy. On sagittal views, evaluation of the subchondral bone of the medial facet is hampered due to partial volume averaging.

The articular cartilage over the patella should be smooth and fairly uniform in thickness. The cartilage of the lateral facet is usually convex on sagittal views. It may have a smoothly rounded convex shape or may appear rather pointed (usually with the apex near the lower one-third). On axial views, the cartilage of the lateral facet is usually slightly concave, while the cartilage of the medial facet is most often slightly convex.

The articular cartilage may show homogeneous intermediate signal intensity on T1-weighted images, or may show a bilaminar appearance with slight increased signal in the superficial quarter to third. This appearance was particularly prominent in disarticulated specimens (Fig. 34.11). The exact nature of the bilaminar appearance is not known, although it would seem to correspond to the division between the vertical and horizontal fibers of the articular cartilage and may relate to relative differences of hydration between these layers.

With standard T2-weighted images, cartilage maintains a low/intermediate signal intensity. Contrast between cartilage and signal-intense synovial fluid is increased, and focal cartilage defects may be seen when fluid filled. Contrast between cartilage and subchondral bone is poor, however.

Gradient-echo pulse sequences have shown great promise in the evaluation of cartilage. Volume acquisition (3D FT) sequences utilizing flip angles between 45 and 75° are useful in demonstrating contrast between intermediate signal cartilage and high-signal synovial fluid. Such images are well suited for the evaluation of surface lesions of cartilage. The relative values of various gradient-echo techniques and magnetization transfer contrast in the evaluation of internal cartilage changes are still under investigation.

The patellar tendon should be homogeneously low in signal throughout. The insertion of the patellar tendon into both the patella and tibial tubercle should be sharp. The patellar tendon forms a V shape with the anterior tibial cortex at its insertion; this V shape is normally filled with prepatellar fat. Any decrease in this fat signal may indicate a patellar tendinitis or traumatic lesion.

Figure 34.11. A pronounced bilaminar appearance of cartilage was found in disarticulated specimens *(arrows)* (TR = 700, TE = 20).

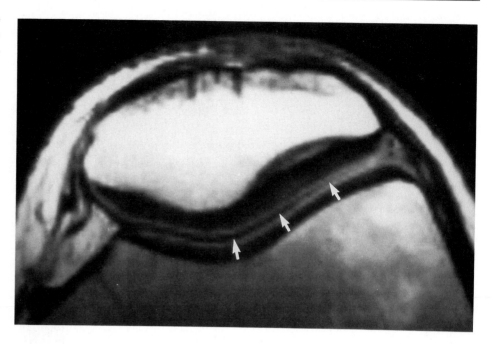

Chondromalacia Patellae

Chondromalacia patellae (CMP) is pathological softening of the patellar cartilage. The etiology of chondromalacia is controversial, although trauma—both repetitive minor trauma and single significant events—and mechanical tracking abnormalities are often cited as causes (20, 34, 38, 39). Chondromalacia patellae has been classified on the basis of macroscopic findings at arthroscopy. The arthroscopic classification by Shahriaree (19) is frequently used (Fig. 34.12). According to this system, Stage I represents softening and swelling of the articular cartilage. Stage II represents blistering of the articular cartilage, producing deformity of the surface. Stage III represents surface irregularity with cartilage fibrillation, without significant exposure of subchondral bone. Stage IV represents ulceration with exposure of significant subchondral bone.

Yulish et al. recently proposed a three-stage classification system of chondromalacia based on MR findings (4). Based on our examination of disarticulated knee specimens, we have adopted a four-stage MR imaging grading system (5), following the Shahriaree arthroscopic grading system (Fig. 34.13).

Stage I: Focal areas of decreased signal intensity on T1-weighted images, usually small, and not extending to or deforming the outer surface.

Stage II: Larger focal areas of decreased signal extending to the cartilage surface but with preservation of a sharp cartilage margin.

Stage III: Focal signal abnormality extending to cartilage surface with loss of sharp cartilage surface margin. Focal or irregular cartilage narrowing may be present, without significant subchondral bone exposure. Decreased signal within the adjacent subchondral bone may be present.

Stage IV: Focal decreased signal extending from subchon-

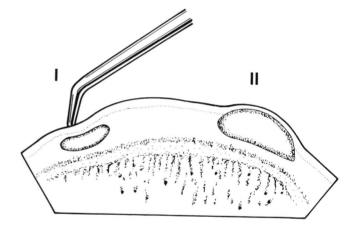

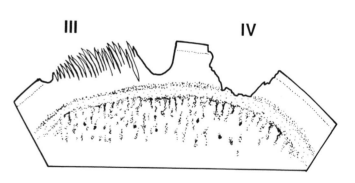

CHONDROMALACIA PATELLAE (Shahriaree)

Figure 34.12. Staging of chondromalacia patellae based on arthroscopic findings. (After Shahriaree H. Contemp Orthopaed 1985;11(5):27–39.)

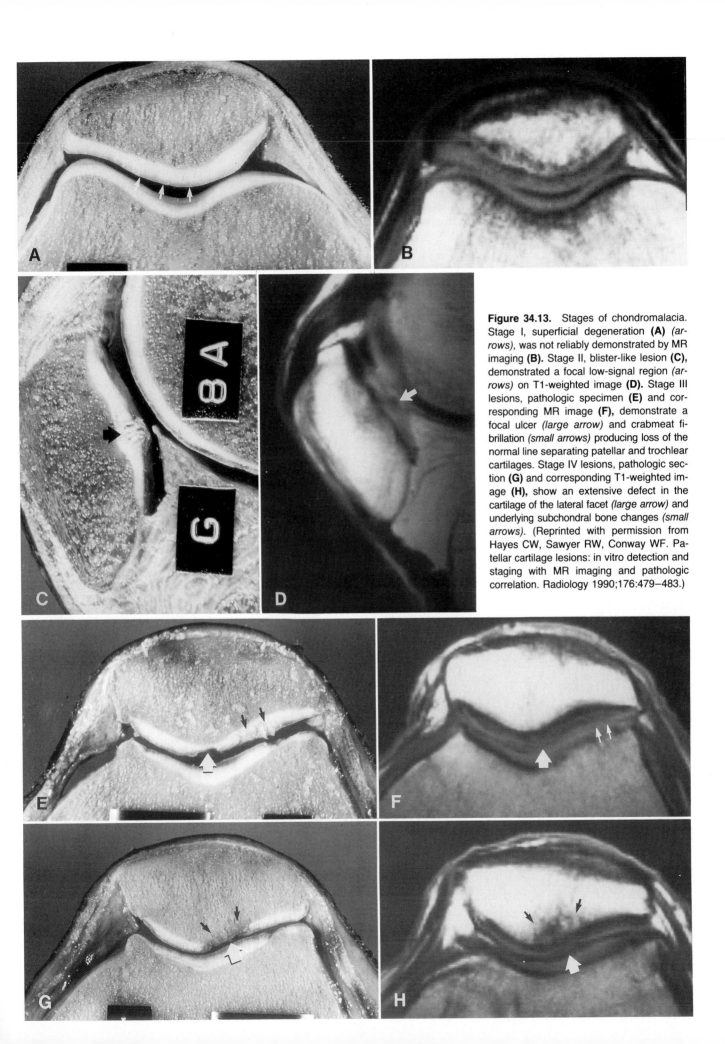

Figure 34.13. Stages of chondromalacia. Stage I, superficial degeneration **(A)** *(arrows),* was not reliably demonstrated by MR imaging **(B)**. Stage II, blister-like lesion **(C)**, demonstrated a focal low-signal region *(arrows)* on T1-weighted image **(D)**. Stage III lesions, pathologic specimen **(E)** and corresponding MR image **(F)**, demonstrate a focal ulcer *(large arrow)* and crabmeat fibrillation *(small arrows)* producing loss of the normal line separating patellar and trochlear cartilages. Stage IV lesions, pathologic section **(G)** and corresponding T1-weighted image **(H)**, show an extensive defect in the cartilage of the lateral facet *(large arrow)* and underlying subchondral bone changes *(small arrows)*. (Reprinted with permission from Hayes CW, Sawyer RW, Conway WF. Patellar cartilage lesions: in vitro detection and staging with MR imaging and pathologic correlation. Radiology 1990;176:479–483.)

dral bone to surface over a significant area or cartilage thinning with exposure of subchondral bone. Decreased signal within subchondral bone is often present at this stage.

Stage I signal changes, focal hypointense regions within the cartilage without distortion of the cartilage surface configuration, are often present in asymptomatic individuals (Fig. 34.14). Whether or not these changes represent the MR imaging equivalent of stage I CMP has not been confirmed. In studies of disarticulated knee specimens, Hayes et al. were unable to detect grade I surface degeneration utilizing T1-weighted spin-echo sequences. The clinical significance of the detection of early (stages I and II) CMP is debatable, since many of these patients are asymptomatic, and therapy is generally conservative.

Stages III and IV of CMP deserve further discussion. In cadaver studies, all lesions rated as stage III and above showed corresponding MR changes on T1-weighted images (5). These changes most commonly show focal or diffuse thinning of the cartilage with irregularity, either not extending to bone (stage III, Fig. 34.15), or with significant subchondral bone exposure (stage IV, Fig. 34.16). In preliminary clinical studies, we have found MR imaging to be 100% sensitive for grade IV changes. Detection of grade III changes is also quite reliable, except on the medial facet

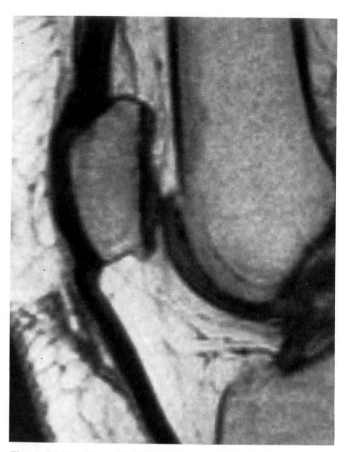

Figure 34.15. Stage III CMP, demonstrating irregular cartilage loss not extending to subchondral bone.

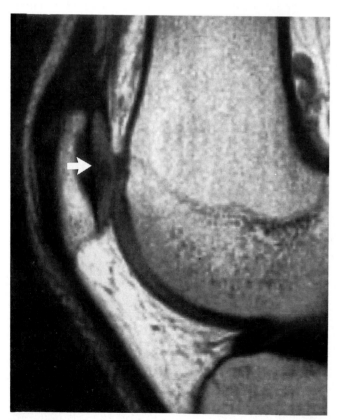

Figure 34.14. A focal area of decreased signal within the articular cartilage on T1-weighted image has been described as a grade I change. Whether this truly represents stage I basal degeneration of CMP is not known. (Reprinted, with permission, from Conway WF, Hayes CW, Loughran T, et al. Cross-sectional imaging of the patellofemoral joint and surrounding structures. RadioGraphics 1991;11:195–217.)

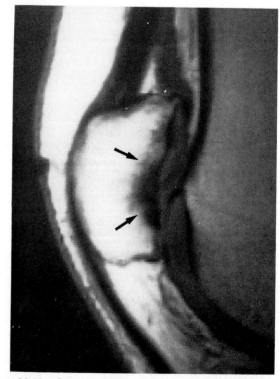

Figure 34.16. Stage IV CMP, with marked inferior patellar cartilage loss, subchondral bone exposure, and subchondral bone changes *(arrows).*

of the patella, where some false positives and false negatives were observed (Hayes et al., unpublished). This is probably due to partial volume-averaging artifacts, which tend to occur on the steeper medial facet compared with the lateral side.

The "crabmeat" appearance of cartilage fibrillation seen grossly in stage III CMP will sometimes produce an obliteration of the normal sharp line between the coapted surfaces of the patellar and trochlear cartilages (Fig. 34.17). While this may be a sensitive indicator of CMP, we have found its specificity to be low, again most likely due to volume-averaging artifacts.

Decreased signal in the subchondral bone may be found in patients with CMP. These findings nearly always are associated with stage III or stage IV chondromalacia. In advanced cases, changes of patellofemoral osteoarthritis include obvious subchondral bone changes, corresponding to sclerosis, excavation of the undersurface of the patella, and significant osteophyte formation.

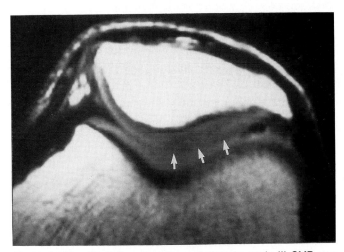

Figure 34.17. Fibrillation of cartilage surface in grade III CMP may obliterate the normal sharp line separating the patella and the trochlear cartilages *(arrows)*.

Traumatic Chondromalacia

Direct trauma to the patella may cause flap-type lesions in the articular cartilage as well as complete chondral or osteochondral fractures. These lesions may produce symptoms of clicking or locking and are generally treated surgically. Magnetic resonance imaging usually shows underlying subchondral bone changes (decreased T1, increased T2) similar to "bone bruises" seen elsewhere in the knee (Fig. 34.18). The subchondral changes are more diffuse than those seen with chronic CMP, and most likely reflect intraosseous edema. Conventional T2-weighted spin-echo sequences and T2*-weighted gradient-echo images are especially useful in determining the extent of the chondral separation.

Patellar Tendon

The patellar tendon is generally homogeneously dark on T1- and T2-weighted sequences. Occasionally, thin, linear, high-signal regions, presumably representing fat, may be seen running through the patellar tendon of asymptomatic individuals. The tendon typically makes sharp insertions on both the tibial tubercule and the inferior pole of the patella, contrasting with the infrapatellar fat pad.

Infrapatellar bursitis is typically seen as an area of low signal on T1-weighted images, between the distal patellar tendon and the anterior tibia, replacing the normal fat in this region (Fig. 34.19). T2-weighted images show corresponding increased signal. Prepatellar bursitis may also be demonstrated as a localized area of decrease signal on T1-weighted images with corresponding increased signal on T2-weighted images in the soft tissues just anterior to the patella.

Symptomatic patellar tendinitis is relatively common and is often accompanied by central mucoid degeneration of the tendon. This disorder is presumably related to overuse and frequently termed "jumper's knee" (40). T1-weighted images easily demonstrate the involved abnormal area of increased signal within the patellar tendon, most com-

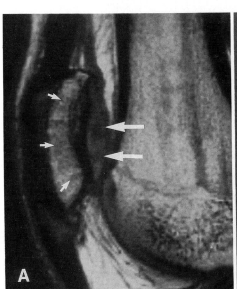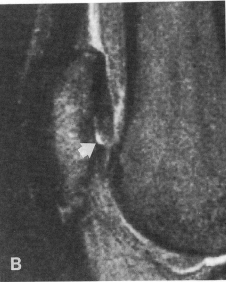

Figure 34.18. Traumatic chondromalacia. T1-weighted image (TR = 700, TE = 20) demonstrates focal areas of decreased signal within the patellar cartilage *(arrows)* as well as decreased marrow signal *(small arrows)* **(A)**. T2-weighted image (TR = 2500, TE = 90) shows fluid extending underneath a chondral fracture **(B,** *arrow)*.

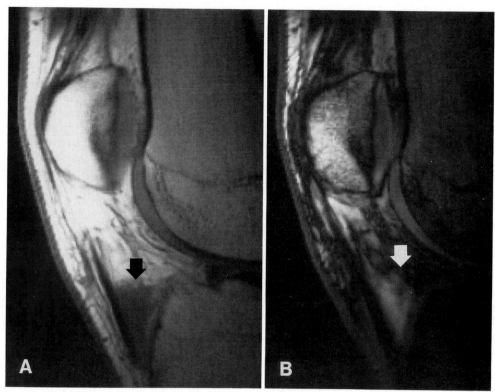

Figure 34.19. Infrapatellar bursitis. T1-weighted image **(A)** and gradient-echo image (FISP 75, TR = 40, TE = 12) **(B)** demonstrate decreased and increased signal, respectively, between patellar tendon and tibia *(arrows)*.

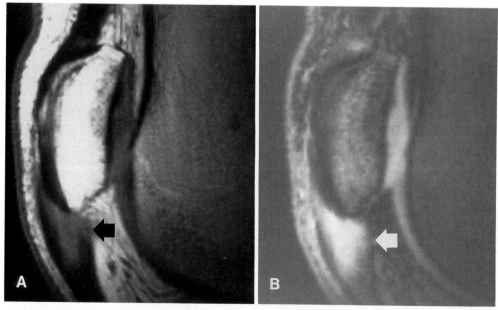

Figure 34.20. Patellar tendinitis. Typical findings of patellar tendinitis include thickening of the tendon and focal areas of increased signal on T1-weighted and gradient-echo images *(arrows)*. (Reprinted, with permission, from Conway WF, Hayes CW, Loughran T, et al. Cross-sectional imaging of the patellofemoral joint and surrounding structures. RadioGraphics 1991;11:195–217.)

monly at the inferior pole of the patella (Fig. 34.20). T2-weighted images show corresponding increased signal in these areas as well. Magnetic resonance imaging is a valuable aid in preoperative planning, since surgical resection of the degenerated area may be considered in refractory cases.

Patellar Tracking and Alignment Abnormalities

The congruence angle is a measurement of lateral subluxation of the patella. Abnormal congruence angles have been associated with chondromalacia patellae. The congruence angle (CA) is formed by a line bisecting the sulcus angle of the trochlea and a second line projecting from the apex of the trochlea through the apex of the patella (Fig. 34.21). The normal congruence angle is $-6° \pm 11°$ (24). Positive congruence angles are associated with patellar instability and CMP (29, 22, 41) (Fig. 34.22). Caution must be observed in applying the above-normal values (derived from radiographic tangential views) to full, or near-full, extension axial MR images. Magnetic resonance imaging is capable of detecting mild degrees of pathologic subluxation that may be undetectable by plain radiographic techniques.

The patellar tilt angle is formed by a line paralleling the lateral facet of the patella and a second line drawn parallel

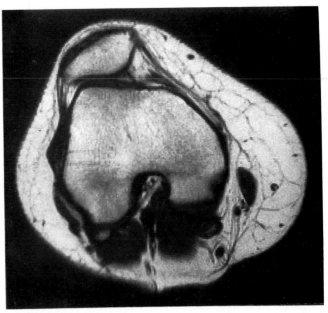

Figure 34.22. Abnormal congruence angle, indicative of lateral subluxation.

to the posterior aspects of the lateral and medial femoral condyles (Fig. 34.23). The normal angle is greater than 8°. Abnormal patellar tilt angles—especially when the CA is normal—may be associated with the excessive lateral pressure syndrome and CMP (26) (Fig. 34.24).

Kinematic MR studies of patellar tracking abnormalities have been described by Shellock et al. (42, 43). Using a special positioning device, the relationship between the patella and femoral trochlear groove may be demonstrated during early stages of knee flexion (between 0 and 30°). Symptomatic patients may show varying degrees of reducing or nonreducing medial or lateral patellar subluxation, frank dislocation, or lateral tilting (Fig. 34.25). Such information may prove very useful in both the preoperative assessment and follow-up of candidates for possible surgical correction of tracking abnormalities.

Numerous variations in the shape of the patella and trochlea have been described. Whether or not certain shapes are apt to produce chondromalacia remains controversial. The Wiberg classification (44) of patellar shape is straightforward and easily adapted to MR images (Fig. 34.26). Wiberg type I patellas show equal-sized medial and lateral facets. Type II patellas have smaller, slightly steeper medial facets. The Wiberg type II patella represents the most common pattern. Wiberg type III patella (small, very steep medial facet) has been associated with recurrent lateral dislocation. Hypoplasia of the lateral trochlear facet has also been implicated in subluxation and dislocation.

Patella alta, an abnormally high patella, has been implicated in both recurrent dislocation and chondromalacia. On sagittal MR images the length of the patellar tendon is easily measured and may be compared with the length of the patella in the midline (Fig. 34.27). According to the Insall-Salvati index (derived from plain radiographs), this ratio should be 1.02 ± 0.13 (39).

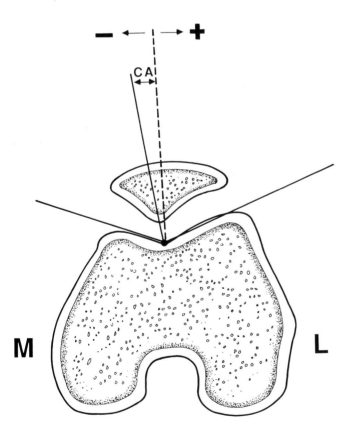

CONGRUENCE ANGLE (CA)

Figure 34.21. Congruence angle.

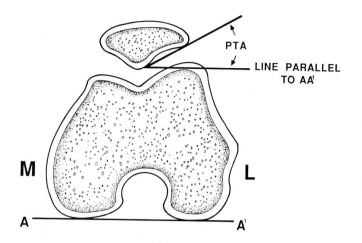

PATELLAR TILT ANGLE (PTA)

Figure 34.23. Patellar tilt angle.

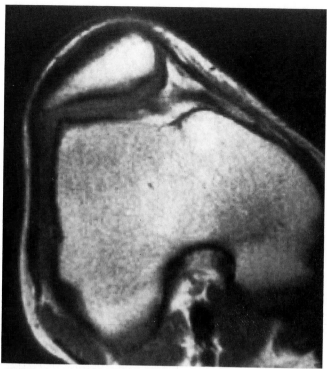

Figure 34.24. An abnormal patellar tilt angle may be associated with the excessive lateral pressure syndrome.

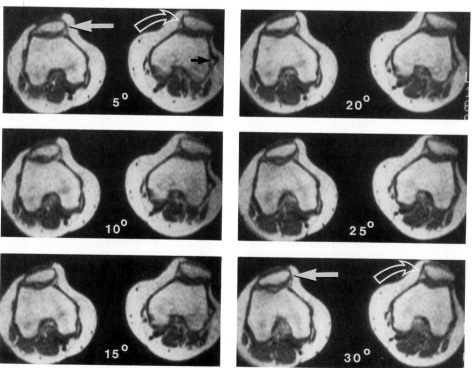

Figure 34.25. Kinematic MR imaging of patellofemoral tracking abnormalities. Consecutive axial images of both knees obtained from 5 to 30° flexion. The left knee shows mild lateral subluxation of the patella throughout the imaged range of motion *(curved arrows)*. The lateral retinaculum *(small arrow)* is lax early during flexion, suggesting that the subluxation is not due to excessive stretching of this structure.

The right knee demonstrates *medial* subluxation that becomes more apparent with greater flexion *(straight arrows)*. (Reprinted with permission from Shellock FG, Mink JH, Deutsch AL, Fox JM. Patellar tracking abnormalities: clinical experience with kinematic MR imaging in 130 patients. Radiology 1989;172:799–804.)

WIBERG I   **WIBERG II**   **WIBERG III**

Figure 34.26. Wiberg classification of patellar shapes. (After Wiberg G. Acta Orthoped Scand 1941;12:319.)

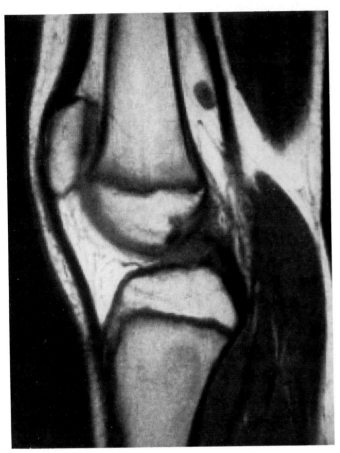

Figure 34.27. Patella alta.

Arthritis

General Considerations

Magnetic resonance imaging has the potential to become the primary imaging modality for the evaluation of both bone and and soft-tissue changes accompanying various adult and pediatric arthritides. Initial detection of early cartilage and synovial changes, staging of the disease, and assessment of progression or therapeutic response are areas in which MR imaging may play a substantial role. There are currently no reports in the literature of the use of MR imaging in a systematic fashion for the evaluation or follow-up of arthritis. Magnetic resonance imaging has been shown to be capable of detecting early synovial changes, as well as small effusions and synovial cysts (1, 45, 46). In our experience, small erosions and subchondral bone changes that may be undetected on plain radiographs may be seen with MR imaging.

At our institution, MR imaging has been used sparingly for the routine evaluation of arthritis, due to the limited availability and cost of the examination. In individual problematic cases, MR imaging may be quite useful, however. In cases of rapid or unexpected progression of symptoms, MR imaging may show unexpected changes such as osteonecrosis, nondisplaced fractures, synovial cysts, focal cartilaginous or bony erosions, and subchondral changes otherwise not visible by plain film technique. Unsuspected meniscal injuries may also be found. T1-weighted spin-echo sequences and conventional T2- or T2*-weighted gradient-echo images are of equal importance in evaluation of the articular cartilage. Both coronal and sagittal planes are useful in evaluating the articular cartilage of the femoral condyles, and to a lesser extent the tibial plateaus. When no significant synovial fluid is present, T1-weighted images may be adequate to detect cartilage defects. When a significant joint effusion is present, however, T2- or T2*-weighted images are essential for their arthrogram effect, to best demonstrate cartilage defects.

One pitfall that must be considered when evaluating cartilage thickness is the effect of chemical shift on the apparent thickness of the cartilage of the femoral condyles and the tibial plateaus. The thickness of cartilage may be under- or overestimated if attention is not paid to the direction of the frequency-encoding gradient (2). Since the artifact occurs along one axis only, rotation of the frequency- and phase-encoding gradients in a repeat sequence will also rotate the direction of the artifact by 90°, thus correcting the artifact in the original plane.

Osteoarthritis

Magnetic resonance imaging is not currently a primary imaging modality in the evaluation of osteoarthritis of the knee, but has proved useful in individual cases for distinguishing osteoarthritis from other sources of pain in the knee. Magnetic resonance imaging may also serve as an accurate evaluation of cartilage status in preoperative cases.

Typical findings seen with MR imaging include focal areas of cartilage thinning, particularly in the posterior aspects of the femoral condyles and along the posterior surface of the patella (Fig. 34.28). Subchondral irregularities with decreased signal on T1-weighted images are commonly found. Small subchondral cyst formation, effusions, and typical marginal spurs are easily identified (Fig. 34.29). In cases of unexplained pain out of proportion to plain radiographic findings, MR imaging may demonstrate unsuspected meniscal tears, osteonecrosis, or radiographically occult fractures.

Rheumatoid Arthritis

Isolated reports and small series demonstrating the MR appearances of rheumatoid arthritis are present in the literature (46, 47), although no large, systematic evaluations have been performed to date. Magnetic resonance imaging may demonstrate early cartilage loss in a tricompartmental distribution, typical of an inflammatory arthritis. Synovial thickening and pannus formation may be detected earlier with MR imaging than by plain radiography

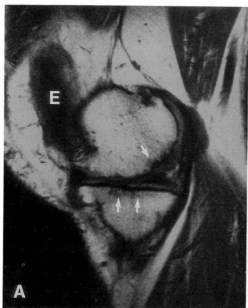
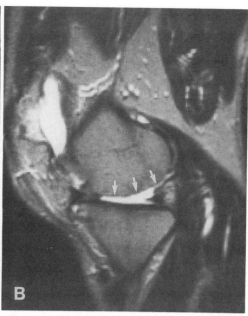

Figure 34.28. Osteoarthritis. T1-weighted image (TR = 700, TE = 20) demonstrates irregularity and focal low-signal areas in the subchondral bone of the femoral condyle and tibial plateau *(arrows)*, effusion *(E)*, and spur formation. T2-weighted image (TR = 2500, TE = 90) demonstrates the severity of the cartilage loss along the femoral condyle *(small arrows)*. This appearance is accentuated by chemical shift artifact.

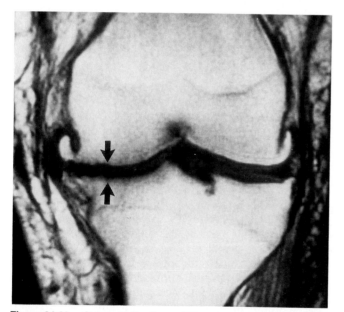

Figure 34.29. Osteoarthritis. Coronal T1-weighted image (TR = 700, TE = 20) demonstrates focal cartilage narrowing *(arrows)* and medial and lateral spurs.

(47). Areas of presumed pannus formation may demonstrate low to intermediate signal characteristics on T1- and T2-weighted images (47; 48) (Fig. 34.30). Additionally, some areas may show nonhomogeneous high signal on T2-weighted images. It has been speculated that low-signal areas on T2 images may represent hemosiderin deposition or fibrosis, while areas of high signal on T2-weighted images may represent focal fluid collections or possibly ac-

tive inflammation (49). A nonspecific sign of synovial reaction—termed the "irregular infrapatellar fat pad sign"—has been described by Stoller et al. (3) (Fig. 34.31). This represents an irregular appearance at the junction of the infrapatellar fat pad and the cartilage of the femoral condyle and may be found in patients with rheumatoid arthritis, hemophilia, hemorrhagic effusions, pigmented villinodular synovitis, and other inflammatory arthritides. Magnetic resonance imaging allows the precise delineation of effusions and synovial cysts as well as evaluation for meniscal and ligamentous injuries that may occur in addition to the underlying disease process.

Bone changes seen with rheumatoid arthritis include marginal and subchondral erosions and intraosseous cysts formation (Figs. 34.32 and 34.33). We have noted on several occasions significant cysts that were undetected by plain radiographic technique. Secondary bone changes such as osteonecrosis or fractures may be detected.

Juvenile Chronic Arthritis

The classification of juvenile chronic arthritis is complex and a thorough discussion of the topic is beyond the scope of this chapter. The precise classification of juvenile chronic arthritis requires attention to clinical history, extraarticular manifestations, laboratory evaluation—including rheumatoid factor and HLA typing—as well as attention to physical and radiographic findings. The true nature of the disease is often not apparent until a significant follow-up time period. In the individual case MR imaging may demonstrate early synovial involvement (nonspecific), while potentially excluding other conditions such as leukemia, neuroblastoma, or osteomyelitis.

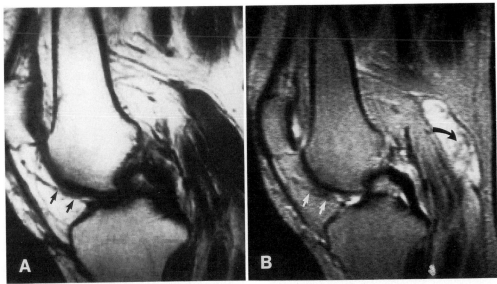

Figure 34.30. Rheumatoid arthritis. T1 (**A,** TR = 700, TE = 20) and T2 (**B,** TR = 2500, TE = 90) images demonstrate a popliteal cyst with heterogeneous signal intensity consistent with synovial hypertrophy or debris within the cyst *(curved arrow)*. Synovial hypertrophy is also present along the anterior aspect of the femoral condyle *(small arrows)*.

Juvenile onset of adult-type rheumatoid arthritis is characterized by positive rheumatoid factor and tends to affect females above the age of 10. The involvement is similar to classic adult rheumatoid arthritis, showing early severe destructive arthritis, subcutaneous nodules, and polyarticular involvement. Periostitis may be present in this form of juvenile chronic arthritis, in distinction to adult rheumatoid arthritis (50).

Magnetic resonance imaging findings in juvenile onset of adult-type rheumatoid arthritis are similar to adult onset findings: pannus formation and synovial thickening, joint effusions and synovial cysts, marginal and subchondral erosions, and secondary subchondral bone changes (Figs. 34.34 and 34.35). Growth disturbances accompanying early onset juvenile chronic arthritis are generally not severe.

Seronegative juvenile chronic arthritis may be divided into systemic disease, polyarticular disease, and pauciarticular or monoarticular disease. The systemic disease (Still's) is characterized by an acute febrile onset, rash, lymphadenopathy, hepatosplenomegaly, and only mild joint involvement. Polyarticular and pauciarticular diseases typically show evidence of growth disturbances, with ballooning of the epiphyses. Evidence of synovial thickening may be found. Bony erosion is a late manifestation, but is well seen with MR imaging (48).

Hemophilia

Patients with hemophilia suffer repeated hemarthroses, which lead to deposition of hemosiderin in the synovial membrane and periarticular tissues. Pannus formation subsequently occurs, with eventual destruction of carti-

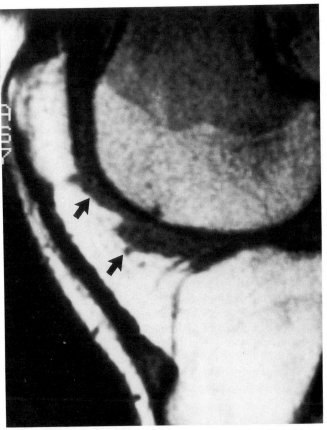

Figure 34.31. Irregular infrapatellar fat pad sign. T1-weighted image (TR = 600, TE = 25) demonstrates an irregular appearance to the junction between the fat pad and condyle, consistent with nonspecific synovial hypertrophy *(arrows)*.

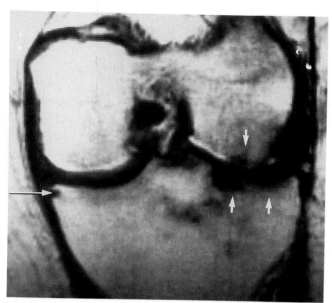

Figure 34.32. Rheumatoid arthritis. Subchondral bone changes *(short arrows)* and small marginal erosions *(long arrow)* may be detected early with MR imaging (TR = 700, TE = 20).

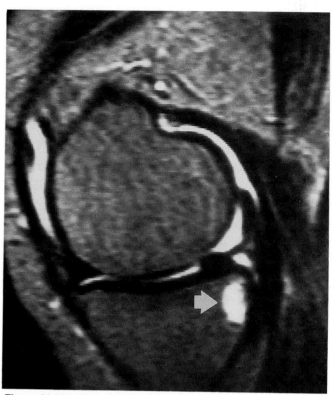

Figure 34.33. Rheumatoid arthritis. T2-weighted image (TR = 2500, TE = 90) demonstrates a moderate-sized subchondral cyst in the posterior aspect of the tibial plateau *(arrow)*.

lage in a pancompartmental distribution. Irregular subchondral bone erosion with eventual deformity, contractures, and ankylosis occur. Epiphyseal overgrowth due to hyperemia is present along with accelerated skeletal maturation.

Magnetic resonance studies may reveal areas of low signal intensity on both T1- and T2-weighted images corresponding to hemosiderin-laden synovium, pannus, and fibrosis (51). In advanced cases, cartilage loss is severe, with marked subchondral bone irregularity and contracture (Fig. 34.36). Widening of the intercondylar notch, a common plain radiographic manifestation, may be visualized on coronal views.

Pigmented Villonodular Synovitis

Pigmented villonodular synovitis (PVNS) is an uncommon lesion arising from the synovial lining of joints. Localized and diffuse forms exist, with the latter form occurring most commonly in the knee. The etiology of PVNS remains uncertain, with some evidence supporting a reactive lesion secondary to intraarticular bleeding and trauma. Pigmented villonodular synovitis derives its name from the pathologic deposition of hemosiderin that occurs in the tissue, giving it a brownish gross appearance. On MR imaging examination PVNS may appear as a localized synovial based mass, or as more diffuse thickening of the synovium surrounding the knee. While parts of the mass may show nonspecific signal changes associated with an admixture of synovial fluid and soft tissue, the lesion may show characteristic areas of hemosiderin deposition, as shown by decreased signal on both T1- and T2-weighted pulse sequences (Fig. 34.37). This appearance was present in seven of seven cases of PVNS reported by Jelinek et al. (52).

Loose Bodies

Loose bodies are most commonly found in patients with preexisting osteoarthritis, or prior osteochondral fracture. Small calcified loose bodies may be extremely difficult to visualize with MR imaging alone, and therefore plain radiographs should accompany any case in which loose bodies are suspected. Many large loose bodies demonstrate well-defined internal areas showing increased signal on T1-weighted images due to marrow fat (Fig. 34.38). Multiple calcified loose bodies will frequently localize in the posterior aspect of the knee, sometimes within a popliteal cyst (Fig. 34.39).

Synovial osteochondromatosis or chondromatosis is an uncommon lesion characterized by multiple cartilaginous and ostocartilaginous loose bodies. The etiology of the disease is uncertain—possible etiologies include trauma, inflammation, osteocartilaginous metaplasia, or benign neoplasia (53). The knee is affected in 50% of cases. The MR image appearance may be variable according to the makeup of the loose bodies. Ossified bodies may show low signal on both T1 and T2 images (3). Noncalcified cartilaginous bodies would be expected to show intermediate signal characteristics typical of cartilage.

Cysts, Plicae

Uncomplicated popliteal cysts are very common in the knee and are consistently found at the posteromedial aspect of the knee joint, between the medial head of the gastroc-

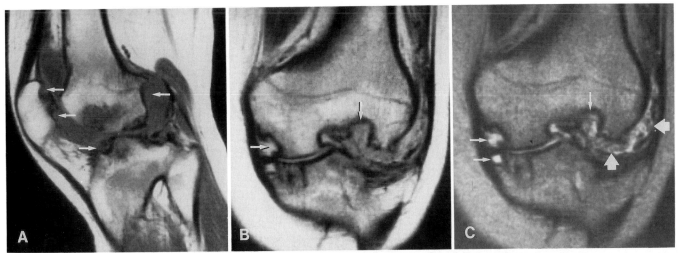

Figure 34.34. Juvenile onset of adult-type rheumatoid arthritis. T1-weighted sagittal **(A)** and coronal **(B)** images (TR = 700, TE = 20) and coronal T2-weighted image (TR = 2500, TE = 90) **(C)** demonstrate effusion, synovial hypertrophy *(short arrows)*, joint space narrowing with subchondral changes, and severe erosive changes *(small arrows)*.

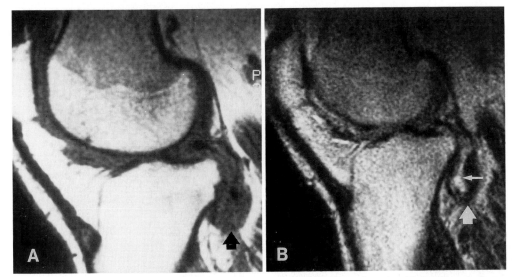

Figure 34.35. Juvenile chronic arthritis. T1-weighted **(A,** TR = 600, TE = 25) and T2-weighted **(B,** TR = 2000, TE = 80) images demonstrate marked synovial hypertrophy anteriorly and posteriorly, showing intermediate to low signal intensity on both sequences. Posteriorly there is synovial thickening *(large arrow)* with a small amount of synovial fluid present *(small arrow).*

nemius muscle and the semimembranosus muscle. Popliteal cysts are common among patients with recurrent effusions, particularly rheumatoid arthritis and chronic internal derangements. Popliteal cysts appear homogeneously hypointense relative to skeletal muscle on T1-weighted images, with increased intensity on T2-weighted images (Fig. 34.40) (1, 45, 46). With rheumatoid arthritis, the appearance may be heterogeneous, secondary to synovial hypertrophy and pannus formation.

Meniscal Cysts

Meniscal cysts are fairly common and are invariably associated with meniscal tears. Previous reports, based on arthrography, found meniscal cysts most commonly associated with the lateral side, although a small series of MR–

demonstrated meniscal cysts showed a medial predominance (54). Meniscal cysts show isointense or hypointense signal compared with skeletal muscle on T1-weighted sequences. T2-weighted images show homogeneous increased signal intensity. The relatively isointense appearance on T1-weighted images may be attributed to the proteinaceous quality of the fluid filling such cysts. T2-weighted images are frequently able to demonstrate the connection of the meniscal cysts to the joint, important from an operative standpoint.

Ganglion Cysts

Ganglion cysts may be found in variable locations around the knee and are not associated with meniscal tears. Ganglion cysts cannot be differentiated from meniscal cysts on

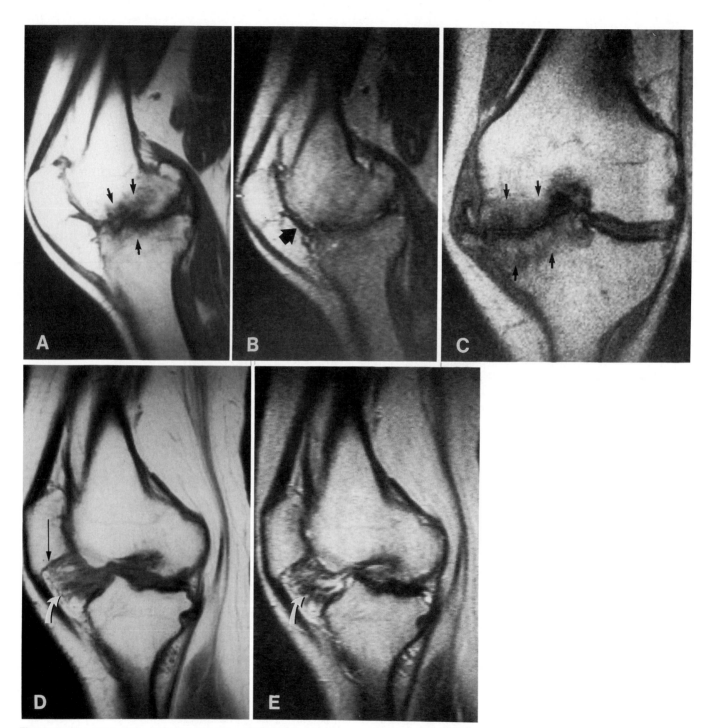

Figure 34.36.   Hemophilia. T1-weighted (TR = 700, TE = 20) and T2-weighted (TR = 2500, TE = 90) sagittal **(A** and **B)** and coronal **(C)** views demonstrate severe cartilage loss involving predominantly the medial compartment of the knee, associated with marked subchondral bone changes *(small arrows)* and osteophyte formation. Very low-signal areas are present anteriorly that may represent hemosiderin-laden synovium or fibrosis *(large arrow)*. T1- and T2-weighted sagittal views of the opposite knee in the same patient **(D** and **E)** demonstrate irregular low-signal regions within the infrapatellar fat pad, all consistent with hemosiderin deposits or fibrosis *(curved arrows)*. "Squaring" of the inferior pole of the patella *(straight arrow)* is also present.

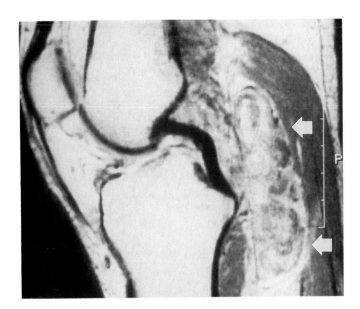

Figure 34.37. Pigmented villonodular synovitis. Intermediate-weighted (TR = 2500, TE = 35) sagittal image demonstrates an oblong region of heterogeneous abnormal signal intensity with some areas of low signal *(arrows)* in the popliteal fossa of the knee. PVNS frequently demonstrates low-signal areas on both T1- and T2-weighted images, secondary to hemosiderin deposits.

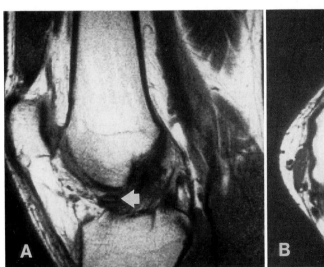

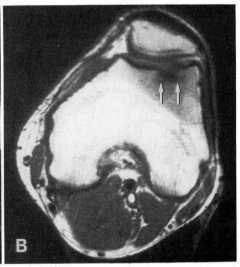

Figure 34.38. Loose body. T1-weighted sagittal image **(A)** shows a corticated loose body between the femoral condyle and tibia *(large arrow)*. Donor site was presumably an osteochondral fracture of the lateral trochlea as demonstrated in **(B)** *(arrows)*.

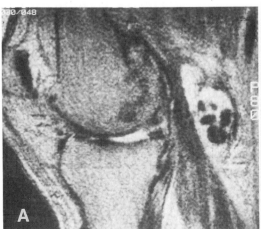

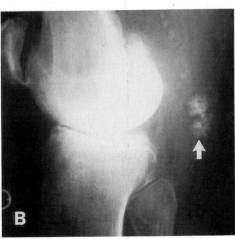

Figure 34.39. Multiple calcified loose bodies are demonstrated within a popliteal cyst on T2-weighted sagittal image (TR = 2150, TE = 80) **(A).** Corresponding lateral plain film **(B)** shows multiple loose bodies in the popliteal fossa *(arrow)*.

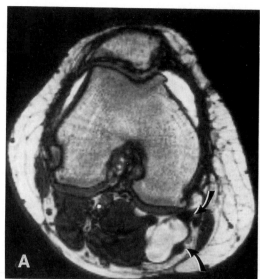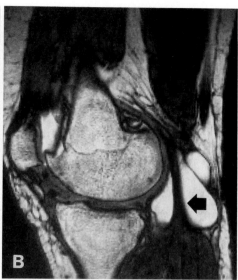

Figure 34.40. Popliteal cyst. Axial **(A)** and sagittal **(B)** gradient-echo images (3D FISP 75, TR = 40, TE = 15) demonstrate a high-intensity lobulated cyst extending from between the medial head of the gastroc-nemius muscle *(large arrow)* and semimembranosus and semitendo-sus tendons *(curved arrows).*

the basis of their signal intensity, but are more apt to show internal septations (54). A connection to the associated joint is generally not found. Ganglion cysts may be entirely intramuscular, and this diagnosis may be entertained when a cystic lesion is seen in such a location. In equivocal cases the use of intravenous Gd will demonstrate the nonenhancing cystic nature of the mass.

Plicae

Plicae are synovial folds within the knee that are due to a failure of involution of normal embryologic synovial septations. Plicae are common normal variants, occurring in 20 to 60% of normal adult knees. Three locations are most common: suprapatellar, mediopatellar, and infrapatellar. The suprapatellar plica is easily visualized on sagittal sequences as a thin, low-signal band extending through the suprapatella bursa, and generally has no clinical significance. A large mediopatellar plica may become inflamed and thickened due to chronic impingement, producing pain, clicking, or locking, mimicking a meniscal tear. The mediopatellar plica is best visualized on axial views as a low-signal linear structure medial to the patella (Fig. 34.41).

Osteonecrosis, Osteochondritis Dissecans, and Related Disorders

Osteonecrosis

Magnetic resonance imaging is extremely sensitive to marrow changes and is therefore excellent for the early detection as well as staging of osteonecrosis and related disorders. Metaphyseal bone infarcts may be idiopathic or associated with steroid use, collagen vascular disease, ethanol abuse, sickle cell disease, or other miscellaneous causes. Bone infarcts typically show a serpiginous low-signal border with a central high-signal region on T1-weighted images, corresponding to a thin rim of reactive bone with central yellow marrow (25) (Fig. 34.42). The length of time

required for a bone infarct to assume this characteristic appearance is currently unknown. Rao et al. described the appearance of acute infarcts in patients with underlying bone marrow disease (55). The authors found focal intramedullary areas of low signal on T1-weighted images and high signal on corresponding T2-weighted images, which they attributed to edema. Since no well-defined low-signal rim was found, this acute appearance could not be distinguished from other infiltrative processes, including infection and neoplasm. The time required for conversion of the acute infarct to the immature (healing) and chronic stages is uncertain. Once the infarct shows a circumscribed low-signal border with central high-signal region on T1-weighted images, the appearance is characteristic.

Spontaneous osteonecrosis of the medial femoral condyle occasionally occurs in middle-aged and elderly individuals, predominantly females. These patients usually experience an abrupt onset of symptoms, frequently mimicking meniscal tears. The lesion may be associated with medial meniscal tears, although the exact cause remains obscure. Magnetic resonance imaging demonstrates the area of osteonecrosis as a focal region of low-signal intensity on T1- and T2-weighted images, predominantly involving the weight-bearing aspect of the medial condyle (Fig. 34.43). The lesion may be associated with subchondral fracture and eventual collapse of the articular surface, which may be visualized with MR imaging.

Osteochondritis Dissecans

Osteochondritis dissecans is a disease of childhood or young adulthood, usually involving the inner aspect of the medial femoral condyle. The exact etiology of osteochondritis dissecans is controversial although most authors favor repeated or acute osteochondral trauma. The radiographic appearance is characterized by a well-defined subchondral bone fragment within an excavated portion

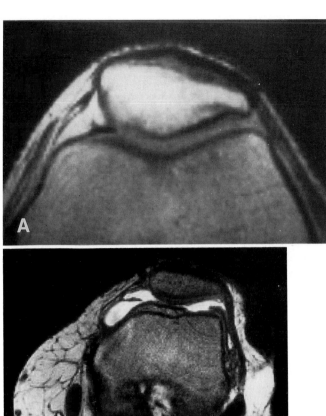

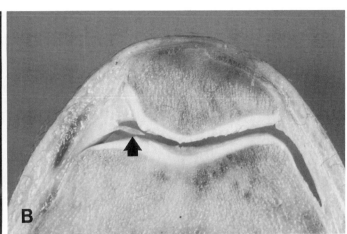

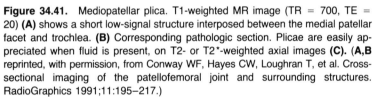

Figure 34.41. Mediopatellar plica. T1-weighted MR image (TR = 700, TE = 20) **(A)** shows a short low-signal structure interposed between the medial patellar facet and trochlea. **(B)** Corresponding pathologic section. Plicae are easily appreciated when fluid is present, on T2- or T2*-weighted axial images **(C)**. (A,B reprinted, with permission, from Conway WF, Hayes CW, Loughran T, et al. Cross-sectional imaging of the patellofemoral joint and surrounding structures. RadioGraphics 1991;11:195–217.)

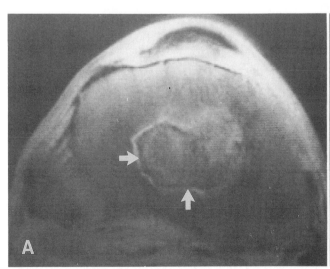

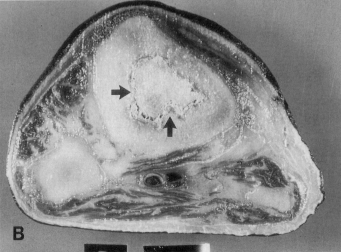

Figure 34.42. Mature bone infarcts show a serpiginous thin low-signal rim on T1-weighted images, with a high-signal center **(A,** *arrows*). The outermost high-signal rim is an artifact of disarticulation. **(B)** Pathologic section from same specimen.

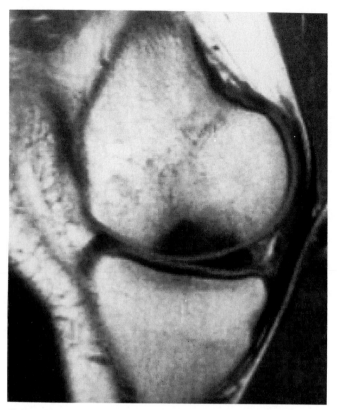

Figure 34.43. Spontaneous osteonecrosis of the medial femoral condyle. T1-weighted image (TR = 700, TE = 20) demonstrates a focal area of decreased signal at the weight-bearing surface of the femoral condyle. There was no evidence of subchondral bone collapse.

of the condyle. The fragment may eventually heal completely or may become detached, becoming an intraarticular loose body.

In determining whether surgical intervention is necessary, the determination of the mechanical stability of the fragment is critical. Both plain film and nuclear medicine appearances are helpful in determining whether the fragment is stable, loose in situ, or grossly loose (56). Magnetic resonance imaging has an advantage over the former modalities by its increased spacial resolution and sensitive evaluation of soft tissues.

Small osteochondral defects on the lateral aspect of the medial femoral condyle are best imaged in the coronal projection. Defects occurring far posteriorly are best seen on sagittal projections. Occasionally, trochlear osteochondral defects are best imaged in the axial projection. Individual lesions are seen as foci of low signal on T1-weighted images. The margins of the adjacent parent bone usually shows a low-signal rim of various thickness. Mesgarzadeh et al. found increased signal on T2-weighted images between the low-signal rim of parent bone and the osteochondral fragment in lesions that were grossly loose or loose in situ (56). Presumably this was due to fluid between the fragment and parent bone. Displacement of fragments from the condyle may be demonstrated. Disruption of the articular cartilage, which could previously be demonstrated only by arthrotomography, may also be seen on T2 or T2* images (Fig. 34.44).

The major differential diagnosis of osteochondritis dissecans is the irregularity of the condyle, which may be seen as a normal variant in children, usually between the

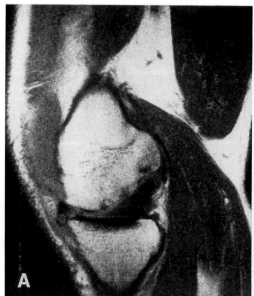

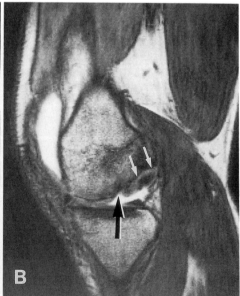

Figure 34.44. Osteochondritis dissecans. T1-weighted (TR = 700, TE = 20) sagittal image demonstrates a focal area of decreased signal in the posterior aspect of the medial femoral condyle **(A)**. Corresponding gradient-echo image (3D FISP 75, TR = 40, TE = 15) clearly demonstrates a large, irregular cartilage defect *(large arrow)*, as well as linear high signal extending between the parent bone and fragment *(small arrows)*, indicative of gross loosening.

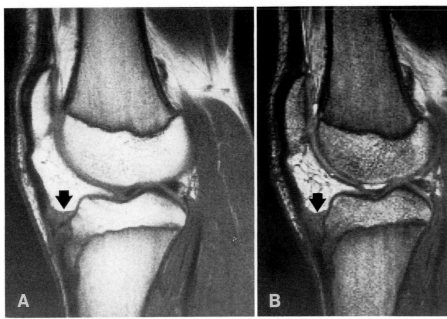

Figure 34.45. Osgood-Schlatter disease. T1-weighted (TR = 700, TE = 20) and T2*-weighted sagittal images (**A** and **B**, respectively) demonstrate filling in of the normal sharp V shape formed by the patellar tendon and anterior tibia with intermediate- to low-signal tissue *(arrows).* There is some increased signal of the patellar tendon at its insertion as well.

ages of 2 and 12. The lesions should show no evidence of loosening, and are characteristically located more posteriorly on the condyle than their pathologic counterparts.

Osgood-Schlatter Disease

Chronic tension from the powerful quadriceps muscles at their insertion on the growing tibial tuberosity is generally regarded as the cause of Osgood-Schlatter disease. The clinically painful swelling at the patellar tendon insertion is neither a true osteonecrosis nor an inflammatory process, but more likely a response to chronic microavulsions. The diagnosis is typically made on the basis of symptoms and physical examination. In these cases MR imaging examination is warranted only to exclude a more serious disease such as infection or tumor, or if the symptoms are atypical.

Plain radiographs may demonstrate soft-tissue swelling over the involved tibial tuberosity along with an indistinct appearance to the inferior margin of the patellar tendon. The normally sharp V shape formed by the patellar tendon and the anterior margin of the tibial tuberosity is rendered unsharp due to replacement of infrapatellar fat by soft-tissue density. Fragmentation of the tibial tuberosity may occur, although this finding, by itself, may represent a normal variant.

On MR imaging, T1-weighted images show replacement of the normal high-signal infrapatellar fat with an area of decreased signal adjacent to the patellar tendon insertion (40, 46) (Fig. 34.45). The tendon itself may show focal areas of increased signal, depending on the degree of associated tendinitis.

Infection

Infections of the appendicular skeleton are well suited to evaluation by MR imaging. T1-weighted images are extremely sensitive to changes within the marrow and the extent of marrow involvement can be readily assessed. Extension of osteomyelitis into the soft tissues may be evaluated with T2-weighted images, also sensitive for soft-tissue abscess (Fig. 34.46). Magnetic resonance imaging may also allow the noninvasive distinction between cellulitis and osteomyelitis or septic arthritis (57) (Fig. 34.47). Specific findings in osteomyelitis are covered elsewhere in this book (see Chapters 13, 37, and 41).

Tumors and Tumor-Like Conditions

The general approach to musculoskeletal tumors has been covered elsewhere in this book (see Chapter 15). Primary bone neoplasms occur most commonly about the knee, and thus will be briefly covered in this section. Likewise, numerous tumor-like conditions occur around the knee and may be a source of confusion in their differential diagnosis.

The knee is a relatively common site for primary soft-tissue tumors, both benign and malignant. Benign tumors such as lipomas, neurofibromas, hemangiomas, and soft-tissue chondromas occur occasionally around the knee (Fig. 34.48). Malignant soft-tissue neoplasms include malignant fibrous histiocytoma, neurofibrosarcoma, liposarcoma, and synovial sarcomas. Magnetic resonance imaging may, in many cases, distinguish between the extraarticular or intraarticular origin of soft-tissue masses and enable differ-

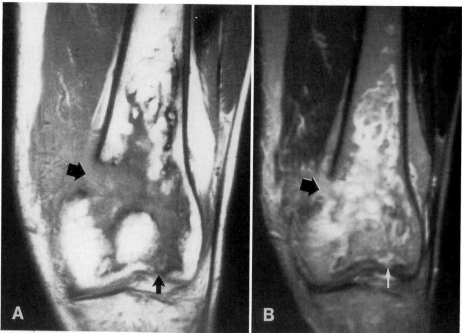

Figure 34.46. Osteomyelitis. T1-weighted (TR = 700, TE = 20) and T2-weighted (TR = 2500, TE = 90) coronal images (**A** and **B,** respectively) show an extensive area of irregular signal changes within the distal shaft and epiphysis of the femur, representing chronic, active osteomyelitis in a patient with preexisting bone infarcts. The large cloaca *(large arrows)* is well demonstrated. Extension to the articular surface is also demonstrated *(small arrows),* although no definite extension into the joint could be demonstrated.

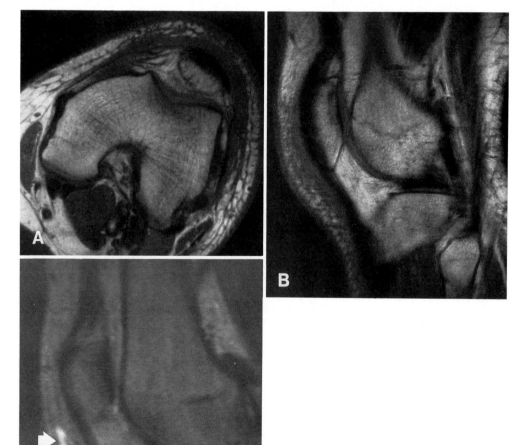

Figure 34.47. Prepatellar cellulitis. Axial **(A)** and sagittal **(B)** T1-weighted images (TR = 700, TE = 20) demonstrate marked irregular decreased signal within the subcutaneous fat around the anterior aspect of the knee. The suprapatellar and infrapatellar fat pads appeared unaffected. Sagittal T2-weighted image (TR = 2500, TE = 90) **(C)** shows a small subcutaneous abscess *(arrow).*

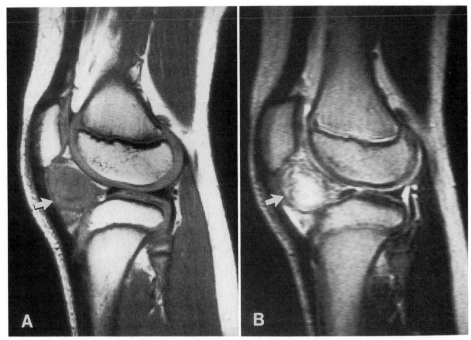

Figure 34.48. Soft-tissue tumor. Sagittal T1-weighted (TR = 700, TE = 20) and T2-weighted (TR = 2300, TE = 90) images (**A** and **B,** respectively) demonstrate a circumscribed mass occupying the infrapatellar fat pad of the knee *(arrows).* The mass showed mildly hyperintense signal relative to skeletal muscle on T1-weighted images, with heterogeneous increased signal on T2-weighted image. Final pathology showed a neurofibroma.

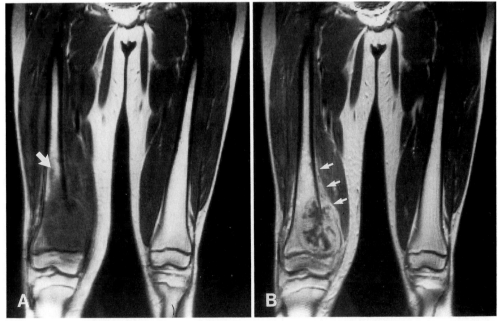

Figure 34.49. Osteosarcoma. Coronal T1-weighted images (TR = 700, TE = 20) pre- and post-Gd administration (**A** and **B,** respectively). Precontrast image best demonstrates intraosseous extent of tumor *(arrow).* Following Gd enhancement, the soft-tissue extent of the lesion is appreciated *(small arrows),* although the intraosseous component becomes isointense with the marrow.

entiation from entities such as PVNS or synovial hypertrophy. Except in selected cases specific tissue characterization and positive differentiation between benign and malignant tumors are usually not possible.

The goal of the MR imaging evaluation of suspected primary bone tumors of the knee is similar to that of tumors elsewhere: *(a)* accurate determination of the intraosseous and extraosseous extent of the lesion *(b)* intracompartmental vs. extracompartmental extension, *(c)* status of the major neurovascular bundles, and *(d)* allowing a reasonable differential diagnosis (Fig. 34.49).

Chondroid tumors deserve special mention. Enchondromas are common in the distal femoral metaphysis and proximal tibia. On the basis of plain films, it may be dif-

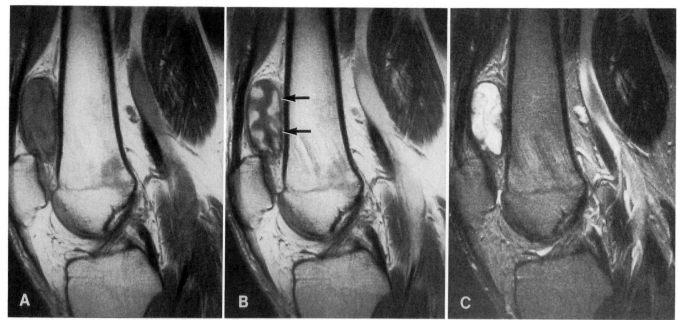

Figure 34.50. Soft-tissue mass. T1-weighted sagittal images (TR = 700, TE = 20) pre- and postintravenous Gd administration (**A** and **B**, respectively) show an ovoid mass in the region of the suprapatellar bursa. No bone involvement is present. A serpiginous rimlike enhancement pattern is demonstrated *(arrows)*. T2-weighted image (TR = 2500, TE = 90) **(C)** shows nearly homogeneous intense increased signal. Both serpiginous enhancement with Gd administration and homogeneous intense signal increase on T2-weighted images have been described in tumors containing hyaline cartilage. Final pathologic diagnosis was a benign soft-tissue chondroma.

ficult to distinguish an enchondroma from a low-grade chondrosarcoma. On the basis of their appearance on T1- and T2-weighted pulse sequences, it is not possible to differentiate benign from malignant chondroid tumors (11). Likewise, the rimlike pattern of enhancement of chondroid matrix seen following intravenous Gd may be present in both benign and malignant cartilaginous tumors (Fig. 34.50). Magnetic resonance imaging may demonstrate areas of endosteal erosion or small areas of cortical breakthrough in the case of chondrosarcoma that are indicative of malignancy. However, one must pay attention to the possibility of chemical shift artifact producing apparent asymmetry of cortical bone, particularly on axial images. Such chemical shift artifact may lead to the erroneous conclusion that bone erosion is present (Fig. 34.4).

Fibrous metaphyseal defects and nonossifying fibromas are lucent lesions frequently occurring eccentrically in the metaphyses of long bones in adolescents, usually near the insertion of a tendon or ligamentous structure. The plain radiographic appearance of these lesions is generally diagnostic by way of their sharp, thinly sclerotic margins, lightly trabeculated bubbly appearance, and classic eccentric location. The lesions are most often encountered incidentally on MR imaging. The MR image appearance frequently shows a well-defined lobulated area of low to intermediate signal intensity with some trabeculation (58). The lesions are located eccentrically in the medullary canal and produce variable cortical thinning or slight bony expansion. Some areas of the lesion may show distinct enhancement following intravenous Gd administration. As with many bone lesions, the diagnosis is based more on morphology and characteristic site of the lesion than its signal characteristics.

References

1. Hartzman S, Reicher MA, Bassett LW, et al. MR imaging of the knee. Part II. Chronic disorders. Radiology 1987;162:553–557.
2. Mink JH, Reicher MA, Crues JV. Magnetic resonance imaging of the knee. New York: Raven Press, 1987.
3. Stoller DW, Genant HK, Helms CA, et al. Magnetic resonance imaging in orthopaedics and rheumatology. Philadelphia: Lippincott, 1989.
4. Yulish BS, Montanez J, Goodfellow DB, et al. Chondromalacia patellae: assessment with MR imaging. Radiology 1987;164:763–766.
5. Hayes CW, Sawyer RW, Conway WF. Patellar cartilage lesions: in vitro detection and staging with MR imaging and pathologic correlation. Radiology 1990;176:479–483.
6. Gylys-Morin VM, Hajek PC, Sartoris DJ, et al. Articular cartilage defects: detectability in cadaver knees with MR. AJR 1987;148:1153–1157.
7. Reiser MF, Bongartz G, Erlemann R, et al. Magnetic resonance in cartilaginous lesions of the knee joint with three-dimensional gradient-echo imaging. Skel Radiol 1988;17:465–471.
8. Spritzer CE, Vogler JB, Martinez S, et al. MR imaging of the knee: preliminary results with a 3DFT GRASS pulse sequence. AJR 1988;150:597–603.
9. Konig H, Sauter R, Deimling M, et al. Cartilage disorders: comparison of spin-echo, CHESS, and FLASH sequence MR images. Radiology 1987;164:753–758.
10. Hajek PC, Baker LL, Sartoris DJ, et al. MR arthrography: anatomic-pathologic investigation. Radiology 1987;163:141–147.
11. Erlemann R, Reiser MF, Peters PE, et al. Musculoskeletal neoplasms: static and dynamic Gd-DTPA-enhanced MR imaging. Radiology 1989;171:767–773.
12. Cohen EK, Kressel HY, Frank TS, et al. Hyaline cartilage-origin bone and soft-tissue neoplasms: MR appearance and histologic correlation. Radiology 1988;167:477–481.
13. Petterson H, Slone RM, Spanier S, et al. Musculoskeletal tumors: T1 and T2 relaxation times. Radiology 1988;176:783–785.
14. Munk PL, Helms CA, Holt RG, et al. MR imaging of aneurysmal bone cysts. AJR 1989;153:99–101.
15. Dick BW, Mitchell DG, Burk L, et al. The effect of chemical shift

misrepresentation on cortical bone thickness on MR imaging. AJR 1988;151:537–538.

16. Deutsch AL, Mink JH, Rosenfelt FP, et al. Incidental detection of hematopietic hyperplasia on routine knee MR imaging. AJR 1989;152:233–236.

17. Carson WG, James SL, Larson RL, et al. Patellofemoral disorders: physical and radiographic evaluation. Part II. Radiographic examination. Clin Orthopaed Relat Res 1984;185:178–186.

18. Dugale TW, Barnett PR. Historical background: patellofemoral pain in young people. Orthoped Clin N Am 1986;17(2):211–219.

19. Shahriaree H. Chondromalacia. Contemp Orthopaed 1985;11(5):27–39.

20. Bentley G, Dowd G. Current concepts of etiology and treatment of chondromalacia patellae. Clin Orthopaed Relat Res 1984;189:209–228.

21. Outerbridge RE, Dunlop JAY. The problem of chondromalacia patellae. Clin Orthoped Relat Res 1975;110:177–196.

22. Lauren CA, Dussault R, Levesque HP. The tangential x-ray investigation of the patellofemoral joint: x-ray technique, diagnostic criteria and their interpretation. Clin Orthopaed Relat Res 1979;144:16–26.

23. Lund F, Nilsson BE. Radiologic evaluation of chondromalacia patellae. Acta Radiol Diagn 1980;21:413–416.

24. Merchant AC, Mercer RL, Jacobsen RH, et al. Radiographic analysis of patellofemoral congruence. J Bone Joint Surg 1974;56A(7):1391–1396.

25. Munk PL, Helms CA, Holt RG. Immature bone infarcts: findings on plain radiographs and MR scans. ARJ 1989;152:547–549.

26. Newberg AH, Seligson D. The patellofemoral joint: 30°, 60°, and 90° views. Radiology 1980;137:57–61.

27. Dye SF, Boll DA. Radionuclide imaging of the patellofemoral joint in young adults with anterior knee pain. Orthoped Clin N Am 1986;17(2):249–262.

28. Ihara H. Double-contrast CT arthrography of the cartilage of the patellofemoral joint. Clin Orthopaed Relat Res 1985;198:50–55.

29. Bovan F, Bellemans MA, Geurts J, et al. The value of computed tomography scanning in chondromalacia patellae. Skel Radiol 1982;8:183–185.

30. Martinez S, Korobkin M, Fondren FB, et al. Diagnosis of patellofemoral malalignment by computed tomography. J Comput Assist Tomogr 1983;7(6):1050–1053.

31. Schutzer SF, Ramsby GR, Fulkerson JP. Computed tomographic classification of patellofemoral pain patients. Orthoped Clin N Am 1986;17(2):235–248.

32. Stanford W, Phelan J, Kathol MH, et al. Patellofemoral joint motion: evaluation by ultrafast computed tomography. Skel Radiol 1988;17:487–492.

33. Ozonoff MB. Pediatric orthopedic radiology. Philadelphia: W. B. Saunders, 1979.

34. Outerbridge RE. The etiology of chondromalacia patellae. J Bone Joint Surg 1961;43B(4):752–757.

35. Outerbridge RE. Further studies on the etiology of chondomalacia patellae. Bone Joint Surg 1964;46B(2):179–190.

36. Delgado-Martins H. A study of the position of the patella using computerized tomography. J Bone Joint Surg 1979;61B(4):443–444.

37. Berquist TH. Imaging of orthopedic trauma and surgery. Chapter 6. In: The knee: Philadelphia: W. B. Saunders, 1986:293–390.

38. Goodfellow J, Hungerford DS, Woods C. Patello-femoral joint mechanics and pathology. 2. Chondromalacia patellae. J Bone Joint Surg 1986;56B(3):291–299.

39. Insall J, Salvati E. Patella position in the normal knee joint. Radiology 1971;101:101.

40. Bodne D, Quinn SF, Murray WT, et al. Magnetic resonance images of chronic patellar tendinitis. Skel Radiol 1988;17:24–28.

41. Moller BN, Krebs B, Jurik AG. Patellofemoral incongruence in chondromalacia and instability of the patella. Acta Orthoped Scand 1986;57:232–234.

42. Shellock FG, Mink JH, Fox JM. Patellofemoral joint: kinematic MR imaging to assess tracking abnormalities. Radiology 1988;168:551–553.

43. Shellock FG, Mink JH, Deutsch AL, Fox JM. Patellar tracking abnormalities: clinical experience with kinematic MR imaging in 130 patients. Radiology 1989;172:799–804.

44. Wiberg G. Roentgenographic and anatomic studies of the patellofemoral joint, with special reference to chondromalacia patellae. Acta Orthoped Scand 1941;12:319.

45. Beltran J, Noto AM, Herman LJ, et al. Joint effusions: MR imaging. Radiology 1986;158:133–137.

46. Burk DL, Kanal E, Brunberg JA, et al. 1.5-T surface-coil MRI of the knee. AJR 1986;147:293–300.

47. Beltran J, Caudill JL, Herman LA, et al. Rheumatoid arthritis: MR imaging manifestations. Radiology 1987;165:153–157.

48. Yulish BS, Lieberman JM, Newman AJ, et al. Juvenile rheumatoid arthritis: assessment with MR imaging. Radiology 1987;165:149–152.

49. Reiser MF, Bogartz GP, Erleman R, et al. Gadolinium-DTPA in rheumatoid arthritis and related diseases: first results with dynamic magnetic resonance imaging. Skel Radiol 1989;18:591–597.

50. Resnick D, Niwayama G. Diagnosis of bone and joint disorders. Philadelphia: W. B. Saunders, 1988.

51. Yulish BS, Lieberman JM, Strandjord SE, et al. Hemophilic arthropathy: assessment with MR imaging. Radiology 1987;164:759–762.

52. Jelinek JS, Kransdorf MJ, Utz JA, et al. Imaging of pigmented villonodular synovitis with emphasis on MR imaging. AJR 1989;152:337–342.

53. Spjut HJ, Dorfman HD, Fechner RE, et al. Tumors of bone and cartilage. Washington, D.C.: Armed Forces Institute of Pathology, 1983.

54. Burk DL, Dalinka MK, Kanal E, et al. Meniscal and ganglion cysts of the knee: MR evaluation. AJR 1988;150:331–336.

55. Rao VM, Fishman M, Mitchell DG, et al. Painful sickle cell crisis: bone marrow patterns observed with MR imaging. Radiology 1986;161:211–215.

56. Mesgarzadeh M, Sapega AA, Bonakdarpour A, et al. Ostochondritis dissecans: analysis of mechanic stability with radiography, scintigraphy, and MR imaging. Radiology 1987;165:775–780.

57. Tang JSH, Gold RH, Bassett LW, et al. Musculoskeletal infection of the extremities: evaluation with MR imaging. Radiology 1988;166:205–209.

58. Ritschl P, Hajek PC, Pechmann U. Fibrous metaphyseal defects: magnetic resonance imaging appearances. Skel Radiol 1989;18:253–259.

CHAPTER 35

Magnetic Resonance Arthrography of the Knee

A. Engel and Paul C. Hajek

Introduction

The attempt to visualize the hyaline cartilage by means of magnetic resonance (MR) imaging represents a challenge to this technique. Comparisons of measurements of the thickness of the cartilage by MR imaging and by visual inspection of the cartilage, or by MR imaging and examination of histological sections of the cartilage, are conspicuous by their absence.

For the entire hyaline cartilage system, assessment of the cartilage surface by MR imaging is necessary both for the analysis of pathological changes (cartilage defects) and so that the MR imaging assessment can be compared with the findings at operation. The findings obtained by MR imaging are hardly comparable, however, because different machines with field strengths from 0.15 to 1.5 T, with different sequences (T1, T2, and gradient echo), have been used, and because of the nonhomogeneous nature of the material investigated—human cartilage in vivo and in vitro, or animal cartilage (1, 2).

The statements made range from: "Articular cartilage is easily visualized in contrast to the neighboring low-intensity cortical bone and menisci" (3), through the laconic observation: ". . . it also allows direct visualization of the articular cartilage . . ." (4) to "It is not possible to distinguish between cortical bone and cartilage" (5). Other authors avoid the cartilage problem altogether, by not dealing with its assessment (6–9). Where a description of the hyaline cartilage is given, the emphasis varies so that attention is focused only on the cartilage surface, the thickness of the cartilage, the boundary between subchondral bone and cartilage, or the behavior of the signal within the cartilage. Buckwalter et al. (10) could detect defects as small

as 1 mm, and Wojtys et al. (11) from 3 mm upward in a cadaver study using 0.35 T. Other authors, however, have observed that small superficial defects can hardly be diagnosed, if at all (12–15). It is not until a chondromalacia stage II is reached (16) that 100% agreement is reported (12, 14).

Almost all authors agree that a joint effusion makes it easier to assess the hyaline cartilage, and in some cases gradient echo makes this possible (2–4, 11, 17, 18). The term "arthrogram effect" is used (12).

This arthrogram effect, whether caused by a joint effusion or by an iatrogenically induced effusion (1, 3), formed the basis for the development of MR arthrography.

In an earlier study (19) the signal behavior and also the T1 and T2 relaxation times of articular cartilage, meniscus, and ligaments were determined and compared with those of intraarticular blood and synovial fluid containing different amounts of albumin (24%, w/v; 12%, w/v; 1%, w/v). Sodium chloride (0.9%, w/v), Urografin (60%), air, and gadolinium-labeled diethylenetriaminepentaacetic acid (Gd-labeled DTPA) (Magnevist; Schering, Berlin, Germany) were also studied in the same way. Out of the different albumin solution concentrations, the one with 12% (w/v) showed the clearest differentiation from articular cartilage on T1-weighted images (Fig. 35.1). Such high albumin concentrations seem to be possible when severe arthritis is present (20). Urografin (60%) and the 0.9% NaCl solution showed equivalent signal behavior, but inadequate contrast. Intraarticular air yielded excellent contrast, especially of the surrounding structures. The accurate recognition of details was limited, however, because of the formation of numerous bubbles. Fresh intraarticular blood yielded inadequate contrast. Thus, except with severe arthritis the various pathological knee joint effusions that may occur do not represent an ideal "biological contrast medium" for MR imaging. However, the ideal conditions were achieved with Gd-DTPA, in that there was excellent contrast in relation to all articular structures in T1-weighted images (Fig. 35.2). In addition, Gd-DTPA did not show in vitro uptake by hyaline cartilage (21).

Gadolinium-DTPA in Cartilage and Synovium

Gadolinium (Gd^{3+}) is described, inter alia, as a trace element, in contrast to the elements that are abundant in

Abbreviations (see also glossary): DTPA, diethylenetriamine pentaacetic acid; FLASH, fast low-angle shot; FISP, fast imaging with steady-state free precession; ICP/MS, inductively coupled plasma quadrupole/mass spectrometer; PPM, parts per million.

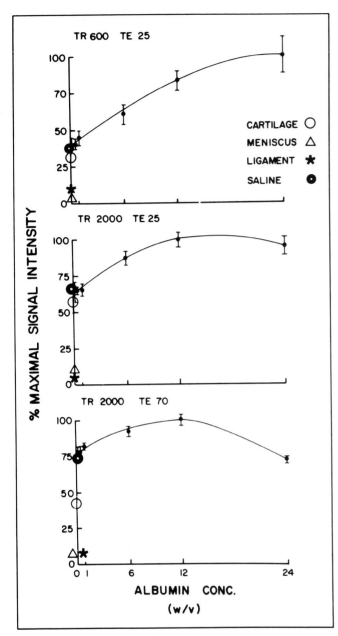

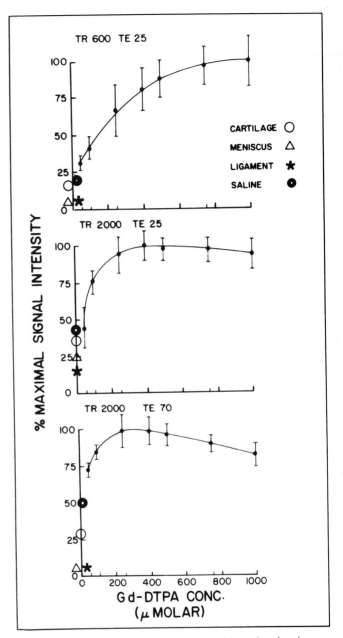

Figure 35.1. Relationship between percentage of maximum signal intensity and various concentrations of human serum albumin in 0.9% (w/v) saline for three different imaging sequences. A value of 100% is defined as highest signal intensity for each specific imaging sequence. For comparison, relative signal intensities of cartilage, meniscus, ligaments, and 0.9% (w/v) saline are plotted on the abscissa. Each point represents the arithmetic mean of five measurements (1 SD). Flattening of curves and a decrease in signal intensity for concentrations above 12% (w/v) result from increasing T2 effect. (Reprinted with permission from Hajek PC, Sartoris DJ, Neumann Ch, Resnick D. AJR 1987;149:97–104.)

Figure 35.2. Percentage of maximum signal intensity of various solutions of Gd-labeled DTPA-dimeglumine in 0.9% (w/v) saline for different imaging sequences. A value of 100% is defined as the highest signal intensity for each specific imaging sequence. Relative signal intensities for cartilage, meniscus, ligaments, and 0.9% (w/v) saline are plotted on the abscissa. Each point represents the arithmetic mean of five measurements (1 SD). Flattening of curves and decrease in signal intensity at high concentrations result from accentuated influence of decreasing T2 values. (Reprinted with permission from Hajek PC, Sartoris DJ, Neumann Ch, Resnick D. AJR 1987;149:97–104.)

the body, such as Ca^{2+}, Cl^-, K^+, Mg^{2+}, P, and S and the principal elements of tissues, C, O, N, and H. Trace elements are increasing in importance in diagnosis and therapy, since a deficiency or excess of these can cause metabolic disorders (22). The essential metabolic actions of a few trace elements are known, but are largely unknown for most of them (23). The methods used for trace element analysis were applied in order to detect any Gd, given by

intraarticular administration, that might still be present in the cartilage and synovial membrane.

Method

After previous administration of the contrast medium to patients who had received a knee endoprosthesis by implantation for osteoarthritis of the knee, pieces of cartilage were taken from two areas of the medial and lateral condyles in accordance with accepted resection practice ($n = 10$ knee joints) 17 to 137 hr following intraarticular injection. The freeze-dried pieces of tissue were subjected to pressure decomposition by the method of T + LG. The analysis of the pieces of cartilage and bone and of the synovial membranes was carried out with an inductively coupled plasma-quadrupole mass spectrometer (ICP/MS) of the plasma-quadrupol type from the Vg/Isotopes Company.

The lowest accurately measurable range for Gd was established as 0.05 ppm. Values above 0.2 ppm were classified as lying above the norm.

Results

The Gd concentrations measured in the different areas of the cartilage showed values which, apart from two borderline values (0.2 and 0.1 ppm), lay exclusively near the lower limit of detection (<0.05 ppm). No relationship was found between the time of the intraarticular administration or the taking of the samples and the concentration of Gd-DTPA or the measured Gd conjugates. This underlines the fact that Gd-DTPA (Magnevist) is eliminated very rapidly via the kidneys. Investigations have shown that after the intravenous administration of 0.1 mol Magnevist/kg, 83% has already been excreted renally after 6 hr. The amount 5 days after the injection is 91%. About 1% of the contrast medium is excreted via the stools (24). Early diffusion of Gd-DTPA-dimeglumine into viable cartilage might happen within the first 2 to 4 hr. The degree of diffusion, however, is very low and does not exceed 0.5 ppm.

The values measured in the synovial membranes were higher than those in the cartilage in all patients, especially in patients aged over 80 years (from 0.07 to 3.4 ppm). Our investigations did not provide any explanation for this, but fibrosis of the capsule and altered permeability might play a role.

Determination of the Optimal Contrast

Concentration and signal intensity studies have shown that the signal intensity increased with increasing concentration of Gd-DTPA (25). Very high concentrations did not result in any further improvement in the signal intensity, but led rather to a decrease. In the investigations of Hajek et al. (19, 26), carried out on cadaver knee joints, 500 μg molar was found to be the lowest concentration that was just about adequate. However, since the amount of free synovial fluid varies in vivo, these two statements about concentration cannot be applied to use in vivo.

Magnetic Resonance Sequences

A 1.5-T superconducting magnet (Magnetom 63; Siemens Co., Erlangen, Germany) was used for the MR investigations. The imaging was carried out by means of a circular polarized surface coil (knee resonator) 20 cm in diameter. Sagittal slices were produced and were supplemented by coronal and transverse section scans in some cases.

The following parameters were used.

T1-weighted spin-echo sequences:
TR = 700 ms, TE = 15 ms; slice thickness, 3 to 4 mm; gap between slices, 0 to 20% of the slice thickness; image matrix used, 256 × 256.
T2-weighted spin-echo sequences;
TR = 2500 ms, TE = 15/90 ms.
Gradient-echo sequences (both two- and three-dimensional (2D and 3D) gradient techniques were used):
FLASH (fast low-angle shot imaging) 2D: TR = 300 ms, TE = 10 ms; flip angle $\alpha = 135°$; effective slice thickness, 1 to 3 mm.
FLASH 3D: TR = 30 ms, TE = 10 ms; flip angle $\alpha = 30$ to 50°; effective slice thickness, 1 to 3 mm.
FISP (fast imaging with steady state free precession) 3D: TR = 30 ms, TE = 10 ms; flip angle $\alpha = 30$ to 50°; effective slice thickness, 1 to 3 mm.

Randomized Study to Establish the Optimal Concentration of Contrast Medium

In a randomized study on 20 patients, a Gd-DTPA solution was given by intraarticular administration in concentration of 1, 2, 4, and 5 mM in vivo (decision of the Ethics Committee of the University of Vienna, minutes of April 12, 1988). Signal behavior in relation to the articular cartilage was then determined on T1-weighted images.

In 1 and 2 mM solutions and occasionally also in 4 mM solutions, there was excellent contrast in the imaging of the articular cartilage. If a fairly large effusion was present there was excessive dilution and thus inadequate contrast with a 1 mM concentration of Gd-DTPA, so that the 2 mM solution was found to be the ideal concentration; 0.5 mM solutions injected twice resulted in only inadequate contrast because of marked dilution phenomena. The 5 mM solution resulted in excessive contrast, leading to spreading of the image in the MR tomogram. Only 2 mM solutions were therefore used for the subsequent investigations.

Magnetic Resonance Arthrography Technique

The technique required to introduce the Gd-DTPA into the knee joint differs only slightly from that currently used for intraarticular injection (27, 28). For the administration of Gd-DTPA (Magnevist) the knee joint is placed in a slightly bent, recumbent position for ventrolateral access or in a hanging position for intrapatellar access. In order to be able to fill the internal cavity of the knee with a small amount of contrast medium dissolved in physiological NaCl solution, compression is applied to the soft tissues 15 to 20 cm above the joint cavity with an elastic bandage (Fig. 35.3). The injection itself is carried out under sterile conditions. The amount of dissolved contrast medium injected amounts to ca. 30 to 40 ml. After the injection, intensive active and passive movements of the knee joint are carried out several times in order to achieve uniform distribution in the joint. This is followed by the MR imaging examination.

When marked synovitis is present, villi or septa may

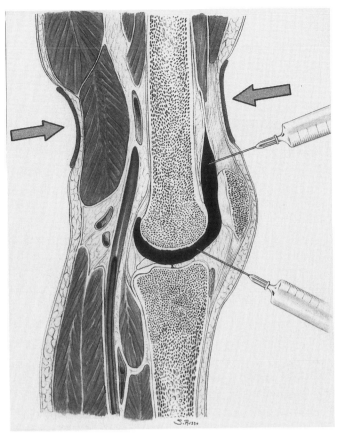

Figure 35.3. Illustration of the injection technique; upper recess reduced in size through compression with an elastic bandage.

prevent uniform distribution of the contrast medium with the use of a ventrolateral access route, so that the articular cartilage is inadequately visualized.

In these cases, but also in severe osteoarthritis of the knee with bone deformities, it is advisable to choose an infrapatellar access route, since after filling of the intercondylar space only a slight degree of movement can distribute the contrast medium sufficiently for visualization of the cartilage.

Clinical Relevance

Normal Cartilage

The macroscopically intact joint surface can be clearly demarcated on T1-weighted and gradient-echo images. The cartilage surface is defined by the contrast medium (Fig. 35.4). The cartilage appears as an area of intermediate signal intensity. This can be demarcated from the subchondral zone on T1-weighted images through the high signal intensity of the bone marrow, which is rich in fat. The zone lying in between, which is of intermediate signal intensity, is made up of calcified and uncalcified cartilage and also of the subchondral bone plate. The zone of calcified cartilage is normally ca. 0.15 mm thick and lies above the uncalcified cartilage, demarcated by the so-called "tidemark." Together with the subchondral bone plate this results in a zone of 0.45 mm. Since this order of magnitude is still less than the pixel size, this zone will make

only an indirect contribution to the signal. Since calcified tissue appears dark on T1 images all this will result in a slight decrease of the signal in this zone compared to the superficial layers of cartilage. On the gradient-echo images the zone of intermediate signal intensity is caused exclusively by uncalcified cartilage tissue. The calcified part of the cartilage, the subchondral plate, and the adjacent structures cannot be differentiated and display a uniform low signal. Healthy cartilage will be visualized as being of approximately the same thickness on T1-weighted and gradient-echo images.

Osteoarthritis

Osteoarthritis of the knee joint is a frequent clinical syndrome, but an advanced age need not necessarily be accompanied by arthritis (29). Casscells (30) was also able to demonstrate this in an investigation on 300 cadaver knee joints.

Not all of the questions concerning the etiology and course of the destruction of cartilage and subchondral bone have been clarified completely yet, but the morphological changes can be classified into stages.

The stage-classification of Otte (31) has been used for the assessment of the histological and MR arthrography findings. Otte distinguishes between four stages:

Stage 1: cartilage fissures close to the surface.
Stage 2: deep fissures extending to the subchondral plate; "cluster" formation.
Stage 3: loss of the uncalcified cartilage, the covering bone plate or remains of cartilage may be present at the base.
Stage 4: opening up of the subchondral medullary space, development of superficial granulation tissue or regenerating fibrocartilage.

This histological cartilage assessment is not directly applicable to the macroscopic assessment, however. The problems here concern the histological stages 1 and 2, because the structural change of the cartilage becomes manifest in a loss of the surface sheen (Table 35.1). The cartilage looks dull and appears of softer consistency than normal cartilage (32). It is only in the advanced histological stage 2 and in stage 3 that circumscribed superficial defects appear macroscopically, seen as areas of roughness, unevenness, or as deeper lesions.

The agreement between the histological and macroscopic findings is greatest in stages 3 and 4, since macroscopically extensive, focal cartilage defects and areas with complete loss of the cartilage, which have been described as "bare bone" (33), are present.

Mild Arthritis

Signs of arthritis include changes in the form of a broadening of the subchondral bone plate, an increase in the zone of calcified cartilage (34), and often a doubling or severalfold increase in the tidemark (35). This therefore makes the cartilage appear thicker on T1-weighted images than on gradient-echo images (Fig. 35.5 *A–D*).

The difference between the cartilage thickness seen on

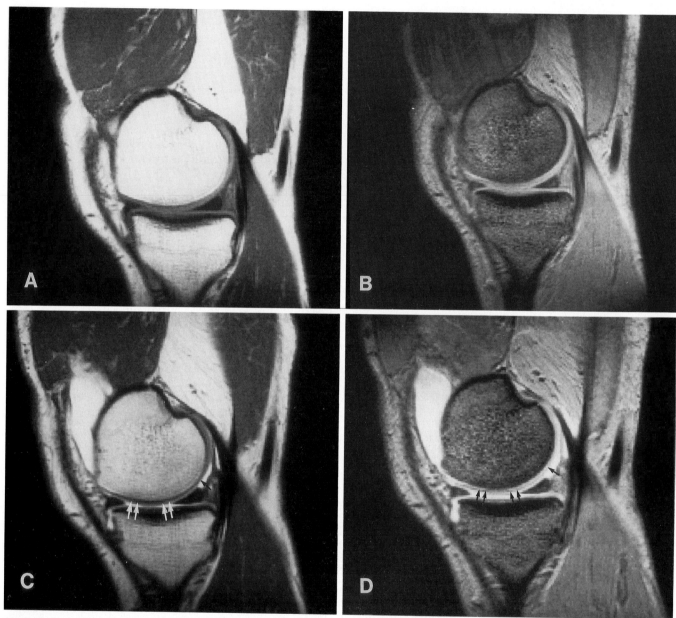

Figure 35.4. MR imaging, spin-echo 700/15 **(A** and **C),** gradient-echo 30/10/30°, FISP 3D **(B** and **D):** normal knee joint, sagittal plane. **(A** and **B)** Before the intraarticular injection of Gd-DTPA; **(C** and **D)** after the injection. The identification of the hyaline articular cartilage (*arrows*) applies to all sections of the joint, as long as sufficient contrast medium is present.

Table 35.1
Assessment of Arthritis[a]

Histological Staging by System of Otte[b]	Macroscopic Staging	Radiographic Staging by System of Jonasch	MR Arthrographic Staging
Stage 1 Development of fine superficial fissures; loss of chondrocytes and proteoglycans	**Stage 1** Loss of surface coloring (dull surface)		**Stage 1** Surface normal Cartilage thickness normal
Stage 2 Fissures extend to subchondral zone "Cluster" formation	**Stage 2** Irregular cartilage surface with deep lesions, circumscribed defects	**Stage 1** Slight narrowing of joint cavity (>2 mm); degenerative loss of roundness and elongation of joint edges	**Stage 2** Surface normal to slightly irregular; difference in cartilage thickness between T1 and gradient-echo images
Stage 3 Loss of uncalcified cartilage; thickening of covering bone plate	**Stage 3** Circumscribed cartilage defects, exposure of subchondral plate ("bare bone")	**Stage 2** Joint cavity >1 mm, marginal rim up to 5 mm wide	**Stage 3** Irregular surface with defects up to "bare bone," residual cartilage; difference between T1 and gradient-echo images
Stage 4 Opening up of subchondral medullary space. Granulation tissue. Regenerative fibrocartilage	**Stage 4** "Bare bone" development of regenerative tissue and fibrous replacement tissue	**Stage 3** Joint cavity <1 mm, marginal rim over 5 mm **Stage 4** Joint cavity <1/2 mm Deformation of the joint surface	**Stage 4** Development of regenerative tissue "Bare bone"

[a]Stage classification.
[b]Adapted from Otte (see Ref. 31).

T1-weighted and gradient-echo images can therefore be interpreted as an indirect sign of structural cartilage changes. The broadening of the zone of calcified cartilage and the severalfold increase in the tidemark occur early during the processes of destruction in the cartilage, and are therefore already found in Otte's stage 1 (Fig. 35.6*A* and *B*).

Mild to Moderate Arthritis (Transitional Stage 2–3) (Fig. 35.7*A*)

In MR arthrography the cartilage is visualized with a smoothly defined surface on both T1-weighted and gradient-echo images. Microscopically detectable fine fissure formations extending to the subchondral zone (Fig. 35.7*B*) do not lead to any detectable changes in the signal of the cartilage. The fibrocartilage, resting on osteophytes (Fig. 35.7*B* and *C*), displays a low signal intensity on T1-weighted and gradient-echo images. Only on gradient-echo images is there a difference in the signal from hyaline cartilage, which shows an intermediate signal intensity (Fig. 35.5).

This possibility of differentiation by MR arthrography allows one to draw conclusions about the degree of severity of the arthritis. This is of therapeutic significance in that the moderately severe form of arthritis is a good indication for surgical procedures to maintain the joint, if mainly only one compartment is affected and if an axial deformity is present.

Severe Arthritis (Stage 3) (Fig. 35.8*A* and *B*)

In MR arthrography severe arthritis becomes manifest by an irregular cartilage surface that is clearly demarcated by the contrast medium (Fig. 35.8*C* and *D*). The broadening of the zone of calcified cartilage and of the subchontral

plate (Fig. 35.8*E*) becomes apparent in an absolute reduction in the cartilage thickness and in the difference between the evaluable cartilage thickness on T1-weighted and gradient-echo images. In our experience the assessment is reliable with a cartilage thickness of less than 1 mm.

Advanced Arthritis (Stages 3 and 4)

In MR arthrography the images in these stages are characterized by the absence of the intermediate signal intensities that are typical of the cartilage. This applies to both T1-weighted and to gradient-echo-sequences (Fig. 35.9*A* and *B*). The assessment of the subchondral zone, up to the demarcation from the subchondral adipose tissue, is best achieved on T1-weighted images. On gradient-echo-images good assessment of the surface is possible and subchondral cysts can also be detected. Osteophytes and degenerative meniscal changes can also be visualized with both methods. The histological section picture shows that besides the absence of cartilage (Fig. 35.9*C*), small areas of cartilage showing severe degenerative changes (Fig. 35.9*D*) and cartilage impressions may be present as well. Regenerative connective tissue is also found in the subchondral space and a very thickened covering plate. The changes described here become manifest on T1-weighted images as a linear hypointense subchondral zone.

It is not possible to differentiate between connective tissue, cartilage, and bone-dense areas on either T1-weighted or gradient-echo images. This is due to the cartilage volumes, which are on the order of magnitude of 1 voxel and the signals of which are reduced through partial volume effects, and also to the fact that this is cartilage that shows marked degenerative changes (few cells, rich in connective tissue). This brings about a signal intensity identical to that of connective tissue and areas of sclerosis or of any

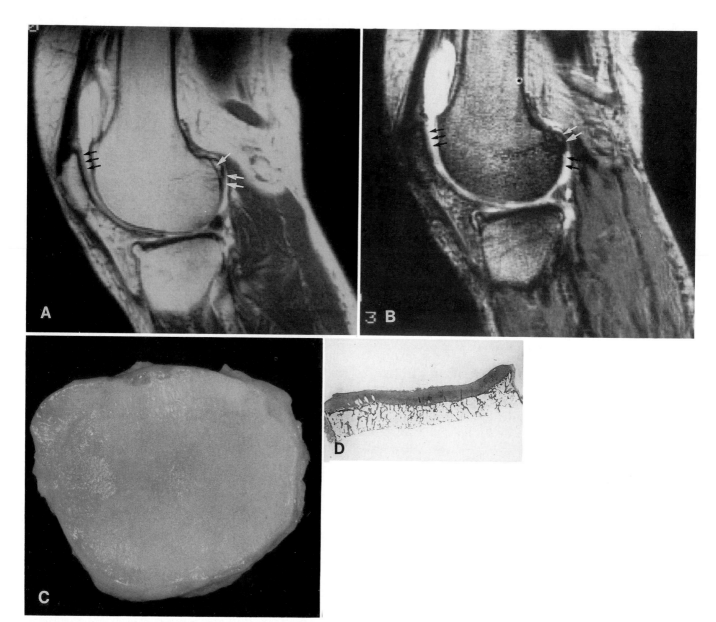

Figure 35.5. Thickened subchondral bone plate and calcified cartilage in mild osteoarthritis. **(A)** MR arthrography (spin-echo 600/15), sagittal plane, posterior section of condyle. Broad hypointense zone (*arrows*) surrounds a signal-rich zone (osteophyte plus fat marrow) (*arrow*). **(B)** MR arthrography (gradient-echo 30/13/35). A broad zone of intermediate signal intensity is visible (*black arrow*). Joint capsule extends up to the osteophyte. The osteophyte is surrounded by a zone of low signal intensity (*white arrows*). The fibrocartilage displays a low signal intensity, the hyaline cartilage an intermediate signal. Patella cartilage appears thicker on the T1-weighted images **(A)** than on the gradient-echo images **(B)** because T1 images show both the calcified and the uncalcified cartilage, whereas gradient-echo images show only the uncalcified cartilage. **(C)** Macrograph of the patellar cartilage surface, which is irregularly structured with slight marginal osteophyte formation. **(D)** Macrograph of the histological section sagittally through the patella.

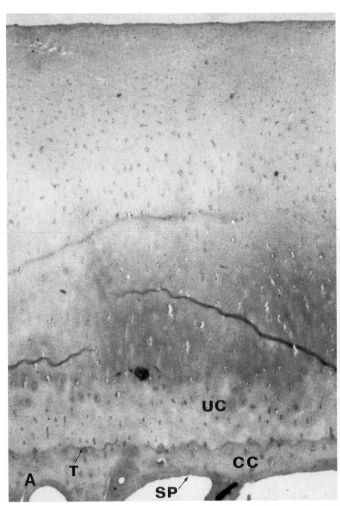

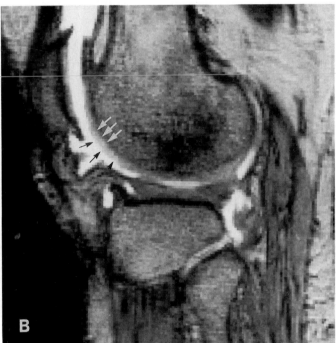

Figure 35.6. Thickening of calcified cartilage and tidal zone in mild osteoarthritis. (**A**) Structure of the hyaline articular cartilage and uncalcified cartilage (*UC*), tidemark (*T*), calcified cartilage (*CC*), and subchondral plate (*SP*) seen on light microscopy. (Staining: hematoxylin-eosin (HE); magnification: ×8.) (**B**) Gradient-echo (200/13/16), sagittal plane of middle section of the lateral condyle. A broad zone of inhomogeneous intermediate signal intensity (hyaline cartilage, *black arrows*) is visible alongside a zone of low signal intensity (subchondral plate plus calcified cartilage, *white arrows*).

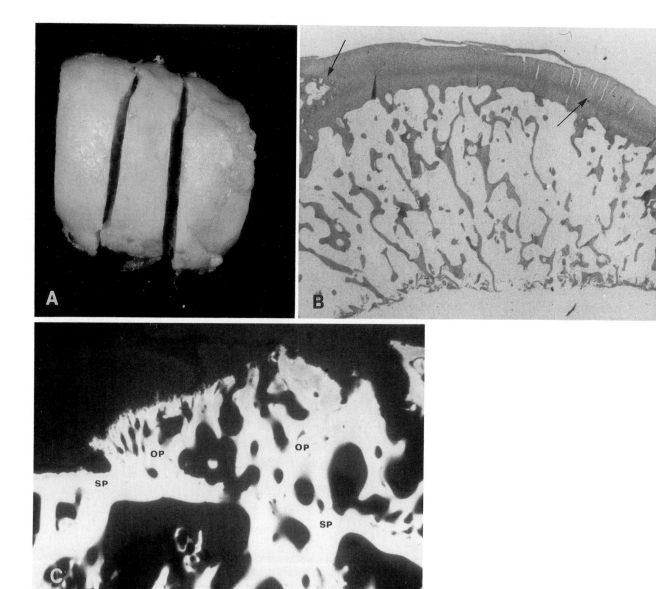

Figure 35.7.   Transitional stage osteoarthritis. **(A)** Micrograph of the lateral posterior section of the condyle: irregular cartilage surface is shown, sometimes with dull, but also with glossy, areas of cartilage. **(B)** Osteophyte in the posterior section of the condyle covered with fibrocartilage (*arrow*), old subchondral plate, development of fine car- tilage fissures (*arrow*) extending to the subchondral plate. (Staining: HE; magnification: macrograph of a histological section.) **(C)** Microra- diograph of the subchondral plate (*SP*), with osteophytic new bone for- mation (*OP*).

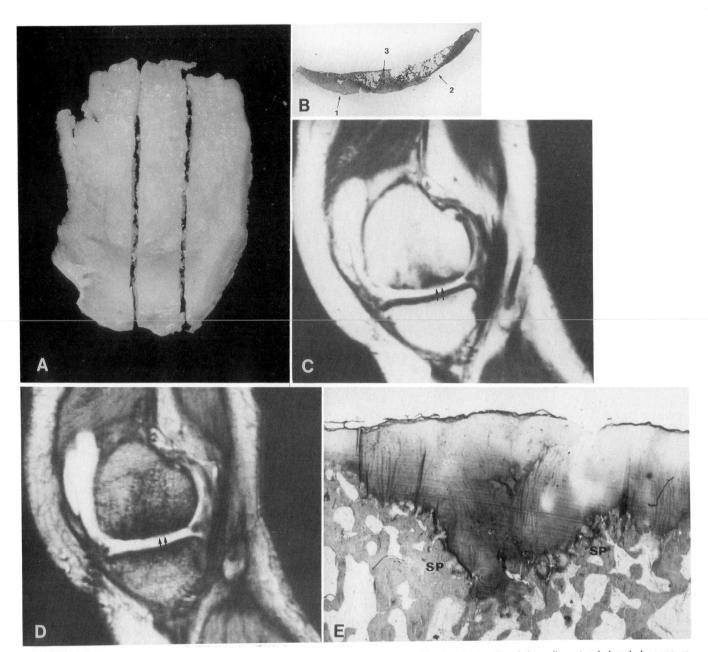

Figure 35.8. Stage 3 osteoarthritis. (**A**) Central section of the medial condyle. Irregular cartilage surface is seen, sometimes with covering bone plate exposed, pseudoregenerative granulation tissue. (Magnification: macrograph.) (**B**) Differences in cartilage thickness are shown [zone with marked degenerative changes (*1*), area with absence of articular cartilage (*2*), cartilage impressions (*3*)]. (Staining: HE; magnification: macrograph of a histological section.) (**C**) MR arthrography (spin-echo) 600/15), sagittal plane. Shown is focal area of cartilage (*arrows*) with signal intensity of the adjacent subchondral space reduced. Demarcation of remaining cartilage surface is somewhat irregular. (**D**) MR arthrography (gradient-echo image, 30/13/35°). Focal area of cartilage is distinctly narrowed, with irregular surface. (**E**) Visible are articular cartilage with few cells, broadening of the subchondral covering plate (*SP*), and fairly high-grade degenerative cartilage changes. (Staining: TB; magnification: ×2.6.)

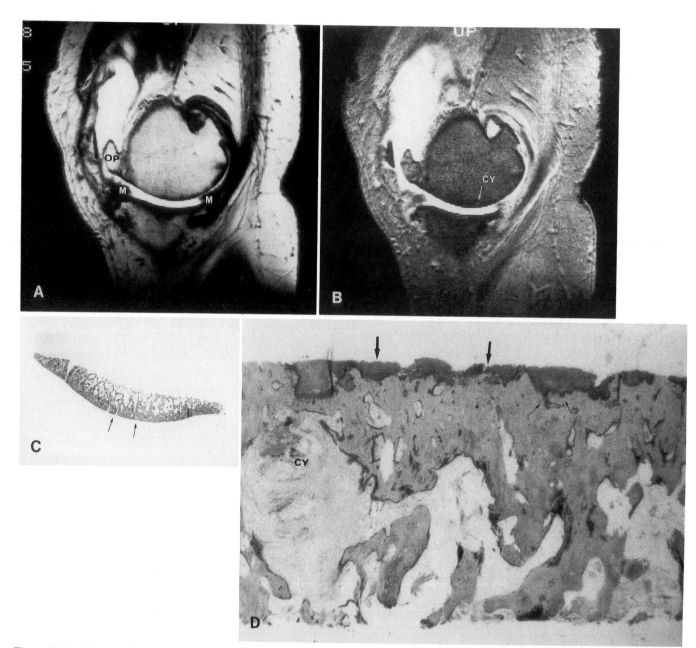

Figure 35.9. Advanced (stages 3 and 4) osteoarthritis. (**A**) MR arthrography (spin-echo 700/15), sagittal plane of middle third of medial condyle: smoothly defined condylar surface without articular cartilage is shown, with irregularly defined linear subchondral zones of low signal intensity, large ventral osteophyte (*OP*) with signal intensity equivalent to fat, and fragmented meniscus (*M*) showing degenerative changes. (**B**) MR arthrography (gradient-echo 30/10/40°): smoothly de-fined condylar surface without articular cartilage is shown, with small subchondral cyst (*CY*). (**C**) Shown is smoothly defined condylar surface without articular cartilage (*arrows*). Magnification: macrograph of the histological section; staining: HE.) (**D**) Cartilage remnants show high-grade degenerative changes, subchondral cyst (*CY*), cartilage impression (*arrows*) surrounded by granulation tissue. (Staining: HE, magnification: ×3.2.)

hemosiderin deposits that may be present. The question of whether smoothed bare bone and/or regeneration tissue are present is of great clinical interest. In the further course of events granulation tissue develops from the medullary space and also pseudoregeneration fibrocartilage tissue on the surface of the bone, denuded of cartilage (former subchondral plate). This "replacement cartilage" can no longer take the load imposed on it (36) so that alloplastic joint replacement is indicated.

Chondromalacia Patellae

Chondromalacia patellae, defined as a localized lesion confined to the articular cartilage (37), cannot be distinguished histologically from osteoarthritis. Neither pathological-anatomical investigations (38–40) nor biomechanical (41) or electron microscopic studies (42) have resulted in any evidence of an independent disease entity. Ficat and Hungerford (43) have pointed out that early detection of the degenerative disease is possible because arthroscopy or arthrotomy permit good assessment of the patellofemoral joint. The only macroscopic difference between chondromalacia patellae and osteoarthritis results from the formation of osteophytes and narrowing of the joint cavity. The stage classifications for chondromalacia patellae, e.g., those of Outerbridge (44), Ficat et al. (45), Bandi (46), Wiles et al. (47), or Otte (31), do not differ therefore, or only slightly, from that for osteoarthritis. Common to all classifications is an assessment of the surface, with Outerbridge additionally including the extent of the surface area.

As already described for osteoarthritis, a limited assessment of cartilage damage is possible by MR arthrography in stage 2 and an unequivocal assessment in stage 3. Intact cartilage surfaces and fine fissures or changes that are already macroscopically detectable (stage 1) do not become accessible, or only indirectly so, through the use of this investigation technique (Table 35.2).

Table 35.2
Comparison between MR Arthrographic Staging and the Macroscopic Staging of Outerbridge[a]

MR Arthrographic Staging	Macroscopic Staging of Outerbridge
Stage 1 Surface normal, cartilage thickness normal	**Stage 1** Local softening and swelling
Stage 2 Surface normal to slightly irregular; difference in cartilage thickness between T1 and GE	**Stage 2** Fissures and fragmentation, lesions up to 1.3 cm in diameter
Stage 3 Irregular surface with defects up to "bare bone"	**Stage 3** As for stage 2, but lesions >1.3 cm in diameter
Stage 4 Development of regenerative tissue "Bare bone"	**Stage 4** Ulcers and erosions down to the bones

[a] Adapted from Outerbridge (see Ref. 44).

The reason for trying out all imaging procedures (48, 49) is the inconclusiveness of clinical diagnosis (50–52). Clinical signs, such as the Zohlen test (53) or Fründ's sign (54), indicate only that cartilage damage is present, but it is not possible to draw conclusions as to its extent and degree of severity. However, the absence of clinical signs does not exclude the possibility that cartilage damage might already be present. Thus the figures for retropatellar cartilage damage in autopsies on patients aged over 30 years are given as 35% by Silfverskjölg (55), as 88% by Schlenzka et al. (51) and as almost 100% by Wiles et al. (47). Grueter (56) found cartilage damage already in 50% of those aged over 30 years and Øwre (57) in 84%. Cases of silent chondromalacia are therefore a frequent occurrence, but an underlying chondromalacia need not always be the cause when positive clinical findings are obtained. The exclusion of diseases such as osteochondritis dissecans, loose joint bodies, synovitis or plica syndrome is therefore just as important as the demonstration of a chondromalacia.

Accurate visualization of the cartilage defect, its location, and extent and also an assessment of the subchondral bone are of decisive importance for the treatment. (Figs. 35.10 and 35.11). The widely differing assessments made of the efficacy of the surgical measures used, in fact, make this absolutely essential (52, 58–65).

It is possible to fulfill all of these requirements for the planning of surgical therapy by means of MR arthrography, with the exception of cases of acute injury to the knee joint, where arthroscopy should primarily be carried out when a hemarthrosis is present. It is also possible to monitor the surgical measures used and to check their effects on the hyaline articular cartilage by means of MR arthrography.

Osteochondritis Dissecans

The assessment of the location and extent of an osteochondritis dissecans and the plan of treatment—surgical or conservative therapy—resulting from this, are of decisive importance for the orthopedic surgeon (43, 66–69). The following questions must primarily be answered when planning surgery:

1. Is a fracture of the cartilage present?
2. Is it a case of partial or total detachment of the fragments?
3. What are the characteristics of the bone layer?

The comprehensive diagnostic therapeutic algorithm of Clanton and De Lee (70) has been used to assess the image information, with the MR arthrography findings being integrated into this scheme. The continuity of the cartilage surface and the degree of detachment of the fragments from the bone layer have been used as the main criteria of assessment for the determination of treatment. This results in a classification of the osteochondritis dissecans into five types:

Type 1: Intact cartilage surface, no line of demarcation, no separation, hypointense subchondral area that cannot be sharply demarcated (Fig. 35.12).

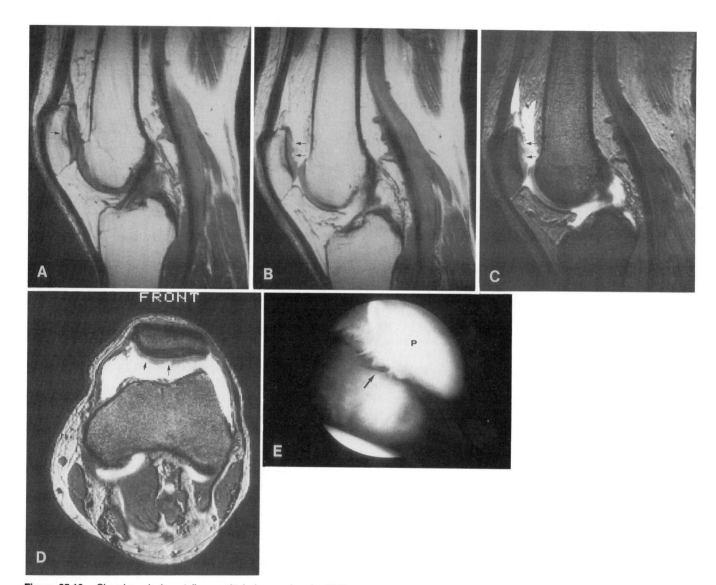

Figure 35.10. Chondromalacia patellae, sagittal plane: spin-echo 700/
15 (**A** and **B**), gradient-echo 30/10/30° (**C**), gradient-echo 30/10/30°,
FISP 3D (**D**) (axial section). (**A**) Articular cartilage of normal appearance, with hypointense linear structure in the subchondral zone of the
patella (*arrow*). (**B–D**) MR arthrography: irregular cartilage surface of
the patella with multiple defects of varying depth (*arrows*). (**E**) Arthroscopy picture: cartilage defect at the patella (*P*) in the form of fissures
and villi (*arrow*).

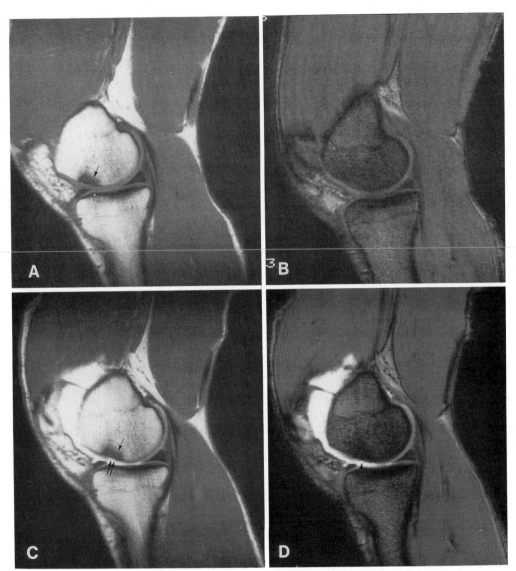

Figure 35.11. MR imaging: Chondromalacia of the medial condyle, focal cartilage defect. Sagittal plane through the medial condyle: (**A** and **C**), spin-echo 700/15; (*B* and *D*), FLASH 300/15/135°. (**A** and **B**) In the transitional region between the ventral and middle third of the medial condyle, circumscribed hypointense area extending into the subchondral cancellous bone (*arrow*) is seen. (**C** and **D**) MR arthrography: cartilage defect extends to the subchondral plate (*arrows, arrowhead.*)

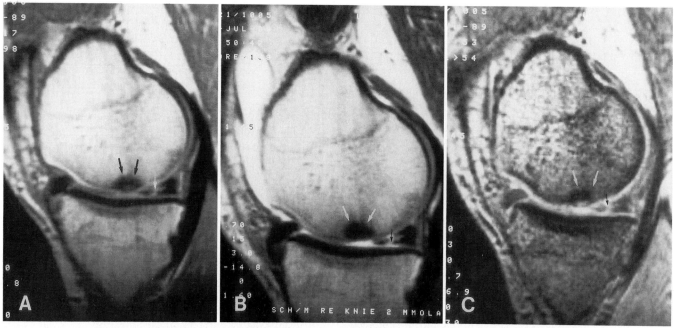

Figure 35.12. Osteochondritis dissecans, type 1. Sagittal plane, medial condyle: (**A** and **B**) spin-echo 700/15; (**C**) gradient-echo 30/10/30°. (**A**) MR imaging: the region of incipient osteochondritis dissecans is characterized by a subarticular zone of low signal intensity (*arrows*). An incidental finding was a complete meniscal lesion in the medial posterior horn. (*small arrow*). (**B** and **C**) MR arthrography of intact cartilage surface with subarticular hypointense area (*arrows*). Entry of contrast medium into the complete meniscal tear is visible (*small arrow*).

Type 2: Intact cartilage surface, line of demarcation, early separation (Fig. 35.13).

Type 3a: Partial cartilage fracture, dislocation of the detached fragments (Fig. 35.14).

Type 3b: Partial cartilage fracture, dislocation of the detached fragments, subchondral cyst formation.

Type 4: Total cartilage fracture, dislocation of the detached fragments (Fig. 35.15).

Type 5: Loose joint body (Fig. 35.16).

Fourteen osteochondral defects of sometimes differing location and also 2 loose joint bodies were classified in accordance with this system.

An assessment of the cartilage surface or cartilage fractures without contrast medium was possible in 50% of the spin-echo images and in 29% of the gradient-echo images (FISP 3D). This is reflected in the percentages for positive agreement in the comparison of the results with and without contrast medium (T1:T1 + Gd = 29%; T1:FISP + Gd = 36%; FISP 3D:T1 + Gd = 36%, FISP 3D + Gd = 64%).

The existence of a cartilage fracture could be unequivocally demonstrated by means of MR arthrography in both T1-weighted and also in FISP 3D images or in the presence of a joint effusion. The best agreement was obtained between T1 + Gd: FISP 3D + Gd at 93%, with the qualitative visualization of the fragments detached from bone being better on T1-weighted images and the cartilage being better visualized with FISP 3D plus contrast medium.

Through MR arthrography a precise assessment of the cartilage can also be included in the staged planning for the surgical treatment of osteochondritis dissecans. When integrated into the algorithm of Clanton and De Lee (70), the picture shown in Table 35.3 is obtained.

Guhl (71) and Ewing and Voto (72) have already pointed out the difficulties of assessing the cartilage surface by arthroscopy in osteochondritis dissecans, and they recommended staining with methylene blue as an aid to the visualization of fine fracture lines or gradations, or reducing the strength of the light to improve the shadow gradation.

Fissure formations in the cartilage and also the extent of the dislocation of cartilage and bone fragments are clearly visualized through the intraarticular use of the contrast medium. The combined view of the cartilage surface, the bone bed of the detached fragments, and adjacent subchondral cancellous bone makes it possible to decide on the treatment—surgical or conservative therapy (relief of pressure and raised shoe edge, such as may be indicated in pediatric osteochondritis dissecans (73)—on the basis of the MR diagnosis system already described. Besides the possibility of determining the optimal surgical procedure (arthroscopy or arthrotomy), and the choice of the fixation material (absorbable pins, osteosynthesis material, bone chip bolting, fibrin bond (74–81) or a cartilage bone transplant (82), MR arthrography also offers the possibility of serial observation.

Plica Syndrome (Shelf Syndrome)

Of all the folds that are ontogenetically possible in the knee joint (plica suprapatellaris, mediopatellaris, and infrapatellaris) the plica mediopatellaris is of the greatest pathological importance. Symptoms of pain induced by the other two plicae have also been reported (83–87).

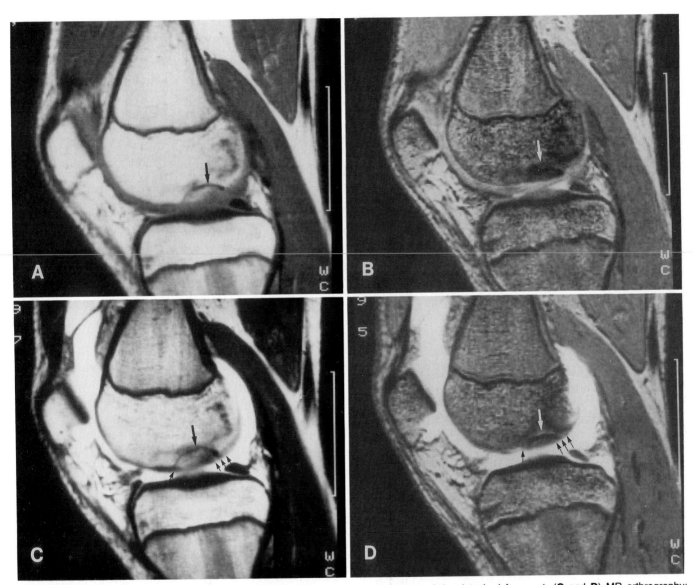

Figure 35.13. Osteochondritis dissecans, type 2. Sagittal plane, medial condyle: (**A** and **C**) spin-echo 700/15; (**B** and **D**) gradient-echo 30/10/30°. (**A**) Distinct demarcation, early separation; (**B**) ill-defined line of demarcation of the detached fragment; (**C** and **D**) MR arthrography: intact hyaline articular cartilage (*small arrows*), demarcation of the detached fragment, early separation (*large arrow*).

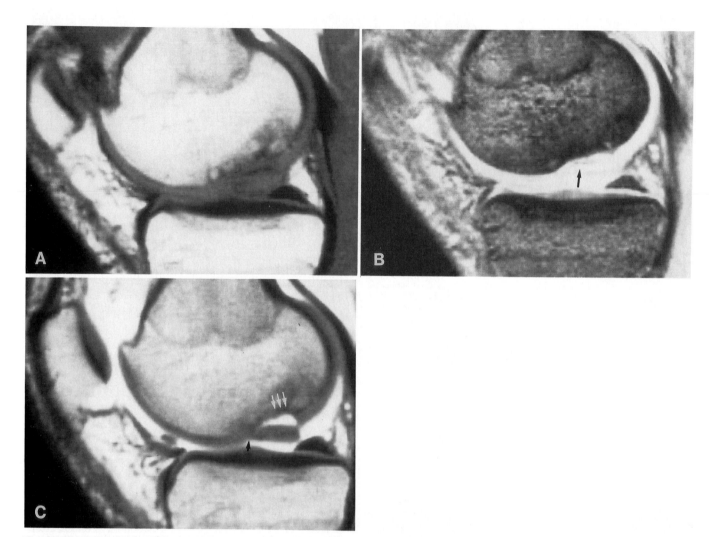

Figure 35.14. Osteochondritis dissecans, type 3a: partial cartilage fracture, with dislocation of the detached fragment. Sagittal plane, medial condyle: (**A** and **C**) spin-echo 700/15; (**B**) gradient-echo 30/10/30°. (**A** and **B**) Formation of a defect in the joint-bearing middle third. The detached fragment is hardly visible on the gradient-echo images (*ar-row*). (**C**) MR arthrography: the detached fragment is demarcated and is joined to the hyaline articular cartilage by a bridge of cartilage (*large arrow*). The hypointense zone in the defect layer (*small arrows*) corresponds to a sclerosis of the subchondral bone.

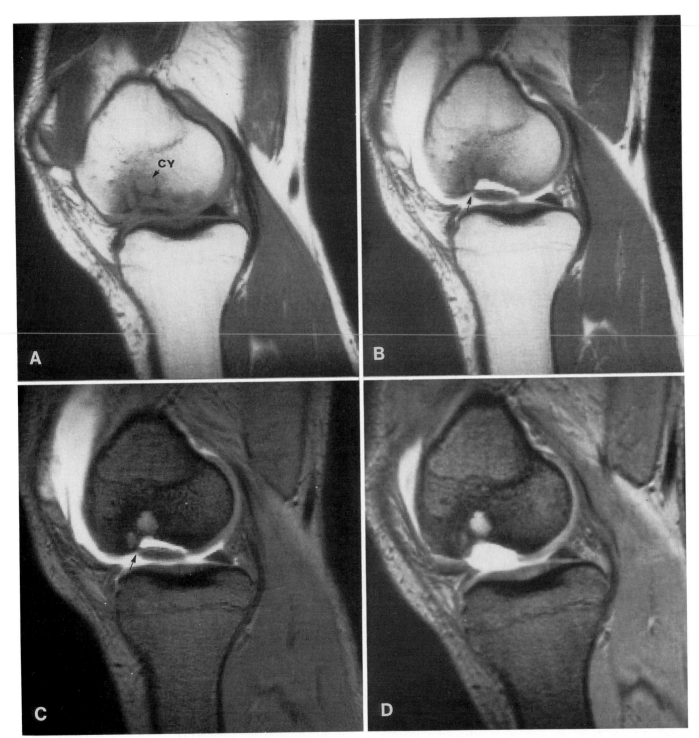

Figure 35.15. Osteochondritis dissecans, type 4. Sagittal plane, medial condyle: (**A** and **B**) spin-echo 700/15; (**C** and **D**) gradient-echo 30/10/30°. (**A**) Visualization of defect: no precise demarcation of detached fragment is visible; cyst formation (*CY*) has occurred in cancellous bone site of fragment. (**B** and **C**) MR arthrography: dislocation of detached fragment has occurred, with complete cartilage fracture apart from a thin bridge of cartilage (*arrow*). (**D**) MR arthrography: site of detached fragment is empty after total cartilage fracture and displacement of fragment.

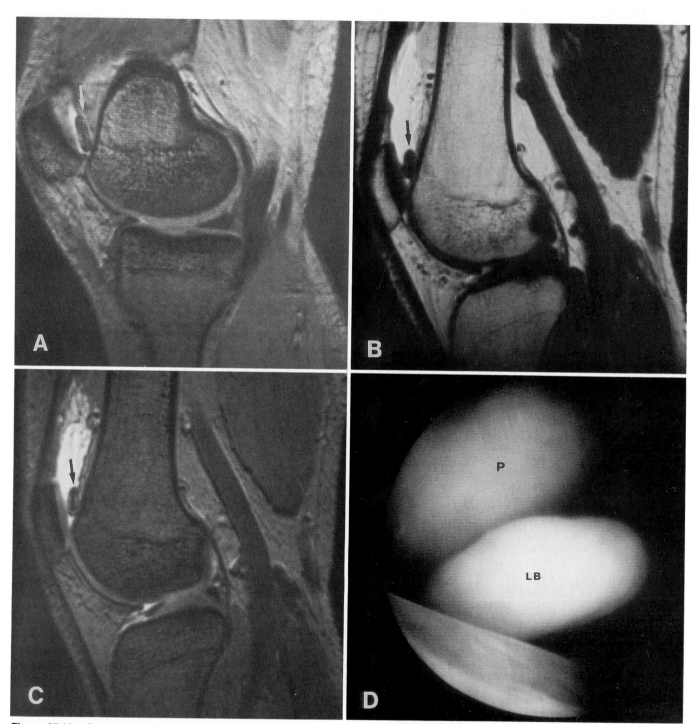

Figure 35.16. Osteochondritis dissecans, type 5 (loose joint body). Sagittal plane: (**A** and **C**) gradient-echo 30/10/30°; (**B**) spin-echo 700/15. Loose joint body lying in the upper recess (*arrow*); (**B** and **C**) MR arthrography; distinct demarcation of the loose joint body, with central ossification surrounded by broad areas of cartilage (*arrow*) (**D**) arthroscopy picture: loose joint body (*LB*), patella (*P*).

Table 35.3
Algorithm for Surgical Treatment of Osteochondritis Dissecans

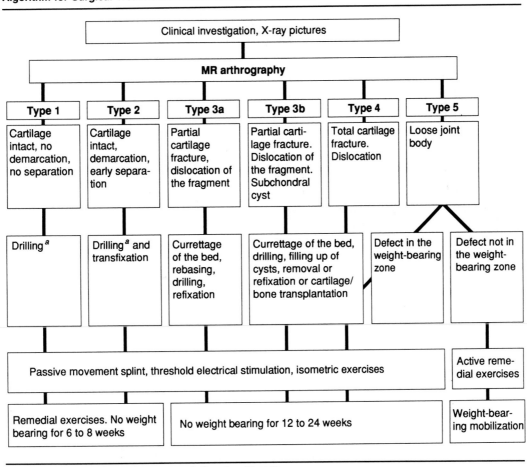

[a]Possibly conservative therapy in young people (type 1 and type 2).
[b]MR arthrography control after 6 months if no metal used for the fixation. Otherwise not until after metal removed.

On average a plica is present in 20% of cases. The data given range from 5 to 55% (83, 84, 86, 88, 89). It is trauma that causes the development of edema, the loss of elasticity, fibrosis, and finally the calcification of the plica. In the case of the plica mediopatellaris, for example, this can lead to the plica coming under tension during flexion, so that it is stretched out over the upper median part of the medial femoral condyle. On increased flexion the plica comes to lie between the condyle and the patella (83–87, 90) so that it can thus become the primary cause for the development of a focal chrondromalacia (91). The form that the plica mediopatellaris can assume has been described by Iino (92).

Clinically the plica syndrome is characterized by diffuse parapatellar or suprapatellar symptoms of pain in the region of the medial patellar margin. In addition snapping may occur on flexion (83, 85) or a pseudo-incarceration may develop. The findings on clinical examination are not significant enough for the unequivocal diagnosis of a plica-syndrome however, nor can a meniscal lesion be excluded (84, 85, 90), or an osteochondromalacia caused by the plica be demonstrated (93).

It is therefore emphasized that a diagnostic arthroscopy should be performed (58, 86, 94–97) but only after conservative therapy has been tried in the form of rest, nonsteroidal antirheumatic drugs, physical therapy and stretching exercises (83, 88). The time when the diagnostic arthroscopy should be performed is given as 4 weeks (93) to 12 weeks (83, 88). In the event of positive arthroscopy findings, either arthroscopic release (98), or total removal of the plica mediopatellaris is recommended (84, 85, 90).

In the context of a comprehensive assessment of the knee joint, good visualization of the plica suprapatellaris (Figs. 35.17 and 35.18) and plica mediopatellaris (Figs. 35.19 and 35.20) is possible by MR arthrography. This is of great importance when it comes to making a diagnosis in cases of unclear pain of the knee joint.

Since, on the one hand, the time specified for carrying out the arthroscopy is relatively late, while on the other the arthroscopy is seen primarily as a diagnostic procedure, MR arthrography can be of considerable help in deciding on the treatment.

It is almost exclusively athletes who are affected by the condition, so that this possibility of obtaining rapid infor-

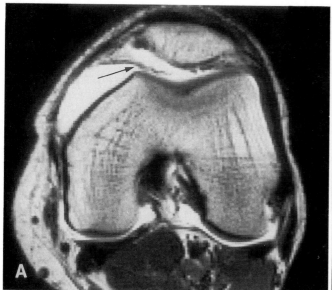 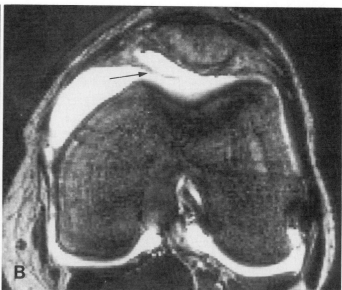

Figure 35.17. MR arthrography: (**A**) spin-echo 700/15; (**B**) gradient-echo 30/10/30° (axial plane). Plica mediopatellaris (*PM*) is stretched out between the patella and the medial condyle. This is type C, according to Iino (92): note the large "sail" that covers the anterior part of the medial condyle (*arrow*).

Figure 35.18. Macrograph of plica mediopatellaris (*PM*).

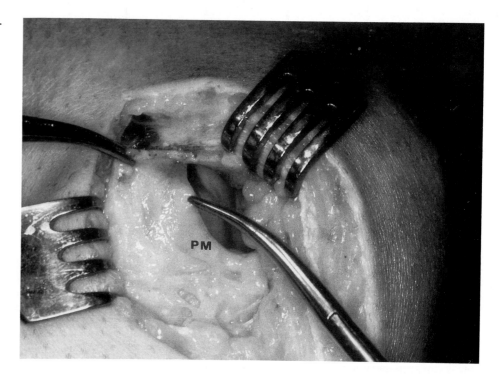

mation with a minimal loss of training time for the athlete is important. The time and form of treatment—surgery or conservative therapy—can thus be controlled and can be included in the planned schedule at the best possible time.

Lesions of the Meniscus

Magnetic resonance imaging has a high degree of accuracy in the diagnosis of lesions of the meniscus (15, 99–101). On T1-weighted images the fibrocartilage of the meniscus appears to be a homogeneous structure with a very low signal intensity (1, 6, 7, 13, 102–106). Any pathological change will result in an increase in the signal intensity on T1-weighted images. Thus meniscal tears appear as linear increases in signal intensity. Cysts and degenerative changes lead to a circumscribed increase in signal intensity inside the meniscus (8, 105–107).

The detection of tears in menisci showing degenerative changes is often problematical, however, since both of these lead to an increase in the signal that can reach that of intraarticular fluid (13).

Through MR arthrography the contours of the menisci can be accurately defined, regardless of whether degenerative changes are present or not (Fig. 35.21*A–D*). Even small superficial tears can be visualized and it can be determined whether a tear extends to the surface. If the contours of the meniscus appear intact, it can be demonstrated that an intrameniscal tear is present. Magnetic resonance arthrography is not the method of first choice among the range of measures available for the clinical diagnosis of a meniscal lesion. It is indicated only for those cases in which clinical examination and nasive MR imaging has not resulted in any definite diagnosis, when arthroscopy is seen as a diagnostic procedure, and for patients in whom arthroscopy is medically contraindicated.

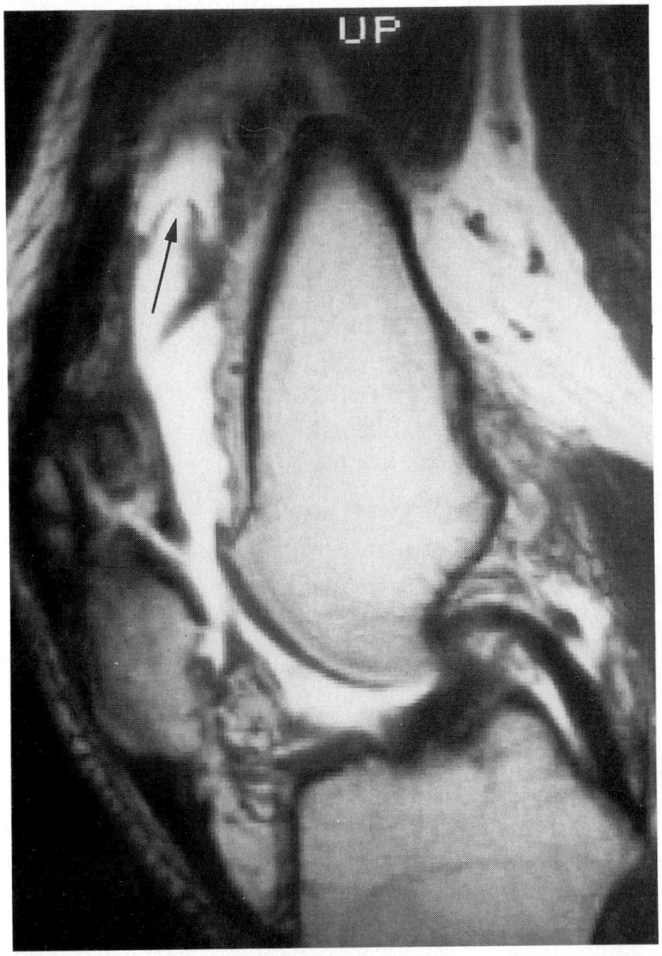

Comparison of Pathological Findings with and without Contrast Medium

Summary of the Magnetic Resonance Findings of Patients with Osteoarthritis of the Knee

The comparison of the findings obtained before and after the administration of the contrast medium underlines the value of MR arthrography for the assessment of the hyaline cartilage and the synovial membranes. In the detail of the images it is possible to distinguish between bare bone, areas with remnants of cartilage, or a focal chondromalacia (108).

The comparison of the sequences used showed that for the assessment of the cartilage, the FISP 3D + Gd images showed the best agreement with the histology.

Summary of the Magnetic Resonance Findings of Patients with Knee Joint Symptoms

The evaluation showed that pathological changes of the cartilage represent the primary indication for MR arthrography (Table 35.4). This is also true when a diagnosis can be made, even without contrast medium, since clinically relevant questions of detail (e.g., the stages of osteochondritis dissecans) (108) can sometimes be answered only by MR arthrography.

Summary of a Prospective Study: Clinical, Magnetic Resonance Arthrography, and Surgical Findings

The clinical, MR arthrography, and surgical findings, obtained in that order, illustrated the problems of making a clinical diagnosis, and also the marked agreement with the findings obtained by arthroscopy or arthrotomy.

Magnetic resonance arthrography (108) possesses great sensitivity and specificity with only a very small percent-

Figure 35.19. MR arthrography: spin-echo 700/15, sagittal plane. Plica suprapatellaris: the suprapatellar septum *(arrow)* shows a central perforation site called the "porta."

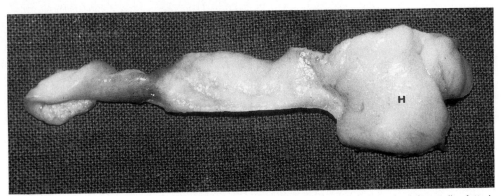

Figure 35.20. Macrograph (resection specimen): plica mediopatellaris extends from the infrapatellar fatty body of Hoffa (*H*) up to the plica suprapatellaris.

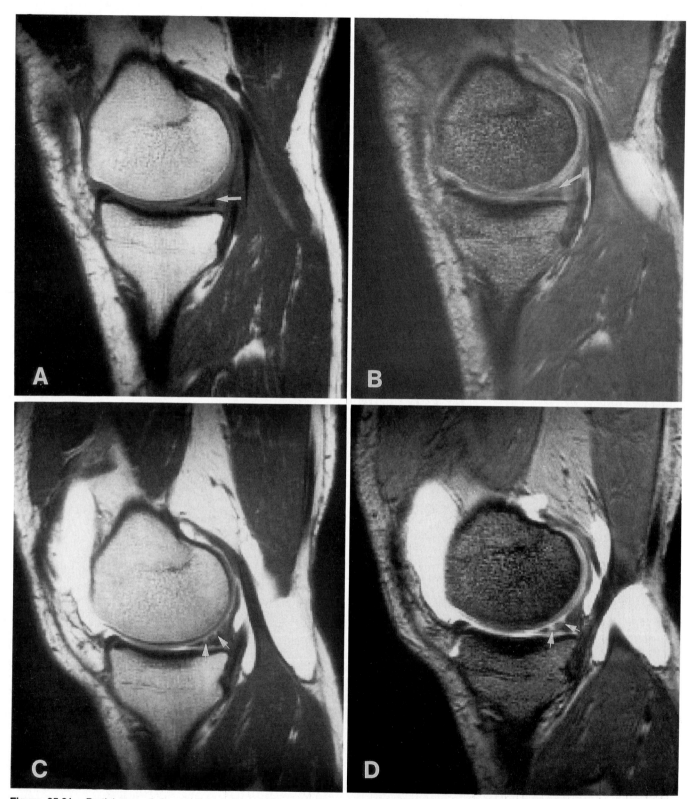

Figure 35.21. Partial tear of the medial meniscus. Sagittal plane through the medial condyle: (**A** and **C**) spin-echo 700/15; (**B** and **D**) FISP-3D, 30/10/30°. **A** and **B** display inhomogeneous signal in the pos- terior horm (*arrow*). **C** and **D** MR arthrography: small meniscal tear (*arrow*) is identified by the contrast medium. Bursal fluid is easily ap- preciated.

Table 35.4
Clinical Use: Detection of Pathological Changes with and without Contrast Medium[a]

	Before Injection		MR Arthrography	
	T1	Gradient-Echo	T1	Gradient-Echo
Chondromalacia (n = 33)	8	8	28	33
Osteochondritis dissecans (n = 10)	10	10	10	10
Loose joint body (n = 5)	4	4	5	5
Meniscal lesion (n = 18)	17	17	18	18
Synovitis (n = 14)	1	1	14	14
Plica syndrome (n = 9)	—	—	9	9
Lesion of the cruciate ligament (n = 4)	4	4	4	4
Bone changes (n = 22)	22	22	22	22

[a] n = 53 patients.

Table 35.5
Agreement between MR Arthrography and Surgical Findings and between Diagnostic Clinical and Surgical Findings

	MR Arthrography/ Surgical Findings (%)	Diagnostic Clinical/ Surgical Findings (%)
Sensitivity	91.6	71.8
Specificity	99.1	92.6
Accuracy	98.5	95.8
False negative findings (%)	1.8	4.7
False positive findings (%)	4.3	39.1

age of false-positive or false-negative findings (Table 35.5). This result is better than that for double-contrast arthrography and high-resolution computed tomography (CT) (109–112). A comparison with the data given for the accuracy rates of MR imaging without contrast medium is difficult, because prospective data are available only for specific regions (meniscus, ligaments).

Conclusions

1. Gadolinium-DTPA is a contrast medium that is suitable for intraarticular use in MR arthrography.
2. Thirty to 40 ml of a 2 mM solution is recommended as the ideal concentration.
3. In this concentration Gd-DTPA brings about optimal contrast of the internal joint structures without inducing any reaction of the synovial membrane.
4. The molecular structure of Gd-DTPA does not prevent it from being taken up into the intercellular fluid space. This process is transitory and reversible. It has been demonstrated that no increased concentration of Gd could be measured in viable human cartilage after 17 hr.
5. The intraarticular administration of the contrast medium makes possible accurate and constantly reproducible imaging of all joint structures and is superior to the examination carried out without the use of the contrast medium. This particularly applies to the imaging of the surface of the articular cartilage.

6. Contrast medium is administered into the knee joint in the same way as with the customarily used injection techniques, but suprapatellar compression is additionally applied.
7. The inadvertent administration of Gd-DTPA into the musculature and the infrapatellar fatty body of Hoffa does not lead to any contrast medium-specific tissue reactions.
8. Magnetic resonance arthrography is characterized by high sensitivity and specificity—comparable with those of arthroscopy.
9. Both the surface and also the osseous parts of the joint can be assessed through this method.
10. Pathological changes an be visualized to a greater extent after the intraarticular administration of contrast medium.
11. Magnetic resonance arthrography is characterized by the fact that it makes good assessment of details possible, this being essential for the planning of treatment.
12. The method makes is possible to answer specific questions relating to treatment.
13. Clinically the use of the method is mainly indicated for the following.
    a. For the diagnosis of uncharacteristic pain in the knee joint.
    b. To determine the degree of severity of pathological cartilage changes.
    c. Possibly to monitor the course of such pathological changes during treatment.
    d. When there are medical contraindications to arthroscopy.

Acknowledgments

My special thanks are due to Prof. Dr. R. Kotz, my academic mentor and head of the Department of Orthopedic Surgery; and to Prof. Dr. H. Imhof, head of the Magnetic Resonance Institute at the Department of Radiodiagnostics, University of Vienna, Austria, who afforded me the possibility to apply the idea of a new imaging technique (magnetic resonance arthrography) to clinical practice.

References

1. Li D. KB, Adams ME, McConkey JP. Magnetic resonance imaging of the ligaments and menisci of the knee. Radiol Clin N Am 1986;24(2):209–227.
2. Beltran J Noto AM, Herman LJ, Mosure JC, Burk JM, Christoforidis AJ. Joint effusions: MR imaging. Radiology 1986;158:133–137.
3. Mink JH, Reicher MA, Cruess JH. Magnetic resonance imaging of the knee. New York: Raven Press, 1987.
4. Hartzmann S, Reicher MA, Bassett LW, Duckwiler GR, Mandelbaum B, Gold RH. MR imaging of the knee. Part III. Chronic disorders. Radiology 1987;162:553–559.
5. Lehner KB, Rechl HP, Gmeinwieser JK, Heuck AF, Lukas HP, Kohl HP. Structure, function, and degeneration of bovine hyaline cartilage: assessment with MR imaging in vitro. Radiology 1989;170:495–499.
6. Gallimore GW Jr, Harms SE. Knee injuries: high resolution MR imaging. Radiology 1986;160:457–461.
7. Kean DM, et al. Nuclear magnetic resonance imaging of the knee: examples of normal anatomy ond pathology. Br J Radiol 1983;56:355–364.

8. Bellon EM, Keith MW, Coleman PE, Shah ZR. Magnetic resonance imaging of internal derangements of the knee. Radiographics 1988;8(1):95–118.

9. Middleton WD, Macrander S, Lawson TL, et al. High resolution surface coil magnetic resonance imaging of the joints: anatomic correlation. Radiographics 1987;7:645–683.

10. Buckwalter KA, Braunstein EM, Wilson MR, Wojtis EM, Martel W. Evaluation of hyaline cartilage of the knee with MR imaging. Radiology 1986;161:239.

11. Wojtys E, Wilson M, Buckwalter K, Braunstein E, Martel W. Magnetic resonance imaging of knee hyaline cartilage and intraarticular pathology. Am J Sports Med 1987;15(5):455–463.

12. Yulish BS, Montanez J, Goodfellow DB, Bryan PJ, Mulopulos GP, Modic MT. Chondromalacia patellae: assessment with MR imaging. Radiology 1987;164:763–766.

13. Steinbrich W, Beyer B, Friedmann G. MR des Kniegelenkes. Fortschr Röntgenstr 1985;143:166–172.

14. Tyrell RL, Gluckert K, Pathria M, Modic MT. Fast three-dimensional MR imaging of the knee: comparison with arthroscopy. Radiology 1988;166:865–872.

15. Polly DW, Callaghan JJ, Sikes RA, McCabe JM, McMahon K, Savory CG. The accuracy of selective magnetic resonance imaging compared with the findings of arthroscopy of the knee. J Bone Joint Surg 1988;70A:192–198.

16. Shahriaree H. Chondromalacia. Contemp Orthoped 1985;11:27–39.

17. Adam G, Bohndorf K, Prescher A, Krasny R, Günther RW. Der hyaline Gelenkknorpel in der MR-Tomographie des Kniegelenkes bei 1,5 T. Fortschr Röntgenstr 1988;148(6):648–651.

18. Adam G, Bohndorf K, Prescher A, Drobnitzky M, Günther RW. Kernspintomographie der Knorpelstrukturen des Kniegelenks mit 3D-Volumen-Imaging in Verbindung mit einem schnellen Bildrechner. Fortschr Röntgenstr 1989;150(1):44–48.

19. Hajek PC, Sartoris DJ, Neumann Ch, Resnick D. Potential contrast agents for MR arthrography: in vitro evaluation and practical observations. AJR 1987;149:97–104.

20. Jessar RA. The study of synovial fluid. In: Hollander JL, McCarty DJ, eds. Arthritis and allied conditions, 8th ed. Philadelphia: Lea & Febiger, 1976:67–81.

21. Engel A, Hamilton G, Hajek P, Fleischmann D. In vitro uptake of ^{153}gadolinium and gadolinium complexes by hyaline articular cartilage. Eur J Radiol 1990;11:104–106.

22. Behne I. Spurenelementanalyse in biologischen Proben. In: Analytiker-Taschenbuch, Vol. 6. von Fresenius UA, ed. Berlin: Springer-Verlag, 1986.

23. Kirchgessner RL. Bedarf und Verwertung von Spurenelementen. In: Spurenelemente—Grundlagen, Ätiologie, Diagnose, Therapie. von Zumkley H, ed. Stuttgart: Thieme Verlag, 1983.

24. Schering FA. Magnevist R, Paramagnetisches Kontrastmittel für die Magnetische Resonanz-Tomographie (MRT). Z Diagn 1988;5–7.

25. Gadian DG, Payne JA, Bryant DJ, Young IR, Carr DH, Bydder GM. Gadolinium-DTPA as a contrast agent in MR imaging—theoretical projections and practical observations. J Comput Assist Tomogr 1985;9:242–251.

26. Hajek PC, Sartoris DJ, Engel A, et al. The effect of intraarticular gadolinium-DTPA on the synovial membrane and cartilage. Invest Radiol 1990;25:179–183. (in press).

27. Vilarrubias JM. Handbuch der Infiltrationen im Bewegungsapparat und bei Sportverletzungen. 1. Deutschsprachige Auflage. U. Bayer: Unas-Verlag, 1987.

28. Möllmann H, Polster J, Knoche H. Technik der intraartikulären Injektion. Munich: Schwarzeck-Verlag, 1977.

29. Forman M, Muller C, Kayaalp O, Kaplan D. Epidemiology of osteoarthritis of the knee [Abstract 102]. Arthr Rheum 1981;24(suppl 4):574.

30. Casscells SW. Gross pathological changes in the knee joint of the aged individual. A study of 300 cases. Clin Orthoped 1978;132:225–232.

31. Otte P. Die konservative Behandlung der Hüft-und Kniearthrose und ihre Gefahren. Dtsch Med J 1969;20:217–225.

32. Harrison MHM, Schajowicz F, Trueta J. Osteoarthritis of the hip: a study of the nature and evolution of the disease. J Bone Joint Surg 1953;35B:598–626.

33. Fassbender HG. Pathologie rheumatischer Erkrankungen. Berlin: Spriner, 1975.

34. Grueter H, Rütt A. Zur Morphologie der in die Koxarthrose einmündenden Hüftgelenkserkrankungen. Z Orthoped 1962;95:401–439.

35. Green WT, Martin GN, Eanes ED, Sokoloff L. Microradiographic study of the calcified layer of articular cartilage. Arch Pathol 1970;90:151–158.

36. Mohr W. Gelenkkrankheiten. Diagnostik und Pathogenese makroskopischer und histologischer Strukturveränderungen. Stuttgart: Thieme Verlag, 1984:180, 235–240.

37. Gschwend N, Bischofberger RJ. Die Chondropathia patellae. Orthoped Praxis 1971;60:562.

38. Collins DH. The pathology of articular and spinal diseases. London: Arnold, 1949.

39. Waisbrod H, Treiman HN. Intra-osseous venography in patellofemoral disorders: J Bone Joint Surg 1980;62B:454.

40. Insall J, Falvo KA, Wise DW. Chondromalacia patellae. A prospective study. J Bone Joint Surg. 1976;58A:1.

41. Shoji H, Granda JL. Acid hydrolases in the articular cartilage of the patella. Clin Orthoped 1974;99:293–297.

42. Paul B. Über die Belastung des Binde-und Stützgewebes, speziell des Gelenkknorpels. Med Sport 1973;13:13–19.

43. Ficat P, Hungerford DS. Disorders of the patellofemoral joint. Paris: Masson, 1977.

44. Outerbridge RE. The etiology of chondromalacia patellae. J Bone Joint Surg 1961;46B:179.

45. Ficat RO, Philippe J, Hungerford DS. Chondromalacia patellae: a system of classification. Clin Orthoped 1979;144:55–62.

46. Bandi W. Verlagerung der Tuberositas tibiae bei Chondromalacia patellae und Femoropatellararthrose. Unfallheilkunde 1975;27:175–186.

47. Wiles PH, Andrews PS, Devas MB. Chondromalacia of the patella. J Bone Joint Surg 1956;38B:95.

48. Outerbridge RE, Dunlop JAY. The problem of chondromalacia patellae. Clin Orthoped 1975;110:177–196.

49. Hvid I, Andersen LI, Schmidt H. Chondromalacia patellae. The relation to abnormal patellofemoral joint mechanics. Acta Orthoped Scand 1981;52:661–666.

50. Fulkerson JP. The etiology of patellofemoral pain in young, active patients: a prospective study. Clin Orthoped 1983;179:129–133.

51. Schlenzka D, Schwesinger G, Plewka F. Zur Chondropathia patellae. Beitr Orthoped Traumatol 1978;25:564–569.

52. Munzinger U, Dubs L, Buchmann R. Das femoropatelläre Schmerzsyndrom. Konservative Behandlung und Resultate 7–10 Jahre nach Maquet-Operation. Orthopäde 1985;14:247–260.

53. Zohlen E. Chondropathia patellae. Über ihre Bedeutung und Wesen. Bruns Beitr Klin Chir 1942;69:174.

54. Fründ H. Traumatische Chondropathie der Patella, ein selbständiges Krankheitsbild Zbl Chir 1926;53:707.

55. Silfverskjöld N. Chondromalacia of the patella. Acta Orthoped Scand 1938;9:214.

56. Grueter H. Untersuchungen zum Patellahinterwandschaden. Z Orthoped 1959;91:486.

57. Øwre A. Chondromalacia Paellae. Acta Chir Scand 1936;77(suppl 41):1–156.

58. Petrick P, Tischer K. Ergebnisse der Behandlung nach Ali Krogius bei der sogenannten habituellen Patellaluxation. Beitr Orthoped Traumatol 1978;25:570–572.

59. Nottage WM, Frazier JK, Anzel SH, Almquist GA, Casperson PC Results of patellar relaignment in patients older than thirty. Clin Orthoped 1981;157:149–152.

60. Hejgaard N, Watt-Boolsen S. The effect of anterior displacement of the tibial tuberosity in idiopathic chondromalacia patellae. Acta Orthoped Scand 1982;53:135–139.

61. DeCesare WF. Late results of Hause procedure for recurrent dislocation of the patella. Clin Orthoped 1979;140:137–143.

62. Lund F, Nilsson BE. Anterior displacement of the tibial tuberosity in chondromalacia patellae. Acta Orthoped Scand 1980;51:679–688.

63. Mynerts R, Loohagen G. Ventralization of the tibial tubercle for patellar pain. Arch Orthoped Trauma Surg 1985;104:49–52.

64. Hirsh DM, Reddy DK. Experience with Maquet anterior tibial tu-

bercle advancement for patellofemoral arthralgia. Clin Orthoped 1974;103:40–45.

65. Noesberger B, Freiburghaus P. Operation nach Elmslie (Indikation, Technik, Ergebnisse). Hefte Unfallheilkunde 1976;127:195–203.

66. Aichroth P. Osteochondritis dissecans of the knee. J Bone Joint Surg 1971;53B:440.

67. König F. Über freie Körper in den Gelenken. Dtsch Z Chir 1887;27(1887–1888):90.

68. Fairbank TJ Knee joint changes after meniscectomy. J Bone Joint Surg 1948;30B:664–670.

69. Axhausen G. Über die osteochondritis dissecans Königs. Klin Wochenschr 1924;3:1057–1060.

70. Clanton TO De Lee JC Osteochondritis dissecans. History, pathophysiology and current treatment concepts. Clin Orthoped 1982;167:50–64.

71. Guhl J. Arthroscopic treatment of osteochondritis dissecans. Clin Orthoped 1982;167:65–74.

72. Ewing JW, Voto SJ. Athroscopic surgical management of osteochondritis dissecans of the knee. Arthroscopy 1988;4:37–41.

73. Brückl R, Rosemeyer, Thiermann G. Behandlungsergebnisse der Osteochondrosis dissecans des Kniegelenkes bei Jugendlichen. Z Orthoped 1982;120:717–724.

74. Dexel M, Jehle U. Resultate der operativen Behandlung der Osteochondrosis dissecans am Kniegelenk. Orthopäde 1981;10:87–91.

75. Gschwend N, Munzinger U, Löhr J. Unsere extraartikuläre Dissekatverschiebung bei Osteochondrosis dissecans des Kniegelenkes. Orthopäde 1981;10:83–86.

76. Cameron HU, Piliar RM, Macnab I. Fixation of loose bodies in joints. Clin Orthoped 1974;100:309–314.

77. Claes L, Burri C, Kiefer H, Mutschler W. Resorbierbare Implantate zur Refixierung von osteochondralen Fragmenten in Gelenkflächen. Akt Traumatol 1986;16:74–77.

78. Ficat RP, Ficat C, Gedeon P, Toussaint JB. Spongialization: a new treatment for diseased patellae. Clin Orthoped 1979;144:74–83.

79. Lütten C, Lorenz H, Thomas W. Refixation bei der Osteochondrosis dissecans mit resorbierbarem Material unter Verlaufsbeobachtung mit der Kernspintomographie (MR). Sportverletzung Sportschaden 1988;2:61–68.

80. Rodegerts, U, Goleissner S. Langzeiterfahrungen mit der operativen Therapie der Osteochondrosis dissecans des Kniegelenkes. Orthroped Praxis 1979;8:612.

81. Wagner H. Operative Behandlung der Osteochondrosis dissecans des Kniegelenks. Z Orthoped 1964;95:333.

82. Menke W, Antoniadis A. Zur operativen Behandlung der Osteochondrosis dissecans des Kniegelenkes. Orthoped Praxis 1986;1:63–66.

83. Hardaker WT, Whipple TL, Bassett FH. Diagnosis and treatment of the plica syndrome of the knee. J Bone Joint Surg 1980;62:221–225.

84. Jackson RW, Marshal DJ, Fujisawa Y. The pathologic medial shelf. Orthoped Clin N Am 1982;13:307–312.

85. Munzinger U, Ruckstuhl J, Scherrer H, Gschwend N. Internal derangement of the knee joint due to pathologic synovial folds: the mediopatellar plica syndrome. Clin Orthoped 1981;155:59–63.

86. Patel D. Arthroscopy of the plicae—synovial folds and their significance. Am J Sports Med 1978;6:217.

87. Tasker T, Waugh W. Articular changes associated with internal derangement of the knee. J Bone Joint Surg (Br) 1982;486–488.

88. Pipkin G. Knee injuries: the role of the suprapatellar plica and suprapatellar bursa in simulating internal derangement. Clin Orthoped 1971;74:161–176.

89. Moller H. Incarcerating mediopatellar synovial plica syndrome. Acta Orthoped Scand 1981;52:357–361.

90. Silver DM. How to deal with plica (shelf) and other synovial impingement syndromes. In: Arthroscopy Clews. New York: P. W. Communications, 1983:2.

91. Schulitz KP, Hille E, Kochs W. The importance of the mediopatellar synovial plica for chondromalacia patellae. Arch Orthoped Trauma Surg 1983;102:37–44.

92. Iino S. Normal arthroscopic findings in the knee joint in adult cadavers. J Jpn Orthoped Assoc 1939;14:467–523.

93. Muse GL, Grana WA, Hollingsworth S. Arthroscopic treatment of medial shelf syndrome. Arthroscopy 1985;1:63–67.

94. Broukhim B, Fox JM, Blazina ME, Pizzo W, Hirsh L. The synovial shelf syndrome. Clin Orthoped 1979;142:135.

95. Richmond JC, McGinty JB. Segmental arthroscopic resection of the hypertrophic mediopatellar plica. Clin Orthoped 1983;178:185–190.

96. Kinnard P, Levesque RY. The plica syndrome. A syndrome controversy. Clin Orthoped 1984;183:141–143.

97. Klein W. The medial shelf of the knee. A follow-up study. Arch Orthoped Trauma Surg 1983;102:67–72.

98. Hoffmann F. Ist die operative Behandlung der hypertrophen Plica mediopatellaris sinnvoll? In: Hofer H, ed. Fortschritte in der Arthroskopie. Stuttgart: Ferdinand Enke Verlag, 1985:93–95.

99. Jung T, Rodriguez M, Augustiny N, Friedrich N, von Schulthess G. 1,5-T-MRI, Arthrographie und Arthroskopie in der Evaluation von Knieläsionen. Fortschr Röntgenstr 1988;148(4):390–393.

100. Schuler M, Naegele M, Lienmann A, et al. Die Wertigkeit der hochauflösenden CT und der Kernspintomographie im Vergleich zu den Standardverfahren bie der Diagnostik von Meniskusläsionen. Fortschr Röntgenstr 1987;146(4):391–397.

101. Stoller DW, Martin C, Cruess JV III, Kaplan L, Mink JH. Meniscal tears: pathologic correlation with MR imaging. Radiology 1987;163:731–735.

102. Silva I, Silver DM. Tears of the meniscus as revealed by magnetic resonance imaging. J Bone Joint Surg 1988;70A:199–202.

103. Reicher MA, Rauschnig W, Gold RH, Basset LW, Lufkin RB, Glen W. High-resolution magnetic resonance imaging of the knee joint: pathologic correlations. AJR 1985;145:903–909.

104. Reicher MA, Hartzman S, Bassett LW, Mandelbaum B, Duckwiller G, Gold RH. MR imaging of the knee. Part 1. Traumatic disorders. Radiology 1987;162:547–551.

105. Cruess JV. Joint imaging. Diagnostic radiology. Magn Reson 1988;175–179.

106. Köppchen J, Fischer H-J, Becker W, Arnold W, Rietig M. Kernspintomographie von Meniskusläsionen. Z Orthoped 1987;125:390–395.

107. Köppchen J, Fischer H-J, Becker W, Arnold W. Meniskusdarstellung in der MR-Tomographie mit Oberflächenspulen. Fortschr Röntgenstr 1987;146(6):617–622.

108. Engel A. Magnetic resonance knee arthrography. Acta Orthoped Scand 1990;60(suppl 239):1–57).

109. Selesnick FH, Noble HB, Bachmann DC, Steinberg FL. Internal derangement of the knee: diagnosis by arthrography, arthroscopy, and arthrotomy. Clin Orthoped 1985;198:26–30.

110. Levinsohn EM, Baker BE Prearthrotomy diagnostic evaluation of the knee: review of 100 cases diagnosed by arthrography and arthroscopy. AJR 1980;134:107–111.

111. Dumas J-M, Edde DJ. Meniscal abnormalities: prospective correlation of double-contrast arthrography and arthroscopy. Radiology 1986;160:435–456.

112. Ghelman B. Meniscal tears of the knee: evaluation by high-resolution CT combined with arthrography. Radiology 1985;157:23–27.

CHAPTER 36

Magnetic Resonance Imaging and Computed Tomography of the Foot and Ankle

Louis H. Wetzel and Mark D. Murphey

Introduction

The human foot is a complex and highly specialized terminal appendage. It must provide the support and stabilization necessary for the habitual upright posture and locomotion that are unique to humans (1). Each foot contains at least 26 bones, 35 articulations, more than 40 discrete ligaments, 20 intrinsic muscles, and 12 long tendons of extrinsic origin. The complex anatomy of the foot and ankle region often limits conventional radiographic evaluation. Sectional imaging, by virtue of its ability to show osseous and soft-tissue structures in a thin section of tissue in various planes, often overcomes the limitations imposed by conventional projection radiography. The purpose of this chapter is to review foot and ankle pathology, which is particularly amenable to evaluation by computed tomography (CT) or magnetic resonance (MR) imaging and the relative merits of each in specific disorders.

Sectional Anatomy

One of the major advantages of sectional imaging of the foot and ankle is the ability to obtain images in multiple planes. Localization and diagnosis of foot lesions by sectional imaging should be based on an understanding of salient anatomic features in the three standard planes. Normal sectional anatomy of the foot and ankle is reviewed in Chapter 8, and there are numerous studies depicting normal foot and ankle anatomy by CT and MR imaging in the literature (2–8).

Technique

Well-positioned high-quality conventional radiographs should always be obtained as the initial radiologic investigation of the foot and ankle. The sectional imaging ex-

amination must be carefully tailored to address specific clinical questions or further investigate abnormalities detected on radiographic or scintigraphic studies. This necessitates a review of the history and any other available imaging studies. Ideally, a brief physical examination should be performed by the physician or technologist performing the imaging examination. All palpable masses or locations of focal pain should be marked for reference. A small metallic object for CT or a gelatin capsule for MR imaging studies should be taped lightly to the skin, since excessive pressure may deform superficial masses. All studies should be monitored to assure that the area of suspected pathology is adequately evaluated and to prescribe additional imaging planes or pulse sequences if necessary.

Computed Tomography

Proper positioning of the extremities simplifies interpretation of sectional imaging studies. Axial CT images are obtained by positioning the feet symmetrically on the scanning table with the patient supine and the soles of the feet perpendicular to the table top. A brace or shoes may be used for reproducible positioning of the feet (9). Direct coronal CT scans are obtained by placing the soles of the feet on the scanning table with the knees flexed and the gantry angled as needed. Direct sagittal CT scanning has also been described (10).

For evaluation of fractures and subtalar joint abnormalities, 1.5- to 3-mm sections in both axial and coronal planes are obtained if the feet are mobile. If the foot cannot be freely positioned because of a cast or splint, then thin axial sections should be obtained as these allow high-quality multiplanar reconstructions. Other disorders, such as soft-tissue tumors or infections, are usually well evaluated with 3- to 5-mm axial or coronal sections. A targeted or magnified field of view technique should be used, just large enough to include both feet in the circle of reconstruction. A high-resolution bone algorithm, available on most current scanners, improves spatial resolution and osseous detail, although depiction of soft tissues is compromised. Intravenous contrast administration may improve identification of vessels and evaluation of tumors.

Magnetic Resonance Imaging

Careful positioning of the feet is particularly important with MR imaging in order to facilitate identification of soft-tissue structures. The patient should be positioned supine with the knees extended and the feet symmetrical. The soles of the feet should be as near perpendicular to the table top as possible. Because some rotation of the foot is inherent with plantar or dorsiflexion the neutral position is preferred. A right-angle foam pad may be used to apply gentle pressure to the balls of the feet, although forced dorsiflexion should be avoided. The feet and toes should be taped lightly to the table or coil, as a reminder and aid to the patient to avoid excessive motion. For most examinations, a standard head coil can be used with the feet positioned near the center of the coil (Fig. 36.1). The use of the head coil has the advantages of simplicity, field homogeneity, good spatial resolution, and wide availability. A small surface coil improves spatial resolution, however, the image intensity is inhomogeneous and imaging of the contralateral extremity is sacrificed.

For localized lesions T1-weighted images in two standard orthogonal planes and T2-weighted images in one of these planes are performed at minimum. For nonspecific complaints such as regional pain, T1-weighted images in all three standard planes are obtained, and the most appropriate plane for T2-weighted images is selected. Both T1- and T2-weighted sequences should be performed in all patients. Standard spin-echo sequences are usually adequate for evaluation of bone marrow and soft-tissue pathology. Gradient-echo pulse sequences may improve visualization of cartilage, which appears as high intensity on these images, but joint fluid may be inseparable from cartilage. With standard spin-echo imaging, 3- to 5-mm sections are obtained with 0- to 1-mm intersection gaps. Thinner sections may be obtained with three-dimensional volume acquisition techniques.

We currently use a superconducting MR imaging system operated at 1.0 T (Siemens Medical Systems, Iselin,

N.J.), and the standard head coil supplied with the unit. The parameters for conventional spin-echo imaging are TR = 500 ms, TE = 17 ms for T1-weighted images, and TR = 2100 ms, TE = 30 and 90 ms for spin-density and T2-weighted images.

Trauma

The foot and ankle are among the most commonly injured structures in the body, and injuries to the ankle alone account for about 10% of all emergency room visits (11). While most acute injuries result from falls, motor vehicle accidents, or athletic injuries, the foot is also uniquely subject to chronic repetitive trauma and stress. Physical and conventional radiographic examinations are adequate for evaluation in the majority of cases. However, soft-tissue injuries and subtle fractures may be overlooked, and geometrically complex or severely comminuted fractures may not be adequately delineated with standard techniques. Radionuclide bone scan and conventional tomography are valuable additional studies but have certain limitations. Bone scan offers limited spatial resolution and specificity, and tomography does not allow evaluation of the ankle and hindfoot in the transverse plane. Subtle or complex osseous injuries are best evaluated by CT. The patient's age, activity, mechanism of injury, and medical history should be considered in evaluation of osseous and soft-tissue trauma.

Complex Ankle Fractures

Distal tibial triplane fractures result from external rotation injury to the closing physis in adolescence (12, 13). Owing to the pattern of growth plate closure, a geometrically complex epiphyseal fracture occurs with fracture compo-

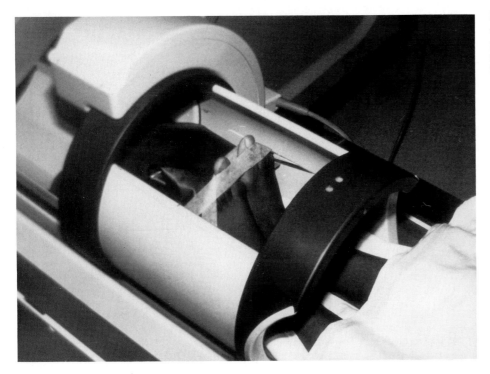

Figure 36.1. MR imaging of the feet in head coil. Head coils are universally available and provide adequate evaluation of most disorders. Careful positioning of the feet facilitates interpretation of the imaging study.

nents in all three orthogonal planes (12–14). Most fractures occur laterally since the anterolateral physis closes last (14). They are difficult to diagnose by plain films and may require open reduction. Axial CT with sagittal and coronal reconstructions accurately demonstrates the fracture fragments, their alignment, and the integrity of the tibial plafond (12, 15) (Fig. 36.2). Fractures are subclassified as two-part or three-part fractures. Evaluation of these features is important before and after attempted closed reduction to assess the need for internal fixation.

Pylon fracture of the ankle results from axial compression of the joint and should be distinguished from trimalleolar fracture caused by rotation injury (16). The pylon fracture usually requires more extensive surgery and has a poorer long-term prognosis. The salient features of the pylon fracture are severe distal tibial comminution and involvement of the tibial plafond. Computed tomography is helpful to evaluate the extent of plafond injury (16). Severe disruption and comminution of the distal tibia may necessitate joint fusion rather than reconstruction (17–19).

Computed tomography has also been advocated for evaluation of trimalleolar fractures (16). Computed tomography may allow more accurate estimation of the size of the posterior malleolar fragment than plain films. If less than 25% of the tibial articular surface is involved, the joint

is considered stable, and posterior internal fixation is usually not necessary (16).

Calcaneal and Other Foot Fractures

Calcaneal fractures are often difficult to evaluate radiographically and complex fractures are routinely evaluated by CT (20–23). Seventy-five percent of calcaneal fractures are intraarticular, involving the posterior subtalar joints, and 25% are extraarticular (20) (Fig. 36.3). The former are produced by downward impaction of the lateral process of the talus into the calcaneal body, producing an oblique vertical shear fracture (24). Increasing severity of calcaneal compression leads to greater comminution and displacement. Management of these fractures is difficult and the precise approach depends on several factors. These include the degree of calcaneal volume loss, size and displacement of the sustentacular fragment, size and location of other fragments, disruption of the posterior articular facet, and impingement or entrapment of the lateral ankle tendons (20, 21) (Fig. 36.4). The CT study should include axial (long axis) and coronal (short axis) scans. Sagittal and three-dimensional reconstructions may enhance depiction of calcaneal or other complex fractures and facilitate communication of findings to the orthopedic surgeon (25, 26).

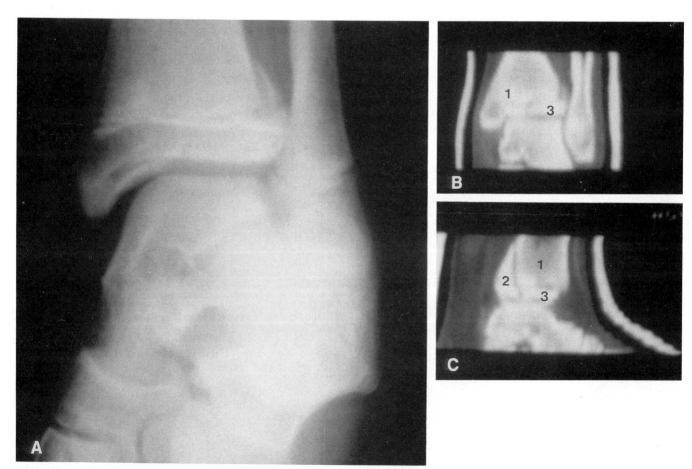

Figure 36.2. Distal tibial triplane fracture. **(A)** Mortise view suggests a Salter-Harris type 2 fracture. **(B)** Coronal and **(C)** sagittal reconstructions reveal a lateral three-fragment variant triplane fracture. Metaphyseal *(1),* metaphyseal-epiphyseal *(2),* and epiphyseal *(3)* fragments are separated by fracture components in the coronal, axial, and sagittal planes. (Courtesy of D. Sartoris, M.D., San Diego, Calif.)

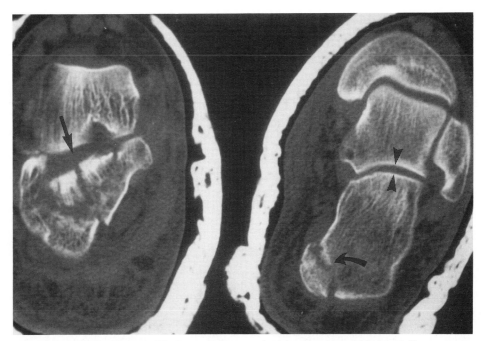

Figure 36.3. Bilateral calcaneal fractures. Coronal CT shows comminuted intraarticular right calcaneal fracture *(straight arrow)* and extraarticular tuberosity fracture on the left *(curved arrow)*. The left subtalar joint *(arrowheads)* is intact. Because many fractures result from falls, as in this patient, bilateral calcaneal fractures are common.

The most common long-term complications are subtalar joint arthritis resulting from joint incongruity and peroneal tendon impingement, which may also be assessed by CT (27).

The value of computed tomography for evaluating complex talar fractures (28) and tarsal-metatarsal fracture-dislocation has also been reported (29). Computed tomography allows accurate assessment of periarticular and intraarticular fracture fragments, extent of subtle dislocation and adequacy of reduction, and may also be helpful in showing nonunion and posttraumatic degenerative arthritis associated with these injuries (30, 31). Computed tomography may also show subtle periarticular fractures in other locations, not evident on plain films, that can be a cause of persistent pain after foot and ankle trauma (32).

Osteochondral Fractures

The talar dome is a common site of impaction fractures resulting from inversion ankle injuries. Fractures involve the medial or lateral aspect of the dome with equal frequency (33–35). Medial lesions are usually deeper, less symptomatic, and tend to heal spontaneously, whereas lateral lesions are usually shallow and are more often displaced. Conventional tomography and CT are useful for radiographic staging of lesions and detection of intraarticular fragments (36). Management of nondisplaced fractures depends on the stability of the osteochondral fragment (37, 38). Conventional arthrotomography and CT arthrography have been used to assess the degree of attachment but require intraarticular contrast injection (36, 39).

Magnetic resonance imaging has been helpful to evaluate the articular cartilage and predict the stability of osteo-chondral fragments (37, 40) (Fig. 36.5). De Smet and colleagues accurately predicted the extent of fragment attachment to the talus with MR imaging in 13 of 14 patients (37). Partially attached fragments showed an incomplete high signal line on T2-weighted images at the fragment/talar interface, whereas unattached fragments had a complete ring of high-signal fluid encircling the fragment. Completely attached fragments had low signal intensity at the interface. Normal cortical irregularity in the posterior aspect of the talus should not be mistaken for an osteochondral fracture (41). The tarsal navicular and metatarsal heads are less frequent sites of osteochondral injuries. While preliminary results are promising, further studies will be needed to determine the precise role and long-term benefits of MR staging of these lesions.

Stress Fractures

Stress fractures are frequent in the foot and ankle and most commonly involve the distal tibia and fibula, calcaneus, navicular bone, and metatarsals. Traditionally, the diagnosis has been established by bone scan and follow-up radiography. Sagittally oriented tarsal navicular stress fractures may be particularly difficult to detect on plain films and conventional tomography (42), but are easily confirmed by axial CT (Fig. 36.6). Occasionally, stress fractures may first be diagnosed on MR imaging as a low-intensity linear or irregular marrow defect on T1-weighted images, usually extending to the cortex (Fig. 36.7). Intraosseous areas of mildly increased intensity may be seen adjacent to the fracture on T2-weighted images. Normal dense trabecular lines of stress in the talus and calcaneus also cause low-intensity linear marrow defects and should not be mistaken for stress fractures.

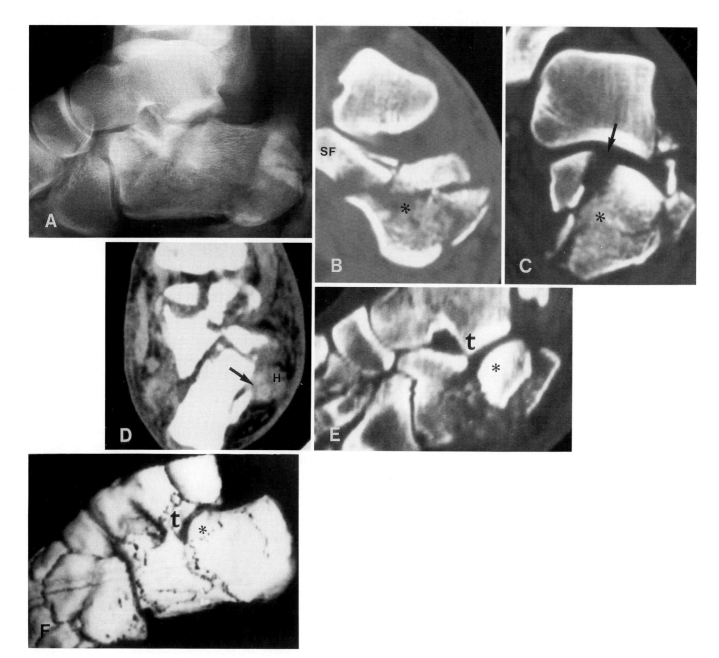

Figure 36.4. Complex calcaneal fracture. **(A)** Lateral view shows comminuted fracture and flattening of calcaneus. **(B and C)** Coronal CT scans show size of nondisplaced sustentacular fragment *(SF)*, intraarticular fracture of posterior subtalar joint *(arrow)*, and central calcaneal volume loss *(asterisks)*. **(D)** Axial CT shows lateral displacement of fragments with impingement on peroneal tendons *(arrow)* and peritendinous hematoma *(H)*. **(E)** Sagittal and **(F)** 3D reconstructions illustrate the mechanism of injury with lateral process of talus *(t)* driven downward into calcaneus. Inferior displacement and rotation of posterior articular fragment *(asterisk)* is demonstrated.

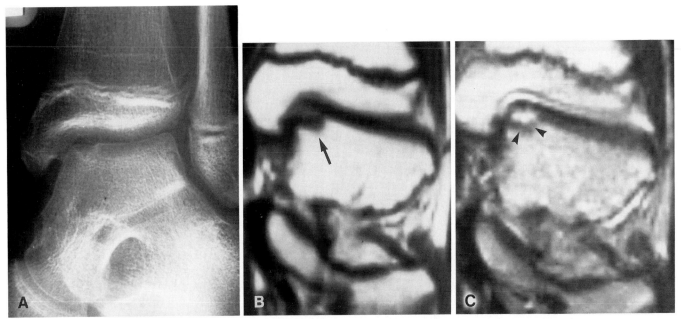

Figure 36.5. Osteochondral fracture of talar dome. **(A)** Mortise view shows nondisplaced medial talar dome lesion. **(B)** Coronal (spin-echo (SE) 500/17) MR image shows low-intensity talar defect *(arrow)*. **(C)** SE 2100/90 image shows high intensity within osteochondral fragment. Low intensity at fragment/talar interface *(arrowheads)* indicates a stable attached fragment.

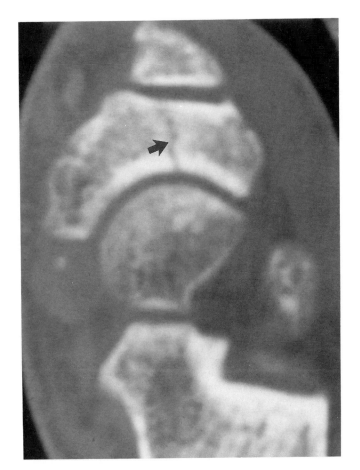

Figure 36.6. Tarsal navicular stress fracture. Axial CT demonstrates sagittally oriented fracture line *(arrow)* that was not evident by conventional radiography in patient with persistent dorsal foot pain. (Courtesy of A. De Smet, M.D., Madison, Wisc.)

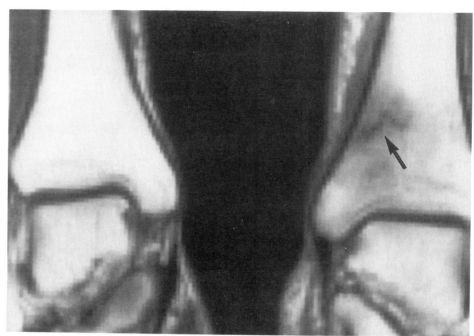

Figure 36.7. Tibial stress fracture. Plain films were normal in middle-aged patient with persistent ankle pain. Coronal SE 500/17 MR image demonstrates unsuspected stress fracture *(arrow)* of the medial meta-physis with surrounding edematous marrow of intermediate intensity. SE 2100/90 images showed mildly increased intensity about fracture.

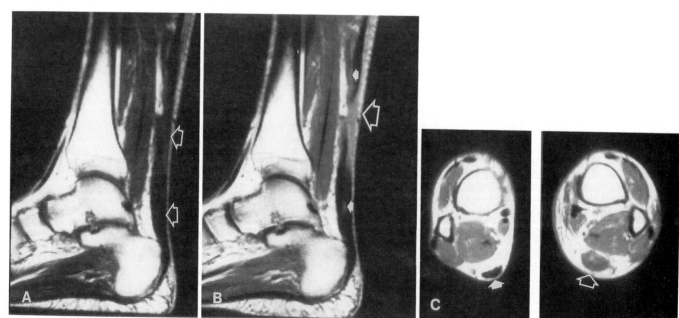

Figure 36.8. Complete Achilles tendon tear. **(A)** Sagittal SE 500/17 MR image shows thickening of entire length of Achilles tendon *(open arrows)*. Discontinuity of tendon is not readily apparent. **(B)** Sagittal SE 2100/30 MR image clearly depicts complete disruption of low-intensity tendon fibers *(open arrow)* just distal to musculotendinous junction with interposed subacute hematoma. Note linear hemorrhage within tendon *(small solid arrows)*. **(C)** Axial SE 500/17 MR image shows thickening and higher intensity of torn tendon *(open arrow)* compared to normal contralateral tendon *(solid arrow)*. The absence of low-intensity tendon fibers at this level is the only evidence of complete tear by axial imaging.

Tendon Injury

Injuries to the ankle tendons may be unrecognized on clinical examination and can lead to chronic instability or deformity. Computed tomography and MR imaging afford noninvasive means of evaluating these injuries. Computed tomography provides accurate evaluation of tendon location and cross-sectional morphology and has been used for diagnosis of tendon dislocations and tears (43–47). Computed tomography is particularly useful when there are associated osseous abnormalities such as periostitis, displaced fracture fragments, or dislocations. Because of superior soft-tissue contrast and multiplanar imaging, MR imaging appears more accurate in assessing the extent of tendon injury (48–51). Tendon ruptures are generally classified as type 1 (partially torn, enlarged tendon), type 2 (partially torn, attenuated tendon), or type 3 (complete rupture with tendon gap). Magnetic resonance allows better assessment of intratendinous hemorrhage or edema, vertical splits, and synovial fluid. Owing to partial volume averaging on axial images, sagittal MR is probably most accurate for diagnosing complete ankle tendon rupture, particularly when there is minimal separation of the torn ends of the tendon. Magnetic resonance may also be use-ful to follow healing of conservatively managed tendon tears.

The Achilles tendon is the strongest, most easily imaged, and most frequently injured tendon of the ankle. It is also the only ankle tendon lacking a synovial sheath. Achilles tendon tears are usually the result of strenuous athletic activities but they also occur in patients with systemic disorders such as rheumatoid arthritis, hyperparathyroidism, gout, systemic lupus erythematosis, and systemic or local steroid therapy. Up to 25% of these injuries may be misdiagnosed clinically (52). Sagittal MR images are helpful to show complete disruption of tendon fibers, which may not be readily appreciated on axial images (Fig. 36.8). Most Achilles tendon tears occur 2 to 3 cm proximal to the calcaneal attachment. Chronic tears show tendon thickening with low intensity on T1- and T2-weighted images (52) (Fig. 36.9). Pain in the region of the Achilles tendon after exercise may also be caused by an accessory soleus muscle (53), which can readily be detected by CT or MR imaging.

Peroneal tendon injury is one of the major complications of intraarticular calcaneal fractures, and CT is useful to detect and categorize these injuries (27). Impingement on the tendons by osseous fragments demonstrated by CT

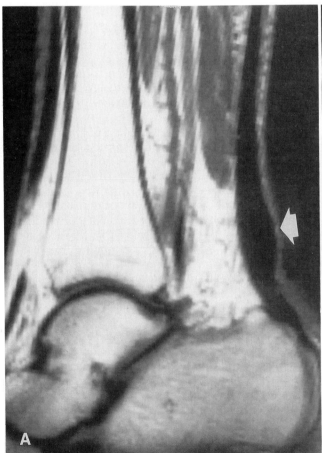
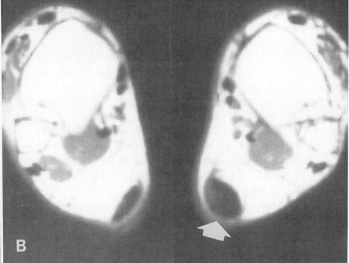

Figure 36.9. Chronic partial Achilles tendon tear. **(A)** Sagittal SE 500/17 MR image shows focal thickening of the distal Achilles tendon *(arrow)*. SE 2100/30 MR image (not shown) confirmed continuity of tendon fibers and showed low intensity within the area of thickening consistent with scarring. **(B)** Axial 500/17 MR image shows thickened low-intensity tendon *(arrow)* as compared to Figure 36.8C.

is an accurate predictor of subsequent development of peroneal tenosynovitis (27). Computed tomography may also demonstrate an absent fibular groove on the lateral malleolus, which predisposes patients to recurrent peroneal tendon subluxation or dislocation (27, 44). Rupture or dislocation of the peroneal tendons is uncommon but may be overlooked as a cause of chronic ankle instability (54). Axial and sagittal MR imaging allows detection of partial or complete tears and tendon sheath fluid. A small amount of normal synovial fluid in the shared tendon sheath of the peroneus longus and brevus tendons should not be mistaken for a vertical tendon tear (41).

The medial ankle tendons are less frequently injured. Of the medial tendons, the posterior tibial tendon is the most commonly ruptured. Rosenberg and coworkers found a high accuracy in detection of tendon rupture by both CT and MR; however, MR was slightly more accurate for classifying the extent of tendon injuries (50). Both CT and MR tended to underestimate the extent of tendon injuries as compared with surgical findings.

Ligament Injury

Ligaments of the ankle may be shown by CT, but there are few reports of CT evaluation of ligamentous ankle injury (46, 55). While early MR imaging of the ligaments was compromised by poor spatial resolution, improved surface coil design and thinner section afforded by 3D volume acquisition techniques have dramatically improved MR depiction of ligamentous anatomy (56–58). Variations in the normal appearance of ankle ligaments have been reported (41). Since most ankle sprains are the result of inversion injuries, reliable evaluation of the lateral ankle ligaments is most often required. In a study of 16 normal ankles, Beltran and colleagues identified the anterior talofibular ligament (the most commonly injured ankle ligament) on axial images in 100% of cases and showed the calcaneofibular ligament in 81% of ankles in the coronal plane (58). The components of the deltoid ligament are best shown in the coronal plane. The talocalcaneal ligament and cervical ligament of the sinus tarsi can be seen on sagittal images. Acute ligamentous tears appear as disruption or absence of the low-intensity ligamentous fibers surrounded by hemorrhage or edema (49). While MR is clearly capable of depicting normal and deranged ligamentous structures, long-range clinical studies will be necessary to determine the efficacy and role of MR in the diagnosis and treatment of ankle sprains.

Infection

Osteomyelitis

The foot is particularly susceptible to infection in all age groups as a result of trauma, plantar puncture wounds, and pressure skin ulcerations. Infection in the diabetic foot is discussed in Chapter 37. In the foot, osteomyelitis usually occurs from direct inoculation of bone with deep penetrating injuries or from local extension of a soft-tissue infection (59, 60). Barefoot children are frequently involved, and calcaneal osteomyelitis may result from diagnostic heel puncture in neonates (61). Plantar infection tends to spread within one of three longitudinally oriented plantar compartments, which may account for development of osteomyelitis at a site distant from a soft-tissue puncture wound (59, 62, 63). Osteomyelitis resulting from hematogenous dissemination is less common in the foot but tends to involve the calcaneus (59) and frequently occurs in the distal tibial metaphysis in children.

Computed tomography can be helpful in the diagnosis of osteomyelitis by showing subtle increased density in the marrow, trabecular lysis, and cortical bone changes when plain film findings are minimal or absent (63–66). Rarely, CT may demonstrate intraosseous gas (67). Computed tomography is probably less sensitive for early osteomyelitis than MR imaging or bone scintigraphy (68). The three-phase bone scan remains the standard and most economical means for diagnosis of osteomyelitis but does not always differentiate bone from soft-tissue infection (69, 70) and may not provide sufficient anatomic detail to guide diagnostic needle aspiration. Previous trauma, surgery, or arthritis also complicates the scintigraphic diagnosis.

At MR imaging, acute osteomyelitis results in an area of decreased intensity on T1-weighted images within the marrow, which shows moderately to markedly increased intensity on T2-weighted images (Fig. 36.10). The MR findings are usually striking in acute pyogenic disease but may be less pronounced in partially treated or subacute infections. Owing to its high sensitivity and spatial resolution, MR imaging appears more accurate than bone scan to confirm or exclude osteomyelitis (69–71). Magnetic resonance imaging is particularly valuable when radiographic and scintigraphic findings are equivocal (Fig. 36.11). Although MR is sensitive for alterations in bone marrow signal intensities associated with osteomyelitis, similar findings may be seen with sterile inflammatory processes such as acute infarction or fracture.

The role of MR imaging in evaluation of chronic osteomyelitis has not been extensively investigated. In one study, foci of active infection in areas of previous chronic osteomyelitis were accurately detected by MR imaging (71). Brodie's abscess is common in the distal tibia (Fig. 36.12) and may involve the tarsal bones in young children (59). Computed tomography is superior to conventional radiography for detection of sequestra, and CT or MR may demonstrate a subcutaneous sinus tract extending from a cortical defect (72) (Fig. 36.13). Localization of these abnormalities is important for surgical planning (59, 72).

Soft-Tissue Infections

Septic arthritis may result from direct spread of infection or from hematogenous dissemination. Direct extension is a more common mechanism in the foot than in other joints (59). *Pseudomonas aeruginosa,* an organism common in the soil, may contaminate plantar puncture wounds and has a propensity to involve cartilage (73, 74). As in other articulations, *Staphylococcus aureus* is the most common cause of foot and ankle joint space infection (59). In infants and adults septic arthritis frequently results from direct extension of osteomyelitis in adjacent bones, whereas hematogenous spread tends to occur in children (59). There is probably little role for CT in diagnosis of septic arthritis.

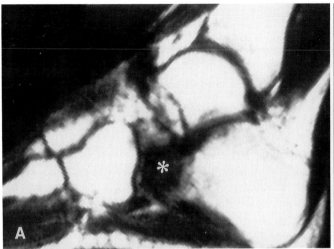

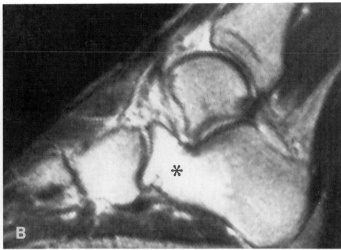

Figure 36.10. Calcaneal osteomyelitis from nail puncture wound. Plain films were normal, but osteomyelitis was suspected by bone scan. MR was requested for localization prior to diagnostic needle aspiration. **(A)** A low-intensity marrow defect is noted in the anterior portion of the calcaneus on sagittal SE 500/17 area *(asterisk)*. **(B)** Sagittal SE 2100/90 image shows high signal intensity from acute osteomyelitis *(asterisk)*.

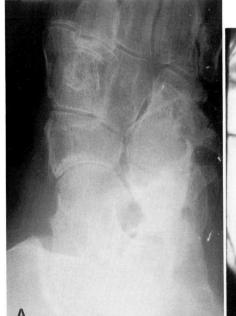

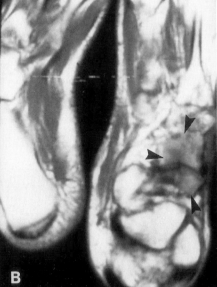

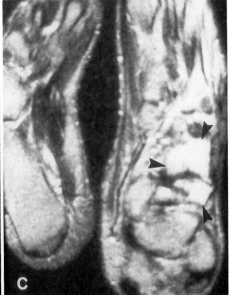

Figure 36.11. Late posttraumatic osteomyelitis. **(A)** Oblique radiograph shows sclerotic deformity of the anterior calcaneus and cuboid resulting from shotgun injury 2 years earlier. Note metallic fragments. **(B)** Axial SE 500/17 image shows decreased marrow intensity within the cuboid and anterior calcaneal remnant *(arrowheads)*. **(C)** Axial SE 2100/90 image demonstrates abnormal high signal intensity within these areas *(arrows)*. Osteomyelitis was confirmed by open biopsy.

Magnetic resonance imaging is more sensitive for detecting joint effusion associated with septic arthritis, but does not distinguish between septic and sterile joint effusions and does not obviate joint aspiration for diagnosis (69, 75). Magnetic resonance imaging may suggest an infected joint when there is extensive surrounding edema, soft-tissue abscess formation, or associated osteomyelitis (69) (Fig. 36.14).

Although cellulitis is usually diagnosed clinically, imaging is often performed to exclude abscess, osteomyelitis, or joint infection. Ill-defined soft-tissue edema representing cellulitis may be detected by CT or MR but is more apparent on T2-weighted MR images. Magnetic resonance is superior for detection of bone or joint involvement. Similarly, synovial fluid associated with septic tenosynovitis may be detected by either method (Fig. 36.15) but is more conspicuous on T2-weighted MR images. Sizeable well-defined abscess cavities are equally well demonstrated by either modality. Magnetic resonance more readily shows small fluid collections in the foot that are not

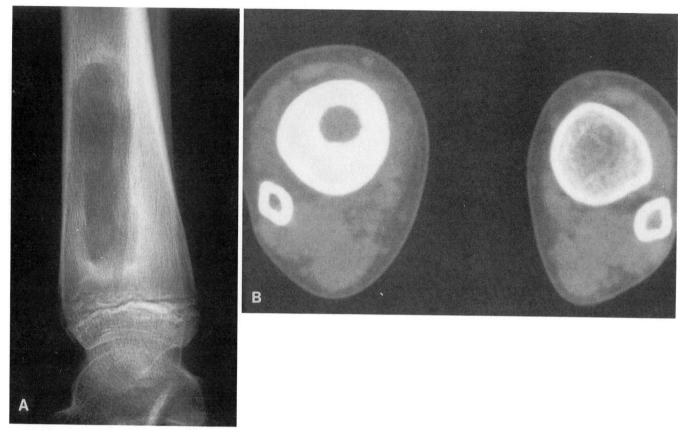

Figure 36.12. Brodie's abscess of distal tibia. **(A)** Lateral film demonstrates lytic lesion with sclerotic borders. **(B)** CT shows lytic area in right tibia and dense circumferential sclerosis. Sequestra were excluded by CT in this case.

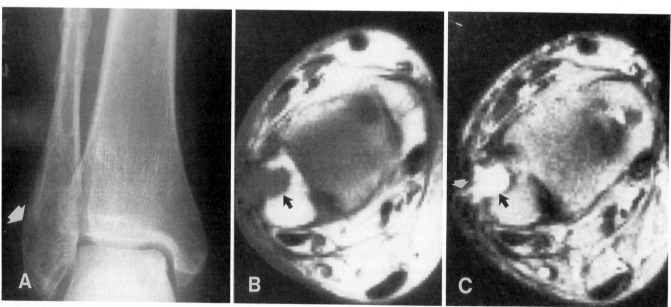

Figure 36.13. Subcortical "pin tract" abscess with draining sinus. **(A)** Frontal radiograph shows deformity of distal fibula from previously internally fixed fracture in patient with draining sinus over lateral malleolus. A vague lucency is noted at site of previous fixation screw *(arrow)*.

(B) Axial SE 500/17 MR image shows low-intensity defect in the lateral fibula *(arrow)*. **(C)** SE 2100/90 MR image shows high intensity of abscess *(black arrow)* extending through lateral cortex and fluid in sinus tract *(white arrow)*.

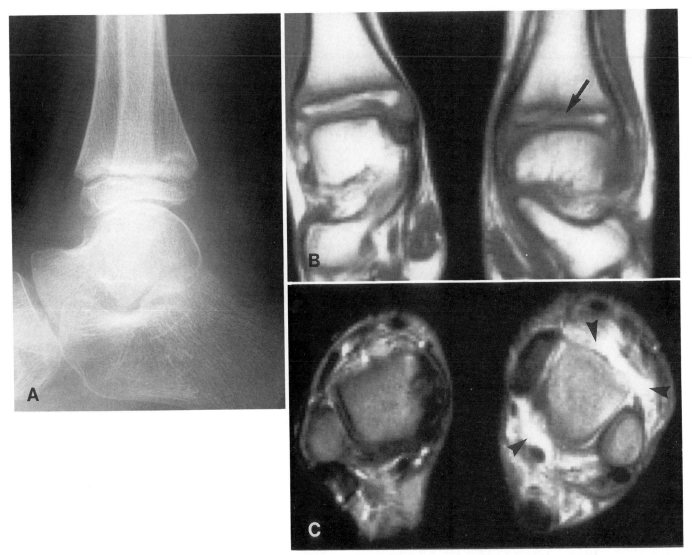

Figure 36.14.   Septic arthritis. **(A)** Lateral radiograph in child with ankle pain and swelling shows increased periarticular density. **(B)** Coronal SE 500/17 MR image demonstrates swelling and inflammation about the left ankle and loss of cartilage thickness. Also note low intensity of left tibial epiphysis *(arrow)* that was infected by direct extension of β-hemolytic streptococcus. **(C)** Axial SE 2100/90 MR image shows joint effusion *(arrowheads)* and periarticular edema.

well seen by CT because of their high density, surrounding edema, or streak artifacts from adjacent bones.

Tumors

Bone and soft-tissue tumors of the foot are uncommon and often present a diagnostic challenge for the radiologist and orthopedic surgeon. As metastases to the foot are extremely rare, most foot tumors are primary lesions (76). Bone tumors are usually detected initially on conventional radiographs, which remain paramount for assessment of tumor aggressiveness and differential diagnosis (77). Soft-tissue tumors of the foot nearly always present as a palpable mass. Plain films are less helpful in evaluation of soft-tissue masses but should still be routinely obtained since detection of soft-tissue calcification or secondary bone involvement may provide important clues to the diagnosis.

Numerous studies have compared CT and MR for evaluation and staging of musculoskeletal tumors (78–83). There is general agreement that MR is more useful for determining intramedullary and soft-tissue extent of tumors and for evaluating joint and neurovascular involvement. Computed tomography is superior for detecting intralesional calcification and may be better for detecting subtle changes in cortical bone. Multiplanar imaging and high soft-tissue contrast make MR imaging particularly advantageous in evaluation of foot tumors owing to the complex anatomy and compact soft tissues of the foot.

Bone Tumors

Considering both the foot and ankle region, benign and malignant bone tumors are more frequent in the distal tibia and fibula than in the foot. The most common benign tumors of the distal tibia and fibula, which occasionally

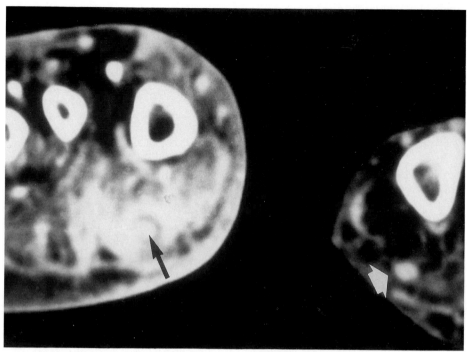

Figure 36.15. Septic tenosynovitis. Coronal CT shows flexor hallucis longus tendon *(black arrow)* surrounded by low-density collar of fluid in tendon sheath and adjacent soft-tissue inflammation. Note normal contralateral tendon *(white arrow).*

involve the foot, are osteochondroma, nonossifying fibroma, and giant cell tumor (84). Giant cell tumor typically extends to the articular cortex, and coronal MR imaging is helpful to assess involvement of the ankle joint (Fig. 36.16). The appearance of giant cell tumor on T2-weighted MR images varies from high to low in signal intensity (85, 86), and we have noted a low intensity in three recurrent tumors. Giant cell tumors in the small bones of the foot are found in younger patients and have a higher rate of recurrence than lesions in other skeletal sites (84) (Fig. 36.17).

Osteoid osteoma usually involves long bones but in the foot appears to have a propensity for the talus (87, 88). They commonly occur in the talar neck and most are subperiosteal in location (87). Such lesions have minimal osseous reaction but may incite a lymphofollicular synovitis mimicking symptoms of arthritis. The tumor nidus may be identified as a lucent or calcified focus by CT and as a high- or low-intensity area on T2-weighted MR images. Cortical lesions show a reactive zone of sclerosis. Medullary lesions usually show less sclerosis, but MR may show a prominent zone of decreased intensity on T1-weighted and increased intensity on T2-weighted images surrounding the nidus resulting from inflammation and reactive hyperemia associated with these vascular lesions (88).

Aneurysmal bone cysts involve the distal tibia and fibula with slightly greater frequency than the foot bones and involve the tarsals, metatarsals, and phalanges in decreasing order of frequency (84). Computed tomography or MR imaging shows an expansile multilocular lesion that occasionally contains multiple fluid/fluid levels of varying density and intensity owing to different aged hemorrhages (89, 90) (Fig. 36.18). This appearance is not specific as fluid/fluid levels have also been found in giant cell tumor, chondroblastoma, fibrous dysplasia, simple cyst, and sarcoma (91).

Enchondroma typically involves the hand and less commonly involves the small bones of the foot, usually the phalanges. Plain films, conventional tomography, or CT often shows central calcification. The diffuse high signal intensity of enchondroma on T2-weighted MR images has been attributed to the hydrophilic nature of the hyaline cartilage matrix of which these lesions are composed (92). Chondroblastomas and chondromyxoid fibromas are rare cartilage tumors that may involve the feet. The former has a predilection for the calcaneus and talus, and the latter frequently involves the calcaneus and metatarsals (84, 93, 94). Chondroblastomas have shown low or mixed low and high signal intensities on T2-weighted MR images owing to matrix calcification (92, 95, 96) while chondromyxoid fibromas have been high in intensity on T2-weighted images (97).

Simple cysts and intraosseous lipomas tend to involve the calcaneus at the junction of its anterior and middle thirds. While the radiographic appearance may be similar, CT or MR can readily distinguish the fluid or fat content of these lesions (98, 99) (Fig. 36. 19). Computed tomography may be more helpful to distinguish intraosseous lipoma from normal rarefaction of trabecular bone in this area (100). When a central calcified nidus is present in lipoma, the plain film is virtually pathognomonic (Fig. 36.19C).

Primary malignant bone neoplasms and lymphoma rarely involve the foot. Only 1 to 3% of osteosarcomas, chondrosarcomas, and Ewing's sarcomas arise in the foot with decreasing frequency distally (84). Computed tomography or

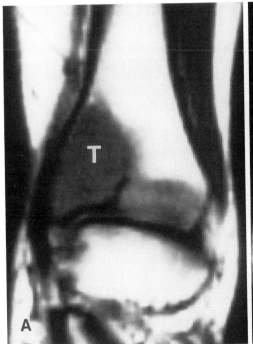

Figure 36.16. Giant cell tumor of distal tibia. **(A)** Coronal SE 500/17 MR image shows low-intensity tumor *(T)* in distal tibia without evidence of ankle joint involvement. **(B)** Axial 2100/90 MR image shows intermediate intensity of lesion *(T)* and well-defined low-intensity rim *(arrowheads)*.

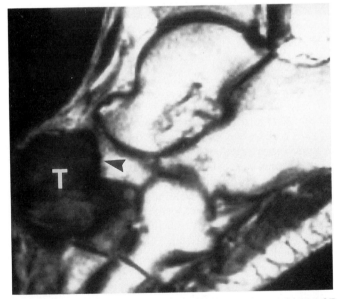

Figure 36.17. Recurrent giant cell tumor of cuneiforms. Sagittal SE 500/17 MR image shows expansile tumor *(T)* with erosion of navicular bone *(arrowhead)*. Tumor had low intensity on SE 2100/90 images (not shown). This was the third recurrence of tumor at this site. Biopsy showed fibrosarcomatous stromal elements, and Syme's amputation was performed.

MR may confirm the aggressive nature of these lesions, but they do not usually help predict the histologic diagnosis. In general, these neoplasms show low signal intensity on T1-weighted images and have intermediate to high signal intensity on T2-weighted images. They are frequently heterogeneous on T2-weighted images owing to varied histologic composition, necrosis, or hemorrhage. Magnetic resonance imaging is routinely performed for staging of tumors, assessment of response to chemotherapy, and for detection of tumor recurrence. For preoperative staging of osteosarcoma in the foot and ankle, the MR study should include T1-weighted images in the long axis of the tibial shaft for detection of skip metastases (101). Computed tomography may prove more helpful for detection of sclerotic osteosarcoma recurrence since the low intensity of such lesions may be difficult to distinguish from scar tissue or heterotopic ossification by MR imaging.

Soft-Tissue Tumors

After the hand and wrist, the foot is the next most common site of ganglion cyst. It usually occurs as a unilocular or multilocular mass over the dorsum of the foot adjacent to extensor tendons but may also be found along the peroneal tendon sheaths (102, 103). The lesions are painful and associated with previous trauma in about half of cases (102). The diagnosis is usually made at physical examination by transillumination of the mass, although obesity or thick fibrous capsule can obscure this finding. The cystic nature of these lesions may be demonstrated by CT or MR, but previous rupture or hemorrhage may alter the typical appearance. Uncomplicated lesions show homogeneous high signal intensity on T2-weighted images and smooth margins (Fig. 36.20). Lesion intensity on T1-weighted images is often higher than that of simple fluid because of a thick gelatinous content (104, 105).

Cavernous hemangiomas are relatively common in the foot. Deep lesions are difficult to diagnose clinically and

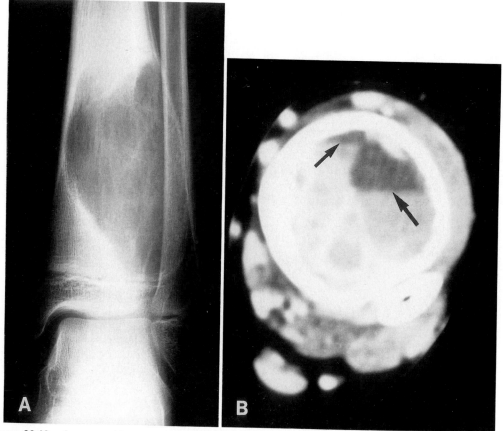

Figure 36.18. Aneurysmal bone cyst of distal tibia. **(A)** Frontal film shows expansile lesion with sclerotic margins and periosteal reaction. **(B)** CT shows fluid/fluid levels *(arrows)* within expansile lesion.

may be confused with soft-tissue sarcomas (106, 107). Plain films are often helpful in diagnosis when phleboliths are seen, but they are found in less than half of lesions (108). Magnetic resonance may suggest the diagnosis by showing a mass of intermediate intensity on T1-weighted images and high signal intensity on T2-weighted images with low-intensity internal septa between vascular spaces (108–112). The lesion margins are usually slightly lobulated or irregular (Fig. 36.21). Small hemangiomas may not show these features (108) (see also Chapter 15).

Plantar fibromatosis typically presents as a nodular subcutaneous growth involving the medial aspect of the plantar aponeurosis and is bilateral in 10 to 25% of cases (113). Magnetic resonance imaging may be helpful to confirm the diagnosis by showing nodular thickening of the plantar aponeurosis with low signal intensity on all pulse sequences due to dense fibrous tissue (112) (Fig. 36.22). In contrast, aggressive fibromatosis is a deep fibromatosis that may involve the muscles of the foot. It has a variable appearance on MR imaging and often shows areas of both high and low signal intensity on T2-weighted images owing to entrapment of muscle by sheets of fibrous tissue (112, 113) (Fig. 36.22*B*).

Since malignant soft-tissue neoplasms of the foot are less common than benign tumors, they may be misdiagnosed clinically and inappropriately managed. If a malignant tumor is initially approached by marginal excision—the standard treatment for benign lesions, rather than by care-

ful preliminary incisional biopsy—then tumor dissemination is more likely (76). Magnetic resonance imaging does not generally distinguish benign from malignant masses (79, 82, 114). However, the location and MR imaging features of certain benign soft-tissue tumors in the foot such as those discussed above may suggest a specific diagnosis (112). Precise delineation of malignant neoplasms is best accomplished by MR imaging and is essential if wide resection and limb-sparing treatment is contemplated.

Synovial sarcoma is the fourth most common soft-tissue sarcoma after malignant fibrous histiocytoma, liposarcoma, and rhabdomyosarcoma (115). Nearly one-fourth of all synovial sarcomas occur in the foot and ankle region, whereas the more common sarcomas rarely or seldom occur there (115–117). Accordingly, synovial sarcoma is probably one of the most common malignant soft-tissue tumors arising in the foot. About one-third show soft-tissue calcification, and 15 to 20% cause secondary bone involvement. These lesions commonly arise in or about the tendons and may show extensive peritendinous growth (Fig. 36.23). They may have infiltrative margins or well-defined borders (104, 114), the latter most likely resulting from slow-growing lesions. They usually appear low in signal intensity on T1-weighted MR images and intermediate to high intensity on T2-weighted images.

Clear cell sarcoma is a rare, deep-seated neoplasm of uncertain histogenesis that arises in intimate association with tendons or aponeuroses and usually affects young

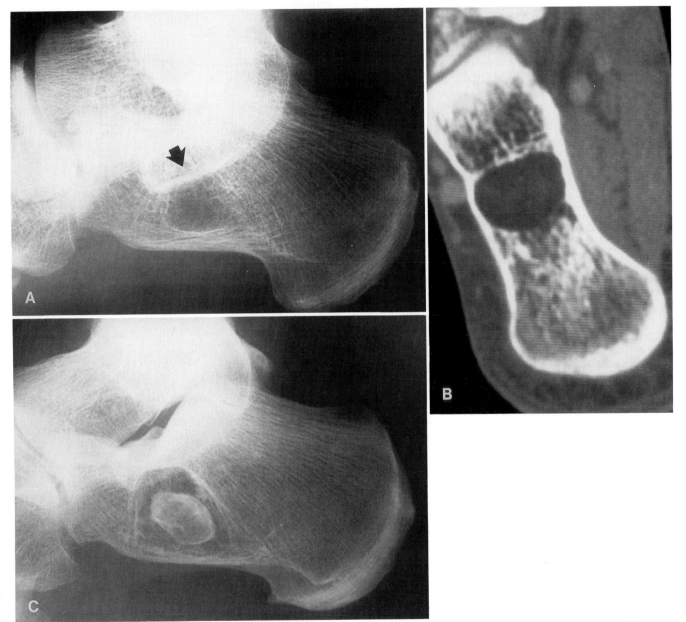

Figure 36.19. Intraosseous lipoma of calcaneus. **(A)** Lateral radiograph shows noncalcified lytic lesion *(arrow)* in patient with vague foot pain. **(B)** Axial CT shows fat-density lesion with faint rim of sclerosis.

(C) Calcified intraosseous lipoma in a different patient. This appearance is considered pathognomonic.

Figure 36.20. Ganglion cyst. Biloculated lesion *(asterisk)* has homogeneous high intensity on SE 2100/90 MR image. Cyst arose from sheath of peroneus tendons *(arrow* shows peroneus longus tendon) just distal to inferior peroneal retinaculum. (From Wetzel LH, Levine E. AJR 1990;155(5):1025–1030. Reproduced with permission.)

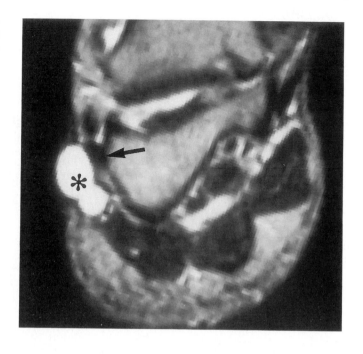

Figure 36.21. Cavernous hemangioma of lateral plantar compartment. Coronal SE 2100/90 MR image shows inhomogeneous tumor *(solid white arrows)* with high intensity, comparable to that of joint fluid *(open arrow).* Note low-intensity internal septa. At surgery, lesion was dissected free of peroneus tendons *(curved arrow)* without functional impairment. (From Wetzel LH, Levine E. AJR 1990;155(5):1025–1030. Reproduced with permission.)

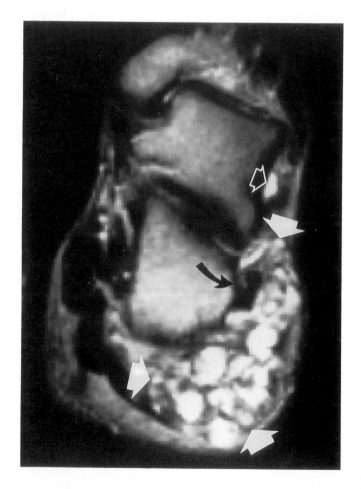

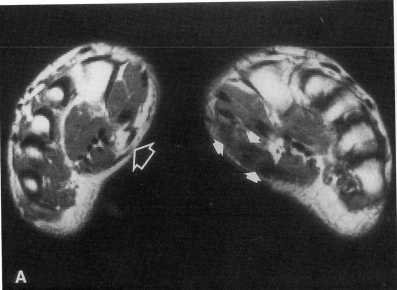
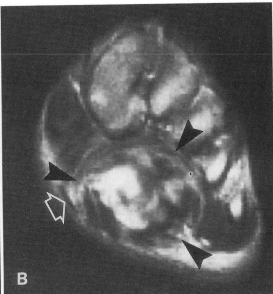

Figure 36.22. Fibromatoses of the foot. **(A)** Plantar fibromatosis. Coronal SE 2100/30 MR image shows low-intensity nodular thickening *(solid arrows)* of medial aspect of left plantar aponeurosis. Lesion remained very low in intensity on an SE 2100/90 image owing to dense fibrous tissue. Note appearance of normal contralateral aponeurosis *(open arrow)*. **(B)** Aggressive fibromatosis in a different patient. Coronal SE 2100/90 MR image shows mixed high- and low-intensity areas within mass *(arrowheads)* in intermediate plantar compartment. Note normal thickness of plantar aponeurosis *(open arrow)* as compared with plantar fibromatosis. (From Wetzel LH, Levine E. AJR 1990;155(5):1025–1030. Reproduced with permission.)

adults (118). The foot and ankle region is the most common site of clear cell sarcoma (43%), and one-fourth of tumors in this region involve the heel, often at the calcaneal attachment of the plantar aponeurosis. In distinction to synovial sarcoma, calcification and secondary bone involvement are uncommon. About half of tumors produce melanin, which causes paramagnetic relaxation enhancement of surrounding tissues, resulting in higher intensity on T1-weighted images and lower intensity on T2-weighted images (112, 119) (Fig. 36.24). Regional lymph node and pulmonary metastases are common, and bone and myocardial metastases also occur. Tumors are also discussed in Chapter 15.

Articular and Periarticular Disorders

Osteoarthritis of the subtalar joints is a frequent complication of intraarticular calcaneal fractures and occasionally results from intraarticular talar fractures or previous surgical fusion of the ankle joint (20, 21, 120). Plain films often do not adequately delineate the extent of subtalar joint disease. The findings of joint space narrowing, articular surface irregularity and depression, subchondral cyst formation and sclerosis, and osteophytes are well shown by CT (Fig. 36.25). Magnetic resonance imaging may directly demonstrate cartilage thinning and subchondral cysts, but sclerosis, small osteophytes, or intraarticular bodies may escape detection by MR.

Rheumatoid arthritis commonly involves the subtalar joints, causing pain, deformity, and gait disturbance. Computed tomography clearly demonstrates soft-tissue swelling, joint space narrowing, and osseous erosions. It has also been helpful to show the extent and plan correction of the pes planovalgus deformity (heel valgus, sustentacular flattening, and inferomedial slippage of the talar head) that occurs in this disorder (121). Magnetic resonance imaging may show cartilage destruction and osseous erosions and in some cases demonstrates these abnormalities when plain film findings are minimal or absent (122, 123) (Fig. 36.26). Synovial hypertrophy appears as areas of low or intermediate signal intensity on T1- and T2-weighted images, although inflamed synovium may show high intensity on T2-weighted images, making distinction from joint fluid difficult (122, 123). Because of its ability to directly image these soft-tissue abnormalities, MR may prove useful for evaluating early disease and monitoring disease progression.

The ankle is the third most common site of hemophilic arthropathy after the knee and elbow (124). Repeated episodes of hemarthrosis result in synovial hypertrophy, subchondral bone destruction and cyst formation, cartilage thinning, and osseous erosions. Synovectomy may slow progression of arthropathy if performed prior to the stage of cartilage destruction (125, 126). Acute hemarthrosis and intramedullary hemorrhage initially show high signal intensity on T1- and T2-weighted images. Their intensity on T1-weighted images decreases over time, becoming intermediate or how in intensity. Focal or diffuse cartilage thinning or loss may be identified, and hypertrophic hemosiderin-containing synovium appears low in intensity on T1- and T2-weighted images (127, 128). Since traditional methods tend to underestimate the extent of articular disease, MR may prove helpful in selecting patients for synovectomy and has also been proposed to monitor effectiveness of clotting factor therapy in acute hemarthrosis (126, 127).

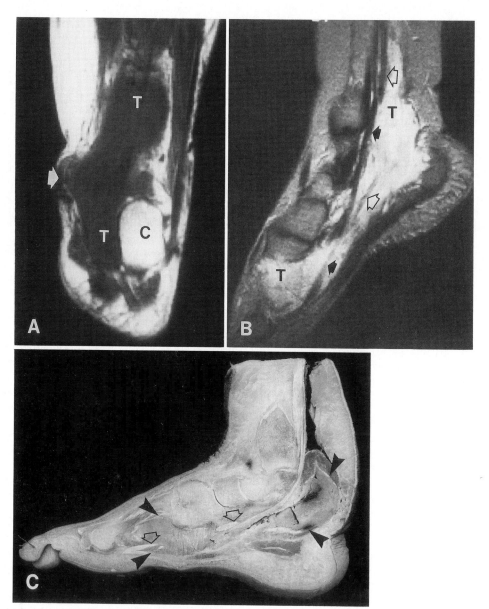

Figure 36.23. Synovial sarcoma. **(A)** Coronal SE 500/17 MR image shows large tumor *(T)* medial to calcaneus *(C)* extending into posterior compartment of leg, which was not detected clinically in obese patient. Small medial component *(arrow)* was palpable, and initial clinical diagnosis was ganglion cyst. **(B)** Sagittal SE 2100/90 MR image reveals longitudinal extent of tumor *(T)* with infiltrative margins and high intensity compared with fat. Note tumor growth along flexor digitorum longus *(solid arrows)* and flexor hallucis longus *(open arrows)* tendons. **(C)** Gross section of specimen confirms extent of neoplasm *(arrowheads)*, which was firmly adherent to long flexor tendons in foot *(open arrows)*. (From Wetzel LH, Levine E. AJR 1990;155(5):1025–1030. Reproduced with permission.)

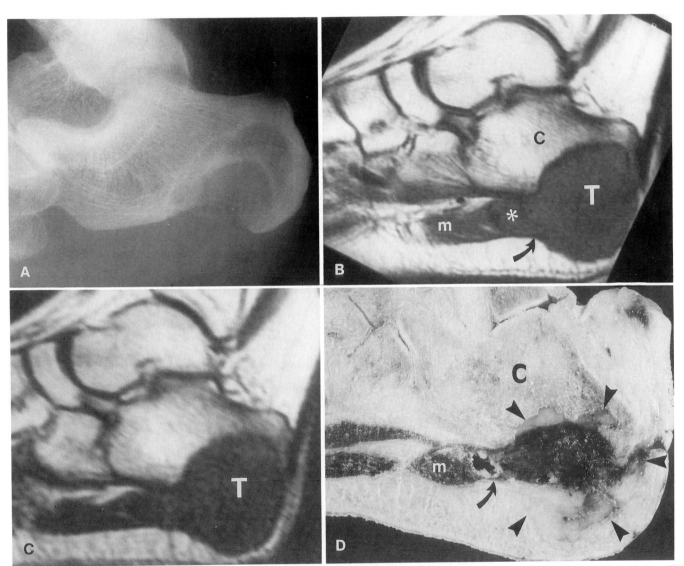

Figure 36.24. Clear cell sarcoma of plantar aponeurosis. **(A)** Lateral radiograph shows large, well-defined erosion of posterior calcaneus. **(B)** Sagittal SE 600/25 MR image reveals oval tumor *(T)* with intensity similar to that of muscle *(m)*, arising near origin of plantar aponeurosis *(curved arrow* shows aponeurosis distal to origin). Note tumor infiltration *(asterisk)* of flexor digitorum brevis muscle *(m)* and erosion of calcaneus *(c)*. Presumptive clinical diagnosis was lipoma of heel. **(C)** Neoplasm *(T)* has low intensity compared with intensity of fat on SE 2000/90 MR image, probably related to its melanin content. **(D)** Gross section just medial to plane of images confirms infiltration *(curved arrow)* of muscle *(m)* by tumor *(arrowheads)*. *c,* Calcaneus. (From Wetzel LH, Levine E. AJR 1990;155(5):1025–1030. Reproduced with permission.)

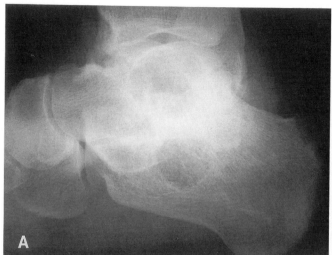
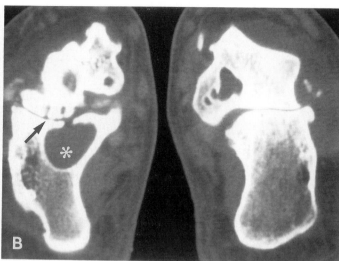

Figure 36.25. Subtalar osteoarthritis. **(A)** Lateral radiograph of right foot shows sclerosis about the subtalar joint with well-defined lucencies in talus and calcaneus. **(B)** Coronal CT shows marked narrowing of the posterior subtalar joints, subchondral sclerosis, and marginal spur-ring. Note large subchondral cyst of right calcaneus *(asterisk)* with superior cortical defect communicating with posterior subtalar joint. Also note depression of articular surface of the right calcaneus *(arrow)* and bilateral talar subchondral cysts.

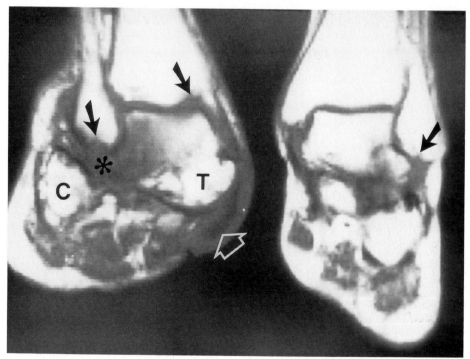

Figure 36.26. Rheumatoid arthritis. Coronal SE 500/17 MR image shows advanced pes planovalgus deformity of the right foot with inferomedial position of talar head *(T)* and lateral dislocation of calcaneus *(C)*. Note pannus formation *(asterisk)*, multiple osseous erosions *(arrows)*, and large synovial cyst *(open arrow).*

Pigmented villonodular synovitis is a proliferative synovial disorder of uncertain etiology. The focal variety frequently involves the hands and feet, and may occur about the ankle or subtalar joints (129). Noncalcified periarticular masses cause pressure erosions with early decrease of cartilage thickness. Magnetic resonance imaging may suggest the diagnosis by showing a periarticular mass of low signal intensity on T1-weighted images and heterogeneous low intensity on T2-weighted images, with a lower intensity peripheral rim owing to hemosiderin deposition (114, 130–133) (Fig. 36.27).

Synovial chondromatosis is most common in the knee, hip, and elbow and infrequently involves the ankle. The diagnosis is usually evident by plain films, which show multiple rounded calcific densities in close proximity to the joint. Computed tomography or MR imaging may be help-

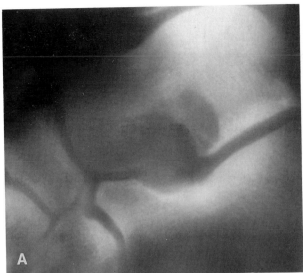

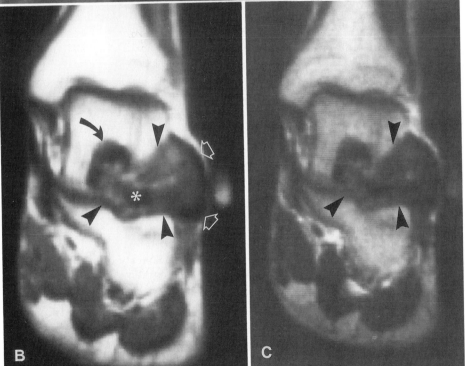

Figure 36.27. Pigmented villonodular synovitis. **(A)** Lateral conventional tomogram shows talar and calcaneal erosions in region of sinus tarsi with preservation of subtalar joint space. **(B)** Coronal SE 2100/28 MR image shows mass *(arrowheads)* surrounding interosseous talocalcaneal ligament *(asterisk)* and causing erosion *(curved arrow)* of talus. Note prominent lateral extension of tumor with hypointense peripheral rim *(open arrows)*. **(C)** Lesion *(arrowheads)* has mildly inhomogeneous, low-intensity signal on SE 2100/90 MR image, reflecting hemosiderin deposition. (From Wetzel LH, Levine E. AJR 1990;155(5):1025–1030. Reproduced with permission.)

ful to confirm the nature, location, and extent of lesions (Fig. 36.28). When chondral fragments are not calcified, MR imaging allows distinction of this disorder from pigmented villonodular synovitis.

Miscellaneous Disorders

Tarsal coalition refers to a fibrous, cartilaginous, or osseous union of the tarsal bones, usually resulting from congenital failure of normal joint formation. The most common sites are the calcaneonavicular and middle subtalar (sustentaculotalar) joints (134). Talonavicular and calcaneocuboid involvement are less common. Symptoms of pain and stiffness often first occur in early adult life, and

spastic flat foot deformity may develop. Plain film findings are often subtle and may be obscured by overlying osseous structures, particularly with middle subtalar joint involvement. Computed tomography is helpful for diagnosis and characterization of these lesions (135–137) (Fig. 36.29). Middle subtalar coalitions are best imaged in the coronal plane, while other forms are evaluated by axial CT. Both feet should be imaged since lesions may be bilateral (134, 138). Computed tomography demonstrates joint space narrowing, cortical irregularity and subcortical sclerosis with fibrous or cartilaginous union, and the extent of osseous union can be assessed. Computed tomography is helpful for surgical planning and detection of associated degenerative disease (139, 140). Magnetic resonance im-

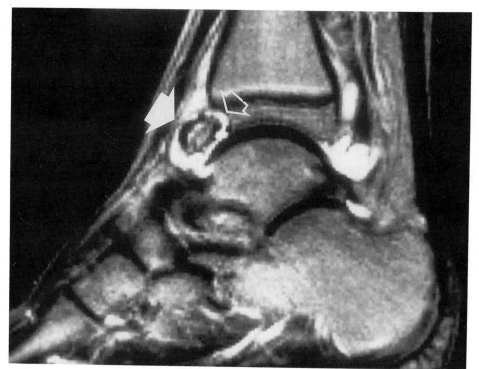

Figure 36.28. Synovial chondromatosis. Sagittal SE 2100/90 MR image shows large joint effusion and osteo-cartilaginous body *(solid arrow)* within joint space. Note erosion of anterior tibia *(open arrow)*.

aging can also demonstrate coalitions, although distinction between the low intensity of sclerosis and fibrosis may be difficult.

Osteonecrosis in the foot is usually on a posttraumatic basis, but widespread involvement may occur in patients with systemic lupus erythematosis (141, 142). The talus is most commonly involved as a result of severe talar neck fractures. Plain films and bone scan are of limited value after trauma, and MR imaging is helpful to confirm or exclude avascular necrosis in this setting (143). Reflex sympathetic dystrophy syndrome frequently involves the foot and may result in swelling, pain, and marked demineralization. In this disorder MR imaging shows only mild alterations in marrow intensity and can be used to exclude infection, which may present with similar symptoms. The practical value of a negative MR examination, e.g., in excluding infection, avascular necrosis, joint effusion, or neoplasm, should not be underestimated.

Magnetic resonance imaging has recently been used to evaluate tarsal tunnel syndrome (144–146). High-resolution images demonstrate the posterior tibial nerve and anatomic structures traversing the tarsal tunnel. Various lesions causing neural compression, such as tumors, tenosynovitis, and posttraumatic fibrosis, may be detected by MR imaging (145). Finally, CT and MR can be very helpful for detection and localization of radiolucent foreign bodies, which commonly involve the plantar aspect of the foot (147–150).

Conclusion

In conclusion, CT and MR imaging often overcome limitations of conventional radiography and may be helpful for evaluation of a wide variety of foot and ankle disorders. Computed tomography shows cortical and trabecular bone detail to best advantage and is less expensive than MR imaging. Magnetic resonance imaging has higher soft-tissue contrast and better multiplanar imaging capability. In general, CT is most helpful for evaluation of complex and subtle osseous trauma. Magnetic resonance imaging appears to be the method of choice for evaluation of soft-tissue trauma, although accurate diagnosis can sometimes be made with CT when MR is not available. Magnetic resonance imaging is useful for evaluation of suspected osteomyelitis when conventional methods are equivocal. Magnetic resonance imaging is generally preferred for staging of bone tumors and is clearly better than CT for detection and staging of soft-tissue tumors. In certain soft-tissue tumors of the foot, MR imaging may often suggest a specific diagnosis. Magnetic resonance imaging allows direct visualization of soft-tissue articular structures, which may improve the diagnosis and management of these conditions. Rapid progress in MR angiography, kinematic MR studies, and MR spectroscopy will undoubtedly provide new and unique means for evaluation of vascular, traumatic, and neoplastic disorders of the foot and ankle.

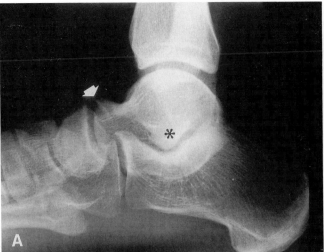

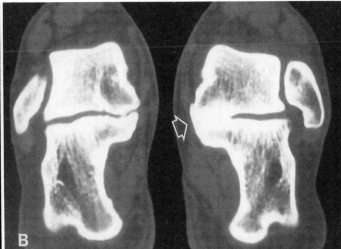

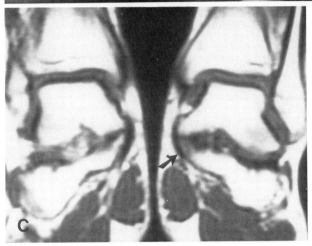

Figure 36.29. Bilateral tarsal coalition of middle subtalar joints. **(A)** Lateral radiograph of left foot shows secondary findings of talar beak *(arrow)* and widening of the lateral process of the talus *(asterisk)*. The middle subtalar facet is not well seen. **(B)** Coronal CT shows complete osseous fusion of left middle subtalar joint *(open arrow)* and fibrocar-tilaginous coalition of the contralateral joint. **(C)** Coronal SE 500/17 MR image shows left osseous coalition with low intensity in the sustentaculum due to sclerosis *(solid arrow)*. Note low intensity at talocalcaneal interface on right consistent with fibrous coalition.

References

1. Warwick R, Williams PL, eds. Gray's anatomy. 35th Br. ed. Philadelphia: W. B. Saunders, 1973:583–584.
2. Seltzer SE, Weissman BN, Braunstein EM, Adams DF, Thomas WH. Computed tomography of the hindfoot. J Comput Assist Tomogr 1984;8(3):488–497.
3. Sartoris DJ, Feingold ML, Resnick D. Axial computed tomographic anatomy of the foot. Part I. Hindfoot. J Foot Surg 1985;24(6):392–412.
4. Solomon MA, Gilula LA, Oloff LM, Oloff J, Compton T. CT scanning of the foot and ankle. 1. Normal anatomy. AJR 1986;146:1192–1203.
5. Heger L, Wulff K. Computed tomography of the calcaneus: normal anatomy. AJR 1985;145:123–129.
6. Hajek PC, Baker LL, Bjorkengren A, Sartoris DJ, Neumann CH, Resnick D. High-resolution magnetic resonance imaging of the ankle: normal anatomy. Skel Radiol 1986;15:536–540.
7. Middleton WD, Macrander S, Lawson TL, et al. High resolution surface coil magnetic resonance imaging of the joints: anatomic correlation. Radiographics 1987;7(4):645–683.
8. Crim JR, Cracchiolo A, Bassett LW, Seeger LL, Soma CA, Chatelaine A. Magnetic resonance imaging of the hindfoot. Foot Ankle 1989;10(1):1–7.
9. Donaldson JS, Poznanski AK, Kieves A. CT of children's feet: an immobilization technique. AJR 1987;148:169–170.
10. Mann FA, Gilula LA. Direct sagittal CT of the foot [letter]. AJR 1989;153(4):886.
11. Hume EL, McKeag DB. Soft tissue ankle injuries: the need for compulsive assessment and therapy. Emerg Med Rep 1984;5:45–52.
12. Feldman F, Singson RD, Rosenberg ZS, Berdon WE, Amodio J, Abramson SJ. Distal tibial triplane fractures: diagnosis with CT. Radiology 1987;164:429–435.
13. Lynn MD. The triplane distal tibial epiphyseal fracture. Clin Orthoped 1972;86:187–190.
14. Kump WL. Vertical fractures of the distal tibial epiphysis. AJR 1966;97:676–681.
15. Karrholm J, Hansson LI, Laurin S. Computed tomography of intraarticular supination-eversion fractures of the ankle in adolescents. J Pediatr Orthoped 1981;1(2):181–187.
16. Mainwaring BL, Daffner RH, Riemer BL. Pylon fractures of the ankle: a distant clinical and radiologic entity. Radiology 1988;168:215–218.
17. Rüedi TP, Allgower M. The operative treatment of intraarticular fractures of the lower end of the tibia. Clin Orthoped 1979;138:105–110.
18. Reckling WF, McNamara GR, De Smet AA. Problems in the di-

agnosis and treatment of ankle injuries. J Trauma 1981;21:943–950.

19. Kellam JF, Waddell JP. Fractures of the distal tibial metaphysis with intra-articular extension: the distal tibial explosion fracture. J Trauma 1979;19:593–601.
20. Guyer BH, Levinsohn EM, Fredrickson BE, Bailey GL, Formikell M. Computed tomography of calcaneal fractures: anatomy, pathology, dosimetry, and clinical relevance. AJR 1985;145:911–919.
21. Heger L, Wulff K, Seddiqi MSA. Computed tomography of calcaneal fractures. AJR 1985;145:131–137.
22. Sartoris DJ, Resnick D. Pictorial analysis—computed tomography of trauma to the ankle and hindfoot. J Foot Surg 1988;27(1):80–91.
23. Forrester DM, Kerr R. Trauma to the foot. Radiol Clin N Am 1990;28(2):423–433.
24. Hammesfahr R, Fleming L. Calcaneal fractures: a good prognosis. Foot Ankle 1981;2:161–171.
25. Leitch JM, Cundy PJ, Paterson DC. Three-dimensional imaging of a juvenile Tillaux fracture. J Pediatr Orthoped 1989;9(5):602–603.
26. Magid D, Michelson JD, Ney DR, Fishman EK. Adult ankle fractures—comparison of plain films and interactive two- and three-dimensional CT scans. AJR 1990;154:1017–1023.
27. Rosenberg ZS, Feldman F, Singson RD. Peroneal tendon injuries: CT analysis. Radiology 1986;161:743–748.
28. Martinez S, Herzenberg JE, Apple JS. Computed tomography of the hindfoot. Orthop Clin N Am 1985;16:481–496.
29. Goiney RC, Connell DG, Nichols DM. CT evaluation of tarso-metatarsal fracture-dislocation injuries. AJR 1985;144:985–990.
30. Kuhlman JE, Fishman EK, Magid D, et al. Fracture nonunion: CT assessment with multiplanar reconstruction. Radiology 1988;167:483–488.
31. Oloff LM, Jacobs AM. Fracture nonunion. Clin Podiatry 1985;2:379–406.
32. Meyer JM, Hoffmeyer P, Savoy X. High resolution computed tomography in the chronically painful ankle sprain. Foot Ankle 1988;8:291–296.
33. Canali ST, Belding RH. Osteochondral lesions of the talus. J Bone Joint Surg Am 1980;62:97–102.
34. Smith GR, Winquist RA, Allan NK, et al. Subtle transchondral fractures of the talar dome: a radiologic perspective. Radiology 1977;124:664–673.
35. Gepstein R, Comforty B, Weiss RE, et al. Closed percutaneous drilling for osteochondritis dissecans of the talus: a report of two cases. Clin Orthoped 1986;213:197–200.
36. Heare MM, Gillespy T, Bittar ES. Direct coronal computed tomography arthrography of osteochondritis dissecans of the talus. Skel Radiol 1988;17:187–189.
37. De Smet AA, Fisher DR, Burnstein MI, Graf BK, Lange RH. Value of MR imaging in staging osteochondral lesions of the talus (osteochondritis dissecans): results in 14 patients. AJR 1990;154:555–558.
38. Mesgarzadeh M, Sapega AA, Bonakdarpour A, et al. Osteochondritis dissecans: analysis of mechanical stability with radiography, scintigraphy, and MR imaging. Radiology 1987;165:775–780.
39. Smith GR, Winquist RA, Allen NK, Northrop CH. Subtle transchondral fractures of the talar dome: a radiological perspective. Radiology 1977;124:667–673.
40. Yulish BS, Mulopulos GP, Goodfellow DB, Bryan PJ, Modic MT, Dollinger BM. MR imaging of osteochondral lesions of talus. J Comput Assist Tomogr 1987;11(3):296–301.
41. Noto AM, Cheung Y, Rosenberg ZS, Norman A, Leeds NE. MR imaging of the ankle: normal variants. Radiology 1989;170:121–124.
42. Pavlov H, Torg JS, Freiberger RH. Tarsal navicular stress fractures: radiographic evaluation. Radiology 1983;148(3):641–645.
43. Rosenberg ZS, Jahss MH, Noto AM, et al. Rupture of the posterior tibial tendon: CT and surgical findings. Radiology 1988;167(2):489–493.
44. Szczukowski M Jr, St. Pierre RK, Fleming LL, Somogyi J. Computerized tomography in the evaluation of peroneal tendon dislocation. Am J Sports Med 1983;11(6):444–447.
45. Keyser CK, Gilula LA, Hardy DC, Adler S, Vannier M. Soft-tis-
sue abnormalities of the foot and ankle: CT diagnosis. AJR 1988;150:845–850.
46. Solomon MA, Gilula LA, Oloff LM, Oloff J. CT scanning of the foot and ankle. 2. Clinical applications and review of the literature. AJR 1986;146:1204–1214.
47. Rosenberg ZS, Feldman F, Singson RD, Price GJ. Peroneal tendon injury associated with calcaneal fractures: CT findings. AJR 1987;149(1):125–129.
48. Rosenberg ZS, Cheung Y, Jahss MH. Computed tomography scan and magnetic resonance imaging of ankle tendons: an overview. Foot Ankle 1988;8(6):297–307.
49. Beltran J, Noto AM, Herman LJ, Lubbers LM. Tendons: high-field-strength, surface coil MR imaging. Radiology 1987;162:735–740.
50. Rosenberg ZS, Cheung Y, Jahss MH, Noto AM, Norman A, Leeds NE. Rupture of posterior tibial tendon: CT and MR imaging with surgical correlation. Radiology 1988;169:229–235.
51. Daffner RH, Riemer BL, Lupetin AR, Dash N. Magnetic resonance imaging in acute tendon ruptures. Skel Radiol 1986;15:619–621.
52. Quinn SF, Murray WT, Clark RA, Cochran CF. Achilles tendon: MR imaging at 1.5T. Radiology 1987;164:767–770.
53. Apple JS, Martinez S, Khoury MB, Nunley JA. Case report 376: accessory (anomalous) soleus muscle. Skel Radiol 1986;15(5):398–400.
54. Abraham E, Sternaman JE. Neglected rupture of the peroneal tendon causing recurrent sprains of the ankle. J Bone Joint Surg 1979;61A:1247–1248.
55. Dihlmann W. Computertomographie des Talocruralgelenkes. Chirurg 1982;53:123–126.
56. Beltran J, Noto AM, Mosure JC, Shamam OM, Weiss KL, Zuelzer WA. Ankle: surface coil MR imaging at 1.5T. Radiology 1986;161(1):203–209.
57. Kneeland JB, Macrandar S, Middleton WD, Cates JD, Jesmanowicz A, Hyde JS. MR imaging of the normal ankle: correlation with anatomic sections. AJR 1988;151:117–123.
58. Beltran J, Munchow AM, Khabiri H, Magee DG, McGhee RB, Grossman SB. Ligaments of the lateral aspect of the ankle and sinus tarsi: an MR imaging study. Radiology 1990;177(2):455–458.
59. Resnick D, Niwayama G. Diagnosis of bone and joint disorders. 2nd ed. Philadelphia: W. B. Saunders, 1988;2525–2610.
60. Whitehouse WM, Smith WS. Osteomyelitis of the feet. Semin Roentgenol 1970;5:367–377.
61. Borris LC, Helleland H. Growth disturbance in the hind foot following osteomyelitis of the calcaneus in the new born. J Bone Joint Surg (Am) 1986;68A:32–35.
62. Grodinsky M. A study of the fascial spaces of the foot and their bearing on infection. Surg Gynecol Obstetr 1929;49:737–751.
63. Sartoris DJ, Devine S, Resnick D, et al. Plantar compartmental infection in the diabetic foot. The role of computed tomography. Invest Radiol 1985;20(8):772–784.
64. Kuhn JP, Berger PE. Computed tomographic diagnosis of osteomyelitis. Radiology 1979;130:503–506.
65. Rosen RC, Anania WC, Chinkes SL, Gerland JS. Utilization of computerized tomography in osteomyelitis of the foot. A case report. J Am Podiatr Med Assoc 1987;77(2):85–88.
66. Williamson BR, Teates CD, Phillips CD, Croft BY. Computed tomography as a diagnostic aid in diabetic and other problem feet. Clin Imaging 1989;13(2):159–163.
67. Ram PC, Martinez S, Korobkin M, et al. CT detection of intraosseous gas—a new sign of osteomyelitis. AJR 1981;137:721–723.
68. Fletcher BD, Scotes PV, Nelson AD. Osteomyelitis in children: detection by magnetic resonance. Radiology 1984;150:57–60.
69. Beltran J, Noto AM, McGhee RB, Freedy RM, McCalla MS. Infections of the musculoskeletal system: high-field-strength MR imaging. Radiology 1987;164:449–454.
70. Unger E, Moldofsky P, Gatenby R, Hartz W, Broder G. Diagnosis of osteomyelitis by MR imaging. AJR 1988;150:605–610.
71. Tang JSH, Gold RH, Bassett LW, Seeger LL. Musculoskeletal infection of the extremities: evaluation with MR imaging. Radiology 1988;166:205–209.
72. Wing VW, Jeffrey RB, Federle MP, et al. Chronic osteomyelitis examined by CT. Radiology 1985;154:171–174.

73. Chusid MJ, Jacobs WM, Sty JR. Pseudomonas arthritis following puncture wounds of the foot. J Pediatr 1979;94:429.
74. Johanson PH. *Pseudomonas* infections of the foot following puncture wounds. JAMA 1968;204:262.
75. Beltran J, Noto AM, Herman LJ, et al. Joint effusions: MR imaging. Radiology 1986;158:133–137.
76. Enneking WF. Musculoskeletal tumor surgery. New York: Churchill Livingstone, 1983:719–740.
77. Keigley BA, Haggar AM, Gaba A, Ellis BI, Froelich JW, Wu KK. Primary tumors of the foot: MR imaging. Radiology 1989;171(3):755–759.
78. Bloem JL, Taminiau AHM, Eulderink F, Hermans J, Pauwels EKJ. Radiologic staging of primary bone sarcoma: MR imaging, scintigraphy, angiography and CT correlated with pathologic examination. Radiology 1988;169:805–810.
79. Petasnick JP, Turner DA, Charters RJ, Citelis S, Zacharias CE. Soft tissue masses of the locomotor system: comparison of MR imaging with CT. Radiology 1986;160:125–133.
80. Pettersson H, Gillespy T, Hamlin DJ, et al. Primary musculoskeletal tumors: examination with MR imaging compared with conventional modalities. Radiology 1987;164:237–241.
81. Wetzel LH, Levine E, Murphey MD. A comparison of MR imaging and CT in the evaluation of musculoskeletal masses. Radiographics 1987;7:851–874.
82. Sundaram M, McGuire MH, Herbold DR. Magnetic resonance imaging of soft tissue masses: an evaluation of fifty-three histologically proven tumors. Magn Reson Imag 1988;6:237–248.
83. Demas BE, Heelan RT, Lane J, Marcove R, Hajdu S, Brennan MF. Soft tissue sarcomas of the extremities: comparison of MR and CT in determining the extent of disease. AJR 1988;150:615–620.
84. McLeod RA. Bone and soft tissue neoplasms. In: Berquist TH, ed. Radiology of the foot and ankle. New York: Raven Press, 1989:247–275.
85. Brady TJ, Gebhardt MC, Pykett IL, et al. NMR imaging of forearms in healthy volunteers and patients with giant cell tumor of bone. Radiology 1982;144:549–552.
86. Hermann SD, Mesgarzadeh M, Bonakdarpour A, et al. The role of magnetic resonance imaging in giant cell tumor of bone. Skel Radiol 1987;16:635–643.
87. Capanna R, Van Horn JR, Ayala A, Picci P, Bettelli G. Osteoid osteoma and osteoblastoma of the talus. Skel Radiol 1986;15:360–364.
88. Yeager BA, Schiebler ML, Wertheim SB, et al. MR imaging of osteoid osteoma of the talus. J Comput Assist Tomogr 1987;11(5):916–917.
89. Hudson TM. Fluid levels in aneurysmal bone cysts: a CT feature. AJR 1987;142:1001–1004.
90. Beltran J, Simon DC, Levy M, Herman L, Weis L, Mueller CF. Aneurysmal bone cysts: MR imaging at 1.5T. Radiology 1986;158:689–690.
91. Tsai JC, Dalinka MK, Fallon MD, Zlatkin MB, Kressel HY. Fluid-fluid level: a nonspecific finding in tumors of bone and soft tissue. Radiology 1990;175:779–782.
92. Cohen EK, Kressel HY, Frank TS, et al. Hyaline cartilage—origin of bone and soft-tissue neoplasms: MR appearance and histologic correlation. Radiology 1988;167:477–481.
93. Kricun ME, Kricun R, Haskin ME. Chondroblastoma of the calcaneus: radiographic features with emphasis on location. AJR 1977;128:613–616.
94. Bloem JL, Mulder JD. Chondroblastoma: a clinical and radiological study of 104 cases. Skel Radiol 1985;14:1–9.
95. Fobben ES, Dalinka ML, Schiebler ML, et al. The magnetic resonance imaging appearance at 1.5 Tesla of cartilaginous tumors involving the epiphysis. Skel Radiol 1987;16:647–651.
96. Harms SE, Greenway G. Musculoskeletal system. In: Stark DD, Bradley WG Jr, eds. Magnetic resonance imaging. St Louis: C. V. Mosby, 1988:1323–1433.
97. Kitamura K, Nibu K, Asai M, et al. Chondromyxoid fibroma of the mastoid invading the occipital bone. Arch Otolaryngol Head Neck Surg 1989;115:384–386.
98. Ketyer S, Brownstein S, Cholankeril J. CT diagnosis of intraosseous lipoma of the calcaneus. J Comput Assist Tomogr 1983;7(3):546–547.
99. Ramos A, Castello J, Sartoris DJ, Greenway GD, Resnick D, Haghighi P. Osseous lipoma: CT appearance. Radiology 1985;157:615–619.
100. Resnick D, Niwayama G. Diagnosis of bone and joint disorders. 2nd ed. Philadelphia: W. B. Saunders, 1988:3782–3786.
101. Wetzel LH, Schweiger GD, Levine E. MR imaging of transarticular skip metastases from distal femoral osteosarcoma. J Comput Assist Tomogr 1990;14(2):315–317.
102. Enzinger FM, Weiss SW. Soft tissue tumors. 2nd ed. St Louis: C. V. Mosby, 1988:923.
103. Resnick D, Niwayama G. Diagnosis of bone and joint disorders. 2nd ed. Philadelphia: W. B. Saunders, 1988:4194–4196.
104. Binkovitz LA, Berquist TH, McLeod RA. Masses of the hand and wrist: detection and characterization with MR imaging. AJR 1990;154:323–326.
105. Weiss KL, Beltran J, Lubbers LM. High-field MR surface-coil imaging of the hand and wrist. Part II. Pathologic correlations and clinical relevance. Radiology 1986;160:147–152.
106. Fergusson ILC. Hemangiomata of skeletal muscle. Br J Surg 1972;59:634–637.
107. Enneking WF. Musculoskeletal tumor surgery, Vol. 2. New York: Churchill Livingstone, 1983:1175–1223.
108. Levine E, Wetzel LH, Neff JR. MR imaging and CT of extrahepatic cavernous hemangiomas. AJR 1986;147:1299–1304.
109. Kaplan PA, Williams SM. Mucocutaneous and peripheral soft-tissue hemangiomas: MR imaging. Radiology 1987;163:163–166.
110. Cohen EK, Kressel HY, Perosio T, et al. MR imaging of soft-tissue hemangiomas: correlation with pathologic findings. AJR 1988;150:1079–1081.
111. Yuh WTC, Kathol MH, Sein MA, Ehara S, Chiu L. Hemangiomas of skeletal muscle: MR findings in five patients. AJR 1987;149:765–768.
112. Wetzel LH, Levine E. Soft-tissue tumors of the foot: value of MR imaging for specific diagnosis. AJR 1990;155(5):1025–1030.
113. Enzinger FM, Weiss SW. Fibromatoses. In: Soft tissue tumors. 2nd ed. St. Louis: C. V. Mosby, 1988:136–163.
114. Kransdorf MJ, Jelinek JS, Moser RP, et al. Soft-tissue masses: diagnosis using MR imaging. AJR 1989;153:541–547.
115. Enzinger FM, Weiss SW. Synovial sarcoma. In: Soft tissue tumors. 2nd ed. St Louis: C. V. Mosby, 1988:659–688.
116. Kearney MM, Soule EH, Ivins JC. Malignant fibrous histiocytoma. Cancer 1980;45:167–178.
117. Owens JC, Shiu MH, Smith R, Hajdu SI. Soft tissue sarcomas of the hand and foot. Cancer 1985;55:2010–2018.
118. Enzinger FM, Weiss SW. Malignant tumors of uncertain histogenesis. In: Soft tissue tumors. 2nd ed. St. Louis: C. V. Mosby, 1988:945–951.
119. Gomori JM, Grossman RI, Shields JA, Augsburger JJ, Joseph PM, DeSimeone D. Choroidal melanomas: correlation of NMR spectroscopy and MR imaging. Radiology 1986;158:443–445.
120. Resnick D, Niwayama G. Diagnosis of bone and joint disorders with emphasis on articular abnormalities. Philadelphia: W. B. Saunders, 1981:1460–1461.
121. Seltzer SE, Weissman BN, Braunstein EM, Adams DF, Thomas WH. Computed tomography of the hindfoot with rheumatoid arthritis. Arthritis Rheum 1985;28(11):1234–1242.
122. Yulish BS, Lieberman JM, Newman AJ, Bryan PJ, Mulopulos GP, Modic MT. Juvenile rheumatoid arthritis: assessment with MR imaging. Radiology 1987;165:149–152.
123. Beltran J, Caudill JL, Herman LA, et al. Characteristics of partial flip angle and gradient reversal MR imaging. Radiology 1987;164:153–157.
124. Resnick D, Niwayama G. Diagnosis of bone and joint disorders with emphasis on articular abnormalities. Philadelphia: W. B. Saunders, 1981:2017–2040.
125. Arnold WD, Hilgartner MW. Hemophilic arthropathy: current concepts of pathogenesis and management. J Bone Joint Surg 1977;59:287–305.
126. Speer DP. Early pathogenesis of hemophilic arthropathy. Evolution of the subchondral cyst. Clin Orthoped 1984;185:250–265.
127. Yulish BS, Lieberman JM, Strandjord SE, Bryan PJ, Mulopulos

GP, Modic MT. Hemophilic arthropathy: assessment with MR imaging. Radiology 1987;164:759–762.

128. Kulkarni MV, Drolshagen LF, Kaye JJ, et al. MR imaging of hemophiliac arthropathy. J Comput Assist Tomogr 1986;10:445–449.

129. Resnick D, Niwayama G. Diagnosis of bone and joint disorders. 2nd ed. Philadelphia: W. B. Saunders, 1988:3925.

130. Jelinek JS, Kransdorf MJ, Utz JA, et al. Imaging of pigmented villonodular synovitis with emphasis on MR imaging. AJR 1989;152:337–342.

131. Spritzer CE, Dalinka MK, Kressel HY. Magnetic resonance imaging of pigmented villonodular synovitis: a report of two cases. Skel Radiol 1987;16:316–319.

132. Kottal RA, Vogler JB, Matamoros A, Alexander AH, Cookson JL. Pigmented villonodular synovitis: a report of MR imaging in two cases. Radiology 1987;163:551–553.

133. Sherry CS, Harms SE. MR evaluation of giant cell tumors of the tendon sheath. Magn Reson Imag 1989;7(2):195–201.

134. Resnick D, Niwayama G. Diagnosis of bone and joint disorders. 2nd ed. Philadelphia: W. B. Saunders, 1988:3563–3574.

135. Herzenberg JE, Goldner JL, Martinez S, Silverman PM. Computerized tomography of talocalcaneal tarsal coalition: a clinical and anatomic study. Foot Ankle 1986;6(6):273–288.

136. Deutsch AL, Resnick D, Campbell G. Computed tomography and bone scintigraphy in the evaluation of tarsal coalition. Radiology 1982;144:137–140.

137. Sarno RC, Carter BL, Bankoff MS, et al. Computed tomography in tarsal coalition. J Comput Assist Tomogr 1984;8:1155–1160.

138. Wheeler R, Guevera A, Bleck EE. Tarsal coalitions: a review of the literature and case report of bilateral dual calcaneonavicular and talocalcaneal coalitions. Clin Orthoped Rel Res 1981;156:175.

139. Marchisello PJ. The use of computerized axial tomography for the evaluation of talocalcaneal coalition. J Bone Joint Surg Am 1987;69:609–611.

140. Scranton PE. Treatment of symptomatic talocalcaneal coalition. J Bone Joint Surg Am 1987;69:533–538.

141. Morris HD. Aseptic necrosis of the talus following injury. Orthoped Clin N Am 1974;5:177–189.

142. Resnick D, Pineda C, Trudell D. Widespread osteonecrosis of the foot in systemic lupus erythematosis: radiographic and gross pathologic correlation. Skel Radiol 1985;13:33–38.

143. Sierra A, Potchen EJ, Moore J, et al. High-field magnetic resonance imaging of aseptic necrosis of the talus. J Bone Joint Surg Am 1986;68:927–928.

144. Neri M, Querin F. CT scan in a study of the normal anatomy of the hindfoot and midfoot. Ital J Orthoped Traumatol 1989;15(4):507–520.

145. Erickson SJ, Quinn SF, Kneeland JB, et al. MR imaging of the tarsal tunnel and related spaces: normal and abnormal findings with anatomic correlation. AJR 1990;155(2):323–328.

146. Zeiss J, Fenton P. Ebraheim N, Coombs RJ. Normal magnetic resonance anatomy of the tarsal tunnel. Foot Ankle 1990;10(4):214–218.

147. Combs AH, Kernek CB, Heck DA. Orthopedic grand rounds. Retained wooden foreign body in the foot detected by computed tomography. Orthopedics 1986;9(10):1434–1435.

148. Nyska M, Pomeranz S, Porat S. The advantage of computed tomography in locating a foreign body in the foot. J Trauma 1986;26(1):93–95.

149. Goldenberg RA, Goldenberg EM, Estersohn HS. Needle localization of foreign bodies using computed tomography. A case report. J Am Podiatr Med Assoc 1988;78(12):629–631.

150. Bauer AR Jr, Yutani D. Computed tomographic localization of wooden foreign bodies in children's extremities. Arch Surg 1983;118:1084.

CHAPTER 37

Magnetic Resonance Imaging of the Diabetic Foot

Timothy E. Moore and William T. C. Yuh

Introduction
Imaging Techniques

Magnetic Resonance Imaging Findings in the Diabetic Foot

Introduction

Clinical Importance of Accurate Diagnosis

Before the successful extraction of insulin from pancreatic tissues by Banting and Best in 1921, and the subsequent introduction of clinically available insulin, most diabetic patients died from ketoacidosis or hyperosmolar nonketotic coma, long before developing neuropathic, vascular, or infective complications. However, these more chronic complications are now major causes of morbidity and mortality in diabetes mellitus (1). Although cerebrovascular accidents, myocardial infarction, and end stage renal disease remain more common causes of death, the changes in the feet are probably secondary only to those in the eyes in contributing to a generalized reduction in quality of life. More in-hospital days are spent treating diabetic foot complications than any other complications of diabetes (2). In the United States alone, there are 20,000 amputations performed each year for vascular insufficiency or osteomyelitis in diabetic patients (3). By far the majority of these are distal lower limb amputations such as toes, metatarsals, and feet.

The changes in diabetic feet can primarily be attributed to neuropathic or vascular factors or to a combination of both. Loss of sensation may result in Charcot joints and spontaneous fractures. However, a more frequent finding is the development of plantar ulcers over areas of weight-bearing bony prominences such as the metatarsal heads, calcaneus, or other bony protuberances in deformed feet. These ulcers are, in turn, more prone to infection because of the effects of ischemia. Soft-tissue infection may extend beyond a local ulcer to form an abscess or develop into widespread cellulitis, which in turn may directly invade adjacent bone, resulting in osteomyelitis. Vascular insufficiency can lead to a situation in which it is almost impossible to heal the soft-tissue lesions without first restoring the vascular supply. The dilemma frequently occurs in which an accurate assessment of the degree of destruction and presence or absence of infection must be made for treatment planning, as there may be insufficient salvageable tissue to warrant vascular reconstitution (4). Combined medical treatment and surgical debridement can result in a better than 50% cure rate for osteomyelitis in the feet of diabetics as long as the diagnosis is made early (5).

Imaging Modalities

Various imaging modalities have been used in the assessment of complications of the diabetic foot. Plain film radiography has been the mainstay and remains a valuable means of documenting major structural changes involving bone and soft tissues (6, 7). Bony changes can usually be depicted in plain films but soft-tissue lesions such as joint effusions, edema, fasciitis, and abscesses may not be as easily appreciated. Active bone marrow infection frequently produces no radiographically detectable changes until the later stages of destruction, sclerosis, or periosteal reaction (8, 9). Subtle radiographic changes may also be obscured by superimposed chronic destructive processes such as neuroarthropathy (10). On the other hand, bony changes can also result from other causes that may not be easily differentiated from osteomyelitis, such as demineralization of bone adjacent to soft-tissue infection or repeated trauma from peripheral neuropathy. These factors contribute to both the false-negative and false-positive rates previously reported (11, 12). Plain films are also useful in the documentation of neuropathic changes, but again the earliest subtle changes of neuropathic disease such as microfractures may not be obvious.

Computed tomography (CT) is an established modality for the diagnosis and management of the musculoskeletal system. As compared to plain films, CT provides improved three-dimensional understanding and much better tissue contrast. However, beam-hardening artifact in the soft tissues adjacent to cortical bone, where the contiguous spread of infection occurs, limits the value of CT in the evaluation of bone marrow pathology. Technetium-99m-labeled methylene diphosphonate, indium-111-labeled leukocyte, and gallium-67-labeled citrate scintigraphy are all well known in the assessment of the diabetic foot, particularly in the detection of infection (10, 13, 14). These are accurate and sensitive techniques in the detection of bone infection, but the lack of spatial resolution makes it difficult to estimate the extent of involvement and to distinguish between bone and soft-tissue changes. In addition, bone marrow pathology cannot be appreciated directly by scintigraphy.

Magnetic resonance (MR) imaging has recently been shown to have higher sensitivity and specificity in the detection of osteomyelitis in diabetic patients (12). With its multiplanar ability, exquisite anatomical detail, excellent soft-tissue contrast, and direct visualization of bone marrow, MR imaging has the potential to provide essential information that may not be available with other modalities, in spite of the fact that early bone changes are not appreciated with MR imaging.

Imaging Techniques

As further knowledge and experience accumulate and technologic advances are made, there are likely to be sig-

nificant changes in MR imaging protocols for the diabetic foot. Standard parasagittal (parallel to the long axis of the foot) T1- and T2-weighted images and axial (perpendicular to the long axis of the foot) T1-weighted images give a combination of good anatomic details in two planes as well as identifying areas of abnormal signal in marrow and soft tissues. However, short τ inversion recovery (STIR) images appear to be more sensitive in the detection of subtle water changes in the bone marrow caused by osteomyelitis (12). T2-weighted images remain more sensitive in the demonstration of soft-tissue abscesses (Fig. 37.1). The protocol in our institution includes 5-mm T1-weighted (TR, 350 to 600; TE, 20 to 26) and STIR (TR, 2000; TI, 125 to 160) images parallel to the long axis of the foot (sagittal) and 7.5- to 10-mm T1-weighted (TR, 350 to 500; TE, 20 to 60) images perpendicular to the long axis of the foot (axial). We use either a 1.5-T (General Electric Signa, Milwaukee, Wisc.) or 0.5-T (Picker International, Highland Heights,

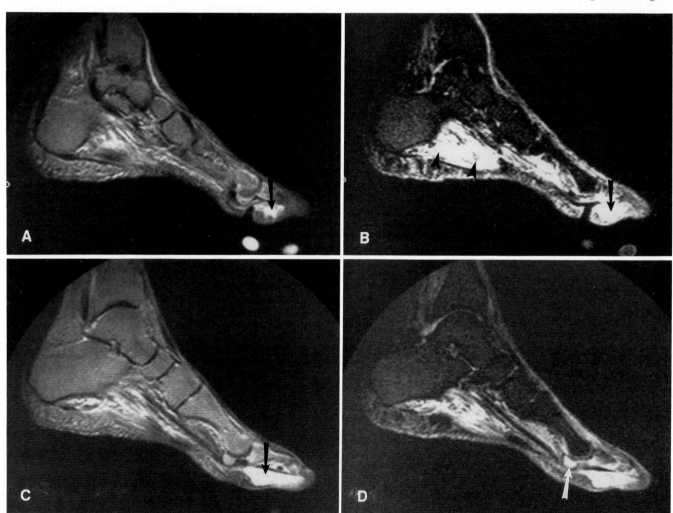

Figure 37.1. Chronic plantar ulcer with adjacent abscess and osteomyelitis of sesamoid bone. **(A)** T2-weighted (TR 2000, TE 100) sagittal image shows a ventral soft-tissue defect (ulcer cavity) overlying the first metatarsal head. Markers (oil beads) indicate site of ulcer. An abscess *(black arrow)* is shown in the soft tissues distal to the base of the ulcer. **(B)** The abscess *(black arrow)* in this STIR (TR 2000, TI 125) sagittal image corresponding to **(A)** is less well defined than in the T2-weighted image **(A)**, because of high-signal background from edema or cellulitis. Note extreme high signal of plantar fascial edema

or cellulitis *(black arrowheads)* seen in this particular sequence. Bone marrow signal of the metatarsal and phalanx is normal in the T1-weighted (not shown), T2-weighted, and STIR images. **(C)** T2-weighted sagittal image lateral to ulcer shows clearly defined abscess *(black arrow).* Note bone marrow signal of sesamoid appears normal. **(D)** STIR sagittal image corresponding to **(C)** shows abnormal signal of sesamoid *(white arrow)* from osteomyelitis, which is clearly seen in this pulse sequence but not demonstrated in the T2-weighted image.

Ohio) superconductive scanner. Studies may be performed with a head coil using a 27- to 30-cm field of view and the foot in a neutral position (15 to 30° external rotation). The matrix size is 192×192, using two excitations (NEX). In view of the necessity of maintaining high signal-to-noise ratios with the small anatomic parts and relative lack of motion artifact, medium (0.5 T) to high (1.5 T) field strengths are probably preferable to lower field strengths. A T1-weighted sequence following intravenous gadolinium (Gd)-labeled diethylenetriaminepentaacetic acid (DTPA) may give additional information, as discussed below (see "Abscesses and Sinus Tracts" and "Osteomyelitis"). An oil-containing marker bead placed over an ulcer or other clinically significant site can serve as a useful pointer to regions of interest (Figs. 37.1 and 37.2).

Magnetic Resonance Imaging Findings in the Diabetic Foot

Edema

Edema consists of excessive soft-tissue fluid, thought to be the result of prolonged weight bearing and repeated minor injuries in the insensitive foot due to peripheral neuropathy.

This is the most common MR imaging finding in diabetic feet. Increased signal intensity can be demonstrated in both T2-weighted and STIR images in the plantar aspect of the foot (Figs. 37.3*B* and 37.4*B*). We believe this represents plantar fascial edema. Increased signal may also be seen, usually to a lesser extent, in the dorsal soft tissues. We believe the plantar edema is caused by a similar mechanism to that which causes plantar ulcer formation, and it seems likely that increased plantar fluid may be the earliest manifestation of diabetic neuroarthropathy before the necrosis (ulceration). We have seen this finding in the ma-

jority of MR imaging examinations of our diabetic patients with or without infection.

The increased plantar soft-tissue fluid is probably not directly related to either arterial or venous insufficiency. Magnetic resonance imaging of vascular insufficiency in nondiabetic patients frequently shows increased fluid predominantly in the dorsal (nondependent) soft tissues rather than the plantar tissues. In these patients, plantar fascial fluid frequently is squeezed away by the weight-bearing regions (Fig. 37.5) whereas in diabetic feet accumulation of fluid in the plantar fascia is the primary site of the underlying process.

Joint Effusions

Joint effusions consist of increased intraarticular fluid thought to be caused by uneven weight-bearing and repeated trauma from peripheral neuropathy.

Small effusions are frequently observed in the diabetic foot with or without infection. These are of unknown etiology, but are probably related to peripheral neuropathic disease. They can be seen anywhere in the foot or ankle but are more common in the ankle and intertarsal joints, where neuropathic bone changes frequently occur but where ulcers and osteomyelitis are less common. Joint effusions are best demonstrated in STIR and T2-weighted images as increased signal delineating the anatomic margins of the affected joints (Figs. 37.3*B* and 37.4*D*).

Deformity

Deformity involves alteration of normal bony configuration caused by uneven weight distribution due to peripheral neuropathy with or without fractures.

Typically, there is collapse of the longitudinal arch of the foot along the vector of the weight-bearing force of the

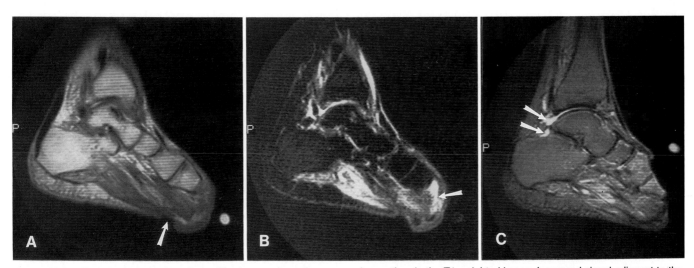

Figure 37.2. Dorsal postoperative ulcer with abscess formation resulting in osteomyelitis. **(A)** T1-weighted (TR 633, TE 20) sagittal image shows transmetatarsal amputation. Marker bead overlies a small, recent unhealed ulcer. Another chronic ulcer is seen in the plantar region *(arrow)*. Marrow signal of the remaining bone is normal. **(B)** STIR (TR 2000, TI 125) sagittal image corresponding to **(A)** shows a sinus tract *(arrow)* linking the ulcer with an abscess. The abscess is in contact with the metatarsal remnant and is better demonstrated in the STIR image than in the T1-weighted image. Increased signal adjacent to the abscess suggests cellulitis. Bone marrow signal is normal. **(C)** STIR (TR 2033, TE 100) sagittal image 16 weeks after **(A)** and **(B)** shows postdebridement changes with exposure of the medial cuneiform and the metatarsal remnant, both of which now demonstrate abnormal signal due to spread of osteomyelitis. Note frequent finding of ankle and subtalar effusions *(arrows)*.

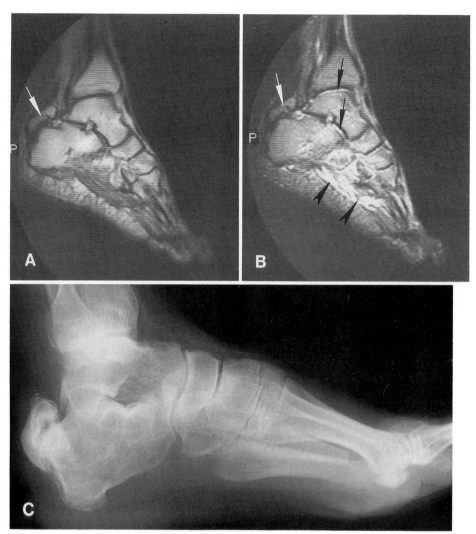

Figure 37.3. Plantar edema, effusions, and bone fragmentation in the insensitive diabetic foot. **(A)** T1-weighted (TR 516, TE 20) sagittal image shows an old spontaneous calcaneal fracture fragment *(white arrow)*. **(B)** T2-weighted (TR 2000, TE 100) sagittal image corresponding to **(A)**. Edema is seen as diffuse high signal intensity predominantly in the region of the plantar fascia, probably due to prolonged weight bearing in the insensitive foot *(black arrowheads)*. Also note scattered ab-

normal high signal in the dorsal aspect, caused by random traumatic episodes due to peripheral neuropathy. Joint effusions are best seen in the ankle and subtalar joints *(black arrows)*. The bone fragment *(white arrow)* has similar intensity to that of normal bone marrow on both T1- and T2-weighted images. **(C)** A lateral radiograph confirms the old calcaneal fracture fragment. Vascular calcification and a fourth metatarsal amputation are also demonstrated.

tibia. The usual sites of weight-bearing bony prominences, such as the metatarsal heads and calcaneus, may be altered and other sites, such as the cuboid bone, assume the lowermost plantar position, bear a disproportionate amount of body weight, and are in turn more prone to ulceration (Fig. 37.6).

Ulcers

Ulcers are focal soft-tissue defects due to necrosis of soft tissues caused by prolonged uneven weight bearing and/or trauma due to peripheral neuropathy.

Ulcers most commonly occur beneath the metatarsal heads and the calcaneus because these sites exert pressure on the underlying soft tissues due to their relatively superficial plantar location (Figs. 37.1, 37.7, and 37.8). Ulcers may also be seen in the midtarsal region when neuroar-

thropathy has resulted in collapse (deformity) of the midfoot (Fig. 37.7). Dorsal ulcers are much less frequently seen and occur mainly at sites of previous surgery (Fig. 37.2). Ulcers do not heal well in diabetes mellitus because of repeated minor trauma and uneven weight distribution, in addition to vascular insufficiency (15). Prolonged exposure to infectious pathogens may result in cellulitis, abscesses, or osteomyelitis. It is not surprising, therefore, to find that the most frequent request for radiologic examination is the differentiation of abscesses and cellulitis from osteomyelitis near a long-standing unhealed ulcer.

Although they are usually obvious clinically, identification of ulcers in MR imaging studies is helpful because it highlights the potential ports of entry of infection to the adjacent bone and soft tissues, and therefore serves as a pointer to the regions most likely to harbor infection. The ulcer craters themselves appear as air- or fluid-containing

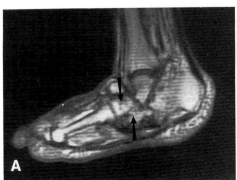
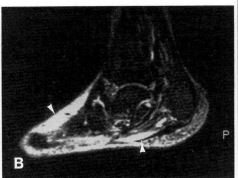
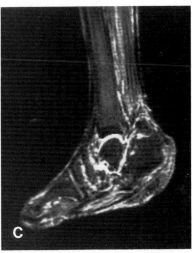

Figure 37.4. Chronic neuropathic disease with acute exacerbation. **(A)** T1-weighted (TR 583, TE 20) sagittal image shows chronic phase with gross deformity of the foot, loss of the longitudinal arch, and fragmentation of the midtarsus. There are regions of inhomogeneous abnormal (decreased) marrow signal *(arrows)* consistent with fracture lines. **(B)** STIR (TR 2233, TI 125) sagittal image shows chronic phase with minimal abnormal signal in the soft tissues and normal bone marrow signal. This is, therefore, unlikely to represent infection. Note dorsal and plantar edema *(arrowheads)*. **(C)** STIR (TR 2000, TI 150) sagittal image shows acute phase. Note significant joint fluid and abnormal signal in the bone marrow and surrounding soft tissues. These findings cannot be easily differentiated from those of an infectious process, in spite of the typical neuropathic location and the absence of adjacent ulcers.

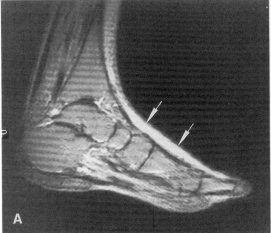
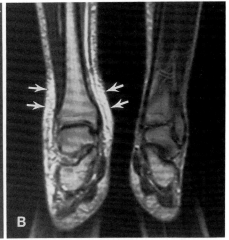

Figure 37.5. Nondependent distribution of edema in vascular insufficiency. **(A)** Arterial insufficiency in a nondiabetic 71-year-old man. In contrast to Figure 37.3*B*, high-signal edema, shown in this T2-weighted (TR 2046, TE 100) sagittal image, is found mainly in the dorsal aspect of the foot *(white arrows)*. **(B)** Venous insufficiency in a nondiabetic 35-year-old man. T2-weighted (TR 200, TE 80) coronal image shows high-signal nondependent edema *(white arrows)* on the abnormal right side around the ankle. Note the relatively normal plantar soft tissue.

defects in the soft tissues continuous with the exterior. Acute ulcers usually show surrounding decreased signal in T1-weighted and increased signal in STIR and T2-weighted images due to adjacent edema or cellulitis (Fig. 37.2). Chronic ulcers are usually surrounded by a region of relatively low signal in STIR, T1-weighted, and T2-weighted images, presumably due to fibrosis (Fig. 37.1).

Cellulitis

Cellulitis is an infiltrating soft-tissue infection, frequently caused by direct exposure to pathogens from adjacent unhealed ulcers.

It is not surprising that cellulitis frequently forms near the site of a chronic unhealed ulcer. Poor wound healing due to vascular insufficiency and hyperglycemia often results in the spread of infection beyond ulcer sites. Because ulcers are usually ventrally located, plantar fasciitis is a common and serious complication.

Cellulitis characteristically appears as a region of diffuse increased signal in STIR and T2-weighted images. In T1-weighted images, the affected region shows signal that is higher than that of muscle but lower than that of fat. Unfortunately, the signal characteristics of edema and cellulitis frequently coexist, and it is often difficult or impossible to differentiate between these two entities, especially

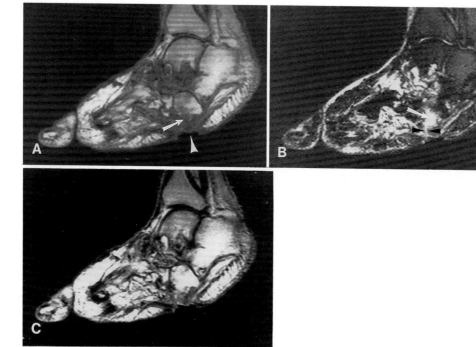

Figure 37.6. Osteomyelitis of the cuboid delineated by Gd-DTPA. **(A)** T1-weighted (TR 600, TE 20) sagittal image shows chronic ulcer *(white arrowhead)* formed as a result of the cuboid becoming a major weight-bearing bone. Note the abnormal signal in the region of the cuboid and surrounding tissues *(white arrow)*. **(B)** STIR (TR 2067, TI 125) sagittal image again shows abnormal signal adjacent to the cuboid *(white arrow)*. It is difficult to determine whether this is located within the cuboid or the adjacent soft tissue, due to the possibility of the partial volume effect or a residual fragment of fractured cuboid. Tarsal bones again show neuropathic changes. Note sinus tract *(black arrowheads)*. **(C)** T1-weighted sagittal image following intravenous Gd-DTPA corresponds to **(A)**. The cuboid is now better delineated. The abnormal signal noted in **(A)** and **(B)** is now confirmed to be located within the cuboid. Therefore osteomyelitis is diagnosed with a high degree of confidence.

Figure 37.7. Osteomyelitis of the calcaneus associated with ulcers and cellulitis. **(A)** T1-weighted (TR 600, TE 20) sagittal image shows a large open ulcer exposing the posterior calcaneus. The infected marrow shows homogeneous abnormal (decreased) signal. Another plantar ulcer is shown in the forefoot *(arrowheads)*. Note an old compression fracture of the navicular with normal signal intensity due to neuroarthropathy. **(B)** STIR (TR 2067, TI 125) sagittal image corresponding to **(A)** shows abnormal (increased) signal in the calcaneus identifying the infected bone marrow. Note cellulitis *(arrows)* tracking forward from the region of osteomyelitis, which cannot be easily differentiated from the frequently coexistent plantar fascial edema.

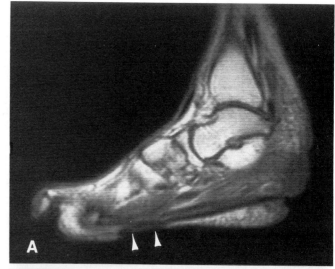

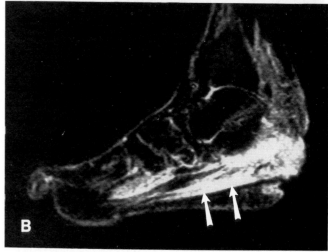

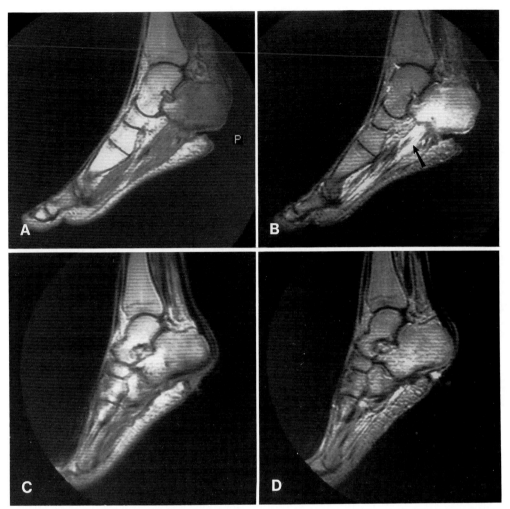

Figure 37.8. Healing osteomyelitis following intravenous antibiotic therapy. **(A)** T1-weighted (TR 616, TE 20) sagittal image shows a large open ulcer exposing much of the calcaneus. Homogeneous abnormal (decreased) signal due to osteomyelitis is seen throughout most of the calcaneal marrow. **(B)** T2-weighted (TR 2000, TE 100) sagittal image corresponding to **(A)** again shows abnormal signal in calcaneal marrow. Note plantar cellulitis *(arrow)* tracking forward from the osteomye-litis, which is not obvious in the T1-weighted image. **(C)** T1-weighted (TR 683, TE 20) sagittal image 16 weeks after **(B)** shows progressive restoration of normal bone marrow signal consistent with healing of osteomyelitis. **(D)** T2-weighted (TR 2016, TE 100) sagittal image corresponding to **(C)** shows abnormal (increased) signal in the calcaneus is now much less intense than in **(B)**.

in the plantar region. Sometimes cellulitis may be identified as abnormal signal tracking from an ulcer or a region of osteomyelitis through the adjacent fascia (Figs. 37.7–37.9).

Abscesses and Sinus Tracts

More focal forms of soft-tissue infection include abscesses and sinus tracts, frequently seen adjacent to ulcers or osteomyelitis, and sometimes coexistent with cellulitis.

Like cellulitis, abscesses frequently develop in the vicinity of ulcers. It is common to see an abscess or sinus tract situated between the ulcer and an adjacent infected bone. This finding is consistent with the well-known mechanism of contiguous spread of infection in the soft tissues of diabetic feet. Due to their proteinaceous content, abscesses typically show focal fluid collections with homogeneous high signal, usually greater than that of the surrounding edema on T2-weighted and STIR images. Actively infected

sinus tracts show low signal in T1-weighted images when compared to adjacent subcutaneous fat and increased signal in T2-weighted and STIR images (Fig. 37.6). Because cellulitis or edema is frequently adjacent, it may be difficult, or even impossible, to identify abscesses and sinus tracts against a background of high signal (Fig. 37.1*B* and *C*). Early trials with Gd suggest that enhancement of abscess walls may help to delineate the extent of abscess formation and the size of the fluid-containing center in some cases (Fig. 37.10).

Osteomyelitis

Osteomyelitis is an infection of bone, typically caused by direct contiguous extension of soft-tissue infection from an unhealed ulcer. Osteomyelitis can also be caused by direct exposure of the bone to pathogens when an ulcer is large enough.

Osteomyelitis is usually a late event following ulcera-

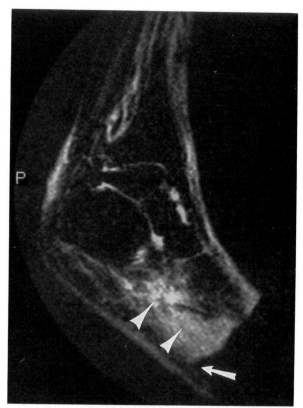

Figure 37.9. Postoperative ulcer due to transmetatarsal amputation, with adjacent cellulitis. STIR (TR 2233, TI 125) sagittal image shows a large open ulcer *(arrow)*. This proven plantar cellulitis *(arrowheads)* tracking posteriorly through the plantar fascia from the ulcer cannot be easily distinguished from the frequently found plantar fascial edema.

tion and cellulitis. The single most common clinical request for imaging the diabetic foot is the detection of osteomyelitis. This may be a search for newly developed or previously undetected osteomyelitis, or documentation of healing or progressing infection. The clinical presentation and physical findings of osteomyelitis and cellulitis are frequently nonspecific. However, the differentiation between osteomyelitis and cellulitis is clinically essential in treatment planning (outpatient oral antibiotics versus 6 weeks of in-patient intravenous therapy).

Normal bone marrow has a characteristic MR signal. Because MR imaging can directly visualize bone marrow, changes in water content within the marrow associated with osteomyelitis can be readily detected without superimposition of other soft-tissue changes. The classic finding in osteomyelitis is abnormal marrow signal emanating from all or a part of a bone(s). This is seen as homogeneous low signal intensity on T1-weighted images and high signal intensity on both STIR and T2-weighted images (Figs. 37.1D and E, 37.7, and 37.8).

Because the detection of infection by MR imaging depends on the demonstration of abnormal water content, it is not always possible to determine clearly whether increased fluid in bone marrow is caused by infection or other pathologic processes such as acute trauma in neuropathic disease. There are, however, indirect indications that can be very helpful. Ulcers, sinus tracts, and abscesses are frequently seen in the soft tissues immediately adjacent to infected bone. Multiple bony fragments without adjacent ulcers or abnormal surrounding soft tissue are more likely to be seen in neuropathic disease. Similarly, lesions in the metatarsal heads and calcaneus with ulceration suggest

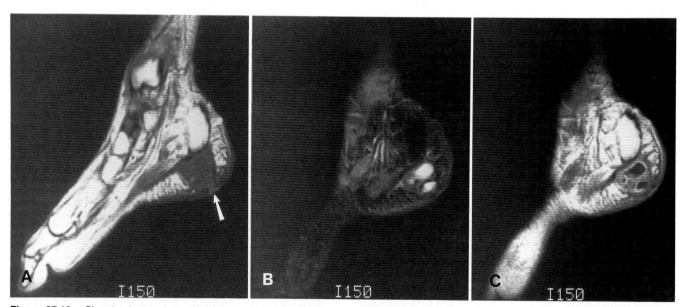

Figure 37.10. Chronic plantar ulcer with abscess formation. **(A)** T1-weighted (TR 600, TE 60) sagittal image shows a rounded soft-tissue mass, with signal intensity close to muscle, adjacent to the calcaneus. A small ulcer with a sinus tract is also shown *(arrow)*. The bone marrow and cortex of the calcaneus appear to be intact. **(B)** STIR (TR 2000, TI 170) sagittal image corresponding closely to **(A)**. Septation is seen within the high-signal abscess. The ulcer has low signal intensity consistent with granulation tissue due to chronic ulceration. Again note the normal appearance of the bone marrow. **(C)** T1-weighted sagittal image following intravenous Gd-DTPA corresponding to **(B)**. The abscess walls and septa are enhanced but there is no enhancement of the fluid-filled cavity. Delineation of the abscess from the bone is also more easily appreciated than in the T1-weighted and STIR images.

osteomyelitis (Figs. 37.1 and 37.7), whereas abnormal marrow signal in the midtarsal and proximal metatarsal regions, in the absence of any evidence of adjacent soft-tissue infection, is more suggestive of neuropathic disease (Fig. 37.4).

Although axial images can usually differentiate cellulitis adjacent to bone from osteomyelitis, difficulties with volume averaging may occur in the detection of osteomyelitis when soft-tissue infection is immediately adjacent to a short bone such as the cuboid or cuneiform. Magnetic resonance imaging is also often unable to differentiate osteomyelitis with extensive bone destruction from cellulitis at a previous amputation site. Because bone marrow shows significant homogeneous enhancement as compared to the adjacent cellulitis, the application of Gd appears, in our early experience, to be a useful way of identifying involved bone in these two problem areas of volume averaging and bony fragmentation (Fig. 37.6).

Neuropathic Osteoarthropathy

Neuropathic osteoarthropathy describes destructive bone and joint changes caused by repeated trauma in the presence of peripheral neuropathy.

Typically, neuroarthropathic changes are first detected as multiple bony fragments with inhomogeneous decreased signal in T1-weighted images and increased signal in T2-weighted and STIR images representing various stages in healing trauma. These high-signal areas noted on T2-weighted images tend to be scattered and are not confluent around the bony fragments. Little or no abnormal signal is seen in T2-weighted or STIR images in surrounding tissue between the bony fragments to suggest cellulitis or osteomyelitis. These changes are often seen in the tarsus and proximal metatarsus and are frequently some distance from ulcers or other indicators of infection such as abscesses and sinus tracts.

The reason for abnormal signal in the bone marrow of neuropathic feet in the absence of infection is uncertain. However, it seems likely that it is related to acute spontaneous microfractures in the major weight-bearing bones such as the calcaneus and talus. Johnson (16), in a series of 118 cases of Charcot hips, found that 50% were preceded by fractures. El-Khoury and Kathol (6) believe that the absence of pain sensation and proprioception leaves the feet without protection from repeated microtrauma in the face of continued activity. This would support the likelihood of multiple acute microfractures and would probably account for the abnormal signal changes that may in time be difficult to differentiate from those of osteomyelitis. However, abnormal bone marrow signal due to increased water content in the acute phase may decrease or disappear in the subacute and chronic phases (Fig. 37.4). In addition, osteomyelitis is frequently associated with ulcers, abscesses, and cellulitis, and therefore confluent abnormal soft-tissue signal should be seen between the bony fragments in the event of osteomyelitis in a Charcot joint.

In the later stages, multiple fractures tend to result in complete collapse of the longitudinal and transverse arches so that the advanced neuropathic foot, even in the absence of infection, may show more scattered abnormal sig-

nal from bone marrow, loss of the normal foot architecture, and gross bone fragmentation (Figs. 37.4 and 37.6). However, the signal intensity and amount of surrounding abnormal tissue should be much less in the neuropathic joint than in the infected one. Old inactive neuroarthropathy may show ununited fractures with low signal fracture lines and adjacent normal marrow signal (Fig. 37.3).

Ischemia

Ischemia, or a deficiency of blood supply, is caused by atherosclerotic changes of arteries. This probably precedes the peripheral neuropathy caused by ischemia to peripheral nerves. Vascular insufficiency and peripheral neuropathy cannot be directly demonstrated by MR imaging, but some of the secondary changes, such as ulcers (see above) and bone infarcts, may be seen.

Despite the frequent clinical and angiographic evidence of severe ischemia in diabetic feet, bone infarcts appear to be an uncommon finding in our series. Only three bones showing infarction have been noted in our experience. Sweet and Madewell (17) list 11 major categories of etiologic factors in the development of bone infarcts but do not include diabetes. We have found normal marrow signal characteristics in the phalanges of toes that were gangrenous. Histologic examination of these phalanges confirmed the presence of normal bone marrow. The large and medium-sized vessels of diabetic patients are affected by atherosclerosis that is morphologically indistinguishable from that found in nondiabetics. However, much confusion surrounds the significance and relevance of small-vessel disease (4). It seems that bone marrow is one of the last tissues to infarct when the vascular supply is compromised. Typically, bone infarcts are best seen in T1-weighted images as areas of low marrow signal (Fig. 37.11). In T2-weighted and STIR images, bone infarction often shows central low signal with peripheral high signal. Older in-

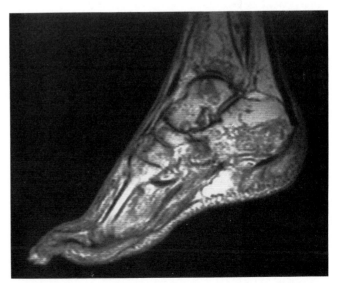

Figure 37.11. Multiple bone infarcts in a 78-year-old diabetic man with peripheral vascular disease. T1-weighted (TR 600, TE 20) sagittal image shows multiple areas of decreased signal in the foot and distal tibia that represent infarcted bone marrow.

farcts, in which marrow has been replaced by calcification and fibrosis, may show low signal with all imaging sequences.

Surgical Changes

Surgical changes include revascularization and surgical procedures, such as major bony amputation and minor soft-tissue debridement.

Reestablished perfusion following revascularization procedures cannot be directly assessed by MR imaging. However, MR imaging is useful in the documentation of healing processes, which are indirectly affected by revascularization. This includes the follow-up of ulcers, soft-tissue infection, and osteomyelitis.

Minor surgical procedures for control of infection or gangrenous tissue include toe and ray amputations, transmetatarsal amputations, debridement, biopsies, and skin grafting. Prior knowledge of any previous surgery is essential information in the interpretation of MR imaging studies of diabetic feet. Because MR imaging consists of multiple slices, it is sometimes surprisingly difficult to define the anatomy when a ray or rays have been amputated or part of a bone excised. Comparison with plain films is very helpful in this situation. Amputation sites are prone to ulcer formation and infection. Magnetic resonance imaging can be useful in documenting the advancing or healing of these postoperative complications (Figs. 37.2 and 37.9). As has been discussed in the section "Osteomyelitis," the differentiation between severe osteomyelitis and adjacent cellulitis at an amputation site can be extremely difficult but may be made easier by using Gd to enhance the bone marrow.

A recent bone marrow biopsy produces increased bone marrow fluid, which may be indistinguishable from the abnormal signal caused by osteomyelitis.

Assessment of Progress

In our institution, MR imaging is frequently used to assess the responsiveness of multiple forms of treatment of the diabetic foot. This includes documentation of healing following revascularization, minor surgical procedures, and conservative measures, such as bed rest, systemic control of hyperglycemia and ketoacidosis, and antibiotic therapy. Figure 37.8 documents healing osteomyelitis.

Acknowledgments

The authors thank Georges Y. El-Khoury, M.D., Phyllis Bergman, and Mary McBride for their help in the writing and preparation of this chapter.

References

1. Garber AJ. Diabetes mellitus. In: Stein JH, ed. Internal medicine. 2nd ed. Boston: Little, Brown, 1987:1997.
2. Gibbons GW. Management of the diabetic foot. Surg Rounds 1986;9:55.
3. Olefsky JM. Diabetes mellitus. In: Wyngaarden JB, Smith LH, eds. Cecil textbook of medicine. 18th ed. Philadelphia: W. B. Saunders, 1988:1361.
4. Corson JD, Jacobs RL, Karmody AM, Leather RP, Shah DM. The diabetic foot. Curr Probl Surg 1986;23:719–788.
5. Peterson LR, Lissack LM, Canter K, Fasching CE, Clabots C, Gerding DN. Therapy of lower extremity infections with ciprofoxacin in patients with diabetes mellitus, peripheral vascular disease, or both. Am J Med 1989;86(6 pt 2):801–808.
6. El-Khoury GY, Kathol MH. Neuropathic fractures in patients with diabetes mellitus. Radiology 1980;134:313–316.
7. Mendelson EB, Fisher MR, Deschler TW, Rogers LF, Hendrix RW, Spies S. Osteomyelitis in the diabetic foot: a difficult diagnostic challenge. Radiographics 1983;3:248–261.
8. Capitanio MA, Kirkpatrick JA. Early Roentgen observations in acute osteomyelitis. AJR 1970;108:488–496.
9. Sherman RS. The nature of radiologic diagnosis in diseases of bone. Radiol Clin N Am 1970;8:227–239.
10. Seldin DW, Heiken JP, Feldman F, Alderson PO. Effect of soft-tissue pathology on detection of pedal osteomyelitis in diabetics. J Nucl Med 1985;26:988–993.
11. Tang JSH, Gold RH, Bassett LW, Seeger LL. Musculoskeletal infection of the extremities: evaluation with MR imaging. Radiology 1988;166:205–209.
12. Yuh WTC, Corson JD, Baraniewski HM, et al. Osteomyelitis of the foot in diabetic patients: evaluation with plain film, ^{99m}Tc-MDP bone scintigraphy, and MR imaging. AJR 1989;152:795–800.
13. Maurer AH, Millmond SH, Knight LC, et al. Infection in diabetic osteoarthropathy: use of indium-labeled leukocytes for diagnosis. Radiology 1986;161:221–225.
14. Sfakianakis GN, Al-Sheikh W, Heal A, Rodman G, Zeppa R, Serafini A. Comparisons of scintigraphy with In-111 leukocytes and Ga-67 in the diagnosis of occult sepsis. J Nucl Med 1982;23:618–626.
15. Jones EW, Peacock I, McLain S, Fletcher E, Edwards R, Finch RG, Jeffcoate WJ. A clinico-pathological study of diabetic foot ulcers. Diabetic Med 1987;4:475–479.
16. Johnson JTH. Neuropathic fractures and joint injuries. Pathogenesis and rationale of prevention and treatment. J Bone Joint Surg (Am) 1967;49A:1–30.
17. Sweet DE, Madewell JE. Pathogenesis of osteonecrosis. In: Resnick D, Niwayama G, eds. Diagnosis of bone and joint disorders, Vol. 5. 2nd ed. Philadelphia: W. B. Saunders, 1988:3188.

CHAPTER 38

Computed Tomography of Degenerative Disorders of the Lumbar Spine

Jaap Schipper

Introduction

The moment computed tomography (CT) became available for studying parts of the body other than the contents of the skull, lumbar spine imaging was revolutionized. Computed tomography for the first time made it possible to obtain direct images of the relationship between bone and soft tissues of the lumbar spine.

Possibly the most overwhelming impact CT has had in lumbar spine imaging involves the evaluation of lumbar root compression. Computed tomography offers a very sensitive noninvasive way of imaging possible lumbosacral root compression caused by degenerative disorders of the intervertebral disk and joints.

Since abnormalities may be detected that are asymptomatic, correlation between CT findings and clinical signs and symptoms remains a very important factor in interpreting CT images of the lumbar spine.

Magnetic resonance (MR) imaging will probably replace CT to a large extent in the evaluation of degenerative disorders of the lumbar spine. Despite this, knowledge of basic CT features of degenerative abnormalities of the lumbar intervertebral disk and facet joints remains a necessity for every general radiologist. This chapter tries to provide some of these basic features, with emphasis on the degenerative intervertebral disk.

Imaging Technique

Patients are examined in the supine position. A scout view is made for proper alignment of the slices. Three to 5-mm slices are obtained in a plane as parallel as possible to the vertebral end plates. This can be achieved by tilting the gantry and minimizing lumbar lordosis by flexing both the knees and hips. Reangulation is usually necessary for

each level and is, due to the limited angle of maximum inclination, still often insufficient at the L5–S1 level. Reformatted images in other planes seldom provide diagnostic information that is not already present on the scans in the axial plane (1). Thinner slices are normally not helpful since they will result in decreased signal:noise ratios. Special attention should be paid to imaging the entire intervertebral foramina in order to exclude the possibility of overlooking intra- or extraforaminal abnormalities (Fig. 38.1). A routine CT examination of the lumbar spine should contain slices through the vertebral bodies and intervertebral disks from the L3 down to the S1 level. It is important not only to obtain slices at the level of the intervertebral spaces but also to visualize the canal in several cuts at the level of the intervertebral spaces but also to visualize the canal in several cuts at the level of the vertebral bodies. This allows the detection of herniated disk fragments that may have migrated cranially or caudally from the interspace level of origin. Scans are made at levels above L3 when clinically indicated.

In cases of previous lumbar back surgery the intravenous administration of contrast agent may be indicated (see "The Postoperative Spine"). The administration of intrathecal contrast material provides excellent delineation of the dural sac (including the nerve root sheets) and its contents (Figs. 38.2 and 38.3). However, this is indicated only in special cases, for instance when the presence of an intradural abnormality is suspected. Hard copies are made at a window width of 600 Hounsfield units (HU) and a level of 150 HU; if necessary additional (bone) settings can be used.

Lumbar Disk Herniation

Lumbar disk herniation (LDH) is a rather common disease: the incidence of LDH per year is about 1.9% for male and 2.2% for female patients in a group of family practices in The Netherlands (2). This corresponds quite well with the 1.5% reported incidence of slipped or ruptured disk in the United States (3). The number of hospital admissions for LDH in The Netherlands is estimated to be approximately 2 per 1000 persons per year, which accounts for up to 2% of the hospital days for all causes (2).

Lumbar disk herniation is defined as a local protrusion of the annulus fibrosus with or without extrusion of nuclear material through a defect in the annulus. Herniated

Abbreviations (see also glossary): HU, Hounsfield unit; LDH, lumbar disk herniation.

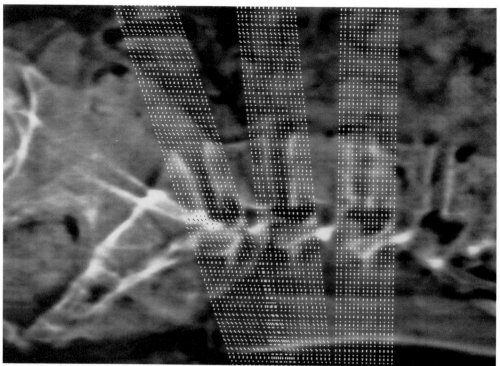

Figure 38.1. Scout view. Note that images are obtained through the entire intervertebral foramina.

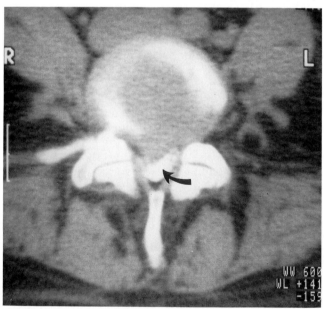

Figure 38.2. Postmyelography CT. Large, right-sided paramedial LDH compresses the contrast-filled dural sac *(arrow)*.

Figure 38.3. Same patient as in Figure 38.2, now displayed in the reconstructed sagittal plane. No additional information is added to the findings in Figure 38.2.

disk material that lies free in the spinal canal is referred to as a sequestered herniation. These disk fragments may still be located anterior to the posterior longitudinal ligament.

Presenting symptoms of patients with LDH are related to compression of a lumbar nerve root by a protrusion of the annulus fibrosus or extruded disk material. This so-called radicular syndrome consists of radiating pain in the area of the sciatic or femoral nerve, which may be accompanied by back pain, feelings of numbness, or even paresis (4). However, physical signs and symptoms alone lack sufficient sensitivity to form the basis for surgical exploration or chemonucleolysis. Moreover, the level and side of the compression are often difficult to determine (5).

Currently, high-resolution CT of the lumbar spine is, together with MR imaging, the principal mode of investigation in the diagnosis of LDH. Although some limitations do exist it offers a highly sensitive and specific, non-invasive and economical examination to patients with symptoms of lumbar nerve root compression probably due to LDH (6, 7).

Computed Tomography Patterns of Lumbar Disk Herniation

The Presence of an Abnormal Epidural Soft-Tissue Mass with Hounsfield Unit Measurements Resembling That of the Intervertebral Disk

These Hounsfield unit measurements will nearly always be higher than those of the dural sac. The presence of such an epidural soft-tissue mass is often very clear (Fig. 38.4). Sometimes, however, the differences in Hounsfield unit measurements are very subtle, so that even large amounts of herniated disk material may be almost completely obscured (Fig. 38.5*A* and *B*). Calcifications may be present in the protruded disk material. The fact that herniated disk material may show late (40 min postinjection) enhancement after the intravenous administration of contrast agent (8) can be helpful, especially in these cases (Fig. 38.6*A* and *B*).

The location of the soft-tissue mass may be classified as medial, paramedial (Fig. 38.4), lateral, intraforaminal (Fig. 38.7), or extraforaminal (Fig. 38.8). The direct relationship with the annulus is often visualized, making the appearance unmistakable (Fig. 38.4). This relation usually remains visible even in the case of large herniations extending cephalad or caudally from the disk space involved. However, it is possible that fragments migrate from the interspace level of origin to the level of an adjacent vertebral body above or below (Fig. 38.9). These fragments are usually still located anteriorly to the posterior longitudinal ligament.

Particular attention should be given to the possibility of an extremely laterally localized LDH (9) (Fig. 38.8). In a retrospective study of noncorresponding radiological and surgical findings these lateral and especially extraforaminal LDHs were missed surprisingly often on CT (10).

Other epidural soft-tissue abnormalities, such as epidural metastases or an epidural abcess, will virtually never give rise to diagnostic difficulties since the presenting symptoms will be different.

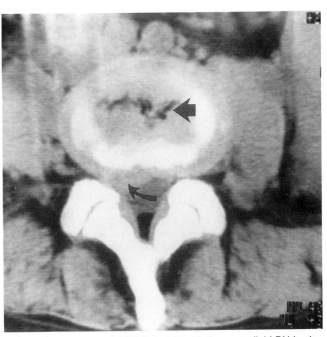

Figure 38.4. Paramedial LDH. A right-sided paramedial LDH is visualized *(curved arrow)*. Note difference in density between intervertebral disk and dural sac. The relation with the disk is obvious in this case. A vacuum phenomenon due to degeneration of the disk is present *(straight arrow)*.

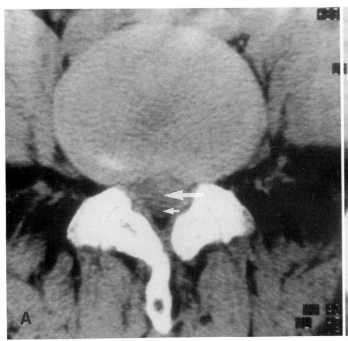

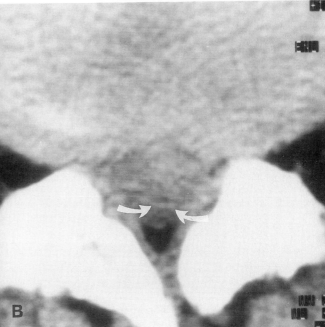

Figure 38.5. **(A)** A large medial LDH is almost completely obscured due to a subtle difference in HU measurement between disk material *(large arrow)* and dural sac *(small arrow)*. **(B)** A detailed view clearly shows the flattened shape of the compressed dural sac *(arrow)*.

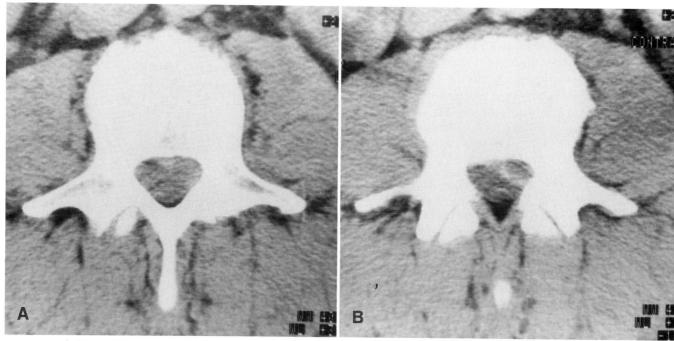

Figure 38.6. Equivocal left-sided LDH **(A)** shows significant peripheral enhancement 40 min after the intravenous administration of contrast **(B)**.

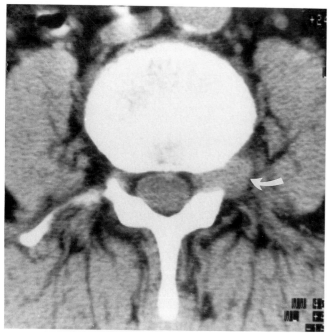

Figure 38.7. Intraforaminal LDH: large left-sided intraforaminal mass *(arrow)*. At surgery a large sequestered disk fragment was found.

Figure 38.8. Extraforaminal LDH: patient with right-sided symptoms of compression of the L5 nerve root. A previous myelogram showed no abnormalities. A large, mostly extraforaminal localized LDH is visualized *(arrow)*. Note that the coronal reconstruction *(below)* adds no additional information to the axial scan.

Obliteration of Epidural Fat

The thecal sac is surrounded on its anterolateral and posterior margin by epidural fat. In the presence of LDH the anterolaterally located fat pads are obliterated and replaced by soft-tissue densities. The obliteration, if caused by LDH, is usually unilateral, in contrast to obliteration caused by hypertrophied facet joints, which is usually bilateral. Another important cause of epidural fat pad obliteration is postsurgical epidural scarring, the features of which are discussed later in this chapter.

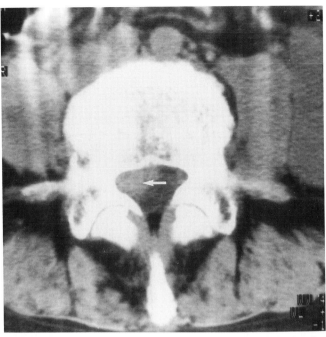

Figure 38.9. Migrated disk fragment: LDH visualized at the level of the vertebral body. Note high density of disk material *(arrow)* compared with the adjacent dural sac.

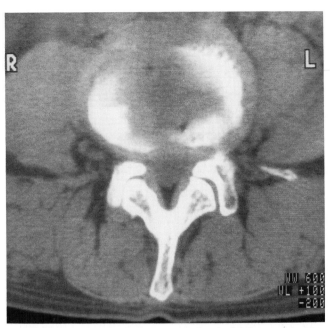

Figure 38.10. Bulging disk at the L3–L4 level: slightly convex contour of the normally concave-shaped L3–L4 disk. Note symmetrical obliteration of the anterolateral epidural fat pads.

Compression and Distortion of the Thecal Sac

The normally triangular shaped thecal sac will be compressed and flattened due to the mass effect generated by extruded disk material.

Displacement of Nerve Roots

A CT display of the actual compression and displacement of a nerve root before it leaves the spinal canal is mostly restricted to the L5–S1 level. This is the major limitation of the CT evaluation of lumbosacral root compression. The relatively large amount of epidural fat on the L5–S1 level enables reliable identification of the nerve root only on this level. On other levels intracanicular root compression is assumed in the presence of other CT signs of LDH fitting the clinical signs and symptoms of radicular compression. However, CT provides excellent visibility of both intra- and extraforaminal nerve root compression on all levels.

The Bulging Disk

In the process of regressive changes of the lumbar intervertebral disk, usually referred to as "degeneration," the nucleus pulposus gradually loses its capacity to bind water. A description of the underlying biochemical changes of this phenomenon is beyond the scope of this text (11). One of the results is that the nucleus pulposus becomes less resistive to the different forces that are applied to it. The ensuing increased stress on the annulus fibrosis will eventually induce morphologic changes within the annulus, and nuclear material may herniate through the weakened fibers. The annulus itself loses its normal shape. The normal configuration of the posterior contour of the annulus

varies from a concave one on the L3–L4 level to a flattened or even slightly convex one on the L5–S1 level. Furthermore, the normal disk margins do not extend cephalad or caudad to the level of the adjacent vertebral endplates.

A bulging lumbar disk is characterized by a smooth, convex, posterior contour of the disk margin that may cause flattening of the dural sac and symmetrical displacement or obliteration of the anterior epidural fat (Fig. 38.10). There may also be a generalized extension beyond the level of the adjacent end plates (12). Due to the strong fibers in the center of the posterior longitudinal ligament sometimes only the lateral disk margins show bulging. Noting the bilateral, symmetrical nature of the abnormality in these cases precludes falsely diagnosing a lateral disk herniation.

A bulging lumbar disk itself will not cause nerve root compression. The diagnosis "bulging disk" without CT features of nerve root compression has characteristically no clinical relevance. This is demonstrated by the fact that it is a common finding on levels that are not associated with corresponding physical signs and symptoms. Recognizing a bulging disk in a patient with symptoms of nerve root compression is important in the presence of other abnormalities that may compromise the lumbar spinal canal. A nerve root in a relatively narrow lateral recess due to hypertrophied articular processes may well become compressed in the presence of a bulging disk (Fig. 38.11). Especially in these complicated cases meticulous correlation of clinical signs with the CT image is necessary. The performance of an additional myelographic study complemented with functional views will often be required to confirm the absence or presence of nerve root compression.

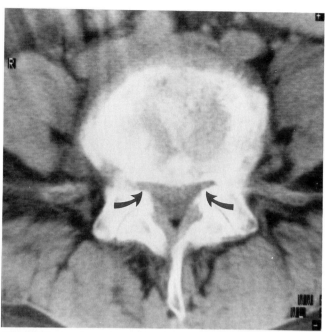

Figure 38.11. Stenosis combined with a bulging disk. Note bilateral stenosis of the lateral recess *(curved arrows)* and the symmetrical convex contour of the bulging disk.

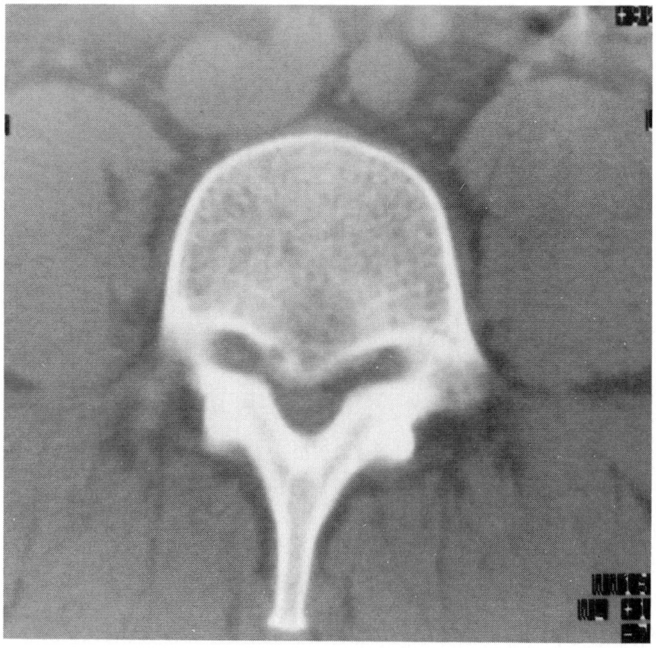

Figure 38.12. Developmental stenosis. Note decreased AP diameter of the spinal canal.

Spinal Stenosis

Stenosis of the spinal canal, causing compression of lumbosacral nerve roots, may be acquired, developmental, or both (the latter being very frequent). Nondegenerative, acquired causes of spinal stenosis, such as Paget's disease or acromegaly, are not described here.

Clinical signs and symptoms may vary, but usually consist of bilateral signs of sciatica and/or subjective symptoms such as burning feelings and sleeping legs.

The two most important noniatrogenic (postoperative) disorders causing spinal stenosis are idiopathic stenosis of the central spinal canal (developmental stenosis) and degenerative stenosis due to stenosis of the lateral recess or degenerative soft-tissue hypertrophy.

The anteroposterior (AP) diameter of the canal can be measured on axial CT scans (13). A distance of less than 12 mm indicates a narrow canal (Fig. 38.12). This measurement, however, is often erroneous. Most important are secondary signs of stenosis, such as the absence of epidural fat at the level of the intervertebral disk and distortion of the dural sac (14). In developmental stenosis the vertebral body and pedicles are often disproportionately large.

Degenerative disorders of the facet joints may give rise to stenosis of the lateral recess caused by bony spur formation at the site of these joints (Fig. 38.13). An AP diameter of the lateral recess of less than 3 mm is considered abnormal (14). Degenerative hypertrophied soft tissue within the spinal canal, notably the ligamentum flava, is another important cause of spinal canal stenosis (15) (Fig. 38.14).

All these forms may occur in a combined form and may be accompanied by a bulging or even herniated disk (Figs. 38.14 and 38.15).

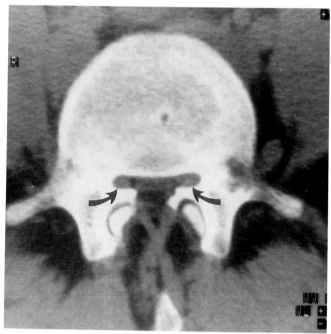

Figure 38.13. Spinal stenosis: Stenosis of the lateral recesses *(arrows)* due to hypertrophied facet joints.

The Postoperative Spine

The patient with recurrent symptoms after back surgery forms one of the major problems encountered in CT imaging of the lumbar spine. Although in about 30% of these patients the origin of the recurrent symptoms will remain unclear (16), CT evaluation of the postoperative spine often provides useful information.

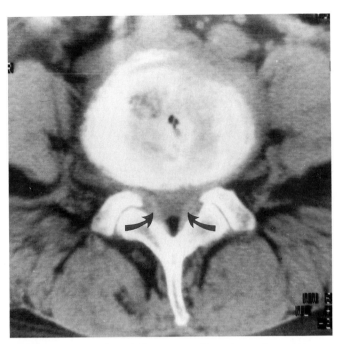

Figure 38.14. Thickened ligamentum flava *(curved arrows)* combined with a bulging disk.

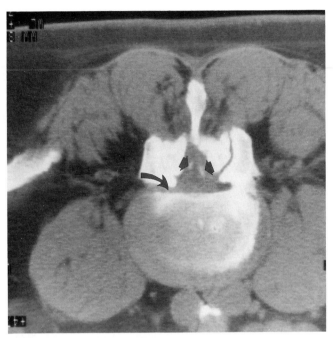

Figure 38.15. Left paramedial LDH in narrow canal. Note calcifications in thickened ligamentum flava *(arrows)* and unilateral obliteration of epidural fat on the left side *(curved arrow).*

The operative site can be determined easily on CT by a surgical defect in the bony arch and the ligamentum flava. Defects caused by fascectomies are also easily recognized. Physiological scarring, recognizible as replacement of the epidural fat by tissue of higher density at the operative site, is almost always present. However, especially after diskectomy and/or fasciectomie, scarring may take place in the anterior parts of the spinal canal and lateral recesses. Differentiating these forms of scarring from recurrent disk herniation can be very difficult. Postoperative scarring (in contrast to recurrent disk herniation) is generally not considered as an indication for intervention. Hence, distinguishing these two entities is very important.

Several features associated with the presence of scar tissue may help to differentiate it from a recurrent disk herniation. First, scar tissue tends to retract the dural sac in the direction of the epidural soft-tissue lesion rather than deform and push away the sac, as can be seen in recurrent herniation (17, 18). Usually precontrast Hounsfield unit measurements of scar tissue are somewhat lower than those of herniated disk material or the disk itself. This finding, however, is too inconsistent to rely on in making the diagnosis. Several reports (19, 20) deal with the importance of intravenous administration of high-dose contrast material (100 ml as a bolus injection). Scar tissue usually shows significant early enhancement (Fig. 38.16*A* and *B*), while the avascular disk material remains virtually nonenhanced (Fig. 38.17*A* and *B*).

Finally, in evaluating the postoperative spine, it is very important not only to search for scarring or recurrent herniation but also to search for possible concomitant findings, most particular stenosis of the lateral recess. The latter can either be due to bony overgrowth after surgery or may already have been present at the initial examination

while both diagnosis and surgery focused erroneously on the disk herniation as the single cause of radiculopathy of the patient.

Correlative Imaging

Ever since the advent of high-resolution CT of the lumbar spine much discussion has involved its possible role in the diagnostic work-up of patients with a radicular compression syndrome. The first CT articles, appearing in 1976 (21) and 1977 (22), merely gave a description of CT findings in patients with LDH. Since that time the question was raised whether CT could replace already existing procedures, such as myelography and epidural phlebography, and in what cases which procedure should be chosen. Many studies have made a comparison between CT and either of these more invasive methods (23, 24). These studies showed sometimes conflicting results. An analysis of the comparative literature that appeared up to 1986 (25) showed no clear difference in the overall diagnostic quality of CT, myelography, and phlebography. Epidural phlebography was especially helpful in the detection of lateral disk herniations. With the introduction of CT, phlebography lost its place in the diagnosis of LDH. Several advantages and disadvantages of both CT and myelography are generally acknowledged. Myelography is an invasive procedure that is likely to reveal the presence or absence of nerve root or cauda equina compression. Its sensitivity in the detection of lateral herniations or herniations at the L5–S1 level is limited. However, the clinical relevance of missing herniations at the L5–S1 level in the absence of myelographic signs of root compression is questionable. Myelography is capable of demonstrating

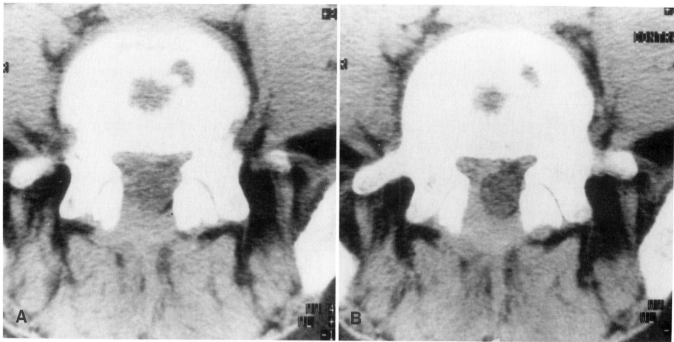

Figure 38.16. Postsurgical epidural scarring: patient with recurrent symptoms after back surgery for LDH. **(A)** Precontrast scan (L4–L5 level). Large epidural mass with almost complete obliteration of the epidural fat pads is evident in postoperative spine. **(B)** Postcontrast scan: significant enhancement of scar tissue. Note difference in Hounsfield unit between mass and dural sac after intravenous administration of contrast.

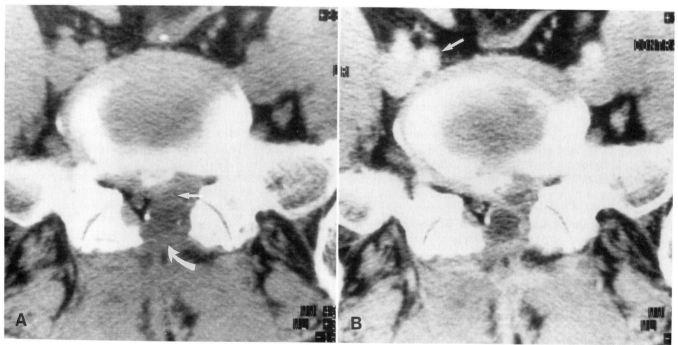

Figure 38.17. Recurrent disk herniation: patient with recurrent symptoms after back surgery for LDH. **(A)** Precontrast scan (L5–S1 level) reveals left-sided epidural mass *(straight arrow)*. Note postsurgical defect in the posterior bony arch *(curved arrow)*. **(B)** Postcontrast scan shows no significant enhancement of extruded disk material. Note contrast-filled iliac veins *(arrow)*.

tumors of the cauda equina, and the L2–L3 level is always visualized. Because it involves lumbar puncture and intrathecal administration of contrast media there is a risk of side effects and hospitalization is usually required. Computed tomography, on the other hand, is a noninvasive procedure that usually reveals the cause of radicular compression and accurately demonstrates lateral and intraforaminal pathologic conditions, as well as L5–S1 herniations. Computed tomography cannot, however, be expected to detect tumors of the cauda equina and the L2–L3 level is not routinely examined. There is seldom a need for contrast administration and hospitalization is not required.

A comparison of the levels of diagnostic accuracy of the two modes of investigation is hampered by the completely different types of imaging for potential lumbosacral nerve root compression. Computed tomography provides a direct axial sectional view of both bony structures and soft tissue, including the disk and the extrathecal part of the nerve root. The intradural part of the nerve root itself, most commonly involved in the etiology of sciatica, however, cannot be adequately imaged. Nerve root compression is assumed in the presence of secondary signs, such as obliteration of epidural fat and indentation of the dural sac. Myelography, on the other hand, gives only an indirect image of soft-tissue lesions in relation to the dural sac and nerve root sheaths but provides "functional" information about the potential root compression caused by these lesions by showing nonfilling of root sheaths and edema of the roots. Myelography is not able to display abnormalities located distally from the contrast-filled subarachnoid space.

A study that investigated the mutual relationship between CT and myelography concluded that if CT is the first investigation in patients suspected of LDH, the number of myelographic procedures can be reduced by two-thirds (23). This study proposes a combined approach, with myelography being reserved for those patients in whom CT provides no adequate explanation for symptoms of nerve root compression or in case of equivocal CT findings. This sequence leads to a very high rate of correctly predicted LDHs (94% in this study population).

Patients with previous back surgery or patients with a narrow, bony spinal canal pose special diagnostic difficulties. Performing both procedures in these patients is often mandatory.

The first reports (26, 27) dealing with MR imaging in the detection of LDH show promising results. The sensitivity of MR imaging in the detection of LDH is probably comparable to that of CT. However, MR imaging lacks most of the above-mentioned drawbacks of CT. Although MR imaging is still more expensive than CT, MR imaging will probably replace CT in the diagnostic work-up of patients with sciatica.

Acknowledgment

J. Th. Wilmink, M.D. (University Hospital, Groningen, The Netherlands) is gratefully acknowledged for his courtesy in providing a large number of the examinations reproduced in this chapter.

References

1. Rosenthal DI, Stauffer AE, Davis KR, Ganott M, Taveras JM. Evaluation of multiplanar reconstruction in CT recognition of lumbar disk disease. AJR 1984;143:169–176.
2. Netherlands Central Bureau of Statistics. Annual survey 1985. Utrecht: Netherlands Central Bureau of Statistics, 1985.
3. Graves EJ. National Hospital discharge survey: annual summary 1987. Vital and health statistics, series 13, No. 99. Hyattsville, Md.: National Center for Health Statistics, 1989.
4. Durning RP, Murphy ML. Lumbar disk disease. Clinical presentation, diagnosis and treatment. Postgrad Med 1986;79:54–74.
5. Hakelius A. Prognosis in sciatica. Acta Orthoped Scand 1970;29:1–76.
6. Carrera GF, Williams AL, Haughton VM. Computed tomography in sciatica. Radiology 1980;137:433–437.
7. Williams AL, Haughton VM, Syvertsen A. Computed tomography in the diagnosis of herniated nucleus pulposus. Radiology 1980;135:95–99.
8. De Santis M, Crisi G, Folchi Vichi F. Late contrast enhancement in the CT diagnosis of herniated lumbar disk. Neuroradiology 1984;26:303–307.
9. Novetsky GJ, Berlin L, Epstein AJ, Lobo N, Miller SH. Extraforaminal herniated disk: detection by computed tomography. Am J Neuroradiol 1982;3:653–655.
10. Slebus FG, Braakman R, Schipper J, van Dongen KJ, Westendorp M. Non-corresponding radiological and surgical diagnoses in patients operated for sciatica. Acta Neurochir (Wien) 1988;94:137–143.
11. Wiesel SW, Bernini P, Rothman RH. The aging lumbar spine. Philadelphia: W. B. Saunders, 1982.
12. Williams AL. The bulging annulus. Radiol Clin N Am 1983;21(2):289–300.
13. Ullrich CG, Binet EF, Sanecki MG. Quantitative assessment of the lumbar spinal canal by computed tomography. Radiology 1980;134:137–143.
14. Lee BCP, Kazam E, Newman AD. Computed tomography of the spine and the spinal cord. Radiology 1987;128:95–102.
15. Beamer YB, Garner JT, Chelden CH. Hypertrophied ligamentum flavum. Clinical and surgical significance. Arch Surg 1973;106:289–292.
16. Naylor A. The late results of laminectomy for lumbar disk disease. J Bone Joint Surg 1974;56:17–29.
17. Braun IF, Lin JP, Benjamin MV, Kricheff II. Computed tomography of the asymptomatic postsurgical spine: analysis of the physiologic scar. AJR 1984;142:149–152.
18. Teplick JG, Haskin ME. CT of the postoperative spine. Radiol Clin N Am 1983;21:395–420.
19. Schubiger O, Valavanis R. CT differentiation between recurrent disk herniation and postoperative scar formation: the value of contrast enhancement. Neuradiology 1980;22:251–254.
20. Yang PJ, Seeger JF, Dzioba RB, et al. High dose IV contrast in CT scanning of the postoperative lumbar spine. Am J Neuroradiol 1986;7:703–707.
21. Di Chiro G. Computed tomography of the spinal cord after lumbar intrathecal introduction of metrizamide (computer assisted myelography). Radiology 1976;120:101–104.
22. Coin CG, Chan YS, Keranen V. Computer assisted myelography in disk disease. J Comput Assist Tomogr 1977;1:398–404.
23. Schipper J, Kardaun JWPF, Braakman R, van Dongen KJ, Blaauw G. Lumbar disk herniation: diagnosis with CT or myelography? Radiology 1987;165:227–231.
24. Bell GR, Rothman RH, Booth RE, et al. A study of computer assisted tomography. Comparison of metrizamide myelography and CT in the diagnosis of herniated lumbar disk and spinal stenosis. Spine 1984;9:552–556.
25. Kardaun JWPF, Schipper J, Braakman R. CT, myelography and phlebography in the detection of lumbar disk herniation: an analysis of the literature. Am J Neuroradiol 1989;10:1111–1122.
26. Modic MT, Masaryk T, Boumphrey F, Goormastic M, Bell G. Lumbar herniated disk disease and canal stenosis: prospective evaluation by surface coil MR, CT and myelography. AJR 1986;147:757–765.
27. Edelman RR, Shoukimas GM, Stark DD, et al. High resolution surface-coil imaging of lumbar disk disease. AJR 1985;144:1123–1129.

CHAPTER 39

Magnetic Resonance Imaging of Degenerative Disorders of the Spine

Michele H. Johnson, S. Howard Lee, and Te Hua Liu

Introduction

Degenerative disorders of the spine represent the major cause of neck and back pain and radiculopathy in the adult population. Nearly 80% of adults will suffer from back pain during their lifetime (1). The spectrum of spinal degenerative disease includes herniated intervertebral disk, facet joint osteoarthritis, uncovertebral joint osteoarthritis, osteophytosis, spondylolisthesis, ligamentous hypertrophy, and degenerative spinal stenosis (acquired). Narrowing of the spinal canal can be central, lateral, or foraminal, depending on the combination of degenerative factors contributing to the stenosis. Herniated disk is a major cause of back and neck pain; however, it should be noted that other pathologic changes may lead to significant pain syndromes.

Magnetic resonance (MR) imaging has become a widely utilized and highly effective, noninvasive technique for the evaluation of the degenerative spine. Technological advances, such as high magnetic field strength systems, surface coils, and faster imaging sequences, allow for high-resolution imaging that can clearly demonstrate normal anatomy as well as a variety of degenerative processes (2, 3).

Imaging Technique and Anatomy

Optimal MR examination of the spine includes images in two planes, usually axial and sagittal. The technical parameters and spatial resolution will vary with manufacturer, field strength, and softward packages. The magnet strengths at our institutions vary from 0.3 to 1.5 T, with Siemens (Iselin, N.J.), General Electric (Milwaukee, Wisc.), and Fonar (Melville, N.J.) represented. Surface coils are routinely utilized to improve the signal:noise ratio.

In general, T1-weighted images (for instance, TR 500,

TE 16 ms) in the sagittal and axial planes provide detailed anatomic information. Axial images are preferably made parallel to the end plates. T2-weighted images (for instance, TR 2000, TE 80 ms) in the sagittal plane may also be obtained using single or multiecho sequences. These images are particularly useful for spinal cord imaging. Gradient-echo (GE) images are often substituted for the T2-weighted images in some protocols. Gradient-echo acquisitions are performed using low flip angles (20 to 30°), displaying a myelographic effect with the disk and vertebrae contrasted by cerebrospinal fluid (CSF). Although scanning time with GE acquisitions is significantly reduced when compared to conventional spin-echo T2-weighted images, the signal:noise ratio is less favorable than in T1- and T2-weighted spin echo images (4–6).

Oblique images or 3D imaging may be added for improved visualization of the neural foramen in the cervical and lumbar regions (7, 8). Coronal views are occasionally useful in the lumbar region. The postsurgical lumbar spine is examined using the imaging parameters discussed above, with the addition of axial and sagittal T1-weighted images following the administration of Gd contrast agent.

On both T1- and T2-weighted images (Fig. 39.1) cortical bone and the longitudinal ligaments are low in signal intensity. The medullary bone compartment of the vertebral bodies is generally of intermediate signal intensity on T1-weighted and low to intermediate intensity on T2-weighted images, although this may vary in the presence of degenerative disk disease (9). The basivertebral plexus and epidural veins are usually hypointense relative to bone marrow on T1-weighted images, while on T2-weighted images they may remain isointense or be hyperintense in signal intensity relative to bone marrow. The CSF is of low signal intensity on T1-weighted and higher signal intensity on T2-weighted images. The CSF demonstrates increasing signal intensity with respect to increasing echo and repetition times (TE and TR, respectively). With GE imaging, CSF signal intensity varies with the flip angle. Decreasing the flip angle results in a relative increase in CSF signal intensity (Fig. 39.2).

The intervertebral disks are biconvex in shape, with a maximum thickness of about one-third of the height of the adjacent vertebral body. The normal disk should not protrude beyond the anterior and posterior margins of the vertebrae. The intervertebral disk is composed of a central nucleus pulposis and a surrounding annulus fibrosis. The

---

Abbreviations (see also glossary): CSF, cerebrospinal fluid; OPLL, ossification of posterior longitudinal ligament; FBSS, failed back surgery syndrome.

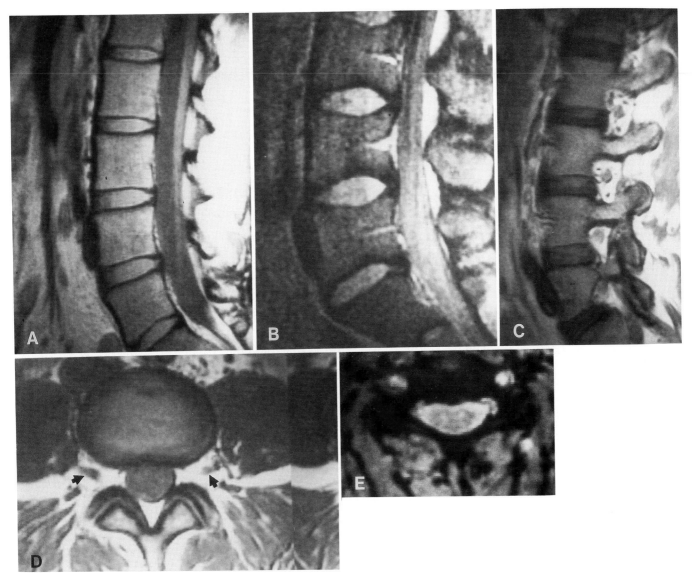

Figure 39.1. Normal lumbar anatomy. **(A)** Sagittal T1-weighted image in the midline demonstrates the cauda equina within the thecal sac. **(B)** Sagittal T2-weighted image in the midline demonstrates increased signal intensity of the intervertebral disk compared to the vertebral bodies. **(C)** Parasagittal T1-weighted image demonstrates nerve roots within the neural foramen surrounded by contrasting fat. **(D)** Axial T1-weighted image through the intervertebral disk also shows the exiting nerve roots *(arrows)*. The facet joints are clearly defined. **(E)** Axial GE image through the disk demonstrates the increased signal intensity of the CSF within the thecal sac.

annulus and the nucleus contain progressively less collagen and more mucoid ground substance (10). The annulus is predominantly composed of fibrocartilage and is thicker and stronger anteriorly than posteriorly. Sharpeys' fibers are collagenous fibers that attach the annulus to the fused epiphyseal ring at the vertebral body end plates. The normal adult intervertebral disk has homogeneous intermediate signal intensity similar to cancellous bone or muscle on T1-weighted images. T2-weighted images allow for the distinction of Sharpey's fibers, which are lower in signal on T2-weighted images, from the remainder of the annulus and from the nucleus pulposis. A low-intensity horizontal band within lumbar disks is a normal finding in patients over the age of 30 (11). This intranuclear cleft (Fig. 39.3) is thought to represent invagination of lamellae of the annulus into the nucleus pulposis. Epidural fat is of

consistently high signal intensity on T1-weighted images and intermediate to high signal intensity on T2-weighted images. On parasagittal and axial sections, lumbar nerve roots of low to intermediate signal intensity are nicely contrasted by fat as they exit the spinal canal in the upper one-third of the intervertebral foramen (Fig. 39.1). Within the dural sac, the normal conus medullaris is at the T12–L2 level (12), and the nerve roots of the cauda equina continue caudally in the posterior compartment bathed in the CSF (Fig. 39.4).

Degenerative Disk Disease

Disk degeneration with an intact annulus is a common finding in individuals over age 40 (1). Biochemical alterations within the disk precede the biomechanical and struc-

Figure 39.2. Normal lumbar anatomy: GE (TR 0.20, TE 11, flip angle 60°) midline image demonstrates increased signal intensity of the CSF contrasted by the signal of the intervertebral disk and vertebrae.

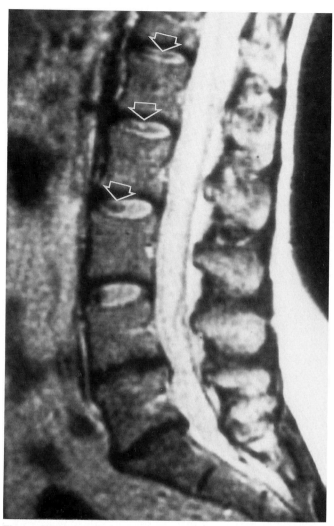

Figure 39.3. Intranuclear cleft and degenerative disk disease. T2-weighted midline sagittal image demonstrates the horizontal intranuclear cleft *(arrows)* within the intervertebral disk spaces (T12–L4). The L4–L5 and L5–S1 disks show decreased signal and decreased intervertebral disk space height. Bulging of the annulus is seen at the L5–S1 level.

tural changes that lead to disk degeneration. Biochemical changes within the nucleus pulposes are related to a decrease in water-binding capacity, disintegration of high-molecular-weight proteoglycans, and an increase in collagen content. By age 60, the water content of the nucleus pulposes has decreased from 90 to 70% (13–15). Decrease in the water content results in a decrease in the signal intensity of the nucleus pulposis on T2-weighted images, as demonstrated in Figure 39.3.

In a series of articles, Yu et al. (13) and Ho et al. (15) describe the changes that occur within the intervertebral disk on aging by examining cadaver specimens of lumbar spines. Normal aging of the intervertebral disk can be described as a range from type 1 (newborn-fibrocartilaginous disk), through type 2 (increasing fibrous content), to type 3, in which the annulus and nucleus are almost indistinguishable by their collagen content. The development of radial tears within the annulus represents an early stage of disk degeneration (type 3B). Recognition of radial tears in the annulus, diminished intervertebral disk space height, and decreased signal intensity within the intervertebral disk constitute the MR findings in the type 3B disk. The severely degenerated disk (type 4) is characterized by marked reduction in the intervertebral disk space height and reduced signal intensity on T2-weighted images (Fig. 39.5), as collagen fibers replace the fibrocartilaginous compo-

nents of the disk. Radial tears and annulus fissures that become fluid filled result in areas of increased signal intensity within the disk on T2-weighted images (10, 16, 17).

The vacuum phenomenon, a familiar sign of degenerative disk disease on CT scans and plain films, is a significantly less reliable sign on MR. Within optimal MR images, a vacuum phenomenon is manifest as a total absence of signal on all pulse sequences. Grenier et al. (18), in a review of 14 cases, suggested looking for decreased signal within the disk as compared to the signal from cortical bone. The vacuum phenomenon in this series was generally associated with other manifestations of degenerative disease of the spine (Figs. 39.6 and 39.10).

The changes in the MR appearance of the lumbar spine in degenerative disk disease are not restricted to those in the disk itself. Changes in the vertebral marrow adjacent to the end plates have been described by Modic as types 1, 2, and 3 (9). The type 1 and type 2 changes may represent a continuum, while the type 3 changes are more dis-

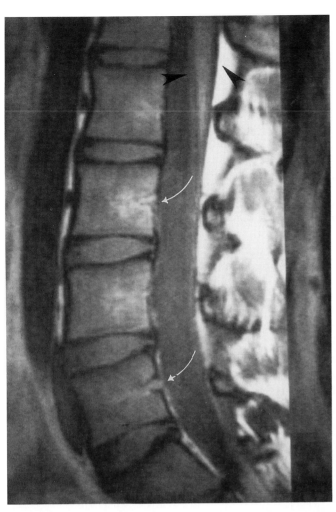

Figure 39.4. Normal conus; basivertebral vein cleft. Sagittal T1-weighted image shows the normal conus medullaris at the T12–L1 level *(arrowheads)*. Note the vertebral cleft for the basivertebral vein *(curved arrows)*.

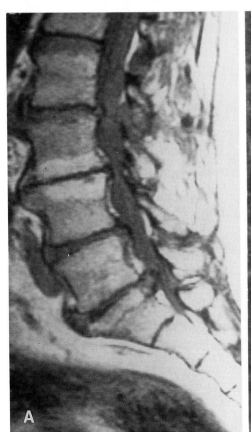
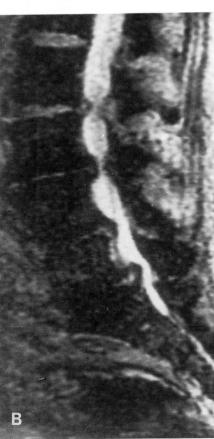

A   **B**

Figure 39.5. Degenerative disk disease. **(A)** Sagittal T1-weighted image demonstrates marked decrease in intervertebral disk space height at the L3–L4, L4–L5, and L5–S1 levels. Disk herniations at the L2–L3, L4–L5, and L5–S1 levels are present. There is increased signal intensity at the end plates adjacent to the L3–L4 intervertebral disk. **(B)** Note the decreased signal intensity within the degenerated disks at L3–L4, L4–L5, and L5–S1 levels on the T2-weighted image. The impingement on the thecal sac from osteophytes at L3–L4, disk herniation at L2–L3, and herniation plus osteophyte at L4–L5 and L5–S1 levels is well demonstrated. The T2-weighted image displays the end plates at L3–L4 as isointense with the remaining vertebral body. This is an example of type 2 marrow change.

Figure 39.6. Vacuum disk phenomenon. **(A)** Marked decrease in signal intensity of the disks at L4–L5 and L5–S1 levels is seen on this sagittal T1-weighted image. Note the degenerative spondylolisthesis of L4 on L5. **(B)** Low signal intensity of the L4–L5 and L5–S1 disks is also seen on the sagittal T2-weighted image associated with marked narrowing of the intervertebral disk spaces. **(C)** A central area of decreased signal intensity is evident on this axial T1-weighted image at L4–L5, although not as apparent as on CT or plain films *(arrows).* **(D)** Lateral plain film demonstrates L4–L5 and L5–S1 vacuum disk phenomenon and spondylolisthesis of L4 on L5.

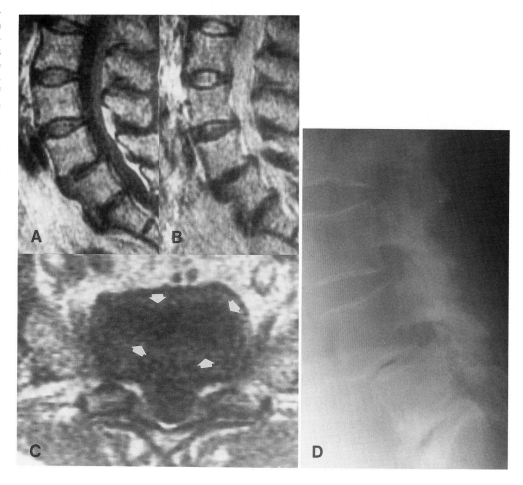

Figure 39.7. Lumbar disk degeneration and type 1 marrow change. A T1-weighted image **(A)** shows a broad area of decreased signal intensity adjacent to the L3–L4 disk. The GE image **(B)** shows mildly increased signal intensity *(small arrows)* in the corresponding area. (Reprinted with permission from Hayes CW, Jensen ME, and Conway WF. Non-neoplastic lesions of vertebral bodies: findings in magnetic resonance imaging. Radiol Soc. N Am 1989;9(5):886).

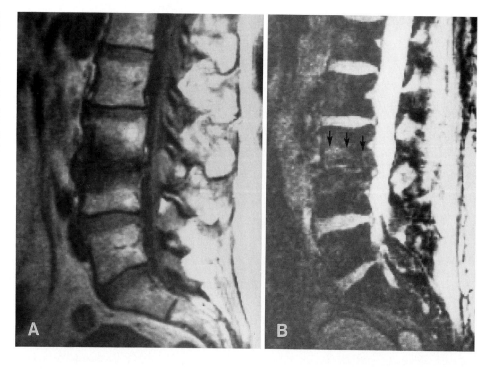

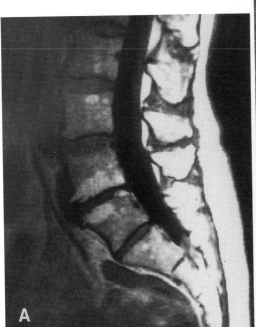
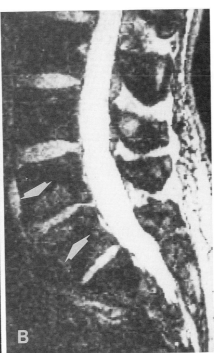

Figure 39.8. Lumbar disk degeneration and type 2 marrow change. A T1-weighted image **(A)** shows a sharply marginated area of increased signal intensity adjacent to a narrowed disk space. The GE image **(B)** shows slightly increased signal intensity *(small arrows).*

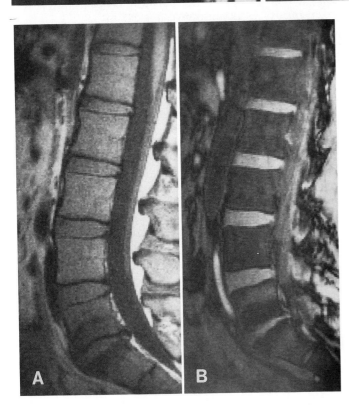

Figure 39.9. Lumbar disk degeneration and type 3 marrow change. The L5–S1 intervertebral disk is reduced in stature secondary to degenerative disk disease. Sagittal lumbar image demonstrates low signal intensity within the marrow adjacent to the end plates at this level on both T1-weighted **(A)** and GE (TR 0.20, TE 11, flip angle 60°) **(B)** images. This is characteristic of type 3 marrow change. Note that the GE image fails to demonstrate the decreased signal intensity within the degenerated L5–S1 disk as well as a true T2-weighted image would (see also Fig. 39.5).

crete. Decreased signal intensity adjacent to the end plates on T1-weighted images, increased signal intensity on T2-weighted images (both from vascularized fibrous tissue as well as fissuring within the vertebral end plates) characterize type 1 changes (Fig. 39.7). The prototype for this appearance may be the postchymopapain-treated disk. Type 2 changes of the end plates are characterized by increased signal intensity on T1-weighted images and isointense to slightly increased signal intensity on T2-weighted images (Fig. 39.8). The increased signal in type 2 changes result from degenerative fatty relacement of the marrow, and may progress from the more acute changes seen with type 1. With type 3 changes, bony sclerosis and the lack of marrow elements within the sclerotic bone may be the etiology of the decrease in signal intensity of the vertebral end plates on both T1-weighted, GE, and T2-weighted images (Fig. 39.9).

Degenerative disk disease can be categorized in a more

Figure 39.10. Bulging annulus at the L5–S1 level. **(A)** Axial T1-weighted image demonstrates concentric extension of the annulus at the L5–S1 level beyond the margins of the end plate without focality *(arrows)*. Note the vacuum phenomenon at this level. **(B)** Sagittal midline T1-weighted image also demonstrates bulging without disk herniation *(arrow)*.

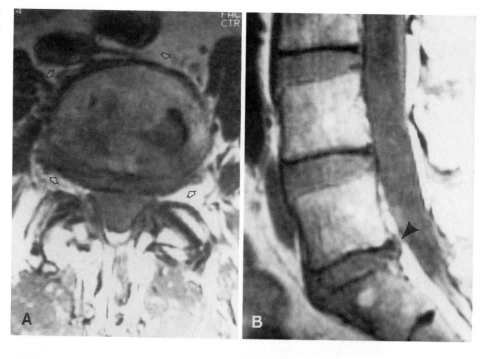

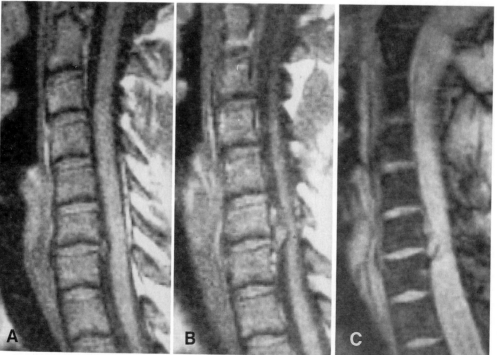

Figure 39.11. Disk herniation: cervical. **(A)** and **(B)** Sagittal T1-weighted images demonstrate a C6–C7 herniated disk. Mild posterior displacement of the cord is identified. **(C)** Sagittal GE image with its myelo-graphic effect demonstrates the herniation by its impression on the thecal sac.

traditional way by the configuration of the intervertebral disk (1, 2, 13). A bulging annulus (bulging disk) is disk degeneration with an intact annulus, which results in the concentric extension of the disk (annulus) beyond the vertebral body end plate (Fig. 39.10). It is generally accepted that a bulging disk implies no focal nerve root compression. A herniated disk is a focal displacement of the nucleus pulposus through a defect in the annulus (Fig. 39.11).

The displaced nuclear material remains connected to the parent nucleus pulposis. Disk herniations may be characterized as midline, posterolateral, or lateral in location (Fig. 39.12).

Masaryk (19) recently described the appearance of annular fibers in the presence of lumbar disk herniation as evaluated by MR. Magnetic resonance can clearly demonstrate the actual status and site of tear in the annulus fi-

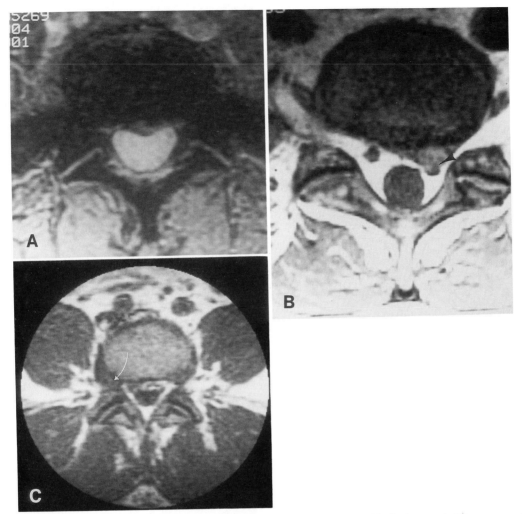

Figure 39.12. Herniated disks: central, posterolateral, and lateral. **(A)** L5–S1 axial GE image demonstrates a central herniated disk indenting the ventral aspect of the thecal sac. **(B)** This axial T1-weighted image through the disk and neural foramen shows a focal disk herniation cen- tral and to the left with displacement of the nerve root *(arrowhead).* **(C)** Axial T1-weighted image demonstrates a predominantly lateral disk herniation with disk material in the neural foramen *(curved arrow)* in addition to the posterolateral component.

brosis. Masaryk used the term *prolapsed disk* to describe a herniation through an annular defect that is incomplete, when a few remaining outer fibers of the annulus (Sharpey's fibers of low signal intensity) contain the prolapsed nucleus (Fig. 39.13). In a prolapsed disk, a fine line of low signal intensity representing the annulus can be identified on T1-weighted images between the prolapsed nucleus and the epidural fat or thecal sac. Identification of the remaining annular fibers may be limited by chemical shift artifact on high field systems.

Extrusion of the disk implies disruption of the remaining annular fibers (Fig. 39.14). Extruded portions of disk material may be of slightly higher signal intensity than that of prolapsed disks. With a sequestered disk, the herniated nuclear material is not contiguous with the remaining intraannular nucleus within the intervertebral disk space. This is often referred to as a free or migrated fragment (20). The fragment may be located above or below the disk space, and anterior or posterior to the posterior longitudinal ligament. They almost invariably require surgical intervention. The sequestered fragment may have a higher signal

intensity than the parent disk on T2-weighted images (Fig. 39.15). Early reparative processes leading to increased water content and neovascularization are postulated as the reason for an increase in signal intensity of up to 70% greater than the parent disk. However, they may have the same signal intensity as the parent disk (Fig. 39.16). Magnetic resonance imaging is most specific in the identification and localization of sequestered disks (2, 15, 16). Magnetic resonance imaging is excellent for the evaluation of multilevel disk disease, particularly with higher level lumbar disks missed due to more limited CT evaluations (Figs. 39.5 and 39.16) (20).

Clinical syndromes may vary, depending on the nerve root compressed by herniations in different locations (14, 21, 22). Displacement of the epidural fat plane, focal deformity of the thecal sac, and displacement of adjacent nerve roots are associated findings. In the cervical and thoracic regions, displacement and/or rotation of the spinal cord within the thecal sac may be secondary manifestations of disk herniation (23, 24) (Figs. 39.11 and 39.17). Rare spontaneous regression of nonsequestered herniated

Figure 39.13. Lumbar disk herniation: prolapsed. Sagittal T1-weighted image demonstrates prolapse of disk material beyond the posterior margins of the vertebral bodies. Some annular fibers remain intact *(arrows)*.

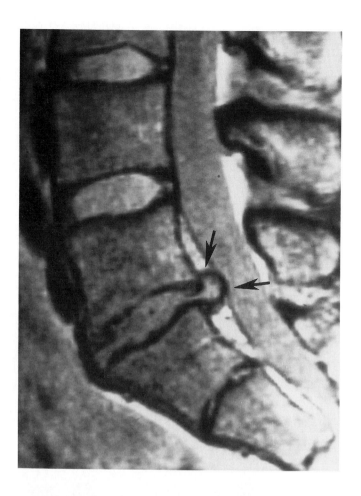

Figure 39.14. Lumbar disk herniation: extruded. Note the loss of the dark line defining the annular fibers around this disk herniation *(arrow)*. This is due to complete disruption of annular fibers at the site of the herniation.

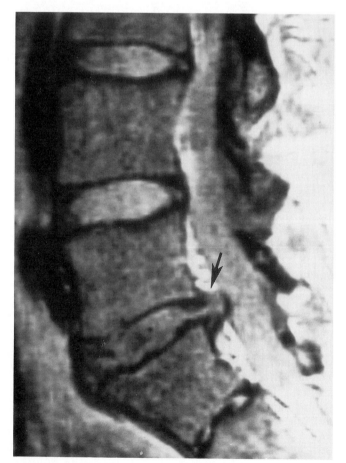

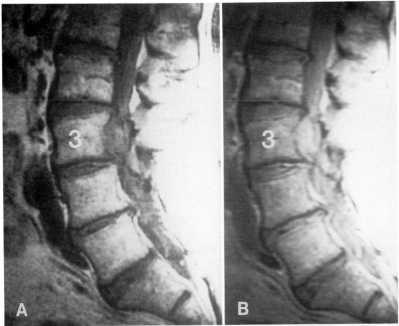

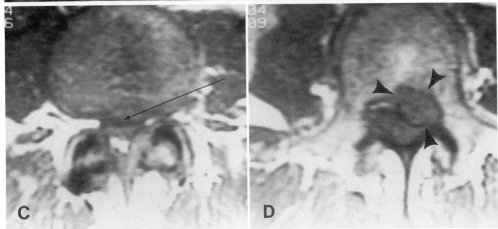

Figure 39.15. Disk herniation: sequestered (migrated) disk. **(A)** T1-weighted image demonstrates a large soft-tissue component behind the L3 vertebral body. This material is higher in signal intensity than the L3–L4 intervertebral disk. The L3–L4, L4–L5, and L5–S1 intervertebral disks are decreased in stature. **(B)** Spin-density image again shows the soft tissue in the canal as increased signal intensity relative to the disk at L3–L4. **(C)** Axial image at the L3–L4 disk level shows a central disk herniation *(arrow)*. **(D)** Axial image at the level of the pedicles of L3 demonstrates a sequestered disk fragment within the canal compressing the sac *(arrows)*. This has migrated cephalad from the L3–L4 level.

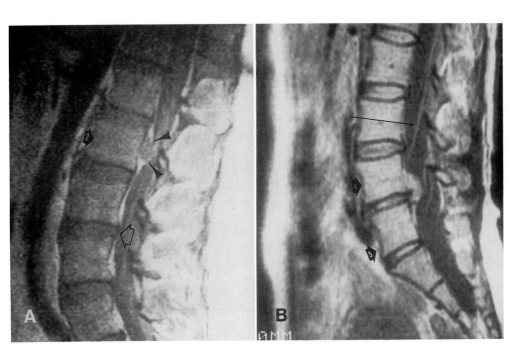

Figure 39.16. Lumbar disk herniation: sequestered (migration). **(A)** Sagittal T1-weighted image demonstrates sequestered fragment behind the L2 vertebra from L1–L2 disk herniation *(arrowheads)*. Note the small L3–L4 disk herniation *(open arrow)*. **(B)** L4–L5 sequestered disk with upward migration of the fragment in another patient. Note the L4–L5 intervertebral disk space height reduction and adjacent deformed nerve roots *(arrow)*.

disk has been demonstrated both by CT and MR (Fig. 39.18) (25).

Although less frequent than those in the lumbar and cervical regions, thoracic disk herniations are also readily identified by MR imaging (20, 26, 27). Thin-section images are usually required because the ventral extradural components of thoracic disk herniations are frequently smaller than their lumbar or cervical counterparts. Compression or indentation of the thecal sac and displacement of the spinal cord within the spinal canal are additional features of thoracic disk disease (Fig. 39.17) (24). When the frequency-encoding gradient is in the anterior to posterior direction, chemical shift artifact arising from fat in the vertebral body marrow may obscure small thoracic disks, resulting in a false-negative examination (28). Enzmann et al. described this finding in two patients. This problem was seen only on T1-weighted images (29). If the frequency-encoding gradient is in the superior to inferior direction in a sagittal T2-weighted image the superior end plate will appear to be different in thickness (either thicker or thinner) than the inferior end plate of the same verte-

bral body, due to chemical shift effects. The amount of chemical shift is directly proportional to imaging field strength and inversely proportional to band width and thus is more prominently seen in high field images (30).

Cervical herniated disks are demonstrated well on T1- and T2-weighted images, as well as on GE images (5, 23, 27) (Figs. 39.11 and 39.21). The signal intensity characteristics of degenerative disks in the cervical region are less predictable than those in the lumbar region. Evaluation of the neural foramen is limited on sagittal images, necessitating axial and sometimes oblique images for assessment (7, 8). Uncovertebral joint degenerative changes may be best demonstrated on axial or oblique views, although prominent osteophytes may be seen on sagittal images (Fig. 39.19). With cervical disk herniations, engorged epidural veins are occasionally seen. Linear high signal above and below intervertebral disk levels may represent epidural venous engorgement (28, 31) (Fig. 39.20).

The assessment of the nature of ventral extradural impressions on the thecal sac in the cervical area frequently involves distinguishing herniated disk from osteo-

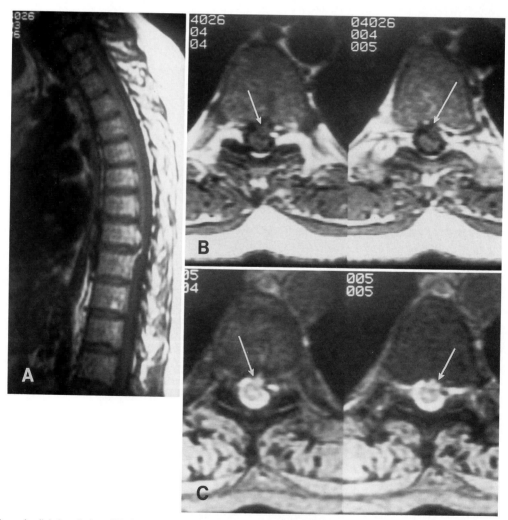

Figure 39.17. Thoracic disk herniation. **(A)** Sagittal T1-weighted image demonstrates a herniation at the T7 level. **(B)** and **(C)**. Axial T1-weighted **(B)** and apial GE images **(C)** show a herniation *(arrows)* central and to the left with slight cord displacement.

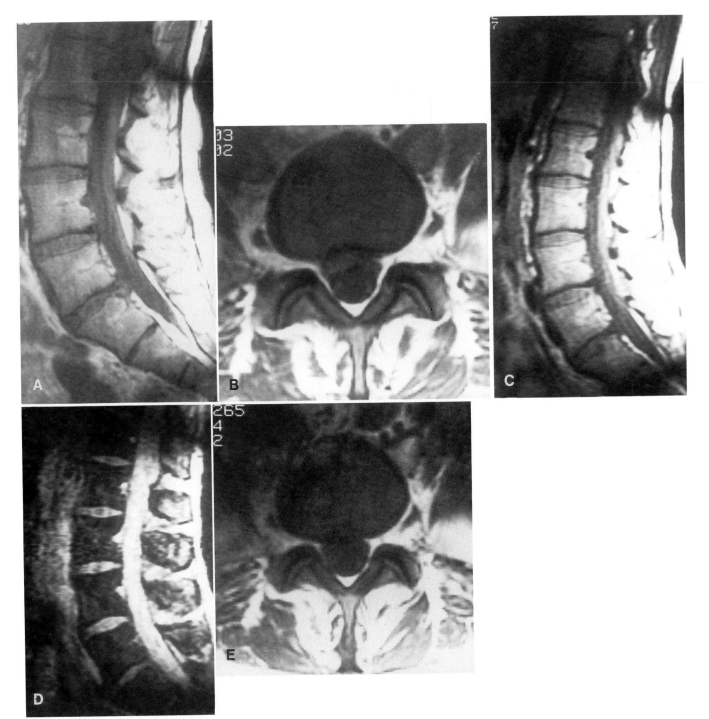

Figure 39.18. Spontaneous regression of lumbar disk herniation. September, 1988: Sagittal **(A)** and axial **(B)** T1-weighted images demonstrate L5–S1 protrusion central and to the right. February, 1990: Repeat examination demonstrates regression of the herniation without surgical treatment. **(C)** T1-weighted sagittal image, **(D)** GE image demonstrate degeneration at L5–S1 level without herniation. Axial T1-weighted image at disk level **(E)** confirms the absence of disk herniation.

phyte. Osteophytes may be high, intermediate, or low in signal intensity, depending on the amount of fatty marrow they contain (5, 23). Gradient-echo images make this delineation easier by accentuating the CSF/soft-tissue interface and by demonstrating osteophyte as low in signal in contrast to the higher signals of CSF and paraspinal fat (5) (Figs. 39.7 and 39.21).

Spinal Stenosis

Spinal stenosis is a reduction in the cross-sectional area of the spinal canal that may be congenital or secondary to degenerative disease (2, 32–34). Both the bony spinal canal and adjacent soft-tissue structures (i.e., intervertebral disk and ligamentous changes) contribute to the development

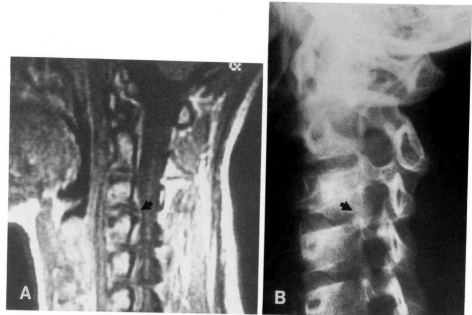

Figure 39.19. Uncovertebral joint degenerative disease. **(A)** Sagittal T1-weighted image demonstrates degenerative spurring of the unco- vertebral joint at the C3–C4 level *(arrow)*. **(B)** Oblique film correlates nicely with MR imaging findings.

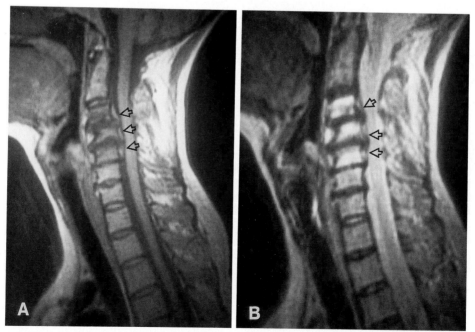

Figure 39.20. Epidural venous plexus engorgement. **(A)** Sagittal T1-weighted image demonstrates a thin band of intermediate signal intensity behind the vertebral levels (C3–C5; *arrows*). This is felt to represent engorgement of the epidural venous plexus. Posterior osteo- phytes causing focal spinal stenosis and degenerative disease are demonstrated at the C3–C4, C4–C5 and C5–C6 levels. **(B)** Sagittal T2-weighted image demonstrates the osteophytes to better advantage. Type 2 marrow change is seen from C3–C5. The intermediate signal intensity band representing the epidural venous plexus engorgement is again seen *(arrows)*.

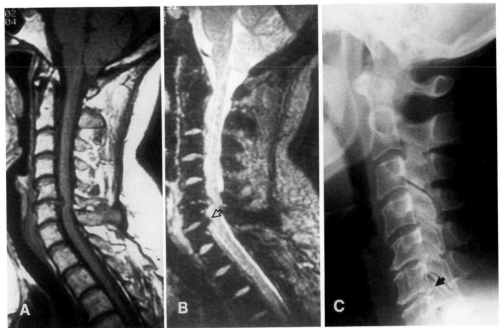

Figure 39.21. Advantages of GE imaging: osteophyte vs. disk. **(A)** T1-weighted image demonstrates diminished intervertebral disk height and disk herniation at C5–C6. **(B)** Sagittal GE image confirms the disk herniation at C5–C6 and demonstrates a ventral impression on the sac at C6–C7, which is lower in signal intensity than the corresponding disk; this is consistent with osteophyte *(arrow)*. **(C)** Plain lateral cervical film demonstrates small osteophytes at the C6–C7 level *(arrow)*.

of stenosis. The presence of spondylolisthesis and other forms of vertebral malalignment may further compromise the spinal canal dimensions with or without the presence of spondylolysis. Spondylolysis may be missed by MR imaging and is better demonstrated by CT and plain films. Narrowing of the anterior-posterior dimension of the bony spinal canal and compromise of the thecal sac in the midline is best seen in sagittal MR images (Fig. 39.22). In a retrospective study of MR images of 13 healthy subjects and 30 patients with lumbar degenerative disease, Grenier et al. found that assessment of foraminal stenosis was best accomplished by review of sagittal MR images (35). The axial images can accurately depict the size of the thecal sac, the status of the intervertebral disk, the ligamentum flavum, and the presence of degenerative facet joint changes and osteophytes (36) (Fig. 39.7). In this respect, MR findings are analogous to those on CT; however, bony spinal stenosis (due to osteophytes, spondylolisthesis, and/or facet joint degenerative disease) is in general best demonstrated by CT. In the MR evaluation of canal stenosis, T1-weighted images were found to be best for the assessment of size and shape of the spinal canal, evaluation of the foramen, and for conus localization (2). The myelographic effect of the increased signal intensity of CSF on T2-weighted images allowed for the best overall assessment of thecal sac encroachment and the presence and location of extradural defects (Fig. 39.22). Uncovertebral joint degenerative changes above or in combination with facet degenerative changes may lead to neural foraminal narrowing in the cervical region. Assessment of cervical foraminal disease generally requires axial images. Oblique images may also be necessary.

Ossification of the posterior longitudinal ligament (OPLL) may cause ventral encroachment on the spinal cord (37). Although OPLL is most commonly identified in the cervical area, the thoracic region may also be affected. Resultant reduction in spinal canal diameter may lead to myelopathy requiring surgical decompression. Distinction of OPPL from herniated disk by MR imaging is possible using T1-weighted, T2-weighted, and GE images (Fig. 39.23) (38). Epidural lipomatosis, a rare cause of symptomatic compression of the thecal sac and contents by an abnormal accumulation of mature fat within the epidural space, can be easily and completely assessed by MR imaging (39) (Fig. 39.24). For congenital varieties of spinal stenosis such as achondroplasia, MR imaging can define the neural structures within the canal as well as the bony deformities seen in these disorders (32–34).

The Postoperative Spine

Magnetic resonance imaging is frequently utilized as the noninvasive imaging method of choice for evaluation of the postoperative spine (40–42). Postoperative changes in the spine commonly identified on MR imaging may include changes in fatty marrow content adjacent to a diskectomy site, loss of posterior elements with laminectomy, and changes in alignment secondary to fusion. Less commonly, dural ectasia, pseudomeningocele, arachnoiditis, the presence of fat grafts, hematoma, gas, or infection are identified within the spinal canal (Fig. 39.25). Epidural fibrosis and recurrent disk herniation are the most common causes of the recurrent pain following lumbar spine surgery (43–46). Accurate evaluation of the causes of the failed back surgery syndrome (FBSS) is critical in order to determine which cases are amenable to surgical therapy. In

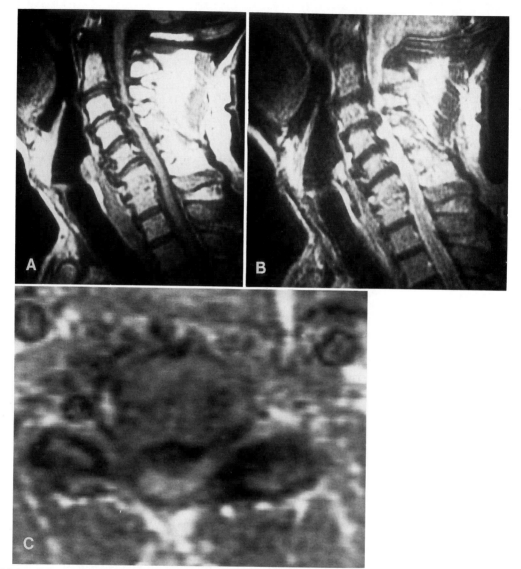

Figure 39.22. Disk herniation and degenerative cervical stenosis. **(A)** Sagittal T1-weighted image demonstrates severe degenerative disk disease (C2–C5) with a reduction in intervertebral disk space height and a large central disk herniation at C4–C5. Note the severe stenosis and the kyphotic deformity at the C6–C7 level where the vertebral bod-ies are fused. **(B)** Sagittal T2-weighted image demonstrates the ste-nosis to better advantage. Reduction in signal intensity of the cervical intervertebral disk, as well as the disk herniations at C3–C4 and C4–C5, are seen. **(C)** Axial T1-weighted image through the C3–C4 inter-vertebral disk demonstrates marked cord compression.

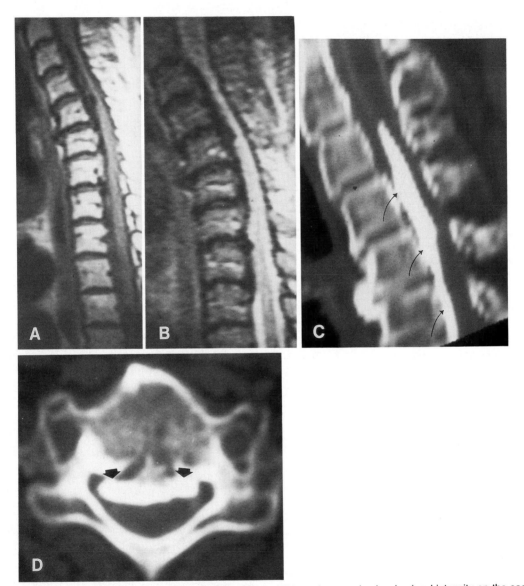

Figure 39.23.   Ossification of the posterior longitudinal ligament. **(A)** Sagittal T1-weighted image demonstrates confluent region of decreased signal intensity in the region of the posterior longitudinal ligament at the C–5 level. Note the compression of the thecal sac. **(B)** This region remains low in signal intensity on the sagittal T2-weighted image. **(C)** Sagittal CT reconstruction postmyelogram demonstrates the thickened and extensively ossified ligament *(arrows)*. **(D)** Axial CT scan demonstrates the ossified posterior longitudinal ligament *(arrows)*.

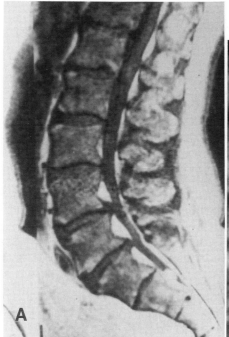
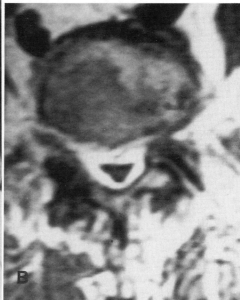

Figure 39.24. Epidural lipomatosis: sagittal **(A)** and axial **(B)** T1-weighted images; note increased amount of epidural fat associated with compromise of the thecal sac. [Reprinted with permisison from Lee SH, Coleman PE, Hahn FJ. Radiol Clin N Am 1988;26(5):963.]

postoperative assessment, plain unenhanced CT was correct in 43% of cases, and intravenous contrast-enhanced CT was correct in 74 to 87% of cases (46). On T1-weighted MR images epidural fibrosis is hyperintense relative to CSF and variable in signal intensity relative to the intervertebral disk (which varies in signal intensity, depending on the degree of disk degeneration) (47). On nonenhanced images, the differences in signal intensity of epidural fibrosis, thecal sac, and recurrent disk are subtle, making a distinction sometimes difficult. Nonenhanced MR imaging was successful in 86% of cases in making the diagnosis of epidural fibrosis (46). The configuration of the scar, with its extension posterior to the thecal sac and often extending to the bony defect, may facilitate diagnosis. Traction effects, such as pulling the thecal sac toward the epidural fibrosis, are typically found (Figs. 39.26 and 39.27). Anterior extension of fibrosis adjacent to the disk may be seen. The distinction between this type of epidural fibrosis and herniation may be particularly difficult without intravenous MR contrast (Figs. 39.26 and 39.27).

The addition of MR contrast techniques to the imaging of the postoperative spine has facilitated the distinction between recurrent herniated disk and epidural fibrosis (Figs. 39.26 and 39.27). Huetfle in 1988 reviewed a series of MR images with and without Gd-diethylenetriaminepenta-acetic acid (DTPA) in 30 postoperative patients, 17 of which had surgical and pathologic correlation at a total of 19 levels (38). In this series, epidural fibrosis demonstrated early heterogeneous enhancement on T1-weighted images. Immediate scanning after contrast administration was recommended in order to accentuate the differentia-

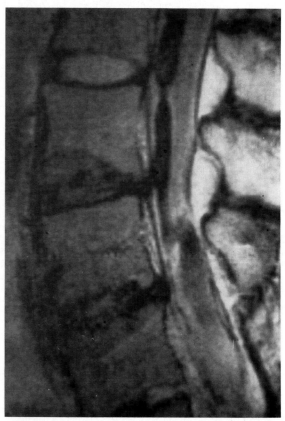

Figure 39.25. Lumbar arachnoiditis. Sagittal spin-density image demonstrates thickened and clumped nerve roots typical of severe arachnoiditis.

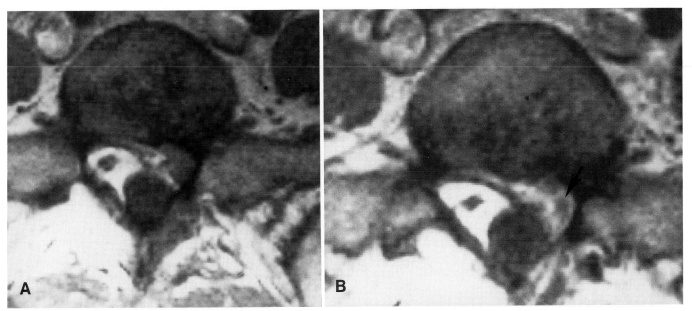

Figure 39.26. Postlaminectomy: scar without recurrent disk. **(A)** Axial T1-weighted image at L5–S1 demonstrates soft-tissue density antero-lateral to the thecal sac. The left nerve root is not distinguishable within this density. **(B)** Axial T1-weighted image after contrast administration demonstrates contrast enhancement throughout the soft-tissue density consistent with postsurgical scar, without recurrent herniated disk. Note the nerve root surrounded by scar *(arrow)*.

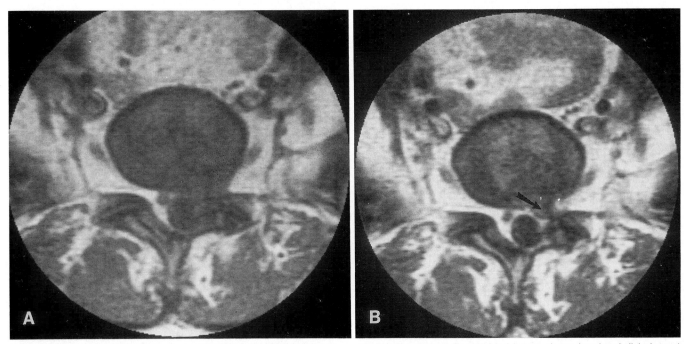

Figure 39.27. Herniated disk and scar. **(A)** Axial T1-weighted image demonstrates soft tissue anterior and lateral *(left)* to the thecal sac. Patient has had a hemilaminectomy on the left, now with recurring pain. **(B)** Enhanced axial T1-weighted image shows herniated disk *(arrow)* surrounded by contrast-enhancing scar.

tion between disk and epidural fibrosis. Herniated disk demonstrated variable enhancement on delayed scans. When enhancement of the disk was identified it was peripheral in location rather than central or diffuse, as is seen with epidural fibrosis. Granulation tissue developing around a chronic disk herniation may be the explanation for the observed peridiskal enhancement (Fig. 39.27) (40).

We have found Gd-enhanced MR to be a most reliable method for the assessment of disk vs. epidural fibrosis in the postoperative patient. This is true on both mid- and high field imaging systems. We have recently had the opportunity to examine a small number of patients using the nonionic MR contrast agent from Squibb (Prohance) (Princeton, N.J.) in the evaluation of disk vs. epidural fibrosis, as a stage 3 trial by the Food and Drug Administration (Fig. 39.27). In this small number of cases, disk was clearly distinguished from epidural fibrosis in all imaging planes. Because this nonionic agent can be administered in a larger dose, evaluation in larger patients may be facilitated.

Conclusion

Magnetic resonance imaging will continue to grow as the primary diagnostic procedure in evaluation of patients with degenerative disease of the spine. The facility with which diagnoses can be made using these techniques is unsurpassed, and should continue to improve with the addition of newer MR technology, both hardware and software. First and later generation MR contrast agents should continue to provide for earlier and more accurate diagnosis of the pathology and lead to a better understanding of the pathophysiology of degenerative disease of the spine.

References

1. Modic MT, Pavlicek W, Weinstein MH, et al. Magnetic resonance imaging of intervertebral disk disease clinical and pulse sequence considerations. Radiology 1984;152:103–111.
2. Modic MT, Masaryk T, Boumphrey F, Goormastic M, Bell G. Lumbar herniated disk disease and canal stenosis: prospective evaluation by surface coil MR, CT and myelography. Am J Neuroradiology 1986;7:709–717.
3. Ross JS, Modic MT, Marsaryk TJ, Carter J, Marcus RE, Bohlman H. Assessment of extradural degenerative disease with Gd DTPA-enhanced MR imaging: correlation with surgical and pathologic findings. Am J Neuroradiology 1989;10;6:1243–1249.
4. Haacke EM. Image behavior: resolution, signal to noise, contrast and artifacts. In: Modic MT, Masaryk TJ, Ross JS, eds. Magnetic resonance imaging of the spine. Chicago: Yearbook Medical Publishers, 1989:22–27.
5. Hedberg MC, Drayer BP, Flom RA, Hodak JA, Bird CR. Gradient echo (GRASS) MR imaging in cervical radiculopathy. Am J Neuroradiology 1988;9:145–151.
6. Watanabe AT, Teitelbaum GP, Lufkin RB, Tsuruda JS, Jinkins JR, Bradley WG. Gradient-echo MR imaging of the lumbar spine: comparison with spin-echo technique. J Comput Assist Tomogr 1990;14(3):410–414.
7. Daniels DL, Hyde JS, Kneel JB, et al. The cervical nerves and foramina: local coil MR imaging. Am J Neuroradiology 1986;7:129–133.
8. Modic MT, Masaryk TJ, Ross JS, Mulopulos GP, Bundschuh CV, Bohlman H. Cervical radiculopathy: value of oblique MR imaging. Radiology 1987;163:227–231.
9. Modic MT, Steinberg PM, Ross JS, Masaryk TJ, Carter JR. Degenerative disk disease: assessment of changes in vertebral body marrow with MR imaging. Radiology 1988;166:193–199.
10. Yu S, Haughton V, Lynch KL, Ho KC, Sether LA. Fibrous structure in the intervertebral disk correlation of MR appearance with anatomic sections. Am J Neuroradiology 1989;10:1105–1110.
11. Aguila LA, Piraino DW, Modic MT, Dudley AW, Duchesneau PM, Weinstein MA. The intranuclear cleft of the intervertebral disk: magnetic resonance imaging. Radiology 1985;155:155–158.
12. Wilson DA, Prince JR. MR imaging determination of the location of the normal conus medullaris throughout childhood. Am J Neuroradiology 1989;10:259–262.
13. Yu S, Haughton VM, Ho PSP, Sether LA, Wagner VN, Ho KC. Progressive and regressive changes in the nucleus pulposis. Radiology 1988;169:93–97.
14. Brown, MD. The pathophysiology of disk disease. Orthoped Clin N Am 1971;2:259–370.
15. Ho PSP, Yu S, Sether LA, Wagner M, Ho KC, Haughton VM. Progressive and regressive changes in the nucleus pulposis. Part 1. The neonate. Radiology 1988;169:87–91.
16. Ross JS, Modic MT, Masaryk TJ. Tears of the annulus fibrosis: assessment with Gd-DTPA-enhanced MR imaging. Am J Neuroradiology 1989;10:(6):1251–1254.
17. Park WM, McCall IW, O'Brien JP, Edin, Webb JK; Fissuring of the posterior annulus fibrosus in the lumbar spine. BJR 1979; 52:382–387.
18. Grenier N, Grossman RI, Schielsler ML, Yeager BA, Goldber HI, Kressel HY. Degenerative lumbar disk disease: pitfalls and usefulness of MR imaging in detection of vacuum phenomenon. Radiology 1987;164:861–865.
19. Masaryk TJ. High resolution MR imaging of sequestered lumbar intervertebral disks. Am J Neuroradiology 1988;9:351–358.
20. Schellinger D, Manz HJ, Vidic BV, Patronas NJ, Deveikis JP, Muraki AS, Abdullah DC. Disk fragment migration. Radiology 199;175:831–836.
21. Jinkins JR, Whittemore AR, Bradley WG. The anatomic basis of vertebrogenic pain and the autonomic syndrome associated with lumbar disk extrusion. Am J Neuroradiology 1989;10:219–231.
22. Grenier N, Gresell JF, Douws C, et al. MR imaging of foraminal lumbar disk herniations. J Comput Assist Tomogr 1990;14(2):243–249.
23. Modic MT, Masaryk TJ, Mulopulos GP, Bundschuh C, Han JS, Bohlman H. Cervical radiculopathy: prospective evaluation with surface coil MR imaging, CT with metrizamide and metrizamide myelography. Radiology 1986;161:753–759.
24. Ross JS, Perez-Reyes N, Masaryk TJ, Bohlman H, Modic MT. Thoracic disk herniation: MR imaging. Radiology 1987;165:511–515.
25. Teplick JG, Haskin M. Spontaneous regression of herniated nucleus pulposus. AJR 1985;145:371–375.
26. Alvarez O, Roque CT, Pampati M. Multilevel thoracic disk herniation: CT and MR studies. J Comput Assist Tomogr 1988; 12(4):649–652.
27. Karnaze MG, Gado MH, Sartor KJ, Hodges FJ. Comparison of MR and CT myelography in imaging the cervical and thoracic spine. AJR 1988;150(2)397–403.
28. Williams MP, Cherryman GR, Husband JE. Significance of thoracic disk herniation by MR imaging. J Comput Assist Tomogr 1989;13(2):211–214.
29. Enzmann DR, Griffin C, Rubin JB. Potential false-negative MR images of thoracic spine in disk disease with switching of phase and frequency-encoding gradients. Radiology 1987;165:635–637.
30. Dwyer JA, Knop RH, Hoult DI. Frequency shift artifacts in MR imaging. J Comput Assist Tomogr 1985;9:16–18.
31. Takahashi M, Sakamoto Y, Miyawaki M, Bussaka H, MR visualization and clinical significance of the anterior longitudinal epidural venous plexus in cervical extra-axial lesions. Compute Med Imaging Grap 1988;12(3):169–175.
32. Wang H, Rosenbaum AE, Reid CS, Zinreich SJ, Pyerito RE. Pediatric patients with achondroplasia: CT evaluation of the cranio cervical junction. Radiology 1987;164:515–519.
33. Fortuna A, Ferrante L, Acqui M, Santoro A, Mastronardi L. Marrowing of thoraco-lumbar spinal canal in achondroplasia. J Neurosurg Sci 1989;32(2):185–196.
34. Epstein BS, Epstein JA, Jones MD. Lumbar spinal stenosis. Radiol Clin N Am 1987;15(2):227–239.
35. Grenier N, Kressel HY, Schiebler ML, Grossman RI, Dalinka MK. Normal and degenerative posterior spinal structures: MR Imaging. Radiology 1987;165:517–525.
36. Weisz GM, Lamond TS, Kitchener PN. Spinal imaging: will MRI replace myelography? Spine 1988;13(1):65–68.
37. Terayama K. Genetic studies on ossification of the posterior longitudinal ligament of the spine. Spine 1989;14(11):1184–1191.
38. Grenier N, Gresell JF, Vital JM, et al. MR or lumbar longitudinal ligaments. Normal anatomy and assessment of disruption in disk disease. Radiology 1989;171:197–205.
39. Quint JD, Boules RS, Sanders WP, Mehta BA, Patel SC, Tiel RL. MRI in epidural lipomatosis. Radiology 1988;169:485.
40. Huetfle MG, Modic MT, Ross JS, et al. Lumbar spine: postoperative MR imaging with Gd DTPA. Radiology 1988;167:817–824.
41. Ross JS, Delamarter R, Hueftle MG, et al. Gadolinium DTPA enhanced MR imaging of the postoperative lumbar spine: time course

and mechanism of enhancement. AJR 1989;152(4):825–834.

42. Ross JS, Masaryk TJ, Modic MT. Magnetic resonance of lumbar arachnoiditis. Am J Neuroradiology 1987;8:885–892.

43. Finnegan WJ, Fenlin JM, Marvel JP, Nardini RJ, Rothman RH. Results of surgical intervention in the symptomatic multiply operated back patient. J Bone Joint Surg 1979;61:1077–1082.

44. Law JD, Lehman RAW, Kirsh WM. Reoperation after lumbar intervertebral disk surgery. J Neurosurg 1978;48:259–263.

45. White AH, ed. Failed back surgery syndrome evaluation and treatment. Spine 1986; 1(1):1–175.

46. Bundschuh CV, Modic MT, Ross JS, et al. Epidural fibrosis and recurrent disk herniation in the lumbar spine: MR imaging assessment. Am J Neuroradiology 1988;9:169–178.

47. Lee SH, Coleman PE, Hahn FJ. Magnetic resonance imaging of degenerative disk disease of the spine. Radiol Clin N Am 1988;26(5):949–964.

Magnetic Resonance Imaging and Computed Tomography of Spinal Trauma

Heinrich Schüller and Maximilian Reiser

Introduction

About 45 spine injuries per million people per year occur in the United States (1–5). Men, 20 to 50 years of age, are most often affected, the main reasons being sports accidents, motor vehicle accidents, and falls from great height.

The most important goal after an accident is the visualization of an existing or imminent cord injury, which accompanies injuries of the spine in 20% of the cases (6). Injuries of the ligaments are common in cases of cord trauma without fractures of the vertebrae.

For therapeutic reasons bone injuries should be exactly defined regarding their location, extent, and form. Analysis of the mechanism of the accident, clinical (especially neurological) examination of the patient, and conventional plain film images render important information and are the basis for further diagnostic procedures.

Technique, results, and value of computed tomography (CT) and magnetic resonance (MR) imaging will be discussed in this chapter.

Computed Tomography Technique

Native Computed Tomography

Transport and positioning of a patient with suspected spine injury must be performed with extreme care (7). Especially when transferring the patient onto the CT table a sufficient number of people must assist. The examination is usually performed in the supine position. Extracorporal metallic devices, e.g., electrocardiogram (ECG) cables, must be removed or repositioned in order that they not produce image artifacts. During the examination of the thoracic or lumbar spine the arms should be outside the examination field, if possible behind the head. In some patients sedation and assisted or controlled respiration is mandatory.

In cases of severe fractures of the cervical spine stabilization or traction may be necessary, which must be sustained during the whole examination. For this purpose nonartifact-producing mechanical traction devices have proved suitable (8).

Contrast-Enhanced Computed Tomography (Myelo-Computed Tomography)

In patients with neurological deficits there may be an indication for lumbar or cervical (C1–C2 or suboccipital) contrast medium injection in order to verify a possible compression of the cord. Five milliliters of a nonionic contrast medium with a content of 170 g/liter iodine are sufficient for complete imaging of the whole subarachnoid space in computed tomography.

Choice of Slices

The slices are chosen from a digital projection radiogram (scanogram, scout view, or topogram according to the CT scanner used). The inclination of the gantry is defined by using the lateral view image and should be parallel to the intervertebral disks. Sagittal projections may be useful in determining the examination area within the thoracic spine. The inclination of the gantry and the field of view should be kept unchanged during the whole examination in order to obtain adequate secondary reformations.

Thickness of Slices

The examination parameters must ensure an acceptable duration of the procedure and diagnostic information, which should be as complete as possible. Slice thickness has a major influence: 1.5-mm slices minimize the partial volume effect, whereas thicker slices reduce the overall scan time. In the craniovertebral region, including C2, a slice thickness of 2 mm or less is recommended; in the other parts of the spine a thickness of 3 to 5 mm is usually sufficient.

Image Documentation

The documentation of the examination must allow the assessment of both the bony and soft-tissue structures by using different windows. Window widths of 800 Houns-

---

Abbreviations: HU, Hounsfield units.

field units (HU) or more are chosen for depiction of bony elements and less than 300 HU for soft-tissue structures.

Digital Image Processing

If the examination is performed with constant gantry inclination and constant field of view in continuous slices, the data can be viewed at as a cube of CT values, which can be used for further calculations, producing images in any chosen plane. Of course the reformatted images can contain only information already present in the original images. They can, however, facilitate the interpretation, especially in cases of subluxation of vertebral bodies or facet joints. Additionally the three-dimensional reconstruction of the spine is a promising development, which allows an even easier interpretation (9).

Magnetic Resonance Imaging Technique

Magnetic Resonance Imaging Examination Technique Planning

The supine position is the most favorable for the patient. The patient should be positioned as securely and comfortably as possible.

All ferromagnetic devices must be removed before entering the MR imaging suite, since they might be moved by the interaction of the static and gradient magnetic fields. Patients with cardiac pacemakers or metal foreign bodies (e.g., bullet splinters in critical anatomic locations, especially intraspinal, intracranial, or intraorbital) must be excluded from the examination (10). Exemptions can be made only in cases of well-known magnetic characteristics of the material (11). In principle, high-frequency and gradient fields can cause heating of the material, but this does not endanger the patient.

In addition, effects of eddy currents may diminish the quality of the image. When in doubt, metallic foreign bodies must be located beforehand by conventional X-ray. Parts of any device for artificial respiration and monitoring that come close to or into the field must be nonferromagnetic in order to prevent image artifacts and risks for the patient.

Superior image quality can be obtained using surface coils or other specially designed coil systems. They allow a better signal:noise ratio as compared to the body coil, better image quality, and shorter examination time. However, examination time is also dependent on the patient's positioning, coil tuning, image reconstruction, as well as interpretation time of the images and the decisions about further proceedings.

Pulse Sequences

Traumatic changes can affect the extradural, intradural, extramedullar, and intramedullar compartment. T1-weighted spin-echo sequences (repetition time (TR) less than 500 ms and echo time (TE) up to 25 ms) allow a good differentiation of the structures in the first two compartments. Cord and cerebrospinal fluid can be distinguished easily.

T2-weighted spin-echo sequences (TR>2000 ms, TE>90 ms) are recommended to assess pathological changes of the spinal cord. Proton-weighted sequences (long TR, short TE) may be helpful in discriminating vertebral bodies and hematomas. T2- and T2*-weighted sequences enable detection of stenosis of the subarachnoid space in a fashion comparable to myelographic imaging. In addition these sequences are suitable for visualization of edema or hemorrhage within the cord and bone marrow.

T2*-weighted gradient-echo sequences allow the acquisition of data in a relatively short time. This technique is highly sensitive for the detection of acute or subacute hemorrhage. Narrowing of the spinal canal is accentuated due to susceptibility effects of bony elements.

Metal implants and even the smallest metal splinters may cause severe image distortion. The longer the acquisition time (TR, averages) the greater the danger of motion artifacts. Gradient-echo sequences are more sensitive to patient movements, but they have the advantage of a shorter examination time. In order to reduce motion artifacts various techniques are available:

1. Reduction of motion (belt or sedating drugs)
2. ECG or respiratory ordered phase encoding
3. Fast imaging technique (gradient echo)
4. Saturation pulses
5. Suitable inclination of the phase-encoding gradient.

Imaging Planes

In contrast to CT, MR imaging allows the free choice of imaging planes. For exact orientation spin-echo sequences with short TR and TE or fast image sequences in the sagittal projection are useful. For detailed anatomical evaluation axial planes are needed. The previous sagittal examination results are used to define the planes of interest. T2- or T2*-weighted sequences in a plane of choice are added for better characterization of the lesion.

Injuries of the Vertebral Column

The main purpose of CT or MR imaging examinations of the traumatized spine is to assess the extent of the injury and the stability of the vertebral column.

Knowing whether a spinal fracture is stable or unstable is crucial for the selection of proper therapeutic regimens. In stable injuries the vertebral column is able to maintain its normal alignment, to provide support for the cranium and body, and to protect the neural elements under normal physiologic stresses (12). In unstable injuries movements of the patient may produce or aggravate bony or neural injury.

The definitions of *stability* and *instability*, however, are still controversial (13). These terms do not necessarily predict prognosis, nor do they define therapeutic regimens or even guide patient management. Although it is useful to define stability for clarifying signs that indicate significant spinal trauma, the terms stability or instability must be used with great care. On the one hand, there is the possibility of missing the correct diagnosis of an unstable fracture (12). On the other hand, "normal physiologic stresses" for injured vertebral columns may be quite different in patients with bed rest compared to those who walk or even work.

To discern between stability and instability of the vertebral column it is best to define three anatomic parts of different static importance. This method is based on the works of Holdsworth (14) and Denis (15). Three columns are defined:

an anterior part, including the disks and the anterior two-thirds of vertebral bodies
a middle part, including the posterior third of a vertebral body, the annulus fibrosus, and the posterior ligament
a posterior part, including the whole dorsal ligament system

If only one of these three columns is injured the fracture is stable. If all three parts are involved the fracture is unstable. Isolated disruptions of the middle part of the vertebral column are very infrequent; therefore in combination with the rupture of the anterior or posterior part the lesion must be regarded as unstable. Diagnostic procedures must therefore focus on the middle column.

Fractures of the Vertebral Bodies

Fractures of the vertebral bodies result from heavy forces exerted onto the vertebral column. Because of the special static characteristics of the vertebral column the lower parts of the cervical spine are most often affected, followed by the thoracolumbar transition. Statistics vary, however, concerning the incidence of fractures of the other segments (12, 16, 17).

Fractures of the Dens Axis

Fractures of the dens account for about 2% of all fractures of vertebral bodies. Hyperflexion as well as hyperextension may lead to a fracture of the dens axis. In hyperflexion trauma ventral dislocation of the dens is found in most cases. A ventral subluxation of C1 or C2 may also be present. Hyperextension trauma, on the other hand, leads to dislocation of the dens dorsally with a dorsal subluxation of C1 or C2. Anderson and D'Alonzo described three types of fractures (18):

Type I (rare): Fractures of the dens axis cranial to the base. The fracture lines usually are oblique. This is a stable fracture that as a rule can be treated conservatively.
Type II (common): Fractures through the base of the dens axis. These fractures are unstable. A risk of pseudarthrosis exists in cases of dislocation of the fragments. Therefore surgical stabilization is the method of choice (Figs. 40.1–40.3).
Type III: Fractures through the base of the axis also affecting the body of the axis. Most of these fractures are stable and can be treated conservatively.

Type II fractures are difficult to detect in CT because the horizontal fracture lines may be parallel to the CT slices and thus not visible due to volume averaging.

Dens fractures are visualized to good advantage on MR imaging because on sagittal or coronal planes a horizontal fracture is clearly depicted.

Fractures of the Body of the Axis

Flexion combined with axial compression is the most common mechanism for fractures of the body of the axis. The fracture line starts at the base of the dens and extends vertically or obliquely into the body of the axis. The oblique fractures result from the same mechanisms of trauma as the Anderson type III fractures.

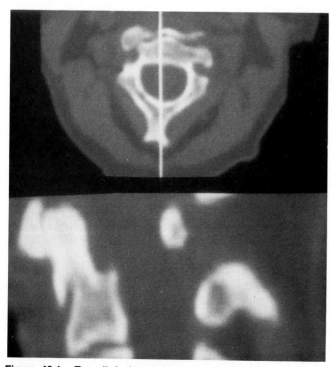

Figure 40.1. Type II Anderson fracture of the dens axis is shown (axial CT image and sagittal secondary reformation).

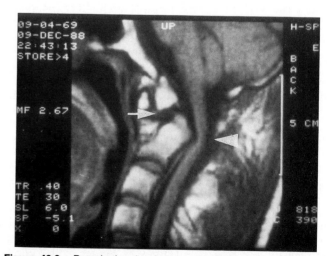

Figure 40.2. Pseudarthrosis of dens axis. Sagittal T1-weighted image: visible are narrowing of spinal canal and compression of the cord *(arrowhead)*. Granulation tissue with intermediate signal intensity *(arrow)* is visible cranial and posterior to the lower part of the second cervical vertebra.

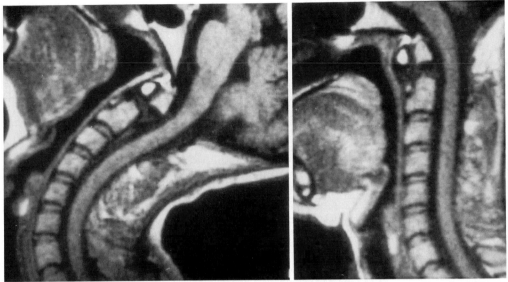

Figure 40.3. Functional images in pseudarthrosis of the dens axis. Retroflexion: minor dislocation of atlas without significant narrowing of the spinal canal. Anteflexion: no dislocation of the vertebrae or narrowing of the spinal canal.

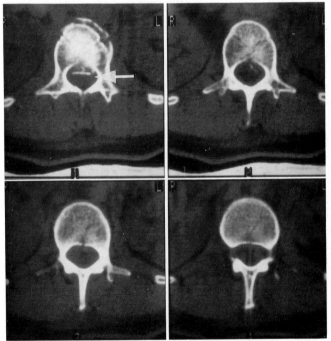

Figure 40.4. Burst fracture of a lower thoracic vertebra: a dislocated fragment narrowing the spinal canal is seen *(arrow)*.

Other Fractures of Vertebral Bodies

Destruction of the trabecular structures within a vertebral body is the consequence of absorption of an external force. If the nucleus pulposus is pushed through the upper or lower end plates into the vertebral body the latter is compressed. The bony fragments are dislocated from the center to the outer parts. Thus dorsally displaced fragments may encroach on the spinal canal and cause a compression of the cord (Fig. 40.4).

Computed tomography visualizes the position of the fragments and the narrowing of the spinal canal can easily be measured. In cases of a mere impression fracture an increase of bone density is found, which may easily be missed on axial slices (Fig. 40.5). Similar to conventional lateral radiograms, alterations of the shape of a vertebral body can easily be recognized in MR, whereas in CT secondary reformations are required in order to depict the vertebral bodies in the coronal or sagittal plane (Figs. 40.6–40.9). Paravertebral hematoma may be the only finding on axial scans indicating minor injuries of the vertebral column.

Using MR imaging, alterations of paravertebral soft tissues as well as bone marrow are visualized to good advantage (Fig. 40.10). Within the vertebral body, high signal intensity is found in T2-weighted images, indicating an increase of fluid content (edema) within the bone marrow. Signal intensities are variable on T1-weighted scans. Edema and acute hemorrhage show low signal intensity whereas subacute hemorrhage results in high signal intensity (Fig. 40.6).

Wedge-shaped compression fractures are predominantly found in the thoracolumbar area. This type of injury results from compressive hyperflexion. The anterior or lateral loss of height is the result of a force exerted on the anterior third of the vertebral body, while the posterior column remains intact. If in addition the middle column has a normal appearance there is no need for CT or MR examination because this fracture may be regarded as stable.

"Chance fracture" was first described by G. Q. Chance in 1948, and is defined as a distraction trauma of the vertebral column (19). This type of fracture was later called a seat-belt fracture, occurring in persons wearing lap seat-belts in sudden deceleration motor vehicle accidents.

Forward flexion of the free part of the vertebral column may lead to different types of lesions (20):

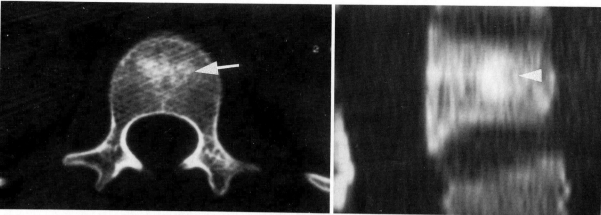

Figure 40.5. Compression fracture of a lower thoracic vertebral body. **(A)** Axial CT scan: cancellous bone *(arrow)* shows increased density. **(B)** Sagittal reformation: wedge-shaped compression of the vertebral body is evident, with increased density in the central portion of the vertebra *(arrowhead)*.

Figure 40.6. Compression fracture of the first lumbar vertebra. **(A)** Two days after trauma: wedge-shaped vertebral body with decreased signal intensity *(arrow)* is evident on the T1-weighted image. **(B)** Two days after trauma: increased signal intensity *(arrowhead)* on a T2-weighted image is due to edema. One year after trauma normal signal intensity on the T1-weighted **(C)** and T2-weighted **(D)** images is evident.

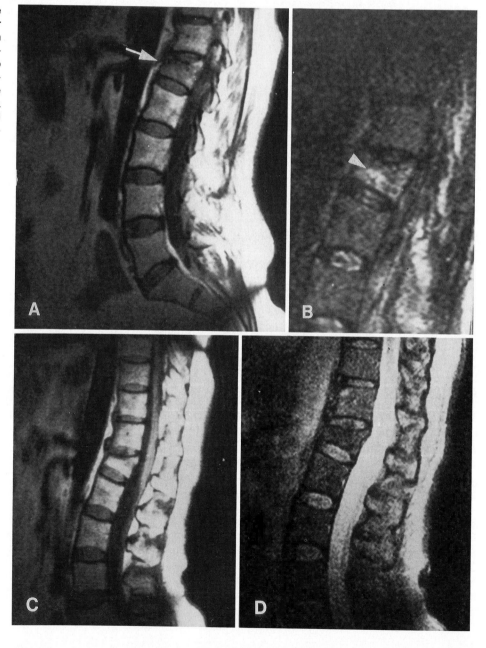

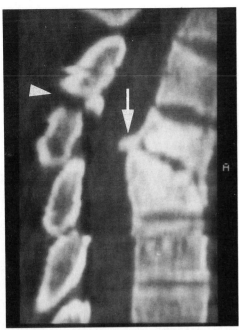

Figure 40.7. Flexion-distraction injury: the sagittal reformation shows an intraspinal fragment *(arrow)* of the vertebral body, widening of the interspinous space *(arrowhead)*, fracture of the articular process, and anterior angulation of the spine.

1. The original Chance fracture (horizontal splitting of the vertebra, including the pedicles and spinal process and sometimes the transverse processes, omitting the anterior ligament)
2. Rupture of all but the anterior ligaments and disk
3. A combination of the above on either one or two adjacent segments.

These fractures are difficult to detect in CT because of the horizontal fracture lines, which may be parallel to the CT plane and thus not visible. They are regarded as an indication for MR imaging.

In cases of recent trauma of the vertebral bodies with intramedullary hematoma a low signal intensity is found on T1-weighted images and even more on T2- or T2*-weighted images. In cases of bone marrow edema T1-weighted images show only a slight decrease in signal intensity whereas T2-weighted images depict a clear increase of signal intensity.

In cases of subacute hemorrhage high signal intensity in both T1- and T2-weighted images is present because of the biochemical changes of hemoglobin (intracellular) into methemoglobin (extracellular). Later, a characteristic decrease of signal intensity, especially on T2- and T2*-weighted images, is observed because of conversion to hemosiderin occurring first in the periphery of the lesion.

Fractures of the Vertebral Arches

Isolated fractures of the vertebral arches are rare. Usually they occur in combination with other lesions of the vertebral column. All of them are instable injuries.

The best known fracture of this type is the so-called Jefferson fracture, resulting from an axial compression to the skull. The axial force is transmitted to the surfaces of the lateral masses of the atlas through the cranium and the occipital condyles. Consequently a symmetrical bilateral fracture of the anterior and posterior arch of the atlas is found.

Asymmetrical fractures, for instance, fractures of the anterior or posterior arches or a combination of unilateral anterior and posterior fractures, are possible (Fig. 40.11). A typical fracture of the axis is the notorious hangman's fracture. Its pathomechanism was first described by Wood-Jones in 1912. It consists of a combination of hyperextension and distraction, leading to a fracture through the pedicles of the axis with anterior dislocation of the vertebral body. A similar type of fracture may be found in car accidents. Although the fracture mechanism is different, hyperextension and axial compression, the result is very similar. Often rupture of the C2–C3 disk in combination with a rupture of the posterior and anterior ligament may be found (21).

Burst injuries may result in a lesion of the vertebral arch. Especially in cases of sagittal fracture lines additional fractures of the arches must be looked for thoroughly. An increased distance of the pedicles may be an important sign on the conventional radiographs.

On axial CT images fractures of the vertebral arches are clearly visualized. Computed tomography shows distance and position of the fragments (Figs. 40.12 and 40.13) as well as a possible additional rotation of the corresponding vertebral body (Fig. 40.14). Diagnostic difficulties may arise from congenital clefts. In contrast to traumatic fissures, however, congenital clefts reveal smooth borders.

Computed tomography is the method of choice for the detection of fractures of vertebral arches because of its superior anatomic resolution of bony structures and the axial imaging plane (Fig. 40.15).

Injuries of the Disks

Trauma of the intervertebral disks is found in up to 30% of all injuries of the vertebral column (22–25). It may occur as isolated injury or combined with fractures and dislocations of vertebral bodies. Anterior, intravertebral, or intraspinal herniations may occur, depending on the direction of dislocation of the herniated disk material (Fig. 40.16). Traumatic herniation may be observed on plain CT if it is localized in the upper cervical and lumbar regions of the vertebral column. Intrathecal contrast enhancement is required in the lower cervical and thoracic region, CT criteria being the same as in nontraumatic disk herniations.

Difficulties may be encountered in interpreting CT images because of an oblique scanning plane in relation to the disk. Compared to CT, MR imaging allows a better distinction of the disk versus surrounding tissue, for example as occurs in an anterior epidural hematoma. The direction of a disk herniation as well as its sequelae may easily be seen in a sagittal image. In contrast to CT, MR imaging does not require contrast enhancement (24).

Injuries of Spinous and Transverse Processes

Fractures of spinous and transverse processes are caused by strong dorsal forces. Abrupt flexion results in oblique

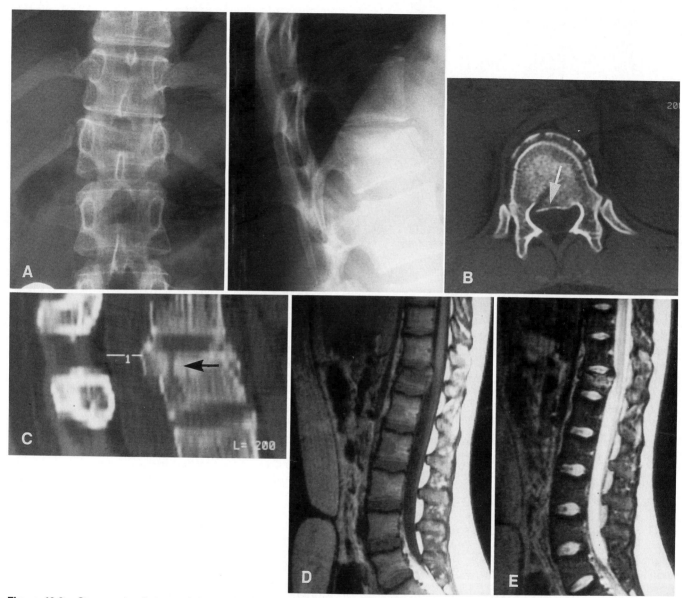

Figure 40.8. Compression fracture of the twelfth thoracic vertebra. **(A)** AP and lateral roentgenogram: wedge-shaped vertebral body with irregular contour of the posterior part is visible. **(B)** The axial CT scan shows narrowing of the spinal canal *(arrow)*. **(C)** On secondary reformation the degree of narrowing is clearly visualized. A fracture line within the vertebral body is visible *(arrow)*. T1-weighted image **(D)** and T2-weighted sagittal image **(E)** 2 days following the trauma (low and high signal intensities in the T1- and T2-weighted images, respectively). The posterior longitudinal ligament is not disrupted.

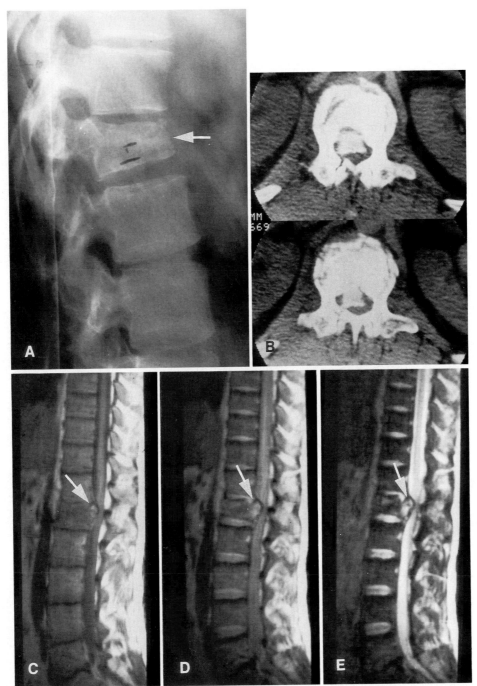

Figure 40.9. Burst fracture of the first lumbar vertebra. **(A)** The lateral roentgenogram: wedge-shaped compression of the vertebral body is shown *(arrow)*. **(B)** The axial CT 2 days after trauma clearly shows displacement of a bony fragment into the spinal canal. Multiple circumferential fracture lines are present. **(C–E)** MR image 7 days after trauma: encroachment of the spinal canal and compression of the spinal cord. High signal intensity *(arrow)* on the T1-weighted **(C)**, spin-density image **(D)**, and more on the T2-weighted image **(E)** within the dorsal part of the vertebral body is due to subacute hematoma.

Figure 40.10. Area of increased signal intensity posterior to C5 *(arrow)* on T1-weighted sagittal image shows the extent of a subligamentous hematoma.

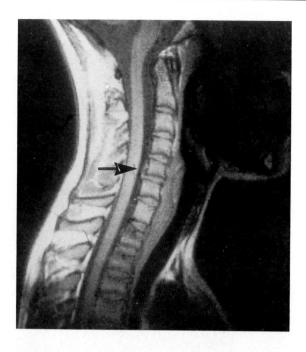

Figure 40.11. Unilateral fracture of the anterior *(arrow)* and posterior *(arrowhead)* arch of C1. Two adjacent slices: note the small fragment torn off by the transverse ligament *(open arrow)*, as well as the lateral displacement of the left portion of C1.

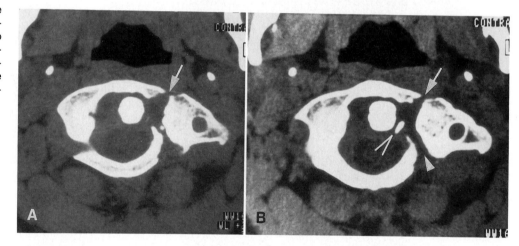

Figure 40.12. Symmetrical and bilateral fracture of the arch of a cervical vertebral body.

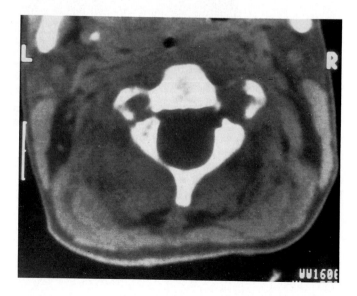

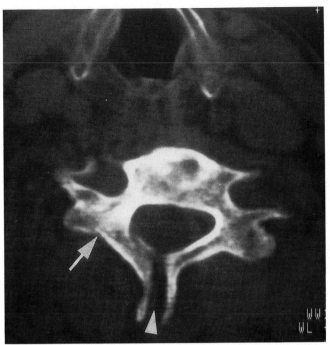

Figure 40.13. Unilateral fracture of arch *(arrow)* and the spinous process *(arrowhead)* of a middle cervical vertebra.

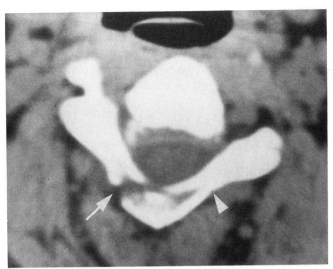

Figure 40.15. Fracture of right lamina *(arrow)* and a possible fracture of the left lamina *(arrowhead)* are visible at C4 level.

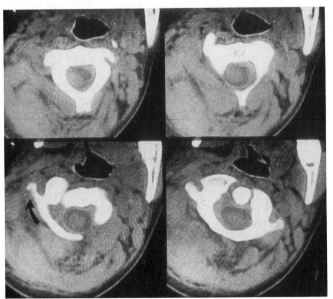

Figure 40.14. Rotational atlantoaxial dislocation *(arrow)*.

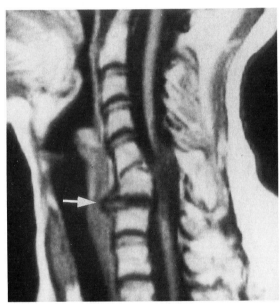

Figure 40.16. Posttraumatic anterior herniation of the C6–C7 cervical disk *(arrow)*: compression fracture of the sixth cervical vertebra with posterior dislocation and encroachment on the spinal canal.

fractures of the spinous processes C6 and C7, and can be found, for instance, in the victims of car accidents. This type of fracture is stable and does not lead to neurologic deficits, because the posterior ligament system remains intact.

Clay shoveler's fracture is a stress fracture of the same spinous processes observed in former times in Australian clay workers as well as in German autobahn-building workers.

Isolated fractures of the transverse processes of the vertebral bodies are rare. These bony injuries are often ac-

companied by paravertebral hematoma. In the thoracolumbar region, a fracture of a transverse process and should focus attention on the possibility of blunt renal trauma.

In most cases fractures of the spinous or transverse processes are sufficiently imaged by conventional radiographs. Computed tomography is the method of choice for additional documentation of hematomas and accompanying lesions of visceral organs.

When renal trauma is suspected, the examination should be performed following intravenous injection of contrast agent. In addition, traumatic injuries accompanying frac-

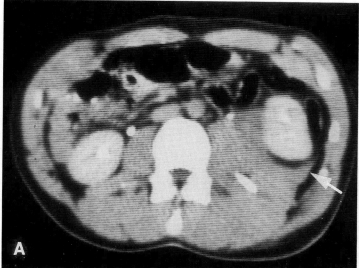

Figure 40.17. CT scan following intravenous contrast enhancement. **(A)** Rib fracture with parirenal hematoma *(arrow)*; **(B)** fracture of the transverse process and avulsion of the left position of the vertebral body and arch *(arrow)* with perirenal hematoma *(arrowheads)*.

tures of transverse processes, such as rib fractures, hematothorax or pneumothorax, and hepatic and splenic lesions, are best detected by CT (Fig. 40.17).

Injuries of Ligaments and Facet Joints

The intervertebral facet joints are injured mostly by hyperflexion, hyperextension, or rotation. In strong hyperflexion of the vertebral column the vertebral body rotates around its anterior rim, leading to a rupture of the dorsal ligament complex (joint capsule, supraspinal, interspinal, and posterior ligament, and ligamentum flavum) and occasionally even the joint process.

In cases of an additional rupture of the ventral and discal ligament complex an anterior subluxation of the vertebral body may occur. This is one of the most instable injuries of the vertebral column.

With even heavier forces acting on the vertebral column, distortion and bilateral locking of the facets may be found. The lower facets of the cranial vertebral body are pushed anteriorly across the upper facets of the lower vertebral body.

In addition, the anterior portions of the vertebral bodies may be compressed, forming the so-called teardrop fractures (teardrop-like form of fragments). This type of fracture results in compression of the cord with a deleterious hemorrhage into the gray substance and subsequent paralysis, even if the compression is only of short duration.

Unilateral joint luxation represents a fracture with fewer neurologic complications and without loss of stability. It occurs after a combination of hyperflexion and rotation. The joint ligaments are torn, but the disk and the posterior ligament remain intact.

Adequate imaging of bony lesions like subluxation or locked joint facets can be achieved by CT with sagittal or oblique secondary reformations. In transient subluxation with spontaneous replacement of the joint-forming bones

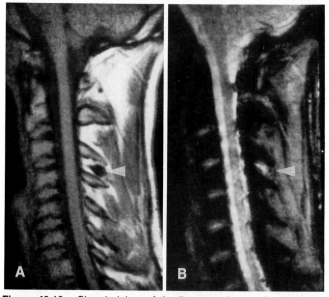

Figure 40.18. Chronic injury of the ligaments in a professional athlete. **(A)** Area of decreased signal intensity is evident on T1-weighted image between the spinous processes of C3 and C4. **(B)** Gradient-echo image shows high signal intensity in the same area, indicative of fluid collection in the interspinal space.

there is no means of imaging bony lesions or ligament avulsion.

When bony lesions are excluded but clinical symptoms suggest ligamentous lesions, additional functional images in flexion and extension position or even in lateral inclination must be performed. Proof of segmental instability or blocking indicates ligamental lesions.

Magnetic resonance imaging enables the visualization of ligament lesion (Fig. 40.18). Within the first weeks after trauma T2-weighted images demonstrate high signal intensity discontinuities of the normally low signal intensity ligaments and joint capsules (24, 26).

Paraspinal Lesions

Paravertebral hematomas are readily identified by CT and MR imaging. Acute and diffuse hemorrhage shows an increase in volume of the involved muscles on CT images. A few hours later fibrin retraction and diffusion of serum into the adjacent tissue lead to an increase in density, which may reach 90 HU.

During the following days and weeks resorption of the blood causes a progressive decrease in density down to 20 HU. Sometimes a stratification phenomenon may be observed, consisting of hyperdense areas in dependent parts due to sedimentation of erythrocytes and hypodense areas in the nondependent parts (27).

Magnetic resonance imaging shows the typical time-dependent phases of hemorrhage as described earlier (cf. "Fractures of the Vertebral Bodies.")

Cord Injuries

Spinal Canal

Intraspinal Bone Fragments

Using CT, intraspinal bone fragments are demonstrated to good advantage. Even small fragments of a size of only 1 or 2 mm can be clearly depicted because of the high density of bone. Furthermore, the extent of the narrowing of the spinal canal can be accurately assessed. If it is difficult to differentiate intra- or extradural localization of a fragment on plain CT, intrathecal contrast enhancement can be used.

For localization of small bony fragments MR imaging is not suitable because of the low signal intensity of compact bone.

Epidural Hematoma

Acute epidural hemorrhage is visualized as hyperdense areas of 50 to 90 HU on CT. It forms either a homogeneous or a speckled ring (28). However, it can also be localized on one side of the spinal canal, usually the side of the fracture, pushing the less hyperdense cord and the subarachnoid space to the opposite side (29).

Epidural hematomas in the lower cervical and in the thoracic spine may be missed in CT if no intrathecal contrast enhancement is employed.

The longitudinal extension of an epidural hematoma is better demonstrated on sagittal MR images than with CT. An additional advantage of MR is its high contrast between hematoma and surrounding structures and its ability to depict the dura. Both facilitate a clear-cut definition of the relevant anatomical structures.

Acute hemorrhage shows low signal intensity on T1- and T2-weighted images in the first hours, which may increase after 24 hr (25).

Tear of the Dura

In traumatic dural tears cerebrospinal fluid effusion will occur. The formation of a cavity filled with cerebrospinal fluid outside the spinal canal is termed *pseudomeningocele*. Myelo-CT is the most accurate modality for the detection of tears of the dura (30).

Using MR imaging and plain CT, larger amounts of cerebrospinal fluid are visualized, evidenced by an extradural hypointense area on CT and an fluid-isointense area on MR imaging, if there is no blood in the area (31). Myelography-like gradient-echo sequences (rapid acquisition with relaxation enhancement, RARE) (32), however, allow a noninvasive and anatomically clear definition of pseudomeningoceles.

Nerve Root Avulsion

Traction injuries of nerve roots are most frequent in the cervical and thoracic area and are rare in the lumbosacral area. They result primarily from motorbike accidents. Lumbosacral nerve root avulsion is sometimes found in severe fractures of the pelvis. Maximal tension leads to a tear of the arachnoid and the dura around the nerve root. The root itself can be partly or completely torn. Torn nerve ends tend to retract, forming an empty space that includes the intervertebral foramen and neighboring areas. These spaces are visualized using myelo-CT (29) or MR imaging (33). Pseudomeningoceles, missing nerve roots, and detection of a cerebrospinal fluid-filled diverticulum are the criteria for the diagnosis of nerve root avulsion (34).

Spinal Cord Concussion

Similar to cerebral lesions posttraumatic lesions of the spinal cord are defined as concussion, contusion, or compression.

Concussion of the spinal cord is a transient disturbance of spinal cord function. It resolves within 48 hr and does not show any pathological or anatomical alterations (35).

Spinal Cord Contusion

Spinal cord contusion includes lesions of the cord without tear or compression. The intensity and duration of the offending force determine the severity of the cord lesion. Therefore findings can vary from small, pointlike hemorrhage and edema to cord distraction with major necrosis (36).

Three stages of the lesions can be distinguished: (a) an early phase of hemorrhage and necrosis, (b) an intermediate stage of resorption and organization, and (c) a final stage of scar formation.

In the early stage microcirculation is essential for the outcome of the injury (37, 38). In the early phase of a hemorrhagic necrosis a pathologically increased permeability of the vessels in the posterior parts of the central gray matter is found. Microhemorrhages also tend to spread centrifugally into the gray matter (39). Three to 4 hr after the trauma edema is evident, also centrifugally spreading, which affects the white matter as well (40). In the following days central necrosis of the cord with softening of the tissue or hemorrhagic necrosis can occur. Hemorrhage can also spread cranially or caudally in the spinal cord, a condition called hematomyelia. Hemorrhage shows its maximum extent within 24 to 48 hr. The following edema has its peak 72 to 144 hr after the traumatic injury. In the days after the end of the necrotizing process resorption and liquefaction is found. This is followed by proliferation of glial cells.

Late sequelae associated with cord contusion include narrowing of the spinal canal with compression of the cord, arachnoid adhesions, syringomyelia, and cord atrophy.

Using CT, edema and increased volume of the cord are difficult to verify because of the small decrease in density. Intramedullary hemorrhage can be assumed if an increase in density of 60 to 90 HU is observed.

Sagittal MR images allow for a precise assessment of the longitudinal extension of pathological alterations within the cord. Intramedullary edema is evidenced by an increase of signal intensity on T2- and T2*-weighted images (Fig. 40.19). Hemorrhagic lesions follow the characteristics of time-dependent changes, which are essentially uniform in various anatomic locations (41). Liquefaction of the hematoma will account for low signal intensity on T1-weighted images and high signal intensity on T2-weighted images.

Spinal Cord Compression

Lasting pressure on the cord can be caused by dislocated fractures or traumatic disk herniation. Spinal compression is aggravated by spondylophytes or hematoma.

The underlying pathological alterations of cord compression can be visualized using CT and MR imaging. Sagittal MR imaging planes allow superior anatomical orientation. However, assessment of the functional deficit is not possible based on the static image of compression because the extent and duration of the force is decisive for the subsequent intramedullary changes. Computed tomography and MR imaging criteria are identical to those of spinal contusion.

Cord Transsection

Complete tears or transsections of the cord are rare events resulting from very strong shear forces, severest flexion-rotation trauma, or direct force such as shooting or stabbing injuries. Following intrathecal contrast enhancement, contrast material is found between both ends of the torn cord, enabling a reliable diagnosis to be made with CT (42). Multiple bony fragments may pose problems for the accurate assessment of the degree of cord injuries (Fig. 40.20).

With MR imaging transsection of the cord is directly depicted as a complete discontinuity of the cord itself (23, 43).

Late Outcome

Late sequelae of trauma to the spine can include *(a)* narrowing of the spinal canal with compression of the cord, *(b)* arachnoid adhesions, *(c)* syringomyelia, and *(d)* cord atrophy.

Narrowing of the Spinal Canal with Compression of the Cord

Stenosis of the spinal canal and compression of the cord can easily be visualized on CT as well as on MR imaging, the latter having the advantage of showing both the spinal canal and the cord simultaneously.

Arachnoid Adhesions

Arachnoid adhesions result from redundant growth and glueing of the soft meninges, which also stick together, mostly as a result of trauma. The causes include trauma of the cord with and without fractures of the vertebral bodies, iatrogenic maneuvers such as punctures, myelography, spinal anesthesia, and operations, or chronic pressure.

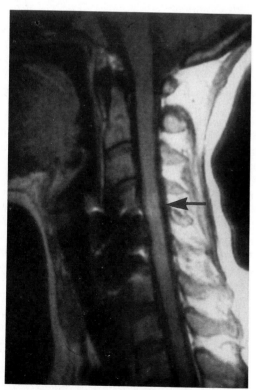

Figure 40.19. High signal intensity on T1-weighted image *(arrow)* following surgery indicates edema of the spinal cord.

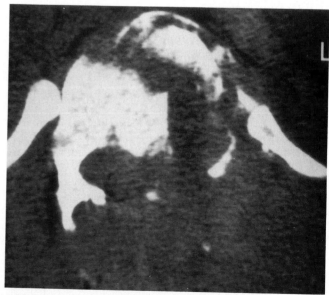

Figure 40.20. Dislocated rotational burst fracture with cord transsection at level T12.

Symptoms of arachnoid adhesions are radicular pain, paresthesia or loss of sensible or tactile features of the skin. In the end paralysis of the extremities occurs. Resection of scar tissue or releasing the cord or nerve roots may reduce symptoms in 50% of the cases (44). Morphological alterations characteristic of arachnoid adhesions are *(a)* adhesions and lumps of the nerve roots, *(b)* contact of nerve roots and meninges, and *(c)* narrowing of the subarachnoid space by soft tissue.

Myelography (45), myelo-CT (46), and MR imaging (47) are well suited for the assessment of these findings. On T1- and T2-weighted MR images the nerve roots can be clearly differentiated from the cerebrospinal fluid, so that adhesions of adjoining nerve roots or sticking together of the meninges are readily visualized. The latter is also referred to as the "empty sac" sign, which is best demonstrated on T2-weighted images. Inflammatory soft-tissue lesions are better demonstrated on T1-weighted images, because soft tissue, adhesions, and cerebrospinal fluid exhibit high signal intensity on T2-weighted images and cannot be reliably differentiated from each other. Both MR imaging and myelo-CT are adequate modalities for imaging of arachnoid adhesions (47).

Arachnoiditis of the thoracic spine can be demonstrated on T1-weighted spin-echo sequences, which reveal a thickening of the leptomeninges with or without adhesions of the cord. An excentric position of the cord within the spinal canal is indicative of cord adhesions (48).

Syringomyelia

The formation of a syrinx, which consists of one or several fluid-filled spaces within the cord, is a late complication of spinal trauma and is found in about 2% of cases. Its etiology is not yet fully understood.

Contusion of the cord with or without hemorrhage and subsequent resorption of the blood or necrotic material may lead to the formation of areas filled with glial cells. Finally these adhesions cause abnormal tension within the already injured cord. The varying pressure of the cerebrospinal fluid, for instance in Valsava's maneuver, coughing, sneezing, or extension movements, may also contribute to syringomyelia (49).

The cystic lesions are usually located cranially to the traumatic lesion. However, they may also be found caudally in rare instances (49, 50). In contrast to dilatations of the central canal, termed *hydromyelia*, syrinx formation is located in an excentric position.

Using nonenhanced CT, hypodense areas may be visible within the cord. Following intrathecal injection of contrast material, diffusion of the contrast material within 2 to 24 hr into the cysts can be found, allowing a superior delineation of the cysts (49, 51, 52). However, delayed CT is an invasive procedure and its specifity is relatively low (51, 53, 54).

For the diagnosis of syringomyelia, MR imaging is the diagnostic procedure of choice (55–57). It is noninvasive and can easily distinguish between myelomalacia and syringomyelia.

On T1-weighted images posttraumatic cysts, syringomyelia, and myelomalacia exhibit low signal intensity. Spin-density images of posttraumatic cysts, however, show relatively low signal isointense with cerebrospinal fluid, whereas malacic tissue is isointense or hyperintense compared to cord tissue. On T2-weighted images both entities are hyperintense and the intensity within the syrinx is identical to that of the cerebrospinal fluid. Low signal intensity on T1-weighted images can also be found in reactive gliosis, which shows, however, high signal intensity on spin-density images (58) (Figs. 40.21 and 40.22).

Cord Atrophy

Posttraumatic cord atrophy is characterized by a localized or more often a diffuse decrease of the cord diameter (Fig.

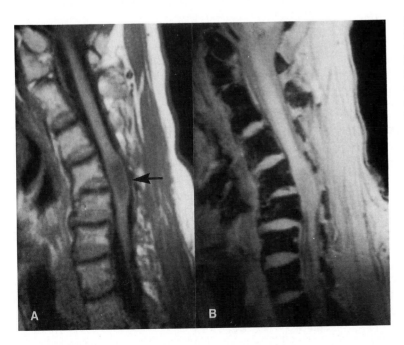

Figure 40.21. Posttraumatic syringomyelia: compression fracture of the fifth cervical vertebra and angulation of the cervical spine. **(A)** Sagittal T1-weighted spin-echo image: central low-intensity area *(arrow)* within the cord is due to syringomyelia. Effacement of the subarachnoid space is due to volume increase of the cord. **(B)** Sagittal gradient-echo image (flip angle 50°): the fluid collection within the cord is faintly visible due to similar signal intensity of cord and syringomyelia.

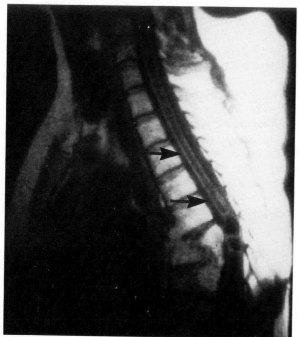

Figure 40.22. Syringomyelia with rounded cavities *(arrowhead)* at the level of the compression fracture is visible, as well as a long syringomyelia cranial to the fracture *(arrows)*.

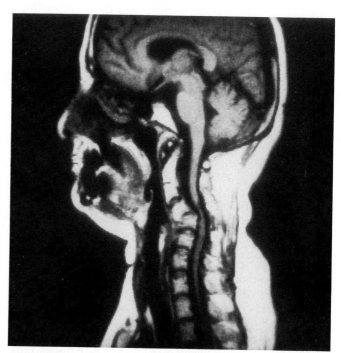

Figure 40.23. Fracture of the fifth cervical vertebra is followed by severe cord atrophy.

40.23). It can be reliably diagnosed by CT and MR imaging (59).

Computed Tomography vs. Magnetic Resonance Imaging

When comparing the merits of CT and MR imaging, it must be emphasized that both modalities have different drawbacks and benefits. In the selection of a specific imaging technique, the acute availability has major importance.

Using CT, bony structures and calcified elements are visualized with unsurpassed sensitivity and detail. It also allows for diagnosis of injuries to the paravertebral soft tissues and organs, e.g., renal or hepatic trauma. In patients with severe trauma, various organs and anatomical areas may be involved. As of yet, CT is the method of choice for the detection of head and abdominal injuries. Therefore CT of the spine can be easily performed in the work-up of acutely traumatized patients together with CT of the head, abdomen, and thorax.

However, plain CT may be insufficient for a variety of pathological conditions. For the assessment of the subarachnoid space, especially in the lower cervical and thoracic areas, intrathecal contrast enhancement may be requested. The restriction to axial imaging planes can also be a major limitation. Therefore longitudinal secondary reformations may be mandatory.

Multiplanar imaging capabilities and high tissue contrast make MR imaging a very valuable tool for the evaluation of the traumatized spine. Large sections of the spine are simultaneously displayed. While the information provided by CT may be reduced by artifacts in the cervicothoracic junction, all parts of the spine are depicted with

similar image quality using MR imaging. In order to achieve adequate imaging quality, a sophisticated technique must be applied, e.g., use of surface coils, saturation pulses, flow compensation, and rotation of phase-encoding gradient. Recently T1- and T2*-weighted gradient-echo techniques have been successfully employed.

Magnetic resonance imaging offers a broad object range: bone marrow, cortical bone, cerebrospinal fluid, spinal cord, and paravertebral soft tissues are displayed with high contrast, when appropriate pulse sequences are used. On T2- and T2*-weighted sequences spinal ligaments can also be evaluated.

Cardiac pacemakers and ferromagnetic foreign bodies are contraindications for MR imaging. In acutely traumatized patients with such inclusions, equipment for monitoring and intensive care must be made without ferromagnetic components.

References

1. Kraus JF, Franti CE, Riggins RS, Richards D, Borhani NO. Incidence of traumatic spinal cord lesions. J Chronic Dis 1975;28:471–492.
2. Bracken MB, Freeman DH, Hellenbrand K. Incidence of acute traumatic hospitalized spinal cord injury in the United States, 1970–1977. Am J Epidemiol 1981;113:615–622.
3. Anderson DW, Mclauren RL, eds. The national head and spinal cord injury survey. J Neurosurg 1980;53(suppl):1–43.
4. Young JS, Burno PE, Bowen AM, et al. Spinal cord injury statistics systems. Phoenix: Good Samaritan Medical Center, 1982.
5. Green BA, Callahan RA, Klose KJ, De La Torre J. Acute spinal cord injury: current concepts. Clin Orthoped 1981;154:125.
6. Leyendecker K, Schirmer M. Traumatische Rückenmarkschädigungen. In: Schirmer M, ed. Querschnittslähmungen. Berlin: Springer-Verlag, 1985.
7. Wales LR, Knopp RK, Morishima MS. Recommendations for

evaluation of the acutely injured cervical spine. A clinical radiologic algorithm. Ann Emerg Med 1980;9:422–428.

8. Stimac GK, Burch D, Livingston RR, Anderson P, Dacey RG. A device for maintaining cervical spine stabilization and traction during CT scanning. AJR 1987;149:345–346.

9. Sartorius DJ, Resnick D. Computed tomography of the spine: an update and review. CRC Crit Rev Diagn Imag 1987;27:271–296.

10. Kelly WM, Payle PG, Pearson JA, et al. Ferromagnetism of intraocular foreign body causing unilateral blindness after MR study. Am J Neuroradiology 1986;7:243–245.

11. Teitelbaum GP, Yee CA, Van Horn DD, Kim HS, Colletti PM. Metallic ballistic fragments: MR imaging safety and artifacts. Radiology 1990;175:855–859.

12. Daffner RH, Deeb ZL, Goldberg AL, Kandabarow A, Rothfus WE. The radiologic assessment of posttraumatic vertebral stability. Skel Radiol 1990;19:103–108.

13. Butt WP. Letters to the editor. Skel Radiol 1990;19:446–447.

14. Holdsworth F. Fractures, dislocations and fracture dislocations of the spine. J Bone Joint Surg 1970;52A:1534–1551.

15. Denis F. The three column spine and its significance in classification of acute thoracolumbar spinal injuries. Spine 1983;8:817–831.

16. Wilcox NE, Stauffer ES, Nickel VL. A statistical analysis of 423 consecutive patients admitted to the spinal cord injury center Rancho Los Amigos Hospital, 1 January 1964 through 31 December 1967. Paraplegia 1970;8:27–35.

17. Reid DC, Saboe L. Spine fractures in winter sports. Sports Med 1989;7:393–399.

18. Anderson LD, D'Alonzo RT. Fractures of the odontoid process of the axis. J Bone Joint Surg 1974;56A:1663–1674.

19. Chance GQ. Note on a type of flexion fracture of the spine. Br J Radiol 1948;21:452–453.

20. Smith WS, Kaufer H. Patterns and mechanisms of lumbar injuries associated with lap seat belts. J Bone Surgery 1969;51A:239–243.

21. Mirvis SE, Young JWR, Lim C, Greemberg J. Hangman's fracture: radiologic assessment in 27 cases. Radiology 1987;163:713–717.

22. Imhof H, Hajek P, Kumpan W, Schratter M, Wagner M. CT in the diagnosis of acute injuries to the spine. Radiology 1986;26:242–247.

23. Kalfas I, Wilberger J, Goldberg A, Prostko ER. Magnetic resonance imaging in acute spinal cord trauma. Neurosurgery 1988;23:295–299.

24. Mirvis SE, Geisler FH, Jelinek JJ, Joslyn JN, Gellad F. Acute cervical trauma: evaluation with 1.5 MR imaging. Radiology 1988;166:807–816.

25. Tarr RW, Drolshagen LF, Kerner TC, Allen JH, Partain CL, James EA. MR imaging of recent spinal trauma. J Comput Assist Tomogr 1987;11:412–417.

26. McArdle CB, Crofford MJ, Mirfakhraee M, Amparo EG, Calhoun JS. Surface coil MR of spinal trauma: preliminary experience. Am J Neuroradiology 1986;7:885–893.

27. Maas R, Gürtler KF. Trauma-induced space occupying lesions. In: Heller M, Jend HH, Genant HK, eds. Computed tomography of trauma. New York: Georg Thieme Verlag, 1986.

28. Coin CG, Pennink M, Ahmad WD, Keranen VJ. Diving type injury of the cervical spine: contribution of computed tomography to management. J Comput Assist Tomogr 1979;3:362–372.

29. Post JD, Seminer DS, Quencer TM. CT-diagnosis of spinal epidural hematoma. Am J Neuroradiology 1982;3:190–192.

30. Morris RE, Hasso AN, Thompson JR, Hinshaw DB, Vu LH. Traumatic dural tears: CT diagnosis using metrizamide. Radiology 1984;152:443–446.

31. Masaryk TJ. Spine Trauma. In: Modic MT, Masary TJ, Ross JS, eds. Magnetic resonance of the spine. Chicago: Year Book Medical Publishers, 1989:214–239.

32. Hennig J, Friedburg G, Ströbel B. Rapid nontomographic approach to MR-myelography without contrast-agents. J Comput Assist Tomogr 1986;10:375–378.

33. Freedy RM, Miller KD, Eick JJ, Granke DS. Traumatic lumbosacral nerve root avulsion: evaluation by MR imaging. J Comput Assist Tomogr 1989;13:1052–1057.

34. Blum U, Friedburg HG, Ott D, Wimmer B. Traction injuries of the brachial plexus: radiographic diagnosis by enhance computed

tomography (CT) and magnetic resonance imaging (MRI). Röntgen Fortschr 1989;151:702–705.

35. Del Bigio MR, Johnson GE. Clinical presentation of spinal cord concussion. Spine 1989;14:37–40.

36. Jellinger K. Neuropathology of cord injuries. In: Vinken PJ, Bruyn GW, eds. Handbook of clinical neurology, Vol. 25, Part I. Amsterdam: Elsevier/North-Holland, 1976:43–121.

37. Dohrmann GJ, Wagner FC, Bucy PC. The microvasculature in transitory traumatic paraplegia. An electron microscopic study in the monkey. J Neurosurg 1971;35:263–271.

38. Ducker TB. Experimental injury of the spinal cord. In: Vinken PJ, Bruyn GW, eds. Handbook of clinical neurology, Vol. 25, Part I. Amsterdam: North-Holland, 1976:9–26.

39. Allen WE, D'Angelo CM, Kier EL. Correlation of microangiography and electrophysiologic changes in experimental spinal cord trauma. Radiology 1974;111:107–115.

40. Yashon D, Bingham WD, Faddoul EM, et al. Edema of the spinal cord following experimental impact trauma. J Neurosurg 1973; 38:693–697.

41. Hackney DB, Asato R, Joseph PM, et al. Hemorrhage and edema in acute spinal cord compression: demonstration by MR imaging. Radiology 1986;161:387–390.

42. Cooper PR, Cohen W. Evaluation of cervical spinal cord injuries with metrizamide myelography-CT scanning. J Neurosurg 1984; 61:281–289.

43. Fasano FJ Jr, Stauffer ES. Traumatic division of the spinal cord demonstrated by magnetic resonance imaging. Report of two cases. Clin Orthoped 1988;233:168–170.

44. Shikata J, Yamamuro T, Iida H, Sugimoto M. Surgical treatment for symptomatic spinal adhesive arachnoiditis. Spine 1989;14:870–875.

45. Jorgensen J, Hansen PH, Steenskov V, Ovesen N. A clinical and radiological study of chronic lower spinal arachnoiditis. Neuroradiology 1975;9:139–144.

46. Simmons JD, Newton TH. Arachnoiditis. In: Newton TH, Potts DG, eds. Computed tomography of the spine and spinal cord. San Anselmo, Calif.: Clavadel Press, 1983:223–229.

47. Ross JS, Masaryk TJ, Modic MT, et al. MR imaging of lumbar arachnoiditis. AJR 1987;149:1025–1032.

48. Ross JS. Inflammatory disease. In: Modic MT, Masaryk TJ, Ross JS, eds. Magnetic resonance imaging of the spine. Chicago: Year Book Medical Publishers, 1989:167–182.

49. Quencer RM, Green BA, Eismont FJ. Posttraumatic spinal cord cysts: clinical features and characterization with metrizamide computed tomography. Radiology 1983;146:415–423.

50. Tator CH, Meguro K, Rowed DW. Favorable results with syringosubarachnoid shunts for treatment of syringomyelia. J Neurosurg 1982;56:517–523.

51. Seibert CE, Dreisbach JN, Swanson WB, Edgar RE, Williams P, Hahn H. Progressive posttraumatic cystic myelopathy: neuroradiologic evaluation. AJR 1981;136:1161–1165.

52. Stevens JM, Olney JS, Kendall BE. Posttraumatic cystic and noncystic myelopathy. Neuroradiology 1985;27:48–56.

53. Gebarski SS, Maynard FW, Gabrielsen TO, Knake JE, Latack JT, Hoff JT. Posttraumatic progressive myelopathy. Radiology 1985; 157:379–385.

54. Quencer RM, Morse BMM, Green BA, Eismont FJ, Brost P. Intraoperative spinal sonography: an adjunct to metrizamide CT in assessment and surgical decompression of spinal cord cysts. AJR 1984;142:593–601.

55. Enzmann DR, O'Donohue J, Rubin JB, Shuer L, Cogen P, Silverberg G. CSF pulsations within nonneoplastic spinal cord cysts. AJR 1987;149:149–157.

56. Shermann JL, Barkovich AJ, Citrin CM. The MR appearance of syringomyelia: new observations. AJR 1987;148:381–391.

57. Gabriel KR, Crawford AH. Identification of acute posttraumatic spinal cord cyst by magnetic resonance imaging: a case report and review of the literature. J Pediatr Orthoped 1988;8:710–714.

58. Fox JL, Wener WL, Drennan DC, Manz HJ, Won DJ, Al-Mefty O. Central spinal cord injury: magnetic resonance imaging confirmation and operative considerations. Neurosurgery 1988;22:340–347.

59. Mawad ME, Hilal SK, Fetell MR, Silver AJ, Ganti SR, Sane P. Patterns of spinal cord atrophy by metrizamide CT. Am J Neuroradiology 1983;4:611–613.

CHAPTER 41

Magnetic Resonance Imaging and Computed Tomography of Infectious Spondylitis

Hassan S. Sharif, Mohammed Y. Aabed, and Maurice C. Haddad

Introduction

Infectious spondylitis is an infection that involves one or more of the different components of the spine. It can be caused by pyogenic, granulomatous, or echinococcal organisms. Vertebral bodies are most frequently affected, but the posterior osseous elements, disks, epidural space, and paraspinal soft tissues can also be either primarily or secondarily involved (1). Mechanisms of infection include blood borne from a distant body source (urinary tract and lungs are the most frequent), spread from a contiguous site, and direct implantation (2). Hematogenous spread results in lodgement of organisms in the vertebral marrow. The venous (Batson's plexus) and arterial systems have both been implicated in carrying the pathogens; but the latter is recognized as the more important method of transmission (1, 3). Involvement of the disk by blood-borne pathogens occurs only in the pediatric age group (below 20 years) because of persistent diskal blood supply (4, 5). In the older patient disk infection is invariably the result of involvement from neighboring vertebrae or soft tissue (1, 3).

The type of causative organisms, incidence, age at onset, mode of presentation, and degree of severity of infectious spondylitis varies greatly from country to country and are closely related to the standards of public health services in the community and the socioeconomical and educational background of the inhabitants (6). In developing countries, for example, granulomatous spine infections are by far more common than those of pyogenic lesions (7–9).

Clinical signs and symptoms of infectious spondylitis are nonspecific. Back pain is a common early symptom frequently associated with localized spinal tenderness and stiffness with or without neurological deficit (1, 2). A high erythrocyte sedimentation rate is invariably present, but the leukocyte count may be normal, elevated, or low, depending on the type of pathogen (10).

Imaging has an important role to play in the overall management of infectious spondylitis, and an ideal modality should be highly sensitive in order to detect early disease and should provide information that will help in

1. Identifying the nature of pathology (infection versus others)
2. Determining the location and extent of involvement
3. Suggesting type of pathogen
4. Guiding biopsy or drainage procedure
5. Deciding on mode of therapy
6. Planning the surgical approach
7. Assessing response to therapy (11).

The value of plain radiography in the overall assessment of infectious spondylitis is limited. Classic features include disk collapse in association with vertebral destruction or sclerosis, a paraspinal mass, and/or angulation deformity (1). Because of the poor contrast and spatial resolution the true extent is difficult to identify and early lesions may not be detected (7).

Bone scintigraphy can detect the presence of vertebral osteomyelitis, but it fails to identify paraspinal and intraspinal soft-tissue involvement (7). Gallium-67 and indium-111 have been advocated as being superior to technetium-99 bone scintigraphy (methylene diphosphonate) for delineating bone and soft-tissue inflammations; however, both suffer from limitations in their sensitivity and specificity.

Computed tomography (CT) is a well-recognized method for evaluating many disease processes affecting the spine and particularly inflammation (12). Computed tomography can accurately assess the bone lesion, paraspinal soft tissue involvement and the degree of cord or nerve root compression (7, 12, 13). Furthermore percutaneous biopsy or drainage procedures are to date best performed under CT guidance (11). Computed tomography, however, has two main limitations: vertebral osteomyelitis can be detected only if it causes alterations in bone density or mor-

Abbreviations (see also glossary): BS, brucellar spondylitis; DTPA, diethylene-triaminepentaacetic acid; CSF, cerebrospinal fluid; TS, tuberculous spondylitis.

phology and the radiation dose incurred by the procedure can be high.

Computed Tomography Technique

Localization of the affected spinal region (cervical, thoracic, lumbar) can be obtained from the relevant clinical signs and symptoms, plain radiographs, or bone scintigraphy. More precise localization is possible using digital anteroposterior and lateral scout views. The following scanning parameters will provide optimum CT assessment of cases with infectious spondylitis.

1. Plain 3-mm high-resolution consecutive slices without gantry angulation through the lesion as localized on the scout views (field of view (FOV) 18 cm). This sequence is aimed at evaluating bony landmarks.
2. Cases with extensive intraspinal and paraspinal spread need further assessment after enhancement with contrast. Precise delineation of soft-tissue involvement, epidural spread, and dural inflammation are best evaluated on enhanced examinations. All cases with tuberculous infection and a few cases of other infectious spondylitis should be reassessed by 3- to 5-mm high-resolution consecutive slices without gantry angulation following a bolus injection of 150 cm^3 of contrast medium (300 mg I/ml).
3. Biopsy and/or drainage procedures can be performed at the same time as the initial examination or can be planned for a later date. The patient should be scanned in the prone position using a large field of view, preferably after intravenous contrast. Fine needle aspiration biopsy can be used initially for isolation of the organism (smear or culture) and a true cut needle gauge (14 to 18 gauge) should be subsequently used. The material obtained should be sent for histological examination and for culture. Trephine needle biopsy can be used to obtain bone specimen and material obtained should also be evaluated by microbiology and histopathology. Where percutaneous drainage of abscesses is required (particularly in tuberculosis), a thoracic catheter (trocar catheter, size 24CH) should be used in order to negotiate the pus. Complete drainage can be achieved in 1 or 2 days. With multiple loculations the position of the catheter can be changed to drain the different locules.
4. If CT follow-up scans are required during the course of therapy, 5- to 10-mm consecutive cuts after contrast enhancement would suffice.
5. All reformatted images should be evaluated on bone and soft-tissue settings.

Magnetic resonance (MR) imaging has become an established method for assessing musculoskeletal infections and in particular those affecting the spine (7, 11, 14, 15). Its advantages include extremely high contrast resolution, high sensitivity for detecting marrow infiltration, multiplanar capability, and absence of ionizing radiation. Invasive procedures, however, cannot be performed under MR imaging guidance. Furthermore an abnormal signal obtained from the lesion is a reflection of the presence of abnormal water content and/or hyperemia and therefore the findings are of limited specificity (11).

Magnetic Resonance Imaging Technique

Optimal MR imaging of a previously undiagnosed patient with suspected spinal infection should include the following sequences:

1. Multisection sagittal images of the affected region of the spine using spin-echo (SE) technique, with a short repetition time (TR) and short echo time (TE) (TR/TE 500/30 ms) and a slice thickness of 3 to 5 mm with four to six excitations.
2. Multisection sagittal images as in sequence 1, but with a long TR/TE (2000/50 to 150) and with two excitations.
3. Multisection sagittal images as in sequence 1, but after intravenous administration of gadolinium (Gd)-labeled diethylenetriaminepentaacetic acid (DTPA) in a dose of 0.2 mmol/kg body weight.
4. Multisection coronal images using the same pulse sequence as in sequence 1 and following the images in sequence 3.
5. Follow-up examinations can be performed using sequences 1 and 2 only.

Cardiac gating is necessary in the thoracic and cervical spine in order to reduce artifacts secondary to cardiac and cerebrospinal fluid (CSF) pulsation.

The following discussion is divided according to the type of offending organisms.

Magnetic Resonance Imaging and Computed Tomography Features of Pyogenic Spondylitis

There has been a major increase in the reported frequency of pyogenic spondylitis in the last decade (1). Elderly diabetic patients in their fifth to seventh decades (males slightly more than females) are most frequently affected, but immunocompromised individuals and drug abusers are also at risk (2). *Staphylococcus aureus* is the commonest organism identified. Other organisms that have been implicated include *Streptococcus viridans, Streptococcus pneumoniae, Escherichia coli, Salmonella, Pseudomonas,* and *Klebsiella* (1). The infection occurs anywhere in the spine but the lumbar region is the area of predilection followed by the thoracic and cervical spine (1, 2).

Two major criteria are essential to establish the diagnosis: first, the recognition of characteristic imaging features of spine infection; and second, the isolation of the offending organism from the blood or the site of infection. Histopathologic changes are nonspecific and show necrotic bone and diffuse inflammatory cellular infiltration (1). The great majority of cases of pyogenic spondylitis respond promptly and adequately to antibiotic therapy, surgical decompression being rarely needed (10).

On CT vertebral changes secondary to the hematogenous form of pyogenic spondylitis occur mainly in the

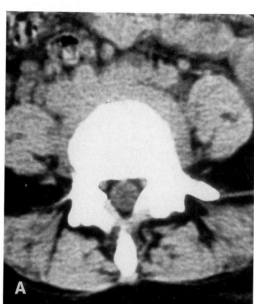

Figure 41.1. **(A)** A 27-year-old woman with lumbar pyogenic spondylitis caused by *Staphylococcus aureus.* Axial section at the superior end plate of L4 shows a circumferential paraspinal soft-tissue mass that has symmetric distribution around the vertebral body. **(B)** The same section as **(A)** on bone settings showing patchy areas of destruction in the end plates. **(C)** Midsagittal reformed image shows the irregularity and marginal sclerosis of the end plates, and the collapse of the intervening disk.

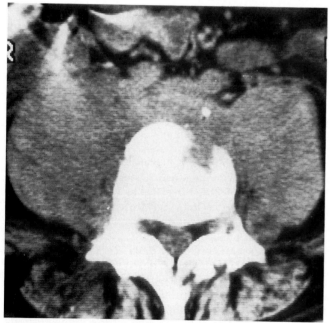

Figure 41.2. A 21-year-old woman with sickle cell anemia presenting with severe low back pain. Axial CT image at the superior end plate of L5 shows partial destruction of the vertebral body anteriorly associated with a large prevertebral soft-tissue mass inseparable from the psoas muscles. *Salmonella typhi* organisms were isolated by fine needle aspiration under CT guidance. (Courtesy of Dr. M. Abdelbagi, Dahran, Saudi Arabia.)

subchondral end plate and range from irregular small erosions to areas of frank destruction (Figs. 41.1 and 41.2) (12). These changes usually involve two consecutive end plates and are associated with disk space collapse (12, 16). Sclerotic margins are not commonly evident in acute infections and their presence indicates attempts at healing (Fig. 41.1) (1). The erosions are clearly seen on axial as well as coronal and sagittal reformed images and they represent areas of intravertebral disk herniation and/or granulation tissue, these however are not specific to pyogenic infections (Fig. 41.1) (7, 13). Vertebral collapse with resultant angulation deformity (gibbus or scoliosis) is extremely rare following pyogenic infections. Associated soft-tissue infection is variable and generally produces granulation tissue rather than abscess (Figs. 41.1 and 41.2) (16). The distribution of granulation tissue is fairly symmetrical around the vertebral body and it extends from one transverse process to the other circumferentially (Figs. 41.1 and 41.2) (16). Epidural extension can occur but usually to a lesser degree than in granulomatous infections (16).

Healing is manifested by new bone formation within affected vertebrae and at their periphery. Exuberant ossification of anterior and posterior spinal ligaments may ultimately lead to complete ankylosis (1).

The MR imaging features of hematogenous pyogenic spondylitis are characteristic (14). On T1-weighted images vertebral osteomyelitis is shown as areas of relatively low signal intensity reflecting the presence of an increase in

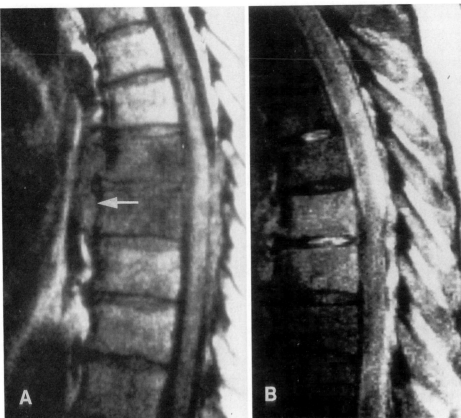

Figure 41.3. **(A)** An adult male with thoracic back pain. Midsagittal MR image (spin echo (SE), TR/TE 500/20 ms) shows diffuse low signal intensity of T7 and T8. Prevertebral granulation tissue can be seen displacing the aorta anteriorly *(white arrow)*. **(B)** Midsagittal section (SE, TR/TE 2000/100 ms) shows diffuse high signal from the bodies of T7 and T8; there is also high signal intensity from the intervening disk in the absence of nuclear cleft. *Staphylococcus aureus* was isolated from the patient's blood. (Courtesy of Dr. M. Modic, Cleveland, Ohio.)

extracellular fluid and/or hyperemia (Fig. 41.3) (14). Intraspinal extension manifests itself as obliteration of the linear echo void area usually found between the posterior vertebral bodies and the cord (the posterior vertebral cortex, posterior spinal ligament, dura and CSF). The abnormal signal is detected in one or more vertebrae: classically two adjacent ones with the intervening disk either remaining normal or reduced in size (Fig. 41.3) (14). On T2-weighted images moderate high signal intensity can be detected from the affected vertebrae and the epidural extension (Fig. 41.3). The proton density image is preferred to appreciate intraspinal spread since the signal from CSF would still be diminished on this pulse sequence. Compared with normal disks, involved disks exhibit marked increased signal intensity, frequently in the absence of a nuclear cleft (Fig. 41.3) (14).

Kramer et al. have recently used Gd-DTPA in pyogenic spondylitis and they demonstrated enhancement of affected marrow and disk (17). We have also used intravenous Gd-DTPA in a few of our cases, again demonstrating diffuse increased signal of the affected vertebrae (becoming isointense with the unaffected) and the disk but not of the soft tissues. The positive aspects of such studies is that they increase confidence of excluding tuberculosis (frequently associated with abscesses that exhibit rim enhancement) and they could also depict meningeal involvement, both of which are difficult to obtain on the long TR/TE sequences. T2-weighted images, however, produce better differentiation between normal and abnormal vertebrae because of the signal loss from the fatty marrow. Fat suppression techniques in conjunction with intravenous contrast media have recently been recommended since they improve visualization of enhanced structures while obliterating the signal from neighboring fatty tissue (Fig. 41.4) (18).

The great majority of cases of infectious spondylitis (also called diskitis) in the pediatric age group is caused by pyogenic organisms (4, 5). The disease is much more difficult to diagnosis clinically than the adult form and the causative organism may be extremely difficult to isolate from blood or the site of the lesion (1, 4, 5). Most of these children respond promptly and adequately to broad-spectrum antibiotics (5). The lumbar spine is the area of predilection, but other spine regions can be involved (4, 5). The CT findings are indistinguishable from those seen in the adult; two vertebral bodies and the intervening disk are usually affected but the paraspinal and intraspinal extension may be less in this age groups (Fig. 41.5). Magnetic resonance imaging features are virtually diagnostic and are also similar to those seen in the adult population (Fig. 41.5).

Vertebral osteomyelitis, compressive radiculopathy, or myelopathy can be caused by direct spread of infection from neighboring structures such as the psoas muscle, pleural cavity, oropharyngeal space, etc. (1, 19). Com-

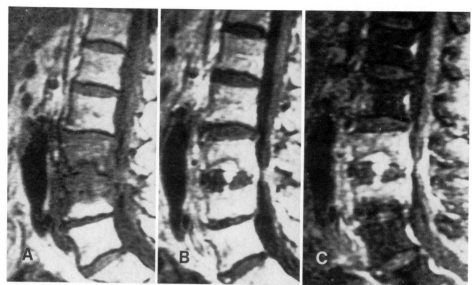

Figure 41.4. **(A)** A 65-year-old man with pyogenic spondylitis at L3–L4. Midsagittal MR image (SE, TR/TE 600/20 ms) shows decreased signal from the entire body of L3, the upper two-thirds of L4, and the prevertebral soft tissues. **(B)** Midsagittal MR image as in **(A)**, taken after intravenous Gd-DTPA administration shows diffuse enhancement of L3 and L4. There is also enhancement of the midportion of the L3–L4 disk. Note that the obtained signal from the affected vertebrae is isointense with the normal ones. **(C)** Midsagittal MR image (SE, TR/TE 600/20/2 ms), taken immediately after the sequence **(B)** and using chemical shift imaging for lipid suppression, demonstrates increased signal from the bodies of L3 and L4, the prevertebral soft tissues, anterior epidural collection, and posterior epidural collection. Note that the signal from the marrow of normal vertebrae and from epidural fat is suppressed. Although the image has more noise than in **(A)** and **(B)**, the vertebral body and disk involvement and the extent of soft-tissue involvement are shown to a better advantage. (Courtesy of Dr. J. H. Simon, Rochester, N.Y.)

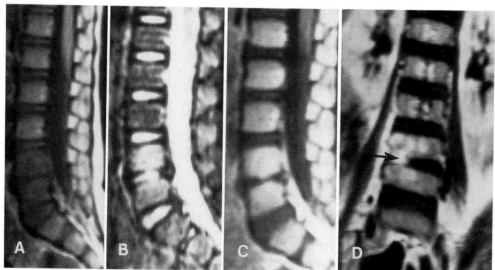

Figure 41.5. **(A)** A 3-year-old girl presenting with abdominal pain and difficulty on walking of a few days' duration. Midsagittal section (SE, TR/TE 500/30 ms) shows decreased signal from the bodies of L4 and L5, with minimal narrowing of the intervening disk. **(B)** Midsagittal image (SE, TR/TE 2000/100 ms) shows diffuse high signal intensity from L4 and L5 bodies and the intervening disk. **(C)** Midsagittal image as in **(A)**, but after intravenous Gd-DTPA administration shows diffuse enhancement of L4 and L5, reaching an isointense level with normal vertebrae. **(D)** Coronal image as in **(C)** shows patchy enhancement of L4 and to a lesser extent of L5. Note also the enhancement of the intervening disk, probably indicating hyperemia *(arrow)*.

puted tomography clearly depicts the true extent of the infection in bone and soft tissue (Fig. 41.6) (19). Postcontrast medium reassessment is essential as abscesses show the characteristic rim enhancement (18). Percutaneous bi-opsy from the soft-tissue component is easy to perform and enables the type of causative organism to be identi-fied early (Fig. 41.6) (19).

Magnetic resonance imaging is ideally suited to evaluate

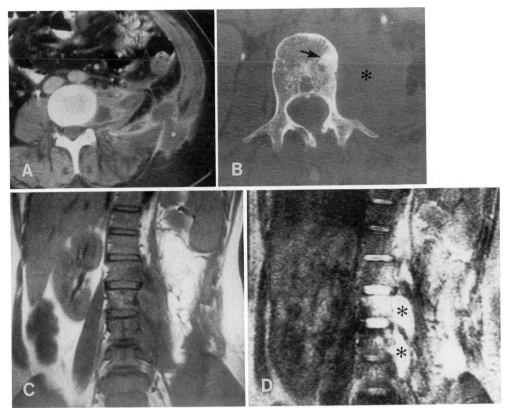

Figure 41.6. **(A)** A 47-year-old man presented with fever and left flank pain of 12 days' duration. Axial postcontrast CT at the level of the inferior end plate of L3 shows a multiloculated abscess involving the left psoas muscle extending into the neural foramen and involving lateral abdominal wall muscles, psoas, paraspinal muscles, and subcutaneous fat *(asterisk)*. *Staphylococcus aureus* was isolated from pus obtained by CT-guided percutaneous biopsy. **(B)** A high-resolution CT axial section at the level of the superior end plate of L3 shows bone sclerosis *(arrow)* on the left side of the body and pedicle in close proximity to the abscess *(asterisk)*. **(C)** Coronal MR image (SE, TR/TE 500/ 30 ms) shows irregularity of the left psoas muscle and a diminished signal from the bodies of L2, L3, and L4. Note the marked thickening of the lateral abdominal wall on the left side as compared to the right. **(D)** Coronal MR image (SE, TR/TE 2000/100 ms) at the same level as in **(C)** shows the marked increase in signal intensity from the infected psoas *(asterisk)* and from the infected vertebral bodies. Note that the disks are still intact. There is also increased signal intensity from the left lateral abdominal wall. (From Simon JH, Szumowski J. Chemical shift imaging with paramagnetic contrast material enhancement for improved lesion depiction. Radiology 1989;171:539–543. Reprinted with permission of Radiology.)

soft-tissue infections that have secondarily affected the spine and spinal canal (Fig. 41.6) (19). Vertebral osteomyelitis, epidural abscesses, and paraspinal spread can be easily assessed (19). Involvement of small bones such as ribs, spinous processes, and iliac bone can be missed because of partial volume averaging from the high signal of surrounding abscess collection (19).

Follow-up of cases with pyogenic spondylitis can easily and conveniently be performed by MR imaging (17). T1- and T2-weighted sagittal images are all that is needed early in therapy. The T2-weighted images may demonstrate a decrease in the high signal noted on the initial examination. The presence of increased signal intensity on T1-weighted images from previously infected vertebrae reflects replacement of cellular marrow by fat and indicates healing (11, 17). Consequently, further follow-up can therefore be done by means of T1-weighted sagittal images only.

Magnetic Resonance Imaging and Computed Tomography Features of Granulomatous Spondylitis

A granulomatous reaction is a nonspecific response of tissues to an antigenic stimulus. Such a reaction has been identified in a wide range of diseases including bacterial, viral, and fungal infections, neoplasia, and autoimmune and idiopathic disorders (1). Granulomatous organisms that have been implicated in spine infections include *Mycobacterium tuberculosis*, *Brucella* species, and fungal disease (11, 20).

Tuberculous Spondylitis

The disease is of worldwide distribution, but the incidence in developing countries is considerably higher than in the United States and Western Europe (7, 9). *Mycobac-*

terium tuberculosis gains access to the osseous vertebral end plate by the hematogenous route with resultant bone destruction and abscess formation (1). The contiguous disk and neighboring vertebrae become secondarily affected by direct spread; distant disks and vertebrae can also be involved by spread via the subligamentous route (1). The latter method is probably responsible for skip lesions, which are common in tuberculous spondylitis (1, 7, 9). Any part of the spine can be involved but the thoracolumbar region is the area of predilection (7).

Age and sex distribution are quite variable in the reported series. Middle-aged adults (with no sex predominance) seem to be most frequently affected, especially those with predisposing conditions such as debilitation, alcoholism, drug addiction, and immunosuppression (1, 6, 7, 9).

Clinically tuberculous spondylitis can be extremely difficult to differentiate from other forms of spinal infection and from other causes of spinal disease (1, 6, 7, 9). Back pain, stiffness, and localized tenderness are the commonest clinical findings, with compressive radiculopathy or myelopathy being more frequent than in pyogenic infections. Routine laboratory investigations do not contribute significantly toward the diagnosis of tuberculous spondylitis. The erythrocyte sedimentation rate can, however, be elevated but is not specific (1, 6). A positive tuberculin test is of limited help, especially in areas of high prevalence (6). The diagnosis is established by isolating the tuberculous bacillus from the site of infection and/or by characteristic histopathological changes (1, 6, 7, 9). Adequate response to antituberculous chemotherapy is another recognized method for establishing the diagnosis.

The treatment of tuberculous spondylitis remains controversial, particularly in patients presenting with minimal or absent neurological deficits (6, 21). Surgery together with antituberculous chemotherapy remain the preferred method in many centers, aimed at draining the abscesses, removing necrotic tissue, decompressing the thecal sac, and fusing the spine (6, 9, 21). In poor-risk patients percutaneous drainage of paraspinal abscesses may help in producing early recovery (11).

Characteristic features of tuberculous spondylitis include thoracolumbar spine predilection, severe vertebral collapse, frequent involvement of posterior osseous elements, large paraspinal abscesses extending beyond the area of vertebra-disk involvement, frequent encroachment on spinal canal and neural elements, and gibbus deformity (7). Multiple noncontiguous vertebral lesions are frequently seen and are caused by subligamentous spread and spread along infected muscle bundles (7).

Pre- and postintravenous contrast medium studies are essential for CT evaluation of tuberculous spondylitis (7). While plain CT is needed for evaluating the bony fragments, especially those that encroach on the spinal canal, the enhanced study is ideal for distinguishing between abscesses and normal tissue (by virtue of rim enhancement) (Fig. 41.7) (7). Affected bone on CT appears osteopenic in the great majority of cases but relative bone sclerosis can be seen in some cases and probably represents devitalized fragments (7). Abscess formation in tuberculosis is a characteristic feature of the infection. It can affect the vertebral body, paraspinal soft tissues, spinal canal, and the poste-

rior elements (Figs. 41.7 and 41.8). Tuberculous lesions affecting nonclassic sites such as the lower lumbar spine can be difficult to differentiate from other infections, even after contrast medium enhancement, especially if granulation tissue rather than abscess formation is associated with it (7).

The pathophysiology of tuberculous spondylitis is best understood by MR imaging (7, 11). Similar to CT, pre- and

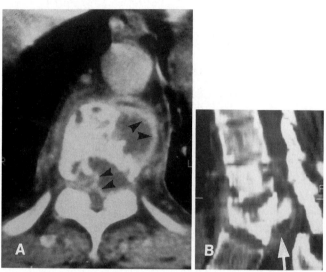

Figure 41.7. **(A)** A 38-year-old woman with tuberculous spondylitis at T10–T11. Postenhanced CT axial section shows patchy destruction of the vertebral body. An abscess with rim enhancement is seen around the vertebral body and encroaching on the spinal canal *(arrowheads)*. **(B)** Sagittal reformed image shows bone destruction with gibbus deformity. There is significant compression on the spinal canal by bone fragments. The enhancing rim represents the inflamed dura *(arrow)*.

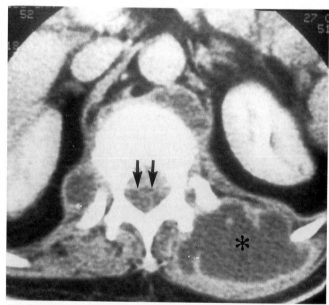

Figure 41.8. A 50-year-old woman with thoracolumbar tuberculous spondylitis. Postcontrast axial image at the level of T12 shows a circumferential abscess with rim enhancement around the vertebral body. Note the encroachment on the spinal canal *(arrows)* and the extension into the erector spinae muscles *(asterisk)*.

postcontrast medium studies are essential for thorough evaluation (11).

As in other hematogenous infections, tuberculosis of the spine starts in the subchondral end plate (1). This creates weakness and improper support for the contiguous disk, which may partially or completely herniate into the affected vertebrae, thus causing spread of infection to the disk and subsequently to neighboring vertebrae (9). Subligamentous spread and along affected muscles are other well-recognized routes (Fig. 41.9) (7, 11). Magnetic resonance imaging in demonstrating these routes is superior to CT (7, 11). An important MR imaging feature of tuberculous spondylitis (TS) is the invariable loss of vertebral body morphology (vertebral contour cannot be identified) (Figs. 41.9 and 41.10). This is in contradistinction to pyogenic or *Brucella* spondylitis, in which vertebral bodies tend to keep their shape despite evidence of osteomyelitis. Encroachment on the spinal canal is also more frequent in TS than in other infections (Fig. 41.10).

The use of Gd-DTPA in TS is highly recommended since it clearly demonstrates abscess rim enhancement in bone and soft tissue, both of which are difficult to visualize on T2-weighted images (Figs. 41.11–41.13) (11). Intraosseous abscesses seem to be a unique feature of TS and their presence can be depicted only on postcontrast MR studies (Figs. 41.11 and 41.12). Partial volume averaging precludes such visualization on CT (11). Similarly, dural involvement can also be appreciated only on enhanced studies, since on T2-weighted images the increased signal from CSF obscures the signal change from the dura (Fig. 41.11) (11).

Involvement of intra- and extraspinal soft tissues with little or no evidence of vertibral osteomyelitis is not unusual in tuberculous spondylitis (18). The psoas muscle is a frequently affected site and the disease may extend along the iliopsoas down to the upper muscles of the thigh (11, 18). Involvement of the ribs and spinous processes can, however, be difficult to identify.

Planning the surgical approach is facilitated by MR imaging mainly because of the clear display of the extent of the active infection and its complications (Fig. 41.13) (11). Posttherapy follow-up can ideally be done by means of MR (11). Only T1- and T2-weighted sequences are required and when the disease is controlled only T1-weighted images are necessary. Increased signal intensity on T1-weighted imagers from previously affected vertebrae indicates healing and has been found to correlate well with symptomatology (Fig. 41.13) (11).

Brucellar Spondylitis

Brucellosis is a zoonosis of worldwide distribution that is caused by small, Gram-negative, nonencapsulated coccobacilli of the genus *Brucella* (7, 13). Of the four species implicated with human infection, *Brucella melitensis*, is the most common, most virulent, most invasive, and the one that is most frequently associated with infectious spondylitis (7, 13). Ingestion of unpasteurized contaminated milk and milk products is the commonest form of transmission in countries where the disease is endemic (7, 8, 13). In the United States and Western Europe, brucellosis is mostly an occupational hazard among animal handlers, abattoir workers, and veterinarians (8). When organisms reach the bloodstream they are phagocytosed by circulating polymorpho-

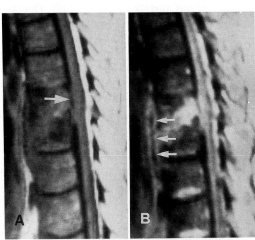
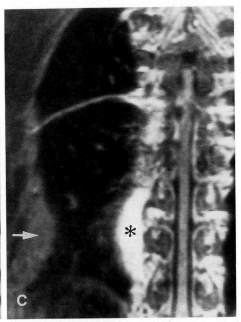

Figure 41.9. **(A)** A 56-year-old-male with tuberculous spondylitis of the midthoracic region. Midsagittal MR image (SE, TR/TE 500/30 ms) shows poor definition of the bodies of T8 and T9. Some obliteration of the epidural space is noted *(arrow)* and there is displacement of the anterior spinal ligament by abscess. **(B)** Midsagittal MR image (SE, TR/TE 2000/100 ms) shows patchy increased signal from the body of T8. There is also increased signal from the anterior component of the abscess *(arrows)*. **(C)** Coronal MR image at the level of the cord (SE, TR/TE 2000/100 ms) shows high signal intensity from the paraspinal abscess *(asterisk)*. There is clear communication between the pleural fluid *(arrow)* and the paraspinal abscess across the minor fissure. Acid-fast bacilli were isolated from both collections.

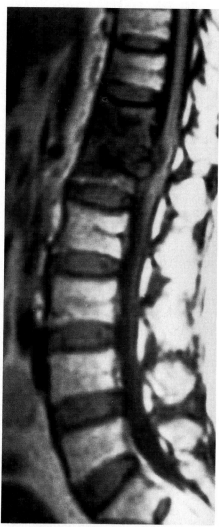

Figure 41.10. A 67-year-old female with tuberculous spondylitis at the T12–L1 level. Midsagittal image (SE, TR/TE 500/30 ms) shows decreased signal from T12 and L1. Note the loss of definition of either vertebral body and the encroachment on the conus medullaris.

nuclear cells and are carried to organs rich in reticuloendothelial cells (liver, spleen, lymph nodes, bone marrow) where they form small miliary granulomata (13). Depending on the virulence of the species, size of the inoculum, and immunity of host, the granulomata may either regress and disappear, or enlarge and cause tissue destruction (13). The musculoskeletal system is most frequently affected and the spine is the commonest site of bone brucellosis (7, 8). Patients with brucellar spondylitis present with back pain with or without other systemic manifestations of brucellosis, neurological deficits being infrequent (22). The diagnosis is established by detecting definite elevation of anti-*Brucella* antibodies and the presence of characteristic imaging features (7, 13). Isolation of the organism from blood culture and/or from tissues removed from the site of involvement is extremely rare and histopathological changes show nonspecific inflammatory changes in the presence or absence of small granulomata (7, 13, 23).

As patients with brucellar spondylitis (BS) respond promptly and adequately to anti-*Brucella* chemotherapy surgical intervention is rarely needed (7, 13).

Brucellar spondylitis is unique among the different forms of infective spondylitis as it can manifest as either focal or diffuse disease (7, 13). In the focal form the osteomyelitis is localized to the anterior aspect of an end plate (classically the superior end plate of L4) at the discovertebral junction (Fig. 41.14) (13, 24). In the diffuse form the infection initially involves an osseous end plate and ultimately extends to involve the entire vertebrae (Fig. 41.14) (13, 24). The disk is secondarily affected and it partially herniates into the weakened end plate. Brucellar spondylitis is slower and less aggressive than tuberculous spondylitis (7, 13). The focal form of the disease is self-limiting and in the great majority of cases seen as an incidental finding on lumbar spine radiographs. On CT, changes of focal brucellar spondylitis include erosions, sclerosis, peripheral vacuum phenomenon, and an anterior osteophyte (Figs. 41.15–41.18) (13). The paraspinal soft tissues and disks are always normal (13, 24). Very rarely may a focal lesion progress to become diffuse (24).

The CT features of the diffuse form of brucellar spondylitis are fairly characteristic (7, 13, 24). Single or multiple well-demarcated end plate defects (simulating Schmorl's nodes) are frequently seen, associated with disk space collapse and sclerotic margins (Fig. 41.19) (7, 13). These are better defined than in pyogenic infections and most probably represent areas of intravertebral disk herniation and/or replacement of bone by granulation tissue (7, 13, 24). Bone sclerosis is more frequently seen in brucellar spondylitis than in pyogenic or tuberculous spondylitis (7). Another finding that seems to be unique to brucellar spondylitis is the relatively common association between disk infection and disk vacuum phenomenon; the latter has been previously reported as being extremely rare in spine infections (Fig. 41.20) (2, 20, 23). The vacuum phenomenon is probably related to the slow nature of the infection, which involves initially the vertebral end plate. As the end plate fails to provide nutritional support to the disk, it becomes devitalized and degenerated (13). Despite the osteomyelitis, vertebral morphology is kept intact and collapse is extremely rare (7, 13, 24). Granulation tissue production rather than abscess formation is the most frequent soft-tissue complication (7, 13, 24). It extends into the paraspinal muscles which on CT appear to have obliterated fat planes (Fig. 41.20) (7, 13, 24). Epidural extension is not uncommon and can be depicted without the need of intravenous contrast (Fig. 41.20) (7, 13).

The commonest site of brucellar spondylitis is the lower lumbar region, but other areas of the spine can be involved and the features are essentially the same (Fig. 41.21) (7, 13, 24). Multiple lesions are not unusual but are probably less frequent than in TS (Fig. 41.21) (7).

On MR imaging the focal and diffuse forms of brucellar spondylitis are clearly displayed (Figs. 41.18, 41.22–41.24) (7, 11, 24). The localized area of osteomyelitis in the focal form is seen as an area of low signal intensity on T1-weighted images and becomes hyperintense on T2-weighted images (Fig. 41.18). Similar but more extensive signal change is noted in the diffuse form of the disease (Figs. 41.22 and 41.24) (7, 11, 24). The disk is invariably

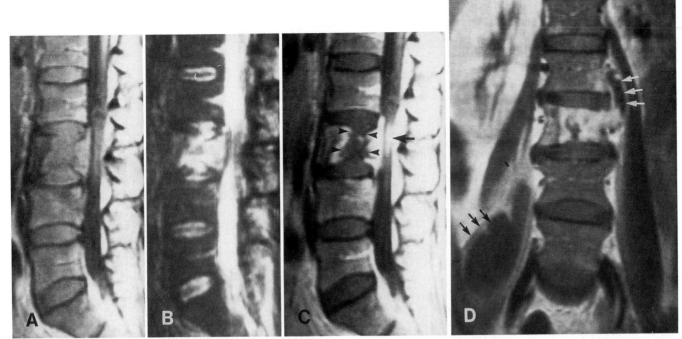

Figure 41.11. **(A)** A 56-year-old man with lower lumbar tuberculous spondylitis. Midsagittal image (SE, TR/TE 500/30 ms) shows an inhomogeneous decreased signal of the body of L3 with some obliteration of the epidural space at the same level. **(B)** Midsagittal image (SE, TR/TE 2000/100 ms) shows an inhomogeneous high signal intensity from L3 and slight compression on the thecal sac. **(C)** Midsagittal MR image as in **(A)** taken after intravenous Gd-DTPA administration shows circumferential enhancement around a central area of necrosis within the body of L3 *(arrowheads);* this represents an osseous abscess. Note the enhancement of the meninges at the same level *(arrow)*, indicating involvement. Disks are intact. **(D)** Coronal image done after sequence **(C)** (SE, TR/TE 500/30 ms) shows heterogeneous enhancement of L3, rim enhancement around a large abscess in the lower aspect of the right psoas muscle, and a smaller abscess medial to the upper portion of the left psoas muscle *(arrows)*. (From Sharif HS, Clark DC, Aabed MY, et al. Granulomatous spinal infection: MR imaging. Radiology 1990;177:101–107. Reprinted with permission of Radiology.)

involved and appears to have reduced height and marked increased signal on the long TR/TE study in the absence of a nuclear cleft (Fig. 41.22) (7, 24). Intravertebral disk herniation can sometimes be clearly visualized (Fig. 41.22). Vertebral collapse is extremely rare. Because granulation tissue exhibits only moderate increase in signal intensity on T2-weighted images, epidural extension is best assessed on T1-weighted images because of obliteration of the linear echo void area anterior to the cord (Figs. 41.22–41.24) (7).

Similar to pyogenic spondylitis, postcontrast studies do not seem to provide additional information to the T2-weighted images, except in demonstrating meningeal involvement and by increasing the confidence of excluding tuberculosis as the etiology (Fig. 41.25) (11).

Follow-up is ideally done by MR. Similar to other spinal infections, the increased signal on T1-weighted images from previously affected vertebrae indicates replacement of the cellular marrow by fat and correlates well with symptomatology (Fig. 41.26) (11).

Fungal Infections in the Spine

Infections of the spine caused by fungus and fungus-like organisms are rare (1). They have been reported with ac-

tinomycosis, nocardiosis, coccidioidomycosis, cryptococcosis, blastomycosis, candidiasis, histoplasmosis, aspergillosis, and maduromycosis (1, 20, 25, 26). Routes of contamination vary from direct implantation by external or internal injuries to hematogenous spread from lesions elsewhere in the body (1, 19, 25, 26). Immunocompromised and debilitated individuals seem to have a higher risk for developing such infections (1).

Some fungi are distributed worldwide and others are endemic only in certain geographical locations (1, 19). The diagnosis is established by isolation of the organism and/or by the presence of characteristic histopathological changes (1, 19, 26). In the event of neurological complications decompressive surgery may be needed, otherwise antifungal therapy can be effective in improving survival and decreasing recurrence rates (25, 26).

Certain organisms may produce a radiological picture that is indistinguishable from tuberculosis (blastomycosis, aspergillosis), i.e., bone destruction, abscess formation, and gibbus deformity; others (actinomycosis, cryptococcoses, coccidioidomycosis, mycetoma) will affect the vertebral bodies by producing patchy destruction or sclerosis but will spare the disks (1). Rib involvement seems to be more frequent in these infections than in the bacterial type of spine infections (1).

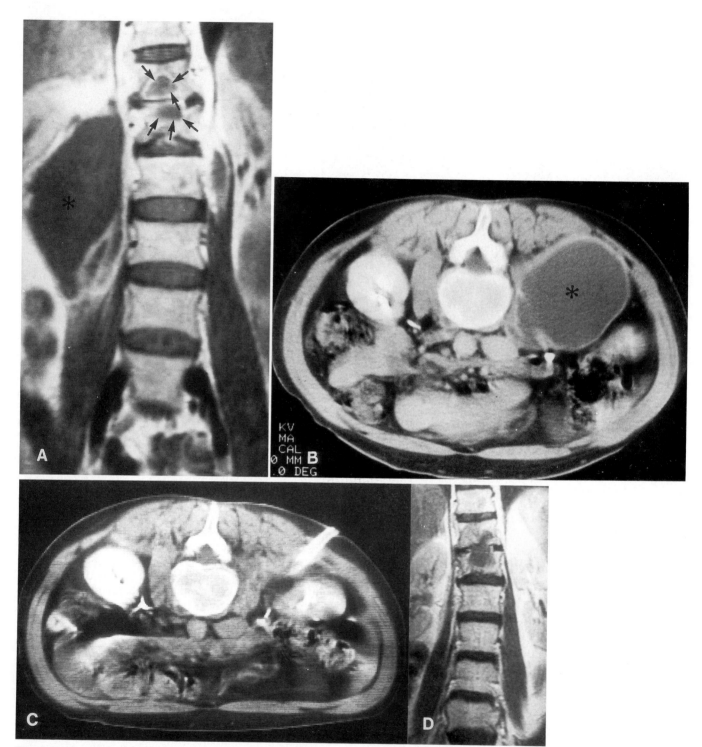

Figure 41.12.   (A) A 64-year-old man with tuberculous spondylitis and severe cardiomyopathy. Coronal MR image (SE, TR/TE 500/30 ms) taken after Gd-DTPA administration demonstrates a large paraspinal abscess on the right *(asterisk)*, a smaller one on the left, and intraosseous abscesses *(small arrows)*. (B) Prone postcontrast CT section at the L2 level shows the large right paraspinal abscess *(asterisk)* displacing the ureter anteriorly. (C) Repeat CT 2 days later than (B) shows the catheter in situ and the complete absence of the abscess. Acid-fast bacilli were isolated from the pus. (D) Repeat coronal post-Gd-DTPA MR image shows complete absence of the psoas abscesses. Residual intraosseous abscesses remain to be seen. (From Sharif HS, Clark DC, Aabed MY, et al. Granulomatous spinal infection: MR imaging. Radiology 1990;177:101–107. Reprinted with permission of Radiology.)

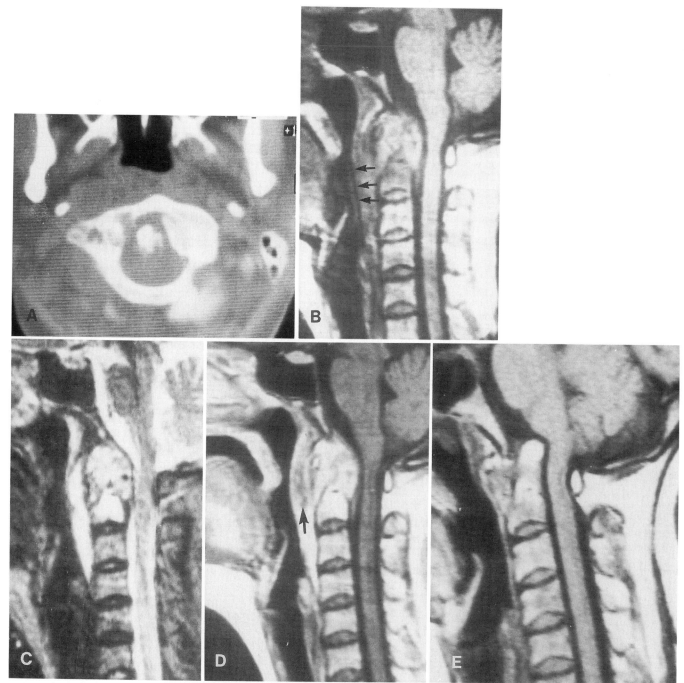

Figure 41.13. **(A)** A 21-year-old woman with tuberculous spondylitis presenting with cervical myelopathy and difficulty in swallowing. Axial CT image at the C1 level shows an irregular odontoid that is dislocated from the anterior arch of the atlas. **(B)** Midsagittal MR image (SE, TR/TE 500/30 ms) shows a large mass involving the odontoid and compressing the cord. There is prevertebral extension of the inflammatory process compressing the air column *(arrows)*. **(C)** Midsagittal MR image (SE, TR/TE 2000/100 ms) shows increased signal from the affected odontoid, the body of C2, and the prevertebral abscess that extends subligamentally down to C4. **(D)** Midsagittal MR image (SE, TR/TE 500/30 ms) taken after Gd-DTPA administration shows diffuse enhancement of C1, C2, and the surrounding inflammatory mass. Note the nonenhancing central region within the abscess, representing pus *(arrow)*. **(E)** Midsagittal MR image (SE, TR/TE 500/30 ms) done 7 months after **(B)** while on antituberculous therapy shows almost complete resolution of the abscess. Note the increased signal in C1 and C2, indicating replacement of cellular marrow by fat. Significant anteroposterior compression of the cord is caused by the odontoid subluxation and the posterior arch of C1. (From Sharif HS, Clark DC, Aabed MY, et al. Granulomatous spinal infection: MR imaging. Radiology 1990;177:101–107. Reprinted with permission of Radiology.)

Figure 41.14. **(A)** Schematics of the two forms of brucellar spondylitis. In focal brucellar spondylitis disks and soft tissues are normal. Changes may include one or more of the following: **(a),** a localized area of bone erosion at the discovertebral junction, **(b)** reactive bone sclerosis, **(c)** a small area of gas (peripheral vacuum phenomenon) entrapped between the vertebral end plate and disk, probably representing soft-tissue destruction, and **(d)** anterior osteophytes (parrot's beak). **(B)** In diffuse brucellar spondylitis, sequential changes include the following: **(a)** the organisms are localized in the superior end plate of the vertebral body *(straight arrows),* similar to the focal lesion; **(b)** the infection spreads throughout the involved vertebrae, and by means of ligamentous *(curved arrows)* and vascular communication, it spreads to involve adjacent vertebrae; and finally **(c)** the osteomyelitis causes bone softening of the osseous end plate, with resultant mechanical instability to the chondral end plate and disk. The disk is secondarily infected and may herniate into the vertebral end plate (Schmorl's nodules) *(arrow).* Granulation tissue may extend into the epidural space. (From Sharif HS, Aideyan OA, Clark DC, et al. Brucellar and tuberculous spondylitis: comparative imaging features. Radiology 1989;171:419–425. Reprinted with permission of Radiology.)

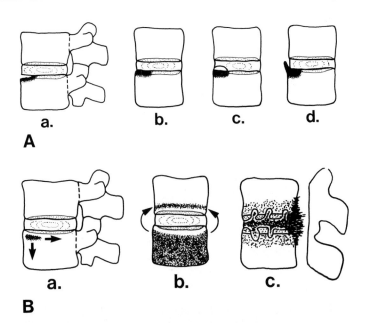

A

B

Figure 41.15. CT example of the focal form of brucellar spondylitis: high-resolution CT scan at the superior end plate of L4 in a 26-year-old woman with spinal brucellosis. Note the erosions in the anterior aspect of the vertebral body *(arrows)* and the normal adjacent soft tissues. (From Madkour MM, Sharif HS, Aabed MY, Al Fayez MA. Osteoarticular brucellosis: results of bone scintigraphy in 140 patients. AJR 1988;150:1101–1105. © Williams & Wilkins, Baltimore, Md.)

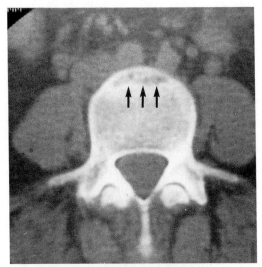

Figure 41.16. CT example of peripheral vacuum phenomenon. Axial CT image at the level of the inferior end plate of L4 of a 38-year-old patient with focal brucellar spondylitis shows a peripheral vacuum phenomenon *(arrow).* Again note the normal surrounding soft tissues.

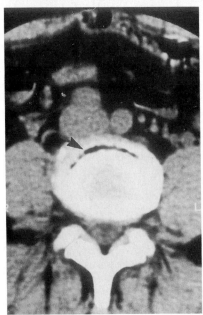

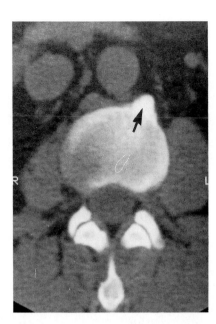

Figure 41.17. CT example of anterior osteophyte formation. Axial image at a lumbar discovertebral junction of a 34-year-old woman with focal brucellar spondylitis shows an anterior osteophyte *(arrow)* with minimal reactive bone sclerosis arising from the body of the vertebrae. The neighboring soft tissues are normal.

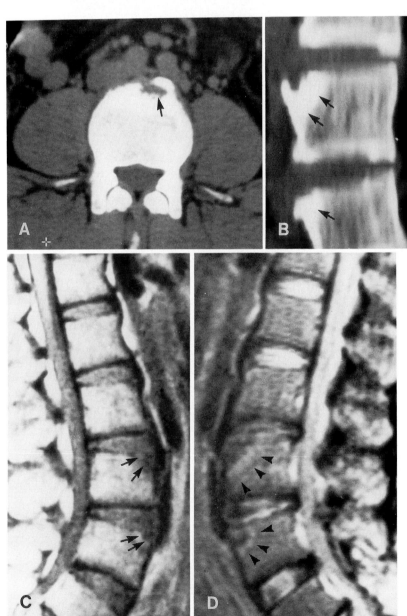

Figure 41.18. **(A)** Focal brucellar spondylitis in the lumbar spine of a 34-year-old man. Axial CT at the superior end plate of L4 shows anterior erosion and localized reactive bone sclerosis *(arrow)*. Note the normal paravertebral soft tissues. **(B)** Midsagittal reformatted image at the level of L4–L5 shows the erosions, sclerosis, and anterior osteophyte in the superior end plates of L4 and L5 (more pronounced at L4; *(arrows)*. **(C)** Midsagittal MR image (SE, TR/TE 500/30 ms) shows areas of decreased signal intensity *(arrows)* in superior end plates of L4 and L5, corresponding to areas of sclerosis noted in **(B)**. **(D)** Midsagittal MR image (SE, TR/TE 2000/100 ms) demonstrates focal increased signal intensity *(arrowheads)* in superior end plates of L4–L5 and L5, corresponding to areas of decreased signal seen in **(C)**. There is moderate loss of signal in L3–L4 and L4–L5 disks. (From Sharif HS, Aideyan OA, Clark DC, et al. Brucellar and tuberculous spondylitis: comparative imaging features. Radiology 1989; 171:419–425. Reprinted with permission of Radiology.)

Figure 41.19. **(A)** Spinal brucellosis in a 54-year-old man. High-resolution CT scan at the level of the superior end plate of L4 shows patchy destruction of the vertebral body. There is obliteration of the fat planes of both psoas muscles and some circumferential granulation tissue around the vertebral body designated by numbers *1* and *2*. **(B)** Sagittal reformatted image demonstrates the extensive bone destruction, particularly in the center of both end plates, into which the disk has most likely herniated. **(C)** Coronal reformatted images on bone settings demonstrating the bone destruction, particularly centrally, as well as a lateral osteophyte, which indicates that the process is long standing. (From Sharif HS, Aideyan OA, Clark DC, et al. Brucellar and tuberculous spondylitis: comparative imaging features. Radiology 1989;171:419–425. Reprinted with permission of Radiology.)

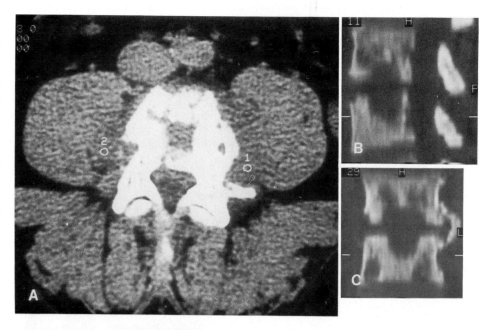
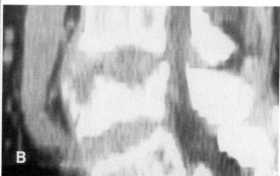

Figure 41.20. **(A)** A 72-year-old man presenting with back ache and paraplegia. Axial CT at the superior end plate of L4 shows marked bone destruction with herniation of bone fragments into the epidural space, evidence of disk gas *(arrow)* and obliteration of the muscle fat planes. Note the paucity of granulation tissue. **(B)** Reformed image in the midsagittal plane shows the severe destruction of the superior end plate of L4 and the associated granulation tissue. Histology of material obtained at surgery showed nonspecific inflammatory reaction.

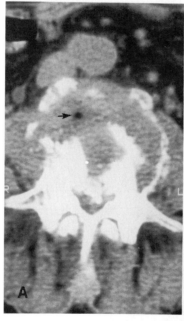

Figure 41.21. **(A)** A 51-year-old woman with brucellar spondylitis of the cervical and thoracic spine. Axial CT of C6 shows destruction of the body with a large central defect *(arrow)*. Irregularities are seen in the anterior aspect of the end plate. **(B)** Axial CT image of T10 shows large and small defects within the vertebral body *(arrow)*. Note the paucity of the soft-tissue component. (From Madkour MM, Sharif HS, Aabed MY, Al Fayez MA. Osteoarticular brucellosis: results of bone scintigraphy in 140 patients. AJR 1988;150:1101–1105. © Williams & Wilkins, Baltimore, Md.)

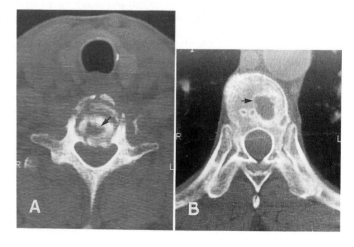

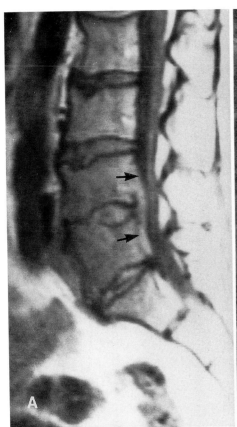
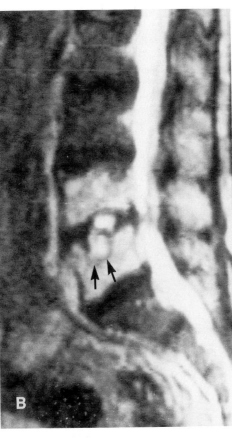

Figure 41.22. **(A)** A 57-year-old man with brucellar spondylitis. Midsagittal MR image (SE, TR/TE 500/30 ms) shows slight diminution in the signal of L4 and L5 vertebral bodies. There is obliteration of the anterior aspect of the L4–L5 disk, as well as obliteration of the epidural space caused by extension of granulation tissue into the spinal canal *(arrows)*. The posterior cortices of L4 and L5 are poorly defined. **(B)** Midsagittal MR image (SE, TR/TE 2000/100 ms) shows diffuse high signal from L4 and L5 and the intervening disk in the absence of a nuclear cleft. Note that the disk has partially herniated into the superior end plate of L5 *(arrows)*.

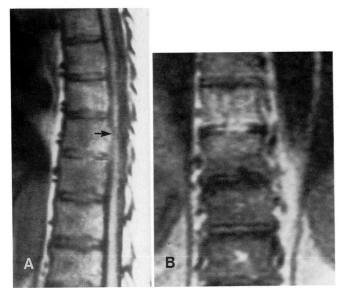

Figure 41.23. **(A)** Brucellar spondylitis in the lower thoracic spine of a 66-year-old man. Midsagittal MR image (SE, TR/TE 500/30 ms) shows a small area of intermediate signal intensity obliterating the signal void region posterior to T9 *(arrow)*, indicating epidural extension. There is also minimal loss of signal in T9 and T10. **(B)** Coronal MR image (SE, TR/TE 2000/50 ms) shows high signal intensity from T9, T10, and the intervening disk. Note the absence of paraspinal soft-tissue abnormality. (From Sharif HS, Aideyan OA, Clark DC, et al. Brucellar and tuberculous spondylitis: comparative imaging features. Radiology 1989;171:419–425. Reprinted with permisison of Radiology.)

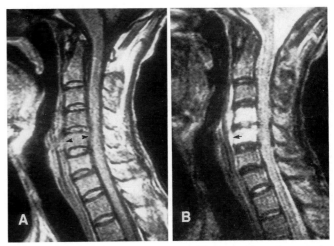

Figure 41.24. **(A)** A 46-year-old man with cervical brucellar spondylitis who presented with quadriparesis. Midsagittal section (SE, TR/TE 500/30 ms) shows interruption of the signal void area in the posterior and anterior aspects of C4 and C5 *(arrowheads)*. The intervening disk is not obliterated. **(B)** Midsagittal section (SE, TR/TE 2000/100 ms) shows diffuse high signal in the bodies of C4 and C5 and in the prevertebral region anterior to C3, C4, and C5 *(arrow)*. Note the obliteration of CSF in the spinal canal at this level.

Figure 41.25. **(A)** Midsagittal MR image (SE, TR/TE 2000/100 ms) shows increased signal from L4 and L5 and the intervening disk in the absence of a nuclear cleft. **(B)** Midsagittal MR image (SE, TR/TE 500/30 ms) taken after Gd-DTPA administration shows diffuse enhancement of both vertebrae and the spondylolisthesis of L4 over L5. Note that there is better contrast between affected and normal vertebrae in **(A)** because of signal loss from marrow and fat. (From Sharif HS, Clark DC, Aabed MY, et al. Granulomatous spinal infection: MR imaging. Radiology 1990;177:101–107. Reprinted with permission of Radiology.)

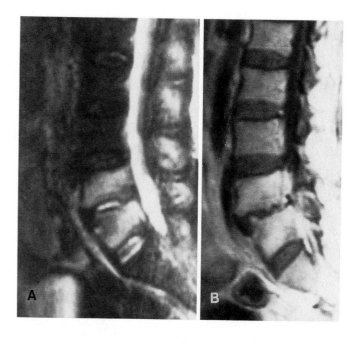

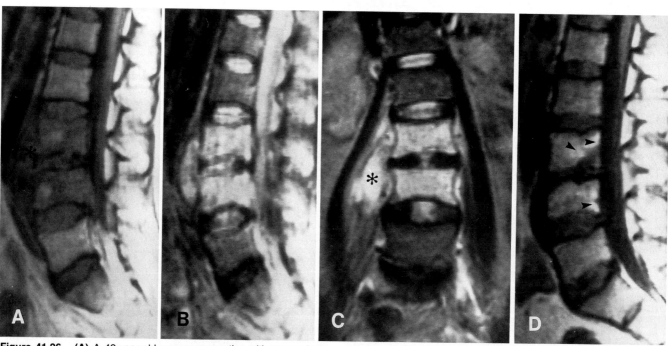

Figure 41.26. **(A)** A 42-year-old woman presenting with severe low back pain. Midsagittal MR image (SE, TR/TE 500/30 ms) shows decreased signal in the bodies of L3 and L4. A prevertebral mass is seen at the same level displacing the abdominal aorta anteriorly *(asterisk)*. **(B)** Midsagittal MR image (SE, TR/TE 2000/100 ms) shows increased signal from L3, L4, and the intervening disk and from the prevertebral mass seen on **(A)**. **(C)** Coronal MR image (SE, TR/TE 2000/50 ms) shows increased signal intensity from affected vertebrae and from the right psoas muscle, which appears to be enlarged *(asterisk)*. The changes in the right psoas suggested abscess and the MR diagnosis was that of TS. The patient had positive serology indicative of active *Brucella* infection. **(D)** Midsagittal MR image (SE, TR/TE 500/30 ms) taken 6 months after anti-*Brucella* chemotherapy. The patient was symptom free. There is a relative increase in signal intensity in L3 and L4, indicating replacement of cellular marrow by fat *(arrows)*. The area of signal void in the posteroinferior aspect of L3 represents increased bone density. (From Sharif HS, Clark DC, Aabed MY, et al. Granulomatous spinal infection: MR imaging. Radiology 1990;177:101–107. Reprinted with permission of Radiology.)

Computed tomography clearly demonstrates the morphological changes associated in the vertebral bodies, spinal canal, and paraspinal soft tissues (19). In a case with long-standing actinomycetoma of the back, extensive destruction of the erector spinae, quadratus lumborum, and psoas muscles was noted (Fig. 41.27). The muscle mass

was replaced by fat and granulation tissue. The inflammatory process extended from the posterior mediastinum at the T4 level down to the coccyx (Fig. 41.27). The gluteii and the upper thigh muscles were also involved and the epidural space was completely obliterated. Although CT has clearly shown the extent of pathology it failed to dem-

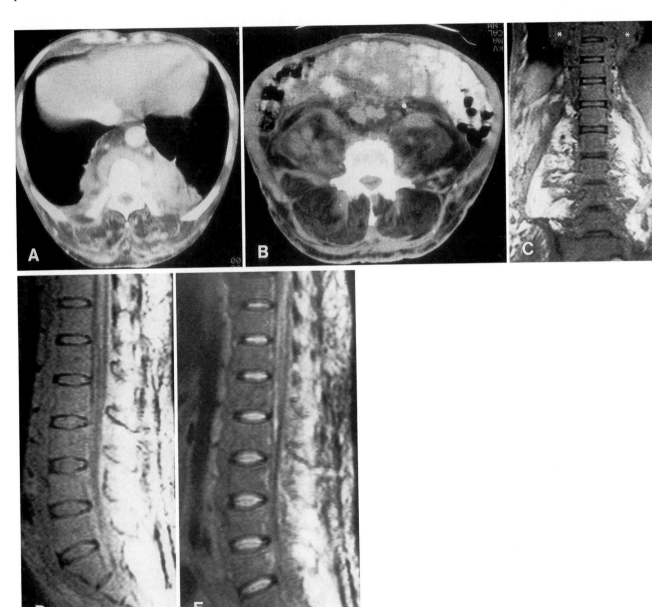

Figure 41.27. **(A)** A 34-year-old man with an extensive mycetoma involving the entire posterior thoracoabdominal wall. Axial postenhanced CT at the level of T11 shows a large posterior mediastinal mass displacing the aorta anteriorly. The mass extends posteriorly to involve the semispinalis dorsi, multifidus, longissimus dorsi, and lower part of the trapezius muscles. The patient was scanned in the prone position because he was unable to lie supine. Note enhancement of the mediastinal mass, erector spinae muscles, and dura. **(B)** Axial postenhanced CT section at the level of L5 shows considerable destruction of the psoas and erector spinae muscles, with involvement of the subcutaneous tissue of the posterior and lateral abdominal walls. Note the extensive epidural extension and the anterior displacement of the bowels. **(C)** Coronal MR image (SE, TR/TE 500/30 ms) shows the posterior mediastinal mass *(asterisk)* extending paraspinally into the abdominal cavity. There is almost complete destruction of the psoas muscles, but the vertebrae and disks appear normal. The spleen is markedly enlarged. **(D)** Midsagittal MR image (SE, TR/TE 500/30 ms) shows displacement of the aorta by an isointense process that represents the exuberant osteophytosis. **(E)** Midsagittal MR image (SE, TR/TE 2000/100 ms) shows almost complete obliteration of the thecal sac, caused by the extensive epidural process, which is of mixed signal intensity. No abnormality is noted in disks and vertebrae. (From Sharif HS, Clark DC, Aabed MY, Aideyan OA, Haddad MC, Mattsson TA. MR imaging of thoracic and wall infections: comparison with other imaging procedures. AJR 1990;154:989–995. © Williams & Williams, Baltimore, Md.)

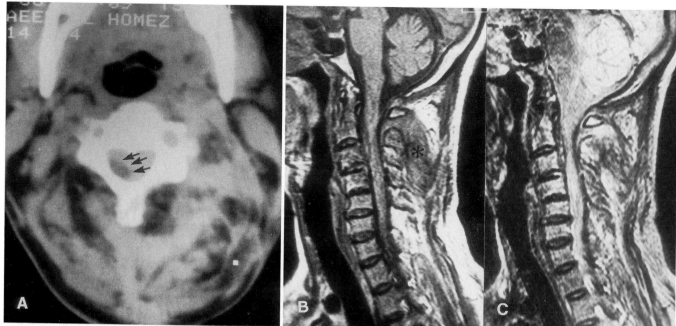

Figure 41.28. **(A)** A 61-year-old man with compressive cervical myelopathy caused by a mycetoma of the posterior elements. Axial postenhanced CT images at the C2 level show an abnormal process involving the posterior muscles of the neck, particularly on the left side, causing patchy enhancement. Severe epidural extension is displacing the thecal sac to the right side and posteriorly *(arrows).* **(B)** Midsagittal MR image (SE, TR/TE 500/30 ms) shows a mass asterisk isointense with muscles arising from soft tissues of the neck posteriorly and in-vading the epidural space between the spinous processes of C1 and C2. The cord is moderately compressed *(arrows).* Moderate signal loss is seen in the bodies of C1 to C4. **(C)** Midsagittal MR image (SE, TR/ TE 2000/50 ms) shows patchy increased signal from the mass seen in **(A)** and increased signal from the bodies of C1 to C4. (From Sharif HS, Clark DC, Aabed MY, et al. Granulomatous spinal infection: MR imaging. Radiology 1990;177:101–107. Reprinted with permission of Radiology.)

onstrate the rib changes noted on the plain radiographs, namely periosteal elevation and patchy erosions (Fig. 41.27).

In a further example of maduromycetoma of the neck, CT demonstrated a large mass arising from the soft tissues posterior to the spinous processes of the cervical spine and showing patchy enhancement. The mass extended epidurally and was responsible for compressing the cord (Fig. 41.28).

Magnetic resonance is able to demonstrate well the diffuse muscle mass destruction and replacement by fat. The inflammatory nature of the process was suggested by the moderate increased signal intensity obtained on the T2-weighted images. The vertebrae and disks appeared normal. Similar to CT, MR imaging failed to demonstrate the extensive rib changes noted on the plain radiographs (Fig. 41.27).

Extension of the inflammatory process into the epidural space was seen more easily on MR imaging than CT (Fig. 41.28).

Magnetic Resonance Imaging and Computed Tomography Features of Echinococcosis of the Spine

Hydatid disease is caused by the parasitic tapeworm *Taenia echinococcus.* In man two main forms have been implicated: *Echinococcus granulosus* and less frequently *Echinococcus multilocularis.* The liver and lungs are the organs most frequently affected: liver 75%, lungs 15%, and the rest of the body 10%. The skeleton is involved only in about 1% of all cases and the spine is the area of predilection (50%) (27).

Because of the rigid nature of bone, the hydatid cysts do not form into the typical spherical shape, but instead permeate and slowly destroy cancellous bone. In the spine, the disease starts in the vertebral body and may spread later to the neural arch and adjacent ribs. If the cortex is breached the cyst may invade the extradural space and cause neurological deficits. The dura always remains intact (27).

Backache with localized tenderness is the main clinical finding frequently associated with compressive myelopathy or radiculopathy. Serologic tests have false-positive and false-negative results with cross-reactions occurring with other parasites (28, 29). Mebendazole (broad-spectrum antihelminthic drug) with or without decompression surgery is recommended for the treatment of these cases. Prognosis is regarded as extremely poor (28).

Spinal echinococcosis is a rare disease and the imaging features of the reported cases do not seem to follow a recognized pattern. There is bone destruction with extension into soft tissues and the epidural space, but the disks are usually not involved. Computed tomography will accurately depict the degree of vertebral body destruction and will also demonstrate the soft-tissue component. Smooth rim enhancement can occur in the wall of the nonosseous cysts, unless complicated by pyogenic superinfection, as a result of which the rim becomes thickened and irregular (30).

On MR imaging the osseous and soft-tissue extent will be depicted on sagittal, coronal, and axial views. The findings can be extremely difficult to differentiate from neoplasia. Posttherapy follow-up is ideally done by MR because of high sensitivity and the lack of ionizing radiation (31, 32).

Differential Diagnosis and Pitfalls

Infectious spondylitis is a difficult disease to diagnose clinically because of the frequent paucity of distinctive signs and symptoms. The great majority of cases have classic presentations and their imaging features are fairly characteristic (CT and MR imaging criteria of the three most common types of infectious spondylitis are summarized in Tables 41.1 and 41.2). Care, however, must be taken for potential pitfalls in atypical presentations, such as the following.

1. Cases of pyogenic spondylitis that are caused by direct spread or direct implantation: These may produce focal vertebral destruction, involvement of posterior elements, and an asymmetric soft-tissue mass; the disks

Table 41.1
Computed Tomography Features of Hematogenous Infectious Spondylitis

	Site of Predilection	Vertebral Body	Disk	Paraspinal Soft Tissue	Epidural Space
Pyogenic infections	Lower lumbar spine	Patchy destruction Minimal sclerosis Localized to end plate Two consecutive vertebrae Vertebral morphology intact No gibbus	Collapsed	Circumferential granulation tissue Patchy enhancement Localized to affected vertebrae	Infrequently involved Moderate: caused by granulation tissue
Tuberculous infections	Thoracolumbar spine	Severe destruction Vertebral morphology lost Sclerosis is not infrequent Gibbus: frequent Multiple sites: frequent	Collapsed Several levels can be involved	Large abscesses Extend paraspinally to distant location Rim enhancement	Frequently involved Severe: caused by abscess and/or bone fragments
Brucellar infections	Lower lumbar spine	Localized end plate destruction: simulating Schmorl's nodes Two consecutive vertebrae Multiple sites: infrequent Vertebral morphology intact No gibbus	Collapsed Disk gas: frequent	Moderate granulation tissue Patchy enhancement	Infrequently involved Moderate: caused by granulation tissue

Table 41.2
Magnetic Resonance Imaging Features of Hematogenous Infectious Spondylitis

	Vertebral Body	Disk	Paraspinal Soft Tissue	Epidural Space
Pyogenic infections	T1: decreased signal in two consecutive vertebrae T2: increased signal Vertebral morphology intact Gd-DTPA: diffuse enhancement	T2: high signal, absent nuclear cleft Gd-DTPA: minimal enhancement	T2: moderate increased signal, from granulation tissue Gd-DTPA: no definite enhancement	T1: obliteration of epidural space Gd-DTPA: meningeal enhancement rare
Tuberculous infections	T1: decreased signal of affected vertebrae T2: diffuse increased signal Vertebral morphology lost Gd-DTPA: rim enhancement demonstrating intraosseous abscesses	T2: high signal, absent nuclear cleft Gd-DTPA: minimal enhancement	T1: abscesses isointense with muscle T2: diffuse high signal from abscess collections Gd-DTPA: rim enhancement around abscesses	T1: obliteration of granulation tissue and bone fragments Gd-DTPA: meningeal enhancement frequent
Brucellar infections	T1: decreased signal in two consecutive vertebrae T2: increased signal, Schmorl's nodes can be seen Vertebral morphology intact Gd-DTPA: diffuse enhancement	As in pyogenic	As in pyogenic	As in pyogenic

may be preserved. The appearances can be indistinguishable from TS or neoplastic processes.

2. Isolated epidural abscesses of any etiology: these may have imaging features similar to any intraspinal space-occupying lesion.

3. Tuberculous spondylitis cases involving the lower lumbar spine and associated with moderate granulation tissue: these have features comparable with BS.

4. Spinal infections caused by more than one pathogen: these may show changes compatible with either infection.

5. Brucellar spondylitis with associated abscess formation: this is difficult to differentiate from tuberculous spondylitis (Fig. 41.26).

6. Unusual fungal infections: these may have changes that are bizarre and difficult to categorize into a specific group (Figs. 41.27 and 41.28).

7. Neoplastic processes involving two consecutive vertebrae and/or intervening disks and resulting in vertebral body necrosis: these can easily be mistaken for being of infective etiology. (Figs. 41.29 and 41.30).

8. An active spondylitis complicating a spinal neoplasm: this would also be difficult to categorize.

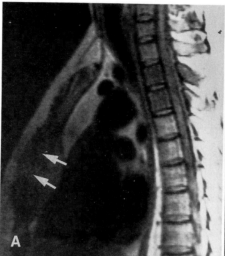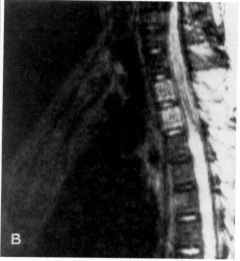

Figure 41.29. (A) A 32-year-old man presenting with fever and acute paraplegia. Surface coil imaging of the cervicothoracic spine shows areas of homogeneous decreased signal intensity in the bodies of T2 and T3. There is also decreased signal from the spinous process of T2. Note the obliteration of the epidural space at the same level, as well as the mass on both sides of the mediastinum. **(B)** Midsagittal section as in **(A)** shows diffuse high signal from the bodies of T2 and T3 and obliteration of the CSF space at the same level. Because of the coincidental involvement of two consecutive vertebral bodies, infectious spondylitis was considered to be high on the list. Incidentally noted was a large mass in the xiphisternum *(white arrows)* in **(A)**. The diagnosis of high-grade lymphoma was made from material obtained by percutaneous biopsy of the sternal mass.

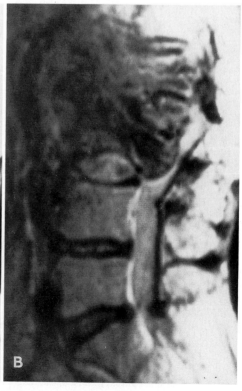

Figure 41.30. (A) A 43-year-old man with extensive multiple myeloma involving the upper lumbar spine. Midsagittal MR image (SE, TR/TE 500/30 ms) shows markedly decreased signal intensity from a large mass completely replacing the bodies of L1 to L3. The disks are also involved. **(B)** Midsagittal MR image (SE, TR/TE 2000/50 ms) shows only moderate increase in the signal intensity of this large mass.

Table 41.3
Diagnostic Criteria of Infectious Spondylitis

	Characteristic Imaging Features	Positive Blood Culture	Positive Serology	Isolation of Pathogen from Site of Involvement[a]	Characteristic Histopathology[a]
Pyogenic infections	Essential criterion	Adults: likely Pediatric age: difficult	—	In case blood culture is negative	—
Tuberculous infections[b]	Essential criterion	Extremely rare	—	Successful in 50% of cases	Essential
Brucellar infections	Essential criterion	Extremely rare	Essential criterion	Extremely rare	—
Fungal infections[c]	Essential criterion	Extremely rare	—	Essential	Essential

[a] Best obtained by CT-guided biopsy or decompression surgery.
[b] Clinical response to antituberculous chemotherapy is another recognized method of diagnosing tuberculous spondylitis.
[c] Fungal infections of the spine are rare, imaging features suggestive of infection can be found in few of the cases.

Conclusion

Infectious spondylitis is an important cause of spinal disease in many parts of the world. Early diagnosis and prompt treatment are essential in order to avoid permanent neurologic damage or spinal deformities. Computed tomography and MR imaging play major roles in the diagnosis and overall management of these disorders (Tables 41.1–41.3).

Despite its numerous advantages in the musculoskeletal system, CT examination suffers from two main limitations when assessing spinal infections.

1. Detection of bone involvement on CT depends mostly on changes in density (osteopenia, sclerosis) and morphology (destruction). Inflammation within the bone in the absence of density and/or morphological alterations cannot be identified. Posttherapy follow-up by CT may not therefore be able to detect early improvement, which can occur before any change in density or structure.
2. The radiation dose incurred by high-resolution CT of the spine can be very high.

Computed tomography, however, is still the method of first choice for guiding percutaneous biopsy and/or drainage procedures in spinal infections. When decompression surgery is needed, CT can be extremely useful in evaluating the osseous components of the spine, especially small bone fragments that may have herniated into the spinal canal.

Magnetic resonance imaging is the most suitable method for the initial evaluation of spine infections. It is cost effective, since as a single study it provides most of the data that otherwise need to be obtained with radiography, scintigraphy, and high-resolution CT. Because of its high sensitivity in detecting bone marrow changes, MR is also ideal for posttherapy follow-up of these lesions; minor alterations in the inflammatory process can be detected with ease. A major limitation of MR imaging is that invasive procedures cannot be performed with ease using currently available units. Further improvement in the technology may soon prove that this can be overcome. Magnetic resonance imaging may also fail to detect with clarity small bone fragments within the spinal canal.

Acknowledgments

We wish to thank the administration of Riyadh al Kharj Hospital (Saudi Arabia) for their support and encouragement; our colleagues in the Radiology Department for their cooperation; the Department of Medical Illustration; Lynette Davies, Evert Blink, and Margaret Campbell for imaging; and Dorothy Davidson for assistance in the preparation of this manuscript.

References

1. Resnick D, Niwayama G. Osteomyelitis, septic arthritis and soft tissue infection: the axial skeleton. In: Resnick D, Niwayama G, eds. Diagnosis of bone and joint disorders. Philadelphia: W. B. Saunders, 1988:2619–2754.
2. Sutton MJ. Pyogenic osteomyelitis of the vertebral body. J Am Osteopath Assoc 1984;83:724–728.
3. Wiley AM, Trueta J. The vascular anatomy of the spine and its relationship to pyogenic vertebral osteomyelitis. J Bone Joint Surg 1959;41B:796–809.
4. Sartoris DJ, Moskowitz PS, Kaufman RA, Ziprkowski MN, Berger PE. Childhood diskitis: computed tomographic findings. Radiology 1983;149:701–707.
5. Fischer GW, Popich GA, Sullivan DE, Mayfield G, Mazat BA, Patterson PH. Diskitis: a prospective diagnostic analysis. Pediatrics 1978;62:543–548.
6. Hsu LCS, Yau ACMC, Hodgson AR. Tuberculosis of the spine. In: Evarts CM, ed. Surgery of musculoskeletal system, Vol. 4. New York: Churchill Livingstone 1983:153–168.
7. Sharif HS, *Aideyan OA*, Clark DC, et al. Brucellar and tuberculous spondylitis: comparative imaging features. Radiology 1989;171:419–425.
8. Lifeso RM, Harder E, McCorkell SJ. Spinal brucellosis. J Bone Joint Surg (Br) 1985;67:345–351.
9. Weaver P, Lifeso RM. The radiological diagnosis of tuberculosis of the adult spine. Skel Radiol 1984;12:178–186.
10. Fredrickson B, Yuan H, Olans R. Management and outcome of pyogenic vertebral osteomyelitis. Clin Orthoped Rel Res 1978;131:160–167.
11. Sharif HS, Clark DC, Aabed MY, et al. Granulomatous spinal infection: MR imaging. Radiology 1990;177:101–107.
12. Burke DR, Brant-Zawadzki M. CT of pyogenic spine infection. Neuroradiology 1985;27:131–137.
13. Madkour MM, Sharif HS, Aabed MY, Al Fayez MA. Osteoarticular brucellosis: results of bone scintigraphy in 140 patients. AJR 1988;150:1101–1105.
14. Modic MT, Feiglin DH, Periuino DW, et al. Vertebral osteomyelitis: assessment using MR. Radiology 1985;157:157–166.
15. Smith AS, Weinstein MA, Mizushima A, et al. MR imaging characteristics of tuberculous spondylitis versus vertebral osteomyelitis. AJR 1989;159:399–405.
16. Van Lom KJ, Kellerhouse LE, Pathria MN, et al. Infection versus

tumour in the spine: criteria for distinction with CT. Radiology 1988;166:851–855.

17. Kramer J, Hajeck CP, Imhof H, Pongracz N, Neuhold A. Vertebral osteomyelitis; MR imaging assessment of early therapy response. Paper presented at the 74th annual meeting of RSNA, Chicago, November, 1988.

18. Simon JH, Szumowski J. Chemical shift imaging with paramagnetic contrast material enhancement for improved lesion depiction. Radiology 1989;171:539–543.

19. Sharif HS, Clark DC, Aabed MY, Aideyan OA, Haddad MC, Mattsson TA. MR imaging of thoracic and abdominal wall infections: comparison with other imaging procedures. AJR 1990;154:989–995.

20. Pritchard DJ. Granulomatous infections of bones and joints. Orthoped Clin N Am 1975;6:1029–1047.

21. Medical Research Council. Sixth report of the Medical Research Council Working Party on tuberculosis of the spine: Five-year assessment of controlled trials of ambularity treatment, debridement and anterior spinal fusion in the management of tuberculosis of the spine. J Bone Joint Surg (Br) 1979;60:163–177.

22. Ariza J, Gudiol F, Valverde J, et al. Brucellar spondylitis: a detailed analysis based on current findings. Rev Infect Dis 1985;7:654–664.

23. Akhtar M. Brucellosis: histopathological features. In: Madkour MM, ed. Brucellosis. London: Butterworths, 1989:59–69.

24. Sharif HS, Madkour MM. Bone and joint imaging. In: Madkour MM, ed. Brucellosis. London: Butterworths, 1989:105–115.

25. Dalinka MK, Dinnenberg S, Greendyke WH, Hopkins R. Roentgenographic features of osseous coccidioidomycosis and differential diagnosis. J Bone Joint Surg 1971;53A:1157–1164.

26. Mariat F. The mycetomas: clinical features, pathology, etiology and epidemiology. Contrib Microbiol Immunol 1977;4:1–39.

27. Charles RW, Grovender S, Maidoo KS. Echinococcal infection of the spine with neural involvement. Spine 1988;13:47–49.

28. Levack B, Kernohan J, Edgar MA, Ransford AO. Observations on the current and future surgical management of the hydatid disease affecting the vertebrae. Spine 1986;11:583–590.

29. Beggs I. The radiology of hydatid disease. AJR 1985;145:639–648.

30. Pau A, Simonetti G, Tortori-Donati P, Turtas S, Viale GL. Computed tomography and magnetic resonance imaging in spinal hydatidosis. Surg Neural 1987;27:365–369.

31. Kaoutzanis M, Anagnostopoulos D, Apostolou A. Acta Neurochir (Wien) 1989;98:60–65.

32. Michael MA, Ciric IC, Tarkington JA. MR imaging in spinal echinococcosis. J Comput Assist Tomogr 1985;9:398–400.

CHAPTER 42

Magnetic Resonance Imaging of Rheumatoid Arthritis of the Cervical Spine

P. R. Algra, F. C. Breedveld, and J. L. Bloem

Introduction

Rheumatoid arthritis (RA) is a chronic inflammatory process of unknown etiology, predominantly involving the synovial joints. The pathologic process in RA is characterized by proliferation of synovial tissue (pannus formation) and destruction of cartilage, ligaments, and bones. The 32 joints in the cervical spine, bearing the head (weighing an average of 14 lb) render this relatively small anatomic area particularly vulnerable to RA (1).

The consequence of this destructive arthritis is that the connections between the different vertebrae and between atlas and skull lose stability, which may lead to subluxation and subsequently to compression of the spinal cord, nerve roots, vertebral arteries, and the medulla or pons. The subluxation occurring most frequently is the anterior atlantoaxial subluxation, with a forward displacement of the atlas and an increased interspace between the dens and the anterior arch of the atlas. In case of destruction of the dens axis and transverse ligament it is possible for the atlas to dislocate backward (2). Destruction of the lateral parts of the cervical spine may cause the dens to protrude into the foramen magnum (3). Lateral subluxations and subaxial subluxation can also be identified.

Involvement of the cervical spine occurs in two-thirds of all patients with longstanding RA (4). Approximately one-third of these patients develop anterior, posterior, lateral, or vertical subluxation (2, 5). Only a minority of the RA patients with a cervical subluxation will develop compressive myelopathy. Once developed this neurological complication tends to progress rapidly. Since the neurological abnormalities can be reversed by immediate sur-

gical therapy, physicians must be aware of clinical signs indicating cord compression.

The clinical expression of RA of the cervical spine varies from absence of symptoms to severe neck and occipital head pains, arm and leg pains, weakness of both upper and lower extremities, loss of deep tendon reflexes, and quadriparesis. Pain is the most frequent symptom, which is aggravated by neck involvement. Sudden death is rare and is caused by medullary compression by the odontoid process (6). Manifestations of cord compression usually occur in the presence of extreme subluxation. The radiologic extent of the subluxations does not correlate with clinical signs of cord compression. The diagnosis of cord compression should be suspected when any neurological syndrome develops. Common presenting features include paresthesias or other sensory disturbances, diminished or absent vibration sense, flexor spasms, urinary retention or other bladder disturbances, and difficulty in walking. However, it is often very difficult to identify clinical signs of cord compression because of the presence of joint deformations.

The outcome of conservative treatment of myelopathy has been uniformly fatal; surgical intervention is the only mode of treatment available. Accurate localization and characterization of the compressive lesion are essential for successful clinical management. Conventional radiology and myelography have not always been sufficient for this purpose.

Magnetic Resonance Imaging Technique

Particularly in the case of patients suffering from rheumatoid arthritis, magnetic resonance (MR) imaging has many advantages over myelography, which is difficult to perform in patients with joint deformities and pain. Many authors favor MR imaging as the imaging method of choice in the diagnostic work-up of the craniovertebral junction (7–14).

The imaging protocol should include T-1 weighted and T2- or T2*-weighted sagittal images. T1-weighted sagittal images of the cervical spine can establish the presence of erosive lesions, destruction of the dens (Fig. 42.1), synovial swelling, pannus formation, and allows evaluation of subluxation of the cervical spine. Both cortical bone and

---

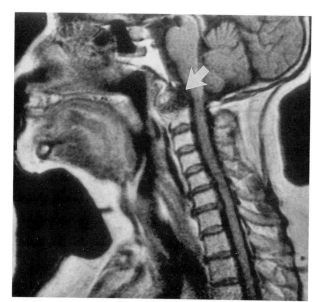

Figure 42.1. Destruction of the dens axis. T1-weighted spin-echo (SE) sagittal image shows complete destruction of the dens axis (*arrow*). No evidence of atlantoaxial subluxation or myelopathy is seen.

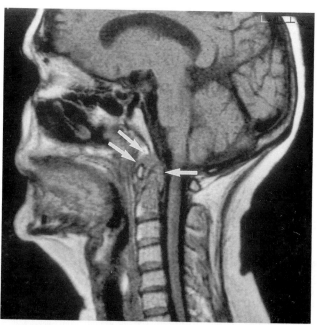

Figure 42.2. Pannus formation around C1–C2. T1-weighted sagittal SE image exhibits a slightly eroded cervical vertebral body with extensive pannus formation of intermediate signal intensity surrounding the odontoid peg (*arrows*).

the cerebrospinal fluid containing subarachnoid space show a very low signal intensity on T1-weighted images. This may cause an overestimation of the diameter of the subarachnoid space, as the former structures cannot be clearly separated on T1-weighted images. T2- and T2*-weighting provide a good contrast between the cortical bone of the vertibral body and the arachnoidal space. T2 weighting makes a reliable estimation of the diameter of the vertebral canal possible (15).

Images can be obtained using a surface coil, head coils, or even body coils. Surface coils and body coils allow imaging during flexion and extension. Because of the central position of the C1–C2 complex, head coils render a very good signal:noise ratio (16).

Images shown in this chapter were obtained with a 0.5- or 0.6-T MR scanner (Phillips Gyroscan, Shelton, Conn., or Technicare Teslacon, General Electric, Milwaukee, Wisc.) using a head coil unless indicated otherwise.

Pannus Formation

Many authors have described the MR imaging characteristics of pannus formation and synovitis (7, 9–11, 16–25). Pannus is usually located around the dens, in the region of the transverse ligament or in the region of the alar ligaments.

Pannus represents an inflammatory response of the synovium with an increase in synovial water content, resulting in increased T1 and T2 relaxtion times (25). However, pannus formation can show both high and low signal intensity on T1- and T2-weighted images (Figs. 42.2 and 42.3). This variability in appearance might be related to the activity of the inflammatory process or the presence of edema (16–18, 21, 22, 25). Active forms of inflammation exhibit a high signal intensity on T2-weighted images as evidenced in experimental models (25). Chronic inflammation can show medium or low signal intensity on both

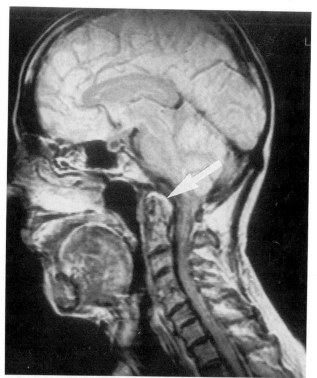

Figure 42.3. Pannus formation shows high signal intensity on moderate T2-weighted SE image (*arrow*). Pannus formation is located around the upper and posterior part of the dens axis.

T1- and T2-weighted images. It has been postulated that the low signal intensity of pannus tissue is related to hemosiderin deposition within the hypertrophied synovium (23). Thus MR imaging may be used in the future to differentiate between acute and chronic inflammation (26).

Pannus formation around the knee in patients suffering from juvenile rheumatoid arthritis shows low to intermediate signal intensity on both T1- and T2-weighted images (27). Sometimes foci of increased signal intensity, most likely due to fluid or inflammation, may be observed on T2-weighted images (21). Images made shortly after intravenous administration of gadolinium (Gd)-labeled diethylenetriaminepentaacetic acid (DTPA) may increase specificity because well-perfused synovial proliferations will enhance (Fig. 42.4), as opposed to fibrosis and intraarticular fluid collections, which do not (20, 28).

The active forms of inflammation occurring in RA of the cervical spine can also be identified with radionuclide studies (29). However, the major component of pannus is fibrous tissue yielding low signal intensity lesions on both T1- and T2-weighted images (17), and which therefore will not be visible on radionuclide examinations.

As pannus tissue occurs in patients with unstable C1–C2 relationships, some authors suggest the high signal intensity on T2-weighted images might be related to formation of fibrous granulation tissue and hypertrophy of connective tissue elements as the abnormal response to chronic stresses and friction (11, 22). Reduction of periodontoid pannus after stable posterior occipitocervical fusion occurs (22).

Metallic wires, used to perform fusion of the atlantoaxial joint in case of atlantoaxial dislocation, cause field inhomogeneities. These inhomogeneities will generate image artifacts such as signal loss and image distortion. For example, stainless steel wires and pins can cause serious image degradation so that adequate analysis of the upper cervical spine is impossible (16). However, as long as the used wires are not very close to the area of interest, these will not necessarily cause serious diagnostic problems (30, 31). Artifacts from the surgical stainless steel fixation material will be confined to the posterior part of the neck and do not interfere with the evaluation of the periodontoid region and the medulla/cervical cord (22).

Magnetic resonance imaging can be used to detect active inflammatory changes within the joint and may thus become a valuable diagnostic technique to monitor treatment (27). Since dynamic Gd-DTPA studies reflect perfusion as well as enlargement of the extracellular fluid compartment, Gd-DTPA has potential for the evaluation of response to medical treatment (20). Longitudinal studies are now being carried out to evaluate the use of MR imaging as a therapy instrument monitor (32). Because of its ability to depict soft-tissue lesions, MR imaging may become a sensitive and objective method for quantitative assessment of the joint changes in rheumatoid arthritis (23).

Rheumatoid diskitis can be assessed by MR imaging as well. Pathological changes in cases of rheumatoid diskitis resemble those of low-grade sponylodiskitis. Soft-tissue masses in the peri- and paraodontoid region can have different etiologies; these masses can be found in cases of inflammatory processes other than rheumatoid arthritis such as polyarteritis nodosa (33), mechanical stress due to degenerative disease, and congenital dysplasia of the dens (11).

Atlantoaxial Subluxation and Compressive Myelopathy

All forms of subluxations (horizontal, vertical, rotational, lateral) can be visualized on sagittal and axial T1-weighted images (34). The relationship of C2 with the foramen magnum is easily appreciated and basal settling of the skull or vertical atlantoaxial subluxation can be clearly visualized by MR imaging (Figs. 42.5 and 42.6).

Magnetic resonance imaging permits precise localization of cord compression and provides information on the nature of the structures that obliterate the arachnoidal space. Medullary compression in cases of vertical settling of the skull and atlas on the axis is readily appreciated on MR images (35–37). Compression of the vertebral artery can be seen on axial images of the neck (34).

The noninvasive character of MR imaging is of particular importance in diagnostic imaging in children. It has been shown that MR imaging is more sensitive than radiography in the detection of loss of cartilage and small

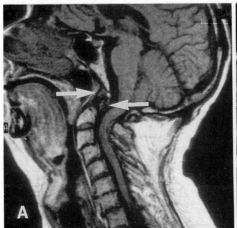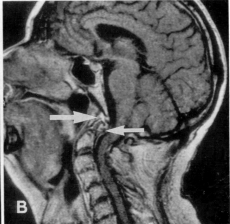

Figure 42.4. Gd-DTPA enhanced images. T1-weighted SE sagittal images before **(A)** and after **(B)** the administration of intravenous Gd-DTPA. Soft-tissue mass at C1–C2 shows marked increase in signal intensity (*arrows*).

erosions. Magnetic resonance imaging can be used also to detect fluid collections, aseptic necrosis, and impingement on the dural sac by involvement of the first and second cervical vertebral articulation (38, 39).

Compression of the medulla and upper cervical cord in patients with severe chronic rheumatoid arthritis is caused not only by atlantoaxial subluxation but frequently by periodontoid pannus also (22).

In compressive myelopathy, intramedullary lesions of high signal intensity on T2-weighted images are depicted by MR imaging (22, 40). Clinical evidence of compressive

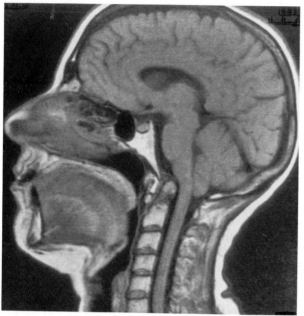

Figure 42.5. Moderate atlantoaxial dislocation. T1-weighted SE sagittal image shows moderate horizontal and vertical subluxation with a slight distortion of the cervical cord.

myelopathy is probably more closely correlated with cord distortion shown by MR imaging than with the extent of both C1–C2 and subaxial subluxation seen on plain radiography (41). The same correlation between MR findings and the degree of myelopathy was also found in a series of patients suffering from atlantoaxial subluxation caused by a variety of diseases (33) and in a series of patients with vertical settling of the skull and atlas on the axis in rheumatoid arthritis (36). In clinical practice, MR imaging evaluation of the rheumatoid cervical spine seems to be particularly useful in selected situations in which myelography was previously indicated, i.e., when demonstration or localization of cord compression is considered necessary (35, 41).

Patients with neurologic symptoms of cord compression will usually show cord or brainstem abnormalities on MR images (40). Anterior compression of the spinal cord caused by the odontoid process, however, can occur without neurological symptoms (35).

An alternative way of establishing the cervical cord status by measuring the cervicomedullary angle. The cervicomedullary angle and the neuraxis configuration can be identified easily on MR imaging. A cervicomedullary angle of less than 135° correlates with brainstem compression, cervical myelopathy, or C2 root pain (42).

Several studies address the comparison of MR imaging with myelography. Usually MR imaging will render information diagnostically equal to myelography. In some cases myelography will afford incomplete visualization of the spinal canal because marked stenosis can impair passage of contrast agent. In these cases MR imaging is particularly useful in imaging the entire length of the vertebral canal as well as the location(s) of stenosis (8).

Atlantoaxial subluxation and other disorders of the cervical spine can occur in juvenile rheumatoid arthritis (43). Cord compression, however, is not very common in the juvenile form of rheumatoid arthritis and although C1–C2

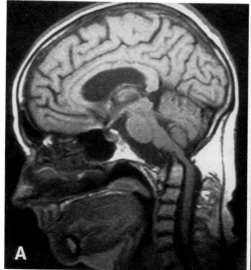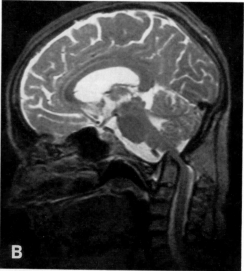

Figure 42.6. Severe atlantoaxial luxation with marked upward displacement of the dens axis. T1-weighted SE sagittal image shows a severely decreased diameter of the cervical cord. Loss of arachnoidal space is more conspicious on T2-weighted image **(B)** than on T1-weighted image **(A)**. Cervicomedullary angle is less than 135° **(B)**. However, there are no signs of acute myelopathy **(B)**.

dislocation has been documented, the incidence of neurological symptoms appears to be very low (38, 39).

Flexion and Extension Studies

Functional studies, such as imaging of the cervical spine in flexion and extension, can be performed during MR examination (Fig. 42.7). Movements of the head and cervical spine are restricted in the bore of the magnet. The use of surface coils or body coils allows dynamic imaging of the cervical spine in flexion and extension. These views allow demonstration of the full range of instability. Patients with atlantoaxial subluxation that does not change with flexion or extension may have extensive pannus tissue (Fig. 42.8) occupying the preodontoid space (10).

Magnetic resonance cine display of flexion and extension maneuvers in the cervical spine has proved useful in the evaluation of cord compression due to bone and joint instability. Condon and Hadley developed techniques that produce quantitative indices of cord deformation and dynamics during flexion and extension (44). A computer program automatically calculates a series of contiguous profiles perpendicular to the cord throughout its length and for each image in the maneuver. Orthogonal polynomial curve-fitting techniques are used to fit these profiles and extract statistical parameters that are quantitative indices of deformation and dynamics. This technique may prove useful to the spinal surgeon in assessing the relative significance of individual lesions and may also be a research tool in the study of cord biomechanics in vivo (44).

Contrast Agents

In cases of inflammatory disease, the albumin content of synovial fluid can rise from 1 to 4% (w/v) to as high as 12%. In general, albumin shortens T1 and T2 relaxation times. An albumin concentration of 12% provides sufficient positive contrast to allow evaluation of cartilage, thereby serving as a biological MR contrast agent on T1-

weighted images (45). It is unclear whether this phenomenon can be useful in the cervical spine.

Postcontrast T1-weighted images are sensitive in detecting periarticular inflammation as a result of T1 shortening by Gd-DTPA (28). In a study in which laboratory animals were used to discriminate inflammatory tissue from normal tissues of intermediate intensity, precontrast T2-weighted images were as effective as postcontrast T1-weighted images (46). Later studies showed a dramatic increase of signal intensity on T1-weighted images after Gd-DTPA administration, which greatly increased the

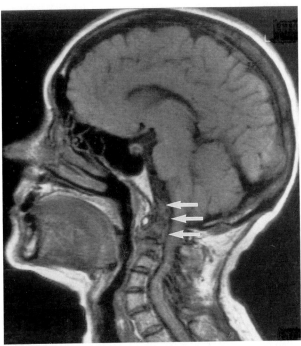

Figure 42.8. T1-weighted sagittal image shows complete destruction of C2. Extensive pannus tissue is present and is seen as a low signal intensity mass extending into the foramen magnum (*arrows*).

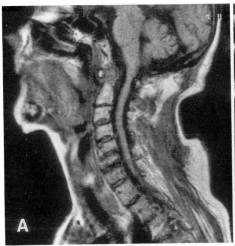
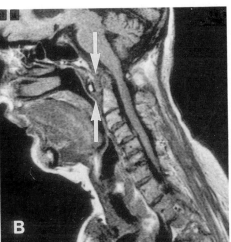

Figure 42.7. Dynamic studies: cervical spine in extension **(A)** and flexion **(B)** Sagittal T1-weighted images obtained with body coil show the unstable relationship between the atlas and the dens axis, resulting in a horizontal atlantoaxial dislocation with the spine in flexion. Although movements of the head and cervical spine are limited to the bore of the body coil, the instability of the C1–C2 relationship is clearly visualized. Pannus formation of intermediate signal intensity is situated around the dens axis (*arrows*).

contrast between pannus and effusion as compared to T2-weighted images (20).

Gd-DTPA contrast-enhanced images can be used to image active pannus tissue, which might be important to monitor therapy. Gd-DTPA contrast-enhanced images can also be helpful in the distinction between joint effusion and pannus formation (28).

References

1. Bland JH. Rheumatoid arthritis of the cervical spine. Bull Rheum Dis 1967;18:471–476.
2. Weissman BW, Aliabadi P, Weinfeld MS, Thomas WH, Sosman JL. Prognostic features of atlanto-axial subluxation in rheumatoid arthritis patients. Radiology 1982;144:745–751.
3. Komusi T, Munro T, Harth M. Radiologic review: the rheumatoid cervical spine. Semin Arthr Rheum 1985;14:187–195.
4. Park WM, O'Neill M, McCall LW. The radiology of rheumatoid involvement of the cervical spine. Skel Radiol 1979;4:1–7.
5. de Carvalho A, Graudal H. The course of atlanto-axial involvement and disc narrowing of the cervical spine in rheumatoid arthritis. Röntgen Fortschr 1981;135:32–37.
6. Mikulowski P, Wollheim FA, Rotmil P, Olsen I. Sudden death in rheumatoid arthritis with atlanto-axial dislocation. Acta Med Scand 1975;198:445–451.
7. Anda S, Nilsen G, Rysland P. Periodontoid changes in rheumatoid arthritis; MRI observations. Report of two cases. Scand J Rheumatol 1988;17:59–62.
8. Masaryk TJ, Modic MT, Geisinger MA, et al. Cervical myelopathy: a comparison of magnetic resonance and myelography. J Comput Assist Tomogr 1986;10:184–194.
9. Modic MT, Weinstein MA, Pavliceck W, Boumphrey F, Starnes D, Duchesneau PM. Magnetic resonance imaging of the cervical spine: technical and clinical observations. AJR 1983;141:1129–1136.
10. Reynolds H, Carter SW, Murtagh FR, Rechtine GR. Cervical rheumatoid arthritis: value of flexion and extension views in imaging. Radiology 1987;164:215–218.
11. Sze G, Brant-Zawadzki MN, Wilson CR, Normanan D, Newton TH. Pseudotumor of the craniovertebral junction associated with chronic subluxation: MR imaging studies. Radiology 1986;161:391–394.
12. Lee BCP, Deck MDF, Kneeland JB, Cahill PT. MR imaging of the cranio cervical junction. Am J Neuroradiology 1985;6:209–213.
13. Wimmer B, Friedburg H, Herning J, Kauffman GW. Möglichkeiten der diagnostischen Bildgebung durch Kernspintomographie. Radiologe 1986;26:137–143.
14. Zacher J, Reiser M. Kernspinntomographie der rheumatischen Halswirbelsäule. Aktuel Rheumatol 1987;12:104–111.
15. Algra PR, Breedveld FC, Vielvoye GJ, Doornbos J, de Roos A. MRI of the cranio-cervical region in rheumatoid arthritis. J Med Imag 1987;1:263–267.
16. Petterson H, Larsson EM, Holtås S, et al. MR imaging of the cervical spine in rheumatoid arthritis. Am J Neuroradiol 1988;9:573–577.
17. Semble EL, Elster AD, Loeser RF, Laster DW, Challa VR, Pisko EJ. Magnetic resonance imaging of the craniovertebral junction in rheumatoid arthritis. J Rheumatol 1988;15:1367–1375.
18. Fezoulidis I, Neuhold A, Wicke L, Seidl G, Eydokimidis B. Diagnostic imaging of the occipito-cervical junction in patients with rheumatoid arthritis. Eur J Radiol 1989;9:5–11.
19. Lipson SJ. Rheumatoid arthritis in the cervical spine. Clin Orthoped Rel Res 1989;239:121–127.
20. Reiser MF, Bongartz GP, Erlemann R, et al. Gadolinium-DTPA in rheumatoid arthritis and related diseases: first results with dynamic magnetic resonance imaging. Skel Radiol 1989;18:591–597.
21. Yulish BS, Lieberman JM, Newman AJ, Bryan PJ, Mulopulos GP, Modic MT. Juvenile rheumatoid arthritis: assessment with MR imaging. Radiology 1987;165:149–152.
22. Larsson EM, Holtås S, Zygmunt S. Pre- and postoperative MR imaging of the craniocervical junction in rheumatoid arthritis. AJR 1989;152:561–566.
23. Beltram J, Caudill JL, Herman LA, et al. Rheumatoid arthritis MR imaging manifestations. Radiology 1987;165:153–157.
24. McAfee PC, Bohlman HH, Han JS, Salvagno RT. Comparison of nuclear magnetic resonance imaging and computed tomography in the diagnosis of upper cervical spinal cord compression. Spine 1986;11:295–304.
25. Terrier F, Hricak H, Revel D, et al. Magnetic resonance imaging and spectroscopy of the periarticular inflammatory soft-tissue changes in experimental arthritis of the rat. Invest Radiol 1985;20:813–823.
26. Neuhold A, Fezoulidis I, Frühwald F, et al. MRI of the occipito-cervical junction in rheumatoid arthritis. Fortschr Röntgenstr 1989;150:413–416.
27. Senac MO Jr, Deutsch D, Bernstein BH, et al. MR imaging in juvenile rheumatoid arthritis. AJR 1988;150:873–878.
28. Kursunoglu-Brahme S, Riccio T, Weisman MH, et al. Rheumatoid knee: role of gadopentetate enhanced MR imaging. Radiology 1990;176:831–835.
29. Pocock DG, Agnew JE, Wood EJ, Bananan EC, Valentine AR. Radionuclide imaging of the neck in rheumatoid arthritis. Rheumatol Rehab 1982;21:131–138.
30. Zygmunt SC, Ljunggren B, Alund M, et al. Realignment and surgical fixation of atlanto-axial and subaxial dislocations in rheumatoid arthritis (RA) patients. Acta Neurochir Suppl (Wien) 1988;43:79–84.
31. Zygmunt SC, Säveland HG, Brattström H, Ljunggren B, Larsson EM, Wollheim F. Reduction of rheumatoid periodontoid pannus following posterior occipito-cervical fusion visualized by magnetic resonance imaging. Br J Neurosurg 1988;2:315–320.
32. Alarcon GS. The use of magnetic resonance imaging (MRI) in patients with rheumatoid arthritis and subluxations of the cervical spine [letter]. Arthr Rheum 1988;31:304.
33. Yamashita Y, Takahashi M, Sakamoto Y, Kojima R. Atlanto-axial subluxation; radiography and magnetic resonance imaging correlated to myelopathy. Acta Radiol 1989;30:135–140.
34. Lignière GC, Montagnani G, Panarace G, Gualdi I. Magnetic resonance imaging for the study of cervical myelopathy in rheumatoid arthritis. Clin Exp Rheumatol 1988;6:343–346.
35. Colée G, Breedveld FC, Algra PR, Padberg GW. Rheumatoid arthritis with vertical atlanto-axial subluxation complicated by hydrocephalus. Br J Rheum 1987;26:56–58.
36. Kawaida H, Sakou T, Morizono Y. Vertical settling in rheumatoid arthritis. Diagnostic value of the Ranawat and Redlund-Johnell methods. Clin Orthoped Rel. Res. 1989;239:128–135.
37. Fuentes JM, Benezech J. Radiologic diagnosis of the craniovertebral joint instability. Radiologie J Cepur 1989;9:71–77.
38. Poznanski AK, Conway JJ, Shkolnik A, Pachman LM. Radiological approaches in the evaluation of joint disease in children. Rheum Dis Clin N Am 1987;13:57–73.
39. Poznanski AK, Glass RBJ, Feinstein KA, Pachman LM, Fischer MR, Hayford JR. Magnetic resonance imaging in juvenile rheumatoid arthritis. Int Pediatr 1988;3:304–311.
40. Aisen AM, Martel W, Ellis JH, McCune WJ. Cervical spine involvement in rheumatoid arthritis: MR imaging. Radiology 1987;165:159–163.
41. Breedveld FC, Algra PR, Vielvoye CJ, Cats A. Magnetic resonance imaging in the evaluation of patients with rheumatoid arthritis and subluxations of the cervical spine. Arthr Rheum 1987;30:624–629.
42. Bundschuh C, Modic MT, Kearny F, Morris R, Deal C. Rheumatoid arthritis of the cervical spine: surface-coil MR imaging. Am J Neuroradiology 1988;9:565–571.
43. Hensinger RN, DeVito PD, Ragsdale CG. Changes in the cervical spine in juvenile rheumatoid arthritis. J Bone Joint Surg (Am) 1986;68:189–198.
44. Condon BR, Hadley DM. Quantification of cord deformation and dynamics during flexion and extension of the cervical spine using MR imaging. J Comput Assist Tomogr 1988;12:947–955.
45. Hajek PC, Sartoris DJ, Neumann CH, Resnick D. Potential contrast agents for MR arthrography: in vitro evaluation and practical observations. AJR 1987;149:97–104.
46. Terrier F, Revel D, Reinhold CE, et al. Contrast enhanced MRI of periarticular soft-tissue changes in experimental arthritis of the rat. Magn Res Med 1986;3:385–396.

CHAPTER 43

Quantitative Computed Tomography for the Assessment of Osteoporosis

Harry K. Genant, Claus-C. Glüer, and Peter Steiger

Introduction

In recent years considerable effort has been expended in the development of methods for quantitatively assessing the skeleton so that osteoporosis can be detected early, its progression and response to therapy carefully monitored, and its risk effectively ascertained. There is not yet, however, a consensus on which method or methods are most efficacious for diagnosing and monitoring the individual patient or for extensive screening of large populations (1). In this regard, the selection of anatomic sites and of methods for quantifying skeletal mass is of considerable current importance.

The skeleton as a whole is composed of about 80% cortical or compact bone and 20% trabecular or cancellous bone (2). The appendicular skeleton is composed of predominantly cortical bone, while the spine is composed of a combination of cancellous bone, predominantly in the vertebral bodies, and compact bone, mostly in the dense end plates and posterior elements. Trabecular bone, because of its high surface:volume ratio, has a presumed turnover rate about eight times that of compact bone, and is highly responsive to metabolic stimuli (2, 3). This high turnover rate in trabecular bone makes it a prime site for detection of early bone loss as well as the monitoring of response to various interventions. The clinical and epidemiological observation that osteoporotic fractures occur first in the vertebral bodies or distal radius, areas of predominantly trabecular bone, substantiates physiologic studies showing a differential early loss from this bone compartment (3).

Numerous methods have been used for quantitative assessment of the skeleton in osteoporosis with variable precision, accuracy, and sensitivity. The first methods to be developed were radiogrammetry (4) and photon absorptiometry (5–8), which measure primarily cortical bone of the peripheral appendicular skeleton. In the past decade, techniques have become available that can quantify bone mineral content in the spine, the site of early osteoporosis. Quantitative computed tomography (QCT) (9–41) provides a measure of purely trabecular bone of the vertebral spongiosum, or other sites, while dual-photon absorptiometry (DPA) and dual X-ray absorptiometry (DXA) (41–48) measure an integral of compact and cancellous bone of the spine, hip, or entire skeleton. The focus of this chapter is on the methodology and clinical application of QCT and its comparison with other methods commonly used to quantitatively assess the skeleton.

Computed Tomography for Bone Mineral Analysis

Computed tomography (CT) has been widely investigated and applied as a means for noninvasive quantitative bone mineral determination (9–41). The usefulness of computed tomography for measurement of bone mineral lies in its ability to provide a quantitative image and, thereby, measure trabecular, cortical, or integral bone, centrally or peripherally. For measuring the spine, the potential advantages of QCT (21, 23, 35) are its capability for precise three-dimensional anatomic localization, providing a direct density measurement, and its capability for spatial separation of highly responsive cancellous bone from less responsive compact bone. The lumbar vertebrae contain substantial amounts of compact bone (60 to 80%), with only part of the spinal mineral being high-turnover trabecular bone (20 to 40%) (29, 49, 50). The sensitivity of a technique measuring an integral of compact and cancellous bone (such as area projection with DPA or DER) may be low compared to QCT due to inclusion of low-turnover compact bone and extraosseous mineral such as osteophytes, sclerosis due to fractures and osteophytosis, or aortic calcification. The selective localization and the direct density measurement provided by QCT permit exclusion of these causes of low sensitivity or error and inclusion of purely trabecular bone.

Quantitative computed tomography has been shown to measure changes in trabecular mineral content in the spine and in the radius and tibia with sensitivity and precision

Abbreviations (see also glossary): QCT, quantitative computed tomography; DPA, dual-photon absorptiometry; DER, dual energy radiography; DXA, dual X-ray absorptiometry; SEQCT, single-energy QCT; DEQCT, dual-energy QCT; ROI, region of interest; SPA, single photon absorptiometry; CCT, combined cortical thickness; FXI, fracture index; BMC, bone mineral content; HU, Hounsfield unit.

(9–41). The extraction of this quantitative information from the CT image, however, requires sophisticated calibration and positioning techniques and careful technical monitoring. Specifically designed, small-scale CT scanners using isotope or X-ray sources have also been developed and applied, principally on a research basis, for measurement of the appendicular trabecular and cortical skeleton (27, 37, 39).

Technical Considerations

Background

Since the discovery of X-rays by Roentgen in 1895, imaging techniques have steadily been refined and made more clinically useful. A major limitation of these techniques is their two-dimensional display of a three-dimensional structure, inevitably leading to a loss of sensitivity due to superposition of all structures along the ray path. This limitation was finally overcome with the advent of computed tomography, which displays the three-dimensional structure of the object slice by slice. The cross-sectional slice under examination is irradiated subsequently from all directions by stepwise rotation of the X-ray source and the detector ($\phi = 0$ to $360°$) around the object. For each step, the projected data $p(x', \varphi)$ are stored digitally. If the section under examination is described by a density function $\mu(x,y)$ representing the linear attenuation coefficient, the projected data at an angle ϕ are given by a linear array of ray-sums p:

$$p(x',\phi) = \int \mu(x,y)\, dy'$$

with x', y' describing the rotated coordinate system.

By inverting this equation we can compute the two-dimensional distribution of $\mu(x,y)$, and by repeating the measurement for adjacent slices we can calculate and display the complete three-dimensional information. Several techniques have been developed for this purpose, namely back projection (51, 52), iterative reconstruction (53, 54), and analytical reconstruction methods such as two-dimensional Fourier reconstruction (53) or filtered back projection (55). Brooks and di Chiro describe the different methods in detail (56) and, in another paper, give a short overview and comparison (57).

If the CT scanner were perfect, the reconstruction algorithm would yield a density function $\mu(x,y)$ that would accurately reflect the distribution of the attenuation and, if the linear attenuation coefficients of the materials within the field of view are known, give quantitative rather than just qualitative information about the density distribution of these structures. Techniques of QCT developed in the last decade succeed in determining the density of a variety of clinically important substances with remaining errors of (or sometimes lower than) the percent level. A major step in overcoming inperfections of real CT scanners has been the introduction of calibration techniques that use calibration phantoms of a known element and density distribution.

Calibration

In determining the density function $\mu(x,y)$ most CT scanners calculate the CT number H rather than the linear attenuation coefficient μ. H, expressed in units of Hounsfield (HU) in honor of the developer of the first clinical CT scanner, is given by the equation

$$H(E) = 1000\left[\frac{\mu(E)}{\mu_W(E_0)} - 1\right]$$

where $\mu(E)$ is the linear attenuation coefficient at the effective energy E and $\mu_W(E_0)$ is the linear attenuation coefficient of water. For $E = E_0$ the CT number of water becomes zero; the CT number of air, -1000. Obviously the Hounsfield number is energy dependent. To determine bone mineral density, the CT number of the patient is compared to a reference standard of known composition and density.

Two different calibration methods are used, simultaneous and nonsimultaneous. Simultaneous calibration was introduced by Cann and Genant (58) and is used in most clinical applications; nonsimultaneous calibration was used with limited success (59) in some early studies (60), and now there is renewed interest in this method (61).

For simultaneous calibration the patient is placed on top of a calibration phantom that has inserts of known mineral density running perpendicular to the plane of the CT slices. The Cann-Genant phantom (Fig. 43.1) (58), which is in use at over 1000 centers worldwide, has cylindrical channels of solute dipotassium hydrogen phosphate (K_2HPO_4) or 50, 100, and 200 mg/cm^3 concentrations along with a water- and fat-equivalent channel (60% ethanol). K_2HPO_4 has attenuation characteristics very similar to those of calcium hydroxyapatite ($Ca_{10}(PO_4)_6(OH_2)$), which closely represents the bulk of the mineral found in bone.

For calibration, a linear regression of the mean CT numbers within the bone equivalent calibration phantom channels, H_{CB}, versus their nominal concentrations MIN_{CB}, is performed for each individual slice:

$$H_{CB} = S_B\, MIN_{CB} + H_W$$

Slope of S_B of bone and intercept H_W, corresponding to the CT number of water, characterize the calibration equation. The linear regression is very good ($r^2 \simeq 0.9999$), regardless of the shape of the object under study (62).

Aqueous solutions, as are used in this phantom, have the potential drawback of limited long-term stability due to the production of gas bubbles, precipitation of the dissolved materials, and impurities (63). Because of these drawbacks, solid-state reference phantoms have been developed. They are totally stable, i.e., their attenuating properties do not change with time, and they are sturdier and more resistant to damage.

Kalender and Süss (62) report on a solid phantom with a small cross-section (2.5 $\times$ 9 cm^2). This phantom incor-

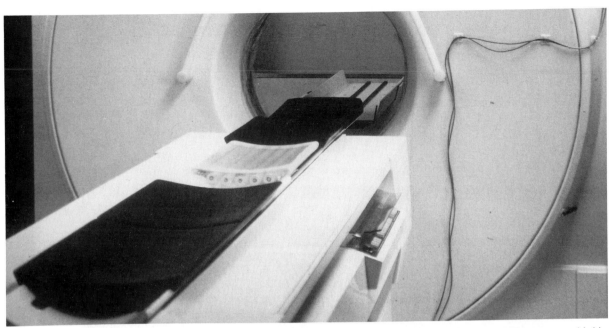

Figure 43.1. Standardized version of Cann-Genant CT mineral calibration phantom (58) in use at over 1000 centers worldwide.

porates only two samples (a 200 mg/cm³ calcium hydrox-yapatite sample and a water-equivalent sample). The density variations of this phantom were found to be better than 0.1%, a tolerance that previously had been difficult to achieve with other solid phantoms (64). Similarly, Arnold (65) has reported on a solid calcium hydroxyapatite-based phantom for simultaneous QCT calibration.

Simultaneous calibration corrects to a large extent for short- and long-term scanner instability. With nonsimultaneous calibration techniques, which are generally used on fourth generation scanners, a quasianthropomorphic tissue-equivalent calibration phantom is scanned before and/or after the patient (66, 67). By mounting one or two additional attenuator rings, the phantom can be matched approximately to the patient's size. Obviously, nonsimultaneous reference phantoms cannot correct for scanner instabilities between the calibration measurements and the subjective choice of attenuator rings is only an approximation of patient size and composition. Moreover, it might lead to reproducibility errors if chosen differently.

Practical Considerations

Quantitative Computed Tomography of the Spine

Vertebral mineral density is determined by measuring representative volumes (about 2 to 3 cm³/vertebra) of purely trabecular bone. The ability to measure that area selectively is advantageous because the turnover rate of trabecular bone is approximately eight times that of cortical bone of the spine (68). The patient is positioned on top of the calibration phantom. To avoid artifacts, a bolus bag is used to fill the air gap between patient and phantom. Using a lateral computed radiograph ("radiographic localizer," "scout view," "topogram") of the lumbar spine and cursor-determined coordinates, the midplanes of typically three to four vertebrae (T12 to L4) are identified. T12 may have

to be excluded depending on the magnitude of superimposed lung tissue. Lordosis is reduced by elevating the knees with a pad. Slices of 8 to 10-mm thickness are taken parallel to the vertebral end plates by tilting the gantry appropriately (Fig. 43.2).

For single-energy QCT measurements, a low-dose, low-energy setting should be chosen if possible. This results in an organ dose of typically 200 to 300 mrem and essentially no gonadal exposure. [This is 1/10 the normal imaging dose per slice, 1/10 the usual number of slices per study, and, therefore, 1/100 the total integral dose of a routine abdominal CT imaging study (18, 34, 69).] The low-voltage setting also provides a relatively high sensitivity of mineral to fat variations (70). Care should be taken to ensure that the patient fits entirely in the field of view and does not move during the exposure. To check for any movement, a second localizer radiograph after the scan can be taken.

For dual-energy QCT measurements, the lower energy setting should be as low as possible, with an optimum effective energy of about 40 keV (71) or a peak voltage setting of 65 kVp (64). Current scanners allow measurements to be taken at about 80 kVp, which corresponds to about 55 keV effective energy. This setting is almost as good as 40 keV (71). The upper energy should be as high as possible (typically 120 or 140 kVp), and the dose should be about equal at both energies (71). For postprocessing DEQCT the organ dose is approximately 600 mrem. Preprocessing DEQCT usually requires higher millamperage settings (59).

Automated evauation optimizes reproducibility and simplifies the operator's task. Recent image evaluation software incorporating contour tracking techniques allows for anatomically adapted and automatically placed regions of interest (ROIs) (72, 73). Calibration is also performed automatically, with the option of operator interference in

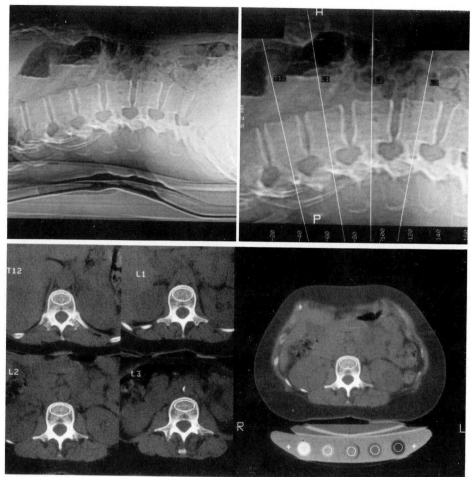

Figure 43.2. QCT spine technique using GE 9800 CT scanner. Lateral scout view provides a rapid and simple localization approach in which the midplane of four vertebral bodies are defined on the video monitor and a single 10-mm thick section is obtained at each level. An oval region of interest, centered in the midvertebral body, is used to determine cancellous bone mineral content (mg/cm³) while circular regions of interest are used to quantify the K_2HPO_4 solutions in the calibration phantom. (Reprinted with permission from Genant HK, Steiger P, Block JE, Ettinger B, Harris ST. Quantitative computed tomography: update 1987 [Editorial]. Calcif Tissue Int 1987;41:174–186.)

case the calibration ROIs are misplaced (73). Calibrated bone mineral density can be averaged over all vertebrae to improve accuracy (70).

Such automated image evaluation packages are already being offered commercially. Moreover, automated determination of the midvertebral CT slice, which has been presented recently (74), may help to further reduce precision errors.

Quantitative Computed Tomography of the Hip

There has been limited research on the clinical utility of QCT of the hip. The complex structural anatomy of the proximal femur makes it difficult to measure specific kinds of skeletal tissue selectively. Recent advances in image processing techniques, however, have opened up the possibility of performing site-specific QCT of the hip (75).

Three different methods of calculating trabecular bone mineral density at the proximal femur have been reported (76, 77). In each, the region to be examined must be scanned contiguously to obtain complete three-dimen-

sional information. Also, in all of the techniques, the standard calibration phantom developed for spine scans is used (58).

Using reformatting techniques, Reiser and Genant (76) calculated images of transverse slices perpendicular to the longitudinal axis of the proximal femur that had been defined by the center of the femoral head and the center of the femoral neck at its smallest width. The CT values of 4 to 8 of these transverse slices of 5-mm thickness were integrated, thus covering 5 to 12 cm³ of bone volume in the femoral neck. The reproducibility of the quantitative measurements was found to be better than 3%.

Results obtained with two other techniques have recently been published by Sartoris et al. (77). From a frontal localization image, they determined the region from the head-neck junction to the lesser trochanter and subsequently scanned it with contiguous 5- to 10-mm slices. The data were analyzed with ROI software much as for quantitative vertebral bone analysis. Several nonoverlapping ROIs were chosen to cover the maximum possible area of trabecular bone in each axial slice. From these data, the average trabecular bone density was calculated.

In the third approach, a threshold pixel value of 200 mg K_2HPO_4/cm^3 was chosen to display the inner and outer contour of the cortical bone areas of each slice. Averaging over all pixels inside these areas yielded a mean value for trabecular bone density.

Comparing short-term reproducibility of the two latter techniques, three scans of the same cadaver yielded a coefficient of variation of 2.2% for the axial slice technique and 0.7% for the histogram analysis technique. The volume of interest was 14 to 19 and 33 to 37 cm^3, respectively.

A comparison of these techniques is difficult because of the limited number of observations. However, the histogram analysis technique has the advantages of reproducibility and simplicity, provided the necessary software is available.

Especially in developing automated techniques for defining volumes of interest, reformatting techniques and three-dimensional representation (78–80) might be helpful in understanding which region of the bone is selected and which is excluded.

Validity of Quantitative Computed Tomography

Functions Describing the Validity of Quantitative Computed Tomography

Quantitative CT has the capability to provide exact results but care must be taken to minimize several possible error sources. The specific functions chosen to describe the validity of the method depend on the question QCT is supposed to answer. Most basic research on the validity of QCT has focused on the question: how well can QCT determine the true bone mineral density of an object? The answer usually is given in terms of two functions: *accuracy* and *precision*. In clinical practice these two functions can be used to describe the performance of QCT both for cross-sectional studies (using QCT to determine the degree of osteopenia) and for longitudinal studies (using QCT to quantify bone loss).

To describe the performance of QCT with respect to other techniques, e.g., its capability to discriminate fractured and nonfractured patients, several concepts have been applied: the *sensitivity* of the techniques can be defined as the ratio of *responsiveness* (e.g., measured rate of bone loss at a given site) to precision of the measurement or it can be used along with *specificity* in the context of *relative (or receiver) operator characteristic (ROC)* analysis.

Finally, there is the question of relevance for screening strategies: is QCT a good risk predictor for bone diseases (e.g., hip fractures)? That answer could be given in terms of *relative risk factors*. Very little has been done so far to relate these three questions. The implicit assumption of most authors, that the best method of determining bone mineral density is also best suited for discriminating diseased and nondiseased patients and, in turn, is the best predictor of risk, has not been tested. Furthermore, it does not follow that the best method for predicting risk is also the best method for studying the causes of bone metabolism. The following discussion is limited to the accuracy and precision of QCT.

Accuracy

The accuracy of a bone mineral measurement technique is defined as the deviation of the results obtained with that technique from the true mineral content, usually expressed as a percentage. For QCT, bone mineral density in milligrams per cubic centimeter of the calibration material (K_2HPO_4 or $Ca_{10}(PO_4)_6(OH)_2$) is usually compared to the results of some "gold standard," typically gravimetric or spectroscopic determination of the mineral in a sample following ashing at ~600°C for 48 to 96 hr. The results of that comparison are correct within 0.5 to 1% if the ROI of the QCT measurement is thoroughly matched with the volume ashed during chemical analyses. If these analyses are not properly done, however, an error of up to 3 to 4% may be introduced, falsely reflecting on the accuracy of QCT (59).

Results of accuracy of SEQCT and DEQCT are summarized in Table 43.1 (10, 20, 36, 75, 81–85). Accuracy data for trabecular bone are usually determined by the coefficient of variation around the regression line of QCT versus ash weight; the assumption is that the average underestimation can be corrected by using scanner-specific normative data. For SE, and to some extent also for DE techniques, the accuracy error depends on the variability of the fat

Table 43.1
Accuracy of Bone Mineral Measurements by Quantitative Computed Tomography

Accuracy (%)	Site	Bone	Technique	Ref.	Year
1	Femur	Cortical	SE (isotope)	Rüegsegger et al. (81)	1974
~6[a]	Tibia/fibula	Cortical	SE	Reich et al. (82)	1976
3–4[a]	Dog bones	Cortical	SE	Posner and Griffiths (83)	1977
1.9	Calcium phantom	Cortical	SE	Genant and Boyd (20)	1977
8	Vertebra	Trabecular	SE	Rohloff et al. (36)	1982
4.3	Distal femur	Trabecular	SE	Adams et al. (9)	1982
26	Vertebra	Trabecular	SE	Laval-Jeantet et al. (84)	1986
10.7	Vertebra	Trabecular	SE	Burgess et al. (85)	1987
13.2	Vertebra	Trabecular	DE	Glüer et al. (70)	1987
9.6	Vertebra	Trabecular	DE	Laval-Jeantet et al. (84)	1986
7.3	Vertebra	Trabecular	DE	Burgess et al. (85)	1987
7	Vertebra	Trabecular	DE	Glüer and Genant (75)	1987

[a] See Goodsitt and Rosenthal (61).

Table 43.2
Reproducibility of Bone Mineral Parameters Measured by Computed Tomography

Bond	Short Term In Vitro (%)	Short Term In Vivo (%)	Long Term In Vitro (%)	Long Term In Vivo (%)	Ref.
Radius	0.2	1.4–2			Isherwood et al. (86)
Radius		2			Rüegsegger et al. (37)
Phantom	1				Genant and Boyd (20)
Vertebra	25				Bradley et al. (87)
Tibia		1.8			Liliequist et al. (88)
Radius	2				Orphanoudakis et al. (31)
Vertebra	1.5		2.8		Cann and Genant (58)
Radius				0.3	Rüegsegger et al. (89)
Radius		0.6			Hangartner and Overton (90)
Vertebra		1.6			Genant et al. (21)
Vertebra	2–3				Firooznia et al. (91)
Radius				0.5	Sashin et al. (92)
Vertebra			2.2		Graves and Wimmer (93)
Vertebra		1.6			Rosenthal et al. (94)
Vertebra		4–5			Meier et al. (95)
Vertebra				0.8	Cann et al. (96)

content of the specimens and the sensitivity of the scanner to fat changes. The data of Burgess et al. (85) and Glüer et al. (70) can be taken as an estimate for a population that covers a wide age range excluding severe osteoporotics.

The physical and physiologic factors affecting accuracy are discussed after the next section and are summarized in Table 43.3.

Precision

The precision of a bone mineral examination technique is defined as the deviation of the outcome of a set of measurements about the expected value. The results of studies on the precision of QCT are summarized in Table 43.2 (20, 21, 31, 37, 58, 86–96). *Precision* and *reproducibility* are terms that are usually used synonymously. Clearly, a precise measurement need not be accurate. Precision can be calculated in two ways.

1. If the bone mineral density of the object is kept unchanged, precision can be determined from n repetitive measurements by using

$$\text{Precision} = \sqrt{\sum_{i=1}^{n} (y_i - \bar{y})^2/(n-1)}$$

where y_i is the outcome of an individual measurement and

$$\bar{y} = \sum_{i=1}^{n} y_i/n \quad \text{(mean of } y_i)$$

2. If bone mineral density can be assumed to have changed in a linear way between subsequent measurements, precision equals the standard error of the estimate of linear regression on the data points, i.e.,

$$\text{Precision} = \sqrt{\sum_{i=1}^{n} (y_i - \hat{y}_i)^2/(n-2)}$$

$$\hat{y}_i = ax_i + b \quad \text{(estimate of } y_i)$$

where a is the slope and b is the intercept of the regression curve.

In view of dosage considerations, repetitive measurements on the same patient may be impossible, and method 2 is preferable for determining in vivo precision. It should be remembered that since there is no way of testing the assumption of linearity, the standard error of the estimate yields just an upper limit of precision.

The long-term reproducibility of a measurement technique defines its clinical utility, since it not only determines an upper limit for accuracy for single measurements. For instance, if a method has a precision of ± 21 mg, the change between two measurements must be 5.7 mg before we can have 95% confidence that the change is real (97). This follows from the law of error propagation: the standard deviation of the difference of two measurements equals $\sqrt{2}$ times the standard deviation of a single measurement and the 95% confidence interval again is larger by about another factor of 2.

The physical and physiological factors that limit precision are described in the following section.

Factors Influencing the Validity of Quantitative Computed Tomography

Bone mineral density as calculated from CT scans differs from the actual density due to several factors. Understanding their origin helps to avoid or limit these error sources. Table 43.3 lists the most prominent factors and indicates whether these factors influence primarily accuracy or precision.

Repositioning

In longitudinal studies it is important to evaluate the same volume of interest in each measurement. Two principal repositioning methods are being reported: computed ra-

Table 43.3
Factors Influencing the Validity of Quantitative Computed Tomography[a]

Factor	Parameter Affected[a]
Technique (SE, DE)	A, P
Quantum noise versus dose	P
Partial volume effects versus resolution	A
Beam-hardening effects	A
Field inhomogeneity	A, P
Scanner stability	P
Interscanner differences	P
Positioning	P
Object size	A
Material composition	A, (P)
Evaluation technique (automated)	P

[a]A, principally influencing accuracy; P, principally influencing precision.

diographs are generally used in clinical environments whereas three-dimensional repositioning software is being developed for research purposes (98). Recently, a method for automated determination of the midvertebral CT slice has been reported (74). Finally, an intermediate approach uses the computed radiograph for longitudinal axis positioning and automated techniques for the definition of the ROI on the axial slices.

In computed radiographic localization systems (scout view, tomogram), lateral radiographs are usually taken for spinal measurements, and anteropoterior radiographs for hip measurements. The operator can address the coordinate system that is logged to the image and uses trackballs to define the positions of the scans to be taken. Generally these systems are accurate to ±2 mm and reproducible to ±1 mm, which results in an error in the definition of the volume of less of than 10% (99). Depending on the homogeneity of the area under study, a 2-mm repositioning longitudinal error again results in a precision error of bone mineral content of 2% for trabecular bone of the vertebrae (58). The error is probably greater at sites of greater longitudinal inhomogeneity, such as the tibia (100), distal femur (10), and radius (90). The new technique of automated determination of the midvertebral CT slice on the localization image promises reduction of errors that are due to operator influence (74).

Whereas the localizer radiograph defines the region to be scanned, the selection of the ROI in the transverse *x–y* plane is performed during the evaluation process and can be corrected if necessary. Automated selection of the ROI eliminates operator influence and is considered to offer better reproducibility, especially if different operators are involved (58, 73, 101). In spinal studies contour tracking algorithms (72, 73, 102) trace the cortical walls of the vertebral body, the spinal canal, and other landmarks to select the optimum (no cortex) position and size (as large as possible (101) of the ROI (Fig. 43.3). For hip studies, no automated procedure has been reported yet. For the spine, a 1-pixel variation in the *x–y* plane introduces a variance of 0.8 HU in the measured area (58). Automated selection of the ROI should yield even better results (59). In studies of the knee bone, mineral density and shape measurements could be reproduced with 0.5 and 1% precision (73). Hence, ROI positioning errors are small compared to the errors introduced by the localizer radiograph.

Automated three-dimensional ROI positioning will further reduce the repositioning error since current scanners have resolving limits of 0.5 to 1 mm in all three axes. If a set of contiguous or overlapping slices is obtained, reconstruction in any plane of the scanned volume can be performed by means of multiplanar reformation techniques (98). These techniques will be helpful for evaluation of nonaxial structures such as the femoral neck (75, 76).

Clinical Applications of Quantitative Computed Tomography

Background and Models of Bone Loss

Quantitative CT has been used to model patterns of vertebral bone diminution with aging, to evaluate skeletal status in health and disease, predict subsequent fracture risk, and to monitor treatment interventions (15, 21, 24, 91, 103). New data from the Osteoporosis Research Group [University of California, San Francisco (UCSF)] based on a study group of 538 healthy women between the ages of 20 and 80 indicate little skeletal involution of spinal ular bone prior to the onset of menopause (Fig. 43.4). Various statistical regressions were performed for the entire population to describe the general pattern of bone loss from the spine; linear, quadratic, cubic, and logarithmic models were found to be equally satisfactory in characterizing this pattern with *R* values of 0.65, 0.67, 0.68 and 0.67, respec-

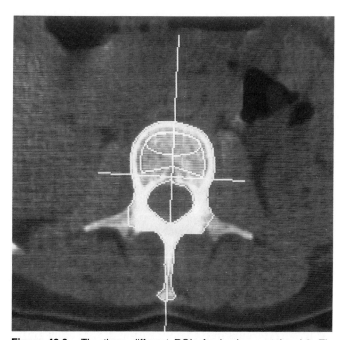

Figure 43.3. The three different ROIs for lumbar vertebra L2. The determination of the ROIs and their statistical analysis take 3 to 4 s of a database VAX 11/780. (Reprinted with permission from Steiger P, Steiger S, Rüegsegger P, Genant HK. Two- and three-dimensional quantitative image evaluation techniques for densitometry and volumetrics in longitudinal studies. In: Genant HK, ed. Osteoporosis update 1987. Berkeley: University of California Press, 1987:171–180.)

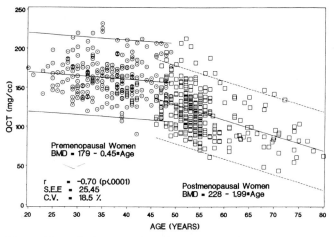

Figure 43.4. Statistical model of bone loss over the lifetime, indicating insignificant premenopausal bone loss, a large decrement at menopause, and rapid postmenopausal loss.

tively. Quantified values were stratified into 5- and 10-year age brackets, and analyzed separately for pre- and post-menopausal women. The 5- and 10-year interval stratification revealed no identifiable bone mineral decrements prior to midlife; significant losses of bone mineral were noted to correspond with the usual time of menopause and to continue into old age. The statistical model of skeletal atrophy is best described as a two-phase regression, consisting of a premenopausal period of skeletal consolidation followed by a period of exponential postmenopausal involution.

Discrimination Capability of Quantitative Computed Tomography

Researchers using QCT have uniformly observed a statistically significant separation between patients with spinal compression deformities and age- and sex-matched comparison subjects without fractures (15, 104). Cann and colleagues (15) found that female osteoporotics had a mean decrement in bone mineral values of 48 mg/cm^3 (39%) compared to normal subjects, whereas males with compression fractures had a mean decrement of 66 mg/cm^3 (50%) compared to men of the same age without fractures. In general, vertebral deformities are absent in individuals with QCT mineral values above 110 mg/cm^3, while almost all patients with bone mineral values below 65 mg/cm^3 have radiographic evidence of vertebral osteoporosis (Fig. 43.5). Although there is significant overlap in patients with intermediate values, it appears that QCT is useful in identifying those patients at increased risk of vertebral fracture. Firooznia and colleagues likewise found a moderate statistical separation between spinal osteoporotics and comparison subjects (104). In a cohort of 96 women with evidence of vertebral compressions, 66% had bone mineral values in the spine below the fifth percentile for age-matched normal subjects, while 85% of these same

subjects had values below the fifth percentile for premenopausal women.

The QCT technique has been used less often to discriminate patients who have suffered spontaneous hip fracture from normal comparison subjects. In one study (18), the trabecular bone mineral content of the lumbar spine was measured in 185 women aged 47 to 84 years with vertebral fracture ($n = 74$), hip fracture ($n = 83$), and both vertebral and hip fracture ($n = 28$). Eighty-seven percent of the vertebral fracture patients, 38% of the hip fracture patients, and 82% of the vertebral and hip fracture patients had spinal bone mineral content (BMC) values below the fifth percentile for healthy premenopausal women and values of 64, 9, and 68% below the fifth percentile for age-matched control subjects, respectively. This study suggests that no preferential loss of spinal trabecular bone occurs in subjects with isolated hip fracture, whereas significant separation can be observed between spine fracture and normal comparison subjects. Further studies utilizing direct hip measurements with QCT need to be undertaken to determine the degree to which bone loss specifically at the hip impacts on the occurrence of fracture.

Quantitative Computed Tomography and Osteoporosis Prophylaxis

Estrogen replacement therapy initiated soon after the cessation of ovarian function is universally accepted as an efficacious means of retarding bone loss and preventing fractures of the spine and hip (105). Quantitative CT has been of great use in chronicling this period of rapid bone loss and substantiating the response to estrogen therapy. Genant and coworkers (21) serially assessed the bone mineral loss in 37 premenopausal women for a duration of 24 months following surgical oophorectomy and determined the dose response for conjugated estrogen therapy in preventing this loss. Without intervention mean annual rates of loss from the spinal trabecular envelope were shown to be on the order of 9% (Fig. 43.6). An estrogen dose of 0.6 mg/day was found to be the lowest effective dose in abbreviating this rapid involution. In a later investigation of 73 women at the time of menopause, slightly slower rates of loss were observed, approximately 5%/year, in subjects on placebo and in those on 1500 mg of daily calcium supplementation (103). However, a dose of 0.3mg/day estrogen combined with 1500 mg daily calcium was found to be as effective as 0.6 mg/day estrogen alone, at least for the first 2 years of observation. In both studies, bone was lost from the appendicular cortical skeleton at a significantly lower rate (1 to 3% annually) than from the vertebral spongiosum.

The effect of estrogen deprivation on younger, premenopausal women has also been examined using QCT. Thirty-eight women (17 to 49 years of age) with hypothalamic and hyperprolactinemic amenorrhea and premature ovarian failure were studied (12). The group with hypothalamic amenorrhea was made up primarily of athletes. Bone mass in the peripheral cortical skeleton was only slightly lower than values for age-matched normal subjects, but spinal trabecular bone was significantly lower (20 to 25%) (Fig. 43.7). A subpopulation of this study included both

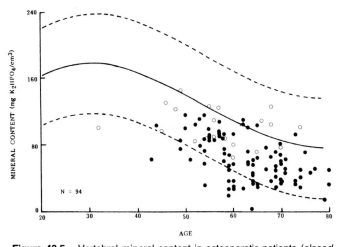

Figure 43.5. Vertebral mineral content in osteoporotic patients (*closed circles*) superimposed on mean and 95% confidence intervals for normal women. (Reprinted with permission from Cann CE, Rutt BK, Genant HK. Effects of extraosseous calcification on vertebral mineral measurement. Paper presented at annual meeting of the American Society of Bone and Mineral Research, Detroit, Mich., April 9–12, 1983.)

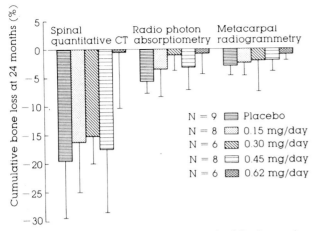

Figure 43.6. Cumulative bone loss 24 months following oophorectomy in 37 women as a function of quantitative technique and estrogen therapy. (Reprinted with permission from Genant HK, Cann CE, Ettinger B, Gordan GS. Quantitative computed tomography of vertebral spongiosa: a sensitive method for detecting early bone loss after oophorectomy. Ann Intern Med 1982;97:699–705.)

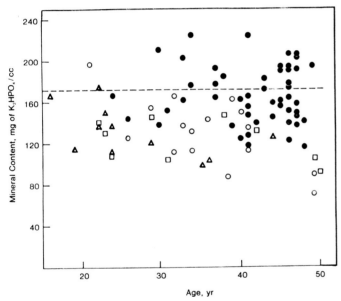

Figure 43.7. Mineral content in lumbar vertebrae for normal controls (*black dots*) and women with amenorrhea from hyperprolactinemia (*squares*), hypothalamic cause (*triangles*), and premature ovarian failure (*circles*). Dashed line is regression with age for controls. K_2HPO_4 indicates potassium phosphate.

amenorrheic and exercise-matched eumenorrheic athletes (106). In this study, eumenorrheic elite athletes had significantly more spinal mineral than sedentary controls, followed in order by amenorrheic elite athletes, amenorrheic casual athletes, and amenorrheic nonathletes. Thus, the exercise effect is apparent in women but can be overridden by the strong effect of estrogen deprivation.

Quantitative Computed Tomography vs. Other Bone Density Techniques

A number of investigators have compared the various techniques of bone mineral assessment to quantify the ef-

fects of metabolic bone diseases on different skeletal compartments, to determine the fracture prediction value of each technique, and to assess its usefulness for describing decrements observed from health to disease. One of the most recent and rigorous of these studies (34) is presented here.

Comparison of Quantitative Computed Tomography and Other Techniques in Normal and Osteoporotic Women

To investigate the associations among the principal methods for the noninvasive measurement of spinal and appendicular bone mass, the UCSF group studied 40 normal early postmenopausal women and 68 older postmenopausal women with osteoporosis (34). The methods of determining bone mineral content included single- and dual-energy QCT and DPA of the lumbar spine, single-photon absorptiometry (SPA) of the distal radius, and combined cortical thickness (CCT) measurements of the second metacarpal shaft. Lateral thoracolumbar radiography was also evaluated, and a spinal fracture index (FXI) was calculated. The correlation coefficients and levels of significance for both patient groups are shown in Tables 43.4 and 43.5. Single-energy QCT showed a good correlation $r = 0.87$) with DPA for early postmenopausal women and a moderate correlation ($r = 0.53$) for postmenopausal osteoporotic women (Fig. 43.8). Dual-energy QCT did not improve these correlations with DPA. Appendicular measurements correlated only modestly among themselves, as well as with the axial measurements by either QCT or DPA. The severity of spinal fracture correlated highly with QCT ($r = -0.91$) but only moderately with DPA ($r = -0.44$) (Fig. 43.9). Reinbold and coworkers (34) also showed a small but significant decrease in the correlation between SEQCT and DEQCT from the younger to the older population (Fig. 43.10). While this decrease is presumably due to age-related increases in marrow fat, generally excellent correlations were observed between DEQCT and SEQCT regardless of the age group. Moreover, it was found that the use of DEQCT failed to improve the correlation with DPA, to reduce the normal biologic variation found with QCT in general, or to enhance the discrimination between normal and osteoporotic women. These facts suggest that DEQCT is, in most circumstances, unnecessary in the assessment of postmenopausal women and that SEQCT, with its lower radiation dose, is adequate. However, when highly accurate assessments are necessary or for special research purposes, the use of DEQCT may be necessary.

Widely differing correlations between QCT and DPA have been reported in apparently similar groups of patients (15, 25, 32, 34, 35, 38, 107, 108). Some reported correlations may be poor due to inadequate sample sizes; however, in our large cohort good correlation ($r = 0.87$ and $r = 0.82$) is shown between SEQCT or DEQCT, respectively, and DPA in early postmenopausal women, and moderate correlations ($r = 0.53$ and $r = 0.42$) in postmenopausal osteoporotic women. The modest correlation observed in osteoporotic women, while statistically significant, shows sufficient dispersion that a reliable prediction

Table 43.4
Pearson Correlation Coefficients for Early Postmenopausal Women (*n* = 40)

	SEQCT	DEQCT	DPA	SPA	SPA/W	CCT
SEQCT		0.9734 0.0001	0.8661 0.0001	0.3824 0.0149	0.4874 0.0014	0.5363 0.0004
DEQCT	0.9734 0.0001		0.8156 0.0001	0.3716 0.0182	0.4345 0.0051	0.5345 0.0004
DPA	0.8661 0.0001	0.8156 0.0001		0.4673 0.0024	0.5396 0.0003	0.5029 0.0009
SPA	0.3824 0.0149	0.3716 0.0182	0.4673 0.0024		0.7307 0.0001	0.3650 0.0206
SPA/W	0.4874 0.0014	0.4345 0.0051	0.5396 0.0003	0.7307 0.0001		0.5466 0.0003
CCT	0.5363 0.0004	0.5345 0.0004	0.5029 0.0009	0.3650 0.0206	0.5466 0.0003	

SPA/W: SPA divided by width.

Table 43.5
Pearson Correlation Coefficients for Postmenopausal Osteoporotic Women (*n* = 68)

	SEQCT	DEQCT	DPA	SPA	SPA/W	CCT	FXI
SEQCT		0.9482 0.0001	0.5310 0.0001	0.2005 0.1012	0.2390 0.0497	0.2285 0.0609	−0.9102 0.0001
DEQCT	0.9482 0.0001		0.4245 0.0003	0.2005 0.1012	0.2133 0.0809	0.2366 0.0521	−0.8817 0.0001
DPA	0.5310 0.0001	0.4245 0.0003		0.3633 0.0023	0.1622 0.1864	0.2522 0.0380	−0.4399 0.0002
SPA	0.2005 0.1012	0.2005 0.1012	0.3633 0.0023		0.4021 0.0007	0.5788 0.0001	−0.0775 0.5316
SPA/W	0.2390 0.0497	0.2133 0.0809	0.1622 0.1864	0.4021 0.0007		0.2522 0.0381	−0.2195 0.0720
CCT	0.2285 0.0609	0.2366 0.0521	0.2522 0.380	0.5788 0.0001	0.2522 0.0381		−0.1720 0.1610
FXI	−0.9102 0.0001	−0.8817 0.0001	−0.4399 0.0002	−0.0775 0.5316	−0.2195 0.0720	−0.1720 0.1610	

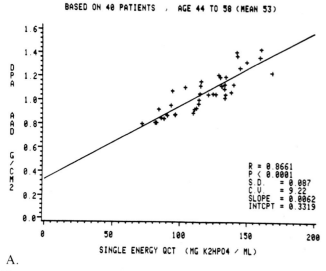

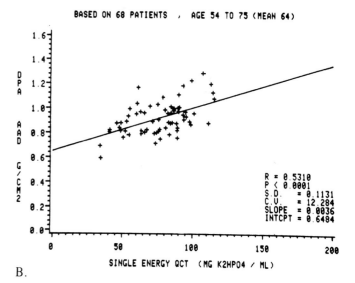

Figure 43.8. Moderately good correlation is shown for DPA versus SEQCT for normal early postmenopausal women **(A)**. Modest correlation is shown for DPA versus SEQCT for postmenopausal osteoporotic women **(B)**.

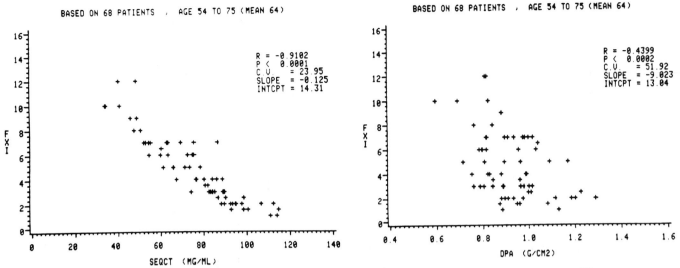

Figure 43.9. High correlation is shown for SEQCT versus spinal fracture index in osteoporotic women (**A**). Modest correlation is shown for DPA versus spinal fracture index, in the same patients (**B**).

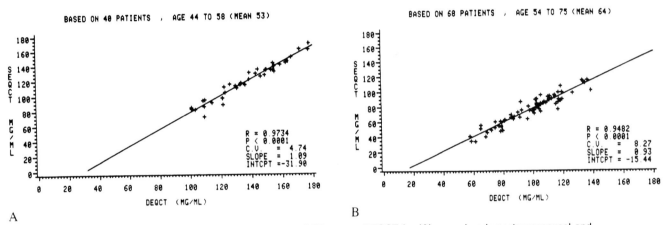

A B

Figure 43.10. High correlation is shown for SEQCT versus DEQCT for (**A**) normal early postmenopausal and (**B**) postmenopausal osteoporotic women.

of QCT value from DPA value or vice versa is precluded for the individual patient.

Some authors (25, 109) have tried to improve the correlation between QCT and DPA by measuring the same bone envelopes with both methods and expressing the quantities measured in dimensionally similar units. The UCSF Osteoporosis Research Group previously found, in a mixed group of 52 patients, that the modest correlation ($r = 0.67$) between the anatomically and dimensionally dissimilar integral DPA (in gm/cm^2) and trabecular QCT (in mg/cm^3) was significantly improved when both techniques measured the same parameters. For example, integral mass (in g) assessed by DPA correlated well ($r = 0.88$) with integral mass as determined by CT; likewise, integral density (in mg/cm^3) assessed by DPA correlated well ($r = 0.84$) with integral density assessed by QCT. Nevertheless, the strongest correlation ($r = 0.91$) was observed between trabecular and integral density when both were measured by

QCT. Thus, the difference observed between QCT and DPA were not explained entirely by anatomic considerations. This finding suggests that residual fundamental differences exist between these techniques, perhaps related to their respective abilities to define bone edges, regions of interest, baseline measurements, and volumes and their respective other sources of error.

Mazess and Vetter (109) used QCT and DPA to examine comparable bone envelopes in 10 excised vertebrae. They found that the mineral mass obtained by integrating all bone voxels by DEQCT correlated highly ($r = 0.97$) with the mass determined by DPA. The mineral density of the total vertebra, including all processes as determined by DPA, correlated highly ($r = 0.87$) with the density as assessed by DEQCT, whereas the correlation with the density of the body was lower ($r = 0.82$) and even lower ($r = 0.79$) with the density of a section from the spongiosum of the anterior vertebral body. In other words, correlations between DPA and QCT weakened when integral density was compared with the more labile trabecular density.

The current study of early postmenopausal and older osteoporotic females showed that the older women have less bone mineral than younger postmenopausal women and that the magnitude of the difference depends on the method used for measurement. The mean differences observed were 35.6% by SEQCT, 28.2% by DEQCT, 13.7% by DPA, 13.7% by SPA, and 15.1% by CCT. The substantial decrement observed by QCT suggests that measurements of spinal trabecular bone density by QCT discriminate those women with spinal osteoporosis from younger postmenopausal normals better than by DPA or of appendicular cortical bone mass by SPA or CCT. This observation is supported by the recent work of Jones et al. (49) showing disproportionate loss of trabecular relative to compact bone in the spine, both in aging and in osteoporosis. Furthermore, two recent studies have compared the differences between young normals and osteoporotics measured by QCT and DPA. Sambrook et al. (38) found a 54% ($p = 0.01$) decrement by QCT and a 23% ($p = 0.05$) decrement by DPA, while Gallagher (19) found a 65% ($p = 0.001$) decrement by QCT and 16% ($p > 0.05$, nonsignificant) decrement by DPA. Finally, Raymakers et al. (33) tested the ability of these two techniques to discriminate between women with crush fractures and age-matched normals. Quantitative CT was slightly better, showing a sensitivity and specificity of 83 and 83%, respectively, compared to 73 and 76% for DPA.

The UCSF group concluded that although measurements of BMC by QCT, DPA, SPA, and CCT are positively correlated, their relationships are not strong enough and their dispersions are too large to predict one measure by another for the individual patient. Single-energy QCT was determined to be adequate, and perhaps preferable, for assessing postmenopausal women. Measurements of spinal trabecular bone density by QCT can also discriminate osteoporotic women from younger normal women more sensitively than measurements of spinal integral bone by DPA or appendicular cortical bone by SPA or CCT.

Recommended Clinical Applications

The question remains whether bone densitometry, by any method, should be recommended widely and become standard medical practice. We know more today about each technique's advantages and shortcomings than we ever have, and there is continuing rapid improvement of current technology and development of exciting new modalities. But in the midst of this development and improvement, and while we await the results of numerous ongoing and future clinical and epidemiologic investigations, we are faced with important management decisions involving the current population of women already afflicted with osteoporosis or at risk for developing it. We must, therefore, rely on sound expert opinion and available technology to guide a clinical approach to patient management.

Previously it was thought that osteoporosis, by clinical definition, could be reduced to a simple distinction between atraumatic fracture and nonfracture. Indeed, this philosophy led some researchers (110–113) to suggest that

bone density, per se, is not a sensitive predictor of fracture risk, because only modest separation was observed between fracture patients and nonfracture comparison subjects of the same age. The suggestion was based, in part, on early densitometry studies (114–116) showing considerable overlap between hip fracture subjects and matched controls. A more recent study, in which the Ward's triangle region of the hip was evaluated using newer techniques, suggests better fracture discrimination (117). Moreover, those recent studies (118, 119) that have assessed the contribution of absolute bone density for fracture risk have shown a strong predictive ability for a variety of techniques. Certainly, many investigators have shown that direct quantitative assessment of the spine by QCT or DPA satisfactorily discriminates spine fracture from nonfracture subjects (34, 104, 120). Additionally, all bone densitometry techniques illustrate a progressive and substantial loss of bone over the life span from all skeletal sites (albeit at different rates) (24, 121, 122), and it is this systematic bone loss that contributes most significantly to fractures in old age. In fact, several studies (15, 123–125) have shown that osteoporotic fractures do not become likely until a certain threshold of bone density is crossed, and that high density values, independent of age, confer significant protection against fracture. Most importantly, these recent findings have convinced many researchers to no longer consider osteoporosis as a fracture/nonfracture dichotomy but rather to redefine osteoporosis as the lower part of a continuum of bone density, with greatest fracture risk among those with lowest absolute density values. Wasnich (126) draws a parallel to the use of blood pressure monitoring to predict stroke risk. Bone mass, he points out, like blood pressure, must be treated as a continuous variable, with fracture, analogous to stroke, as a primary outcome. Indeed, most individuals with hypertension have not suffered a stroke, yet they are nonetheless at great risk. Similarly, because of the probabilistic nature of fractures, many individuals with low bone density have not experienced a fracture, but neither should they be considered normal or free of osteoporosis.

Although controversy exists with regard to the appropriate use of bone densitometry (111, 112, 126–129), we currently consider a number of clinical applications to be valid: (*a*) assessment of patients with metabolic diseases known to affect the skeleton, (*b*) assessment of perimenopausal women for the initialization of estrogen replacement or experimental therapy, (*c*) establishing a diagnosis of osteoporosis or assessing its severity in the context of general clinical care, and (*d*) monitoring the efficacy of treatment interventions or the natural course of disease (130, 131). Only the first of these, however, could be said to be supported by all investigators in the field. Nonetheless, a growing body of literature supports wider applications for the individual patient (126–128). We realize, of course, that certain prerequisites must be achieved before these recommendations could be realistically followed.

Knowledge about the proper use and interpretation of bone densitometry studies and an understanding of appropriate medical interventions are not universal among physicians, nor is instrumentation and technical performance of bone density studies of uniformly high quality.

Indeed, this deficiency of medical and technical expertise is the principal deterrent to widespread implementation of our recommended clinical applications at this time. However, given the current impetus to disseminate information about osteoporosis, to make newer instrumentation more readily available, and to limit the cost of these techniques, we anticipate that our recommendations may soon become standard medical practice.

References

1. National Institutes of Health. Osteoporosis consensus conference. JAMA 1984;252:199–802.
2. Snyder W. Report of the task group on reference man. Oxford: Pergamon Press, 1975.
3. Frost HM. Dynamics of bone remodeling. In: Frost HM, ed. Bone biodynamics. Boston: Little, Brown, 1964:315–334.
4. Garn SM. The earlier gain and later loss of cortical bone. Springfield: Charles C. Thomas, 1970.
5. Cameron JR, Mazess RB, Sorenson JA. Precision and accuracy of bone mineral determination by direct photon absorptiometry. Invest Radiol 1968;3141–145.
6. Mazess RB, Cameron JR. Bone mineral content in normal U.S. whites. In: Mazess RB, ed. Proceedings of the international conference on bone and mineral measurement, October 1973, Chicago, Ill. Washington, D.C.: US Government Printing Office, 1974. DHEW publication No. (NIH) 75-683.
7. Sorenson JA, Mazess RB. Effects of fat on bone mineral measurements. In: Cameron JR, ed. Proceedings of the bone measurement conference (US Atomic Energy Commission (USAEC) Conference 700515). Washington, D.C.: USAEC, 1970:255–262.
8. Wahner HW, Eastell R, Riggs BL. Bone mineral density of the radius: where do we stand? J Nucl Med 1985;26(11):1339–1341.
9. Adams JE, Pullan BR, Adams PH, et al. Dual energy computed tomography (CT) and the estimation of bone mass. J Comput Assist Tomogr 1982;6:204.
10. Adams JE, Chen SZ, Adams PH, Isherwood I. Measurement of trabecular bone mineral by dual energy computed tomography. J Comput Assist Tomogr 1982;6:601–607.
11. Burgess AE, Colborne B, Zoffmann E. Vertebral bone mineral content. Proceedings of the 5th international workshop on bone and soft tissue densitometry, Bretton Woods, N.H., October 14–18, 1985.
12. Cann CE, Martin MC, Genant HK, et al. Decreased spinal mineral content in amenorrheic women. JAMA 1984;251:626.
13. Cann CE, Genant HK. Cross-sectional studies of vertebral mineral content using quantitative computed tomography. J Comput Assist Tomogr 1982;6:216.
14. Cann CE, Rutt BK, Genant HK. Effects of extraosseous calcification on vertebral mineral measurement. Paper presented at annual meeting of the American Society of Bone and Mineral Research, Detroit, Mich., April 9–12, 1983.
15. Cann CE, Genant HK, Kolb FO, Ettinger BF. Quantitative computed tomography for prediction of vertebral fracture risk. Bone 1985;6:1–7.
16. Cann CE, Ettinger B, Genant HK. Normal subjects versus osteoporotics: No evidence using dual energy computed tomography for disproportionate increase in vertebral marrow fat. J Comput Assist Tomogr 1985;9:617–618.
17. Faul DD, Couch JL, Cann CE, Boyd DP, Genant HK. Composition-selective reconstruction for mineral content in the axial and appendicular skeleton. J Comput Assist Tomogr 1982;6:202.
18. Firooznia H, Rafii M, Golimbu C, Schwartz MS, Ort P. Trabecular mineral content of the spine in women with hip fracture: CT measurement. Radiology 1986;159:737–740.
19. Gallagher JC, Gogar D, Mahoney P, McGill J. Measurement of spine density in normal and osteoporotic subjects using computed tomography: relationship of spine density to fracture threshold and fracture index. J Comput Assist Tomogr 1985;9:634.
20. Genant HK, Boyd D. Quantitative bone mineral analysis using dual energy computed tomography. Invest Radiol 1977;12(6):545–551.
21. Genant HK, Cann CE, Ettinger B, Gordan GS. Quantitative computed tomography of vertebral spongiosa: a sensitive method for detecting early bone loss after oophorectomy. Ann Intern Med 1982;97:699–705.
22. Genant HK, Boyd DP, Rosenfeld D, Abols Y, Cann CE. Computed tomography. In: Cohn SH, ed. Non-invasive measurements of bone mass and their clinical application. Boca Raton: CRC Press, 1981;121–149.
23. Genant HK, Cann CE, Boyd DP, Kolb FO, Ettinger B, Gordan GS. Quantitative computed tomography for vertebral mineral determination. In: Frame B, Potts JT, eds. Clinical disorders of bone and mineral metabolism: proceedings of the Frances and Anthony D'Anna memorial symposium (international congress series No. 617), Detroit, Mich., May 9–13, 1983. Amsterdam: Excerpta Medica, 1983:355–359.
24. Genant HK, Cann CE, Pozzi-Mucelli RS, Kantner AS. Vertebral mineral determination by QCT: clinical feasibility and normative data. J Comput Assist Tomogr 1983;7:554.
25. Genant HK, Powell MR, Cann CE, Stebler B, Rutt BK, Richardson ML, Kolb FO: Comparison of methods for in vivo spinal bone mineral measurement. In: Christiansen C, Arnaud CD, Nordin BEC, Parfitt AM, Peck WA, Riggs BL, eds. Osteoporosis. Proceedings of the Copenhagen International symposium on osteoporosis, Cophenhagen, Denmark, June 3–8, 1984, Copenhagen: Aalborg Stiftsbogtrykkeri, 1984:97–102.
26. Genant HK, Cann CE, Ettinger B, et al. Quantitative computed tomography for spinal mineral assessment. J Comput Assist Tomogr 1985;9:602–604.
27. Hangartner TN, Overton TR. The Alberta gamma CT system. J Comput Assist Tomogr 1983;6:1156.
28. Laval-Jeantet AM, Cann CE. Roger B, Allant P. A postprocessing dual energy technique for vertebral CT densitometry. J Comput Assist Tomogr 1984;8(6):1164–1167.
29. Laval-Jeantet AM, Jones CD, Bergot C, et al. Comparison of bone loss from spongiosa and from compact vertebral bone in the aging process and in osteoporotics. Proceedings of the 5th international workshop on bone and soft tissue using computed tomography, Bretton Woods, N.H., October 14–18, 1985.
30. Mack LA, Hanson JA, Kilcoyne RF, et al. Correlation between fracture index and bone densitometry by CT and dual photon absorptiometry. J Comput Assist Tomogr 1985;9:635–636.
31. Orphanoudakis SC, Jensen PS, Rauschkolb EN, Lang R, Rasmussen H. Bone mineral analysis using single energy computed tomography. Invest Radiol 1979;14:122–130.
32. Powell MR, Kolb FO, Genant HK, et al. Comparison of dual photon absorptiometry and quantitative computed tomography of the lumbar spine in the same subjects. In Frame B, Potts JT, eds. Clinical disorders of bone and mineral metabolism. Amsterdam: Excerpta Medica, 1983:58–61.
33. Raymakers JA, Hoekstra O, Van Puten J, Kerkhoff H, Duursma SA. Osteoporotic fracture prevalence and bone mineral mass measured with CT and DPA. Skel Radiol 1986;15:191.
34. Reinbold WD, Genant HK, Reiser UJ, Harris ST, Ettinger B. Bone mineral content in early-postmenopausal and postmenopausal osteoporotic women: comparison of measurement methods. Radiology 1986;160(2):469–478.
35. Richardson ML, Genant HK, Cann CE, Ettinger B, Gordan GS, Kolb FO, Reiser UJ: Assessment of metabolic bone diseases by quantitative computed tomography. Clin Orthoped 1985;185:224–238.
36. Rohloff R, Hitzler H, Arndt W, Frey W. Vergleichende Messungen des Kalksalzgehaltes spongioeser Knochen mittels Computertomographie und J-125-Photonen-Absorptionsmethode. In: Lissner J, Doppman JL, eds. CT' 82. Konstanz: Schnetztor Verlag, 1982:126–130.
37. Rüegsegger P, Elsasser U, Anliker M, Gnehm H, Kind HP, Prader A. Quantification of bone mineralization using computed tomography. Radiology 1976;121:93–97.
38. Sambrook PN, Bartlett C, Evans R, Hesp R, Kaltz D, Reeve J. Measurement of lumbar spine bone mineral: a comparison of dual-photon absorptiometry and computed tomography. Br J Radiol 1985;58:621.
39. Stebler BG, Rutt BK, Hosier K, Cann CE, Boyd DP, Genant HK. Signal system and data acquisition system for multielement ger-

manium detectors. Proceedings of CT Densitometry Workshop. J Comput Assist Tomgr 1985;9:610–611.

40. Cann CE. Quantitative bone mineral analysis using dual energy computed tomography. Radiology 1987;152:257–261.

41. Dalen N, Lamke B. Bone mineral losses in alcoholics. Acta Orthoped Scand 1976;47:469–471.

42. Krolner B, Pors-Nielsen S. Measurement of bone mineral content of the lumbar spine: theory and application of a new two-dimensional dual photon attenuation method. Scand J Clin Lab Invest 1980;40:485–487.

43. Madsen M, Peppler W, Mazess RB. Vertebral and total body bone mineral content by dual photon absorptiometry. Calcif Tissue Res 1976;2:361–364.

44. Nilas L, Borg J, Gotfredsen A, Christiansen C. Comparison of single and dual photon absorptiometry in postmenopausal bone mineral loss. J Nucl Med 1985;26:1257–1262.

45. Peppler WW, Mazess RB. Total body bone mineral and lean body mass by dual-photon absorptiometry. Calcif Tissue Int 1981;33:353.

46. Riggs BL, Wahner HW, Dunn WL, Mazess RB, Offord KP, Melton LJ III., Differential changes in bone mineral density of the appendicular and axial skeletal with aging. J Clin Invest 1981;67:328–335.

47. Roos B, Rosengren B, Skoldborn H. Determination of bone mineral content in lumbar vertebrae by a double gamma-ray technique. In Cameron JR, ed. Proceedings of the bone measurement conference: USAEC Conf-700515. Springfield, Va.: Clearinghouse for Federal Scientific and Technical Information, National Bureau of Standards, US Department of Commerce, 1970:243–254.

48. Wahner HW, Dunn WL, Mazess RB, et al. Dempster: Dual-photon Gd-153 absorptiometry of bone. Radiology 1985;156:203–206.

49. Jones CD, Laval-Jeantet AM, Laval-Jeantet MH, Genant HK. Importance of measurement of spongious vertebral bone mineral density in the assessment of osteoporosis. Bone 1987;8(4):201–206.

50. Nottestad SY, Baumel JJ, Kimmel DB, Recker RR, Heaney RP. The proportion of trabecular bone in human vertebrae. J Bone Min Res 1987;2:221–229.

51. Kuhl DR, Edwards RQ. Image separation radioisotope scanning. Radiology 1963;80:653–661.

52. Kuhl DE, Hale J, Easton WL. Tranmission scanning: a useful adjunct to conventional emission scanning for accurately keying isotope deposition to radiographic anatomy. Radiology 1966;87:278–284.

53. Bracewell RN. Strip integration in radioastronomy. Aust J Phys 1956;9:198–217.

54. Gordon R, Bender R, Herman GT. Algebraic reconstruction techniques (ART) for three-dimensional electron microscopy and x-ray photography. J Theor Biol 1970;29:471–481.

55. Bracewell RN, Riddle AC. Inversion of fan-beam scans in radio astronomy. Astrophys J 1967;150:427–434.

56. Brooks RA, di Chiro G. Principles of computer assisted tomography (CAT) in radiographic and radioisdotopic imaging. Phys Med Biol 1976;21(5):689–732.

57. Brooks RA, di Chiro G. Theory of image reconstruction in computed tomography. Radiology 1975;117:561–572.

58. Cann CE, Genant HK. Precise measurement of vertebral mineral content using computed tomography. J Comput Assist Tomogr 1980;4(4):493–500.

59. Cann CE. Quantitative computed tomography for bone mineral analysis: technical considerations. In: Genant HK, ed. Osteoporosis update 1987. Berkeley: University of California Press, 1987:131–145.

60. Abols Y, Genant HK, Rosenfeld D, Boyd DP, Ettinger B, Gordon GS. Spinal bone mineral determination using computerized tomography in patients, control and phantoms. In Mazess RB, ed. Proceedings of the fourth international conference on bone measurement (NIH 80-1928). Washington, D.C.: US Government Printing Office, 1979.

61. Goodsitt MM, Rosenthal DI. Quantitative computed tomography scanning for measurement of bone and bone marrow fat content: a comparison of single and dual energy techniques using a solid synthetic phantom. Invest Radiol 1987:22;799–810.

62. Kalender WA, Süss C. A new calibration phantom for quantitative computed tomography. Med Phys 1987;9:816–819.

63. Reiser U, Heuck F, Faust U, Genant HK. Quantitative Computertomographie zur Bestimmung des Mineralgehalts in Lendenwirbeln mit Hilfe eines Festkoerper-Referenzsystems. Biomed Technol 1985;30:187–188.

64. Zamenhof RGA. Optimization of spinal bone density measurement using computerized tomography. In: Genant HK, ed. Osteoporosis update 1987. Berkeley: University of California Press, 1987:145–169.

65. Arnold B. Solid phantom for QCT-bone mineral analysis. Proceedings of the 7th international workshop on bone densitometry, Palm Springs, Calif., September 17–21, 1989.

66. Computerized Imaging Reference Systems: CIRS model IV lumbar reference simulator technical manual. Norfolk, Virginia: CIRS, 1986.

67. Computerized Imaging Reference Systems. CIRS model V femoral neck reference simulator technical manual. Norfolk, Virginia: CIRS 1987.

68. Report of the Task Group on Reference Man. (ICRP Publication No. 23) Oxford: Pergamon Press, 1975.

69. Cann CE: Low-dose CT scanning for quantitative spinal mineral analysis. Radiology 1981;140:813–815.

70. Glüer CC, Steiger PW, Block JE, Genant HK. Precision studies of quantitative computed tomography of the proximal femur. Radiology 1987;165(P):297.

71. Talbert AJ, Brooks RA, Morgenthaler DG. Optimum energies for dual-energy computed tomography. Phys Med Biol 1980:25(2):261–269.

72. Sandor T, Kalender WA, Hanlon WB, Weissman BN, Rumbaugh C. Spinal bone mineral determination using automated contour detection: application to single and dual energy CT. SPIE Medical Imag Instrument 1985;555:188–194.

73. Steiger P, Steiger S, Rüegsegger P, Genant HK. Two- and three-dimensional quantitative image evaluation techniques for densitometry and volumetrics in longitudinal studies. In: Genant HK, ed. Osteoporosis update 1987. Berkeley: University of California Press, 1987:171–180.

74. Kalender WA, Brestowsky H, Felsenberg D. Bone mineral measurements: automated determination of the midvertebral CT section. Radiology 1988;168:219–221.

75. Glüer CC, Genant HK. Quantitative computed tomography of the hip. In: Genant HK, ed. Osteoporosis update 1987. Berkeley: University of California Press, 1987:187–195.

76. Reiser UJ, Genant HK. Determination of bone mineral content in the femoral neck by quantitative computed tomography. Washington, D.C.: Radiological Society of North America, 1984.

77. Sartoris DJ, Andre M, Resnick C, Resnick D. Trabecular bone density in the proximal femur: quantitative CT assessment. Radilogy 1986;160:707–712

78. Steiger P, Rüegsegger P, Felder M. Three-dimensional evaluation of bone changes in joints of patients who have rheumatoid arthritis. J Comput Assist Tomogr 1985;9(3):622–623.

79. Sartoris DJ, Resnick D, Bielecki D, Andre M, Gershuni D, Meyers M. A technique for multiplanar reformation and three-dimensional analysis of computed tomographic data: application to adult hip disease. J Can Assoc Radiol 1986;37:69–72.

80. Gillespie JE, Isherwood I. Three-dimensional anatomic images from computed tomographic scans. Br J Radiol 1986;59:289–292.

81. Rüegsegger P, Niederer P, Anliker M. An extension of classical bone mineral measurements. Ann Biomed Eng 1974;2:194.

82. Reich NE, Seidelmann FE, Tubbs RR. Determination of bone mineral content using CT scanning. AJR 1976;127:593.

83. Posner I, Griffiths HJ. Comparison of CT scanning with photon absorptiometric measurement of bone mineral content in the appendicular skeleton. Invest Radiol 1977;12:545.

84. Laval-Jeantet AM, Roger B, Bouysse S, Bergot C, Mazess RB. Influence of vertebral fat content on quantitative CT density. Radiology 1986;159:463–466.

85. Burgess AE, Colborne B, Zoffman E. Vertebral trabecular bone: comparison of single and dual-energy CT measurements with chemical analysis. J Comput Assist Tomogr 1987;11:506–515.

86. Isherwood I, Rutherford RA, Pullan BR, Adams PH. Bone mineral estimation by computer assisted transverse axial tomography. Lancet 1976; II:712–715.

87. Bradley JG, Huang HK, Ledley RS. Evaluation of calcium con-

centration in bones from CT scans. Radiology 1978;128;103–107.

88. Liliequist B, Larsson SE, Sjogren I, Wickman G, Wing K. Bone mineral content in the proximal tibia measured by computed tomography. Acta Radiol Diagn 1979;20:957–966.

89. Rüegsegger P, Anliker M, Dambacher M. Quanitification of trabecular bone with low dose computed tomography. J Comput Assist Tomogr 1981;5:384–390.

90. Hangartner TN, Overton TR. Quantitative measurement of bone density using gamma-ray computed tomography. J Comput Assist Tomogr 1982;6:1156.

91. Firooznia H, Golimbu C, Rafii M, et al. Quantitative computed tomography assessment of spinal trabecular bone: age-related regression in normal men and women. J Comput Assist Tomogr 1984;8:91–97.

92. Sashin D, Sternglass EJ, Sandler RB, et al. The development and evaluation of a CT technique for measurement of the density of cortical bones in the appendicular skeleton. J Comput Assist Tomogr 1983;7:552.

93. Graves VB, Wimmer R. Long term reproducibility of quantitative computed tomography vertebral mineral measurements. Comput Tomogr 1985;9:73–76.

94. Rosenthal DI, Ganott MA, Wyshak G, Slovik DM, Doppelt SH, Neer RM. Quantitative computed tomography for spinal denisty measurement factors affecting precision. Invest Radiol 1985;20:306–310.

95. Meier DE, Orwoll ES, Jones JM. Marked disparity between trabecular and cortical bone loss with age in healthy men. Ann Intern Med 1984;101:605–12.

96. Cann CE, Henzi M, Burry K, et al. Reversible bone loss is induced by GnRH agonists. Presented at the 68th annual meeting of the Endocrine Society, Anaheim, Calif., June 25–27, 1986.

97. Heaney RP, Recker RR. Distribution of calcium absorption in middle-aged women. Am J Clin Nutr 1986;43:299–305.

98. Cann CE, Heller M, Skinner HB. A functional presentation format for 3-D bone images. Radiology 1984;153(P):311.

99. Cann CE. Quantitative CT applications: comparison of current scanners. Radiology 1987;162:257–261.

100. Helms CA, Cann CE, Brunelle FO, Gilula FA, Chafetz N, Genant HK. Detection of bone marrow metastases using quantitative CT. Radiology 1981;140:745.

101. Banks LM, Stevenson JC. Modified method of spinal computed tomography for trabecular bone mineral measurements. J Comput Assist Tomogr 1986;10:463–467.

102. Kalender WA, Klotz E, Süss C. Vertebral bone mineral analysis: an integrated approach with CT. Radiology 1987;164:419–423.

103. Ettinger B, Genant HK, Cann CE. Postmenopausal bone loss is prevented by treatment with low-dosage estrogen with calcium. Ann Intern Med 1987;106:40–45.

104. Firooznia H, Golimbu C, Rafii M, et al. Quantitative computed tomography assessment of spinal trabecular bone in osteoporotic women with and without vertebral fractures. J Comput Assist Tomogr 1984;8:99–103.

105. Ettinger B, Genant HK, Cann CE. Long-term estrogen replacement therapy prevents bone loss and fractures. Ann Intern Med 1985;102:319–324.

106. Marcus R, Cann CE, Madvig P, et al. Menstrual function and bone mass in elite women distance runners. Ann Intern Med 1985;102:158–163.

107. Kilcoyne RF, Hanson JA, Ott SM, Mack L, Chesnut CH. Vertebral bone mineral content measured by two techniques of computed tomography and compared with dual photon absorptiometry. J Comput Assist Tomogr 1984;8:1164–1167.

108. Ott SM, Chesnut CH, Hanson JA, Kilcoyne RF, Murano R, Lewellen TK. Comparison of bone mass measurements using different diagnostic techniques in patients with postmenopausal osteoporosis. In: Christiansen C, Arnaud CD, Nordin BEC, Parfitt AM, Peck WA, Riggs BL, eds. Osteoporosis Vol. 1. Proceedings of the Copenhagen international symposium on osteoporosis, Copenhagen, Denmark, June 3–8, 1984. Copenhagen: Aalborg Stiftsbogtrykkeri, 1984:93–96.

109. Mazess RB, Vetter J. The influence of marrow on measurement of trabecular bone using computed tomography. Bone 1985;6:349–351.

110. Aitken JM. Relevance of osteoporosis in women with fractures of the femoral neck. Br J Med 1984;288:1084–1085.

111. Cummings SR, Black D. Should perimenopausal women be screened for osteoporosis. Ann Intern Med 1986;104:817–823.

112. Hall FM, Davis MA, Baran DT. Bone mineral screening for osteoporosis. N Engl J Med 1987;316:212–214.

113. Ott, SM. Should women get screening bone mass measurements? Ann Intern Med 1986:104:874.

114. Bohr H, Schaadt O. Bone mineral content of femoral bone and the lumbar spine measured in women with fracture of the femoral neck by dual photon absorptiometry. Clin Orthoped 1983;179:240–245.

115. Krolner B, Pors Nielsen S. Bone mineral content of the lumbar spine in normal and osteoporotic women: cross-sectional and longitudinal studies. Clin Sci 1982;62:329–336.

116. Riggs BL, Wahner HW, Seeman E, et al. Changes in bone mineral density of the proximal femur and spine with aging: differences between the postmenopausal and senile osteoporosis syndromes. J Clin Invest 1982;70:716–718.

117. Mazess RB, Barden H, Ettinger M, Schultz E. Bone density of the radius, spine, and proximal femur in osteoporosis. J Bone Min Res 1988;3:13–18.

118. Ross PD, Wasnich RD, MacLean CJ, Vogel JM. Prediction of individual lifetime fracture expectancy using bone mineral measurements. In: Christiansen C, Johansen JS, Riis BJ, eds. Osteoporosis 1987, Vol. 1. Proceedings of the international symposium on osteoporosis, Copenhagen, Denmark, September 27–October 2, 1987. Copenhagen: Osteopress ApS 1987:288–293.

119. Melton LJ, Kan SH, Wahner HW, Riggs BL. Lifetime fracture risk: an approach to hip fracture risk assessment based on bone mineral density and age. J Clin Epidemiol 1988;41(10):985–994.

120. Mazess RB, Barden HS, Ettinger M, et al. Spine and femur density using dual-photon absorptiometry in US white women. Bone Min 1987;2:211–219.

121. Garn SM, Rohmann CG, Wagner B. Bone loss as a general phenomenon in man. Fed Proc 1967;6:1729–1736.

122. Mazess RB. On aging bone loss. Clin Orthoped 1982;165:239–252.

123. Ross PD, Wasnich RD, Vogel JM. Detection of prefracture spinal osteoporosis using bone mineral absortiometry. J Bone Min Res 1988;3(1):1–11.

124. Melton LJ III, Wahner HW, Richelson LS, O'Fallon WM, Riggs BL. Osteoporosis and the risk of hip fracture. Am J Epidemiol 1986;124:254–261.

125. Odvina CV, Wergedal JE, Libanati CR, Schulz EE, Baylink DJ. Relationship between trabecular vertebral body density and fractures: a quantitative definition of spinal osteoporosis. Metabolism 1988;37(3):221–228.

126. Wasnich RD. Fracture prediction with bone mass measurements. In: Genant HK, ed. Osteoporosis update 1987. Berkeley: University of California Press, 1987:95–101.

127. Riggs BL, Wahner HW. Bone densitometry and clinical decision-making in osteoporosis. Ann Intern Med 1988;108(2):293–295.

128. Riis BJ, Christiansen C. Measurement or spinal or peripheral bone mass to estimate early postmenopausal bone loss? Am J Med 1988;84:646–653.

129. Slemenda CW, Johnston C. Bone mass measurement: which site to measure. Am J Med 1988;84:643–645.

130. Genant HK, Block JE, Steiger P, Glüer CC, Ettinger B, Harris ST. Appropriate use of bone densitometry. Radiology 1989;170:817–822.

131. Genant HK, Steiger P, Block JE, Ettinger B, Harris ST. Quantitative computed tomography: update 1987 [Editorial]. Calcif Tissue Int 1987;41:174–186.

Physical Principles of Magnetic Resonance Imaging and Computed Tomography

Joost Doornbos

Introduction

The physical backgrounds of computerized tomography (CT) and magnetic resonance (MR) imaging have been described elsewhere thoroughly and in full detail (1–4). In this appendix the physical background and the technique of both methods will be reviewed in a comparative way.

On the one hand CT and MR imaging have many common features, e.g., their tomographic nature and the exquisite anatomical detail. The body is usually imaged slice by slice. Each slice consists of many small volume elements (voxels, three-dimensional); these voxels are displayed as picture elements (pixels, two-dimensional). On the other hand, however, the modalities display differences that are even more striking than their resemblance (e.g., tissue contrast manipulation).

The following sections summarize the CT and MR imaging characteristics of some relevant practical topics in tomographic imaging.

Image Creation and Display

Computed Tomography

A CT image is created by computer reconstruction of data acquired from projections of X-ray attenuation in the slice of interest. The image characteristics in CT are based on the specific X-ray absorption in different tissues. A narrow X-ray beam is used to intersect only one slice of the patient's body (Fig. A.1). The X-ray source is stepwise rotated around the patient. A detector ring detects the transmitted radiation. Up to 1000 views (projections, one-dimensional strings of numbers) are stored in the computer memory and subsequently an image can be calculated from these views. This calculation (reconstruction) is performed by a computer using the so-called "filtered back projection" algorithm (Fig. A.2). Subsequent slices are imaged after

translation of the patient table. Slice orientation is almost always in the axial plane. In some instances, e.g., in the hand and foot, direct sagittal and coronal images are obtained. Moreover, data from axial slices may be used to generate two-dimensional images in other planes (this is called reformatting), or to create three-dimensional views. To obtain reformatted images of high quality, thin axial slices should be acquired, preferably with incremental table translations less than the slice thickness.

Magnetic Resonance Imaging

The phenomenon of nuclear magnetism is used to obtain images of the human body. Atomic nuclei with an odd number of protons or neutrons (e.g., the hydrogen atom, which is present in water and fat molecules) possess a magnetic moment. These moments (also called *spins*) usually have random orientation in space (Fig. A.3*A*). In the presence of a magnetic field a net alignment of the spins will occur (Fig. A.3*B*). The image slice is selected by shap-

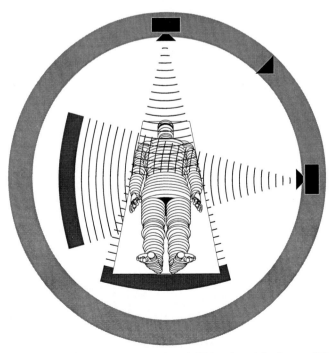

Figure A.1. Schematic presentation of CT imaging. A fan beam of X-rays intersects the patient. Projections of the image slice are sampled by the detector ring.

Abbreviations (see also glossary): DTPA, diethylenetriaminepentaacetic acid; GE, gradient echo; RF, radio frequency; SE, spin echo; TE, echo time; TR, repetition time.

ing the magnetic field with field gradients. The orientation of the spins in the selected slice can be perturbed by radio waves at a specific (resonance) frequency (Fig. A.3C). After a pulse of radio-frequency (RF) energy has been applied with a transmitter coil near the patient, the perturbed spins will realign with the static magnetic field (Fig. A.3D). This process of realignment is called relaxation and it is governed by two tissue-specific parameters: the relaxation times T1 (or longitudinal relaxation time) and T2 (or transverse relaxation time). During realignment the spins emit radio energy, which is received with an antenna coil. Spatial in-

formation about the slice of interest is present in the received radio signal.

Previous to every RF pulse, the slice of interest is subjected to magnetic field gradients of a predefined nature. In this way the properties of spins become dependent on their spatial localization. The strength of the predefined gradient fields is increased in a stepwise manner before every RF pulse. Similar to CT, up to 256 different views (projections) are recorded in this way (Fig. A.4). An image may now be reconstructed from these views via the two-dimensional Fourier transform (2D FT)) algorithm. Because images of other slices are obtained by manipulation of the magnetic field, there is no need to change the patient position. Figure A.5 gives an overview of the electromagnetic spectrum: X-rays are located in the high-energy region whereas the radio waves applied in MR imaging are found in the low-energy region of the electromagnetic spectrum.

Image Contrast

Computed Tomography

Image contrast in CT is based on interaction between X-rays and electrons of tissue traversed. The process of X-ray attenuation is described by photoelectric absorption and Compton interaction. Thus quality of X-rays, tissue electron density, and spatial tissue distribution in the image slice determine the X-ray attenuation. For example, cortical bone shows up bright, whereas lung tissue has a low density in the image. Image contrast in CT can be manipulated by administration of contrast agents to locally reduce or augment density (e.g., air or iodinated chemical compounds, respectively).

Magnetic Resonance Imaging

Image contrast generation in MR imaging is dependent on several specific tissue characteristics: (*a*) the density of mobile protons (mainly in water and fat molecules) and (*b*) the nuclear magnetic relaxation times (T1 and T2) of these protons.

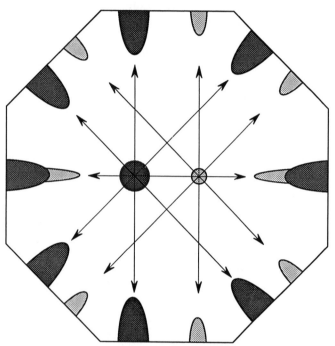

Figure A.2. A series of up to 1000 CT projections is processed by the computer and "back projected" to form the image. In this example eight projections of an object consisting of two parts are shown.

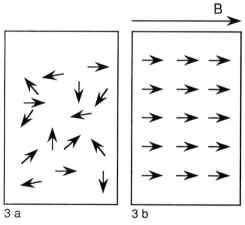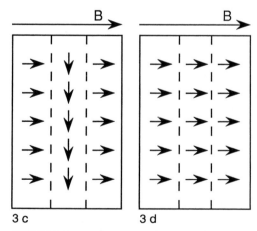

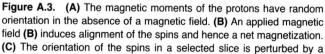

Figure A.3. **(A)** The magnetic moments of the protons have random orientation in the absence of a magnetic field. **(B)** An applied magnetic field **(B)** induces alignment of the spins and hence a net magnetization. **(C)** The orientation of the spins in a selected slice is perturbed by a radio-frequency pulse. The nature, extent, and time evolution of this perturbation determine the image content. **(D)** The relaxation processes realign the spins with the external magnetic field.

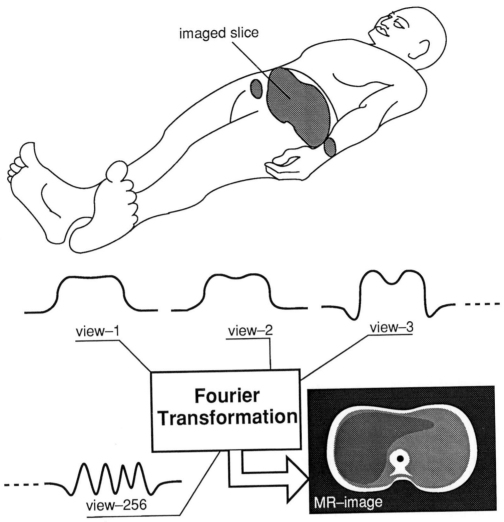

Figure A.4.   In MR imaging up to 256 views are processed by the Fourier transform algorithm in a computer to reconstruct the image.

Depending on the imaging method used, each of these three parameters may be emphasized in the image. Therefore MR images are usually classified as proton density, T1-weighted, or T2-weighted images. The differences in tissue brightness are caused by differences in proton density and T1 and T2 relaxation times among tissues.

Contrast in MR imaging can be influenced to a large degree by variation of the time interval between the successive RF pulses (the repetition time, TR), the duration of the RF pulse (pulse angle), and by variation of the time interval between application of the RF pulse and the recording of the radio signal from the patient (the echo time, TE). As TR and TE determine the T1 and T2 weighting of the image, respectively, TR may be regarded as a "T1 knob" and TE as a "T2 knob." A generally used MR method is the spin-echo (SE) technique. Spin-echo MR images obtained with a TR of 600 ms or shorter and a TE of 20 ms or shorter are called T1-weighted images, images obtained with a TR of 2000 ms or longer and a TE of 20 ms are called spin-density images, and a TR of 2000 ms or longer in combination with a TE of 100 ms or longer results in T2-weighted images. Another frequently used method is

the gradient-echo (GE) technique. In this method the RF pulse angle, TR, and TE are the contrast-generating parameters. Application of a low pulse angle (e.g., 20°) will result in a proton density and T2*-weighted GE image (T2* is the apparent T2), whereas a high pulse angle (e.g., 90°) may be applied to obtain a T1-weighted GE image, or T2*-weighted image if TR is very short (steady state).

Blood flow is an additional contrast-determining factor in imaging of vessels. Furthermore, contrast in MR imaging can be altered by administration of contrast agents that influence the magnetic behavior of tissues.

Gadolinium-labeled diethylenetriaminepentaacetic acid (Gd-DTPA) is an example of a contrast agent that decreases the relaxation times of certain tissues and thereby changes image contrast. Gd-DTPA has pharmacokinetic properties similar to the conventional iodinated Roentgen contrast agents. The degree of decrease of T1 and T2 relaxation times is dose dependent. In clinical practice the administered dose is 0.1 to 0.2 mmol Gd-DTPA-kg body weight. Well-perfused tissue (e.g., viable tumor) will have a high signal intensity of postcontrast T1-weighted images due to T1 shortening by Gd-DTPA. Gd-DTPA in the clin-

Type of radiation

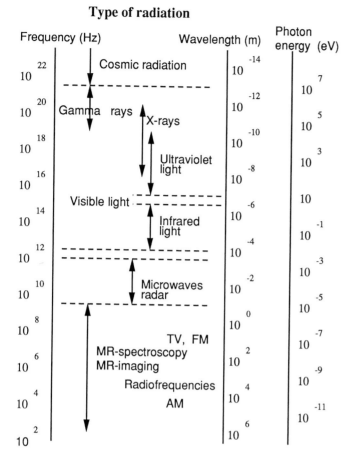

Figure A.5. The electromagnetic spectrum. The wavelengths, frequencies, and photon energies are displayed. Magnetic resonance imaging radio waves are found in the low-energy part, X-rays are located in the high-energy part of the spectrum.

ical dose will not considerably shorten T2. High concentrations of Gd-DTPA, as present in the most dependent part of the urinary bladder after administration of Gd-DTPA, cause shortening of T2.

Tissue Characterization

Quantitative tissue characterization in CT and MR is feasible but the results must be regarded with caution. Image brightness in CT is given in Hounsfield specific absorption units (HU). In MR signal intensity is given in arbitrary units, but it is possible to calculate the proton density and tissue relaxation times from MR images.

The variation in these quantitative data can be large between patients and scanner types. The numerical values should therefore not be used as absolute diagnosis indicators. However, Hounsfield units and MR relaxation times may be used as adjunctive parameters in the diagnostic procedure.

Image Characteristics

A CT or MR image is characterized by several technical factors, such as slice thickness (e.g., 1 to 10 mm), resolution (e.g., 128×256 up to 512×512 pixels) and field of view

Table A.1
Magnetic Resonance Imaging Intensity of Normal Musculoskeletal Tissues

Tissue Type	T1-Weighted MR imaging	T2-Weighted MR imaging
Cortex	Very low	Very low
Red marrow	Intermediate	Intermediate
Yellow marrow	High	Intermediate
Muscle	Intermediate	Low
Fat	High	Intermediate to high
Fibrous cartilage	Low	Low
Hyaline cartilage	Intermediate	High
Ligaments/tendons	Very low	Very low

(10 to 40 cm). The brightness and contrast (windowing) of the images displayed at the computer screen can be controlled by the operator. Especially in CT, correct windowing is important to obtain information from very bright (bone) or very dark (lung) structures. Table A.1 displays MR tissue characteristics of various normal musculoskeletal tissues in T1- and T2-weighted images, respectively.

Scanner Types

Computed Tomography

In CT scanning scanners are categorized by generation. Currently in use are third and fourth generation scanners with a rotating and a fixed detector ring, respectively (Figs. A.6 and A.7). The rotation is usually restricted to 360°, after which the direction of rotation is reversed. A fifth generation scanner type with continuous rotation instead of back and forth rotation (enabling very fast scanning) has been recently introduced. Scanners that use an electron beam instead of the mechanically moving X-ray tube are used in some institutions in cardiovascular applications.

Magnetic Resonance Imaging

Magnetic resonance imagers are generally classified according to the field strength (0.06 to 2.0 T) of the magnet.

Machine performance is to some degree dependent on generation or field strength. However, a wealth of other technical factors also determines the quality of an imaging system.

Artifacts

Computed Tomography

Apart from hardware maladjustments and patient motion, artifacts in CT generally arise from interfaces between substances with strongly different density (e.g., air-tissue interface, metallic clips). The artifacts show up as bright streaks in the image. Another well-known problem is the beam-hardening artifact. This causes areas of diminished density and broad streaks. The beam-hardening artifact originates from the selective absorption of low-energy X-rays in the body. The remaining high-energy rays are less attenuated and tissues near dense regions like bony structures may thus show up with artifactual density.

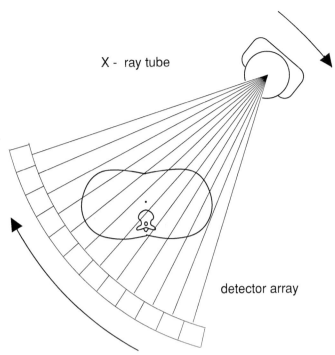

Figure A.6. Schematic drawing of the X-ray source and detector in a third generation CT scanner. Both the source and the detector move around the patient. The detector is a curved array of detector units.

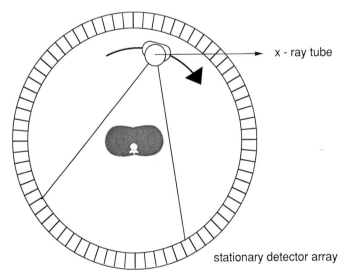

Figure A.7. Schematic drawing of the X-ray source and detector in a fourth generation CT scanner. Only the X-ray source moves around the patient. Detection is performed in the circular array of detectors.

Magnetic Resonance Imaging

Even when MR images are obtained in a technically correct way and the apparatus has been well adjusted, image artifacts may remain present. During the relatively long acquisition time physiologic motion (cardiac and vascular pulsations, breathing) influences the image. Motion artifacts show up as image blurring and ghost images. Image acquisition with electrocardiogram triggering and artifact reduction by software adjustments are applied to obtain

artifact-free images. Metallic objects like surgical clips and prosthetic implants may distort the local magnetic field due to a local change in magnetic susceptibility and thus may cause image degradation. (5) In general, the presence of orthopedic implants does not pose safety hazards for the patient: deflection forces are very small or absent and no significant heating of the implant is observed.

Iron present in hematoma may influence the relaxation times and the local magnetic susceptibility and hence the image intensity. Depending on the age of the hematoma its aspect will change (see Chapter 15). Artifacts due to local changes in the magnetic susceptibility are observed especially in gradient-echo images.

Another source of artifacts is the varying fat content of tissues. Fatty tissue is somewhat shifted in the images as compared to tissues mostly made up of water. This misregistration (chemical shift artifact) shows up at interfaces in the frequency-encoding direction; for instance, a dark border line may show up at one side of the cortex, whereas the interface at the other side will appear bright. A similar phenomenon is observed at the intervertebral disks in sagittal views.

When the chosen field is too small, parts of the body outside this field of view may be backfolded on top of the desired image (aliasing).

Advantages

Computed Tomography

High anatomical resolution is obtained, characterization of small lesions of cortical bone is possible, motion artifacts are avoided by breath-hold scanning, and relatively good soft-tissue contrast is present.

Magnetic Resonance Imaging

Magnetic resonance imaging yields a high soft-tissue contrast, anatomical resolution is adequate, image contrast can be manipulated by the user, usually no contrast agent is required, imaging planes are freely selectable, the technique is noninvasive, and no ionizing radiation is used.

Disadvantages

Computed Tomography

In CT the patient radiation dose must be considered; this may limit, for instance, the number of slices or the slice thickness. The choice of imaging planes is restricted. Since soft-tissue contrast is not always sufficient for diagnosis, administration of a contrast agent is often required. Artifacts due to the presence of bony structures and metallic objects may degrade the image.

X-ray tube heating and computer data processing speed may prohibit very fast scanning.

Magnetic Resonance Imaging

At present MR imaging is an expensive and time-consuming imaging modality. The scan speed of MR imaging is limited by the inherent low signal strength from the proton spins and by the capacity of the gradient field generators. The availability of MR imaging units is still limited in some countries.

Artifacts due to physiologic motion and metallic objects may degrade the image. Small calcifications are not well delineated.

Acknowledgment

The illustrations were prepared by Jan Beentjes.

References

1. Berland LL, Practical CT. New York: Raven Press, 1986.
2. Lee JKT, Sagel SS, Stanley RJ, eds. Computed body tomography. New York: Raven Press, 1989.
3. Stark DD, Bailey WG, eds. Magnetic resonance imaging. St. Louis: C. V. Mosby, 1988.
4. Wells, PNT, ed. Scientific basis of medical imaging. Edinburgh: Churchill Livingstone, 1982.
5. Shellock FG, Crues JV. High-field-strength MR imaging and metallic biomedical implants: An ex vivo evaluation of deflection forces. AJR 1988;151:389–392.

Glossary

Acquisition matrix Number of independent data samples in each direction. In two-dimensional Fourier transform (2D FT) imaging it is the number of samples in frequency and phase-encoding directions (e.g., 256×246 to 128×256).

Acquisition time Time required to carry out an MR imaging procedure comprising only the data acquisition time. The acquisition time equals the product of repetition time (TR), number of signals averaged (NEX or NSA), and the number of different phase-encoding steps.

Aliasing (wrap-around artifact) Consequence of sampling in which any components of the signal that are at a higher frequency than the Nyquist limit (this frequency is equal to one-half the sampling rate) will be folded in the spectrum so that they appear to be at lower frequency. In Fourier transform imaging, this can produce an apparent wrapping around to the opposite side of the image of a portion of the object that extends beyond the edge of the reconstructed region.

Angular frequency (ω) Frequency of oscillation or rotation.

Angular momentum A vector quantity given by the vector product of the momentum of a particle and its position vector. In the absence of external forces, the angular momentum remains constant, with the result that any rotating body tends to maintain the same axis of rotation. When a torque is applied to a rotating body in such a way as to change the direction of the rotation axis, the resulting change in angular momentum results in precession. Atomic nuclei possess an intrinsic angular momentum referred to as spin, measured in multiples of Planck's constant.

Artifacts False features in the image produced by the imaging process.

Attentuation Reduction of (electric) power, due to passage through a medium or electrical component.

Averaging Combining more than one acquisition matrix to increase the signal:noise ratio (SNR).

B_0 A symbol for the constant magnetic (induction) field in an NMR system.

B_1 A symbol for the radio-frequency magnetic induction field used in an NMR system.

Bandwidth A range of frequencies (e.g., contained in a signal or passed by a signal processing system).

Bloch equations Phenomenological "classical" equations of motion for the macroscopic magnetization vector. They include the effects of precession about the magnetic field (static and RF) and the T1 and T2 relaxation times.

Boltzmann distribution Describes the relative number of particles in two particular energy states with corresponding energies in a system of particles in thermal equilibrium that are able to exchange energy. The small excess of nuclei in the lower energy state is the basis of the net magnetization and the resonance phenomenon.

BOSS (bimodal out-of-slice saturation) A presaturation motion artifact suppression technique. *See also* Saturation and Saturation pulse.

Capacitance (C) Electrical capacity, determines the energy stored in a capicitor. Also determines the charge on a capicitor that is maintained at a given voltage.

Car-Purcell *See* CP sequence.

Car-Purcell-Meiboom-Gill *See* CPMG sequence.

Chemical shift Change of Larmor frequency of a given nucleus when bound in different sites in a molecule, due to magnetic shielding of the nucleus by nearby electron orbitals. The amount of chemical shift in hertz is proportional to the magnetic field strength.

Chemical shift artifact Image artifact of apparent spatial offset of regions with different chemical shifts (fat and water) along the direction of the frequency-encoding gradient.

Chemical shift selectives. *See* CHESS.

CHESS (chemical shift selective) Imaging process that makes use of frequency differences to differentiate fat from water.

CNR (contrast:noise ratio) The ratio of the absolute difference in intensities between two regions to the level of fluctuations in intensity due to noise.

Coherence Maintenance of a constant phase relationship between rotating or oscillating waves or objects. Loss of phase coherence of spins (dephasing) results in a decrease in the transverse magnetization and hence a decrease in the MR signal.

Coil Single or multiple loops of wire (or other electrical conductor) designed either to produce a magnetic field from current flowing through the wire or to detect a changing magnetic field by voltage induced in the wire.

Contrast The relative difference of signal intensities in two adjacent regions.

Contrast:noise ratio *See* CNR.

CP (Carr-Purcell) sequence Pulse sequence of a 90° RF pulse followed by repeated 180° pulses to produce a train of spin echoes; useful for measuring T2.

CPMG (Car-Purcell-Meiboom-Gill) sequence Modification of CP sequence with 90° phase shift in the rotating frame of reference between the 90° pulse and the subsequent 180° pulses in order to reduce accumulating effects of imperfections in the 180° pulses.

Adapted by Johan L. Bloem, M.D. with permission from American College of Radiology. Glossary of MR terms. 2nd ed. Reston, Va.: American College of Radiology, 1986.

Cryostat An apparatus for maintaining a constant low temperature (as by means of liquid helium). Requires a vacuum chambers to help with thermal isolation.

CT Computed tomography.

DAC Digital-to-analog converter.

Decoupling Techniques to avoid interactions between coils, such as separate transmitting and receiving coils.

Dephasing Loss of phase coherence. *See also* Coherence.

Diamagnetic A substance that will slightly decrease a magnetic field when placed within it. Its magnetization is oppositely directed to the magnetic field, i.e., with a small negative susceptibility.

2D FT imaging Two-dimensional Fourier transform imaging. *See* FT.

3D FT imaging Three-dimensional Fourier transform imaging. *See* FT.

Display matrix The number of pixels (usually 256 or 512) along each image axis.

Dixon technique An imaging technique that can resolve the chemical shifts of lipid and water. In- and out-of-phase images are acquired to resolve the two constituents. *See also* In-phase image and Out-of-phase image.

DSA Digital subtraction angiography.

DVI Digital vascular imaging.

Echo planar imaging *See* EPI.

Eddy currents Electric currents induced in a conductor by a changing magnetic field or by motion of the conductor through a magnetic field. One of the sources of concern about the potential hazard to subjects in very high magnetic fields or rapidly varying gradient or main magnetic fields. Can be a practical problem (image degradation) in the cryostat of superconducting magnets.

Electron paramagnetic resonance *See* ESR.

Electron spin resonance *See* ESR.

EPI (echo planar imaging) Magnetic resonance technique of planar imaging in which a complete planar image is obtained from one slice-selective excitation pulse. The free induction decay (FID) is observed while periodically switching the Y (phase-encoding) gradient field in the presence of an X (frequency-encoding) gradient field. The Fourier transform of the resulting spin-echo train can be used to produce an image of the excited plane.

EPR Electron paramagnetic resonance. *See* ESR.

ESR (electron spin resonance) Magnetic resonance phenomenon involving unpaired electrons, e.g., in free radicals. Frequencies are much higher (i.e., microwave range) than corresponding NMR frequencies in the same static magnetic field.

Excitation Putting energy (in the form of an RF pulse) into the spin system; if a net transverse magnetization is produced an NMR signal can be observed.

Exorcist *See* ROPE.

FFE Fast field echo, a gradient-echo (GE) image sequence. *See* Gradient echo and Gradient-echo imaging.

FID (free induction decay) If transverse magnetization of the spins is produced, e.g., by a 90° pulse, a transient NMR signal will result that will decay to zero with a characteristic time constant T2*; this decaying signal is the FID.

Field of view *See* FOV.

FISP Fast imaging with steady state free precession. *See* GRASS and SFP.

FLASH (fast low-angle single-shot imaging) A fast gradient-echo imaging technique (also called spoiled Grass) that uses a spoiler gradient to destroy transverse magnetization. Flip angles (excitation pulses) of less than 90° are used, leaving a portion of the longitudinal magnetization unperturbed. Repetition time (TR) may be reduced to less than 100 ms. Signal depends mainly on T1 (high-flip angle) or proton density (low flip angle).

Flip angle Angle of rotation of the macroscopic magnetization vector produced by a radio-frequency pulse, with respect to the direction of the static magnetic field ($\mathbf{B}_0$).

Flow-related enhancement Increase in signal intensity that may be seen for flowing blood with some MR imaging techniques due to washout of saturated spins from the imaging region.

Flow void Signal loss due to phase changes, reflecting rapid flow.

Fourier transform *See* FT.

Fourier transform imaging Magnetic resonance imaging techniques in which at least one dimension is phase encoded by applying variable gradient magnetic fields along that dimension before "reading out" the NMR signal with a magnetic field gradient perpendicular to the variable gradient. The Fourier transform is then used to reconstruct an image from the set of encoded NMR signals.

FOV (field of view) Size of the area that is imaged. It is the product of pixel dimensions and acquisition matrix.

Frequency encoding Encoding the distribution of sources of MR signals along a direction by detecting the signal in the presence of a magnetic field gradient along that direction so that there is a corresponding gradient of resonance frequencies along that direction.

FT (Fourier transform) A mathematical procedure to separate the frequency components of a signal from its amplitudes as a function of time or vice versa. The Fourier transform is used to generate the spectrum from the free induction decay or spin echo in pulse MR techniques.

G (Gauss) Unit of magnetic flux density in the CGS system. The earth's magnetic field is between 0.5 and 1 G. The currently preferred (SI) unit of magnetic flux density is the tesla (T) (1 T = 10,000 G).

Gd-DTPA (gadolinium-labeled diethylenetriaminepentaacetic acid) Paramagnetic MR contrast agent.

GE (gradient echo) *See* Gradient echo and Gradient-echo imaging.

GMR (gradient moment refocusing) A gradient moment-nulling motion artifact reduction technique. *See also* Gradient moment nulling.

Gradient The amount and direction of the rate of change in space of some quantity, such as magnetic field strength.

Gradient coil Current-carrying coil designed to produce a desired gradient magnetic field.

Gradient echo Spin echo produced by reversing the direction of the magnetic field gradient or by applying balanced pulses of magnetic field gradients before and after a refocusing RF pulse so as to cancel out the position-dependent phase shifts that have accumulated due to the gradient. In the latter case the gradient echo is generally adjusted to be coincident with the RF spin echo. *See also* Gradient-echo imaging.

Gradient-echo imaging Rapid image acquisition using gradient reversal instead of 180°-refocusing pulse, excitation pulses of less than 90°, and short repetition times (TR). Gradient-echo images are more sensitive than standard spin-echo images to magnetic field inhomogeneity. *See also* FFE, FISP, FLASH, and GRASS.

Gradient magnetic field A magnetic field that changes in strength in a certain given direction.

Gradient moment nulling Correction of motion-induced phase errors with gradient magnetic fields. *See also* GMR, MAST, and Saturation pulse.

GRASS Gradient-recalled acquisition in the steady state. *See also* SFP and FISP.

GRE (gradient-recalled echo) *See* Gradient echo and Gradient-echo imaging.

G_x, G_y, G_z Symbols for gradient magnetic field. Used with subscripts to denote spatial direction component of gradient, i.e., direction along which the field changes.

Gyromagnetic ratio (γ) The ratio of the magnetic moment to the angular momentum of a particle. This is constant for a given nucleus.

HYBRID A hybrid technique between spin-echo and echo planar imaging (EPI), in which several lines are gathered in one pass.

Inductance Results in a voltage across a coil when the current in the coil changes with time. It also determines the energy stored in the coil.

Induction Transference of electric or electromagnetic force without direct contact.

In-phase image Spin-echo image acquired when lipid and water signals are together (in phase). The 180° pulse is at TE/2. *See also* Dixon technique and Out-of-phase image.

IR (inversion recovery) An RF pulse sequence in which a 180°-inverting pulse is used prior to the 90°-detection pulse. In practice, IR sequences are in fact IR spin-echo (SE) sequences because a 180°-refocusing pulse is also used.

Inversion recovery. *See* IR.

Inversion time *See* TI.

Larmor equation States that the frequency of precession of the magnetic nuclear moment is proportional to the magnetic field, $\omega_0 = \gamma\ B_0$.

Larmor frequency (ω_0) The frequency at which magnetic resonance can be induced. For protons the Larmor frequency is 42,58 MHz/T.

Lattice By analogy to NMR in solids, the magnetic and thermal environment with which nuclei exchange energy in longitudinal relaxation.

Longitudinal magnetization (M_z) Component of the macroscopic magnetic magnetization vector along the static magnetic field. Following excitation by an RF pulse, M_z will approach its equilibrium value M_0, with a characteristic time constant T1.

Longitudinal relaxation Return of longitudinal magnetization to its equilibrium value after excitation; requires exchange of energy between the nuclear spins and the lattice.

M_0 Equilibrium value of the magnetization; directed along the direction of the static magnetic field. Proportional to spin density *(N)*.

Macroscopic magnetization vector Net magnetic moment per unit volume (a vector quantity) of a sample in a given region, considered as the integrated effect of all the individual microscopic nuclear magnetic moments.

Magnetic field (H or B) The region surrounding a magnet (or current-carrying conductor) is endowed with certain properties. One is that a small magnet in such a region experiences a torque that tends to align it in a given direction. Magnetic field is a vector quantity; the direction of the field is defined as the direction that the north pole of the small magnet points when in equilibrium. A magnetic field produces a magnetizing force on a body within it. Although the dangers of large magnetic fields are largely hypothetical, this is an area of potential concern for safety limits. Formally, the forces experienced by moving charged particles, current-carrying wires, and small magnets in the vicinity of a magnet are due to magnetic induction **(B),** which includes the effects of magnetization, while the magnetic field **(H)** is defined so as not to include magnetization. However, both **B** and **H** are often loosely used to denote magnetic fields.

Magnetic moment A measure of the net magnetic properties of an object or particle. A nucleus with an intrinsic spin will have an associated magnetic dipole moment, so that it will interact with a magnetic field (as if it were a tiny bar magnet).

Magnetic susceptibility Measure of the ability of a substance to become magnetized.

Magnetization The magnetic polarization of a material produced by a magnetic field. *See also* Macroscopic magnetization vector.

MAST Motion artifact suppression technique. *See also* Gradient moment nulling.

Matrix *See* Acquisition matrix and Display matrix.

MR (magnetic resonance) *See* NMR.

NEX Number of excitations. *See also* NSA.

NMR (nuclear magnetic resonance) The absorption or emission of electromagnetic energy by nuclei in a static magnetic field, after excitation by a suitable radio-frequency magnetic field. The peak resonance frequency is proportional to the magnetic field, and is given by the Larmor equation. Only nuclei with a nonzero spin exhibit NMR.

NMR imaging Creation of images of objects such as the body by use of the nuclear magnetic resonance phenomenon. The immediate practical application involves imaging the distribution of hydrogen nuclei (protons) in the body. The image brightness in a given region is usually dependent jointly on the spin density and the relaxation times, with their relative importance determined by the particular imaging technique employed. Image brightness is also affected by motion such as blood flow.

NMR signal Electromagnetic signal in the radio-frequency range produced by the precession of the transverse magnetization of spins. The rotation of the transverse magnetization induces a voltage in a coil, which is amplified and demodulated by the receiver; the signal may refer only to this induced voltage.

NSA (number of signals averaged) Number of signals averaged together to determine each distinct position-encoded signal to be used in image reconstruction. *See also* NEX.

Nuclear magnetic resonance *See* NMR, NMR imaging, and NMR signal.

Nuclear spin An intrinsic property of certain nuclei (odd number of neutrons and/or protons) that gives them an associated characteristic angular momentum and magnetic moment.

Nutation A displacement of the axis of a spinning body away from the simple cone-shaped figure that would be traced by the axis during precession. In the rotating frame of reference, the nutation caused by a radio-frequency pulse appears as a simple precession, although the motion is more complex in the stationary frame of reference.

Out-of-phase image Image obtained when lipid and water signals are opposed (out of phase). This is achieved by decentering the 180° pulse in a spin-echo sequence. *See also* Dixon technique.

Paramagnetic A substance with a small but positive magnetic susceptibility.

Partial saturation *See* PS.

Permanent magnet A magnet the magnetic field of which originates from permanently magnetized material.

Phase In a periodic function (such as rotational or sinusoidal motion), the position relative to a particular part of the cycle.

Pixel Acronym for a picture element; the smallest discrete part of a digital image display (two dimensional).

Precession Comparatively slow gyration of the axis of a spinning body so as to trace out a cone; caused by the application of a torque tending to change the direction of the rotation axis, and continuously directed at right angles to the plane of torque. The magnetic moment of a nucleus with spin will experience such a torque when inclined at an angle to the magnetic field, resulting in precession at the Larmor frequency.

PS (partial saturation) Magnetic resonance imaging technique applying repeated RF pulses with repetition times (TR) less or equal to T1.

Pulse sequences Set of RF (and/or gradient) magnetic field pulses and time spacings between these pulses; used in conjunction with magnetic field gradients and NMR signal reception to produce MR images.

Q (quality factor) Applies to any electrical circuit component; most often the coil Q determines the overall Q of the circuit. Inversely related to the fraction of the energy in an oscillating system lost in one oscillation cycle. Q is inversely related to the range of frequency over which the system will exhibit resonance. It affects the signal:noise ratio, because the detected signal increases proportionally to Q while the noise is proportional to the square root of Q. The Q of a coil will depend on whether it is unloaded (no patient) or loaded (patient).

Quenching Loss of superconductivity of the current-carrying coil that may occur unexpectedly in a superconductive magnet. As the magnet becomes resistive, heat will be released that can result in rapid evaporization of liquid helium in the cryostat.

Radiofrequency *See* RF.

RARE (rapid acquisition with relaxation enhancement) Phase encoding between 180° pulses.

Receiver Portion of the MR apparatus that detects and amplifies RF signals picked up by the receiver coil.

Receiver coil Coil of the RF receiver, picks up the MR signal.

Relaxation times After excitation the spins will tend to return to their equilibrium distribution, in which there is no transverse magnetization and the longitudinal magnetization is at its maximum value and orientated in the direction of the static magnetic field. It is observed that in the absence of applied RF, the transverse magnetization decays toward zero with a characteristic time constant T2, and the longitudinal magnetization returns toward the equilibrium value (M_0) with a characteristic time constant T1.

Refocusing Restoration of phase coherence. *See also* Coherence, Gradient echo, and SE.

Resistive magnet A magnet the magnetic field of which originates from current flowing through an ordinary (nonsuperconducting) conductor.

Resonance A large-amplitude vibration in a mechanical or electrical system caused by a relatively small periodic stimulus with a frequency at or close to a natural frequency of the system; in NMR apparatus, resonance can refer to the NMR itself or to the tuning of the RF circuitry.

RF (radio frequency) Wave frequency of electromagnetic radiation intermediate between auditory and infrared.

RF (radio-frequency) coil Coil used for transmitting RF pulses and/or receiving NMR signals.

RF (radio-frequency) pulse Brief burst of RF magnetic field delivered to object by RF transmitter. An RF pulse near the Larmor frequency will result in rotation of the macroscopic magnetization vector in the rotating frame of reference. The amount of rotation will depend on the strength and duration of the RF pulse.

ROPE (respiratory ordered phase encoding) ROPE and exorcist are respiratory motion compensation systems.

Rotating frame of reference A frame of reference (with corresponding coordinate systems) that is rotating about the axis of the static magnetic field B_0 (with respect to a stationary frame of reference) at a frequency equal to that of the applied RF magnetic field, B_1. Although B_1 is a rotating vector, it appears stationary in the rotating frame, leading to simpler mathematical formulations.

Saturation A nonequilibrium state in MR, in which equal numbers of spins are aligned against and with the magnetic field, so that there is no net magnetization.

Saturation pulse A frequency-selective motion artifact suppression pulse that excites and thereby rephases or saturates a defined area. *See also* Gradient moment nulling.

SE (spin echo) Reappearance of an NMR signal after the free induction decay (FID) has apparently died away, as a result of the effective reversal of the dephasing of spins (refocusing) by technique such as specific RF pulses, e.g., Carr-Purcell (CP) sequence (RF spin echo), or pairs of magnetic field gradient pulses (gradient echo), applied in times shorter than or in the order of T2. Unlike RF spin echoes, gradient echoes will not refocus phase differences due to chemical shifts or inhomogenities of the magnetic field.

SE (spin echo) image Any of many MR imaging techniques in which the spin echo is used rather than the free induction decay (FID). The standard SE sequence consists of a 90° excitation pulse followed by a 180° rephasing pulse.

SFP (steady state free precession) Excitation in which strings of RF pulses are applied rapidly and repeatedly with short interpulse times compared to T1 and T2. FISP and GRASS are SFP sequences that refocus in the phase-encoding direction. A true FISP sequence refocuses in both read and phase encoding.

SI Signal intensity.

SNR (signal:noise ratio) Used to describe the relative contributions to a detected signal of the true signal and random superimposed signals (noise).

Spatial resolution The smallest distance between two points in the object that can be distinguished as separate details in the image. Depends on slice thickness and pixel dimensions.

Spin The intrinsic angular momentum of an elementary particle, or system of particles such as a nucleus, that is also responsible for the magnetic moment; or, a particle or nucleus possessing such a spin. The spins of nuclei have characteristic fixed values.

Spin density *(N)* The density of resonating spins in a given region; one of the principal determinants of the strength of the NMR signal from the region.

Spin echo *See* SE.

SSFP Steady state free precession. *See* SFP.

Statistical formulas TP, true positive; FP, false positive; TN, true negative; FN, false negative:

$$\text{Sensitivity} = \frac{\text{TP}}{(\text{TP} + \text{FN})}$$

$$\text{Specificity} = \frac{\text{TN}}{(\text{FP} + \text{TN})}$$

$$\text{Accuracy} = \frac{(\text{TP} + \text{TN})}{(\text{TP} + \text{TN} + \text{FP} + \text{FN})}$$

$$\text{Positive predictive value} = \frac{\text{TP}}{(\text{TP} + \text{FP})}$$

$$\text{Negative predictive value} = \frac{\text{TN}}{(\text{FN} + \text{TN})}$$

$$\text{Prevalence} = \frac{(\text{TP} + \text{FN})}{(\text{TP} + \text{TN} + \text{FP} + \text{FN})}$$

95% confidence limits (interval): a range of values selected in such a way that there is a specified probability (95%) of including the true value.

Steady state free precession *See* SFP.

STIR (Short τ (TI) inversion recovery) Inversion recovery technique combining short inversion time with magnitude reconstruction, thus adding T1- and T2-dependent contrasts. STIR imaging also suppresses signal from fat.

Superconducting magnet A magnet the magnetic field of which originates from current flowing through a superconductor. Such a magnet must be enclosed in a cryostat.

Superconductor A substance the electrical resistance of which essentially disappears at temperatures near absolute zero.

Susceptibility *See* Magnetic susceptibility.

Surface coil A small RF receiver coil, which is placed close to the surface of the body and over the region of interest.

T1 Spin-lattice or longitudinal relaxation time; the characteristic time constant for spins to tend to align themselves with the external magnetic field.

T2 Spin-spin or transverse relaxation time; the characteristic time constant for loss of phase coherence among spins oriented at an angle to the static magnetic field, due to interactions between the spins, with resulting loss of transverse magnetization and NMR signal.

T2* (apparent T2) Observed time constant of the free induction decay (FID) due to loss of phase coherence among spins oriented at an angle to the static magnetic field, commonly due to a combination of magnetic field inhomogeneities, $\Delta\mathbf{B}$, and spin-spin transverse relaxation with resultant more rapid loss in transverse magnetization and NMR signal.

τ Denotes different time delays, such as T1, between RF pulses.

TE (echo time) Time between middle of 90° pulse and middle of spin-echo production.

Tesla (T) The preferred (SI) unit of magnetic flux density. One tesla is equal to 10,000 G (gauss), the older (CGS) unit.

TI (Inversion time) In inversion recovery, the time between the middle of the inverting (180°) RF pulse and the middle of the subsequent (90°) excitation pulse.

Torque The effectiveness of a force setting a body into rotation. It is a vector quantity given by the vector product of the force and the position vector where the force is applied.

TR (repetition time) The period of time between the beginning of a pulse sequence and the beginning of the succeeding pulse sequence.

Transverse magnetization ($\mathbf{M}_{xy}$) Component of the macroscopic magnetization vector at right angles to the static magnetic field ($\mathbf{B}_0$). Precession of the transverse magnetization at the Larmor frequency is responsible for the detectable NMR signal.

Two-dimensional Fourier transform imaging (2D FT) A form of sequential plane imaging using Fourier transform imaging.

Volume imaging (3D FT) Imaging techniques in which NMR signals are gathered from the whole object volume to be imaged at once, with appropriate encoding pulse RF and gradient sequences to encode positions of the spins. Many sequential plane imaging techniques can be generalized to volume imaging, at least in principle.

Voxel Volume element; the element of three-dimensional space corresponding to a pixel, for a given slice thickness.

Bibliography

American College of Radiology. Glossary of MR terms. 2nd ed. Reston, Va.: American College of Radiology, 1986.

Fletcher RH, Fletcher SW, Wagner EH. Diagnostic test. In: Clinical epidemiology, the essentials. Baltimore: Williams & Wilkins, 1984:41–59.

Gelfand DW, Ott DJ. Methodologic considerations in comparing imaging methods. AJR 1985;144:1117–1121.

Haacke EM. Special editorial. Magn Reson Imag 1988;6:353–354.

Levy TL, Shellock FG, Cruess JV III. Glossary. In: Stark DD, Bradley WG Jr. Magnetic Resonance Imaging. St. Louis: C. V. Mosby, 1988:1453–1473.

INDEX